Ecological Biomarkers

Indicators of Ecotoxicological Effects

Ecological Biomarkers

Indicators of Ecotoxicological Effects

Edited by

Claude Amiard-Triquet
Jean-Claude Amiard
Philip S. Rainbow

CRC Press
Taylor & Francis Group
Boca Raton London New York

CRC Press is an imprint of the
Taylor & Francis Group, an **informa** business

Cover photograph courtesy of Olivia Fossi Tankoua.

CRC Press
Taylor & Francis Group
6000 Broken Sound Parkway NW, Suite 300
Boca Raton, FL 33487-2742

Printed in the United States of America on acid-free paper
Version Date: 20121112

International Standard Book Number: 978-1-4398-8017-3 (Hardback)

Library of Congress Cataloging-in-Publication Data

Ecological biomarkers : indicators of ecotoxicological effects / edited by Claude Amiard-Triquet, Jean-Claude Amiard, and Philip S. Rainbow.
pages cm
Includes bibliographical references and index.
ISBN 978-1-4398-8017-3 (hardcover : alk. paper)
1. Biochemical markers. 2. Indicators (Biology) 3. Environmental toxicology. I. Amiard-Triquet, C. II. Amiard, J. C. III. Rainbow, P. S.

QH438.4.B55E26 2013
571.9'5--dc23 2012029269

Visit the Taylor & Francis Web site at
http://www.taylorandfrancis.com

and the CRC Press Web site at
http://www.crcpress.com

Contents

Preface

Aims and Scope

The biomarker concept was initially developed with the medical purpose of the early diagnosis of pathological status and for use in mammalian toxicology. At the beginning of the 1990s, ecotoxicologists became interested in the concept, which stimulated important debate, for instance, at the 2nd European Conference on Ecotoxicology organized by the Society of Ecotoxicology and Environmental Safety (SECOTOX) in Amsterdam in 1992. In 1994, Depledge proposed a definition that is still authoritative today: "A biochemical, cellular, physiological or behavioural variation that can be measured in tissue or body fluid samples or at the level of whole organisms that provides evidence of exposure to and/or effects of, one or more chemical pollutants (and/or radiations)."

In the United States, the Clean Water Act is the primary federal law governing water pollution. Because of its statutory responsibilities, the US Environmental Protection Agency has developed a strategy to improve monitoring and assessment of environmental risk in aquatic ecosystems at local, state, regional, and national scales. In this framework, the Environmental Monitoring and Assessment Program (EMAP) has substantially advanced the scientific basis for monitoring the condition of aquatic ecosystems. The EMAP strategy includes physicochemical indicators in sediments and the water column and, for biological indicators, mainly responses at the level of the community. The Water Framework Directive (WFD) promulgated by the European Parliament and Council is the chosen way forward to maintain or improve the quality of European aquatic environments. In this aim, it is necessary to attain a good status of these waters. This good status is based on both the chemical and the ecological status of the water masses. The chemical status is considered "good" when the concentrations of chemicals in the medium are below the limits defined in EC's regulations. The characterization of the ecological status of water masses is mainly based on the composition and abundance of certain plant and animal taxa. The failure of the WFD to recognize a role for biomarkers in this context is regrettable as is their limited use in the EMAP strategy.* By neglecting biomarkers, both regulatory bodies ignore a category of biological tools well known to be precocious and sensitive indicators of the degradation of organism health. Effects at the community level allow an ecotoxicological assessment after severe environmental degradation has already occurred, thus leading to expensive remediation processes, whereas biomarkers have an interesting potential as predictive tools usable much earlier in any environmental degradation process.

Ecological analyses recommended in the EMAP or the WFD are useful to describe differences between sites, differently impacted by anthropogenic pressure, or to reveal temporal changes when historical records are available. However, ecological approaches are of no help in determining the origin of such changes, whereas so-called "specific" biomarkers can contribute to answering this type of question. Some biomarkers are currently used for the implementation of the OSPAR Convention for the Protection of the Marine Environment of the

* USEPA, July 2002. EMAP research strategy. Report EPA 620/R-02/002.

Northeast Atlantic, such as those for metal-specific biological effect monitoring (e.g., metallothionein, δ-amino levulinic acid dehydratase inhibition in blood [ALA-D]) and PAH-specific biological effect monitoring (e.g., cytochrome P4501A, DNA adducts).

Chemical data needed to fulfill the requirements of the WFD or the EMAP strategy may be useful to predict the potential effects on living organisms but only if the dose–effect relationship is well established. Predicted No Effect Concentrations can be derived from laboratory toxicological tests, but the main limit of this practice is that toxicity data are nearly always determined for individual chemicals, whereas in real life numerous molecules or classes of molecules coexist in waters with the possibility of multiple interactions. Among these toxic compounds (including numerous persistent organic pollutants), many are not yet analytically accessible or are analyzable only at exorbitant cost. Thus, it is necessary to develop other strategies to assess the degree to which a given ecosystem is impacted or not by toxic contaminants. In attempting to fulfill this aim, "generalist" biomarkers can reveal the integrated ecotoxicity of complex mixtures, particularly physiological markers linked to the growth and reproduction of organisms.

At the end of the 1990s, several books established the state of the art of biomarker methodology, such as *Use of Biomarkers for Environmental Quality Assessment*, published by Science Publishers, Enfield, USA, in 2000 (Lagadic, Caquet, Amiard and Ramade, eds.). However, as mentioned above, the use of biomarkers remains comparatively marginal in ecological risk assessment. Several reasons may be responsible for this. In the first issue of the journal *Ecotoxicology* (1992), Cairns pointed to one of them, termed the "signal-to-noise ratio." If the natural variation of a given biomarker is weak in the absence of chemical stress, the change induced by chemical stress will be easily detectable. On the other hand, significant natural variation in a biomarker has the potential to conceal—at least partly—a stress-induced additional variation. However, the question of such confounding factors (season, age, sex, etc.) is not peculiar to the methodology of biomarkers and has been mastered (using adapted sample strategies and statistical treatments) in the framework of Mussel Watch programs, based on the monitoring of pollutant concentrations in biological matrices. A second reason for the lack of wider take-up of the use of biomarkers appeared when it became clear that several biomarkers previously considered specific (e.g., decrease of AChE activity in the presence of organophosphate pesticides and carbamates) were also found to be responsive to other molecules (metals, algal toxins) or other forms of stress. Lastly, both specific and generalist biomarkers are determined at the individual or suborganismal level. Does a change that affects a few biological macromolecules, some cells, or a few individuals within a population have any ecological significance that would allow the prediction of deleterious effects at higher levels of biological organization, namely, the population, community, and ultimately the ecosystem?

Over the past decade, the importance of developing biomarkers with added ecological value has been recognized. Subsequent to the publication of our first book, *Les biomarqueurs dans l'évaluation de l'état écologique des milieux aquatiques*, published by Lavoisier, Paris, in 2008 (Amiard and Amiard-Triquet, eds.), it is time to revisit those biological responses that are the most ecologically relevant in order to diagnose degradation of the health status of an aquatic environment well before it becomes unmanageable. The literature reviewed in this book supports the efficacy of the use of lysosomal biomarkers, immunotoxicity effects, behavioral disturbances, energy metabolism impairments, endocrine disruption measures, and genotoxicity as all indicative of probable toxic effects at higher biological levels. These biomarkers thus provide a real possibility of delivering the holy grail—an easily measured biomarker at a simple level of biological organization that is predictably linked to a potentially ecologically significant effect at higher levels of biological organization. This book provides the burning torch to light our way in this quest.

Editors

Claude Amiard-Triquet is a research director in the CNRS (French National Research Center) based at the University of Nantes, France. She earned the degree of DSc in 1975 for her research in radioecology at the French Atomic Energy Commission. Dr. Amiard-Triquet's topics of research interest include metal ecotoxicology, biomarkers, and, more recently, emerging contaminants (endocrine disruptors, nanoparticles). As the head of multidisciplinary research programs, she has managed research collaborations between specialists in organic and inorganic contaminants and chemists and biologists involved in studies from the molecular to ecosystem levels, with a constant concern for complementarity between fundamental and applied research. Dr. Amiard-Triquet regularly acts as an expert for the assessment of scientific proposals (e.g., the European Framework Program for Research and Development, the International Foundation for Science, and the Sea Grant Administration, Oregon State) and is also in demand as a referee for a dozen or so international journals. She has authored or co-authored more than 180 research papers and has authored 27 chapters in various books. Dr. Amiard-Triquet has also co-authored one book, *La Radioécologie des Milieux Aquatiques*, with J.C. Amiard and co-edited three books: *L'Évaluation du Risque Écologique à l'Aide de Biomarqueurs* with J.C. Amiard, *Environmental Assessment of Estuarine Ecosystems: A Case Study* with P.S. Rainbow, and *Tolerance to Environmental Contaminants* with P.S. Rainbow and M. Roméo. She has given or contributed to about 100 presentations at international meetings.

Jean-Claude Amiard is a research director in the CNRS based at the University of Nantes, France. He was an associate professor at the University of Quebec at Rimouski from 1994 to 2010. He earned his DSc degree in 1978 from the University Pierre and Marie Curie, Paris. He has directed 16 PhD theses and contributes to master's teaching in several French and foreign universities. In 2011, he has gathered all this teaching material into a book, *Risques chimiques environnementaux. Méthodes d'évaluation et impacts sur les êtres vivants.* He acts as an expert for governmental organizations in charge of health security Agence nationale de sécurité sanitaire de l'alimentation, de l'environnement et du travail (ANSES) or information on nuclear activities Association nationale des comités et commissions locales d'information (ANCCLI), and in this framework, he has co-edited a book, *Le tritium, actualité d'aujourd'hui et de demain*, with S. Gazal. Previously, he has co-authored and co-edited two books on biomarkers with L. Lagadic, T. Caquet, and F. Ramade and one book, *L'Évaluation du Risque Écologique à l'Aide de Biomarqueurs*, with C. Amiard-Triquet. His research activities have focused on the fate and effects of trace metals in marine and estuarine ecosystems, on the tolerance of organisms to chronic exposure to contaminants, and, more recently, on the application of biomarkers to the assessment of ecotoxicity of emerging contaminants. He has published more than 130 papers in peer-reviewed journals, 90 papers in national journals or congress proceedings, and 32 book chapters or books. He has participated in 140 national and international congresses.

Philip Rainbow is the head of the Department of Zoology at the Natural History Museum, London, leading a staff of more than 100 working scientists. He earned a PhD (1975) and a DSc (1994) from the University of Wales. Dr. Rainbow was appointed (1994) to a personal chair in the University of London, where he was head of the School of Biological

Sciences at Queen Mary (1995–1997) and is now a visiting professor. He has taught Metals in the Marine Environment at Queen Mary for more than a decade. Professor Rainbow has served as a member of the Natural Environment Research Council (NERC) Marine Science Peer Review Committee, NERC Peer Review College, the Council of the Linnean Society of London, and the Advisory Committee of the Darwin Initiative (DEFRA, UK Government). He has been an editor of the *Journal of Zoology* and is on the editorial boards of *Environmental Pollution, Marine Environmental Research,* and the *Journal of the Marine Biological Association UK*. In 2002, Dr. Rainbow was invited to give the Kenneth Mellanby Review Lecture by the journal *Environmental Pollution* at the Society of Environmental Toxicology and Chemistry annual meeting at Salt Lake City, Utah. He has more than 210 peer-reviewed publications, six co-edited books, and two co-authored books. The first (*Biomonitoring of Trace Aquatic Contaminants* with D.J.H. Phillips) went to two editions. The second, co-authored with Professor Sam Luoma, *Metal Contamination in Aquatic Environments: Science and Lateral Management*, was published in 2008 by Cambridge University Press and has now been issued in paperback. Dr. Rainbow's recent research has focused on the factors affecting the bioavailability of trace metals to aquatic invertebrates from both solution and the diet and the biodynamic modeling of trace metal bioaccumulation.

Contributors

Jean-Claude Amiard
CNRS, Université de Nantes
Mer, Molécule, Santé, EA 2160
Nantes, France

Claude Amiard-Triquet
CNRS, Université de Nantes
Mer, Molécule, Santé, EA 2160
Nantes, France

Franck André Atienzar
Responsable unité de toxicologie *in vitro*
UCB SA
Braine-l'Alleud, Belgium

Doris W. T. Au
Department of Biology and Chemistry
City University of Hong Kong
Kowloon, Hong Kong

Michel Auffret
Institut Universitaire Européen de la Mer
LEMAR UMR CNRS 6539
Plouzané, France

Brigitte Berthet
ICES and Université de Nantes
Mer, Molécule, Santé, EA 2160
Nantes, France

Pauline Brousseau
INRS–Institut Armand-Frappier
Laval, Quebec, Canada

Arnaud Chaumot
IRSTEA - UR "Milieux aquatiques, écologie et pollutions"
Laboratoire D'écotoxicologie
Lyon, France

Michael W. L. Chiang
Department of Biology and Chemistry
City University of Hong Kong
Kowloon, Hong Kong

Tracy K. Collier
Oceans and Human Health, NOAA
Bainbridge Island, Washington

Carole Cossu-Leguille
Université Paul Verlaine de Metz
CNRS UMR 7146 Laboratoire des Interactions Ecotoxicologie, Biodiversité, Ecosystèmes (LIEBE)
Metz, France

Mickaël Couderc
Université de Nantes
Mer, Molécule, Santé, EA 2160
Nantes, France

Odile Dedourge-Geffard
Université Reims Champagne Ardenne
Unité Interactions Animal-Environnement EA4689
UFR Sciences Exactes et Naturelles
Reims, France

Françoise Denis
Université du Maine - Muséum National d'Histoire Naturelle
Département Milieux et Peuplements Aquatiques, UMR 5178 "BOME"
Concarneau, France

Cyril Durou
CEHTRA
Sainte Eulalie, France

Michel Fournier
INRS–Institut Armand-Frappier
Laval, Québec, Canada

Héloïse Frouin
Institute of Ocean Sciences (Fisheries and Oceans Canada)
Sidney, British Columbia, Canada

François Gagné
Section Recherche sur les Écosystèmes Fluviaux
Direction de la Recherche pour la Protection des Écosystèmes Aquatiques
Science et Technologie de l'Eau, Environnement Canada
McGill, Montréal, Québec, Canada

Jeanne Garric
IRSTEA, Laboratory of Ecotoxicology and Biology
Lyon, France

Alain Geffard
Université de Reims Champagne Ardenne
Unité Interactions Animal-Environnement EA 4689
UFR Sciences Exactes et Naturelles
Reims, France

Olivier Geffard
IRSTEA, Laboratory of Ecotoxicology and Biology
Lyon, France

Laure Giambérini
Université Paul Verlaine de Metz
CNRS UMR 7146 Laboratoire des Interactions Ecotoxicologie, Biodiversité, Ecosystèmes (LIEBE)
Metz, France

Matthew J. Gubbins
Marine Scotland Science, Marine Laboratory
Aberdeen, Scotland

Martial Huet
Université de Bretagne Occidentale
Institut Universitaire Européen de la Mer
LEMAR UMR CNRS 6539
Plouzané, France

Jean Laroche
Université de Bretagne Occidentale
Institut Universitaire Européen de la Mer Laboratoire des Sciences de l'Environnement Marin
LEMAR UMR CNRS 6539
Plouzané, France

Sébastien Lemière
Maître de conférences des universités
Université des Sciences et Technologies de Lille
Laboratoire "Ecologie numérique et Ecotoxicologie"
Villeneuve d'Ascq, France

Reinier M. Mann
Hydrobiology, Consulting Company
Auchenflower, Queensland, Australia

Justine Marchand
Université du Maine (Le Mans)
Mer, Molécule, Santé, EA 2160
Le Mans, France

Christophe Minier
Laboratory of Ecotoxicology
University of Le Havre
Le Havre, France

Michael N. Moore
European Centre for Environment and Health
Peninsula College of Medicine and Dentistry
Universities of Exeter and Plymouth
Truro, England

Catherine Mouneyrac
CEREA, Université Catholique de l'Ouest
Université de Nantes, Mer, Molécule, Santé, EA 2160
Nantes, France

Frédéric Palais
Université Reims Champagne Ardenne
Unité Interactions Animal-Environnement EA4689
UFR Sciences Exactes et Naturelles
Reims, France

Alexandre Péry
INERIS, Unité Modèles pour l'Écotoxicologie et la Toxicologie
Verneuil-en-Halatte, France

Stéphane Pillet
Research Institute of the McGill University Health Center
Montreal, Quebec, Canada

Philip S. Rainbow
Department of Zoology
The Natural History Museum
London, England

François Rodius
Maître de Conférences des universités
Université Paul Verlaine Metz
CNRS UMR 7146
Metz, France

Michèle Roméo
Chargée de Recherche INSERM, Université de Nice Sophia-Antipolis
Faculté des Sciences, EA ECOMERS
Nice, France

Susanna Sforzini
Department of Science and Technological Innovation (DiSIT)
University of Piemonte Orientale "A. Avogadro"
Alessandria, Italy

Paul Somerfield
Plymouth Marine Laboratory
Plymouth, England

Paule Vasseur
Université Paul Verlaine de Metz
CNRS UMR 7146 Laboratoire des Interactions Ecotoxicologie, Biodiversité, Ecosystèmes (LIEBE)
Metz, France

Aldo G. Viarengo
Department of Science and Technological Innovation (DiSIT)
University of Piemonte Orientale "A. Avogadro"
Alessandria, Italy

Benoît Xuereb
Laboratory of Ecotoxicology
University of Le Havre
Le Havre, France

1

Introduction

Claude Amiard-Triquet and Jean-Claude Amiard

CONTENTS

Anthropogenic activities are responsible for the environmental input of many classes of chemicals through industrial sources, domestic and urban effluents, and diffuse sources linked to agriculture. The main categories of contaminants include both organic [petroleum hydrocarbons, polychlorobiphenyls (PCBs), pesticides, etc.] and inorganic (metals and nonmetallic elements) compounds. These compounds were studied as soon as ecotoxicology appeared as a specific branch of environmental studies, whereas emerging contaminants have become a topic of concern more recently, even though some of them have been present in the environment for years. Emerging contaminants include pharmaceutical and care products, alkylphenols, brominated flame retardants, perfluorinated organic compounds, and nanoparticles.

Depending on their physical characteristics, three main categories may be distinguished among chemical wastes: solids, liquids, and gases. Each category corresponds to one of the compartments of our physical environment: lithosphere, hydrosphere, atmosphere. However, it is impossible to describe chemicals entering our environment as continental, aquatic, or atmospheric contaminants since many exchanges occur between these compartments. Whatever the point of entrance of a given substance into the environment, an important fraction may be carried over what may be a significant distance as a result of water and air circulation. As a consequence, even polar environments are not spared, and in a charismatic species such as the polar bear, increasing levels of persistent organic pollutants are well documented, with possible ecotoxicological effects at the population level (Letcher et al. 2010).

Even if contaminants are distributed on a worldwide scale, dilution in air or water masses increases with distance from the contamination source. This contamination gradient is the primary factor controlling contaminant uptake into organisms (Figure 1.1). Environmental conditions influence the transformation of many chemicals through chelation, hydrolysis, photodegradation, biodegradation, etc. However, some degradation products of contaminants are not less toxic than the initial molecule, sometimes being even more noxious.

Many toxicants are able to cross biological membranes but these membranes and associated structures can act as barriers to contaminant entry (Figure 1.1). For instance, metal speciation and therefore dissolved metal bioavailability may be modified through ligand secretion into the external medium or by precipitation of dissolved metals as microcrystals of metal sulfides onto the cell surface. Secretion of exudates by a variety of organisms (bacteria, plants, animals) can involve a great variety of compounds. Subtle changes in the charge and types of reactive groups in such secretions can interfere markedly with

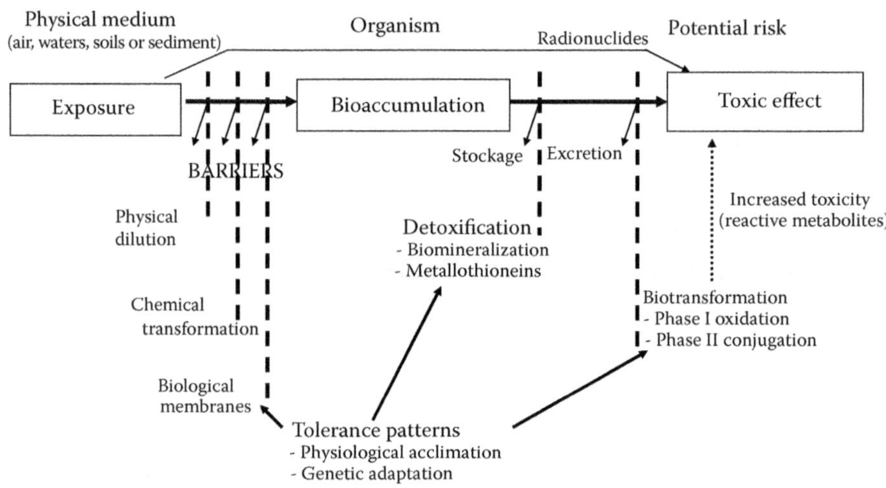

FIGURE 1.1
The ecotoxicology triad.

their metal binding characteristics and consequently the biological uptake of the metal. Another mechanism of limiting contaminant uptake is the existence of impervious extracellular barriers such as cuticles, integuments, tests, shells, and scales that contribute to reduce the cell epithelial surface available to contribute to transepithelial transport (for details, see Mason and Jenkins 1995).

Once incorporated into an organism (Figure 1.1), contaminants can be either stored in tissues or excreted. Storage in intra- or extracellular compartments does not necessarily result in a toxic effect in organisms. For instance, metal detoxification is efficient in numerous organisms. It may be based on the synthesis of metallothioneins (MTs), a family of metalloproteins able to sequester metals via metal binding to their constituent thiol groups, thus blocking any interference between the metals and enzymes that would otherwise result in subsequent enzymatic activity impairments. MT induction is the most common toxic metal defense mechanism in vertebrates. It is also present in most biological taxa (Amiard et al. 2006), but among invertebrates, the major mode of metal detoxification is metal biomineralization in various types of cellular inclusions (Marigomez et al. 2002). It is only when the metal-binding capacity of these ligands is overwhelmed that metal toxicity can occur.

On the contrary, processes responsible for excretion are not systematically free of noxious effects on organisms. Biotransformation of certain organic pollutants [polycyclic aromatic hydrocarbons (PAHs), PCBs] is organized into two phases (Figure 1.1). Phase I reactions consist of oxidation, reduction, and hydrolysis processes. Phase II enzymes serve to link metabolites from phase I with endogenous substrates, increasing their water solubility and thereby facilitating their excretion. However, phase II biotransformation sometimes leads to reactive metabolites, the interactions of which with cellular macromolecules can engender toxicity (Roméo and Wirgin in Amiard-Triquet et al. 2011). Biotransformation is followed by phase III leading to the elimination of metabolites by the multixenobiotic transport system (Damiens and Minier in Amiard-Triquet et al. 2011).

The activity of biotransformation enzymes (such as cytochrome P450 enzymes, including ethoxyresorufin *O*-deethylase involved in phase I; glutathione *S*-transferase involved in phase II) or MT concentrations are some examples of biomarkers that have been proposed to assess the exposure of organisms to contaminants present in their environment (Chapter 2). In addition to inducing MT synthesis or activating cytochrome P450 enzymes, metals, PCBs, and PAHs can increase oxidative stress by increasing the concentrations of reactive oxygen species naturally present in organisms. Cytotoxicity can occur, including lipid peroxidation and DNA damage, but the degree of such damage depends on the efficiency of enzymatic (superoxide dismutase, catalase, glutathione peroxidase, etc.) and nonenzymatic defenses. If DNA damage induced by metabolites resulting from contaminant biotransformation is not adequately repaired by specialized nuclear enzymes, this can lead to an erroneous expression of the genome, including the activation of oncogenes, which constitutes the first step of the transformation of a normal cell in a tumoral cell (Newman and Clements 2008).

As an indicator of neurotoxicity effects, acetylcholinesterase (AChE) activity has been initially considered a specific biomarker of exposure to organophosphate and carbamate pesticides. More recently, however, other groups of chemicals present in the marine environment including metals, detergents, hydrocarbons, and also cyanobacterium toxins have been shown to affect AChE activity (Table 4.1).

This lack of biomarker specificity poses a problem for environmental management. Although biomarkers are able to reveal the presence of contaminants, and subsequent changes in the biology of organisms, any lack of specificity in their response reduces the likelihood of precise targeting of a particular contaminant, thereby affecting management decisions to reduce contamination and its impacts. To date, only a few biomarkers seem really specific: δ-amino levulinic acid dehydratase inhibition in blood able to reveal lead contamination, bile fluorescent compounds for petroleum hydrocarbons (Anderson and Lee 2006), and imposex in gastropod mollusks in response to TBT contamination (Chapter 9). However, less specific biomarkers are also interesting environmental management tools as general responses to the degradation of environmental conditions, and they are still important in assessing the health status of a given medium exposed to chronic or acute (e.g., oil spill) pollution pressure. Among these biomarkers, stress proteins, which contribute to cellular protection and are highly conserved throughout evolution from bacteria to humans, can provide information on a large spectrum of environmental stress (Newman and Clements 2008). Histological alterations generally result from the integration of biochemical and physiological changes that may be caused by various chemical contaminants (Newman and Clements 2008). Until now, no immune response specific for a given contaminant has been described, but this category of biomarkers is useful in detecting effects linked to simultaneous exposure to multiple contaminants (Fournier et al. 2005). Lastly, a variety of nonspecific biomarkers are important because they are involved in growth and development and contribute to the success of reproduction with possible ecological consequences on population sustainability and ecosystem functioning when key species are impacted. To aggregate the benefit of specific, less specific, and general biomarkers, it is generally recommended to date to use biomarkers in a battery for ecological risk assessment, as recommended, for instance, by Anderson and Lee (2006) and Thain et al. (2008) in oil spill risk assessment (Chapter 2).

Classically, biomarkers have been classified as biomarkers of exposure, effect, and susceptibility (Manahan 2003). However, the definitions of these classes vary depending on different authors (Chapter 2). So, certain ecotoxicologists prefer the terminology proposed by De Lafontaine et al. (2000), contrasting biomarkers of defense (Chapter 3) and biomarkers of damage (Chapters 4–6).

Biomarkers of defense include MTs, phase I, II, and III enzymes evoked above, as well as antioxidant defenses (Regoli et al. in Amiard-Triquet et al. 2011) and stress proteins

(Mouneyrac and Roméo in Amiard-Triquet et al. 2011). These defense mechanisms have a positive impact on the health of biota, allowing the survival of organisms in a degraded environment. In highly contaminated zones, many plant and animal species are indeed able to cope with the presence of potentially toxic substances (Amiard-Triquet et al. 2011). On the other hand, development of tolerance through physiological acclimation and genetic adaptation can induce energy and fitness costs (Mouneyrac et al. in Amiard-Triquet et al. 2011).

Biomarkers of damage reveal more or less severe biological impairments, potentially responsible for detrimental effects on reproduction or even survival. The importance of toxic effects depending on the degree of environmental contamination is quantified using a dose–effect relationship. The lowest doses do not induce any noxious effect, but with increasing doses biological impairments are progressively enhanced. The theoretical dose–effect relationship is depicted in Figure 1.2 for different levels of biological organization. The curve is limited to the domain of low doses to show the first observed effects or initial effects. At the molecular level, the initial effect is observed at a dose X_1 that is lower than the dose X_2 able to induce a cellular effect, this in turn being lower than X_3, acting at the tissue level. The same argument can be expanded to the level of organs, individuals, populations, etc. This scheme highlights that the lower the level of biological organization, the more sensitive the biological response will be. The rationale for this is quite evident: if only a few molecules have suffered a toxicant effect, cell functioning will not be significantly disturbed; if only a few cells are no longer functional within a whole organ, the function of this organ will still be efficient.

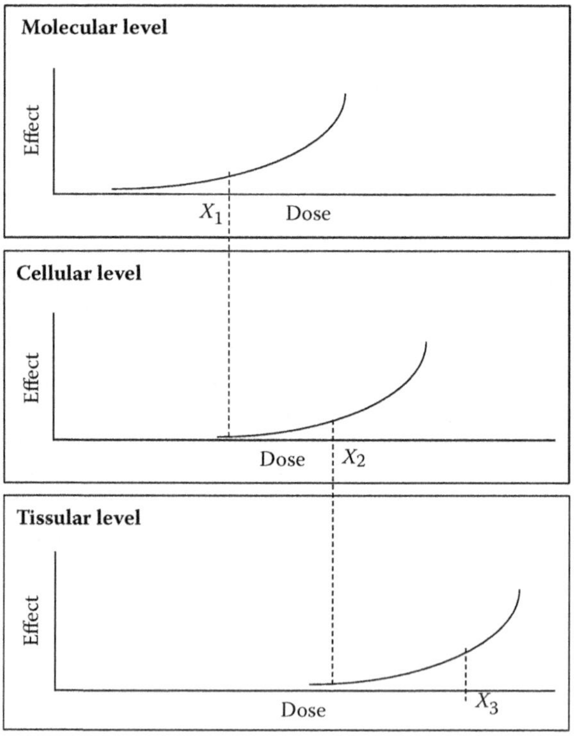

FIGURE 1.2
Biomarkers of damage: progression of the dose–effect relationship according to the level of biological organization.

Because responses of biomarkers of damage at the lowest levels of biological organization are so sensitive, they would appear to have the potential to be particularly useful in a management scheme to prevent any pollution effect. However, because organisms have very efficient mechanisms of regulation and repair, the use of such low level biomarkers brings with it a serious risk of a false positive if they are used as a warning signal for impairments at the level of communities or ecosystems. This is even more true for biomarkers of defense since this type of biological response shows that the organisms are coping actively with environmental degradation.

To put more ECO into ECOtoxicology, Chapman (2002) recommends the use of biological models more representative of the communities or ecosystems under examination than organisms classically used in biomonitoring programs or laboratory tests. It is generally admitted that protecting the most sensitive species within an ecosystem results in the protection of the whole community. This notion of susceptibility is not so simple. Reproduction and development of juveniles are commonly used as endpoints when assessing interspecific susceptibility to chronic toxicity, because these life traits are considered equally relevant in all species. This hypothesis was tested in two nematode species exposed to copper (Kammenga and Riksen 1996). Despite juvenile survival, duration of juvenile and reproduction periods, and daily reproduction rate being more affected in one species, fitness (which was defined by these authors as the population growth rate) was identically reduced in both species.

Species most commonly used as biological models in ecotoxicology are representative of the water column, whereas it is well established that sediments and soils are the main stores for a large majority of contaminants entering the environment. The choice of the most relevant species for the determination of biomarkers will be discussed in Chapter 7, considering the different objectives of conservation programs: ecosystem functioning, biodiversity integrity, survival of charismatic species, etc.

Responses to pollutants at different levels of biological organization are depicted in Figure 1.3 in the case of fish, considering the latency between exposure and the occurrence of the effect on the *X* axis, and the degree of ecological relevance on the *Y* axis. Molecular effects that are the most sensitive (Figure 1.3) are also the most precocious. On the other hand, they are mainly toxicological tools for which ecological relevance is poor. In contrast, population or community responses are obviously relevant to assess the "good ecological status" or "ecological integrity" of water masses [United States' Clean Water Act (CWA), 1972; European Community Water Framework Directive (WFD), 2000], but effects at these levels become significant only after severe environmental degradation has already occurred, thus leading to expensive remediation processes.

An extreme case provides a striking illustration of the magnitude of remediation problems: the experiences of the Minamata Bay project in Japan (Hosokawa 1993). A chemical factory released mercury into this bay from 1932 to 1968, leading to the death of 900 people among more than 2000 affected patients as a result of seafood contamination. The remediation project commenced in 1977 and was completed in 1990 after 1.5 million m^3 of Hg-contaminated sediment had been treated by careful dredging and confined reclamation at a total cost of 48,500 millions yen (equivalent to 650 millions).

Is it possible to reconcile the benefits of biochemical markers and ecological responses? It may be seen in Figure 1.3 that processes involved in reproduction include a set of responses from the molecular level leading to consequences of reproductive success on the sustainability of populations in ecosystems impacted by anthropogenic activities. Although it is excessive to consider that the pursuit of toxicological endpoints other than those concerned with reproduction is likely to be a wasted effort (Tannenbaum 2005), it is

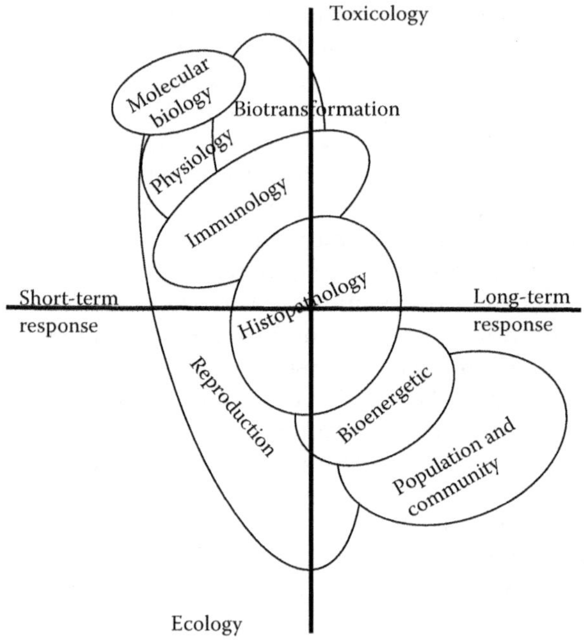

FIGURE 1.3
Latency between exposure of fish to pollutants and the occurrence of effects at different levels of biological organization. (After Adams, S.M. et al., *Mar. Environ. Res.*, 28, 459–464, 1989.)

evident that reproductive success is key for environmental conservation. The impairments at infra-individual and individual levels that can most probably affect the success of reproduction are depicted in Figure 1.4. These include endocrine disruption (Chapters 8 and 9), behavioral changes (Chapter 10), energy disturbances (Chapters 11 and 12), and genetic responses either adaptive or detrimental (Chapters 13 and 14).

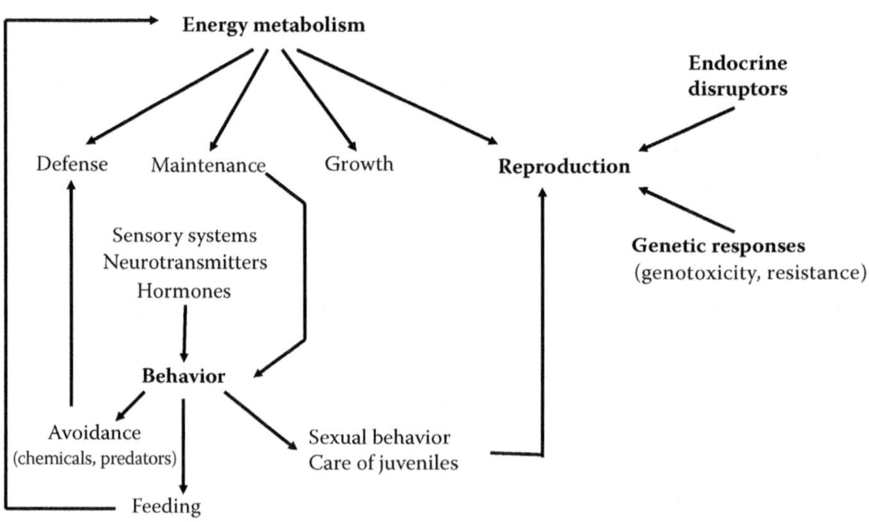

FIGURE 1.4
Linkage between effects of contaminants from molecular to population levels via the success of reproduction.

The problem of endocrine disruption was first realized because of the disastrous ecotoxicological effect of tributyltin (TBT), a compound used in antifouling paints. TBT-mediated imposex (for details, see Chapter 9) has been observed in more than 195 species of prosobranch gastropods worldwide (Sternberg et al. 2010). Subsequent population depletion of such gastropods has been observed in harbors and marinas where many individual snails were presenting morphological symptoms of imposex. In the case of the dogwhelk *Nucella lapillus*, population-level effects on other species (barnacles, fucoid seaweeds, hermit crabs) belonging to the same ecological community would be attributable to such a population drop in the affected gastropods (Bryan and Gibbs 1991).

Endocrine glands and the hormones they secrete are not only indispensable to the success of reproduction but are also involved in the development of organisms, their growth, and their behavior. However, most scientific research, particularly in fish, focuses on interactions between pollutants and male and female sexual hormones (Chapters 8 and 9). A peculiar topic of concern is that the effects of endocrine disruptors on reproduction are typically subtle, occurring at low doses, in the absence of any other appearance of toxicity. The spatial distribution of endocrine-disrupting chemicals, particularly steroid estrogens and nonylphenols, is related to the discharge of domestic and industrial wastewaters everywhere in the world (Jugan et al. 2009; Bertin et al. 2011; Gong et al. 2011; Tetreault et al. 2011). The presence of intersex (male gonads invaded with oocytes) individuals is increasingly documented in bivalves and fish. Natural or xenoestrogens could be a contributory factor in the induction of intersex (Baroiller and D'Cotta 2001; Langston et al. 2007). However, it is still unclear if intersex can have consequences on the production of progeny (Chapters 8 and 9).

A wide variety of anthropogenic, waterborne contaminants can also affect the hypothalamic–pituitary–thyroid axis and its role in development and reproduction as recently reviewed in teleost fish and amphibians (Blanton and Specker 2007; Carr and Patiño 2011). Impairment of thyroid functioning can influence behavior as neurotoxic effects such as the inhibition of neurotransmitters (AChE, serotonin) have also been observed (Figure 1.4). Many aspects of behavior can be affected (Dell'Omo 2002; Amiard-Triquet 2009; Hellou 2011): avoidance of predators or contaminated sediment or other habitat, contributing to the defense and survival of organisms; location of sexual partners and care of juveniles indispensable to reproductive success; feeding behavior and prey capture important for acquiring energy. Thus, behavioral ecotoxicology is potentially useful to link biochemical impairments to population effects (Chapter 10).

The success of reproduction is clearly linked to the relative energy allocation of an organism to defense against exposure to chemical stressors, basal metabolism, growth, and reproduction. Organisms obtain their energy from ingested food. For predators, the impairment of foraging activity can lead to a shift toward easily accessible food such as detritus, the energy value of which may be lower. Chemical contaminants can also influence food assimilation through the impairment of digestive enzyme activity. Lastly, prey species can be susceptible to environmental contamination, thus leading to decreased food availability for predators (Chapter 11).

Energy analysis can reveal a disequilibrium in energy balance associated with toxic or more general stress. Different energy parameters can be used as biomarkers of pollutant effects (Chapter 12). These parameters can be linked to macroscopic criteria representative of maintenance and growth (condition indices, size, or biomass increase, etc.) or reproduction (gonadosomatic index, egg production, offspring number, etc.). For ecological risk assessment, it is necessary to determine to what extent populations may be affected when such adverse effects are revealed (loss of their ecosystem function or even local extinction). Models that can allow extrapolation from individual- and suborganismal-level responses to the population level have been reviewed (Maltby et al. 2001). Among those, dynamic energy budget models combined with demographic models have been well developed (Charles et al. 2009).

Exposure to chemicals can lead to DNA damage (Figure 1.5), the consequences of which may be limited by DNA repair (Peterson and Côté 2004). Mutations frequently have toxic effects, including carcinogenesis, and when affecting germinal tissues, they are inheritable and can also affect future generations, provided that the offspring are viable and able to survive and reproduce. In fact, impairments of germinal cells often result in embryo lethality or early death of the progeny. From an ecological point of view, it is questionable if these precocious deaths can impact the fate of populations (Manahan 2003; Newman and Clements 2008). In some cases, mutations can confer a selective advantage leading to the selection of resistant genotypes. Biomarkers of exposure to genotoxic pollutants are reviewed in Chapter 13, and Vasseur et al. explore the relationships between genotoxicity and population effects.

Chronic exposure to chemicals can exert a selection pressure leading to the presence of resistant genotypes in organisms living in impacted areas. The acquisition of tolerance is particularly well documented for pesticide-exposed insects (Hemingway et al. 2004), but other classes of contaminants (metals, PAHs, PCBs) can be responsible for the predominance of resistant genetic patterns in bacteria (Nies 1999), plants (Frérot et al. in Amiard-Triquet et al. 2011), invertebrates (Nevo et al. 1984), and vertebrates (Athrey et al.

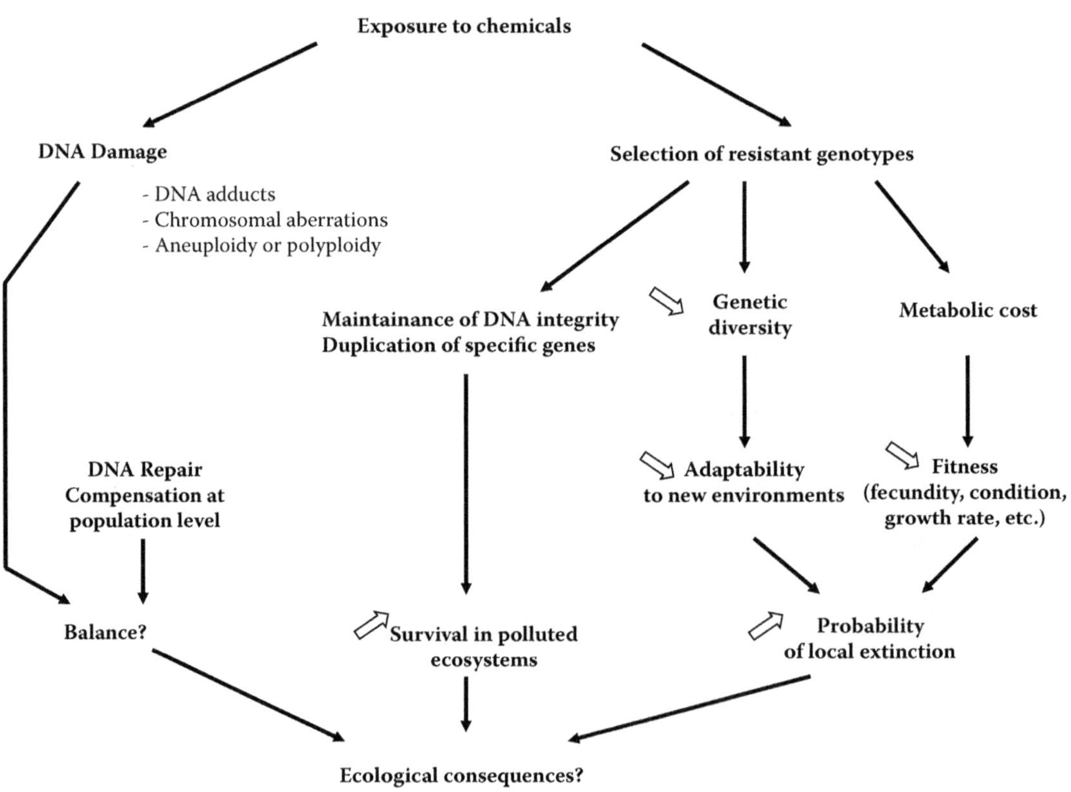

FIGURE 1.5
Genetic responses to chemical exposure: DNA damage versus selection of resistant genotypes.

2007; Romeo and Wirgin in Amiard-Triquet et al. 2011). In contaminated areas, an increased frequency of resistant genotypes has often been reported, allowing the maintenance of DNA integrity associated with the duplication of specific genes (Figure 1.5). However, negative consequences of being resistant may be observed, such as decreased fitness and decreased adaptability to new environments or stressors, thus increasing the probability of local extinction (Chapter 14).

Biomarkers are available as crucial tools in ecotoxicology, because they can be used as early warning signals of environmental change before the onset of irreversible damage at the population level. Syntheses published at the turn of the century (Lagadic et al. 1997, 1998; Garrigues et al. 2001) suggested that scientists were then ready to transfer the methodology of biomarkers to end users in charge of environmental biomonitoring. A decade later, certain biomarkers are used to assess the health status of aquatic environments in different parts of the world (Chapter 15). However, this use is generally limited to a relatively small number of more or less specific biomarkers, the worst counterexample being the WFD—a very important regulation aiming at the protection of aquatic environments from the river source to the seashore—which totally ignores the use of biomarkers despite the efforts of European scientists to demonstrate the relevance of biomarkers as tools for the implementation of the WFD (Allan et al. 2006; Hagger et al. 2008; Sanchez and Porcher 2009). Independently of regulatory frameworks, many important studies have demonstrated "the usefulness of applying a large array of various combined biomarkers at different levels of biological organization, in assessing the toxic effects of a mixture of pollutants in a natural aquatic environment" (Huadi River, a tributary of the Pearl River, China) (He et al. 2011). In the Bay of Cadiz, biomarkers determined in caged clams *Ruditapes philippinarum* allowed assessment of chemical exposure and sediment quality (Ramos-Gómez et al. 2011). In the Río Champotón (southwestern Mexico), a set of biomarkers determined in a native fish *Astyanax aeneus* was shown to be a sensitive and effective tool for identifying periods of environmental conditions adverse to fish health (Trujillo-Jiménez et al. 2011).

Several problems contributing to limit the use of biomarkers have been recognized: the problem of confounding factors (e.g., Thain et al. 2008; Martínez-Gómez et al. 2010), the question of a reference site, and the lack of ecological relevance (Forbes et al. 2006). The problem of confounding factors was well conceptualized by Cairns (1992). When a biological parameter is highly fluctuating, the occurrence of a stress may be concealed by natural fluctuations. On the other hand, when background values are relatively stable, any change due to contamination factors is easily revealed (Figure 1.6). As already mentioned by Kalman et al. (2010), "The question of confounding factors is well mastered in biomonitoring programs based on the determination of contaminants in the tissues of bioaccumulators such as the bivalves used in the 'Mussel Watch'–type programs." The literature indicates that the same natural factors are at work in the case of biomarkers (Thain et al. 2008). Consequently, in the objective of using a peculiar species as a model for the determination of biomarkers, it is still indispensable to determine the natural fluctuations, as exemplified for worms (Kalman et al. 2010), bivalves (Burgeot et al. 2010; Fossi Tankoua et al. 2011), and fish (Sanchez et al. 2008). Temporal surveys provide significant advantages over spot sampling techniques, allowing the assessment of pollution trends responsible for population changes while providing data on background levels that would be of great use in case of a future accident, as often experienced for oil spills (Martínez-Gómez et al. 2010).

For many aspects of environmental monitoring, our present state of knowledge and the insufficiency of background data available mean that the use of a reference site for

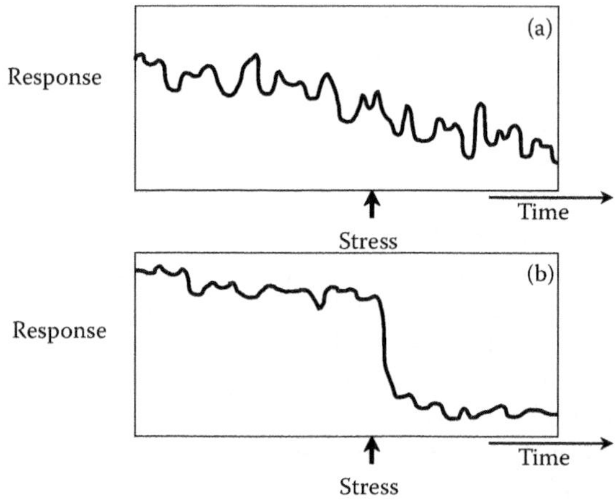

FIGURE 1.6
Relative importance of natural fluctuations of a biomarker response compared to stress-induced response. (a) Highly variable background masking stress response. (b) Background relatively stable allowing significant variation due to stress. (After Cairns, J. Jr., *Ecotoxicology*, 1, 3–16, 1992.)

comparison is essential. However, to date, with the worldwide dispersion of contaminants evoked above, pristine areas have disappeared and, at best, reference sites can be chosen in only a few places that remain comparatively clean. To choose a reference site, geographical proximity and similarity in terms of temperature, granulometry, and organic content of sediment, salinity regime (in estuaries), etc., are mandatory to mitigate the importance of confounding factors. This is not an easy task, as described, for instance, in estuaries (Amiard-Triquet and Rainbow 2009). Potential reference estuaries with low perceived anthropogenic pressure are generally small, whereas the human activities responsible for the presence of many chemicals in the environment have historically developed on the banks of larger main watercourses. This does provide a potential problem when trying to eliminate comparative differences resulting from hydrodynamic differences between the estuaries under comparison. Even in the less fluctuating conditions of a freshwater biomonitoring program, the interpretation of fish biomarker results is strongly influenced by the selected reference system (Sanchez et al. 2010).

The addition of more than one reference site into any comparative study, however superficially attractive, has significant resource implications. Associated with the need for temporal surveys instead of spot sampling techniques and the development of the need to analyze a battery of biomarkers (Chapter 2), methodology involving biomarkers is not always as initially claimed: sensitive, simple, and cost-effective. Even despite this complexification, the biomarker methodology to be proposed to end users—although efficient in assessing chemical exposure, sediment quality, and the toxic effects of mixed pollutants—still fails at predicting chemical risk at supra-individual levels (Forbes et al. 2006). The development of an integrated indicator framework using biological effect techniques remains key to improve the risk assessment of contaminants in aquatic ecosystems (Thain et al. 2008).

Since pioneering papers (Atrill and Depledge 1997; Clements 2000) underlined the importance of targeting links between levels of biological integration, certain research groups have focused their attention on the cascading effects of interrelated biomarkers that can be linked to important biological processes and for which changes can be interpreted (Amiard-Triquet and Rainbow 2009; Ankley et al. 2010; Taylor and Maher 2010; Mouneyrac and Amiard-Triquet, accepted). Ecologically relevant biomarkers such as lysosomal integrity (Chapter 5), immunotoxicity (Chapter 6), endocrine disruption (Chapters 8 and 9), behavior (Chapter 10), energy metabolism (Chapters 11, 12), and genomic biomarkers (Chapters 13, 14) appear to be promising candidates to fill the gap existing between suborganismal and organismal responses to stress and effects occurring at higher levels of biological organization.

The main objective of the present book is to review biomarker research that examines the effects of contaminants using an integrative approach. In order to improve the predictive value of biomarkers, special attention will be devoted to biological responses that can be observed at infra-individual or individual levels (early and sensitive warning signals) but have a serious potential to reveal threats at supra-individual levels (population, community, ecosystem). For each category of biomarkers (biochemical, physiological, behavioral, etc.), their usefulness for predictive (e.g., effects of different nanoparticles in aquatic organisms, Koelher et al. 2008; Li et al. 2009; Galloway et al. 2010; Ringwood et al. 2010; Tedesco et al. 2010; Buffet et al. 2011) or retrospective (e.g., adverse effects of pharmaceuticals in wild fish; Sanchez et al. 2011) risk assessment of emerging contaminants will be considered. The final aim is to contribute to the search for a conceptual framework to support the assessment of the health status of aquatic ecosystems.

References

Adams, S.M. et al. 1989. The use of bioindicators for assessing the effects of pollutant stress on fish. *Mar. Environ. Res.* 28: 459–464.

Allan, I.J. et al. 2006. A "toolbox" for biological and chemical monitoring requirements for the European Union's Water Framework Directive. *Talanta* 69: 302–322.

Amiard, J.C. et al. 2006. Metallothioneins in aquatic invertebrates: Their role in metal detoxification and their use as biomarkers. *Aquat. Toxicol.* 76: 160–202.

Amiard-Triquet, C. 2009. Behavioral disturbances: The missing link between sub-organismal and supra-organismal responses to stress? Prospects based on aquatic research. *Hum. Ecol. Risk Assess. Behav. Ecotoxicol.* (Special issue) 15: 87–110.

Amiard-Triquet, C., and P.S. Rainbow. 2009. Conclusions. In *Environmental Assessment of Estuarine Ecosystems. A Case Study*, ed. C. Amiard-Triquet and P.S. Rainbow, 323–348. Boca Raton, FL: CRC Press.

Amiard-Triquet, C., P.S. Rainbow, and M. Roméo. 2011. *Tolerance to Environmental Contaminants*. Boca Raton, FL: CRC Press.

Anderson, J.W., and R.F. Lee. 2006. Use of biomarkers in oil spill risk assessment in the marine environment. *Human Ecol. Risk Assess.* 12: 1192–1222.

Ankley, G.T. et al. 2010. Adverse outcome pathways: A conceptual framework to support ecotoxicological research and risk assessment. *Environ. Toxicol. Chem.* 29: 730–741.

Athrey, N.R.G., P.L. Leberg, and P.L. Klerks. 2007. Laboratory culturing and selection for increased resistance to cadmium reduce genetic variation in the least killifish, *Heterandria formosa*. *Environ. Toxicol. Chem.* 26: 1916–1921.

Atrill, M.J., and M.H. Depledge. 1997. Community and population indicators of ecosystem health: Targeting links between levels of biological organization. *Aquat. Toxicol.* 38: 183–197.

Baroiller, J., and H. D'Cotta. 2001. Environment and sex determination in farmed fish. *Comp. Biochem. Physiol.* 130C: 399–409.

Bertin, A., P.A. Inostroza, and R. Quinones. 2011. Estrogen pollution in a highly productive ecosystem off central-south Chile. *Mar. Pollut. Bull.* 62: 1530–1537.

Blanton, M.L., and J.L. Specker. 2007. The hypothalamus–pituitary–thyroid (HPT) axis in fish and its role in fish development and reproduction. *Crit. Rev. Toxicol.* 37: 97–115.

Bryan, G.W., and P.E. Gibbs. 1991. Impact of low concentrations of tributyltin (TBT) on marine organisms: A review. In *Metal ecotoxicology, Concepts and Applications*, ed. M. Newman and A.W. McIntosh, 323–361. Chelsea, MI: Lewis Publishers.

Buffet et al. 2011. Behavioural and biochemical responses of two marine invertebrates *Scrobicularia plana* and *Hediste diversicolor* to copper oxide nanoparticles. *Chemosphere* 84: 166–174.

Burgeot, T. et al. 2010. Acetylcholinesterase: Methodology development of a biomarker and challenges of its application for biomonitoring. ICES CM 2010/F: 25.

Cairns, J. Jr. 1992. The threshold problem in ecotoxicology. *Ecotoxicology* 1: 3–16.

Carr, J.A., and R. Patiño. 2011. The hypothalamus–pituitary–thyroid axis in teleosts and amphibians: Endocrine disruption and its consequences to natural populations. *Gen. Comp. Endocr.* 170: 299–312.

Chapman, P.M. 2002. Integrating toxicology and ecology: Putting the "eco" into ecotoxicology. *Mar. Pollut. Bull.* 44: 7–15.

Charles, S. et al. 2009. Matrix population models as relevant modelling tools in ecotoxicology. In *Ecotoxicology Modelling*, ed. J. Devillers, 261–298. Dordrecht: Springer.

Clements, W.H. 2000. Integrating effects of contaminants across levels of biological organization: An overview. *J. Aquat. Ecosyst. Stress Recov.* 7: 113–116.

De Lafontaine, Y. et al. 2000. Biomarkers in zebra mussels (*Dreissena polymorpha*) for the assessment and monitoring of water quality of the St Lawrence River (Canada). *Aquat. Toxicol.* 50: 51–71.

Dell'Omo, G. 2002. *Behavioural Ecotoxicology*. Chichester, UK: Wiley.

Forbes, V.E., A. Plamqvist, and L. Bach. 2006. The use and misuse of biomarkers in ecotoxicology. *Environ. Toxicol. Chem.* 25: 272–280.

Fossi Tankoua, O. et al. 2011. Potential influence of confounding factors (size, salinity) on biomarker tools in the sentinel species *Scrobicularia plana* used in monitoring programmes of estuarine quality. *Environ. Sci. Pollut. Res.* 18: 1253–1263.

Fournier, M. et al. 2005. Biomarqueurs immunologiques appliqués à l'écotoxicologie. *Bull. Soc. Zool. Fr.* 130: 333–351.

Galloway, T. et al. 2010. Sublethal toxicity of nano-titanium dioxide and carbon nanotubes in a sediment dwelling marine polychaete. *Environ. Pollut.* 158: 1748–1755.

Garrigues, P. et al. 2001 *Biomarkers in Marine Organisms: A Practical Approach*. Amsterdam: Elsevier Science.

Gong, J. et al. 2011. Occurrence of endocrine-disrupting chemicals in riverine sediments from the Pearl River Delta, China. *Mar. Pollut. Bull.* 63: 556–563.

Hagger, J.A. et al. 2008. Application of biomarkers for improving risk assessments of chemicals under the Water Framework Directive: A case study. *Mar. Pollut. Bull.* 56: 1111–1118.

He, X. et al. 2011. Assessment of typical pollutants in waterborne by combining active biomonitoring and integrated biomarkers response. *Chemosphere* 84: 1422–1431.

Hellou, J. 2011. Behavioural ecotoxicology, an "early warning" signal to assess environmental quality. *Environ. Sci. Pollut. Res.* 18: 1–11.

Hemingway, J. et al. 2004. The molecular basis of insecticide resistance in mosquitoes. *Insect Biochem. Mol. Biol.* 34: 653–665.

Hosokawa, Y. 1993. Remediation work for mercury contaminated bay – experiences of Minamata bay project, Japan. *Water Sci. Technol.* 28: 339–348.

Jugan, M.L. et al. 2009. *In vitro* assessment of thyroid and estrogenic endocrine disruptors in wastewater treatment plants, rivers and drinking water supplies in the greater Paris area (France). *Sci. Total Environ.* 407: 3579–3587.

Kalman, J. et al. 2010. Assessment of the influence of confounding factors (weight, salinity) on the response of biomarkers in the estuarine polychaete *Nereis diversicolor*. *Biomarkers* 15: 462–469.

Kammenga, J.E., and J.A.G. Riksen. 1996. Comparing differences in species sensitivity to toxicants: Phenotypic plasticity versus concentration–response relationships. *Environ. Toxicol. Chem.* 15: 1649–1653.
Koelher, A. et al. 2008. Effects of nanoparticles in *Mytilus edulis* gills and hepatopancreas—A new threat to marine life? *Mar. Environ. Res.* 66: 12–14
Lagadic, L. et al. 1997. *Biomarqueurs en écotoxicologie. Aspects fondamentaux.* Paris: Masson.
Lagadic, L. et al. 1998. *Utilisation de biomarqueurs pour la surveillance de la qualité de l'environnement.* Paris: Lavoisier Tec & Doc.
Langston, W.J., G.R. Burt, and B.S. Chesman. 2007. Feminisation of male clams *Scrobicularia plana* from estuaries in Southwest UK and its induction by endocrine-disrupting chemicals. *Mar. Ecol. Prog. Ser.* 333: 173–184.
Letcher, R.J. et al. 2010. Exposure and effects assessment of persistent organohalogen contaminants in arctic wildlife and fish. *Sci. Total Environ.* 408: 2995–3043.
Li, H. et al. 2009. Effects of waterborne nano-iron on medaka (*Oryzias latipes*): Antioxidant enzymatic activity, lipid peroxidation and histopathology. *Ecotox. Environ. Saf.* 72: 684–692.
Maltby, L. et al. 2001. Linking individual-level responses and population-level consequences. In *Ecological variability: Separating Natural from Anthropogenic Causes of Ecosystem Impairment*, ed. D.J. Baird and G.A. Burton, 27–82. Pensacola, FL: Society of Environmental Toxicology and Chemistry (SETAC).
Manahan, S.E. 2003. *Toxicological Chemistry and Biochemistry*. Boca Raton, FL: Lewis Publishers.
Marigomez, I. et al. 2002. Cellular and subcellular distribution of metals in molluscs. *Microsc. Res. Technol.* 56: 358–392.
Martínez-Gómez, C. et al. 2010. A guide to toxicity assessment and monitoring effects at lower levels of biological organization following marine oil spills in European waters. *ICES J. Mar. Sci.* 67: 1105–1118.
Mason, A.Z., and J.D. Jenkins. 1995. Metal detoxification in aquatic organisms. In *Metal Speciation and Bioavailability in Aquatic Systems*, ed. A. Tessier and D.R. Turner, 479–608. Chichester: Wiley.
Mouneyrac, C., and C. Amiard-Triquet. 2011. Biomarkers of ecological relevance. In *Encyclopedia of Aquatic Ecotoxicology*, ed. C. Blaise and J.F. Férard. Berlin: Springer (accepted).
Nevo, E., R. Ben-Shlomo, and B. Lavie. 1984. Mercury selection of allozymes in marine organisms: Predictions and verification in nature. *Proc. Natl. Acad. Sci. U. S. A.* 81: 1258–1259.
Newman, M.C., and W.H. Clements. 2008. *Ecotoxicology. A Comprehensive Treatment*. Boca Raton, FL: CRC Press.
Nies, D.H. 1999. Microbial heavy-metal resistance. *Appl. Microbiol. Biotechnol.* 51: 730–750.
Peterson, C.L., and J. Côté. 2004. Cellular machineries for chromosomal DNA repair. *Genes Dev.* 18: 602–616.
Ramos-Gómez, J. et al. 2011. Biomarker responsiveness in different tissues of caged *Ruditapes philippinarum* and its use within an integrated sediment quality assessment. *Environ. Pollut.* 159: 1914–1922.
Ringwood, A.H. et al. 2010. The effects of silver nanoparticles on oyster embryos. *Mar. Environ. Res.* 69: S49–S51.
Sanchez, W, and J.M. Porcher. 2009. Fish biomarkers for environmental monitoring within the Water Framework Directive of the European Union. *TRAC-Trends Anal. Chem.* 28: 150–158.
Sanchez, W. et al. 2008. Assessment of seasonal variability of biomarkers in three-spined stickleback (*Gasterosteus aculeatus* L.) from a low contaminated stream: Implication for environmental biomonitoring. *Environ. Int.* 34: 791–798.
Sanchez, W. et al. 2010. Comparison of two reference systems for biomarker data analysis in a freshwater biomonitoring context. *Environ. Int.* 36: 377–382.
Sanchez, W. et al. 2011. Adverse effects in wild fish living downstream from pharmaceutical manufacture discharges. *Environ. Int.* 37: 1342–1348.
Sternberg, R.M. et al. 2010. Environmental-endocrine control of reproduction in gastropods: Implications for the mechanisms of tributyl-induced imposex in prosobranchs. *Ecotoxicology* 19: 4–23.
Tannenbaum, L.V. 2005. A critical assessment of the ecological risk assessment process: A review of misapplied concepts. *Integr. Environ. Assess. Manag.* 1: 66–72.

Taylor, A.M., and W.A. Maher. 2010. Establishing metal exposure–dose–response relationships in marine organisms: Illustrated with a case study of cadmium toxicity in *Tellina deltoidalis*. In: *New Oceanography Research Developments: Marine Chemistry, Ocean Floor Analyses and Marine Phytoplankton*, ed. L. Martorino and K. Puopolo, 1–57. Hauppayge, NY: Nova Science Publ.

Tedesco, S. et al. 2010. Oxidative stress and toxicity of gold nanoparticles in *Mytilus edulis*. *Aquat. Toxicol.* 100: 178–186.

Tetreault, G.R. et al. 2011. Intersex and reproductive impairment of wild fish exposed to multiple municipal wastewater discharges. *Aquat. Toxicol.* 104: 278–290.

Thain, J. E., A.D. Vethaak, and K. Hylland. 2008. Contaminants in marine ecosystems: Developing an integrated indicator framework using biological-effect techniques. *ICES J. Mar. Sci.* 65: 1508–1514.

Trujillo-Jiménez, P. et al. 2011. Assessing environmental conditions of the Río Champotón (México) using diverse indices and biomarkers in the fish *Astyanax aeneus* (Günther, 1860). *Ecol. Indic.* 11: 1636–1646.

2

History of Biomarkers

Michèle Roméo and Laure Giambérini

CONTENTS

2.1 Context

Although knowledge of the existence of a link between biological dysfunction and the environment is very old, as testified by writings dating from more than 2000 years ago (Hippocrates, translated by Littré 1861), serious consideration of pollution by both society and scientists is a more recent phenomenon. Rachel Carson, fighting against the unreasonable use of organochlorine pesticides and their effects on living organisms, in her book *Silent Spring* (Carson 1962), can be considered a pioneer for ecotoxicological studies. After a period when the effects of the dispersion of chemical compounds into the environment tended to be evaluated *a posteriori* and possibly corrected, a will to carry out evaluations *a priori* was essential in the last quarter of the twentieth century. Until the end of the 1980s, monitoring of the environment was based on conventional chemical methods of variable significance (chromatography, spectrophotometry, electrochemistry, radiochemistry, etc.), generally leading to the measurement of concentrations of chemical substances considered to be dangerous, in water, sediments, and organisms living in coastal ecosystems.

Although such chemical analyses are essential to identify concentration trends of contaminants (increase, plateau, or reduction) in the environment, they do not provide information about the real impact of the pollutant on its final target—the living organism. It is apparent then that this physicochemical assessment is insufficient to evaluate the health of a complex medium, with a mixture of contaminants potentially leading to the phenomena of synergy and antagonism. The concept of biological monitoring, based on the study of the biological response of organisms to pollutants, termed biomarkers, is today well established. The characterization of these biomarkers can constitute an early warning system before the further deterioration of the structure and function of an organism, and particularly before all the population or the ecosystem is disturbed. This concept is not new: it is the principle of diagnosis in human medicine, founded on the detection of symptoms likely to reveal a disease (Lafaurie et al. 1992).

2.2 Definition

In the past nearly 30 years, several definitions of biological markers have been published. The historical development of the biomarker approach has been closely related to advances in medicine and biology of vertebrates [National Research Council (NRC) 1987]. Biological markers were classified as exposure, effect, and susceptibility biomarkers. Moreover, in the publications of the NRC (1987, 1989), the authors highlighted that biological markers can be simultaneously used for biological monitoring and for monitoring of health. According to McCarthy and Shugart (1990), "biological markers are measurements at the molecular, biochemical, or cellular level in either wild populations from contaminated habitats or in organisms experimentally exposed to pollutants that indicate that the organism has been exposed to toxic chemicals, and the magnitude of the organism's response to the contaminant."

The definition was generalized by Depledge (1994): a biomarker is "a biochemical, cellular, physiological or behavioral change which can be measured in body tissues or fluids or at the level of the whole organism that reveals the exposure at/or the effects of one or more chemical pollutants." In September 1994, the journal *Ecotoxicology* presented four reviews on the role of the biomarkers in environmental assessment, as carried out with invertebrates (Depledge and Fossi 1994), vertebrates (Peakall and Walker 1994), terrestrial plants (Ernst and Peterson 1994), and populations and communities of invertebrates (Lagadic et al. 1994). These articles were required by the European Foundation for Science (ESF) to understand to what extent biomarkers could be used to evaluate environmental damage and to formulate possible rules to control any such damage.

Finally, Van Gestel and Van Brummelen (1996) attempted a redefinition of the terms biomarkers, bioindicators, and ecological indicators, by calling on previous work published in *Ecotoxicology* in 1994 when Lagadic et al. (1994) made a clear distinction between biomarkers and bioindicators and restricted the use of the term "biomarker" to the sublethal biochemical changes resulting from individual exposure to xenobiotics. However, this reductionist definition was not generally accepted (Van der Oost et al. 2005; Allan et al. 2006), with many scientists voicing their concern about not neglecting responses (e.g., physiological, behavioral) that could be used in risk assessments involving a change in scale of biological organization from the individual to the population. According to Van Gestel and Van Brummelen (1996), a biomarker is defined as any biological response to an environmental chemical contaminant at the infra-individual level, measured in an organism or its products (urine, feces, hair, feathers, etc.), indicating a change compared to the normal state and which cannot be detected in a healthy organism. The term bio-indicator should be restricted to an organism providing information on the environmental conditions of its habitat by its presence or its absence or its behavior. The concept of specific biomarkers (responding to metal pollutants, or to organics or to any defined pollutant) led to the definition of damage and defense biomarkers put forward by De Lafontaine et al. (2000). From the 1970s, great developments in biochemistry and molecular toxicology made it possible to progress quickly in our knowledge of the mechanisms of the toxicity of xenobiotics, mainly with mammalian models. Thereafter, significant specific biochemical effects were highlighted in species exposed to some contaminants, particularly in birds, fish, and mollusks considered as being of ecological interest. The majority of the examples in this chapter concern the aquatic environment, particularly the marine environment, which is the final receptacle of chemical pollutants.

Well-known biomarkers, which have been recognized in laboratory and environmental studies, have been called "core biomarkers" (Pampanin et al. 2006). Such core biomarkers include the stability of the lysosomal membrane (measured by the neutral red retention time), inhibition of acetylcholinesterase (AChE) activity, metallothionein (MT) concentration, ethoxyresorufin *O*-deethylase (EROD), and the fluorescent metabolites of the bile [fluorescent aromatic compound (FACs)].

2.3 Defense Biomarkers

2.3.1 Ethoxyresorufin *O*-Deethylase

Payne and Penrose (1975) were among the first to report elevated cytochrome P450–dependent monoxygenase activity in fish from petroleum-contaminated areas. The first biomarker that gained international recognition was consequently the enzymatic activity of EROD, an isoenzyme cytochrome P4501A termed as CYP1A. EROD belongs to the group of CYP enzymes that are the main enzymes responsible for the metabolism of certain endogenous compounds (hormonal and membrane steroids, biliary acids, vitamin D, fatty acids, prostaglandins, and pheromones) and nonpolar xenobiotics, including the metabolism of many environmental toxic chemicals and carcinogens (Nebert 1994). CYPs are enzymes referred to as mixed function oxidases (MFOs) (Di Giulio et al. 1995). Klingenberg (1958) and Garfinkel (1958) described successively a pigment present in the microsomal fraction from mammalian liver, which, in its reduced form, fixes carbon monoxide and absorbs at 450 nm. The denomination "P450 cytochrome" was proposed by Omura and Sato (1964), who showed that this pigment is a hemoprotein with molecular mass ranging from 43 to 60 kDa. For the first time, Estabrook et al. (1963) demonstrated the involvement of this hemoprotein in a reaction of monoxidation: the hydroxylation of 17α-hydroxyprogesterone. CYPs are found to be associated with membranes in the endoplasmic reticulum or mitochondria of different tissues: liver, lung, kidney, intestine, etc. (Stegeman and Hahn 1994). They catalyze the oxidation of a substrate RH (an organic compound that becomes hydroxylated) by inserting one atom of molecular oxygen, whereas the second atom is reduced to water following the equation:

$$RH + O_2 + NADPH + H^+ \rightarrow ROH + NADP^+ + H_2O$$

This reaction constitutes the first phase (phase I) of the biotransformation of organic compounds that causes hydrophobic compounds to become more water soluble.

The *de novo* synthesis of P450 proteins by organisms termed as "induction" leads to increased enzymatic activity. Induction has been well known for 40 years in humans and other mammals, more recently in fish and plants, and of late in invertebrates (Stegeman and Hahn 1994). The induction of cytochrome P450 isoenzymes responds to exposure to xenobiotics by way of a selective, receptor-mediated stimulation of the CYP1A gene transcription rate, resulting in increased levels of specific mRNA, new synthesis of cytochrome P450 isoenzymes, and an increase in their catalytic activities (e.g., EROD for CYP1A). The receptor that mediates the regulation of the CYP1A gene expression is known as the AH (aryl hydrocarbon) receptor (AHR) (Poland and Glover 1975; Guengerich 1993). Studies have demonstrated that activation of the AHR pathway is necessary for benzo[*a*]pyrene (B[*a*]P)-induced hepatic carcinogenicity in mice (Shimizu et al. 2000), and 2,3,7,8-tetrachlorodibenzo-*p*-dioxin (TCDD) and polychlorobiphenyl (PCB) induced early life stage toxicities in fish (Antkiewicz et al. 2006). The functioning of the AHR pathway in fishes is almost identical to that in mammals, except that fish have two or more forms of AHR (AHR1 and AHR2) due to genome duplication events (Hahn 2002). After diffusing into the cell, the xenobiotic binds to a protein complex in the cytoplasm consisting of AHR, a dimer of heat-shock protein 90 (Hsp90), p23, and ZAP2 (also known as ARA9 and AIP) (Figure 2.1). Upon ligand binding, ZAP2 is released, exposing the nuclear localization signal on AHR and

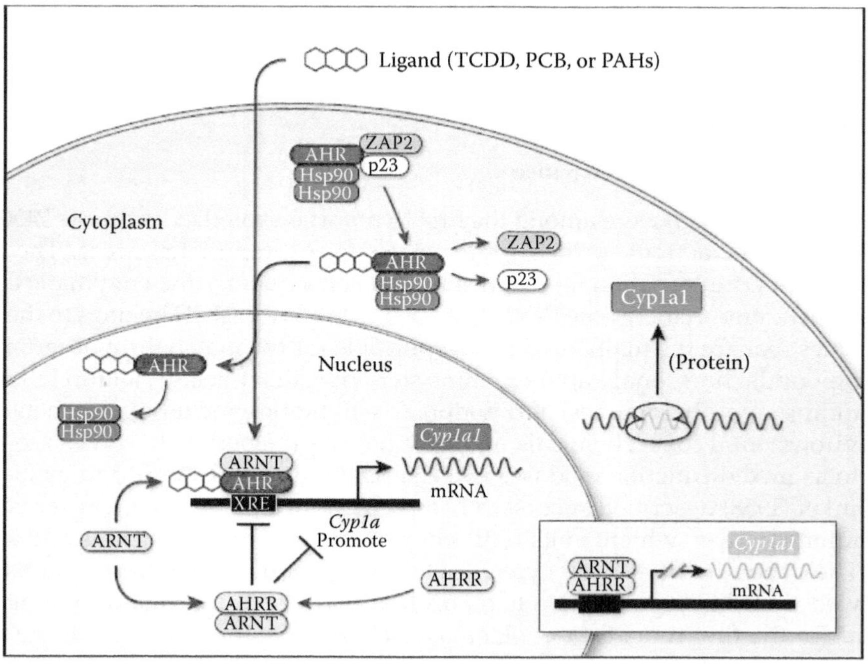

FIGURE 2.1
Functioning of the AHR (aryl hydrocarbon receptor) pathway in fishes. After diffusing into the cell, the xenobiotic binds to a protein complex in the cytoplasm consisting of AHR, Hsp90, p23, and ZAP2. Upon ligand binding, ZAP2 is released leading to translocation of AHR from the cytoplasm to the nucleus. Within the nucleus, Hsp90s are released, and AHR heterodimerizes with the Aryl Receptor Nuclear Translocator (ARNT). The AHR–ARNT complex then binds to multiple enhancer elements in the promoter region of responsive genes in the AHR battery such as CYP1A. (From Figure 8.2 of Roméo, M., Wirgin, I.I., in C. Amiard-Triquet, P.S. Rainbow, and M. Roméo, *Tolerance to Environmental Contaminants*, CRC Press, Boca Raton, 175–208, 2011. With permission.)

leading to translocation of AHR from the cytoplasm to the nucleus. Within the nucleus, Hsp90s are released, and AHR heterodimerizes with another protein, the Aryl Receptor Nuclear Translocator (ARNT). The AHR–ARNT complex then binds to multiple enhancer elements in the promoter region of responsive genes in the AHR battery such as CYP1A.

The P450 enzymes, involved in the detoxification of xenobiotics, are slightly expressed under normal physiological conditions, but are on the other hand strongly inducible: their content or their activity is increased in response to one or more exogenic molecules. The biological advantage of this induction process by xenobiotics is generally to amplify their metabolic degradation. Nelson regularly publishes a review of P450 cytochromes according to their families and subfamilies (drnelson.uthsc.edu/CytochromeP450.html). As of February 2009, more than 8100 distinct CYP gene sequences have already been known. The nomenclature used for cytochrome P450s is based on sequence homology (Nebert and Nelson 1991): two cytochrome P450s belong to the same family when their peptide sequence presents more than 45% amino acid homology and to the same subfamily if the homology is higher than 55%. The abbreviation CYP (cytochrome P450 gene) is completed with a number representing the family, then a letter indicating the subfamily (e.g., CYP4A), and a last number when there are several genes within the same subfamily (e.g., CYP4A1, CYP4A2). Conventionally, genes are written in italics *CYP1A1* (Goksøyr and Förlin 1992), whereas mRNA and proteins are in capitals. Nelson (1998) has developed a classification scheme where CYP families are classified into CLANS, that is, clusters of higher order groupings of P450 families.

They are ubiquitous proteins, the presence of which was demonstrated in plants and animals, from bacteria to mammals. P4501A1 enzymes (in particular, EROD measured in fish) may be induced by compounds sterically analogous to dioxin such as aromatic hydrocarbons, polychlorinated biphenyls, and polychloroazobenzenes. The first work on EROD and other P450 enzymes as biomarkers was completed on freshwater and marine fish livers (Addison 1984; Addison and Payne 1987; Flammarion et al. 1998). Polycyclic aromatic hydrocarbons (PAHs) induce P4501A1 in all fish considered by different authors from agnathans to teleosts and selachians (Stegeman 1987; Andersson and Nilsson 1989).

CYP1As are induced by PAHs, coplanar PCBs, polychlorinated dibenzodioxins, and polychlorinated dibenzofurans (Goksøyr and Förlin 1992), which are pollutants of the 3-methylcholanthrene type and are now considered AH receptor agonists. Three enzyme activities, EROD, ethoxycoumarin *O*-deethylase, and arylhydrocarbon (B[*a*]P) hydroxylase are largely specific in their response to these compounds. Many PAHs are both inducers and substrates for CYP1A. In contrast, coplanar PCBs, although often good inducers, are frequently poor substrates for CYP1A (Di Giulio et al. 1995). In their review, Goksøyr and Förlin (1992) reported that CYP2B is induced by coplanar PCBs (phenobarbital type), CYP3A by endogenous steroids, and CYP4A by endogenous fatty acids and xenobiotics such as phthalates and peroxisome proliferators (Simpson 1997). Therefore, members of the cytochrome P450 family of monoxygenases can metabolize and often produce more toxic forms from (see below) a wide variety of endogenous molecules and xenobiotics.

In contrast to fish, the presence of the AH receptor is not confirmed in mollusks. The cytochrome P450 pathway in PAH metabolism in mussels is low compared to the radical manner which leads to the formation of quinones. However, the existence of a CYP1A-like gene in mussels (Wootton et al. 1995) justifies research into the mechanisms of activation and detoxification already identified in fish. The capacity to metabolize *in vitro* B[*a*]P into derived diol, quinone, and phenol was demonstrated in the mussel *Mytilus galloprovincialis* (Michel et al. 1993). The activity of B[*a*]P hydroxylase BPH, measured in the digestive gland of this mussel (measurement based on the production of phenol metabolites resulting from

B[*a*]P oxidation), proved to be a biomarker of exposure to PAHs (Akcha et al. 2000). In some cases, the biotransformation can induce processes of carcinogenesis, mutagenesis, and toxicity. For example, B[*a*]P is metabolized (7,8-epoxidation, then 9,10-epoxidation) into a mutagenic compound, the (+)-anti-B[*a*]P, 7*R*,8*S*-diol-9S, 10*R*-epoxide, which is able to bind in a covalent manner to DNA and leads to the formation of adducts (Vermeulen 1996; Akcha et al. 1999).

2.3.2 Fluorescent Aromatic Compounds in Fish Bile

The exposure of fish to crude oils containing PAHs causes an increase in FACs in the bile (Aas et al. 2000; Gagnon and Holdway 2000). When the exposure takes place through the food chain, PAHs are absorbed, transported to the liver where they are converted into more water-soluble metabolites, and are excreted in the bile (Varanasi et al. 1995; Lee 2002). Laboratory studies show that the depuration period after exposure lasts several weeks, suggesting that an increased concentration in FACs in bile reflects a relatively recent exposure to PAHs (Huggett et al. 2003). Crude oils with PAHs with two to three rings are very different in their FACs in bile compared to pyrogenic hydrocarbons with four to six nonsubstituted rings. This is why it is difficult to link the induction of CYP1A and the increased concentrations of FACs in the bile to a specific source of PAHs. However, the concentration of FACs in the bile constitutes a fast and practical tool that clearly shows the extent of exposure to PAHs in the framework of biomonitoring: they thus constitute a "relevant" biomarker (Lehtonen et al. 2006).

2.3.3 Phase II Enzymes

Conjugation intervenes in the metabolism of xenobiotics, either following the reactions of oxidation (phase I), or directly on molecules bearing hydroxylated, thiol, or carboxylic groups. These reactions, also called phase II reactions, are catalyzed by membrane or cytosolic enzymes functioning with various cofactors (glutathione, sulfates, glucuronic acid). The enzymes responsible for these conjugations are glutathione *S*-transferases (GSTs), UDP-glucuronosyl-transferases (UDPGTs), and sulfotransferases. The activities of phase II enzymes are lower in fish (Gregus et al. 1983) than in higher vertebrates. In the fish *Platycephalus bassensis*, exposed to a mixture of PCBs, UDPGT activities significantly increase as do cytochrome P450 enzymes (Brumley et al. 1995), whereas the exposure of trout *Salmo gairdneri* to various polychlorinated phenols causes a reduction in UDPGT activities (Castren and Oikari 1987). GSTs are enzymes whose activity is used as a biomarker of organic substance exposure, especially in mollusks, where EROD activity is not routinely measured (Cajaraville et al. 2000). GSTs represent an important enzyme family whose function is to combine reduced glutathione (GSH) with electrophilic compounds by formation of a thioether bridge (Foureman 1989). The products are then metabolized in mercapturates that are excreted in the bile or the urine. GST activity increases in exposed organisms according to the xenobiotic concentration in the medium.

In fish, contradictory results have been reported (Van Veld and Lee 1988). However, several authors have shown that glutathione transferases are involved in the detoxification of many chemical pollutants: hydrocarbons, organochlorine insecticides, and PCBs (Monod et al. 1988; George 1994). In mollusks, GST activity is used with more success than in fish as a biomarker of exposure to these substances (in the marine environment: Fitzpatrick et al. 1997; Hoarau et al. 2001; and for freshwater bodies: Boryslawskyj et al. 1988; Robillard et al. 2003). GSTs play an additional role in the detoxification process, being used as transporting molecules that increase the bioavailability of lipophilic compounds to the phase I enzymes [such as mixed function oxygenases (MFOs)]. They therefore reduce, by covalent linkage to electrophilic compounds, the probability of these compounds binding to other cellular macromolecules such as DNA (Van Veld et al. 1987).

2.3.4 Phase III Enzymes

Surprisingly, after phase II, it was generally considered that the xenobiotics were "detoxified" and no longer considered. However, accumulation of the metabolites that may result in cell injury and their excretion, occurring during phase III of biotransformation, is of particular importance (Damiens and Minier 2011). Phase III includes detoxification enzymes involved in the elimination from the cell of phase I and II products (metabolites) by transmembrane transport carried out by P-glycoproteins (PGPs) or by multidrug resistance–associated proteins (MRPs) (Gottesman and Pastan 1993). By now, it has been realized that transport systems are just as important as the previously known processes (Leslie et al. 2005; Cascorbi 2006). Phase III proteins, involved in the modulation of exit from the cell, are involved in key processes that result in the modulation of toxicological effects, and the multixenobiotic transport system is considered a system governing intracellular contaminant bioavailability. Membrane proteins MRPs are part of the large family of ABC (ATP binding cassette) transporters present in prokaryote and eukaryote cells. These ABC transporters have almost all the same architecture, with two binding domains of ATP located in the cytoplasm, and two hydrophobic regions inserted in the plasma membrane.

The first PGP was discovered in 1976 (Juliano and Ling 1976) in the context of resistance to multiple chemotherapy, and was named MDR (multidrug resistance protein). It transported a large number of compounds with different structures and modes of action—hence, the idea was presented that if different organisms live, grow, and reproduce in contaminated environments, they must have mechanisms allowing them to be resistant. Kurelec (1992) showed that resistance to many xenobiotics (multixenobiotic resistance MXR) has similarities with MDR. MXR proteins are found throughout the tree of life. Kurelec (1992) has reviewed MXR proteins in aquatic organisms. The wide taxonomic distribution of these proteins and their induction in the presence of xenobiotics show their importance in the nonspecific defense of organisms (Tutundjian and Minier 2002). How MXRs expel pollutants is not yet well known. Some models assume that removal is carried out by an enzyme called "flippase," which would capture the substrates at the inner leaflet of the membrane and translocate them to the outer leaflet (Tutundjian and Minier 2002). Minier et al. (1993) showed that mussels *Mytilus edulis* and *M. galloprovincialis* and oysters *Crassostrea gigas* express proteins immunologically similar to mammalian MDR proteins. Moreover, there is a relationship between their expression levels and the level of environmental contamination. Parallel to these studies, Kurelec et al. (1995) showed that the MXR system of the gastropod mollusk *Monodonta turbinata* could be induced by treatment with hydrocarbons.

Competition studies for transport increased our knowledge of the substrates involved. The possibility for *M. edulis* to expel pesticides such as triazines has been demonstrated (Minier and Moore 1998). Results have enabled the description of the phenomenon of resistance that is present in aquatic organisms and is expressed when they are exposed to compounds such as organochlorine pesticides, PCBs, and PAHs (Kurelec et al. 1995; Galgani et al. 1996; Eufemia and Epel 2000). There are also xenobiotics that inhibit MDR; they are called "chemosensitizers," and their presence induces an increase in concentrations of pollutants in the body with subsequent damage (Smital and Kurelec 1998).

2.3.5 Metallothioneins

MTs are nonenzymatic proteins with a low molecular weight (12–15 kDa), high cysteine content, heat stability, and no aromatic amino acids. The thiol groups (–SH) of cysteine residues enable MTs to bind particular trace metals. The first MT was found in equine renal cortex (Margoshes and Vallee 1957). MTs or MT-like proteins have since been reported in many vertebrates including many species of fish (reviewed by Hamilton and Mehrle 1986), and in aquatic invertebrates (reviewed by Amiard et al. 2006) such as echinoderms (Riek et al. 1999), mollusks (Amiard-Triquet et al. 1998; Bebianno and Langston 1998; Bebianno et al. 2003) and their larvae (Damiens et al. 2004), and crustaceans (Roesijadi 1992), but also in terrestrial invertebrates (Dallinger 1996). In aquatic species, MT concentrations were measured mainly in tissues involved in the uptake, storage, and excretion of metals such as gills, digestive glands, and kidneys, but also in muscular and nervous tissues. Fowler et al. (1987) defined three classes of MT according to the location of cysteine residues in the amino acid sequences. Class I includes MTs of vertebrates and MTs with a closely similar structure (mollusks, crustaceans). Class II includes MTs whose structure does not resemble that of class I (*Drosophila*, sea urchins, nematodes, fungi, cyanobacteria), and finally the third class includes the nonprotein MTs, synthesized from glutathione such as phytochelatins, present in plants.

Several reviews have synthesized the research completed mainly in aquatic species concerning the structure and the functions of MTs as well as the progress of assay techniques (Roesijadi 1992, 1996; Roméo et al. 1997; Cosson and Amiard 2000; Cosson 2000; Isani et al. 2000; Amiard et al. 2006). MTs whose behavior is related to the chemistry of thiol groups assume many biological functions and even if some remain under discussion, in general, authors agree on the participation of MTs in the homeostasis and detoxification of essential metals such as zinc and copper and in the detoxification of nonessential metals such as cadmium and mercury. Studies have also shown MT involvement in protection mechanisms against oxidative stress, apoptosis, and growth regulation of nervous cells (Cavaletto et al. 2002).

In vertebrates as well as in invertebrates, MT levels differ according to species and tissues. They are generally higher in the gills and digestive gland in mollusks (Baudrimont et al. 1997). The concentrations vary in different tissues not only according to the developmental stage, age, sex, size, and nutritional status of an organism, but also according to their gonadic development under hormonal control (Hamza-Chaffai et al. 1995, 1999; Leung and Furness 2001; Bebianno et al. 2003; Riggio et al. 2003; Leiniö and Lehtonen 2005). If the organism is exposed to a very high metal concentration, MT synthesis can be inhibited, as demonstrated by George et al. (1992).

MT synthesis is mainly induced by metals (essential or not) such as Cu, Zn, Cd, Hg, and Ag but also to a lesser extent by organic compounds such as some pesticides or antibiotics. The great variability of induction and the various abiotic or biotic factors influencing MT synthesis lead to contradictory results in the literature, which have been discussed in a review relating to the role of MTs in invertebrates and their use as biomarkers (Amiard et al. 2006).

For about the past 20 years, many studies carried out in laboratory conditions and *in situ* have shown the potential of increased concentrations in MTs for use as biomarkers of exposure to contaminant metals. Currently in ecotoxicological studies carried out in terrestrial and aquatic environments, their measurement may be integrated into a multibiomarker approach so *inter alia* mitigating for the presence of other inducers than metals.

2.3.6 Enzymatic and Nonenzymatic Antioxidant Defenses

In biological systems, reactive oxygen species (ROS) are continuously produced by several mechanisms involving exo- or endogenous compounds such as xenobiotics (Di Giulio et al. 1989; Livingstone et al. 1990; Winston and Di Giulio 1991). In aerobic organisms, they are part of basal cellular metabolism such as cellular respiration or phagocytosis activity (Cossu et al. 1997; Valavanidis et al. 2006). Their production is also a result of the activity of different oxidative enzymes such as tryptophan dioxygenase, xanthine oxidase, and cytochrome P450 reductase that produce superoxide anions, and guanyl cyclase and glucose oxidase, which are able to generate hydrogen peroxide.

Moreover, chemical pollutants are important producers of ROS. The xenobiotics known for their redox properties such as quinones, transition metals, diazoïc staining, bipyridyl herbicides, and nitric aromatic compounds induce the formation of superoxide radicals.

The imbalance in favor of ROS production instead of their neutralization by antioxidant systems corresponds to oxidative stress. At the cellular level, it results in the alteration and more particularly in the oxidation of components such as DNA, proteins, and lipids and in a total disturbance of the redox balance (e.g., ratios GSH/GSSG and $NADH/NAD^+$). Its cytotoxic effects are expressed by structural and functional perturbations such as enzymatic inhibition, protein damage, lipid peroxidation, inflammatory processes, and apoptosis (Figure 2.2).

During evolution, aerobic organisms have developed antioxidant defense mechanisms whose main function is to block off and to deactivate ROS. The extent of oxidative damage is directly related to the efficiency of antioxidant systems occurring in the different species. The systems are composed of a suite of cytosolic enzymes [mainly superoxide dismutases (SODs), peroxidases, catalases], reducing molecules of low molecular weight (glutathione, ascorbates, urates) and several liposoluble vitamins (α-tocophérol, β-carotene).

Among enzymatic antioxidant systems, SODs correspond to a metallo-enzyme family (containing Cu, Zn, Fe, or Mn) known to convert superoxide anion in hydrogen peroxide, H_2O_2. Among peroxidases, glutathione peroxidase (GPx), depending or not on selenium,

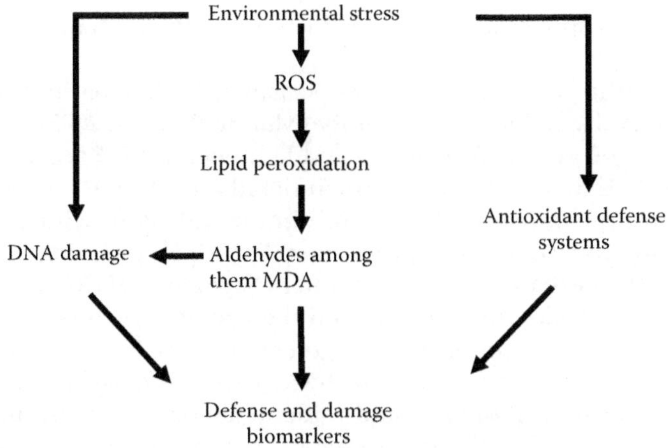

FIGURE 2.2
Environmental stress in organisms could generate ROS able to induce damage to membrane lipids and DNA molecules but also to antioxidant defenses. The cellular damage and the induction of defense systems could be used as defense or damage biomarkers.

uses reduced glutathione (GSH) to reduce different types of peroxides. Its enzymatic activity is related to that of glutathione reductase that generates GSH from the oxidized form of glutathione (GSSG). Catalases are hemoproteins occurring in peroxisomes and act by decomposing H_2O_2 into H_2O and O_2.

Nonenzymatic antioxidant systems are mainly formed by compounds of low molecular weight showing reducing functions or the ability to trap free radicals. In the first category, glutathione in its reduced form is considered the universal detoxificant (Vasseur and Leguille 2004). This triptide is an important antioxidant in eukaryote and prokaryote species. It acts as an electron donor directly able to inactivate several types of ROS. It also constitutes a substrate for enzymatic activity of GPx. Low levels of cellular GSH usually make the cells more sensitive to pro-oxidant compounds. The liposoluble vitamins E and A occurring in the cell membrane are able to capture some ROS as the superoxide anion or the hydroxyl radical right from their formation and then avoid the effects of oxidative stress.

Under stress conditions, the activity of antioxidant defense systems could be induced or inhibited. Usually, induction is interpreted as an adaptation of organisms faced by environmental disturbances, whereas inhibition reflects the toxic effect of pollutants and indicates cell damage (Cossu et al. 2000; Vasseur and Cossu-Leguille 2003). The measurement of antioxidant enzymes could give an indication of the organism's antioxidant status and could be used as a biomarker of oxidative stress. More generally, the assessment of the components of the antioxidant defense systems occurring among animals in different tissues, represents a nonspecific biomarker of the adverse effects of xenobiotics (Valavanidis et al. 2006). In the past decade, this assessment has been used more widely given the general ability of tissues to eliminate different forms of ROS as measured by the total oxyradical scavenging capacity (TOSC) method developed by Regoli et al. (2002a). This method presents advantages that provide to the organism or tissue in an integrated approach:

- A general view of the antioxidant status that could only be obtained with difficulty by the individual measurement of one or several components of the antioxidant systems;
- An antioxidant response against a specific kind of ROS (Monserrat et al. 2007).

The systems of antioxidant defense show seasonal variations in relation to temperature, reproductive cycle, and food availability (Manduzio et al. 2005) in different mollusk and fish species (Regoli et al. 2002b; Leiniö and Lehtonen 2005; Bocchetti and Regoli 2006; Ansaldo et al. 2007). Usually, the maximum antioxidant activities are recorded in spring. They decrease during summer and reach minimum values in winter. The variations of antioxidant systems are conversely proportional to lipid peroxidation, explaining the increased sensitivity of organisms during winter (Niyogia et al. 2001).

Over the two past decades, the literature on the use of antioxidant system response as a defense biomarker has been important (Regoli et al. 2011). In this framework, numerous invertebrate and vertebrate, marine, and freshwater species have been used as sentinels to evaluate the effects of several organic and mineral xenobiotics both under experimental and natural conditions. Today, these biochemical responses are associated with those at other levels of biological organization in species belonging to different trophic levels in a multibiomarker approach required to obtain an integrated evaluation of contaminant impact (Beliaeff and Burgeot 2002; Orbea et al. 2002; Roberts and Oris 2004; Aït Alla et al. 2006; Damiens et al. 2007).

2.3.7 Heat Shock Proteins

Heat shock proteins (Hsps) are ubiquitous proteins, widely conserved throughout the evolution of eukaryotes. They are named according to their apparent molecular weight using sodium dodecyl sulfate-polyacrylamide gel electrophoresis (SDS-PAGE) (Schlesinger et al. 1982; Atkinson and Walden 1985; Moromoto et al. 1990), in particular HSP 40, 60, 70, and 90. The Hsp of lower molecular weight (8 kDa) is called ubiquitine. Cellular response to stress was reported for the first time by Ritossa (1962), who observed Hsp induction in the case of a very significant temperature rise, hence their name. Hsps are now called stress proteins because they are overexpressed in response to a certain number of physical and chemical factors including anoxia (Spector et al. 1986), salinity stress (Ramagopal 1987), metals (Hammond et al. 1982; Caltabiano et al. 1986), xenobiotics (Sanders 1990), and oxidative stress in general (Freeman et al. 1999).

Some Hsps are constitutive; for example, Hsp 60 and 70 are involved in the homeostasis of proteins under normal conditions while playing a protective and repairing role in the event of environmental stresses (Rothman 1989; Welch 1990). Stress proteins have a capacity to repair proteins harmed by stress or to eliminate them when they cannot be repaired any further. They work as molecular "chaperones," accompanying, monitoring, and protecting other proteins (Frydman 2001; Hartl and Hayer-Hartl 2002). They can act in the posttranslational spatial configuration of proteins and intervene in the transfer of proteins to the mitochondria, and in the induction and control of apoptosis (Craig et al. 1994; Creagh et al. 2000). Stress proteins and the genes that code for them have been sequenced in many organisms. Because of their sensitivity to environmental pollutants such as metals, several researchers quantified Hsp 60 and 70 in the bivalve sentinel species *M. edulis* (Sanders et al. 1991, 1994; Brown et al. 1995; Werner and Hinton 1999). Hsp levels reflect the physiological state of the animal.

Another group of proteins, that of glucose-regulated proteins (GPRs), has been discovered (Welch 1990; Hightower 1993). GPRs have very strong analogies with Hsps.

2.4 Damage Biomarkers

2.4.1 AChE Activity

The inhibition of cholinesterase activity can be regarded as one of the first biomarkers proposed in environmental monitoring studies, since its development in human medicine as an index of exposure to neurotoxins, in particular organophosphates from war gases, goes back several decades. For many authors, the measurement of AChE activity is the best marker of contamination by organophosphorous pesticides and carbamates (Holland et al. 1967; Coppage and Braidech 1976; Galgani and Bocquené 1989; Day and Scott 1990). Cholinesterases are enzymes that catalyze the hydrolysis of esters of choline more quickly than other esters. In vertebrates, two cholinesterases have been identified: AChE (EC 3.1.1.7) and butyrylcholinesterase (EC 3.1.1.8, BuChE). AChE is inhibited by excess of substrate but BuChE is not. In spite of the limited number of genes apparently involved, ChEs present a large variety of molecular forms including globular (monomer, dimer, tetramer) and asymmetric forms (from 4 to 12 subunits with a collagen tail). At least eight forms of AChEs are found with a different oligomeric organization, solubility, and mode of membrane anchorage in vertebrates (Mora et al. 1999). Some studies suggest that a polymorphism of ChEs may exist for mollusks. Indeed, two distinct ChEs differentiated by their solubility and their sensitivity toward organophosphates have been found in the oyster *C. gigas* (Bocquené et al. 1997). In some biomonitoring studies, it is not clear whether only AChEs or also pseudocholinesterases are able to hydrolyze the substrate (acetylthiocholine) used; thus, authors should choose to use the nonspecific term of cholinesterases when presenting biological monitoring results.

Measurements carried out on dab (the flatfish *Limanda limanda*) along a 360-km transect in the North Sea (Galgani et al. 1992) showed important inhibitions of various types of cholinesterases. This effect, mainly observed in animals coming from near the coast, is due to compounds carried from the estuaries of the Elba and Weser rivers. The identification of the inhibiting compounds of ChEs nevertheless remains delicate, and it is not possible to definitely conclude that organophosphorous and carbamates are the only chemicals responsible for the observed inhibition effects on ChEs in the various marine compartments. The chemical data on these products are scarce, and marine organisms are subjected permanently to the effects of complex mixtures of contaminants. Payne et al. (1996) wonder whether AChE activity is an old biomarker with a new future. Indeed, these authors show that an inhibition of AChE activity could be associated with an induction of EROD activity in the livers of trout (*Salmo trutta*) and flounders (*Pleuronectes americanus*) caught in an area contaminated with pulp mill effluents.

Contaminants other than pesticides can inhibit AChE activity. Leiniö and Lehtonen (2005) report inhibition of AChE by metals, detergents, and algal toxins. These authors conclude that the inhibition of AChE activity can be regarded as a marker of the physiological state of the animals. Moreover, Pfeifer et al. (2005) emphasize that AChE activity in mussels *Mytilus* sp. collected from Baltic Sea is negatively correlated with salinity. The abiotic parameters of the environment thus need to be taken into account as with other biomarkers when performing biological monitoring.

2.4.2 Vitellogenin

Biomarkers of endocrine disruption are used more and more since many studies have shown that the reproduction of fish is very sensitive to chemical pollutants. Among the chemical compounds reaching the aquatic environment, the first endocrine disruptor compounds (EDCs) were those acting as estrogens by their capacity to mimic the natural estrogen, estradiol, thus causing a feminizing action on organisms. The general term of EDCs now includes molecules of very varied structure and origin (PCBs, tributyltins, or natural phytoestrogens coming from the metabolism of soya or clover). The incidence of fish hermaphroditism close to wastewater treatment plants in the United Kingdom (Purdom et al. 1994) led to a study of the "estrogenicity" of the effluents of the treatment plants. Ethynylestradiol, a synthetic estrogen used as contraceptive, is involved in these effects (Purdom et al. 1994). Human natural estrogens (17β-estradiol, estriol, and estrone) and their conjugates, excreted in urine and feces, contribute to estrogenicity (Larsson et al. 1999). Another chemical molecule is nonylphenol, used as an intermediate in the industrial production of nonylphenol ethoxylates, a large group of nonionic surfactants widely used in plastics, latex paints, household and industrial detergents, and paper and textile industries (Lee 2002). However, according to Soto et al. (1995), EDCs mimic not only the sex steroid hormones estrogens but also androgens, by binding to hormone receptors or influencing cell signaling pathways; they block, prevent, and alter hormonal binding to hormone receptors or influence cell signaling pathways; they alter production and breakdown of natural hormones and modify levels and function of hormone receptors.

When exposed to estrogens and "mimetic estrogens," the liver synthesizes vitellogenin (VTG), a lipoglycophosphoprotein (which is a precursor of yolk egg reserves) specific to females, regardless of the age of fish. VTGs are high-density (300–600 kDa, according to species) glycolipophosphoproteins having Ca and Zn ligands and are considered to have similar characteristics in vertebrates, such as fish (Nagler et al. 1987), and invertebrates, particularly mollusks (Blaise et al. 1999). The "estrogen mimics" exert a feminizing action, thus concerning *a priori* more male individuals with VTG induction, oocyte and oviduct presence in the testes, fecundity decrease, modification of the sex ratio, and reduction in the secondary sexual characters in the male (Tyler and Routledge 1998).

However, field measurements of effects on the reproduction of fish are far from clear; a full demonstration of any effect on fecundity and reproduction, size, or structure of fish populations indeed requires field investigations that are time consuming and spatially limited. The feasibility of the measurement of VTG and the interpretation of histological slides of gonads of male fish collected from French rivers was studied in the chub (*Leuciscus cephalus*) (Flammarion et al. 2000). First results have been followed by a large-scale field experiment with this species. Measurements have demonstrated moderate but significant VTG induction in chub collected downstream from large towns in France (Paris or Lyon). Iwanowicz et al. (2009) evaluated the reproductive status of smallmouth bass (*Micropterus dolomieu*) in the upper Potomac River and its tributaries. They noted the presence of immature female germ cells (oocytes) in the testes of some of the male fish. Further evidence of endocrine disruption occurred when the authors detected the presence of VTG in the blood of male fish. In addition to the effects on male fish, a substantial decrease in VTG in females also suggested endocrine disruption. At present, VTG is considered a biomarker of endocrine disruption in fish and some mollusks. In the freshwater mussel (*Elliptio complanata*), VTG concentrations in hemolymph and gonad increase after exposure to effluents from wastewater treatment plant (Gagné et al. 2001).

2.4.3 Lysosomal Membrane Stability

It is known that lysosomes play a significant role in the catabolism of cellular compounds, the intracellular transport of macromolecules, and the storage of metals (Viarengo et al. 1984) and of organic contaminants (Moore 1988). The lysosomal membrane is weakened in the liver or digestive gland of animals subjected to pollution. It is very difficult to evaluate the molecular changes affecting the permeability of the lysosomal membrane. Analyses of this permeability require extremely purified preparations of lysosomal membrane and their study at a molecular level (see Chapter 5). An easier way to evaluate this parameter is to examine whether its physiological function is changed or destroyed following an exposure to pollutants. Cytochemistry is the relevant tool that links descriptive morphology and biochemistry to observe such pathological modifications. This technique was used successfully to estimate the integrity of the lysosomal membrane by visualizing the hydrolytic enzymes inside the lysosome, and it proved to be a fast and sensitive research tool to evaluate the effects of different xenobiotics (Pellerin-Massicotte and Tremblay 2000). This unspecific response intervenes in all cellular types from fungi to vertebrates. Viarengo et al. (1995) showed that a short-term exposure to pollutants in micromolar amounts (ionic copper Cu^{2+}, dimethylbenzoanthracene, and Aroclor 1254) reduced the stability of the lysosomal membrane (LMS) of the digestive gland of mussels *M. galloprovincialis*. Broeg et al. (2002) studied LMS in livers of the flounder (*Platichthys flesus*) from the North Sea; the lysosomal membrane was affected in fish from the Elba river between 1995 and 1999 but then recovered its integrity in 2000. On the other hand, fish from the Eider river or around Helgoland, which are located farther from pollution sources (DDT and PCB) than the Elba river, showed a decrease in the integrity of lysosomal membrane that has been constant between 1995 and 2000. The authors suggest that the fish populations not continuously exposed to anthropogenic stress have a lower potential or take longer time to recover a good physiological state.

2.4.4 Thiobarbituric Acid Reactive Substances

Deficiency of antioxidant defense systems to eliminate an excess of ROS could induce different types of cellular damage, of which the most widely studied is the peroxidation of lipids (Figure 2.2), able to induce structural and chemical alterations of cellular membranes (Livingstone et al. 1990; Winston and Di Giulio 1991; Vasseur and Cossu-Leguille 2003; Valavanidis et al. 2006). The process of lipid peroxidation involves a chain of reactions leading to the breakdown of polyunsaturated fatty acids that are relatively sensitive to oxidative reactions. Their degradation induces the formation of various compounds such as lipid alcoxyl radicals, ketones, alkanes, epoxides, and aldehydes. Among them, malondialdehyde (MDA) is both the most important and the most studied. Most of these compounds are toxic and mutagenic. The peroxidation of lipids could be initiated by hydroxyl radicals particularly in reactions catalyzed by transition metals (Viarengo et al. 1990; Valavanidis et al. 2006; Almeida et al. 2007).

The effects of lipid peroxidation can be assessed at the different steps of the lipid breakdown: at the initial phase (conjugated diene), intermediate phase (lipid hydroperoxides), or final phase [substances (TBARS) reactive with thiobarbituric acid (TBA) considered as MDA-like peroxides]. This test based on the use of these substances mainly reveals the formation of MDA by colorimetric or fluorimetric methods. Because TBA can react with compounds other than MDA, the results are usually expressed as TBARS concentrations (Knight et al. 1988; Pannuzio and Storey 1998; Durou et al. 2007).

The levels of MDA and TBARS have been used as markers of oxidative stress indicating the peroxidation of cellular membranes in numerous marine and freshwater invertebrate and vertebrate species. They can be influenced by different environmental parameters such as salinity and temperature in bivalves (Damiens et al. 2004) or in fish and can increase 20-fold in goldfish (*Carassius auratus*) exposed to a temperature elevation of 14°C (Lushchak and Bagnyukova 2006). In different populations of the same species, the levels of TBARS can show seasonal variations. In the estuarine polychaete (*Nereis diversicolor*), no variations were observed in the Seine estuary (Durou et al. 2007), but higher levels were recorded in January and October at different Moroccan sites (Aït Alla et al. 2006). In bivalves, no TBARS variations were observed in *Mytilus* sp. (Shaw et al. 2004; Bocchetti and Regoli 2006), whereas their concentrations were maximum in *Perna viridis* during spawning in May despite a strong activation of antioxidant systems (Wilhelm Filho et al. 2001). In marine bivalves, other environmental factors such as tidal cycles can influence lipid peroxidation, which increases during emersion (Durand et al. 2001; Almeida et al. 2005). On the contrary, these phases of immersion/emersion did not induce variations of TBARS in the gastropod *Littorina littorea*, whose antioxidant systems neutralize ROS formation during the aerial phase (Pannuzio and Storey 1998).

Moreover, numerous studies conducted during the past two decades in marine and freshwater media have shown that the levels of lipid peroxidation can be affected by environmental pollutants belonging to different classes of a different nature (Cossu et al. 2000; Giguère et al. 2003; Roméo et al. 2003; Aït Alla et al. 2006; Damiens et al. 2007).

2.4.5 DNA Damage

As reported above, ROS continuously produced in aerobic organisms when not neutralized may cause deleterious cellular effects such as lipid peroxidation described in the previous paragraph, protein breakdown, or DNA base oxidation (Figure 2.2). The preservation of DNA molecule integrity is critical for all living organisms, and they possess efficient protective systems for their genetic material.

Between the first contact of a xenobiotic with the DNA molecule and a potential mutation, an event sequence is produced beginning with the direct or indirect formation of DNA adducts. The secondary modifications of DNA produced can be induced by an oxidative stress and correspond to a single- or double-strand breakdown, an increase of its repair level or base oxidation. When DNA disturbances become permanent, they can induce an alteration of cellular functions and uncontrolled proliferation leading to carcinogenesis. Finally, when the contaminant impact is observed during cell division, it can produce a mutation transmitted to future generations (Møller and Wallin 1998; Burcham 1999; Valavanidis et al. 2006; Almeida et al. 2007; Hwang and Kim 2007; Monserrat et al. 2007 and references quoted by these authors).

The detection and quantification of DNA damage allow its use as a biomarker of genotoxicity under acute or chronic conditions (Chapter 13). Usually, stress conditions induce cellular disturbances in organisms and an increase in DNA damage. Most of the recent published studies are focused on DNA damage induced by oxidative stress.

DNA oxidation generates different modified bases of which 8-oxo-7,8-dihydro-2′-deoxiguanosine (8-oxodGuo), produced by the reaction between oxygen and guanine, are the most measured in aquatic organisms by high-performance liquid chromatography. Other oxidized bases can be studied such as thymine glycol, 5-hydroxymethyluracil, formylamidopyrimidine, and 8-hydroxydeoxyadenine (Martinez et al. 2003; Hwang and Kim 2007).

The Comet test (SCG or single cell gel electrophoresis) is a quantitative technique, quick and visual, to measure DNA strand breakdown in eukaryote cells (Devaux et al. 1997; Burlinson et al. 2007). The method is based on migration during electrophoresis of damaged DNA from the nucleus, forming an impression of a comet, the head of which corresponds to the cell nucleus with intact DNA, whereas the tail is formed by the cut DNA strands. Recent modifications of this test specifically reveal the oxidized DNA bases (Hwang and Kim 2007).

Other DNA damages assessed as genotoxicity biomarkers involve the DNA adducts formed by the nucleotides on which the chemical mutagens are fixed (^{32}P postlabeling) and the mutation quantified at the chromosomal level by the micronucleus test (Monserrat et al. 2007).

More recent molecular biology techniques of DNA amplification (random amplified polymorphic DNA) or polymerase chain reaction have been used to assess the direct effects of xenobiotics on DNA, and also the genetic diversity of studied populations. Actually, these techniques still lack reproducibility and only with difficulty allow the separation of the two mechanisms (Atienzar and Jha 2006).

An increasing number of aquatic and terrestrial ecotoxicological studies include the measurement of different forms of DNA damage in order to evaluate the genotoxicity of physical and chemical environmental stress on plants or animals, whether vertebrates or invertebrates (Flammarion et al. 2002; Gagné et al. 2002; Charissou et al. 2004; Radetski et al. 2004; Almeida et al. 2005; Cadet et al. 2005; Gagné et al. 2006; Nigro et al. 2006; Toyooka and Ibuki 2007; Almeida et al. 2007).

2.5 Multibiomarker Approach

The multibiomarker approach to evaluate the environmental quality of water is recommended by all specialists in ecotoxicology for the biological monitoring of the pollution of the environment henceforth. However, a long way had to be traveled before this point was reached, as discussed below.

At the University of Oslo in Norway, in August 1986 there took place a practical workshop on the biological effects of the pollutants under the auspices of the Group of Experts on the Effects of Pollutants (GEEP) of the Intergovernmental Oceanographical Commission of UNESCO. A special publication of the journal *Marine Ecology Progress Series* (volume 46, 1988) was devoted to the results of this workshop (GEEP Workshop). The workshop, according to Bayne et al. (1988a), had several goals: (1) to evaluate methods covering a broad spectrum from molecular approaches (biochemical level) to cellular and physiological processes (levels of the cell and whole organism) to the structure of communities of benthic organisms (community level); (2) the participants were to be researchers working on these subjects and interested in the measurement of the impact of pollution; (3) biological samples have to be taken from a site known for its pollution gradient according to a very precise protocol of sampling and analysis, and carried out during the workshop; (4) the participants to the workshop should follow a rigorous statistical model, that is, without knowing the ranking of sites along the pollution gradient; (5) the biological analyses carried out throughout the workshop were to be supplemented by meticulous chemical analyses in order to evaluate the relationship between the levels of contamination and the biological responses. The collected material consisted of mussels (*M. edulis*), crabs (*Carcinus maenas*), winkles (*L. littorea*), and flounders (*P. flesus*), as well as sediments. The Frier and Langesund fjords of the south of Norway were selected as sites of interest because they showed a chemical gradient of contamination from the bottom of the Frier fjord to the bay of Langesund. In the conclusions of the GEEP workshop, Bayne et al. (1988b) emphasized the development of biochemical measurements responding to specific organic pollutants: PAHs and PCBs (P450 enzymes) or metals (MTs). These authors concluded that measurements of EROD activity in the flounder *P. flesus* give the clearest and most sensitive response to the gradients of organic pollution. Later, an international (European) program, Biological Effects of Environmental Pollution (BEEP) in Marine Coastal Ecosystems, 2001–2004, was established with the aim of validating and intercalibrating a battery of biomarkers of contaminant exposure and effects in selected indicator species in the Mediterranean, the North Atlantic, and the Baltic Seas. One of the main goals of the program was to set up a network of biomarker researchers around Europe and to assess the applicability of biomarkers for different regions and species in the surrounding sea areas (Lehtonen et al. 2006). The selected biomarkers were specific biomarkers (EROD, MT, AChE inhibition, FACs) but also histochemical biomarkers of toxic effects such as neutral red accumulation showing a disturbed lipid metabolism or "general health" biomarkers, reflecting cytotoxicity LMS and immunotoxicity [acid phosphatase activity of macrophage aggregates (M-ACT) and macrophage aggregate size (M-AREA)] as well as mutagenic damage (frequency of micronuclei); they were measured in flounder (*P. flesus*), eelpout (*Zoarces viviparus*), and blue mussel (*Mytilus* spp.).

De Kock and Kramer (1994) developed the concept of active biomonitoring based on comparing the chemical and/or biological properties of samples collected from one population that, after randomization and translocation, have been exposed to different environmental conditions at monitoring sites. On the other hand, passive biomonitoring consists in analyzing (pollutant concentrations and biomarkers for instance) samples collected from the field (see also Chapter 7).

Field experiments always give a series of results that have to be statistically or hierarchically treated and integrated with environmental data to find the main sources producing a change in the measured biomarkers whatever the type of monitoring (active or passive) used. Authors use several types of treatments: principal component analysis (PCA, already used in many ecological studies), integrated biomarker response (IBR), and the expert system.

Roméo et al. (2003) established a comparison between resident and transplanted mussels along the NW Atlantic coast (France). Mussels (*M. galloprovincialis)* were collected in June (after 4 months' caging) and October (after 8 months' caging). A PCA was performed with the chemical (metal concentrations; unfortunately, measured PAH and PCB concentrations in mussels could not be included in PCA) and biochemical (catalase, GST and AChE activities, and TBARS level) data. The evaluations of the resident and transplanted mussels collected in June allowed them to be separated into three groups: resident mussels from La Rochelle with high metal and TBARS levels, resident mussels from Baie de L'Aiguillon with a very high condition index, and resident mussels from Fier d'Ars (less polluted site) and transplanted mussels at La Rochelle and Baie de L'Aiguillon with low TBARS and AChE activities. Strong seasonal variation from June to October of all parameters was noted. Mussels transplanted to La Rochelle appeared to be the most "polluted" in their pollutant concentrations and biochemical responses; moreover, the La Rochelle site had the highest concentration of organics in sediments of all sites. The choice of Fier d'Ars as a reference site may be questionable because some of the biomarker responses of the mussels were higher than expected there, although pollutants in mussels and sediment were present at the lowest concentrations measured. PCA presents, according to Guerlet (2007), several advantages: the possibility of bringing together the biological and physicochemical data without the latter influencing the profile of the PCA (illustrative variables); possible application without any *a priori* information on the gradient of stress; reduced effect on discriminative power in the case of addition of redundant parameters.

Beliaeff and Burgeot (2002) have established a simple method of summarizing biomarker responses, the IBR, which simplifies their interpretation in biomonitoring programs. They worked with two species belonging to different phyla, the mussel *M. edulis* and the flounder *P. flesus.* They underlined that the selection of an appropriate battery of biomarkers (such as GST, catalase, and AChE activities measured on mussels; EROD and AChE activities as well as DNA adducts on flounders) can avoid false-negative responses obtained with a single biomarker and allow information to be summarized in the form of a multivariate data set. Damiens et al. (2007) determined pollutant concentrations and biomarker levels in transplanted mussels (*M. galloprovincialis*) and established IBR. Three experiments of 1 month's caging at sea (NW Mediterranean Sea, France) were conducted in 2004 and 2005. Pollutant concentrations, displayed as star plots, were compared to IBR star plots. Visualization was thus possible between sites, and there was a correlation between the copper gradient measured in the transplanted mussels and IBR variation. In 2004 (Figure 2.3), the agreement between the copper gradient and the PCB gradient measured in caged mussels and IBR variation was good, whereas the PAH gradient did not seem to

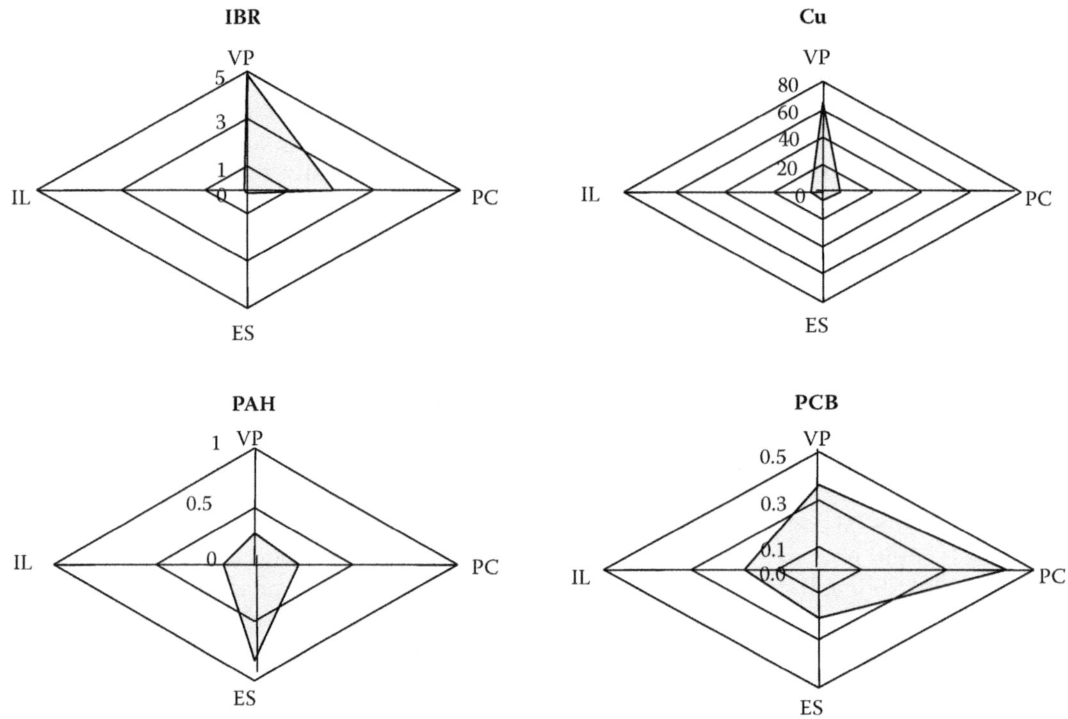

FIGURE 2.3
Integrated biomarker response (IBR) and pollutant star plots: IBR, copper, polycyclic aromatic hydrocarbon (PAH $\mu g \cdot g^{-1}$), and polychlorobiphenyl (PCB $\mu g \cdot g^{-1}$) star plots in mussels *Mytilus galloprovincialis* transplanted in spring 2004 at four sites in the Bay of Cannes (NW Mediterranean Sea, France): VP (old harbor), PC (Canto harbor), ES (mouth of the Siagne River), and IL (Lérins Island). (Adapted from Damiens, G. et al., *Chemosphere*, 66, 574–583, 2007.)

contribute to IBR, demonstrating that the chosen biomarkers did not respond to PAHs. In 2005, IBR (not presented in Figure 2.3) showed that other contaminants, not measured by the authors, might be present at exposed stations compared to the reference station.

According to Broeg and Lehtonen (2006), due to its mathematical basis, the IBR becomes more robust when the number of biomarkers increases. However, according to Guerlet (2007), several inconveniences can limit the use of this tool: a potentially significant influence of the order of the biomarkers on the value of the IBR, the impossibility of its application without *a priori* information on the stress gradient because of the fluctuating character of the types of responses of biomarkers (inhibition or antagonism), and overestimation of the stress as a result of redundancy of the responses integrated into the IBR.

Yeom and Adams (2007) have developed an aquatic ecosystem health index, based on the sum of all star-plot areas over several levels of biological organization to reflect an integrative and holistic assessment of stressors on ecosystem health and identify those levels of biological organization that have the greatest response to environmental stressors.

Dagnino et al. (2007) proposed an expert system that utilizes a suite of biomarker tests measured in marine mussels to translate complex biological responses into a relatively simple, easy to understand, and objective evaluation of the changes in the physiology of an organism induced by pollutants. Their classification was developed using a battery of nine biomarkers at different levels of biological organization, cell, tissue, and organism.

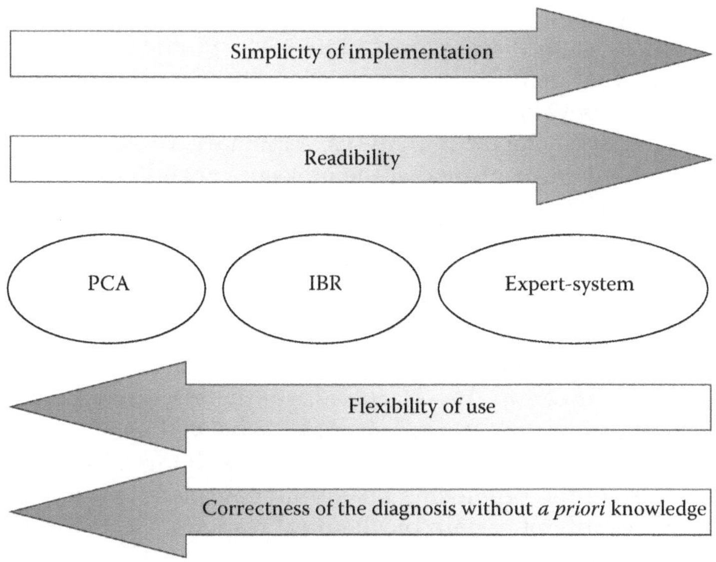

FIGURE 2.4
Synthetic comparison of three tools of integration of the responses of the battery of biomarkers in aquatic organisms. (Adapted from Guerlet, E., PhD thesis, University of Metz, France, 2007.)

The authors describe the profile of biomarkers (MTs, CAT, GST, AChE) along a gradient of pollution. The expert system selects as a guide parameter the biomarker that shows the highest sensitivity to stress, and interprets the other data in light of the alteration level reached by the guide parameter. More precisely, Viarengo et al. (2007), on the basis of the work of Dagnino et al. (2007), proposed a two-tier approach to assess the level of pollutant-induced stress syndromes in sentinel organisms. The LMS assessed either by neutral red retention or by a histochemical technique, provides a robust Tier 1 screening biomarker for environmental impact assessment. Tier 2, constituted by biomarkers of genotoxicity and by biomarkers revealing an exposure (MTs, AChE, EROD, MXR, transport activity, etc.), is used only for animals (mussels) sampled at sites in which LMS changes are evident, and there is no mortality. Then, the above-mentioned expert system is used. However, Guerlet (2007) notes that there is no parallel integration of the physicochemical data, and that for this tool an overestimation of the effects is also observed when there is redundancy between biomarkers.

Figure 2.4, adapted from Guerlet (2007), synthesizes the use of PCA, IBR, and the expert system in the integration of the battery of biomarkers in aquatic organisms. The comparison between different ways of treating the data shows that from PCA to the expert system, simplicity of implementation and readability increase. On the contrary, flexibility of use and correctness of diagnosis without *a priori* knowledge decrease.

2.6 Conclusions

It is known that toxicity resulting from pollutant exposure appears at the subcellular level before being observed at individual or population level. The relevant use of biomarkers rests on their feasibility within the framework of *in situ* studies and on a good knowledge of the risks for the ecosystem (Flammarion et al. 1998). Studies carried out over the past 30 years tend to show that monitoring of pollutant effects by measurement of biomarker responses in organisms is valid, especially if a battery of biomarkers is analyzed on the same sample. In contrast to chemical analyses, a biomarker response reflects the physiological state of an organism, examined at the molecular, cellular, or individual organism level. However, in spite of the acquired knowledge (laboratory experiment and field collection programs or active biomonitoring), certain points deserve to be underlined: (1) more chemical analyses are necessary to validate future biomarkers; (2) the sampling strategy of species of interest still can be improved; (3) comparisons between large geographical areas can be skewed because the biomarker response in some organisms varies, for example, along a gradient of salinity or because of seasonal variations in temperature, and of the physiological processes linked to these factors (assimilation, growth, and reproduction). Environmental conditions of each studied site have to be well known: (1) it is necessary to know the basic levels of the biomarkers according to the changes in temperature, salinity, and sexual maturation in the organisms taken into consideration in a given area; (2) an excess of pollutants can inhibit certain biochemical responses (e.g., EROD activity or MT level), just as a mixture of various pollutants.

Novel methods, in particular (eco)toxicogenomics and (eco)toxicoproteomics, provide integrated approaches to combine the responses of well-established core biomarkers in response to pollutants. The recent cloning of multiple genes in microalgae (Simon et al. 2008; Hutchins et al. 2010), but also in other species belonging to different phyla, has revealed several novel features of their transcriptional response, and recent progress in proteomics indicates that proteome modifications are useful to evaluate the effects of water pollution (Manduzio et al. 2005; Amelina et al. 2007). Profiles of differentially expressed genes can also be obtained via transcriptomics studies that have been developed considerably in recent years. Gornati et al. (2004) reported the coding sequences of Hsp70 and Hsp90 and a partial sequence of heat shock constitutive protein (HSC70) in the fish *Dicentrarchus labrax*. According to Geist et al. (2007), exposure of the striped bass (*Morone saxatilis*) to the pyrethroid insecticide esfenvalerate had tissue-specific effects on the transcription of HSP70, HSP90, and CYP1A1. The authors concluded that stress response at the transcriptome level is a more sensitive indicator for esfenvalerate exposure at low concentrations than swimming behavior, growth, or mortality. Dowling and Sheehan (2006) have demonstrated that proteomics could be a route to identification of toxicity targets in environmental toxicology. Relationships between the induction of responses, sensitivity to pollutants, and the possible consequences for exposed individuals and populations must be characterized; rapid development of genomics and proteomics tools is promising in this respect. Moreover, more and more work is being carried out with nonmodel organisms, and gene and protein sequences are increasing in databases, demonstrating the possibility of using organisms from different phyla according to their sensitivity to environmental pollutants.

Even if some biomarkers do not permit the assessment of ecological risks, they nevertheless give complementary and relevant information compared to chemical analyses because they take into account the bioavailability of chemical pollutants and not only their total concentration. Authorities in charge of environmental problems are speaking in terms of the Precautionary Principle and, in the absence of contrary evidence, any detection of a biomarker response (e.g., EROD activity, Lopez-Barea 1994, quoted in Flammarion et al. 1998) could be regarded as a signal of a potential risk for living organisms.

References

Aas, E. et al. 2000. PAH metabolites in bile, cytochrome P4501A and DNA adducts as environmental risk parameters for chronic oil exposure: A laboratory experiment with Atlantic cod. *Aquat. Toxicol.* 51:241–58.

Addison, R.F. 1984. Hepatic mixed function oxidase (MFO) induction in fish as a possible biological monitoring system. In: *Contaminant Effects on Fisheries*, ed. V.W. Cairns, P,V, Hodson, J.O. Nriagu, 51–60. Toronto: Wiley.

Addison, R.F., and J.F. Payne. 1987. Assessment of hepatic mixed function oxidase induction in winter flounder (*Pseudopleuronectes americanus*) as a marine petroleum pollution monitoring technique, with an appendix describing practical field measurements of MFO activity. *Can. Tech. Rept. Fish Aquat. Sci.* no. 150.

Aït Alla, A. et al. 2006. Tolerance and biomarkers as useful tools for assessing environmental quality in the Oued Souss estuary (Bay of Agadir, Morocco). *Comp. Biochem. Physiol.* 143C:23–9.

Akcha, F. et al. 1999. Relationship between kinetics of benzo[*a*]pyrene bioaccumulation and DNA binding in the mussel *Mytilus galloprovincialis*. *Bull. Environ. Contam. Toxicol.* 62:455–62.

Akcha, F. et al. 2000. Enzymatic biomarker measurement and study of DNA adduct formation in benzo[*a*]pyrene-contaminated mussels, *Mytilus galloprovincialis*. *Aquat. Toxicol.* 49:269–87.

Allan, I.J. et al. 2006. A "toolbox" for biological and chemical monitoring requirements for the European Union's Water Framework Directive. *Talanta* 69:302–22.

Almeida, E.A. et al. 2005. Oxidative stress in digestive gland and gill of the brown mussel (*Perna perna*) exposed to air and re-submersed. *J. Exp. Mar. Biol. Ecol.* 318:21–30.

Almeida, E.A. et al. 2007. Oxidative stress in *Perna perna* and other bivalves as indicators of environmental stress in the Brazilian marine environment: Antioxidants, lipid peroxidation and DNA damage. *Comp. Biochem. Physiol.* 146A:588–600.

Amelina, H. et al. 2007. Proteomics-based method for the assessment of marine pollution using liquid chromatography coupled with two-dimensional electrophoresis. *J. Proteome Res.* 6:2094–104.

Amiard, J.C. et al. 2006. Metallothioneins in aquatic invertebrates: Their role in metal detoxification and their use as biomarkers. *Aquat. Toxicol.* 76:160–202.

Amiard-Triquet, C. et al. 1998. Metallothionein in Arctic bivalves. *Ecotoxicol. Environ. Saf.* 41:96–102.

Andersson, T., and E. Nilsson. 1989. Characterization of cytochrome P450 dependent activities in hagfish, dogfish, perch and spectacle caiman. *Comp. Biochem. Physiol.* 94B:99–105.

Ansaldo, M., H. Sacristán, and E. Wider. 2007. Does starvation influence the antioxidant status of the digestive gland of *Nacella concinna* in experimental conditions? *Comp. Biochem. Physiol.* 146C:118–23.

Antkiewicz, D.S., R.E. Peterson, and W. Heideman. 2006. Blocking expression of AHR2 and ARNT1 in zebra larvae protects against cardiac toxicity of 2,3,7,8-tetrachlorodibenzo-*p*-dioxin. *Toxicol. Sci.* 94:175–82.

Atienzar, F.A., and A.N. Jha. 2006. The random amplified polymorphic DNA (RAPD) assay and related techniques applied to genotoxicity and carcinogenesis studies: A critical review. *Mutat. Res.* 613:76–102.

Atkinson, B.G., and D.B. Walden (eds). 1985. *Changes in Eukaryotic Gene Expression in Response to Environmental stress*. New York: Academic Press.

Baudrimont, M. et al. 1997. Bioaccumulation and metallothionein response in the Asiatic clam (*Corbicula fluminae*) after experimental exposure to cadmium and inorganic mercury. *Environ. Toxicol. Chem.* 16:2096–105.

Bayne, B.L., K.R. Clarke, and J.S. Gray. 1988a. Background and rationale to a practical workshop on biological effects of pollutants. *Mar. Ecol. Prog. Ser.* 46:1–5.

Bayne, B.L. et al. 1988b. An overview of the GEEP workshop. *Mar. Ecol. Prog. Ser.* 46:235–43.

Bebianno, M.J. et al. 2003. Metallothionein concentrations in a population of *Patella aspersa*: Variation with size. *Sci. Total Environ.* 301:151–61.

Bebianno, M.J., and W.J. Langston. 1998. Cadmium and metallothionein turnover in different tissues of the gastropod *Littorina littorea*. *Talanta* 46:301–13.

Beliaeff, B., and T. Burgeot. 2002. Integrated biomarker response: A useful tool for ecological risk assessment. *Environ. Toxicol. Chem.* 21:1316–22.

Blaise, C. et al. 1999. Determination of vitellogenin-like properties in *Mya arenaria* hemolymph (Saguenay Fjord, Canada): A potential biomarker for endocrine disruption. *Environ. Toxicol.* 14:455–65.

Bocchetti, R., and F. Regoli. 2006. Seasonal variability of oxidative biomarkers, lysosomal parameters, metallothioneins and peroxisomal enzymes in the Mediterranean mussel *Mytilus galloprovincialis* from Adriatic Sea. *Chemosphere* 65:913–21.

Bocquené, G., A. Roig, and D. Fournier. 1997. The molecular forms of acetylcholinesterase from the common oyster (*Crassostrea gigas*). *FEBS Lett.* 407:261–6.

Boryslawskyj, M. et al. 1988. Elevation of glutathione *S*-transferase activity as a stress response to organochlorine compounds, in the freshwater mussel, *Sphaerium corneum*. *Mar. Environ. Res.* 24:101–4.

Broeg, K., and K.K. Lehtonen. 2006. Indices for the assessment of environmental pollution of the Baltic Sea coasts: Integrated assessment of a multi-biomarker approach. *Mar. Pollut. Bull.* 53:508–22.

Broeg, K., A. Köhler, and H. Van Westernhagen. 2002. Disorder and recovery of environmental health monitored by means of lysosomal stability in liver of European flounder (*Platichthys flesus* L.). *Mar. Environ. Res.* 54:569–73.

Brown, D.C., B.P. Bradley, and M. Tedengren. 1995. Genetic and environmental regulation of HSP70 expression. *Mar. Environ. Res.* 39:181–4.

Brumley, C.M. et al. 1995. Validation of biomarkers of marine pollution exposure in sand flathead using Aroclor 1254. *Aquat. Toxicol.* 31:249–62.

Burcham, P.C. 1999. Internal hazards: Baseline DNA damage by endogenous products of normal metabolism. *Mutat. Res.* 443:11–36.

Burlinson, B. et al. 2007. Fourth international workgroup on genotoxicity testing: Results of the in vivo comet assay workgroup. *Mutat. Res.* 627:31–5.

Cadet, J., E. Sage, and T. Douki. 2005. Ultraviolet radiation-mediated damage to cellular DNA. *Mutat. Res.* 571:3–17.

Cajaraville, M.P. et al. 2000. The use of biomarkers to assess the impact of pollution in coastal environments of the Iberian Peninsula: A practical approach. *Sci. Total Environ.* 247:295–311.

Caltabiano, M.M. et al. 1986. Induction of 32 and 34-kDa stress proteins by sodium arsenic, heavy metals, and thiol-reactive agents. *J. Biol. Chem.* 261:13381–6.

Carson, R. 1962. *Silent Spring*. Boston: Houghton Mifflin; re-published by Mariner Books (2002).

Cascorbi, I. 2006. Role of pharmacogenetics of ATP-binding cassette transporters in the pharmacokinetics of drugs. *Pharmacol. Ther.* 112:457–73.

Castren, M., and A. Oikari. 1987. Changes of the liver UDP-glucuronosyltransferase activity in trout (*Salmo gairdneri* Rich.) acutely exposed to selected aquatic toxicants. *Comp. Biochem. Physiol.* 86C:357–60.

Cavaletto, A. et al. 2002. Effect of hydrogen peroxide on antioxidant enzymes and metallothionein level in the digestive gland of *Mytilus galloprovincialis*. *Comp. Biochem. Physiol.* 131C:447–55.

Charissou, A.M., C. Cossu-Leguille, and P. Vasseur. 2004. Relationship between two oxidative stress biomarkers, malondialdehyde and 8-oxo-7,8-dihydro-29-deoxyguanosine, in the freshwater bivalve *Unio tumidus*. *Sci. Total Environ.* 322:109–22.

Coppage, D.L., and T. Braidech. 1976. River pollution by anticholinesterase agents. *Wat. Res.* 10:19–24.

Cosson, R.P. 2000. Bivalve metallothioneins as a biomarker of aquatic ecosystem pollution by trace metals: Limits and perspectives. *Cell. Mol. Biol.* 46:295–309.

Cosson, R.P., and J.C. Amiard. 2000. Use of metallothionein as biomarkers of exposure to metals. In *Use of Biomarkers for Environmental Quality Assessment*, ed. L. Lagadic et al., 79–111. Enfield, NH: Science Publishers.

Cossu, C. et al. 1997. Biomarqueurs du stress oxydant chez les animaux aquatiques. In *Biomarqueurs en écotoxicologie*, ed. L. Lagadic et al., 149–63. Paris: Masson.

Cossu, C. et al. 2000. Antioxidant biomarkers in freshwater bivalve *Unio tumidus* exposed to different pollution profiles. *Ecotoxicol. Environ. Saf.* 45:106–21.

Craig, E.A., J.S. Weissman, and A.L. Horwich. 1994. Heat shock proteins and molecular chaperones: Mediators of protein conformation and turnover in the cell. *Cell* 78:365–72.

Creagh, E.M., R.J. Carmody, and T.G. Cotter. 2000. Heat shock protein 70 inhibits caspase-dependent and -independent apoptosis in Jurkat T cells. *Exp. Cell. Res.* 257:58–66.

Dagnino, A. et al. 2007. Development of an expert system for the integration of biomarker responses in mussels into an animal health index. *Biomarkers* 12:155–72.

Dallinger, R. 1996. Metallothionein research in terrestrial invertebrates: Synopsis and perspectives. *Comp. Biochem. Physiol.* 113C:125–33.

Damiens, G., and C. Minier. 2011. The multixenobiotic transport system: A system governing intracellular contaminant bioavailability. In: *Tolerance to Environmental Contaminants*, ed. C. Amiard-Triquet, P.S. Rainbow, and M. Roméo, 229–46. Boca Raton, FL: CRC Press.

Damiens, G. et al. 2004. Evaluation of biomarkers in oyster larvae in natural and polluted conditions. *Comp. Biochem. Physiol.* 138C:121–8.

Damiens, G. et al. 2007. Integrated biomarker response index as a useful tool for environmental assessment evaluated using transplanted mussels. *Chemosphere* 66:574–83.

Day, K.E., and I.M. Scott. 1990. Use of acetylcholinesterase activity to detect sublethal toxicity in stream invertebrates exposed to low concentrations of organophosphate insecticides. *Aquat. Toxicol.* 18:101–14.

De Kock, W.C., and K.J.M. Kramer. 1994. Active biomonitoring (ABM) by translocation of bivalve molluscs. In *Biomonitoring of Coastal Waters and Estuaries*, ed. K.J.M. Kramer, 51–84. Boca Raton, FL: CRC Press.

De Lafontaine, Y. et al. 2000. Biomarkers in zebra mussels (*Dreissena polymorpha*) for the assessment and monitoring of water quality of the St Lawrence River (Canada). *Aquat. Toxicol.* 50:51–71.

Depledge, M.H. 1994. The rational basis for the use of biomarkers as ecotoxicological tools. In *Nondestructive Biomarkers in Vertebrates*, ed. M.C. Fossi and C. Leonzio, 271–95. Boca Raton, FL: Lewis Publishers.

Depledge, M.H., and M.C. Fossi. 1994. The role of biomarkers in environmental assessment (2). Invertebrates. *Ecotoxicology* 3:161–72.

Devaux, A., M. Pesonen, and G. Monod. 1997. Alkaline comet assay in rainbow trout hepatocytes. *Toxicol. In Vitro* 11:71–9.

Di Giulio, R.T. et al. 1989. Biochemical responses in aquatic animals: A review of determinants of oxidative stress. *Environ. Toxicol. Chem.* 8:1103–23.

Di Giulio, R.T. et al. 1995. Biochemical mechanisms of contaminant metabolism, adaptation, and toxicity. In *Fundamentals of Aquatic Toxicology*, 2nd ed., ed. G. Rand, 523–61. Bristol, PA: Taylor & Francis.

Dowling, V.A., and D. Sheehan. 2006. Proteomics as a route to identification of toxicity targets in environmental toxicology. *Proteomics* 6: 5597–5604.

Durand, F., F. Peters, and D.R. Livingstone. 2001. Effect of intertidal compared to subtidal exposure on the uptake, loss and oxidative toxicity of water-born benzo[*a*]pyrene in the mantle and whole tissues of the mussel, *Mytilus edulis* L. *Mar. Environ. Res.* 54:271–4.

Durou, C. et al. 2007. Biomonitoring in a clean and a multi-contaminated estuary based on biomarkers and chemical analyses in the endobenthic worm *Nereis diversicolor. Environ. Pollut.* 66:402–11.

Ernst, W.H.O., and P.J. Peterson. 1994. The role of biomarkers in environmental assessment (4). Terrestrial plants. *Ecotoxicology* 3:180–92.

Estabrook, R.W., D.Y. Cooper, and O. Rosenthal. 1963. The light reversible carbon monoxide inhibition of steroid C-21 hydroxylase system of the adrenal cortex. *Biochem. Zeit.* 338:741–55.

Eufemia, N.A., and D. Epel. 2000. Induction of multixenobiotic defense mechanism (MXR), P-glycoprotein, in the mussel *Mytilus californianus* as a general cellular response to environmental stresses. *Aquat. Toxicol.* 49:89–100.

Fitzpatrick, P.J. et al. 1997. Assessment of a glutathione *S*-transferase and related proteins in the gill and digestive gland of *Mytilus edulis* (L), as potential organic pollution biomarkers. *Biomarkers* 2:51–6.

Flammarion, P., J. Garric, and G. Monod G. 1998. Utilisation de l'activité enzymatique EROD chez les poissons des hydrosystèmes continentaux. In *Utilisation de biomarqueurs pour la surveillance de la qualité de l'environnement*, ed. L. Lagadic, T. Caquet, J.C. Amiard, and F. Ramade, 57–75. Paris: Lavoisier Tec & Doc.

Flammarion, P. et al. 2000. Induction of fish vitellogenin and alterations in testicular structure: Preliminary results of estrogenic effects in chub (*Leuciscus cephalus*). *Ecotoxicology* 9:127–35.

Flammarion, P. et al. 2002. Multibiomarker responses in fish from the Moselle river (France). *Ecotoxicol. Environ. Saf.* 51:145–53.

Foureman, G.L. 1989. Enzymes involved in metabolism of PAH by fishes and other aquatic animals: Hydrolysis and conjugation enzymes (or phase II enzymes), In *Metabolism of Polycyclic Aromatic Hydrocarbons in the Aquatic Environment*, ed. U. Varanasi, 185–202. Boca Raton, FL: CRC Press.

Fowler, B.A. et al. 1987. Nomenclature of metallothioneins. In *Metallothionein II*, ed. J.H.R. Kagi and Y. Kojima, 19–22. Basel: Birhauser-Verlag.

Freeman, M.L. et al. 1999. On the path to the heat shock response: Destabilisation and formation of partially folded protein intermediates, a consequence of protein thiol formation. *Free Radic. Biol. Med.* 26:737–45.

Frydman, J. 2001. Folding of newly translated proteins *in vivo*: The role of molecular chaperones. *Annu. Rev. Biochem.* 70:603–47.

Gagné, F. et al. 2001. Evaluation of estrogenic effects of municipal effluents to the freshwater mussel *Elliptio complanata*. *Comp. Biochem. Physiol.* 128C:213–25.

Gagné, F. et al. 2002. Biomarker study of a municipal effluent dispersion plume in two species of freshwater mussels. *Environ. Toxicol.* 17:149–59.

Gagné, F. et al. 2006. Health status of *Mya arenaria* bivalves collected from contaminated sites in Canada (Saguenay Fjord) and Denmark (Odense Fjord) during their reproductive period. *Ecotoxicol. Environ. Saf.* 64:348–61.

Gagnon, M.M., and D.A. Holdway. 2000. EROD induction and biliary metabolite excretion following exposure to the water accommodated fraction of crude oil and to chemically dispersed crude oil. *Arch. Environ. Contam. Toxicol.* 38:70–7.

Galgani, F., and G. Bocquené. 1989. A method for routine detection of organophosphorous and carbamates in sea water. *Environ. Technol. Lett.* 10:311–22.

Galgani, F., G. Bocquené, and Y. Cadiou. 1992. Evidence of variation in cholinesterase activity in fish along a pollution gradient in the North Sea. *Mar. Ecol. Prog. Ser.* 91:77–82.

Galgani, F. et al. 1996. Interaction of environmental xenobiotics with a multixenobiotic defense mechanism in the bay mussel *Mytilus galloprovincialis* from the coast of California. *Aquat. Toxicol.* 15:325–31.

Garfinkel, D. 1958. Studies on pig liver microsomes. Enzyme and pigment composition of different microsomal fractions. *Arch. Biochem. Biophys.* 77:493–509.

Geist, J. et al. 2007. Comparisons of tissue-specific transcription of stress response genes with whole animal endpoints of adverse effect in striped bass (*Morone saxatilis*) following treatment with copper and esfenvalerate. *Aquat. Toxicol.* 5:28–39.

George, S.G. 1994. Enzymology and molecular biology of phase II xenobiotic-conjugating enzymes in fish. In *Aquatic Toxicology*, ed. D.C. Malins and G.K. Ostrander, 37–85. Boca Raton, FL: Lewis.

George, S.G. et al. 1992. Metallothionein induction in cultured fibroblasts and liver of a marine flatfish, the turbot *Scophtalmus maximus*. *Fish Physiol. Biochem.* 10:43–54.

Giguère, A.Y. et al. 2003. Steady-state distribution of metals among metallothionein and other cytosolic ligands and links to cytotoxicity in bivalves living along a polymetallic gradient. *Aquat. Toxicol.* 64:185–200.

Goksøyr, A., and L. Förlin. 1992. The cytochrome P-450 system in fish, aquatic toxicology and environmental monitoring. *Aquat. Toxicol.* 22:287–311.

Gornati, R. et al. 2004. Rearing density influences the expression of stress-related genes in sea bass (*Dicentrarchus labrax*, L.). *Gene* 341:111–8.
Gottesman, M.M., and I. Pastan. 1993. Biochemistry of multidrug resistance mediated by the multidrug transporter. *Annu. Rev. Biochem.* 62:385–427.
Gregus, Z. et al. 1983. Hepatic phase I and phase II biotransformations in quail and trout: Comparison to other species commonly used in toxicity testing. *Toxicol. Appl. Pharmacol.* 67:430–441.
Guengerich, F.P. 1993. Cytochrome P450 enzymes. *Am. Sci.* 81:440–7.
Guerlet, E. 2007. Utilisation de biomarqueurs cellulaires chez plusieurs espèces d'invertébrés pour l'évaluation de la contamination des milieux dulçaquicoles. Thèse de Doctorat, Université de Metz.
Hahn, M.E. 2002. Aryl hydrocarbon receptors: Diversity and evolution. *Chem-Biol. Interact.* 141:131–60.
Hamilton, S.J., and P.M. Mehrle. 1986. Metallothionein in fish: Review of its importance in assessing stress from metal contaminants. *Trans. Am. Fish. Soc.* 115:596–609.
Hammond, G.L., Y.K. Lai, and C.L. Market. 1982. Diverse forms of stress lead to new patterns of gene expression through a common and essential metabolic pathway. *Proc. Natl. Acad. Sci. U. S. A.* 79:3485–8.
Hamza-Chaffai, A. et al. 1995. Physicochemical forms of storage of metals (Cd, Cu and Zn) and metallothionein like proteins in gills and liver of marine fish from the Tunisian coast: Ecotoxicological consequences. *Comp. Biochem. Physiol.* 111C:329–41.
Hamza-Chaffai, A., J.C. Amiard, and R.P. Cosson. 1999. Relationship between metallothioneins and metals in a natural population of the clam *Ruditapes decussatus* from Sfax Coast: A non-linear model using Box–Cox transformation. *Comp. Biochem. Physiol.* 123C:153–63.
Hartl, F.U., and M. Hayer-Hartl. 2002. Molecular chaperones in the cytosol: From nascent chain to folded protein. *Science* 295:1852–8.
Hightower, L.E. 1993. A brief perspective on the heat-shock response and stress proteins. *Mar. Environ. Res.* 35:IN1–2.
Hoarau, P. et al. 2001. Differential induction of glutathione *S*-transferases in the clam *Ruditapes decussatus* exposed to organic compounds. *Environ. Toxicol. Chem.* 20:523–9.
Holland, H.T., D.L. Coppage, and P.L. Butler. 1967. Use of fish brain acetylcholinesterase to monitor pollution by organophosphorous pesticides. *Bull. Environ. Contam. Toxicol.* 2:156–62.
Huggett, R.J. et al. 2003. Biomarkers in fish from Prince William Sound and the Gulf of Alaska. *Environ. Sci. Technol.* 37:4043–51.
Hutchins, C.M. et al. 2010. Transcriptomic signatures in *C. reinhardtii* as Cd biomarkers in metal mixtures. *Aquat. Toxicol.* 100:120–7.
Hwang, E.S., and G.H. Kim. 2007. Biomarkers for oxidative stress status of DNA, lipids, and proteins *in vitro* and *in vivo* cancer research. *Toxicology* 229:1–10.
Isani, G. et al. 2000. Metallothioneins (MTs) in marine molluscs. *Cell. Mol. Biol.* 46:311–30.
Iwanowicz, L.R. et al. 2009. Reproductive health of bass in the Potomac, USA, drainage: Part 1. Exploring the effects of proximity to wastewater treatment plant discharge. *Environ. Toxicol. Chem.* 28:1072–83.
Juliano, R.L., and V. Ling. 1976. A surface glycoprotein modulating drug permeability in Chinese hamster ovary cell mutants. *Biochim. Biophys. Acta* 455:152–62.
Klingenberg, M. 1958. Pigment of rat liver microsomes. *Arch. Biochem. Biophys.* 75:376–86.
Knight, J.A., R.K. Pleper, and L. McClellan. 1988. Specificity of the thiobarbituric acid reaction: Its use in studies of lipid peroxidation. *Clin. Chem.* 34:2433–8.
Kurelec, B. 1992. The multixenobiotic resistance mechanism in aquatic organisms. *Crit. Rev. Toxicol.* 22:23–43.
Kurelec, B. et al. 1995. Induction and reversion of multixenobiotic resistance in the marine snail *Monodonta turbinata*. *Mar. Biol.* 123:305–12.
Lafaurie, M. et al. 1992. Indicateurs biochimiques de contamination de l'environnement marin. *Analusis Mag*. 20:M27–33.
Lagadic, L., T. Caquet, and F. Ramade. 1994. The role of biomarkers in environmental assessment (5). Invertebrate populations and communities. *Ecotoxicology* 3:193–208.
Larsson, D.G.J. et al. 1999. Ethinyloestradiol—an undesired fish contraceptive? *Aquat. Toxicol.* 45:91–7.

Lee, R.F. 2002. Bioavailability, biotransformation, and fate of organic contaminants in estuarine animals. In *Coastal and Estuarine Risk Assessment*, ed. M.C. Newman, M.H. Roberts, and R.C. Hale, 97–126. Boca Raton, FL: Lewis Publishers.

Lehtonen, K.K. et al. 2006. The BEEP project in the Baltic Sea: Overview of results and outline for a regional biological effects monitoring strategy. *Mar. Pollut. Bull.* 53:508–22.

Leiniö, S., and K.K. Lehtonen. 2005. Seasonal variability in biomarkers in the bivalves *Mytilus edulis* and *Macoma balthica* from the northern Baltic Sea. *Comp. Biochem. Physiol.* 140C:408–21.

Leslie, E.M., R.G. Deeley, and S.P. Cole. 2005. Multidrug resistance proteins: Role of P-glycoprotein, MRP1, MRP2, and BCRP (ABCG2) in tissue defense. *Toxicol. Appl. Pharmacol.* 204:216–37.

Leung, K.M.Y., and R.W. Furness. 2001. Survival, growth, metallothionein and glycogen levels of *Nucella lapillus* (L.) exposed to sub-chronic cadmium stress: The influence of nutritional state and prey type. *Mar. Environ. Res.* 52:173–94.

Littré, E. 1861. *Oeuvres complètes d'Hippocrate (1839–1861)*. Paris: Baillière.

Livingstone, D.R. et al. 1990. Oxyradical production as a pollution-mediated mechanism of toxicity in the common mussel, *Mytilus edulis* L. and other molluscs. *Funct. Ecol.* 4:415–24.

Lushchak, V.I., and T.V. Bagnyukova. 2006. Temperature increase results in oxidative stress in goldfish tissues: 1. Indices of oxidative stress. *Comp. Biochem. Physiol.* 143C:30–5.

Manduzio, H. et al. 2005. Proteome modifications of blue mussel (*Mytilus edulis* L.) gills as an effect of water pollution. *Proteomics* 5:4958–63.

Margoshes, M., and B.L. Vallee. 1957. A cadmium protein from equine kidney cortex. *J. Am. Chem. Soc.* 79:4813–4.

Martinez, G.R. et al. 2003. Oxidative and alkylating damage in DNA. *Mutat. Res.* 544:115–27.

McCarthy, J.F., and L.R. Shugart. 1990. *Biomarkers of Environmental Contamination*. Boca Raton, FL: Lewis Publishers.

Michel, X.R. et al. 1993. Effects of benzo[*a*]yrene, 3,3′,4,4′-tetrachlorobiphenyl and 2,2′,4,4′,5,5′-hexachlorobiphenyl on the xenobiotic-metabolizing enzymes in the mussel (*Mytilus galloprovincialis*). *Aquat. Toxicol.* 27:335–44.

Minier, C., and M.N. Moore. 1998. Calcein accumulation in mussel blood cells. *Mar. Environ. Res.* 46:425–8.

Minier, C., F. Akcha, and F. Galgani. 1993. P-glycoprotein expression in *Crassostrea gigas* and *Mytilus galloprovincialis* in polluted water. *Comp. Biochem. Physiol.* 106B:1029–36.

Møller, P., and H. Wallin. 1998. Adduct formation, mutagenesis and nucleotide excision repair of DNA damage produced by reactive oxygen species and lipid peroxidation product. *Mutat. Res.* 410:271–90.

Monod, G., A. Devaux, and J.L. Rivière. 1988. Effects of chemical pollution on the activities of hepatic xenobiotics metabolizing enzymes in fish from the River Rhone. *Sci. Total Environ.* 73:189–201.

Monserrat, J.M. et al. 2007. Pollution biomarkers in estuarine animals: Critical review and new perspectives. *Comp. Biochem. Physiol.* 146C:221–34.

Moore, M.N. 1988. Cytochemical responses of the lysosomal system and NADPH-ferrihemoprotein reductase in molluscan digestive cells to environmental and experimental exposure xenobiotics. *Mar. Ecol. Prog. Ser.* 46:81–9.

Mora, P., X. Michel, and J.F. Narbonne. 1999. Cholinesterase activity as potential biomarker in two bivalves. *Environ. Toxicol. Pharmacol.* 7:253–60.

Moromoto, R., A. Tissières, and C. Georgopoulos. 1990. *The Role of the Stress Response in Biology and Disease*. Cold Springs Harbor: Cold Spring Harbor Laboratory.

Nagler, J.J. et al. 1987. Serum phosphoprotein phosphorus and calcium levels as reproductive indicators of vitellogenin in highly vitellogenic mature female and oestradiol-injected immature rainbow trout (*Oncorhyncus mykiss*). *Can. J. Zool.* 65:2421–5.

Nebert, D.W. 1994. Drug-metabolizing enzymes in ligand-modulated transcription. *Biochem. Pharmacol.* 47:25–37.

Nebert, D.W., and D.R. Nelson. 1991. P-450 gene nomenclature based on evolution. *Methods Enzymol.* 206:3–11.

Nelson, D.R. 1998. Metazoan cytochrome P450 evolution. *Comp. Biochem. Physiol.* 121C:15–22.

Nigro, M. et al. 2006. Cellular biomarkers for monitoring estuarine environments: Transplanted versus native mussels. *Aquat. Toxicol.* 77:339–47.
Niyogia, S. et al. 2001. Antioxidant enzymes in brackishwater oyster, *Saccostrea cucullata* as potential biomarkers of polyaromatic hydrocarbon pollution in Hooghly Estuary (India): Seasonality and its consequences. *Sci. Total Environ.* 281:237–46.
National Research Council (NRC). 1987. Biological markers in environmental health research. *Environ. Health Perspect.* 74:3–9.
National Research Council (NRC). 1989. *Biological Markers in Reproductive Toxicology*. Washington, DC: National Academic Press.
Omura, T., and R. Sato. 1964. The carbon monoxide pigment of microsomes. *J. Biol. Chem.* 239:2379–85.
Orbea, A. et al. 2002. Antioxidant enzymes and peroxisome proliferation in relation to contaminant body burdens of PAHs and PCBs in bivalve molluscs, crabs and fish from the Urdaibai and Plentzia estuaries (Bay of Biscay). *Aquat. Toxicol.* 58:75–98.
Pampanin, D.M. et al. 2006. Background for the BEEP Stavanger workshops: Biological effects on marine organisms in two common, large, laboratory experiments and in a field study. Comparison of the value (sensitivity, specificity, etc.) of core and new biomarkers. *Aquat. Toxicol.* 78S:S1–4.
Pannuzio, T.M., and K.B. Storey. 1998. Antioxidant defenses and lipid peroxidation during anoxia stress and areobic recovery in marine gastropod *Littorina littorea*. *J. Exp. Mar. Biol. Ecol.* 221:277–92.
Payne, J., and W.R. Penrose. 1975. Induction of aryl hydrocarbon (benzo[*a*]pyrene) hydroxylase in fish by petroleum. *Bull. Environ. Contam. Toxicol.* 14:112–6.
Payne, J.F. et al. 1996. Acetylcholinesterase, an old biomarker with a new future? Field trials in association with two urban rivers and a paper mill in Newfoundland. *Mar. Pollut. Bull.* 32:225–31.
Peakall, D.B., and C.H. Walker. 1994. The role of biomarkers in environmental assessment (3). Vertebrates. *Ecotoxicology* 3:173–9.
Pellerin-Massicotte, J., and R. Tremblay. 2000. Lysosomal fragility as cytological biomarker. In *Use of Biomarkers for Environmental Quality Assessment*, ed. L. Lagadic et al., 229–46, Enfield, NH: Science Publishers.
Pfeifer, S., D. Schiedek, and J.W. Dippner. 2005. Effect of temperature and salinity on acetylcholinesterase activity, a common pollution biomarker, in *Mytilus* sp. from the south-western Baltic Sea. *J. Exp. Mar. Biol. Ecol.* 320:93–103.
Poland, A., and E. Glover. 1975. Genetic expression of aryl hydrocarbon hydroxylase by 2,3,7,8-tetrachloro dibenzo-*p*-dioxin: Evidence of a receptor mutation in genetically non-responsive mice. *Mol. Pharmacol.* 11:389–98.
Purdom, C.E. et al. 1994. Estrogenic effects of effluents from sewage treatment works. *Chem. Ecol.* 8:275–85.
Radetski, C.M. et al. 2004. Evaluation of the genotoxic, mutagenic and oxidant stress potentials of municipal solid waste incinerator bottom ash leachates. *Sci. Total Environ.* 333:209–16.
Ramagopal, S. 1987. Salinity stress induced tissue specific proteins in barley seedlings. *Plant Physiol.* 84: 324–31.
Regoli, F. et al. 2002a. Oxidative stress in ecotoxicology: From the analysis of individual antioxidants to a more integrated approach. *Mar. Environ. Res.* 54:419–23.
Regoli, F. et al. 2002b. Seasonal variations of susceptibility to oxidative stress in *Adamussium colbecki*, a key bioindicator species for the Antarctic marine environment. *Sci. Total Environ.* 289:205–11.
Regoli, F., M. Benedetti, and M.E. Giulani. 2011. Antioxidant defenses and acquisition of tolerance to chemical stress. In *Tolerance to Environmental Contaminants*, ed. C. Amiard-Triquet, P.S. Rainbow, and M. Roméo, 153–73. Boca Raton, FL: Taylor & Francis.
Riek, R. et al. 1999. NMR structure of the sea urchin (*Strongylocentrotus purpuratus*) metallothionein MTA. *JMB* 291:417–28.
Riggio, M. et al. 2003. Changes in zinc, copper and metallothionein contents during oocyte growth and early development of the teleost *Danio rerio* (zebrafish). *Comp. Biochem. Physiol.* 135C: 191–6.

Ritossa, F. 1962. A new puffing pattern induced by heat shock and DNP in *Drosophila*. *Experientia* 18:571–3.

Roberts, A.P., and J.T. Oris. 2004. Multiple biomarker response in rainbow trout during exposure to hexavalent chromium. *Comp. Biochem. Physiol.* 138C:221–8.

Robillard, S., G. Beauchamp, and M. Laulier. 2003. The role of abiotic factors and pesticide levels on enzymatic activity in the freshwater mussel *Anodonta cygnea* at three different exposure sites. *Comp. Biochem. Physiol.* 135C:49–59.

Roesijadi, G. 1992. Metallothioneins in metal regulation and toxicity in aquatic animals. *Aquat. Toxicol.* 22:81–114.

Roesijadi, G. 1996. Metallothioneins and its role in toxic metal regulation. *Comp. Biochem. Physiol.* 113C:117–23.

Roméo, M. et al. 1997. Metallothionein determination in the liver of the sea bass *Dicentrarchus labrax* treated with copper and B[*a*]P. *Mar. Environ. Res.* 44: 275–284.

Roméo, M. et al. 2003. Multimarker approach in transplanted mussels for evaluating water quality in Charentes, France, coast areas exposed to different anthropogenic conditions. *Environ. Toxicol.* 18:295–305.

Roméo, M., and I.I. Wirgin. 2011. Biotransformation of organic contaminants and the acquisition of tolerance. In *Tolerance to Environmental Contaminants*, ed. C. Amiard-Triquet, P.S. Rainbow, and M. Roméo, 175–208. Boca Raton, FL: CRC Press.

Rothman, J.E. 1989. Polypeptide chain binding proteins: Catalysts of protein folding and related processes in cells. *Cell* 59:591–601.

Sanders, B.M. 1990. Stress proteins: Potential as multitiered biomarkers. In *Environmental Biomarkers*, ed. L. Shugart and J. McCarthy, 165–91. Chelsea, MI: Lewis Publishers.

Sanders, B.M. et al. 1991. Relationships between accumulation of a 60 kDa stress protein and scope for growth in *Mytilus edulis* exposed to a range of copper concentrations. *Mar. Environ. Res.* 31:81–7.

Sanders, B.M. et al. 1994. Specific cross-reactivity of antibodies raised against two major stress proteins, stress 70 and chaperonin 60 in diverse species. *Environ. Toxicol. Chem.* 13:1241–9.

Schlesinger, M.J., M. Ashburner, and A. Tissières. 1982. *Heat Shock from Bacteria to Man*, 1–44. Cold Springs Harbor, NY: Cold Spring Harbor Laboratory.

Shaw, J.P. et al. 2004. Seasonal variations in cytochrome P450 immunopositive protein levels, lipid peroxidation and genetic toxicity in digestive gland of the mussel *Mytilus edulis*. *Aquat. Toxicol.* 67:325–36.

Shimizu, Y. et al. 2000. Benzo[*a*]pyrene carcinogenicity is lost in mice lacking the aryl hydrocarbon receptor. *Proc. Natl. Acad. Sci. U. S. A.* 97:779–82.

Simon, D.F. et al. 2008. Global expression profiling of *Chlamydomonas reinhardtii* exposed to trace levels of free cadmium. *Environ. Toxicol. Chem.* 27:1668–75.

Simpson, A.E.C.M. 1997. The cytochrome P450 4 (CYP4) family. *Gen. Pharmacol. Vasc. Syst.* 28:351–9.

Smital, T., and B. Kurelec. 1998. The chemosensitizers of multixenobiotic resistance mechanism in aquatic invertebrates: A new class of pollutants. *Mutat. Res.* 399:43–53.

Soto, A.M. et al. 1995. The E-SCREEN assay as a tool to identify estrogens: An update on estrogenic environmental pollutants. *Environ. Health Perspect.* 103:113–22.

Spector, M.P. et al. 1986. Global control in *Salmonella typhimurium*: Two dimensional electrophoretic analysis of starvation, anaerobiosis, and heat shock-inducible proteins. *J. Bacteriol.* 168:420–4.

Stegeman, J.J. 1987. Cytochrome P450 isozymes and monoxygenase activity in aquatic animals. *Environ. Health Perspect.* 71:87–95.

Stegeman, J.J., and M.E. Hahn. 1994. Biochemistry and molecular biology of monoxygenases: Current perspectives on forms, functions, and regulation of cytochrome P450 in aquatic species. In *Aquatic Toxicology: Molecular, Biochemical and Cellular Perspectives,* ed. G.K. Ostrander and D. Malins, 87–206. Boca Raton, FL: Lewis Publishers.

Toyooka, T., and Y. Ibuki. 2007. DNA damage by coexposure to PAHs and light. *Environ. Toxicol. Pharmacol.* 23:256–63.

Tutundjian, R., and C. Minier. 2002. Les protéines de résistance multiple et leur exploitation pour la biosurveillance chez les organismes aquatiques. *Regard Biochim.* 4:37–49.

Tyler, C.R., and E.J. Routledge. 1998. Natural and anthropogenic environmental oestrogens: The scientific basis for risk assessment. Oestrogenic effects in fish in English rivers with evidence of their causation. *Pure Appl. Chem.* 70:1795–804.

Valavanidis, A. et al. 2006. Molecular biomarkers of oxidative stress in aquatic organisms in relation to toxic environmental pollutants. *Ecotoxicol. Environ. Saf.* 64:178–89.

Van der Oost, R., C. Porte-Visa, and N.W. Van den Brink. 2005. Biomarkers in environmental assessment. In *Ecotoxicological Testing of Marine and Freshwater Ecosystems: Emerging Techniques, Trends, and Strategies,* ed. P.J. Den Besten and M. Munawar, 87–152. Boca Raton, FL: Taylor & Francis.

Van Gestel, C.A.M., and T.C. Van Brummelen. 1996. Incorporation of the biomarker concept in ecotoxicology calls for a redefinition of terms. *Ecotoxicology* 5:217–25.

Van Veld, P.A., and R.F. Lee. 1988. Intestinal glutathione S-transferase activity in flounder *Platichthys flesus* collected from contaminated and reference sites. *Mar. Ecol. Prog. Ser.* 46:61–3.

Van Veld, P.A. et al. 1987. Glutathione S-transferase in intestine, liver and hepatic lesions of mummichog. *Fish Physiol. Biochem.* 9:369–76.

Varanasi, U. et al. 1995. Assessment of oil spill impacts on fishery resources: Measurement of hydrocarbons and their metabolites, and their effects in important species. State/Federal Natural Resources Damage Assessment Final Report, NRDA Project Subtidal 7. Seattle, WA: NOAA (National Oceanic and Atmospheric Administration)/NFMS (National Marine Fisheries Service).

Vasseur, P., and C. Cossu-Leguille. 2003. Biomarkers and community indices as complementary tools for environmental safety. *Environ. Int.* 28:711–7.

Vasseur, P., and C. Leguille. 2004. Defense system of benthic invertebrates in response to environmental stressors. *Environ. Toxicol.* 19:433–6.

Vermeulen, N.P.E. 1996. Role of metabolism in chemical toxicity. In *Cytochromes P450: Metabolic and Toxicological Aspects,* ed. C. Ionnides, 29–53. London: CRC Press.

Viarengo, A. et al. 1984. Possible role of lysosomes in the detoxification of copper in the digestive gland cells of metal-exposed mussels. *Mar. Environ. Res.* 14:469–70.

Viarengo, A. et al. 1990. Heavy metal effects on lipid peroxidation in the tissues of *Mytilus galloprovincialis* Lam. *Comp. Biochem. Physiol.* 97C:37–42.

Viarengo, A. et al. 1995. Stress on stress response: A simple monitoring tool in the assessment of a general stress syndrome in mussels. *Mar. Environ. Res.* 39:245–8.

Viarengo, A. et al. 2007. The use of biomarkers in biomonitoring, a two-tier approach assessing the level of pollutant-induced stress syndrome in sentinel organism. *Comp. Biochem. Physiol.* 146C:281–300.

Welch, W.J. 1990. The mammalian stress response: Cell physiology and biochemistry of stress proteins. In *The Role of Stress Response in Biology and Disease,* ed. R. Moromoto, A. Tissières, and C. Georgopoulos. Cold Springs Harbor, NY: Cold Spring Harbor Laboratory.

Werner, I., and D.E. Hinton. 1999. Field validation of hsp70 stress proteins as biomarkers in Asian clam (*Potamocorbula amurensis*): Is downregulation an indicator of stress? *Biomarkers* 4:473–84.

Wilhem Filho, D.W. et al. 2001. Seasonal changes in antioxidant defenses of the digestive gland of the brown mussel (*Perna perna*). *Aquaculture* 203:149–58.

Winston, G.W., and R.T. Di Giulio. 1991. Prooxidant and antioxidant mechanisms in aquatic organisms. *Aquat. Toxicol.* 19:137–61.

Wootton, A.N. et al. 1995. Evidence for the existence of cytochrome P450 gene families (CYP1A, 3A, 4A, 11A) and modulation of gene expression (CYP1A) in the mussel *Mytilus spp. Mar. Environ. Res.* 39:21–6.

Yeom, D.H., and S.M. Adams. 2007. Assessing effects of stress across levels of biological organization using an aquatic ecosystem health index. *Ecotoxicol. Environ. Saf.* 67:286–95.

3

Biomarkers of Defense, Tolerance, and Ecological Consequences

Claude Amiard-Triquet, Carole Cossu-Leguille, and Catherine Mouneyrac

CONTENTS

3.1 Introduction

Tolerance may be defined as the ability of organisms to cope with stress, either natural (such as temperature changes, salinity variations, oxygen level fluctuations, and plant toxins) or anthropogenic, resulting from chemical input of many different classes of contaminants into the environment. Tolerance resulting from physiological acclimation acquired during the course of the life of an organism exposed to sublethal concentrations of contaminants is not inheritable. However, tolerance leading to a genetic adaptation in response to selection pressure in populations exposed to toxicants may be transmitted to the progeny. Resistance is frequently used in the scientific literature as a synonym for tolerance. Several authors have tried to clarify these terms, for example, Lotts and Stewart (1995) and Morgan et al. (2007), but the definitions they proposed were strongly different, and none of them is currently generally adopted. Nevertheless, the use of the term resistance is usually preferred by authors interested in the genetic basis of an organism's ability to survive in a contaminated environment.

Responses to chemical stress may be assessed using the methodology of biomarkers and specifically in the case of tolerance, the so-called biomarkers of defense (De Lafontaine et al. 2000). These biomarkers were developed on the basis of research on a variety of biochemical processes [metallothionein (MT) or stress protein induction, enhanced activities of biotransformation enzymes, antioxidative defenses, etc.] involved in the ability of organisms to cope with the presence of contaminants such as metals, polycyclic aromatic hydrocarbons (PAHs), polychlorobiphenyls (PCBs), etc. in their medium.

In addition to the intrinsic relative sensitivity characteristic of different species to a contaminant (see Chapter 7), it is well established that within the same species, populations chronically exposed to chemical contaminants in their medium are often more able to cope with chemical stress than "naïve" individuals originating from cleaner sites. The best known examples include bacterial resistance to antibiotics, the tolerance of terrestrial plants to metals (Frérot et al. in Amiard-Triquet et al. 2011), and the resistance of insects to pesticides (Ghanim and Ishaaya in Amiard-Triquet et al. 2011).

Tolerance appears primarily as beneficial for environmental conservation because it contributes to the protection of biodiversity, thus allowing normal functioning of ecosystems. However, some mechanisms involved in tolerance can have less positive consequences in the longer term, such as the production of carcinogenic metabolites during the biotransformation of organic pollutants, the reduced performance of some resistant genotypes, or the energy cost of being tolerant. Lastly, in polluted ecosystems, tolerance may be responsible for high body burdens of toxicants in certain prey species with a subsequent risk of trophic transfer or biomagnification in food webs. Thus, it is necessary to assess carefully the health and ecological consequences of tolerance.

3.2 Tolerance to Chemical Stress in Chronically Exposed Populations

Species either tolerant or susceptible to pollution have been recognized in numerous taxonomic groups (Chapter 7). In this section, we will focus on data about the relative susceptibility of populations originating from natural environments that are comparatively contaminated or as clean as practically possible (reference sites). Results obtained with experimental populations exposed in the laboratory over several generations will also be taken into account. Tolerance appears as a widespread phenomenon, particularly well documented for metals (Table 3.1), but a number of studies have also reported tolerance to organic contaminants (Table 3.2).

A relationship between the origin of phytoplankton strains and their tolerance to metals originating from mining activities (Cu or Zn), industrial effluents, and PCBs has been established, whereas several laboratories have developed resistant strains by exposing them to sublethal doses of other organic contaminants (Cosper et al. 1987 and literature cited therein; Takamura et al. 1989).

In the freshwater crustacean *Daphnia magna*, tolerance was induced over successive generations exposed in the laboratory to different metals (Bossuyt and Janssen 2004b and literature cited therein), whereas *Ceriodaphnia dubia* reared in a metal-depleted medium showed an abnormal sensitivity to metals (Muyssen and Janssen 2002). As soon as the second generation of daphnia was obtained from herbicide (molinate)-exposed parents, longevity was increased and reproduction improved (Sánchez et al. 2004). Ethyl parathion also induced a certain tolerance (Barata et al. 2001). On the contrary, exposure over several generations to another insecticide (diazinon) induced an increased susceptibility; young daphnia obtained from parents exposed to an acaricide (tetradifon) or an industrial effluent showed an increased susceptibility to these contaminants (in Sánchez et al.

TABLE 3.1

Metal Tolerance in Organisms Chronically Exposed to Metal Pollution in the Field or Preexposed in the Laboratory

Taxon	Species	Element	Reference
Ciliate	*Uronema nigricans*	Hg	Berk et al. 1978
Microalgae	Many different species	Cd, Cu, Zn	Takamura et al. 1989
Microalga	*Scenedesmus acutus*	Cr, Cd, Cu	Twiss et al. 1993; Corradi et al. 1995; Torricelli et al. 2004; Gorbi et al. 2006
Microalga	*Scenedesmus* sp.	Hg	Capolino et al. 1997
Microalga	*Chlorella* sp.	Cd	Kaplan et al. 1995
Microalga	*Gomphonema parvulum*	Zn	Ivorra et al. 2002
Microalga	*Pseudokirchneriella subcapitata*	Cu	Bossuyt and Janssen 2004a
Microalga	*Amphidinium caterii*	Fluoride	Antia and Klut 1981, in Cosper et al. 1987
Macroalga	*Stigeoclonium tenue*	Zn	Pawlik-Skowrońska 2003
Macroalgae	*Ectocarpus silicosus* *Fucus vesiculosus*	Cu	Review by Bryan 1984
Nematodes	Estuarine communities	Cu	Millward and Grant 1995
Bryozoan	*Bugula neritina*	Cu	Piola and Johnston 2006
Annelid	*Limnodrilus hoffmeisteri*	Cd, Ni	Klerks and Levinton 1989
Annelid	*Sarganophilus pearsei*	Hg	Vidal and Horne 2003
Annelid	*Nereis diversicolor*	Cd, Cu, Zn	Ait Alla et al. 2006 and literature quoted therein; Burlinson and Lawrence 2007
Bivalve	*Macoma balthica*	Cu	Luoma et al. 1983
Bivalve	*Scrobicularia plana*	Zn	Amiard 1991
Bivalve	*Ostrea edulis*	Cu, Zn	Bryan et al. 1987
Bivalve	*Crassostrea gigas* (larvae)	Cu	Damiens et al. 2006
Bivalve	*Mytilus edulis*	Hg	Roesijadi et al. 1982
Bivalve	*Mytilus edulis* (embryos)	Cu	Hoare et al. 1995
Crustacean	*Daphnia* sp.	Cd, Cu, Hg, Ni, Zn	Bossuyt and Janssen 2004b; Lopes et al. 2004; Tsui and Wang 2005; Lopes et al. 2005, 2006; Haap and Kohler 2009
Crustacean	*Acartia clausi*	Cd, Cu	Moraitou-Apostolopoulou and Verriopoulos 1979; Luoma et al. 1983
Crustacean	*Tisbe holothuriae*	Cd, Co, Cr	Review by Bryan 1984; Miliou et al. 2000
Crustacean	*Artemia salina*	Cu	Review by Bryan 1984
Crustacean	*Gammarus duebeni*	Zn	Jones and Johnson 1992
Crustacean	*Gammarus pulex*	Cd, Zn	Naylor et al. 1990; Stuhlbacher and Maltby 1992
Crustacean	*Asellus aquaticus*	Zn	Naylor et al. 1990
Crustacean	*Asellus meridianus*	Cu, Pb	Review by Bryan 1984
Crustacean	*Platynympha longicaudata*	Cd, Cu, Mn, Pb, Zn	Ross et al. 2002
Crustacean	*Palaemonetes pugio*	Hg	Kraus et al. 1988
Crustacean	*Carcinus maenas*	Zn	Review by Bryan 1984

(*continued*)

TABLE 3.1 (Continued)

Metal Tolerance in Organisms Chronically Exposed to Metal Pollution in the Field or Preexposed in the Laboratory

Taxon	Species	Element	Reference
Crustacean	*Eriocheir sinensis*	Cd	Silvestre et al. 2006 and literature quoted therein
Insect	*Chironomus tentans* (larvae)	Mixture (Cd, Cr, Zn) Cd	Wentsel et al. 1978 Postma and Davids 1995
Insect	*Chironomus riparius* (larvae)	Cd, Zn	Miller and Hendricks 1996; Groenendijk et al. 2002
Insects	*Hydropsyche* spp. *Baetis* spp.	Cu	Cain et al. 2004
Insects	*Hydropsyche betteni* (larvae)	Zn	Balch et al. 2000
Fish	*Fundulus heteroclitus*	Methylmercury	Burnett et al. 2007
Fish	*Heterandria formosa*	Cd	Xie and Klerks 2004
Fish	*Catostomus commersoni*	Cd, Cu	Duncan and Klaverkamp 1983; Munkittrick and Dixon 1988
Fish	*Salmo gairdnerii*	Zn	Bradley et al. 1985
Fish	*Oncorhynchus mykiss*	Cd, Cu, Zn	In McGeer et al. 2000; Chowdhury et al. 2004
Fish	*Gobio gobio*	Cd	Knapen et al. 2004
Fish	*Gambusia affinis*	Cd	Annabi et al. 2009

2004). In another crustacean, the isopod *Platynympha longicaudata*, field exposure to metal-rich effluents from a smelter functioning since 1889 had induced an enhanced tolerance to experimental metal exposure in comparison with populations from reference sites as well as a significant decrease in genetic diversity (Ross et al. 2002). In the crab *Eriocheir sinensis*, preexposure to cadmium induced an increased tolerance to an acute subsequent exposure. This is partly due to MT induction but also involved disulfide bond protection, and enhancement of cell antioxidant capacity and protein degradation potential (Silvestre et al. 2006).

In the fish *Heterandria formosa*, an experimental selection of tolerance to cadmium was carried out over eight generations, leading to a three times longer survival to acute exposure and a reduction of genetic variation (Xie and Klerks 2004; Athrey et al. 2007). In an area in the North Ontario impacted by mining activities (Ag, Cu, Zn), despite metal concentrations shown to be toxic under other conditions, fertilization rate and gamete quality were not impaired in the fish *Catostomus commersoni*. Larvae from the contaminated site also showed an increased tolerance during the period of reliance on yolk reserves, but this effect was no longer observed as soon as individuals began eating (Munkittrick and Dixon 1988). In fish *Melantaenia nigrans* exposed to copper in their environment for more than 40 years, 96 h EC_{50}s were 8.3 times higher than those in controls. Reduced copper uptake by gills and the selection of less sensitive allozymes (AAT-1 and GPI-1) could explain this tolerance (Gale et al. 2003). In several estuaries along the Atlantic coast of North America, the Atlantic killifish (*Fundulus heteroclitus*) and the Atlantic tomcod (*Microgadus tomcod*) are resistant to organic chemicals including PCBs, PCDDs, and PAHs. Mechanisms responsible for tolerance have given rise to numerous studies, which have been recently reviewed (Romeo and Wirgin in Amiard-Triquet et al. 2011; Wirgin et al. 2011). Differential tolerance

TABLE 3.2

Tolerance in Organisms Chronically Exposed to Organic Chemicals in the Field or Preexposed in the Laboratory

Taxon	Species	Contaminant class	Molecule	Reference
Microalga	*Asterionella japonica*	PCB		Cosper et al. 1984
Microalga	*Ditylum brightwellii*	PCB		Cosper et al. 1984
Microalgae	*Asterionella glacialis* *Thalassiosira nordenskioldii*	PCB PCB		Cosper et al. 1988
Phytoplankton	Microplankton, nanoplankton		Tributyltin	Petersen and Gustavson 1998
Microalgae	Phytoplankton communities	Biocide in antifouling paint	4,5-Dichloro-2-*n*-octyl-isothiazoline-3-one	Larsen et al. 2003
Microalgae	Microphytobenthos	Herbicide	Isoproturon	Schmitt-Jansen and Altenburger 2005a
	Periphyton	Herbicides	Atrazine, prometryn, isoproturon	Schmitt-Jansen and Altenburger 2005b
Microalgae	Phytoplankton community	Herbicide	Atrazine	Seguin et al. 2002
Microalga	*Ditysphaerium pulchellum*	Herbicide	Monuron	Bernarz 1981, in Cosper et al. 1987
Microalga	*Chlorella protothecoides*	Organophosphorous insecticide	Methyl parathion	Saroja-Subbaraj and Bose 1983, in Cosper et al. 1987
Cyanophyceae	*Microcystis aeruginosa*	Pesticide	Dinitrophenol	Genoni et al. 2001
Cyanophyceae	*Anabaena variabilis*		Hydroxylamine	Jain et al. 1967, in Cosper et al. 1988
Nematodes		PAHs		Carman et al. 1995
Annelids	*Nereis virens*	PAHs		Lewis and Galloway 2008
Annelids	*Monopylephorus rubroniveus*	PAHs	Fluoranthene	Weinstein et al. 2003
Crustaceans			Diesel	Carman et al. 2000
Crustaceans	*Daphnia magna*	Organophosphorous insecticide	Ethyl parathion	Barata et al. 2001
Crustaceans	*Daphnia magna*	Herbicide	Molinate	Sánchez et al. 2004
Crustaceans	*Daphnia magna*	Pesticides	Toxaphene, carbaryl	Kashian 2004; Coors et al. 2009
Crustaceans	*Daphnia magna*	Pharmaceuticals	17α-Ethinylestradiol faslodex	Clubbs and Brooks 2007
Crustaceans	*Hyalella curvispina*	Organophosphorous insecticide	Azinphosmethyl	Anguiano et al. 2008
Fish	Several species of minnows	Residual chlorine		Lotts and Stewart 1995
Fish	*Fundulus heteroclitus*	TCDD, PCBs, PAHs		Burnett et al. 2007
Fish	*Microgadus tomcod*	PAHs	B[*a*]P	Sorrentino et al. 2004
Fish	*Microgadus tomcod*	PCB, PCDD		Yuan et al. 2006
Fish	*Menidia menidia*	Dioxin-like compounds	PCB 126	Roark et al. 2005
Amphibians	Toad embryos	Pesticides		Anguiano et al. 2001

in subsequent generations coming from field-collected populations in comparatively polluted and clean sites was recently reviewed by Johnston (in Amiard-Triquet et al. 2011). Evidence is reported for copepods (exposed to metals, Co, Cr), daphnids (with Cd, Cu, or a pesticide), chironomid larvae (Cd), bryozoans (Cu), gastropods (Cd, Pb, Zn), and fish (Cd, PCB, pro-oxidant *t*-butyl hydroperoxide).

Within a given population, certain individuals have an inherent ability to cope better with the presence of chemical contaminants in their environment. Studying microalgal responses to a petroleum spill, Carrera-Martínez et al. (2010, 2011) have shown that crude oil-resistant mutants had arisen through rare spontaneous mutations that had occurred before crude oil exposure in the field or in the laboratory. Resistant mutants were enough to assure the survival of microalgal species exposed to oil spills. In the crab *Carcinus maenas*, Depledge et al. (1995) have shown that specimens with naturally low concentrations of proteins in their hemolymph were more susceptible when exposed to copper. In shrimps *Palaemonetes pugio* exposed to chromium (VI) or to fluoranthene, individuals that were heterozygous for the glucose phosphate isomerase allozyme, involved in energy metabolism, survived longer and had less overall mortality than the homozygous genotype (Harper-Arabie et al. 2004). In eels *Anguilla anguilla* exposed to an herbicide thiocarbamate or to an organophosphate insecticide, survival was improved for individuals able to adapt their glutathione metabolism to respond to oxidative stress (Peña-Llopis et al. 2001, 2003).

Co-tolerance may occur when organisms that have been exposed to one toxicant, but not to another one, become tolerant to both of them. Co-tolerance occurs most probably for compounds that have similar chemical structures and activities and share common tolerance mechanisms. Co-tolerance may arise also because genes for resistance to, or transformation of, different contaminants are found on the same mobile genetic element such as a plasmid or a transposon, thus eliciting co-tolerance to contaminants that are unrelated structurally or functionally (Top and Springael 2003; Wright et al. 2008). Examples of co-tolerance between toxicants have been provided in recent reviews for microbes including bacteria, phytoplankton, and periphyton (Tlili and Montuelle; Amiard-Triquet and Roméo, both in Amiard-Triquet et al. 2011). Other microalgal examples have been reported involving different metals and also different organic compounds such as PCBs and DDT (Cosper et al. 1987 and literature quoted therein; Takamura et al. 1989). Such studies are scarce for animal species. However, Brown (1978) has shown the ability of copper-tolerant freshwater isopods *Asellus meridianus* to detoxify lead by storing this metal in intracellular structures involved in copper accumulation. Xie and Klerks (2003) have shown that *Heterandria formosa* (a fish species) that had acquired cadmium resistance in the course of experimental exposure over six generations had also become tolerant to copper. More frequent are studies dealing with cross resistance between pollutants and more natural factors such as temperature, which is important in the context of global warming (http://www.citeulike.org/user/dortsjennifer/tag/crossresistance). The induction of heat shock proteins (HSPs) by environmental factors and cross-tolerance with metals and organics have been recently reviewed (Mouneyrac and Roméo in Amiard-Triquet et al. 2011). The estuarine fish *F. heteroclitus* resident in a harbor highly contaminated with PCBs, evolved tolerance to these chemicals, possibly involving mechanisms that minimize the immunosuppressive effects of a bacterial pathogen *Vibrio harveyi* (Nacci et al. 2009). Likewise, parasitized individuals of the freshwater bivalve *Pisidium amnicum* had an increased tolerance toward pentachlorophenol (Heinonen et al. 2001). Such phenomena may have great ecological significance since most impacted sites are subjected to multiple pollutions. Co-tolerance between different classes of toxicants or between toxicants and natural stress factors can act as a confounding factor complicating the interpretation of biomarker data.

3.3 Biomarkers of Defense

Biomarkers of defense reveal mechanisms that allow aquatic organisms to cope with the presence of pollutants in their environment, at least when they remain at "reasonable" levels, but with an energy cost.

3.3.1 Mechanisms of Defense against Metals

The relative efficiency of different mechanisms of defense used by organisms exposed to chemical stress, governs the interindividual, interpopulational, or interspecific variability of tolerance. Strategies to prevent contaminant toxicity include the limitation of bioaccumulation (controlled uptake, increased excretion) and, when the chemical compound is internalized, its storage in nontoxic physicochemical form (Mason and Jenkins 1995; Marigomez et al. 2002; Amiard et al. 2006; Perales-Vela et al. 2006; Sigel et al. 2009).

MTs and related sulfur-rich chelators are recognized as important in metal ion homeostasis owing to their metal binding capacity. In addition, MT antioxidant properties are frequently evoked (Falfushynska et al. 2012) even though several conflicting experimental studies about the antioxidant protection conferred by MTs have been reported (Moreau et al. 2008 and literature cited therein). These authors have shown that different isoforms of MT, present in different taxa from bacteria to mammals, exhibit different properties. A recent book has been devoted to these ligands in many different taxa including vertebrates and invertebrates from marine and freshwater ecosystems (Sigel et al. 2009). In vertebrates, MTs are considered the major ligand for metal detoxification. In fish originating from a site polluted for decades by Cd and Zn, increased resistance to Cd in acute toxicity tests by comparison with "naïve" individuals was probably attributable to liver MT induction (Knapen et al. 2004). Similarly, MTs were involved in resistance to Cd acquired over several generations in laboratory contaminated fish *Heterandria formosa* but, at the maximum, 26.5% of bioaccumulated Cd was associated with MTs, indicating that a large fraction of this metal was not detoxified by this means (Xie and Klerks 2004). In invertebrates, different detoxification processes can be activated in response to metal stress. In different species and different populations within the same species (depending on their adaptation to contaminated environments), the respective roles of MTs and biomineralization of metals as metal-rich granules (MRG) may be more or less important (Wallace et al. 1998; Berthet et al. 2003; Mouneyrac et al. 2003). As exemplified in zebra mussels from clean and polluted (Cd, Cu, Zn) field locations, in more polluted specimens the contributions of MRGs and MTs become more important, but metal detoxification was not sufficient to prevent metal binding to low molecular weight (LMW) proteins (Voets et al. 2009). In another freshwater bivalve (*Pyganodon grandis*) translocated from a control site to a contaminated site, the cytosolic distribution of Cd in the gills was strongly modified, and the presence of Cd bound to LMW compounds was associated with toxicity symptoms including lipid peroxidation, decreased condition index and delayed growth (Couillard et al. 1995). According to the findings of Ivanina et al. (2008) on *Crassostrea virginica* exposed to cadmium, MT expression may provide sufficient protection against Cd-induced damage to intracellular proteins in the digestive gland. In contrast, Cd detoxification mechanisms appear to be insufficient to fully prevent protein damage in gill cells, thus necessitating induction of HSPs as a secondary line of cellular defense. Gills appear to be Cd-sensitive tissues in oysters, with possible important implications for impaired oxygen uptake contributing to energy misbalance. In crustaceans, the saturation of the detoxification capacity of MTs could be responsible for behavioral impairments in the presence of excess Cd (Wallace and Estephan 2004 and literature quoted therein). In two clones of *Daphnia magna* exposed to cadmium over several generations, MT concentration had a critical role in coping with chemical stress, leading to significant differences in survival (Guan and Wang 2006). In the oligochaete worm *Tubifex tubifex* and the dipteran *Chironomus riparius* exposed to Cd, above a MT concentration threshold (14 and 20 nmol g^{-1}, respectively), compensatory mechanisms were no longer efficient, and impairments of reproduction (*T. tubifex*) or growth (*C. riparius*) were observed (Gillis et al. 2002). From a practical point of view, the saturation of MTs as a defense mechanism poses a problem for the use of MT as a biomarker since very different levels of exposure can induce identical responses (Amiard-Triquet and Roméo in Amiard-Triquet et al. 2011).

In algae, phytochelatins (also termed class III MTs) and other intracellular ligands are produced in response to metal exposure (Perales-Vela et al. 2006). Phytochelatin induction is highly variable depending on species. Species that produce few phytochelatins could cope with metal toxicity by relying on biomineralization of metals in polyphosphate bodies (Ballan-Dufrançais et al. 1991; Le Faucheur et al. 2006).

Mechanisms involving increased metal excretion have been reviewed by Mason and Jenkins (1995). More recently, the role of multixenobiotic resistance (MXR) (see Section 3.3.5) has attracted increasing attention.

3.3.2 Antioxidative Defenses

The pros and cons of using responses to oxidative stress as biomarkers have been recently reviewed (Regoli et al. in Amiard-Triquet et al. 2011; Abele et al. 2012). Toxic effects of pollutants such as PAHs, PCBs, metals, or pesticides often depend on their capacity to increase the cellular levels of reactive oxygen species (ROS). When ROS production exceeds antioxidant defenses, oxidative stress leading to transient or permanent cellular effects at the protein, lipid, or DNA levels can occur. The increase or the reduction in ROS levels induced by pollutants depends on the balance between pro- and antioxidant systems. Indeed, aerobic organisms have developed antioxidant defense systems that enable them to cope with endogenous as well as exogenous ROS production. Among the most widely studied parameters are, on the one hand, activities of enzymes such as superoxide dismutases (SOD), catalase, glutathione peroxidases (GPx) or glutathione reductase (GRd), and, on the other hand, LMW antioxidants such as reduced glutathione (GSH) and vitamins E (α-tocopherol), B (β-carotene), or C (ascorbate). The procedures for carrying out evaluation of antioxidant defenses have been recently reviewed (Abele et al. 2012).

In aquatic environments, numerous studies have shown that antioxidant defense systems represent biomarkers that are able to reveal the early effects of xenobiotics that exert their toxicity via oxidative stress (Viarengo et al. 2007; Regoli et al. in Amiard-Triquet et al. 2011; Abele et al. 2012). Utilization of molecular biomarkers is widely accepted to be the most appropriate approach for early diagnostic of chemical pollution. Depending on the duration and the intensity of the pro-oxidative toxic exposure, antioxidant defense systems can be induced only during the first phase of the response of organisms to xenobiotics. No variation at all or a transient response suggests adaptive or compensatory mechanisms in organisms chronically exposed to pollutants (Regoli and Principato 1995; Fernández et al. 2010).

The dose-dependent increase in GPx activity in gastropod mollusks (*Austocochlea porcata*) exposed to different crude oil concentrations in the laboratory highlighted that these organisms display a compensatory adaptive response. The response was confirmed under field conditions, where an increase in GPx activity was measured after 96 h of exposure of the gastropods to crude oil fractions, and activities returned to levels close to those of controls after 2 weeks of exposure (Reid and MacFarlane 2003). This transient GPx activity response highlights that *A. porcata* can adapt to stress conditions. Significantly higher levels of GR, GPx, and GST measured in gills of *Mytilus galloprovincialis* chronically exposed to metals seem to constitute a specific adaptation in gills to prevent and/or repair metal-induced damage in cellular components, as no signs of lipid peroxidation were observed (Fernandez et al. 2010).

Regoli et al. (in Amiard-Triquet et al. 2011) consider that analyses of antioxidants can be profitably integrated with the measurement of total oxyradical scavenging capacity (TOSC), which quantifies overall cellular resistance toward different ROS. Compared to individual defense biomarkers, TOSC is less sensitive but has a greater prognostic value since an impaired capability to neutralize ROS has been associated with the onset of various forms of oxidative damage such as lysosomal alterations and genotoxic damage.

Falfushynska et al. (2011, 2012) observed strong differences in the ability of two populations of gibel carp (*Carassius auratus gibelio*) originating from control or high polluted sites to withstand additional toxic metal (copper or manganese) or pesticide (thiocarbamate or tetrazine) exposure. The authors highlighted that fish from the polluted area mobilized both antioxidant defense and biotransformation systems more effectively than control fish, despite lower antioxidant defense activities and greater lipid peroxidation damage. These peculiarities could be the result of the adaptation to prolonged life in a toxic environment. Meyer et al. (2003) demonstrated that larval first- and second-generation (F1 and F2) offspring of killifish (*Fundulus heteroclitus*) originating from a site highly contaminated with PAHs, metals, and pentachlorophenol displayed higher resistance when exposed in the laboratory to *t*-butyl hydroperoxide than F1 larvae of control killifish. Such resistance could be explained by high antioxidant activity levels transmittable to offspring. However, although the resistance and the adaptation of *F. heteroclitus* exposed to contaminated sediments can be explained by higher GPx, GRd, and SOD activity levels and higher glutathione production rates in exposed adult killifish as compared to control ones, none of these parameters appears to play a role in acquired resistance. Indeed, only higher basal levels of glutathione and manganese SOD were measured in F1 and F2 larvae of killifish from the contaminated site as compared to the levels measured in control F1 larvae in the absence of any exposure to xenobiotics. Thus, Meyer et al. (2003) showed that, in *F. heteroclitus* chronically exposed to high pollutant levels, up-regulated antioxidant defenses play a role in both short-term (physiological) and heritable (multigenerational) tolerance of the toxicity of these pollutants, as antioxidant defense capacities could be transmitted to offspring and lead to long-term genetic adaptation and to resistance acquired over generations. Comparative studies of different populations of *F. heteroclitus* with different physiological tolerances to pollutants have established that neither the level of gene expression nor the level of DNA polymorphisms was well conserved, because of the heterogeneity of the stress factors involved coupled with the genetic variation of the populations (Whitehead et al. 2011). These results suggest that the differential survival of chronically exposed populations results from genetic adaptation rather than physiological acclimation.

Antioxidant defenses vary depending on the season, the nutrient load, and the reproductive cycle of vertebrate and invertebrate aquatic organisms, and it has been established that antioxidant activity is usually highest in spring and lowest in winter. Organisms may therefore be more sensitive to ROS during winter. Indeed, oxidative stress is also a seasonal phenomenon, and a drop in temperature usually induces an increase in oxidative stress in organisms (Viarengo et al. 1998; Sheehan and Power 1999; Abele et al. 2002). However, compensation phenomena are also possible. Thus, Borković et al. (2005) showed that mussels *Mytilus galloprovincialis* sampled in winter and in spring from areas impacted by industrial and urban wastewaters displayed higher SOD and GPx activities in winter, which suggests a rearrangement of metabolic cellular components to compensate for environmental fluctuations and cope with the pollutant load.

The strategy developed by the amphipod crustacean *Gammarus roeseli* against oxidative stress seems to differ with gender with higher levels of catalase and GPx in females. Moreover, GPx activities fluctuate with oocyte maturation with high levels in previtellogenic oocytes and in early ovaries. Higher MDA levels were also measured in males than in females (Sroda and Cossu-Leguille 2011). This could be related to lower antitoxic capacities in males, but may also be a result of sex-specific biochemical composition in polyunsaturated fatty acids known to be potential targets of ROS in males, which are higher than that in females (Maazouzi et al. 2008).

In periods of food deprivation, Guderley et al. (2003) showed a 3-fold increase in catalase activity in cod (*Gadus morhua*) livers, whereas GPx activity decreased. Under conditions of unfavorable nutrient resources, organisms therefore appear to set up an energy strategy that favors low energy-consuming enzymes: indeed, catalase requires neither a cofactor nor energy for its activity, whereas glutathione peroxidase uses reduced glutathione and NADPH (Janssens et al. 2000). Indeed, mobilizing energy reserves could increase the sensitivity of aquatic organisms to ROS-induced damage, but maintaining or even increasing antioxidant activity contributes to their tolerance to stress. In order to cope with stress conditions, aquatic organisms maintain their antioxidant systems at high levels, and these systems in turn have metabolic priority over other physiological functions such as weight gain or reproduction (Wilhelm Filho et al. 2005).

Antioxidant defense systems are biomarkers that can be used to diagnose individually the effects of oxidative stress-induced damage and constitute early warning systems for possible damage at the ecosystem level. However, in order to use them as predictive elements at the individual and community levels, it is necessary to establish the link between antioxidant defenses and individual health indicators such as weight gain, growth, energy reserves, or metabolic functions (Depledge et al. 1995). Indeed, establishing correlations between antioxidant defenses measured in individuals and their health indicators is essential to define the relevance of these biomarkers for predicting possible effects at the population level (Figure 3.1).

Ferrari et al. (2007) showed that the decrease in reduced GSH contents during the exposure of juvenile rainbow trout (*Oncorhynchus mykiss*) to sublethal concentrations of carbaryl and azinphos-methyl was linked to an increase in fish mortality. Conversely, an increase in GSH levels was reported to enable marine bivalves exposed to organophophorous pesticides (Peña-Llopis et al. 2002) or copper (Hoare et al. 1995) to tolerate these pollutants. In an area impacted by metals (Ni, Cr, Fe), Tsangaris et al. (2007) showed significant correlations between glutathione peroxidase response and energy allocation to growth and reproduction [Scope for Growth (SfG), see Chapter 12] in the mussel *Mytilus galloprovincialis*. Exposing mussels to metals in the laboratory yielded similar results, suggesting that the organisms' health degradation could be due to metal-induced ROS production. This correlation between an early biochemical biomarker (GPx) and a health degradation biomarker (SfG) can be interpreted as evidence for the potential of using GPx to predict effects at the population level.

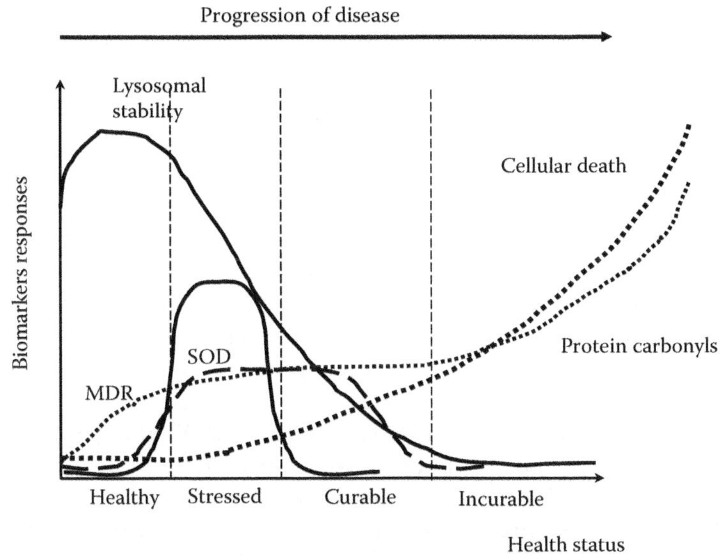

FIGURE 3.1
Theoretical diagram of the conceptual links between biomarkers and health status of individuals in the context of "effects at the population level" prediction. (Adapted from Allen, J.J., Moore, M.N., *Mar. Environ. Res.*, 58, 227–232, 2004.)

3.3.3 Phases I and II Enzymes

Phase I enzymes, such as 7-ethoxyresorufin *o*-deethylase (EROD), and phase II enzymes, such as GST (glutathione-*S*-transferase), are usually considered as defense biomarkers (cf. Chapter 2), involved in the detoxification of organic compounds (Newman and Unger 2003). Yet the activity of phase I cytochrome P450–dependent enzymes can trigger the activation of the initial compounds, especially of PAHs, whose subsequent metabolites can cause cellular damage by binding to biological macromolecules such as DNA and various proteins. The induction of cytochrome P4501A (CYP1A) by nonmetabolized halogenated aromatic hydrocarbons can induce the production of ROS. Resistance to various organic contaminants (PCBs, PCDDs, PAHs) in fish populations living in highly contaminated sites is linked to the absence of CYP1A induction (Romeo and Wirgin in Amiard-Triquet et al. 2011). Various hypotheses have been proposed to explain this resistance, such as high GST activity (Armknecht et al. 1998). Indeed, in response to exposure to 1-chloro-2,4 dinitrobenzene, GST expression and activity in resistant fish (*Fundulus heteroclitus*) from a contaminated estuary were respectively four times and twice as high as in fish from a reference site. Antioxidant defenses (Meyer et al. 2003; see Section 3.3.2) and MXR (Cooper 1999; see Section 3.3.5) have also been suggested. Paetzold et al. (2009) suggested that in multixenobiotic-resistant killifish (*F. heteroclitus*) populations liver coordinated up-regulation of phase I and II enzymes associated with ABC transporters (ABCC2 and ABCG2) may confer contaminant resistance to organisms. Moreover, the resistance and the altered CYP1 phenotype observed in a population of chronically PAH-exposed killifish may be explained by blocking AhR2 expression, leading to protection of organisms from the teratogenicity of PAH in exposed embryos (Wills et al. 2010).

Nutrient resources or the quality of food resources can have consequences on the total energetic budget of organisms with possible effects on metabolization capacities. A 3- to 7-week-long fasting period in rainbow trout *Oncorhynchus mykiss* led to a modification of their metabolization capacities with a decrease in EROD and GST activities and an increase in UDP-glucuronosyl transferase activity (Bloom et al. 2000). These decreases in enzyme activities are considered to be a strategy of the fish that lowers energy costs to deal with stress-induced energy demands.

3.3.4 Stress Proteins

In response to cellular stress, so far the only known universal system is the induction of a protein family called stress proteins (HSP 90 or stress 90, HSP 70 or stress 70, chaperonin 60, stress proteins with low molecular weight: 16–24 kDa), which has been highly conserved through evolution (Feige et al. 1996; Sonna et al. 2002; Gross 2004). These stress proteins are able to repair those proteins damaged by stress, or eliminate them when they cannot be further repaired. They act as molecular "chaperones," supporting, monitoring, and protecting other proteins (see reviews by Frydman 2001; Hartl and Hayer-Hartl 2002; Wang et al. 2004). Moreover, the induction of stress proteins is maintained over time, making them relevant for use as biomarkers (Bierkens 2000).

Initially, HSPs were given this name as their synthesis is induced when cultured cells or whole organisms are exposed to elevated temperature. Among HSPs, the HSP70 family members are the most investigated for their characterization and induction in response to numerous environmental stressors in a range of species (Morimoto et al. 1992; Clark and Peck 2009). Currently, literature data provide numerous examples of stress protein induction in various animal, plant, and bacteria species, in response to an exposure to environmental or chemical stress, although a few counterexamples have been reported (see reviews by De Pomerai 1996; Bierkens 2000; Mukhopadhyay et al. 2003). Assuming that stress proteins play a protective role against a wide variety of stress agents, is their induction in response to a specific stress linked to the development of tolerance to any subsequent stress? The first example demonstrated both *in vivo* and *in vitro* was that of "thermo-tolerance," defined as the ability of a cell or an organism to resist heat stress after exposure to a sublethal heat shock. It has been clearly established that the induction threshold of HSP is correlated with the stress level experienced by species in their natural habitat; reflecting the significance of the "thermal history" of a particular species throughout its evolution, and suggesting that HSPs are ecologically relevant for use by a species to improve its tolerance to heat stress (Fangue et al. 2006 and literature cited therein).

In addition, examples of "cross tolerance" to various stresses acquired after a heat shock have been observed. For example, this happens to be the case in daphnia (*Daphnia magna*), which exhibit tolerance after exposure to a usually lethal dose of malathion following heat pretreatment (Bond and Bradley 1997). In mussels (*Mytilus edulis*), heat pretreatment involves an induction in HSP 70 concentrations and increased resistance to cadmium (Tedengren et al. 2000). In organisms living in environments subjected to natural or chemical stress, the role played by stress proteins in the acquisition of tolerance to an additional stress may vary according to the species and/or population. In oysters *Crassostrea virginica* originating from three sites differing in their thermal regimes, overall HSP and MT patterns were similar in oysters from the three geographically distant populations (Ivanina et al. 2009). HSP levels were lower in Cd-exposed organisms than in their control counterparts during heat stress, suggesting that both stressors may have partially suppressed the cytoprotection up-regulation of molecular chaperones. Synergistic interactions between the effects of metals and heat could lead to a reduced tolerance to heat in metal-exposed organisms (Sokolova and Lannig 2008). However, mussels (*M. edulis*) adapted to low salinity levels in the Baltic Sea—at the limits of their geographical distribution—had lower

HSP 70 levels than mussels from the North Sea, and were more sensitive to cadmium exposure (Brown et al. 1995). Similarly, in another study carried out on the same species (*M. edulis*), Tedengren et al. (1999) demonstrated that Baltic Sea mussels were more sensitive in their physiological response and survival when exposed to contaminants, compared with populations originating from the North Sea. Can the differences between these two populations be explained by environmental factors or genetic differences in their ability to synthesize HSP 70? Juvenile specimens from the Baltic Sea were translocated into the North Sea for 1 month, and then exposed to copper under laboratory conditions. The results revealed that the differences in physiological performance between the two populations can be mainly explained by environmental factors, even though lower levels of HSP induction in Baltic Sea mussels were reported compared to those from the North Sea. Pyza et al. (1997) compared the HSP 70 levels between centipedes (*Lithobius mutabilis*) from a reference site or from sites differing in their level of Pb or Zn contamination. No differences in HSP levels were observed between the centipedes from the contaminated and reference sites, and between sites with different contamination levels. The authors concluded that tolerance acquisition through HSP induction is only possible up to a certain degree and is specific to each species. HSP levels cannot increase indefinitely because the cost of HSP induction is higher than its benefits (Eckwert et al. 1997; Pyza et al. 1997), a feature actually not specific to HSP but which can be found for all proteins. Moreover, there are significant variations in the responses among HSP classes and isoforms that are overexpressed according to inducer agent, species, and even within a species, and consequently their potential use as biomarkers is questionable (De Pomerai 1996; Pyza et al. 1997; Yamashita et al. 2004; Ojima et al. 2005). It has been suggested that the modulation of HSP mRNA expression, highlighting several HSP families or isoforms, could help to ensure phenotype flexibility in response to environmental fluctuations (Hofmann and Somero 1995; Tomanek 2002, 2005; Tomanek and Somero 2002). De Wit et al. (2008) observed in adult zebrafish (*Danio rerio*) exposed to the flame-retardant tetrabromobisphenol-A differential expression of genes, and the most obvious response was an up-regulation of HSP 70 genes, indicating that responses at the genome level can provide information about effects on the proteome.

3.3.5 Multixenobiotic Resistance

MXR has been termed in reference to a homologous phenomenon, the multidrug resistance (MDR) observed in cancer cells. MDR was linked to the presence of transport proteins responsible for the efflux of chemotherapeutic drugs. ATP- binding cassette (ABC) transporters can efflux many drugs, contaminants such as metals (Cd, Zn), pesticides (Buss and Callaghan 2008), PCBs, PAHs, etc. (reviewed by Damiens and Minier in Amiard-Triquet et al. 2011). MXR has been detected in many marine and freshwater organisms including sponges, worms, gastropods and bivalves, crustaceans, fish, and amphibians (Bard 2000). Partial or complete cloned sequences of ABC genes in mollusks, echinoderms, fish, and amphibians are now available from the Swiss-Prot Database (Damiens and Minier in Amiard-Triquet et al. 2011). Studying the transcriptional expression of some ABC transporters in Nile tilapia (*Oreochromis niloticus*) after exposure to benzo(a)pyrene, Costa et al. (2012) have shown that mRNA expression was up-regulated for ABCC2 in gill (up to 16-fold) and ABCG2 in liver (up to 2-fold) and proximal intestine (up to 7-fold). From a review of *in vitro* and *in vivo* studies, various authors have highlighted that ABC-like efflux activity is related to the concentration of the toxic compound, and that MXR activity—as an inducible mechanism—could be a suitable biomarker of exposure to environmental contaminants. An extensive survey performed at 43 sites along the French coast showed clearly that ABC protein expression in bivalves was related to xenobiotic exposure (Minier et al. 2006a). As already mentioned for other biomarkers of defense (see Section 3.3.1 and Figure 3.1), significant linear relationships exist between ABCB1 protein expression in mussels *Mytilus galloprovincialis* from the French coast and the body burdens of contaminants, up to a concentration limit of ca. 1.2 mg Cd kg^{-1} dw cadmium and 1 mg PCB kg^{-1} dw. This could indicate that the mussels were then relying on an increased transport activity or on another defense mechanism. Alternatively, the organisms' health might have already been compromised so that they were unable to further intensify their MXR defense mechanism. The protective role of MXR also showed a limit in the freshwater mussel *Dreissena polymorpha*, for, in the Seine estuary in France downstream of Rouen (390,000 inhabitants), a decrease in lysosomal stability and a reduction in condition index were observed despite increased levels of MXR proteins (Minier et al. 2006b).

The protective role of MXR proteins may be hampered by exposure to so-called chemosensitizers (synthetic musk fragrances studied by Luckenbach et al. 2004; emerging contaminants, natural substances produced by certain invasive species studied by Smital et al. 2004; and others reviewed by Bard 2000). At environmentally realistic doses, they are able to inhibit the normal functioning of the MXR system, thus enhancing the accumulation of xenobiotics that are normally transported from the cell. The role of chemosensitizers as environmental pollutants and the ecotoxicological consequences of transporter inhibition have been highlighted (Bard 2000). Because biotransformation activities (phases I and II) are generally not observed in early development stages, Damiens and Minier (in Amiard-Triquet et al. 2011) suggest that embryos may rely on other defense mechanisms such as the ABC system, which appears as a first line of defense, and that inhibition of MXR activity may have dramatic consequences.

3.4 Ecological Consequences of Tolerance

3.4.1 Conservation of Biodiversity

In a number of cases, defense responses are called upon only for a limited period, for instance, when an animal is able to avoid exposure after it has detected the presence of contaminants (Chapter 10). This type of response is interesting for the conservation of a population in the case of a short-term pollution (accident, occasional discharge, possibly cyclic discharges). Thus, Lotts and Stewart (1995) have shown a temporary acclimation to residual chlorine in several species of minnows, enabling the presence of the fish in areas where concentrations are generally considered lethal. In the fish *Catostomus commersoni* living in metal-contaminated lakes, tolerance provided to larvae by a maternal yolk factor disappeared when larvae began feeding, 24 days after hatching (Munkittrick and Dixon 1988). At the other extreme, genetic adaptation to chemical stress is responsible for the transmission of tolerance to the progeny (Chapter 14), and in this case, the protective effect will last in the long term.

Moving from tolerance at the populational level to the intrinsic relative insensitivity of each species, there is evidence that acute contamination resulting from accidents can cause the local extinction of sensitive species. This is particularly well documented in the case of oil spills that can cause selective mortality of the benthic meiofauna (Ernst et al. 2009; Martínez-Colon et al. 2009) and the macrofauna (Gomez-Gesteira and Dauvin 2005). Similarly, in a given environment, increasing chronic contamination will lead to the local extinction of sensitive species, followed by that of less sensitive species. The new community as a whole is more tolerant to the toxicant responsible than another community, initially identical, but which has never been exposed to this toxicant. This interspecific variability of tolerance is the basis of the pollution-induced community tolerance (PICT) concept proposed by Blanck et al. (1988). PICT has been demonstrated in many studies of microbial communities (reviewed by Tlili and Montuelle in Amiard-Triquet et al. 2011), and nematodes (Millward and Grant 1995, 2000). Considering macrofauna, in a river impacted by mining, Cain et al. (2004) have shown that insect species that incorporate metals in nondetoxified form were rare or absent from the most contaminated areas, whereas tolerant species equipped with efficient mechanisms of detoxification were present along the whole watercourse. Depending on the ecological role of tolerant species in the community, such community-level effects can manifest themselves in various ways (Fleeger et al. 2003). If the sensitive species is a host or a prey, its extinction will lead to a depletion of the populations of its symbionts or predators (Figure 3.2). Population modeling of cod larvae shows their high sensitivity to loss of zooplankton prey, for example, after an oil spill (Stige et al. 2011). On the contrary, if the sensitive species is a competitor or a predator of a tolerant species, the latter will be favored.

Among organisms able to cope with chemical stress, some might be keystone species with important roles in ecosystem functioning. Thus, resistant bacteria will be able to maintain their role in biogeochemical cycling of nutrients. By using these nutrients, primary producers at the base of food webs will function normally, and so on (Chapter 7). However, in certain environments where the level of natural stress is high, the number of species is restricted even in the absence of any pollutant impact. In estuarine waters, the

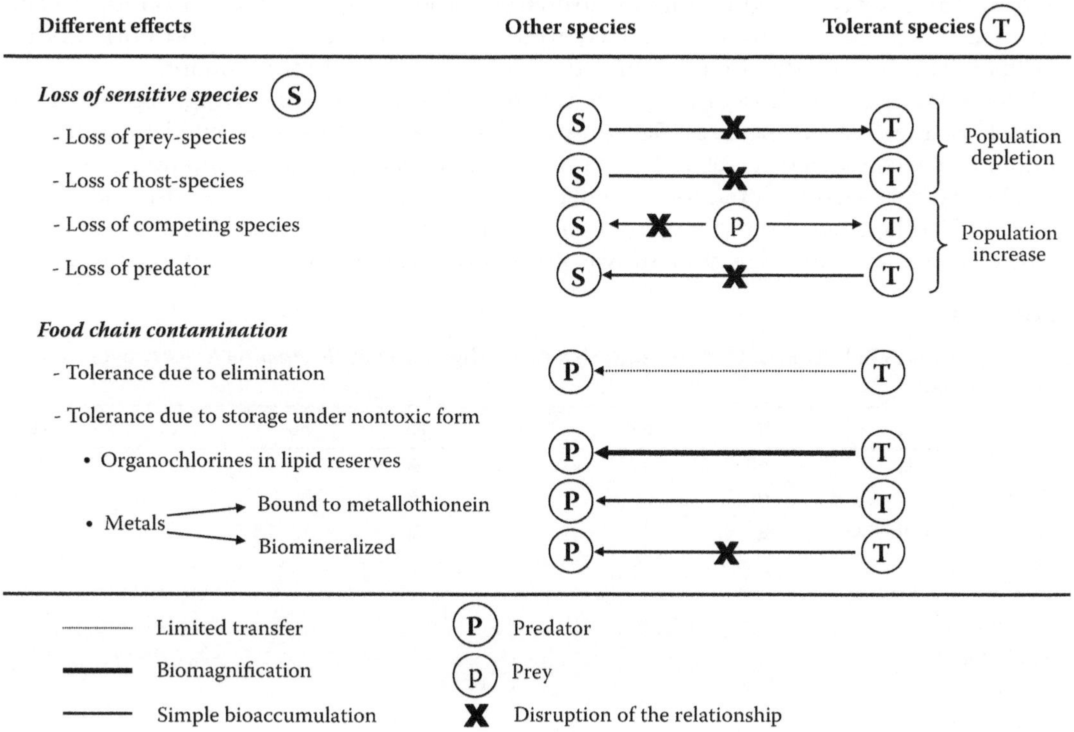

FIGURE 3.2
Community effects of tolerance. (Modified after Moore, N.W., *Advances in Ecological Research*, Academic Press, New York, 1967.)

number of species is reduced, reflecting the number of species able to adapt to low and variable salinity levels and thus survive (McLusky and Elliott 2004). Thus, in estuaries that are among the most polluted areas worldwide, the extinction of a small number of species would be sufficient to hamper ecosystem functionality.

3.4.2 Cost of Tolerance

Living organisms have many defense mechanisms against toxicants present in their environment. The ensuing metabolic cost and physiological stress that can be observed in individuals can have subsequent impacts on populations (Mouneyrac et al. in Amiard-Triquet et al. 2011). This hypothesis of physiological cost also has implications for the evolution of resistance to chemical stress, whether it is a fixed or an inducible response (Calow 1991). Since 1991, this reference has been quoted in scores of articles to support many observations showing an increase in the metabolic rate of organisms exposed to various stress factors (e.g., Canli 2005; Smolders et al. 2005; Guan and Wang 2006; Muyssen et al. 2006; Lannig et al. 2006; Alonzo et al. 2006; Wiegand et al. 2007) inducing, for example, the synthesis of MTs, HSPs, biotransformation enzymes, and antioxidant mechanisms. Rowe (1998) emphasizes that an increase in metabolic rate is similarly observed in species belonging to taxa widely separated phylogenetically (crustaceans, amphibians, reptiles) in response to combustion waste rich in metals, suggesting a general response to metals. Literature data (see subsection 3.2) show that this phenomenon affects other taxa and other types of chemical contaminants. However, when physiological disturbances (in oxygen consumption, energy reserves, condition index, growth, reproduction, etc.) do occur in organisms exposed to chemical stress, it is not easy to distinguish precisely the contribution of the cost of tolerance from the direct costs of the toxic effects of the contaminant.

Interestingly, research on the freshwater fish *Heterandria formosa* highlights various aspects of the cost of tolerance (Xie and Klerks 2004 and literature cited therein). The authors have conducted selective breeding experiments over eight generations, exposing specimens from a field population to high doses of cadmium (Table 3.3). Third- and fourth-generation offspring (F3 and F4) from cadmium-adapted lines were born smaller than control specimens, and size at birth was positively correlated to survival in this species.

TABLE 3.3

Consequences of Selection of Cadmium-Resistant Freshwater Fish *Heterandria formosa* over Eight Generations

	F2	F3	F4	F5	F6	F7	F8
Resistance to Cd	⇗	⇗	⇗	⇗	⇗	⇗	⇗
Cross-resistance to Cu	⇗	⇗		⇗			
Heat resistance at 38°C	⇘	⇘		⇘			
Size at birth		⇘	⇘				
Lifetime fecundity						⇘ (−18%)	
Mean brood size						⇘ (−13%)	
Female life span						⇘ (−7%)	
Time to first brood						⇗ (+6%)	
MT induction							⇗
Cd uptake	⇘	⇘					

Source: Xie, L., and Klerks, P.L., *Environ. Toxicol. Chem.*, 23, 1499–1503, 2004 (and quoted literature therein). With permission.

Note: Empty cells correspond to investigations that were not carried out in all generations.

Moreover, cadmium-tolerant F3 and F4 specimens were less resistant to heat, even at temperatures naturally observed in summer in their habitat. Without exposure to cadmium, F7 specimens displayed numerous life history traits that were negatively influenced by tolerance in comparison to control specimens. Likewise, although they were more tolerant of acute Cu contamination, larvae of *Catostomus commersoni* spawned from adults living in contaminated lakes were hatched at smaller size, grew less, and showed a lower survival rate than those spawned from adults living in a comparatively healthy habitat (Munkittrick and Dixon 1988). The authors hypothesized that this altered condition could be caused by the cost of synthesis of protective proteins. The literature quoted by Xie and Klerks (2004) also shows a lowered fecundity in cadmium-tolerant *Drosophila*, a shorter lifetime in mercury-resistant fish, and a longer period of growth in insecticide-resistant mosquitoes. Investigation of defense systems in fish (lower level of cadmium intake in F3 and F4 offspring, production of MT observed in F8 offspring) has led the authors to consider that the deteriorating life history traits (see Table 3.3) could be caused by a change in energy allocation.

Metal concentrations in the environment, apart from anthropogenic sources, vary depending on the geographical area and specific location within this area, according to season and water supply from feeder watercourses. Organisms are typically able to maintain intracellular concentrations of essential metals in the range of optimal concentrations thanks to homeostasis, regardless of external concentrations. Various observations have been published that demonstrate that when this homeostasis occurs, animals are not subjected to stress (Van Tilborg and Van Assche 1998). These observations no longer apply when considering metals with no vital functions or to xenobiotics, but the dose–response relationship generally has a sigmoid shape and the no observed effect level (NOEL) can be accepted as a valid concept for many contaminants. In several cases, there is apparently no significant physiological cost for various insects and acarid mites resistant to pesticides (quoted by Xie and Klerks 2004). In two isopod crustaceans able to survive in an area impacted by smelting works, it has been shown that differing strategies were implemented, involving (in *Oniscus asellus*) or not involving (in *Porcellio scaber*) an energy cost (Schill and Köhler 2004). In polychaete worms, *Nereis diversicolor*, laboratory exposure to silver or copper induces a higher production of mucus in individuals adapted in the field to chronic metal pollution compared to individuals from a control site (Mouneyrac et al. 2003). Metal stress also induces mucus secretion in mussels (*Mytilus edulis*) or fish (*Oncorhynchus mykiss*) (see Wicklum and Davies 1996). In freshwater invertebrates, mucus secretion contributes significantly to the energy budget, representing 13% to 32% of absorbed energy (see Wicklum and Davies 1996). In the marine gastropod *Patella vulgata*, mucus production accounts for 23% of the energy acquired through food ingestion (Davies et al. 1990). This cost of mucus production in a gastropod is more important than the total cost of locomotion in a reptile or a mammal of the same size (Denny 1980, in Leung et al. 2000). Consequently, these authors consider that mucus production in cadmium-exposed gastropods *Nucella lapillus* is linked to changes observed at the level of energy metabolism (decreased rate of oxygen consumption and glycogen concentration).

As regards the cost of tolerance, the NOEL is all the lower when chemical stress combines with nonchemical stress, more particularly with those stressors affecting energy metabolism—that is, temperature and, to a greater extent, food availability as discussed above. Thus, in the rainbow trout (*Oncorhynchus mykiss*), long-term exposure (100 days) to low concentrations of metals (3 μg Cd L^{-1}, 75 μg Cu L^{-1}, or 250 μg Zn L^{-1}) involves three types of successive responses: damage, repair, and acclimatization. When the rainbow trout can get enough food, there was no effect of metal exposure on growth, but copper exposure generated increased food intake, lower swimming speed, and high oxygen consumption, thus involving a metabolic cost (McGeer et al. 2000 and quoted literature). It is noteworthy that chronic exposure to xenobiotics does not systematically involve increased acquisition of tolerance in populations, as shown by the reduction of diversity commonly observed in contaminated environments. Theory suggests that individuals tolerant to one particular type of stress may have reduced performance when confronted with another stressor. The cost of resistance, which can be associated with physiological acclimatization as well as genetic adaptation, could originate from increased allocation of energy and resources to defense mechanisms. However, other processes have also been reported in literature, such as an alteration in the function of some protein targets or a reduction of physiological plasticity or evolution (Meyer and Di Giulio 2003 and literature cited therein). Indeed, in the cyanophycean *Microcystis aeruginosa* the acquisition of tolerance to dinitrophenol reduces variability in growth when the blue green bacterium is subsequently exposed to a concentration gradient of this molecule (Genoni et al. 2001). In a PCB-resistant strain of the marine diatom *Ditylum brightwellii*, growth in the presence of this contaminant is better than that of a sensitive strain. In other diatoms (*Asterionella glacialis, Thalassiosira nordenskioldii*), the growth of resistant clones originating from contaminated estuaries is enhanced by the addition of PCB in the culture medium. Similar observations were made in the case of polynuclear hydrocarbons with low molecular weight. Nevertheless, in *D. brightwellii*, resistance to PCB reduces tolerance to lower salinity and nitrogen restriction, but increases tolerance to lower temperatures (Cosper et al. 1987). These findings corroborate previous research on terrestrial plants or bacterial strains resistant to antibiotics, revealing that resistant organisms are favored in the presence of the toxin, but in contrast are at a disadvantage in its absence (Cosper et al. 1988 and literature cited therein). In F1 and F2 offspring of fish (*Fundulus heteroclitus*) exposed for decades to a mixture of contaminants (mainly creosote) in the field, there was enhanced sensitivity to photodegradation products of anthracene and fluoranthene, and to hypoxia (Meyer and Di Giulio 2003).

3.4.3 Contamination of Food Webs

Tolerance is responsible for the survival of organisms in polluted environments, but tolerant individuals/populations/species may constitute contaminated links in food webs. This risk is more or less critical, depending on the physiological mechanisms used by organisms along a food chain to cope with chemical exposure: particularly elimination or storage (Figure 3.2). The influence of tolerance on the trophic transfer of contaminants has been recently reviewed (Amiard-Triquet and Rainbow in Amiard-Triquet et al. 2011).

If the metal tolerance mechanism of an invertebrate involves increased storage detoxification, there is a real risk of increased trophic transfer. In Cu-resistant bacteria *Vibrio* sp., important bioaccumulation of this metal was observed. In the presence of these bacteria, the larvae of the bivalve *Argopecten purpuratus* accumulated Cu to very high levels. Thus, bacterial copper accumulation could be very significant in marine environments, increasing copper transfer at the base of marine food chains (Miranda and Rojas 2006). The ecotoxicological significance of trophic transfer has been documented in some species. Thus, decapod crustaceans *Palaemonetes varians* fed on metal-rich Restronguet Creek polychaetes *Nereis diversicolor* showed significant mortality (Rainbow et al. 2006). Zebrafish *Danio rerio* also fed on Restronguet Creek *N. diversicolor* in the laboratory showed reduced reproductive outputs, attributed by the authors to the trophic transfer of arsenic from these worms

(Boyle et al. 2008). Even if metal detoxification by biomineralization does not guarantee the "transfer of metal detoxification along marine food chains" according to the expression of Nott and Nicolaidou (1990), it is a factor limiting the risk of transfer. The physicochemical form of metals in their prey clearly influences subsequent trophic transfer, but the pattern varies between food items, consumers, and metals. From the different studies synthesized by Rainbow et al. (2011), it may be concluded that what is trophically available to one predator (feeding on one prey type) is not necessarily trophically available to another (taxonomically separated) predator even if feeding on the same prey, given the variability between animal digestive systems (Figure 3.3).

The ecotoxicological risk is greater for metals that have organometallic forms such as methylmercury, which is prone to biomagnify in aquatic food chains as dramatically demonstrated by the Minamata disaster. Biomagnification is defined as an increase in contaminant concentration from one trophic level to the next owing to accumulation from food.

Biomagnification is also well documented for persistent organic contaminants such as dichlorodiphenyltrichloroethane (DDT), PCBs, and PBDEs. Hydrophobicity is an important chemical property favoring biomagnification in biota but it is not the whole story, and despite being hydrophobic, PAHs are not biomagnified. The fate of organic contaminants in the food web depends on a set of biological mechanisms including (1) mucus production; (2) induction of MXR that, by limiting bioaccumulation in prey species, reduces contaminant transfer to predators (Section 3.3.5); (3) biotransformation based on phases I and II enzymes, which favor excretion (Section 3.3.3) but with side effects linked to the presence of intermediate reactive metabolites. These genotoxic/carcinogenic metabolites may be responsible for a transfer of toxicity in the food chain, even in the absence of biomagnification. Studies involving PAH-contaminated polychaetes fed to juvenile English sole or mussels contaminated with hydrocarbons released into the field after the oil spill of the tanker *Erika* fed to mammals provide examples of a transfer of toxicity between successive trophic levels (Amiard-Triquet and Rainbow in Amiard-Triquet et al. 2011).

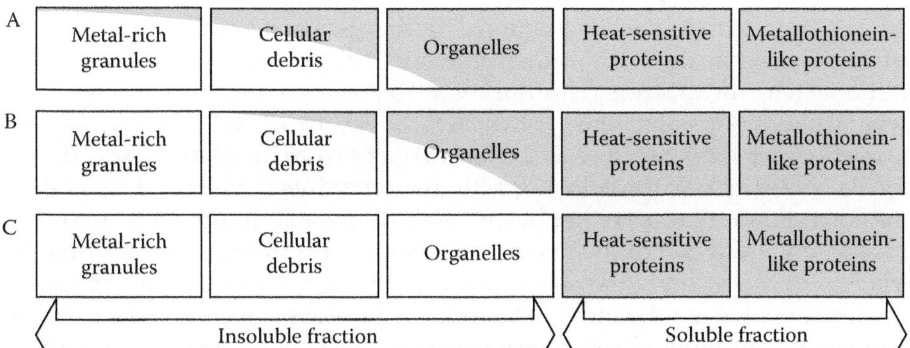

FIGURE 3.3
Fractionation of metal accumulated in prey into five components. (After Wallace, W.G. et al., *Mar. Ecol. Prog. Ser.*, 249, 183–197, 2003.) (a) Highlighted areas covering all five fractions to some degree represent metal accumulated in prey trophically available to a neogastropod mollusk (Cheung and Wang 2005; Rainbow et al. 2007). (b) Highlighted areas (from four fractions) represent metal accumulated in prey trophically available to a predator with weaker digestive powers than a neogastropod mollusk. (c) Highlighted areas (from two fractions) represent metal accumulated in prey trophically available to a planktonic copepod filtering phytoplankton (Reinfelder and Fisher 1991). (From Rainbow, P.S. et al., *Environ. Pollut.*, 159, 2347–2349, 2011. With permission.)

3.5 Conclusions

Considering the general aim of this book, we need to conclude this chapter on biomarkers of defense in terms of their utility as tools for the assessment of the impact of chemical stress on populations and ecosystems. The ecological importance of tolerance and underlying defense mechanisms depends on the extent of this phenomenon. Many species belonging to most taxonomic groups are able to develop tolerance toward the major classes of contaminants to which they have been chronically exposed (Tables 3.1 and 3.2), and sometimes to other compounds thanks to cross-tolerance. Thus, it seems that many species can cope with chemical stress in their environment, so contributing to the conservation of biodiversity and normal or subnormal functioning of the ecosystem.

However, this positive interpretation of the information provided by biomarkers of defense needs to be moderated. First, it is questionable whether the literature accurately reflects the field situation, since all authors have experienced that negative results are not as easily publishable as positive results. A number of counterexamples have shown an increased sensitivity of the progeny of exposed parents (Bervoets et al. 1996; Villarroel et al. 2000). Second, logistical constraints have limited scientific work to species that are easy to collect in the field and keep in the laboratory. Because interspecific variations of sensitivity are important (Chapter 7), the risk assessment of chemicals based on a small number of species may be seriously biased. If species selected as test organisms argue for a reduction of logistical constraints, it is because they are often tolerant to the nonchemical stress generated by laboratory conditions. Athrey et al. (2007) have shown a loss of genetic variation resulting from maintaining populations of fish *Heterandria formosa* in the laboratory. These authors underline that the potential for loss of genetic variation in laboratory populations must be taken into consideration when extrapolating from laboratory to natural populations. For sentinel species collected from the wild where they experience important variations of natural factors (in the intertidal zone, in estuaries), it has been thought that this tolerance to natural stress could spread to tolerance to chemical stress, leading to an undervaluation of risk in field situations. More recently, it has been established that, on the contrary, species at the limit of their tolerance to natural stress are more sensitive to any additional (chemical) stress (Hummel et al. in Amiard-Triquet et al. 2011).

For several biomarkers of defense (MXR, SOD in Figure 3.2, MT, HSP), it has been shown that the relationship between dose and effect deviates from linearity for severe contamination. Equal concentrations (proteins) or activities (enzymes) of biomarkers of defense can therefore correspond to doses either below or above the maximum value for induction. In the first case, the induction of the defense mechanism is efficient in protecting the organisms, whereas in the second case, the induction is thwarted and toxic effects can occur.

The protective value of tolerance mechanisms must not be overvalued since this chapter has documented a number of secondary negative effects of being tolerant: (1) energy cost leading to changes in energy allocation with a risk of cascading effects from individuals to populations (Chapter 12), whatever the origin of tolerance, either physiological acclimation of individuals or inheritable genetic adaptation (Chapter 14); (2) formation of metabolites, which may be more toxic than parent compounds (carcinogenic, generating oxidative stress); (3) increased sensitivity to another type of stress such as photosensitivity or temperature, which may be crucial considering the reduction of the ozone layer or global warming. The status of fish populations in highly contaminated estuaries of the east coast of North America is a good illustration of the difficulty of deciding upon the beneficial role of tolerance (Romeo and Wirgin in Amiard-Triquet et al. 2011). Paradoxically, high prevalence of hepatic neoplasms has been observed in two populations—killifish *Heterandria formosa* from the Elizabeth River and tomcod *Microgadus tomcod* from the Hudson River, United States—resistant, respectively, to PAHs and PCBs. Resistant hepatocytes are able to proliferate vigorously, resist cytotoxicity, and exhibit unusual patterns of gene induction (phases I and II enzymes). Thus, it has been hypothesized that hepatic neoplasia may provide a complementary mechanism for impacted populations to persist in highly contaminated environments (considering that cancer is postreproductive in these fish models), but at the cost of an altered population age structure. Cascading effects at higher levels of biological organization depend on the role of impacted populations in the structure of communities and the functioning of ecosystems.

Another ecosystem aspect must be considered to assess the protective value of biomarkers of defense. Tolerance allows the persistence of organisms in highly contaminated environments but perhaps at the cost of contaminant transfer in food webs, which is particularly worrying for those compounds prone to biomagnification (methylmercury, DDT, PCBs) or in the case of toxicity transfer between successive links (PAHs). If we include our own species in the ecosystem, tolerance may be considered more detrimental than beneficial, responsible for the development of bacterial strains coresistant to chemicals and antibiotics so important in medicine or the development of insect populations resistant to pesticides, thus requiring insecticides with novel modes of action (Ghanim and Ishaaya in Amiard-Triquet et al. 2011).

Tolerance clearly poses a problem for risk assessment in contaminated ecosystems. In regulations for the control of water quality based on toxicity tests, the protective effect of acclimation is never taken into account. This can generate positive false results since living organisms are present in areas where contaminant concentrations are so high that they are not expected to be able to survive. On the other hand, laboratory tests carried out with tolerant strains or populations can lead to negative false results, with underestimated toxicological parameters. Consequently, environmental assessment must not be limited to a chemical approach, which only makes it possible to verify that environmental standards are enforced, but must also include a biological constituent involving a multibiomarker approach (Chapter 2).

References

Abele, D. et al. 2002. Temperature-dependence of mitochondrial function and production of reactive oxygen species in the intertidal mud clam *Mya arenaria*. *J. Exp. Biol.* 205:1831–41.

Abele, D., J.P. Vázquez-Medina, and T. Zenteno-Savín. 2012. *Oxidative Stress in Aquatic Ecosystems*. Chichester, UK: Wiley-Blackwell.

Ait Alla, A. et al. 2006. Tolerance and biomarkers as useful tools for assessing environmental quality in the Oued Souss estuary (Bay of Agadir, Morocco). *Comp. Biochem. Physiol.* 143C:23–9.

Allen, J.J., and M.N. Moore. 2004. Environmental pronostics: Is the current use of biomarkers appropriate for environmental risk evaluation? *Mar. Environ. Res.* 58:227–32.

Alonzo, F. et al. 2006. Effects of chronic internal alpha irradiation on physiology, growth and reproductive success of *Daphnia magna*. *Aquat. Toxicol.* 80:228–36.

Amiard, J.C. 1991. Réponses des organismes marins aux pollutions métalliques. In *Réactions des êtres vivants aux changements de l'environnement*, pp. 197–205. Paris: Piren, CNRS.

Amiard, J.C. et al. 2006. Metallothioneins in aquatic invertebrates: Their role in metal detoxification and their use as biomarkers. *Aquat. Toxicol.* 76:160–202.

Amiard-Triquet, C., P.S. Rainbow, and M. Roméo, eds. 2011. *Tolerance to Environmental Contaminants.* Boca Raton, FL: CRC Press.

Anguiano, O.L., A. Caballero de Castro, and A.M. Pechen de D'Angelo. 2001. The role of glutathione conjugation in the regulation of early toad embryos' tolerance to pesticides. *Comp. Biochem. Physiol.* 128C:35–43.

Anguiano, O.L. et al. 2008. Enhanced esterase activity and resistance to azinphosmethyl in target and nontarget organisms. *Environ. Toxicol. Chem.* 27:2117–23.

Annabi, A. et al. 2009. Comparative study of the sensitivity to cadmium of two populations of *Gambusia affinis* from two different sites. *Environ. Monit. Assess.* 155:459–65.

Armknecht, S.L., S.L. Kaattari, and P.A. Van Veld. 1998. An elevated glutathione *S*-transferase in creosote-resistant mummichog (*Fundulus heteroclitus*). *Aquat. Toxicol.* 41:1–16.

Athrey, N. R. G., P. L. Leberg, and P. L. Klerks. 2007. Laboratory culturing and selection for increased resistance to cadmium reduce genetic variation in the least killifish, *Heterandria formosa. Environ. Toxicol. Chem.* 26:1916–21.

Balch, G.C. et al. 2000. Weight loss and net abnormalities of *Hydropsyche betteni* (caddisfly) larvae exposed to aqueous zinc. *Environ. Toxicol. Chem.* 19:3036–43.

Ballan-Dufrançais, C., C. Marcaillou, and C. Amiard-Triquet. 1991. Response of the phytoplancton alga *Tetraselmis suecica* to copper and silver exposure: Vesicular metal bioaccumulation and lack of starch bodies. *Biol. Cell.* 72:103–12.

Barata, C. et al. 2001. Biochemical factors contributing to response variation among resistant and sensitive clones of *Daphnia magna* Straus exposed to ethyl parathion. *Ecotoxicol. Environ. Saf.* 49:155–63.

Bard, S.M. 2000. Multixenobiotic resistance as a cellular defense mechanism in aquatic organisms. *Aquat. Toxicol.* 48:357–89.

Berk, S.G. et al. 1978. Effects of ingesting mercury-containing bacteria on mercury tolerance and growth rates of ciliates. *Microb. Ecol.* 4:319–30.

Berthet, B. et al. 2003. Accumulation and soluble binding of Cd, Cu and Zn in the polychaete *Hediste diversicolor* from coastal sites with different trace metal bioavailability. *Arch. Environ. Contam. Toxicol.* 45:468–78.

Bervoets, L. et al. 1996. Evaluation of effluent toxicity and ambient toxicity in a polluted lowland river. *Environ. Pollut.* 91:333–41.

Bierkens, J.G. 2000. Applications and pitfalls of stress-proteins in biomonitoring. *Toxicology* 153: 61–72.

Blanck, H., S.Å. Wängberg, and S. Molander. 1988. Pollution-induced community tolerance–A new ecotoxicological tool. In *Functional Testing of Aquatic Biota for Estimating Hazards of Chemicals*, ed. J. Cairns Jr. and J.R. Pratt, 219–30. Philadelphia, PA: JR STP 988 American Society for Testing and Materials.

Bloom, S., T.B. Andersson, and L. Förlin. 2000. Effects of food deprivation and handling stress on head kidney 17α-hydroxyprogesterone 21-hydroxylase activity, plasma cortisol and the activities of liver detoxification enzymes in rainbow trout. *Aquat. Toxicol.* 48:265–74.

Bond, J.A., and B.P. Bradley. 1997. Resistance to malathion in heat-shocked *Daphnia magna. Environ. Toxicol. Chem.* 16:705–12.

Borković, S.S. et al. 2005. The activity of antioxidant defence enzymes in the mussel *Mytilus galloprovincialis* from the Adriatic Sea. *Comp. Biochem. Physiol.* 141C:366–74.

Bossuyt, B.T.A., and C.R. Janssen. 2004a. Long-term acclimation of *Pseudokirchneriella subcapitata* (Korshikov) Hindak to different copper concentrations: Changes in tolerance and physiology. *Aquat. Toxicol.* 68:61–74.

Bossuyt, B.T.A., and C.R. Janssen. 2004b. Influence of multigeneration acclimation to copper on tolerance, energy reserves, and homeostasis of *Daphnia magna* Straus. *Environ. Toxicol. Chem.* 23:2029–37.

Boyle, D. et al. 2008. Natural arsenic contaminated diets perturb reproduction in fish. *Environ. Sci. Technol.* 42:5354–60.

Bradley, R.W., C. Duquesnay, and J.B. Sprague. 1985. Acclimation of rainbow trout, *Salmo gairdneri* Richardson, to zinc: Kinetics and mechanism of enhanced tolerance induction. *J. Fish. Biol.* 27:367–79.

Brown, B. E. 1978. Lead detoxification by a copper-tolerant isopod. *Nature* 276:388–90.

Brown, D.C., B.P. Bradley, and M. Tedengren. 1995. Genetic and environmental regulation of HSP70 expression. *Mar. Environ. Res.* 39:181–4.

Bryan, G.W. 1984. Pollution due to heavy metals and their compounds. In *Marine Ecology*, Kinne, O., Ed. New York: Wiley, pp. 1289–431.

Bryan, G.W. et al. 1987. Copper, zinc and organotin as long-term factors governing the distribution of organisms in the Fal Estuary in Southwest England. *Estuaries* 10:208–19.

Burlinson, F.C., and A.J. Lawrence. 2007. Development and validation of a behavioural assay to measure the tolerance of *Hediste diversicolor* to copper. *Environ. Pollut.* 145:274–8.

Burnett, K.G. et al. 2007. *Fundulus* as the premier teleost model in environmental biology: Opportunities for new insights using genomics. *Comp. Biochem. Physiol.* 2D:257–86.

Buss, D.S., and A. Callaghan. 2008. Interaction of pesticides with p-glycoprotein and other ABC proteins: A survey of the possible importance to insecticide, herbicide and fungicide resistance. *Pestic. Biochem. Physiol.* 90:141–53.

Cain, D.J., S.N. Luoma, and W.G. Wallace. 2004. Linking metal bioaccumulation of aquatic insects to their distribution patterns in a mining-impacted river. *Environ. Toxicol. Chem.* 23:1463–73.

Calow, P. 1991. Physiological costs of combating chemical toxicants: Ecological implications. *Comp. Biochem. Physiol.* 100C:3–6.

Canli, M. 2005. Dietary and water-borne Zn exposures affect energy reserves and subsequent Zn tolerance of *Daphnia magna. Comp. Biochem. Physiol.* 141C:110–6.

Capolino, E. et al. 1997. Tolerance to mercxury chloride in *Scenedesmus* strains. *BioMetals* 10:85–94.

Carman, K.R. et al. 1995. Experimental investigation of the effects of polynuclear aromatic hydrocarbons on an estuarine sediment food web. *Mar. Environ. Res.* 40:289–318.

Carman, K.R., J.W. Fleeger, and S.M. Pomarico. 2000. Does historical exposure to hydrocarbon contamination alter the response of benthic communities to diesel contamination? *Mar. Environ. Res.* 49:255–78.

Carrera-Martínez, D. et al. 2010. Microalgae algae response to petroleum spill: An experimental model analysing physiological and genetic response of *Dunaliella tertiolecta* (Chlorophyceae) to oil samples from the tanker *Prestige. Aquat. Toxicol.* 97:151–9.

Carrera-Martínez, D. et al. 2011. Adaptation of microalgae to a gradient of continuous petroleum contamination. *Aquat. Toxicol.* 101:342–50.

Chowdhury, M.J., E.F. Pane, and C.M. Wood. 2004. Physiological effects of dietary cadmium acclimation and waterborne cadmium challenge in rainbow trout: Respiratory, ionoregulatory, and stress parameters. *Comp. Biochem. Physiol.* 139C:163–73.

Clark, M.S., and L.S. Peck. 2009. HSP70 heat shock proteins and environmental stress in Antarctic marine organisms: A minireview. *Mar. Genomics* 2:11–18.

Clubbs, R.L., and B.W. Brooks. 2007. *Daphnia magna* responses to a vertebrate estrogen receptor agonist and an antagonist: A multigenerational study. *Ecotoxicol. Environ. Saf.* 67:385–98.

Cooper, P.S. 1999. Altered expression of the xenobiotic P-glycoprotein in liver and liver tumours of mummichog *Fundulus heteroclitus* from a creosote-contaminated environment. *Biomarkers* 4:48–58.

Coors, A. et al. 2009. Land use, genetic diversity and toxicant tolerance in natural populations of *Daphnia magna*. *Aquat. Toxicol.* 95:71–9.

Corradi, M.G. et al. 1995. Chromium-induced sexual reproduction gives rise to a Cr-tolerant progeny in *Scenedesmus acutus*. *Ecotoxicol. Environ. Saf.* 32:12–18.

Cosper, E.M., C.F. Wurster, and R. George Rowland. 1984. PCB resistance within phytoplankton populations in polluted and unpolluted marine environments. *Mar. Environ. Res.* 12:209–23.

Cosper, E.M. et al. 1987. Induced resistance to polychlorinated biphenyls confers cross-resistance and altered environmental fitness in a marine diatom. *Mar. Environ. Res.* 23:207–22.

Cosper, E.M., C.F. Wurster, and M.F. Bautista. 1988. PCB-resistant diatoms in the Hudson River estuary. *Estuar. Coast. Shelf Sci.* 26:215–26.

Costa, J. et al. 2012. Gene expression analysis of ABC efflux transporters, CYP1A and GSTa in Nile tilapia after exposure to benzo(*a*)pyrene. *Comp. Biochem. Physiol.* 155C:469–82.

Couillard, Y. et al. 1995. Field transplantation of a freshwater bivalve, *Pyganodon grandis*, across a metal contamination gradient: II. Metallothionein response to Cd and Zn exposure, evidence for cytotoxicity, and links to effects at higher levels of biological organization. *Can. J. Fish. Aquat. Sci.* 52:703–15.

Damiens, G. et al. 2006. Metal bioaccumulation and metallothionein concentrations in larvae of *Crassostrea gigas*. *Environ. Pollut.* 140:492–9.

Davies, M.S., S.J., Hawkins, and H.D. Jones. 1990. Mucus production and physiological energetics in *Patella vulgata* L. *J. Molluscan Stud.* 56:499–503.

De Lafontaine, Y. et al. 2000. Biomarkers in zebra mussels (*Dreissena polymorpha*) for the assessment and monitoring of water quality of the St Lawrence River (Canada). *Aquat. Toxicol.* 50:51–71.

Depledge, M.H., A. Aagard, and P. Györkös. 1995. Assessment of trace metal toxicity using molecular, physiological and behavioural biomarkers. *Mar. Pollut. Bull.* 31:19–27.

De Pomerai, D.I. 1996. Heat-shock proteins as biomarkers of pollution. *Hum. Exp. Toxicol.* 15:279–85.

De Wit, M. et al. 2008. Molecular targets of TBBPA in zebrafish analysed through integration of genomic and proteomic approaches. *Chemosphere* 74:96–105.

Duncan, D.A., and J.F. Klaverkamp. 1983. Tolerance and resistance to cadmium in white suckers (*Catostomus commersoni*) previously exposed to cadmium, mercury, zinc, or selenium. *Can. J. Fish. Aquat. Sci.* 40:128–38.

Eckwert, H., G. Alberti, and H.R. Köhler. 1997. The induction of stress proteins (hsp) in *Oniscus asellus* (Isopoda) as a molecular marker of multiple heavy metal exposure: I. Principles and toxicological assessment. *Ecotoxicology* 6:249–62.

Ernst, S.R. et al. 2009. Benthic foraminiferal response to experimentally induced Erika oil pollution. *Mar. Micropaleontol.* 61:76–93.

Falfushynska, H.I. et al. 2011.Various responses to copper and manganese exposure of *Carassius auratus gibelio* from two populations. *Comp. Biochem. Physiol.* 154C:242–53.

Falfushynska, H.I., L.L. Gnatyshyna, and O.B. Stoliar. 2012. Population-related molecular responses on the effect of pesticides in *Carassius auratus gibelio*. *Comp. Biochem. Physiol.* 155C:396–406.

Fangue, N.A., M. Hofmeister, and P.M. Schulte. 2006. Intraspecific variation in thermal tolerance and heat shock protein gene expression in common killifish, *Fundulus heteroclitus*. *J. Exp. Biol.* 209:2859–72.

Feige, U. et al. 1996. In *Stress-Inducible Cellular Responses*, ed. U. Feige, R.I. Morimoto, I. Yahara, B.S. Polla. Basel: Birkhaüser Verlag, 492 pp.

Fernández, B. et al. 2010. Antioxidant responses in gills of mussel (*Mytilus galloprovincialis*) as biomarkers of environmental stress along the Spanish Mediterranean coast. *Aquat. Toxicol.* 99:186–97.

Ferrari, A. et al. 2007. Effects of carbaryl and azinphos methyl on juvenile rainbow trout (*Oncorhynchus mykiss*) detoxifying enzymes. *Pest. Biochem. Physiol.* 88:134–42.

Fleeger, J.W., K.R. Carman, and R.M. Nisbet. 2003. Indirect effects of contaminants in aquatic ecosystems. *Sci. Total Environ.* 317:207–33.

Frydman, J. 2001. Folding of newly translated proteins *in vivo*: The role of molecular chaperones. *Annu. Rev. Biochem.* 70:603–47.

Gale, S.A. et al. 2003. Insights into the mechanisms of copper tolerance of a population of black-banded rainbowfish (*Melanotaenia nigrans*) (Richardson) exposed to mine leachate, using $^{64/67}$Cu. *Aquat. Toxicol.* 62:135–53.

Genoni, G.P. et al. 2001. Complex dynamics of adaptation in a nonaxenic *Microcystis* culture: 1. Effects of dinitrophenol on population growth. *Ecotoxicol. Environ. Saf.* 48:235–40.

Gillis, P.L. et al. 2002. Cadmium-induced production of a metallothionein-like protein in *Tubifex tubifex* (Oligochaeta) and *Chironomus riparius* (Diptera) correlation with reproduction and growth. *Environ. Toxicol. Chem.* 21:1836–44.

Gomez-Gesteira, J.L., and J.-C.Dauvin. 2005. Impact of the Aegean Sea oil spill on the subtidal fine sand macrobenthic community of the Ares-Betanzos Ria (Northwest Spain). *Mar. Environ. Res.* 60:289–316.

Gorbi, G. et al. 2006. Differential responses to Cr(VI)-induced oxidative stress between Cr-tolerant and wild-type strains of *Scenedesmus acutus* (Chlorophyceae). *Aquat. Toxicol.* 79:132–9.

Groenendijk, D. et al. 2002. Dynamics of metal adaptation in riverine chironomids. *Environ. Pollut.* 117:101–9.

Gross, M. 2004. Emergency services: A bird's eye perspective on the many different functions of stress proteins. *Curr. Protein Pept. Sci.* 5:213–23.

Guan, R., and W.X. Wang. 2006. Comparison between two clones of *Daphnia magna*: Effects of multigenerational cadmium exposure on toxicity, individual fitness, and biokinetics. *Aquat. Toxicol.* 76:217–29.

Guderley, H. et al. 2003. Metabolic priorities during starvation: Enzyme sparing in liver and white muscle of Atlantic cod, *Gadus morhua* L. *Comp. Biochem. Physiol.* 135A:347–56.

Haap, T., and H.R. Köhler. 2009. Cadmium tolerance in seven *Daphnia magna* clones is associated with reduced hsp70 baseline levels and induction. *Aquat. Toxicol.* 94:131–7.

Harper-Arabie, R.M. et al. 2004. Protective effects of allozyme genotype during chemical exposure in the grass shrimp, *Palaemonetes pugio*. *Aquat. Toxicol.* 70:41–54.

Hartl, F.U., and M. Hayer-Hartl. 2002. Molecular chaperones in the cytosol: From nascent chain to folded protein. *Science* 295:1852–8.

Heinonen, J., J.V.K. Kukkonen, and I.J. Holopainen. 2001. Temperature- and parasite-induced changes in toxicity and lethal body burdens of pentachlorophenol in the freshwater clam *Pisidium amnicum*. *Environ. Toxicol. Chem.* 20:2778–84.

Hoare, K., A.R. Beaumont, and J. Davenport. 1995. Variation among populations in the resistance of *Mytilus edulis* embryos to copper: Adaptation to pollution? *Mar. Ecol. Prog. Ser.* 120:155–61.

Hofmann, G., and G. Somero. 1995. Evidence for protein damage at environmental temperatures: Seasonal changes in levels of ubiquitin conjugates and hsp70 in the intertidal mussel *Mytilus trossulus*. *J. Exp. Biol.* 198:1509–18.

Ivanina, A.V., A.S. Cherkasov, and I.M. Sokolova. 2008. Effects of cadmium on cellular protein and glutathione synthesis and expression of stress proteins in eastern oysters, *Crassostrea virginica* Gmelin. *J. Exp. Biol.* 211:577–86.

Ivanina, A.V., C. Taylor, and I.M. Sokolova. 2009. Effects of elevated temperature and cadmium exposure on stress protein response in eastern oysters *Crassostrea virginica* (Gmelin). *Aquat. Toxicol.* 91:245–54.

Ivorra, N. et al. 2002. Metal-induced tolerance in the freshwater microbenthic diatom *Gomphonema parvulum*. *Environ. Pollut.* 116:147–57.

Janssens, B.J. et al. 2000. Reduced enzymatic antioxidative defense in deep-sea fish. *J. Exp. Biol.* 203:3717–25.

Jones, M.B., and I. Johnson. 1992. Responses of the brackish-water amphipod *Gammarus duebeni* (Crustacea) to saline sewage. *Neth. J. Sea Res.* 30:141–7.

Kaplan, D. et al. 1995. Cadmium toxicity and resistance in *Chlorella* sp. *Plant Sci.* 109:129–37.

Kashian, D.R. 2004. Toxaphene detoxification and acclimation in *Daphnia magna*: Do cytochrome P-450 enzymes play a role? *Comp. Biochem. Physiol.* 137C:53–63.

Klerks, P.L., and J.S. Levinton. 1989. Rapid evolution of metal resistance in a benthic oligochaete inhabiting a metal-polluted site. *Biol. Bull.* 176:135–41.

Knapen, D. et al. 2004. Resistance to water pollution in natural gudgeon (*Gobio gobio*) populations may be due to genetic adaptation. *Aquat. Toxicol.* 67:155–65.

Kraus, M.L., J.S. Weis, and P. Weis. 1988. Effects of mercury on larval and adult grass shrimp (*Palaemonetes pugio*). *Arch. Environ. Contam. Toxicol.* 17:355–63.

Lannig, G., A.S. Cherkasov, and I.M. Sokolova. 2006. Temperature-dependent effects of cadmium on mitochondrial and whole-organism bioenergetics of oysters (*Crassostrea virginica*). *Mar. Environ. Res.* 62:S79–S82.

Larsen, D.K. et al. 2003. Long-term effect of Sea-Nine on natural coastal phytoplankton communities assessed by pollution induced community tolerance. *Aquat. Toxicol.* 62:35–44.
Le Faucheur, S. et al. 2006. Thiols in *Scenedesmus vacuolatus* upon exposure to metals and metalloids. *Aquat. Toxicol.* 80:355–61.
Leung, K.M.Y., A.C. Taylor, and R.W. Furness. 2000. Temperature-dependent physiological responses of the dogwhelk *Nucella lapillus* to cadmium exposure. *J. Mar. Biol. Assoc. U.K.* 80:647–60.
Lewis, C., and T. Galloway. 2008. Genotoxic damage in polychaetes: A study of species and cell-type sensitivities. *Mutat. Res./Genet. Toxicol. Environ. Mutagen* 654:69–75.
Lopes, I., D.J. Baird, and R. Ribeiro. 2004. Genetic determination of tolerance to lethal and sublethal copper concentrations in field populations of *Daphnia longispina*. *Arch. Environ. Contam. Toxicol.* 46:43–51.
Lopes, I., D.J. Baird, and R. Ribeiro. 2005. Resistance to metal contamination by historically-stressed populations of *Ceriodaphnia pulchella*: Environmental influence versus genetic determination. *Chemosphere* 61:1189–97.
Lopes, I., D.J. Baird, and R. Ribeiro. 2006. Genetic adaptation to metal stress by natural populations of *Daphnia longispina*. *Ecotoxicol. Environ. Safety* 63:275–85.
Lotts, J.W., and A.J. Stewart. 1995. Minnows can acclimate to total residual chlorine. *Environ. Toxicol. Chem.* 14:1365–74.
Luckenbach, T., I. Corsi, and D. Epel. 2004. Fatal attraction: Synthetic musk fragrances compromise multixenobiotic defence systems in mussels. *Mar. Environ. Res.* 58:215–9.
Luoma, S.N. et al. 1983. Variable tolerance to copper in two species from San Francisco Bay. *Mar. Environ. Res.* 10:209–22.
Maazouzi, C. et al. 2008. Chronic copper exposure and fatty acid composition of the amphipod *Dikerogammarus villosus*: Results from a field study. *Environ. Pollut.* 156:221–6.
Marigomez, I. et al. 2002. Cellular and subcellular distribution of metals in molluscs. *Microsc. Res. Technol.* 56:358–92.
Martínez-Colon, M., P. Hallock, and C. Green-Ruíz. 2009. Strategies for using shallow-water benthic foraminifers as bioindicators of potentially toxic elements: A review. *J. Foram. Res.* 39:278–99.
Mason, A.Z., and K.D. Jenkins. 1995. Metal detoxication in aquatic organisms. In *Metal Speciation and Bioavailability in Aquatic Systems*, ed. A. Tessier and D.R. Turner, 479–608. Chichester: Wiley.
McGeer, J.C. et al. 2000. Effects of chronic sublethal exposure to waterborne Cu, Cd or Zn in rainbow trout: 1. Iono-regulatory disturbance and metabolic costs. *Aquat. Toxicol.* 50:231–43.
McLusky, D.S., and M. Elliott. 2004. *The Estuarine Ecosystem; Ecology, Threats and Management,* 3rd ed. Oxford: Oxford University Press.
Meyer, J.N., and R.T. Di Giulo. 2003. Heritable adaptation and fitness costs in killifish (*Fundulus heteroclitus*) inhabiting a polluted estuary. *Ecol. Appl.* 13:490–503.
Meyer, J.N. et al. 2003. Antioxidant defenses in killifish (*Fundulus heteroclitus*) exposed to contaminated sediments and model prooxidants: Short-term and heritable responses. *Aquat. Toxicol.* 65:377–95.
Miliou, H. et al. 2000. Influence of life-history adaptations on the fidelity of laboratory bioassays for the impact of heavy metals (Co^{2+} and Cr^{6+}) on tolerance and population dynamics of *Tisbe holothuriae*. *Mar. Pollut. Bull.* 40:352–59.
Miller, M.P., and A.C. Hendricks. 1996. Zinc resistance in *Chironomus riparius*: Evidence for physiological and genetic components. *J. North Am. Benthol. Soc.* 15:106–16.
Millward, R.N., and A. Grant. 1995. Assessing the impact of copper on nematode communities from a chronically metal-enriched estuary using pollution-induced community tolerance. *Mar. Pollut. Bull.* 30:701–6.
Millward, R.N., and A. Grant. 2000. Pollution-induced tolerance to copper of nematode communities in the severely contaminated Restronguet Creek and adjacent estuaries, Cornwall, United Kingdom. *Environ. Toxicol. Chem.* 19:454–61.
Minier, C. et al. 2006a. Multixenobiotic resistance protein expression in *Mytilus edulis, Mytilus galloprovinciallis* and *Crassostrea gigas* from the French coasts. *Mar. Ecol. Prog. Ser.* 322:155–68.

Minier. C. et al. 2006b. A pollution-monitoring pilot study involving contaminant and biomarker measurements in the Seine estuary, France, using zebra mussels (*Dreissena polymorpha*). *Environ. Toxicol. Chem.* 25:112–9.

Miranda, C.D., and R. Rojas. 2006. Copper accumulation by bacteria and transfer to scallop larvae. *Mar. Pollut. Bull.* 52:293–300.

Moore, N.W. 1967. A synopsis of the pesticide problem. In *Advances in Ecological Research*, vol. 4, pp. 75–129. New York: Academic Press.

Moraitou-Apostolopoulou, M., and G. Verriopoulos. 1979. Some effects of sub-lethal concentrations of copper on a marine copepod. *Mar. Pollut. Bull.* 10:88–92.

Moreau, J.L. et al. 2008. Metal binding and antioxidant properties of chimeric tri- and tetra-domained metallothioneins. *Biochimie* 90:705–16.

Morgan, A.J., P. Kille, and S.R. Stürzenbaum. 2007. Microevolution and ecotoxicology of metals in invertebrates. *Environ. Sci. Technol.* 41:1085–96.

Morimoto, R.I., K.D. Sarge, and K. Abravaya. 1992. Transcriptional regulation of heat shock genes. *J. Biol. Chem.* 267:21987–90.

Mouneyrac, C. et al. 2003. Physico-chemical forms of storage and the tolerance of the estuarine worm *Nereis diversicolor* chronically exposed to trace metals in the environment. *Mar. Biol.* 143:731–44.

Mukhopadhyay, I. et al. 2003. Heat shock response: hsp70 in environmental monitoring. *J. Biochem. Mol. Toxicol.* 17:249–54.

Munkittrick, K.R., and D.G. Dixon. 1988. Evidence for a maternal yolk factor associated with increased tolerance and resistance of feral white sucker (*Catostomus commersoni*) to waterborne copper. *Ecotoxicol. Environ. Saf.* 15:7–20.

Muyssen, B.T.A., and C.R. Janssen. 2002. Tolerance and acclimation to zinc of *Ceriodaphnia dubia*. *Environ. Pollut.* 117:301–6.

Muyssen, B.T.A., K.A.C. De Schamphelaere, and C.R. Janssen. 2006. Mechanisms of chronic waterborne Zn toxicity in *Daphnia magna*. *Aquat. Toxicol.* 77:393–401.

Nacci, D. et al. 2009. Evolution of tolerance to PCBs and susceptibility to a bacterial pathogen (*Vibrio harveyi*) in Atlantic killifish (*Fundulus heteroclitus*) from New Bedford (MA, USA) harbor. *Environ. Pollut.* 157:857–64.

Naylor, C., L. Pindar, and P. Calow. 1990. Inter- and intraspecific variation in sensitivity to toxins: The effects of acidity and zinc on the freshwater crustaceans *Asellus aquaticus* (L.) and *Gammarus pulex* (L.). *Water Res.* 24:757–62.

Newman, M.C., and M.A. Unger. 2003. *Fundamentals of Ecotoxicology*. Boca Raton, FL: Lewis Publishers.

Nott, J.A., and A. Nicolaidou. 1990. Transfer of metal detoxification along marine food chains. *J. Mar. Biol. Assoc. U.K.* 70:905–12.

Ojima, N., M. Yamashita, and S. Watabe. 2005. Quantitative mRNA expression profiling of heat-shock protein families in rainbow trout cells. *Biochem. Biophys. Res. Commun.* 329:51–7.

Paetzold, S.C. et al. 2009. Up-regulation of hepatic ABCC2, ABCG2, CYP1A1 and GST in multi-xenobiotic-resistant killifich (*Fundulus heteroclitus*) from the Sydney Tar Ponds, Nova Scotia, Canada. *Mar. Environ. Res.* 68:37–47.

Pawlik-Skowrońska, B. 2003. Resistance, accumulation and allocation of zinc in two ecotypes of the green alga *Stigeoclonium tenue* Kutz. coming from habitats of different heavy metal concentrations. *Aquat. Bot.* 75: 189–98.

Peña-Llopis, S. et al. 2001. Glutathione-dependent resistance of the European eel *Anguilla anguilla* to the herbicide molinate. *Chemosphere* 45:671–81.

Peña-Llopis, S., M.D. Ferrando, and J.B. Peña. 2002. Impaired glutathione redox status associated with decreased survival in two organophosphate-poisoned marine bivalves. *Chemosphere* 47:485–97.

Peña-Llopis, S., M.D. Ferrando, and J.B. Peña. 2003. Fish tolerance to organophosphate-induced oxidative stress is dependent on the glutathione metabolism and enhanced by *N*-acetylcysteine. *Aquat. Toxicol.* 65:337–60.

Perales-Vela, H.V. et al. 2006. Heavy metal detoxification in eukaryotic microalgae. *Chemosphere* 64:1–10.
Petersen, S., and K. Gustavson. 1998. Toxic effects of tri-butyl-tin (TBT) on autotrophic pico-, nano-, and microplankton assessed by a size fractionated pollution-induced community tolerance (SF-PICT) concept. *Aquat. Toxicol.* 40:253–64.
Piola, R.F., and E.L. Johnston. 2006. Differential tolerance to metals among populations of the introduced bryozoan *Bugula neritina*. *Mar. Biol.* 148:997–1010.
Postma, J.F., and C. Davids. 1995. Tolerance induction and life cycle changes in cadmium-exposed *Chironomus riparius* (Diptera) during consecutive generations. *Ecotoxicol. Environ. Saf.* 30:195–202.
Pyza, E. et al. 1997. Heat shock proteins (Hsp70) as biomarkers in ecotoxicological studies. *Ecotoxicol. Environ. Saf.* 38:244–51.
Rainbow, P.S. et al. 2006. Trophic transfer of trace metals from the polychaete worm *Nereis diversicolor* to the polychaete *N. virens* and the decapod crustacean *Palaemonetes varians*. *Mar. Ecol. Prog. Ser.* 321:167–81.
Rainbow, P.S., S.N. Luoma, and W.X. Wang. 2011. Trophically available metal—A variable feast. *Environ. Pollut.* 159:2347–9.
Regoli, F., and G. Principato. 1995. Glutathione, glutathione-dependent and antioxidant enzymes in mussels, *Mytilus galloprovincialis*, exposed to metals in different field and laboratory conditions: Implications for a proper use as biochemical biomarkers. *Aquat. Toxicol.* 31:143–64.
Reid, D.J., and G.R. MacFarlane. 2003. Potential biomarkers of crude oil exposure in the gastropod mollusc *Austocochlea porcata*: Laboratory and manipulative field studies. *Environ. Pollut.* 126:147–55.
Roark, S.A. et al. 2005. Population genetic structure and tolerance to dioxin-like compounds of a migratory marine fish (*Menidia menidia*) at polychlorinated biphenyl-contaminated and reference sites. *Environ. Toxicol. Chem.* 24:726–32.
Roesijadi, G. et al. 1982. Enhanced mercury tolerance in marine mussels and relationship to low molecular weight, mercury-binding proteins. *Mar. Pollut. Bull.* 13:250–3.
Ross, K. et al. 2002. Genetic diversity and metal tolerance of two marine species: A comparison between populations from contaminated and reference sites. *Mar. Pollut. Bull.* 44:671–9.
Rowe, C.L. 1998. Elevated standard metabolic rate in a freshwater shrimp (*Palaemonetes paludosus*) exposed to trace element-rich coal combustion waste. *Comp. Biochem. Physiol.* 121A:299–304.
Sánchez, M., E. Andreu-Moliner, and M.D. Ferrando. 2004. Laboratory investigation into the development of resistance of *Daphnia magna* to the herbicide molinate. *Ecotoxicol. Environ. Saf.* 59:316–23.
Schill, R.O., and H.R. Köhler. 2004. Energy reserves and metal-storage granules in the hepatopancreas of *Oniscus asellus* and *Porcellio scaber* (Isopoda) from a metal gradient at Avonmouth, UK. *Ecotoxicology* 13:787–96.
Schmitt-Jansen, M., and R. Altenburger. 2005a. Toxic effects of isoproturon on periphyton communities—A microcosm study. *Estuar. Coast. Shelf Sci.* 62:539–45.
Schmitt-Jansen, M., and R. Altenburger. 2005b. Predicting and observing responses of algal communities to photosystem: II. Herbicide exposure using pollution-induced community tolerance and species-sensitivity distributions. *Environ. Toxicol. Chem.* 24:304–12.
Seguin, F. et al. 2002. A risk assessment of pollution: Induction of atrazine tolerance in phytoplankton communities in freshwater outdoor mesocosms, using chlorophyll fluorescence as an endpoint. *Water Res.* 36:3227–36.
Sheehan, D., and A. Power. 1999. Effects of seasonality on xenobiotic and antioxidant defence mechanisms of bivalve molluscs. *Comp. Biochem. Physiol.* 123C:193–9.
Sigel, A., H. Sigel, and R.K.O. Sigel, eds. 2009. *Metallothionein and Related Chelators*. Cambridge, UK: RSC Publishing.
Silvestre, F. et al. 2006. Differential protein expression profiles in anterior gills of *Eriocheir sinensis* during acclimation to cadmium. *Aquat. Toxicol.* 76:46–58.
Smital, T. et al. 2004. Emerging contaminants—pesticides, PPCPs, microbial degradation products and natural substances as inhibitors of multixenobiotic defense in aquatic organisms. *Mutat. Res. Fund. Mol. Mech. Mutat.* 552:101–17.

Smolders, R., M. Baillieul, and R. Blust. 2005. Relationship between the energy status of *Daphnia magna* and its sensitivity to environmental stress. *Aquat. Toxicol.* 73:155–70.

Sokolova, I.M., and G. Lannig. 2008. Interactive effects of metal pollution and temperature on metabolism in aquatic ectotherms: Implications of global climate change. *Clim. Res.* 37:181–201.

Sonna, L.A. et al. 2002. Effects of heat and cold stress on mammalian gene expression. *J. Appl. Physiol.* 92:1725–42.

Sorrentino, C. et al. 2004. B[*a*]P-DNA binding in early life-stages of Atlantic tomcod: Population differences and chromium modulation. *Mar. Environ. Res.* 58:383–8.

Sroda, S., and C. Cossu-Leguille. 2011. Seasonal variability of antioxidant biomarkers and energy reserves in the freshwater gammarid *Gammarus roeseli*. *Chemosphere* 83:538–44.

Stige, L.C. et al. 2011. Environmental toxicology: Population modeling of cod larvae shows high sensitivity to loss of zooplankton prey. *Mar. Pollut. Bull.* 62:395–8.

Stuhlbacher, A., and L. Maltby. 1992. Cadmium resistance in *Gammarus pulex* (L.). *Arch. Environ. Contam. Toxicol.* 22:319–24.

Takamura, N., F. Kasai, and M.M. Watanabe. 1989. Effects of Cu, Cd and Zn on photosynthesis of freshwater benthic algae. *J. Appl. Phycol.* 1:39–52.

Tedengren, M. et al. 1999. Heavy metal uptake, physiological response and survival of the blue mussel (*Mytilus edulis*) from marine and brackish waters in relation to the induction of heat-shock protein 70. *Hydrobiologia* 393:261–9.

Tedengren, M. et al. 2000. Heat pretreatment increases cadmium resistance and Hsp 70 levels in Baltic sea mussels. *Aquat. Toxicol.* 48:1–12.

Tomanek, L. 2002. The heat-shock response: Its variation, regulation and ecological importance in intertidal gastropods (*genus Tegula*). *Integr. Comp. Biol.* 42:797–807.

Tomanek, L. 2005. Two-dimensional gel analysis of the heat-shock response in marine snails (*genus Tegula*): Interspecific variation in protein expression and acclimation ability. *J. Exp. Biol.* 208:3133–43.

Tomanek, L., and G.N. Somero. 2002. Interspecific and acclimation-induced variation in levels of heat-shock proteins 70 (hsp70) and 90 (hsp90) and heat-shock transcription factor-1 (HSF1) in congeneric marine snails (*genus Tegula*): Implications for regulation of hsp gene expression. *J. Exp. Biol.* 205:677–85.

Top, E., and D. Springael. 2003. The role of mobile genetic elements in bacterial adaptation to xenobiotic organic compounds *Curr. Opin. Biotechnol.* 14:262–9.

Torricelli, E. et al. 2004. Cadmium tolerance, cysteine and thiol peptide levels in wild type and chromium-tolerant strains of *Scenedesmus acutus* (Chlorophyceae). *Aquat. Toxicol.* 68:315–23.

Tsangaris, C., E. Papathanasiou, and E. Cotou. 2007. Assessment of the impact of heavy metal pollution from a ferro-nickel smelting plant using biomarkers. *Ecotoxicol. Environ. Saf.* 66:232–43.

Tsui, M.T.K., and W.X. Wang. 2005. Influences of maternal exposure on the tolerance and physiological performance of *Daphnia magna* under mercury stress. *Environ. Toxicol. Chem.* 24:1228–34.

Twiss, M.R., P.M. Welbourn, and E. Schwärtzel. 1993. Laboratory selection for copper tolerance in *Scenedesmus acutus* (Chlorophyceae). *Can. J. Bot.* 71:333–8.

Van Tilborg, W.J.M., and F. Van Assche. 1998. Homeostatic regulation defines a stress-free concentration band for essential elements relevant for risk assessment. *SETAC Eur. News* 9:7–8.

Viarengo, A., D. Abele-Oeschger, and B. Burlando. 1998. Effects of low temperature on prooxidant process and antioxidant systems in marine organisms. In *Cold Ocean Physiology*, ed. H.O. Pörtner and R.C. Playle, 213–35. Cambridge: Cambridge University Press.

Viarengo, A. et al. 2007. The use of biomarkers in biomonitoring: A 2-tier approach assessing the level of pollutant-induced stress syndrome in sentinel organisms. *Comp. Biochem. Physiol.* 146C:281–300.

Vidal, D.E., and A.J. Horne. 2003. Mercury toxicity in the aquatic oligochaete *Sparganophilus pearsei*: II. Autotomy as a novel form of protection. *Arch. Environ. Contam. Toxicol.* 45:462–7.

Villarroel, M.J. et al. 2000. Effects of tetradifon on *Daphnia magna* during chronic exposure and alterations in the toxicity to generations pre-exposed to the pesticide. *Aquat. Toxicol.* 49:39–47.

Voets, J. et al. 2009. Differences in metal sequestration between zebra mussels from clean and polluted field locations. *Aquat. Toxicol.* 93:53–60.

Wallace, W.G., and A. Estephan. 2004. Differential susceptibility of horizontal and vertical swimming activity to cadmium exposure in a gammaridean amphipod (*Gammarus lawrencianus*). *Aquat. Toxicol.* 69:289–97.

Wallace, W.G., B.G. Lee, and S.N. Luoma. 2003. Subcellular compartmentalization of Cd and Zn in two bivalves. I. Significance of metal-sensitive fractions (MSF) and biologically detoxified metal (BDM). *Mar. Ecol. Progr. Ser.* 249:183–97.

Wallace, W.G., G.R. Lopez, and J.S. Levinton. 1998. Cadmium resistance in an oligochaete and its effect on cadmium trophic transfer to an omnivorous shrimp. *Mar. Ecol. Prog. Ser.* 172:225–37.

Wang, W.B. et al. 2004. Role of plant heat-shock proteins and molecular chaperones in the abiotic stress response. *Trends Plant Sci.* 9:244–52.

Weinstein, J.E., D.M. Sanger, and A.F. Holland. 2003. Bioaccumulation and toxicity of fluoranthene in the estuarine oligochaete *Monopylephorus rubroniveus*. *Ecotoxicol. Environ. Saf.* 55:278–86.

Wentsel, R., A. McIntosh, and G. Atchison. 1978. Evidence of resistance to metals in larvae of the midge *Chironomus tentans* in a metal contaminated lake. *Bull. Environ. Contam. Toxicol.* 20:451–5.

Whitehead, A. et al. 2011. Functional genomics of plasticity and local adaptation in killifish. *J. Hered.* 102:499–511.

Wicklum, D., and R.W. Davies. 1996. The effects of chronic cadmium stress on energy acquisition and allocation in a freshwater benthic invertebrate predator. *Aquat. Toxicol.* 35:237–52.

Wiegand, C. et al. 2007. Bioaccumulation of paraquat by *Lumbriculus variegatus* in the presence of dissolved natural organic matter and impact on energy costs, biotransformation and antioxidative enzymes. *Chemosphere* 66:558–66.

Wilhelm Filho, D. et al. 2005. Effect of different oxygen tensions on weight gain, feed conversion, and antioxidant status in piapara, *Leporinus elongatus* (Valenviennes, 1847). *Aquaculture* 244:349–57.

Wills, L.P. et al. 2010. Characterization of the recalcitrant CYP1 phenotype found in Atlantic killifish (*Fundulus heteroclitus*) inhabiting a Superfund site on the Elizabeth River, VA. *Aquat. Toxicol.* 99:33–41.

Wirgin, I. et al. 2011. Mechanistic basis of resistance to PCBs in Atlantic tomcod from the Hudson River. *Science* 331:1322–5.

Wright, M.S. et al. 2008. Influence of industrial contamination on mobile genetic elements: Class 1 integron abundance and gene cassette structure in aquatic bacterial communities. *ISME J.* 2:417–28.

Xie, L., and P.L. Klerks. 2003. Response to selection for cadmium resistance in the least killifish, *Heterandria formosa*. *Environ. Toxicol. Chem.* 22:313–20.

Xie, L., and P.L. Klerks. 2004. Fitness cost of resistance to cadmium in the least killifish (*Heterandria formosa*). *Environ. Toxicol. Chem.* 23:1499–503.

Yamashita, M., K. Hirayoshi, and K. Nagata. 2004. Characterization of multiple members of the HSP70 family in platyfish culture cells: Molecular evolution of stress protein HSP70 in vertebrates. *Gene* 336:207–18.

Yuan, Z. et al. 2006. Evidence of spatially extensive resistance to PCBs in an anadromous fish of the Hudson River. *Environ. Health Perspect.* 114:77–84.

4

Molecular and Histocytological Biomarkers

Jean-Claude Amiard and Claude Amiard-Triquet

CONTENTS

4.1 Introduction

By addressing biomarkers of damage (de Lafontaine et al. 2000), we reach a new stage in the ecotoxicology triad of exposure—bioaccumulation—effect. Exposure and bioaccumulation are actually far from always inducing toxic effects since various mechanisms allow organisms to cope with the presence of contaminants in their medium, at least so long as the degree of exposure remains moderate (cf. Chapter 3).

Currently, linking damage at infra-individual and individual levels to population-level effects potentially leading to local extinction is a major aim of ecotoxicological research. Indeed, impairments are frequently observed at the level of the individual organism, but only some specimens may be affected or these impairments are only transitional and the individual can recover totally or at least enough to be able to reproduce.

The main biomarkers of damage are molecular biomarkers (cortisol, markers of oxidative stress and lipid peroxidation, neurotransmitters, particularly acetylcholinesterase (AChE), vitamins such as retinol), biomarkers of genotoxicity (notably DNA adducts, micronucleus, and comet assays), subcellular and cellular biomarkers (lysosomal stability, immunotoxicological responses), cytological alterations, notably carcinogenesis, and physiological responses (metabolism impairments, imposex, survival of aquatic animals in air, etc.).

In this chapter, we will not review all of these biomarkers of damage because some of them are the subject of a particular chapter in this book, but will concentrate specifically on certain molecular and histocytological biomarkers of damage.

4.2 Molecular Biomarkers

4.2.1 Cortisol

The question of endocrine disruption is well developed in Chapters 8 and 9. In this chapter, we consider only investigations devoted to cortisol, a biomarker of damage frequently used in ecotoxicological monitoring.

Cortisol is a corticosteroid hormone synthesized in fish by interrenal tissue in response to a stimulation by ACTH (adrenocorticotropic hormone). In fish, the induction of plasma cortisol has been observed in response to general stress (handling, capture) or after exposure to acute chemical stress (Hontela 2000 and literature cited therein). In immature female rainbow trout (*Oncorhynchus mykiss*) intraperitoneally injected with vegetable oil containing polycyclic aromatic hydrocarbons (PAHs) (β-NF or BaP at 10 mg kg^{-1}), Tintos et al. (2008) observed increased levels of plasma cortisol, and this response was accompanied by metabolic changes (increased glucose and lactate levels in plasma, increased glycogenolysis and gluconeogenesis in liver with both PAHs, stimulated amino acid catabolism in liver of β-NF–treated individuals).

On the other hand, several studies provide contradictory conclusions. In juvenile Atlantic salmon *Salmo salar* exposed just before the parr–smolt transformation to 1 or 10 µg PCBs L^{-1} (PCB mixture Aroclor 1254), plasma cortisol was reduced by 58% in response to exposure to either concentration. In addition, plasma triiodothyronine was reduced by 35–50%, and fish treated with the higher dose of A1254 also exhibited a 50% decrease in gill Na^+,K^+-ATPase activity and a 10% decrease in plasma chloride levels in freshwater. Exposure to A1254 in the freshwater environment can inhibit preparatory adaptations that occur during smolting, thereby reducing marine survival and sustainability of salmon populations (Lerner et al. 2007). In another fish, the brown bullhead *Ameiurus nebulosus*, exposed to the polychlorobiphenyl (PCB) mixture, Aroclor 1248 (via intraperitoneal injection), cortisol was significantly lower in concentration as was the thyroid hormone, T3 (Iwanowicz et al. 2009). In rainbow trout exposed to dietary Aroclor 1254 (10 mg kg^{-1} body mass/day) for 3 days, PCB exposure did not modify the acute stressor-induced plasma cortisol, glucose, and lactate responses (Wiseman and Vijayan 2011). A field study in Ria de Aveiro (Portugal) has shown that the fish *Liza aurata* at PAH-contaminated (Vagos) and mercury-contaminated (Laranjo) sites displayed low cortisol and high glucose as well as high lactate levels, but no clear relation was found between stress and thyroidal responses (Oliveira et al. 2011).

Körner et al. (2008), examining concomitantly the gene expression of the estrogen receptor beta-1 (ERβ-1) and the glucocorticoid receptor (GR) in the liver of ethinylestradiol-exposed fish, showed no treatment-related alterations. In line with observed constant bile cortisol concentrations, their data did not indicate corresponding stress-related effects on hepatic vitellogenin production. Jørgensen et al. (2001) investigated the responses to stress in 2-(chlorophenyl)-2-(4-chlorphenyl)-1,1-dichloroethane (*o'p*-DDD) exposed (given a single, oral dose of 75 mg *o'p*-DDD kg^{-1} fish) and unexposed Arctic char *Salvelinus alpinus*. No effects of *o'p*-DDD were observed on post-stress hormone secretion (i.e., peak post-stress plasma ACTH and cortisol levels).

According to Hontela (2000), at that time there was very little information available on the cortisol status of fish chronically exposed to sublethal chemical stress in their medium, despite the biological importance of cortisol that is implicated directly or indirectly (interactions with other hormones such as thyroid hormones, reviewed by Peter 2011) in the regulation of growth, reproduction (Milla et al. 2009), and resistance to disease, which are vital functions, potentially impaired by chemicals. For instance, in the lake trout *Salvelinus namaycush*, combinations of environmental contaminants (mercuric chloride or Aroclor 1254) and cortisol interact to produce a greater toxicity than that of the environmental contaminant alone. Hence, stressors that lead to increased cortisol production may increase the toxicity of mercury and Aroclor 1254 to lake trout thymocytes (Miller et al. 2002). Pre-exposure to copper and atrazine resulted in the abolition of an acute cortisol post-stress in the freshwater fish *Prochilodus lineatus* (Nascimento et al. 2012) and the rainbow trout *Oncorhynchus mykiss* (Tellis et al. 2012) exposed to other stressors (air exposure or confinement). In trout, there was no Cu accumulation in the hypothalamus-pituitary-interrenal axis (HPI axis) suggesting this was not a direct toxic effect of Cu on the cortisol regulatory pathway and the ability of the fish to maintain ion and carbohydrate homeostasis was maintained. Tellis et al. (2012) suggest that this effect on cortisol may be a strategy to reduce costs during the chronic stress of Cu exposure, and not endocrine disruption as a result of toxic injury. However, Nascimento et al. (2012) suggest that *P. lineatus* suffering an impaired cortisol stress response may not be able to respond to any additional stressors.

The response of cortisol has been used by Hontela's (2000) team to evaluate the functional integrity of the hypothalamo–hypophysio–interrenal axis in fish living in contaminated environments. Cortisol failure (with addition of low levels of plasma thyroxin) was detected in mature males and females and immature yellow perch *Perca flavescens* and northern pike *Esox lucius* in the Saint Lawrence River by comparing reference and contaminated (PCBs, PAHs, Cd, Hg) sites. Cortisol depletion was observed by the same team in both species in a river impacted by a kraft paper mill. Lockhart et al. (1972 in Hontela 2000) reported lower levels of plasma cortisol and glucose in pike originating from a mercury-contaminated lake compared to fish from a reference lake. Cortisol and glucose levels appeared as responsive stress biomarkers in a field study using the barbel (*Barbus bocagei*) and the carp (*Cyprinus carpio*) collected in the Tagus River (Iberian peninsula) at a reference site and nine sampling sites selected on the basis of whether various human activities and hydrographic characteristics were present (Carballo et al. 2005).

Less information is available for cortisol in other taxa. However, the review by Letcher et al. (2010) on effect assessment of persistent organohalogen contaminants in arctic wildlife and fish reports that organochlorine (OC) pesticides combined with PCBs and their interactions could account for more than 25% of the variation in plasma cortisol concentrations in polar bears. Cortisol concentration in East Greenland polar bears was found at significantly higher concentrations in historical hair samples (1892–1927; n = 8) relative to recent ones (1988–2009; n = 88). In addition, there was a linear time trend in cortisol concentration of the recent samples, with an annual decrease of 2.7% but there were no obvious correlations between hair cortisol and hair POP concentrations (Bechshøft et al. 2012). Thus, cortisol in polar bear hair appears to be a relatively unspecific biomarker of their contamination by persistent organic pollutants (POPs) but as a relevant biomarker of general stress.

4.2.2 Oxidative Stress and Lipid Peroxidation

According to Sies (1991), oxidative stress may be defined as "a disturbance in the prooxidative–antioxidant balance in favor of the former, leading to potential damage." The presence of free radicals and reactive species of oxygen (ROS) in biological systems and their mode of action are well established in biology and medicine (Halliwell and Gutteridge 2007), and have been recently reviewed in aquatic ecosystems (Abele et al. 2012). Oxidative stress is induced by a wide range of environmental factors including UV stress, oxygen shortage, pathogen invasion, presence of symbionts, cyanobacterial toxins such as microcystin, contaminants such as transition metal ions (Fe, Cu, Cr, Hg, As), pesticides (insecticides, herbicides, fungicides), oil, and related contaminants (Blokhina et al. 2003; Lushchak 2011; Abele et al. 2012). For emerging contaminants, oxidative stress is recognized as a main effect of nanoparticles on biota (Moore 2006; Klaine et al. 2008; Canesi et al. 2011). In addition, natural factors such as temperature and salinity may enhance the production of ROS (Lushchak 2011). Consulting Google Scholar in February 2012 with search terms "aquatic" and "oxidative stress" yielded 20,800 occurrences, whereas the search terms "marine" and "oxidative stress" showed 39,600 occurrences. Rapid browsing of this mass of data shows at least that nearly all taxa are affected.

Cellular responses to oxidative stress include adaptation, damage, repair, senescence, and death (Halliwell and Gutteridge 2007). Oxidative stress gives rise to antioxidant defenses that provide a number of biomarkers of defense (Chapters 2 and 3), but when defenses are overwhelmed, oxidative damage is observed, providing biomarkers of damage. ROS induce modification of lipids, proteins, and nucleic acids. Assessing lipid and protein oxidation is classically used in environmental studies (Chapter 2; Lushchak 2011). Malondialdehyde (MDA), an oxidative by-product of lipid peroxidation, is commonly used as a biomarker of oxidative damage. It is classically detected through spectrophotometric detection of the thiobarbituric acid–MDA derivative, but this has been criticized for its lack of specificity (Chapter 2). More accurate methods (high-performance liquid chromatography or gas chromatography coupled to UV–Vis, fluorescence, and mass spectrometry detectors) have been recently reviewed (Miyamoto et al. in Abele et al. 2012). Another possibility lies in the direct analysis of various radical species. Evaluation of oxidative DNA damage in aquatic organisms has also been well developed (Abele et al. 2012), using several damage parameters (Chapter 13).

Because environmental conditions (oxygen level, UV intensity, temperature, salinity, diet) are recognized as inducers of oxidative stress in aquatic organisms (Blokhina et al. 2003; Lushchak 2011; Miyamoto et al. in Abele et al. 2012), particular attention must be paid to natural fluctuations that can interfere with contamination effects, acting as confounding factors in the interpretation of biomarkers of oxidative damage (Chapter 2). Seasonal and reproductive cycles, which are often accompanied by changes in membrane lipid composition, uptake of fatty acids for energy supply, or changes in antioxidant defenses, are known sources of natural changes in MDA levels (Miyamoto et al. in Abele et al. 2012). Organ-specific and age effects must also be taken into account to avoid misinterpretation, as exemplified in the case of mercury-induced peroxidative damage in bivalves (Ahmad et al. 2011).

4.2.3 Markers of Genotoxicity

Markers of genotoxicity such as DNA adducts, micronucleus, and Comet assay tests are well-documented biomarkers of damage (Chapter 13). Recently, several authors have focused their research on the reproductive consequences of paternal genotoxin exposure in aquatic organisms (Lewis and Galloway 2009; Lacaze et al. 2010; Devaux et al. 2011). DNA damage to sperm was observed in freshwater crustaceans (*Gammarus fossarum*) and fish (*Salmo trutta, Salvelinus alpinus*) and in marine polychaetes (*Arenicola marina*) and bivalves (*Mytilus edulis*) exposed to the model genotoxicant methyl methane sulfonate and/or to the PAH benzo[*a*]pyrene (B[*a*]P). No effect occurred on fertilization success, but severe developmental abnormalities were observed in freshwater fish and marine invertebrates. Prolonged effects were observed in *S. trutta* such as increased mortality (×3) after 2 months, and increased malformations after 1 year (Devaux et al. 2011). These findings are in agreement with field observations reported for herring *Clupea pallasi* after the accident involving the tanker *Exxon Valdez* (for details, see Chapter 13).

4.2.4 Cholinesterases

The majority of insecticides currently in use are organophosphorous, carbamate, and synthetic pyrethroid compounds. Organophosphorous (OP) insecticides produce toxicity by inhibiting cholinesterase enzymes in the nervous system. Monitoring of AChE inhibition has been widely used in terrestrial and freshwater aquatic systems as an indicator of OP exposure and effects (reviews by Galgani and Bocquené 2000; Fulton and Key 2001).

Impairments of AChE activity lead to the accumulation of acetylcholine in neural junctions, responsible for an overstimulation of the peripheral nervous system. The inhibition of AChE activity can have important effects on individuals, including lethal effects in the short term if cholinesterase inhibition exceeds a threshold of about 70% in fish brain. Selected species, however, appear capable of tolerating much higher levels (90%) of brain AChE inhibition. Less drastic inhibition can also have clear repercussions on behavior: sublethal effects on stamina have been reported for some estuarine fish in association with brain AChE inhibition levels as low as 50% (Fulton and Key 2001).

4.2.4.1 AChE Activity Changes Induced by Laboratory or Field Exposure

More recent studies have provided new evidence of the effects of cholinesterase-inhibiting pesticides both in the laboratory (Table 4.1) and in the field (Table 4.2).

In addition to OP pesticides and carbamates, exposure to other classes of contaminants (metals, petroleum, detergents, complex mixtures) as well as natural toxins can inhibit AChE activity (Table 4.1). Thus, AChE inhibition has been proposed for consideration as a generalist biomarker, representative of the physiological status of an organism (Leiniö and Lehtonen 2005).

A dose-additive inhibition of Chinook salmon (*Oncorhynchus tshawytscha*) AChE activity by mixtures of OP and carbamate insecticides has been described by Scholz et al. (2006). Because both classes of contaminants are concomitantly present in water bodies, a relevant risk assessment must not be focused individually on each of them, a practice that would lead to an underestimation of potential risk. This topic has been recently reviewed for invertebrates, and numerous examples of additive, synergistic, but also antagonistic effects have been registered (Domingues et al. 2010).

Two scallops—the Antarctic *Adamussium colbecki* and the Mediterranean *Pecten jacobaeus*—differ widely in AChE molecular forms. However, the presence of inhibitor-sensitive

TABLE 4.1
Influence of Laboratory Exposure to Contaminants on AChE and Behavior in Different Aquatic Species

Mode of Contamination	Molecule	Zoological Taxon	Species	AChE Inhibition	Behavioral Impairment	Reference
Pesticide	Chlorpyrifos	Amphibian	*Rana sphenocephala*	Inhibition (whole body) (maximum inhibition, 43%)	–	Widder and Bidwell 2006
Pesticides	Carbofuran Molinate	Fish (larvae)	*Pimephales promelas*	Inhibition at higher concentrations (pool)	Swimming capacity reduced Sensibility to electric shock increased	Heath et al. 1997
Pesticide	Carbofuran	Fish (larvae)	*Oreochromis niloticus*	Inhibition (pool)	Swimming speed Attacks to prey	Pessoa et al. 2011
Pesticide	Carbaryl	Fish (larvae)	*Oncorhynchus mykiss*	Inhibition (50%) in brain	Swimming speed decreased	Beauvais et al. 2001
Pesticides	Malathion Diazinon	Fish	*Oncorhynchus mykiss*	Inhibition in brain	Swimming speed and distance decreased	Brewer et al. 2001
Pesticide	Chlorpyrifos	Fish (juvenile)	*Oncorhynchus kisutch*	Inhibition in brain and muscle	Swimming, feeding	Sandahl et al. 2005
Pesticide	Endosulfan	Fish	*Channa punctata*	Inhibition	Surfacing activity, distance traveled enhanced	Gopal et al. 1985
Pesticides	Malathion Diazinon	Fish (larvae)	*Oncorhynchus mykiss*	Inhibition in brain	Swimming speed decreased	Beauvais et al. 2000
Pesticides	Mixture (diazinon, chlorpyrifos, malathion, carbaryl, carbofuran)	Fish	*Oncorhynchus tshawytscha*	Inhibition in olfactory tissues	–	Scholz et al. 2006
Pesticides	Carbofuran Deltamethrin	Fish	*Tinca tinca*	Inhibition in brain No inhibition in brain	–	Hernandez-Moreno et al. 2010
Pesticide	Diazinon	Fish	*Morone saxatilis* × *M. chrysops*	Inhibition in brain	Time to capture prey increased	Gaworecki et al. 2009
Pesticide	Malathion	Insect (larvae)	*Hydropsyche slossonae*	Inhibition (pool)	Anomalies on capture nets	Tessier et al. 2000
Pesticide	Deltamethrine	Insect (larvae)	*Chironomus xanthus*	Inhibition in head	Feeding rates decreased	Moreira-Santos et al. 2005

Pesticide	Methamidophos	Crustacean	*Litopenaeus vannamei*	Inhibition in muscle and eye	Locomotory time increased No effect on feeding rate	Garcia-de la Parra et al. 2006
Pesticides	Dimethoate Pirimicarb	Crustacean	*Daphnia magna*	–	Immobility	Andersen et al. 2006
Pesticide Parathion	Chlorpyrifos	Malathion	*Acephate*	Propoxur	Crustacean *Daphnia magna* inhibition (pool)	Printes and Callaghan 2004
Pesticide	Methyl paraoxon	Crustacean	*Daphnia magna*	Inhibition (pool)	–	Duquesne 2006
Pesticide	Methyl paraoxon	Crustacean	*Daphnia magna*	Inhibition (pool)	Swimming activity increased Filtration activity decreased	Duquesne and Küster 2010
Pesticide	Acephate	Crustacean	*Daphnia magna*	Inhibition (pool)	–	Printes et al. 2008
Pesticide	Atrazine	Crustacean	*Tigriopus brevicornis*	Inhibition (pool)	–	Forget et al. 2003
Pesticides	Chlorpyrifos Methomyl	Crustacean	*Gammarus fossarum*	Inhibition (pool)	Feeding rate Locomotion activity	Xuereb et al. 2009a
Pesticide	Deltamethrin	Crustacean	*Penaeus monodon*	Inhibition in muscle	–	Tu et al. 2012
Pesticides	Carbofuran Malathion	Mollusk (larvae)	*Crassostrea gigas*	Inhibition (pool)	–	Damiens et al. 2004
Pesticides	Parathion Chlorpyrifos Malathion Acephate Propoxur	Crustacean	*Daphnia magna*	Inhibition (pool)	–	Printes and Callaghan 2004
Pesticides	Carbofuran Lindane	Mollusk	*Murex trunculus*	Inhibition (whole body)	–	Roméo et al. 2006
Pesticide	Chlorpyrifos	Mollusk	*Corbicula fluminea*	Inhibition (whole body)	Capacity to burrow reduced	Cooper and Bidwell 2006
Pesticides	Methydathion Chlorpyrifos Diazinon IBP	Mollusk	*Ruditapes philippinarum*	Inhibition in adductor muscle	–	Choi et al. 2011
Pesticide	Chlopyrifos	Mollusk	*Potamopyrgus antipodarum* *Valvata piscinalis*	Inhibition (whole body) No inhibition for *Valvata piscinalis*	–	Gagnaire et al. 2008

(continued)

TABLE 4.1 (Continued)

Influence of Laboratory Exposure to Contaminants on AChE and Behavior in Different Aquatic Species

Mode of Contamination	Molecule	Zoological Taxon	Species	AChE Inhibition	Behavioral Impairment	Reference
Pharmaceutical	Propranolol	Mollusk	*Mytilus galloprovincialis*	Inhibition in gills	Inhibited feeding rate	Solé et al. 2010
Pharmaceutical	Acetaminophen	Mollusk	*Mytilus galloprovincialis*	Inhibition in gills (58%)	Increased feeding rate	Solé et al. 2010
Metal, Pesticide	Cu, Cu + Malathion	Mollusk	*Mytilus edulis*	Inhibition in gills	–	Lehtonen and Leiniö 2003 Leiniö and Lethonen 2005
Metal, Pesticide	Cu, Cu + Malathion	Mollusk	*Macoma balthica*	Inhibition in foot tissue	Low siphon activity	Lehtonen and Leiniö 2003
Metal	Cr (VI)	Mollusk	*Mytilus galloprovincialis*	Inhibition	–	Guilhermino et al. 1998
Metal	Pb U	Mollusk	*Corbicula* sp.	Inhibition	–	Labrot et al. 1996
Metal	Cu	Mollusk	*Mytilus edulis* *Patella vulgata*	No effect in hemolymph Increase in hemolymph	–	Brown et al. 2004
Metal	Cu	Crustacean	*Carcinus maenas*	No effect in hemolymph	–	Brown et al. 2004
Metal	Cu, Zn, Cd, Hg	*In vitro*		Inhibition	–	Frasco et al. 2005
Detergents	Dodecyl benzyl sulfonate Sodium dodecyl sulfate Mixture of domestic detergents	Mollusk	*Mytilus galloprovincialis*	Inhibition	–	Guilhermino et al. 1998
Organic pollutants	HAP PCB	Mollusk	*Mytilus galloprovincialis*	Inhibition (soft tissues)	–	Damiens et al. 2007
Toxin	Cyanobacterium toxin	Mollusk	*Macoma balthica*	Inhibition in foot tissue	Low siphon activity	Lehtonen et al. 2003
Mixture	Metals (As, Cu, or Cd) and pesticides (carbofuran, dichlorvos, or malathion)	Crustacean	*Tigriopus brevicornis*	Inhibition in whole body (>50%) Synergy (except Cd): inhibition >65%	–	Forget et al. 1999

TABLE 4.2

Influence of Field Exposure to Contaminants on AChE and Behavior in Different Aquatic Species

Zoological Taxon	Species	AChE Inhibition	Behavioral Impairment	Reference
Amphibian	*Hyla regilla*	Inhibition in brain and tongue	–	Sparling et al. 2001
Fish	*Platichthys flesus*	Inhibition in muscle	–	Kirby et al. 2000
Fish	*Geophagus brasiliensis*	Inhibition in muscle	–	Linde-Arias et al. 2008a
Fish	*Oreochromis niloticus*	Inhibition in muscle	–	Linde-Arias et al. 2008b
Fish	*Gasterosteus aculeatus*	Inhibition in muscle	–	Sanchez et al. 2008
Fish	*Platichthys flesus*	Inhibition in muscle (1)	–	Kopecka and Pempkowiok 2008
Fish	*Salmo trutta*	No inhibition in brain Inhibition in muscle (two sexes)	–	Payne et al. 1996
Fish	*Pleuronectes americanus*	Inhibition in muscle of females No inhibition in muscle of males	–	Payne et al. 1996
Crustacean	*Daphnia magna*	Inhibition (pool) (2)	Feeding rate decreased	Barata et al. 2007
Crustacean	*Carcinus aestuarii*	Inhibition in gills No inhibition in hemolymph	–	Ricciardi et al. 2010
Crustacean	*Procambarus clarkii*	Inhibition in digestive gland	–	Vioque-Fernández et al. 2009
Mollusk	*Mytilus edulis*	Inhibition (soft tissues)	–	Devier et al. 2005
Mollusk	*Mytilus edulis*	Inhibition in gills (3)	–	Burgeot et al. 2010
Mollusk	*Mytilus galloprovincialis*	Inhibition in gills	–	Tsangaris et al. 2010
Mollusk	*Ruditapes philippinarum*	Inhibition in adductor muscle (2)	–	Choi et al. 2011
Mollusk	*Cerastoderma glaucum*	Tissue-dependent response		Jebali et al. 2011
Mollusk	*Scrobicularia plana*	No inhibition in digestive gland	–	Solé et al. 2009
Mollusk	*Scrobicularia plana*	No inhibition (soft tissue)	Burrowing kinetics decreased	Fossi Tankoua et al. 2010
Mollusk	*Donax trunculus*	Inhibition in digestive gland	–	Tlili et al. 2010
Annelid	*Nereis diversicolor*	Inhibition (whole body)	–	Solé et al. 2009
Annelid	*Nereis diversicolor*	Inhibition (whole body)	Post-feeding rates decreased	Fossi Tankoua et al. 2010

Note: Main contaminants: (1) confounding factors (temperature or/and contamination); (2) pesticides; (3) oil spill.

AChE forms only in the gills of the two bivalves could be the consequence of particular adaptive features in these filter feeding organisms (Romani et al. 2006). The interpretation proposed by these authors is that AChEs located in the gills must react first with toxic compounds as a protection for other AChEs involved in neurotransmission. The resistance of AChE forms to modern pesticides could be considered a preadaptation of a common origin resulting from the development of resistance to natural marine neurotoxins.

In vertebrates, two isoforms occur—AChE, the main function of which is the rapid hydrolysis of the neurotransmitter acetylcholine, and butyrylcholinesterase (BChE; or pseudocholinesterase), which has no known specific natural substrate, although it is able to hydrolyze acetylcholine. The sensitivity of different ChEs differs greatly, as shown in the three-spined stickleback (*Gasterosteus aculeatus*) after exposure to the OP insecticide parathion-ethyl (Wogram et al. 2001). After exposure to 1 mg L^{-1} parathion, BChE activity was significantly decreased in liver (~60%) and axial muscle (~30%), whereas its decrease in gills (~30%) was not significant. No effects on BChE activity were observed with 0.1 and 0.01 mg L^{-1} parathion. AChE activity remained unaffected at all parathion concentrations used. Similarly, Monteiro et al. (2005) highlight the fact that different forms of ChE existing in fish have different sensitivities to cholinesterase-inhibiting compounds. Thus, with ChE properties differing between species, several authors are happy to characterize the type of enzyme present in the species studied in order to interpret this biomarker correctly (Scaps et al. 1996; Kristoff et al. 2006; Gagnaire et al. 2008; Jebali et al. 2011).

Oliveira et al. (2007) have examined brain AChE in 20 fish species from the coast of Rio de Janeiro state, Brazil, as a possible pesticide biomarker in marine environmental monitoring. The enzyme sensitivity to methyl paraoxon, shows that *Paralonchurus brasiliensis* and *Genidens genidens*—belonging to the super-order Acanthopterygii, which includes more recently evolved species—are more sensitive than *Merluccius hubbsi* and *Percophis brasiliensis*—belonging to the super-order Paracanthopterygii, which includes the more ancient bony fish species. These authors suggest a possible evolutionary linkage for AChE sensitivity to methyl paraoxon. Interspecific differences in the responses of ChEs to environmental pressure are well illustrated by the studies of Solé et al. (2009) and Fossi Tankoua et al. (2010), who have determined biomarkers including AChE in the bivalve *Scrobicularia plana* and the polychaete *Nereis diversicolor* collected from the same sites at the same dates. Both studies carried out independently in Spain and France concluded that the polychaete was highly responsive, whereas the bivalve was of no help in distinguishing sites according to different degrees of contamination by cholinesterase-inhibiting compounds.

In addition to being inhibited by different xenobiotics, AChE activity may also be influenced by natural factors. In a recent review, Burgeot et al. (2010) explain that an increase in water temperature significantly affects the expression of AChE activity, because temperature can change the activity of the enzymes by changing the protein conformation and the catalytic efficiency or binding capacity. The literature provides numerous examples of the influence of temperature on AChE activity and as a corollary, temporal variations have been observed in different species (Kopecka and Pempkowiak 2008; Burgeot et al. 2010). Seasonal variations can also result from physiological changes as exemplified by Xuereb et al. (2009b), who report that significant differences in AChE activity were observed between female amphipod crustaceans depending on gonadal and embryonic development. In estuarine species, salinity is an important factor influencing AChE expression, for instance, in polychaetes (Scaps and Borot 2000), copepods (Cailleaud et al. 2007), and bivalves (Fossi Tankoua et al. 2011). In addition to salinity effects, changes in AChE levels were observed during the tidal cycle and between surface and bottom-living copepods related to variations in hydrophobic organic contaminant concentrations (Cailleaud et al. 2009). Body size (or weight and age) has been also recognized as a confounding factor for instance in polychaetes (Durou et al. 2007), amphipods (Xuereb et al. 2009b), and bivalves (Fossi Tankoua et al. 2011). In crustaceans, the life cycle stage must also be taken into account (Hoguet and Key 2007).

Nevertheless, most of these confounding factors may be controlled with an appropriate sampling strategy and mastered by using a careful evaluation of sources of fluctuations. Evidence is provided by the series of data obtained during a 2-year survey following the wreck of the tanker *Erika* in the Loire estuary, France, which allowed the determination of the background response level of the AChE in mussels (*Mytilus galloprovincialis*) and the evaluation of the neurosuppressive effects of oil spillage on the mussels (Burgeot et al. 2010). A model of classification was designed from these results, which seems to be very promising for future monitoring initiatives in the Coordinated Environmental Monitoring Programme (CEMP) (monitoring under the OSPAR Joint Assessment and Monitoring Programme where the national contributions overlap and are coordinated through adherence to commonly agreed monitoring guidelines, quality assurance tools, and assessment tools) and in the European Marine Strategy Framework Directive.

4.2.4.2 Linking Neurotoxic Effects and Behavioral Impairments

Among physiological mechanisms inducing behavioral impairments (Chapter 10), the inhibition of neurotransmitters is well documented in aquatic organisms as a result of many studies dealing with the toxic effects of pesticides (Tables 4.1 and 4.2).

In the endobenthic worm *Hediste (Nereis) diversicolor*, exposure to contaminated sediments (both in the laboratory and in *in situ* tests) induced a depletion of food uptake whereas AChE activity was not affected (Moreira et al. 2006). In addition to sublethal effects on stamina in some estuarine fish in association with brain AChE inhibition levels reported by Fulton and Key (2001), temporary loss of hierarchy in food uptake (in the trout *Salvelinus fontinalis*), behavioral deficiency (in the Mediterranean fish *Serranus scriba*), and increased vulnerability to predation (in the Atlantic salmon *Salmo salar*) have been reported as consequences of exposure to cholinesterase-inhibiting insecticides (Zinkl et al. 1991). In the freshwater fish *Channa punctata*, exposure to the neurotoxin endosulfan induced decreases in AChE activity and concentrations of serotonin (5-HT) associated with changes in surfacing behavior (Gopal et al. 1985). Dopamine was also affected, differently depending on the level and the duration of exposure.

There are suspicions that contaminants, other than pesticides, which cause neurotoxicity, could also alter different aspects of behavior. Commonly used pharmaceuticals (the β-adrenergic receptor blocker propranolol or the anti-inflammatory drug paracetamol) alter gill AChE activity (and other biochemical responses) and feeding rate in mussels but at doses not likely to be encountered in the marine environment (Solé et al. 2010).

4.2.4.3 Linking AChE Activity Inhibition and Population Effects

Because cholinesterase-inhibiting pesticides disrupt neuromuscular signaling, reduction in performance seems to be a logical outcome of this biochemical disruption at the organism level (Hopkins and Winne 2006). Several studies have examined fitness-related traits, growth and reproduction impairments, and survival in aquatic organisms exposed to such pesticides but effects on AChE activity were only implicit, not measured (Andersen et al. 2006; Hopkins and Winne 2006). More interestingly, several studies examined concomitantly effects at different levels of biological organization in order to highlight implications for population dynamics (Duquesne 2006; Gaworecki et al. 2009; Duquesne and Küster 2010).

In the hybrid striped bass (*Morone saxatilis* × *M. chrysops*), diazinon exposure inhibited brain AChE activity at all concentrations tested, whereas only the medium and high treatment groups showed impairment of prey capture. Gaworecki et al. (2009) concluded that sublethal exposure to AChE-inhibiting substances may decrease the ecological fitness of hybrid striped bass, a situation that has been described for another species (*Fundulus heteroclitus*) in field conditions (Weis et al. 2001). It may be also noted that the more sensitive response of the biochemical marker provides a predictive assessment of the potential risks associated with diazinon exposure.

In *Daphnia magna*, Duquesne (2006) observed that above a threshold concentration of 2.2 $\mu g\ L^{-1}$ paraoxon-methyl, inhibition of ChE activity was accompanied by effects on survival, reproduction, and body size, and a reduced population growth rate was also reported. In a complementary study, Duquesne and Küster (2010) showed that ChE and swimming activities were significantly affected at lower exposure concentrations (1.0 and 0.7 $\mu g\ L^{-1}$, respectively) than filtration activity, which had the same response threshold (1.5 $\mu g\ L^{-1}$) as physiological responses (use of energy reserves and body size). Despite a high potential for the affected parameters to recover, these authors consider that "the effects of pesticides can propagate through biological systems and possibly induce long-term effects at higher levels of biological organisation."

The pesticides currently used have been preferred to OC pesticides particularly because they are less persistent. Pollution incidents in the aquatic environment often occur as pulses. Thus, it is important to integrate into risk assessments the influence of exposure duration on the effects of pesticides. In *D. magna*, it seems that the longer the exposure, the weaker the recovery (Andersen et al. 2006; Duquesne 2006). A review by Sánchez-Hernandez (2001) indicates that recovery duration varies from 3 to 28 days in different vertebrate and invertebrate species. In the copepod *Tigriopus brevicornis*, recovery from pesticide exposure was nearly complete within 14 days (Forget et al. 2003).

Sparling et al. (2001) evoked a link between the contamination of aquatic media by pesticides and the decline of numerous amphibian populations across the world. They mention that when AChE activity was >2 $\mu mol\ min^{-1}\ g^{-1}$, populations of the frog *Hyla regilla* showed good health status, whereas the health status turned bad when AChE activity was <1.7 $\mu mol\ min^{-1}\ g^{-1}$. However, although neurotoxicity of pesticides can contribute to amphibian decline, other causes may be also involved (Chapter 9).

4.2.5 Retinol

Retinol (vitamin A) and its biologically active derivatives [retinoids, most notably retinoic acids (RAs)] are essential compounds for several basic physiological functions such as growth, cellular differentiation, and reproduction. Food provides a regular amount of retinoids by means of precursors such as β carotene. Cellular and tissue needs are fulfilled by retinol, which is the major form present in blood. Because an excess of retinol in tissues may be toxic, surplus amounts are stored in the liver as retinol esters. A schematic representation of the metabolism of retinoids has been proposed by Inoue et al. (2010). In the nucleus of target cells, retinoids bind to RA receptors (RAR) and retinoid X receptors (RXRs). In the basal state, the RAR/RXR heterodimer is bound to a nuclear receptor corepressor; then, binding to the ligand results in the release of corepressors and recruitment of coactivators (Figure 4.1). This permits the transcriptional activation of target genes via specific RA response elements as described by Inoue et al. (2010).

Different classes of environmental pollutants have been shown to interfere with retinoid physiology through effects on retinoid content and gene transcription level, retinoid

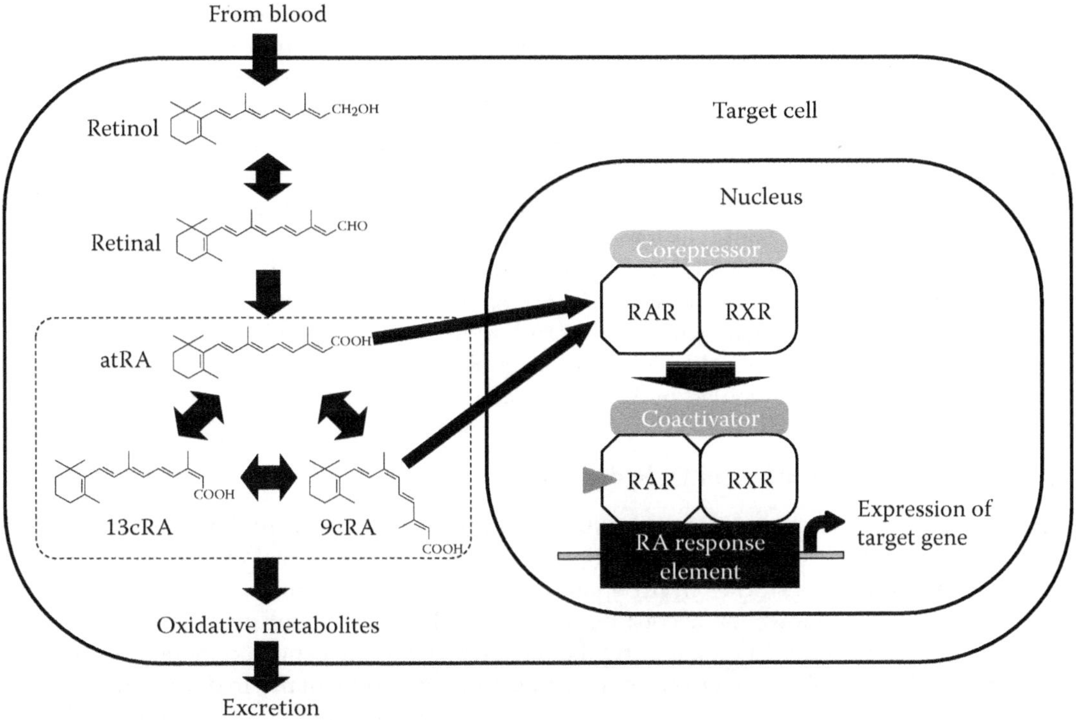

FIGURE 4.1
Schematic representation of the metabolism of retinoids. (After Inoue, D. et al., *J. Health Sci.*, 56, 221–230, 2010. With permission.)

receptors, disruption of retinoid metabolism, or transport (Rolland 2000; Novák et al. 2008; Inoue et al. 2010; Letcher et al. 2010; Chen et al. 2012). They include alkylphenolic compounds with various alkyl groups, pesticides, polychlorinated dioxins, polychlorinated biphenyls, polybrominated diphenyl ethers, polycyclic aromatic compounds, and other organic pollutants as well as environmental complex mixtures such as pulp mill–produced compounds or contaminated sediment extracts.

Retinol was proposed as a biomarker of contaminant-related toxicity as early as 2000 in marine mammals (Simms and Ross 2000). However, Simms and Ross recommended a cautionary approach to avoid misinterpretations associated with the effects of confounding factors. A recent paper by Routti et al. (2010) reinforces this recommendation since it showed that contaminant levels and their relationships with physiological or endogenous variables can be highly confounded by molting/fasting status. The sampling matrix must also be chosen adequately, as demonstrated by Bechshøft et al. (2011), who have shown that retinol was not uniformly distributed in the kidney of polar bears (*Ursus maritimus*). Despite these limitations, associations between blubber PCB concentrations and plasma retinol concentrations as well as concentrations of stored retinol in blubber were established in harbor seals (*Phoca vitulina*) from the Pacific coast of British Columbia (Canada) and Washington state (USA) (Mos et al. 2007). According to these authors, retinol concentrations and retinoic acid receptor α (RARα) expression levels can therefore represent relevant and sensitive biomarkers of PCB-associated toxic effects in toxicological studies of marine mammals. In ringed seals (*Phoca hispida baltica*) and gray seals (*Halichoerus grypus*) living in the multipolluted

Baltic Sea and suffering from pathological impairments, retinyl palmitate levels showed a negative correlation with POP loads (Nyman et al. 2003). These authors proposed that the depletion of vitamin A stores may be used as a potential effect biomarker. On the other hand, PCBs did not affect the retinol status in polar bears (Braathen et al. 2004), and no significant relationships were noted between serum, liver, and blubber retinol concentrations, and serum and blubber OC concentrations in the bowhead whale (*Balaena mysticetus*) (Rosa et al. 2007). In the first case, the PCB concentrations were nevertheless high enough to affect five thyroid hormone variables in female polar bears. In the second case, it may be because bowhead whales have relatively low concentrations of OCs, a threshold effect that has also been documented in gray heron (*Ardea cinerea*) hatchlings (Jenssen et al. 2001).

Retinoids have been also extensively studied in fish (see review by Rolland 2000), amphibians (e.g., Boily et al. 2009; Mann et al. 2009), and birds (e.g., Champoux et al. 2006) including Arctic birds reviewed by Letcher et al. (2010).

4.2.6 δ-Amino Levulinic Acid Dehydratase

The presence of anthropogenic lead in aquatic systems is largely due to the burning of leaded fuels, metal smelters, and mining activities. The effect of the removal of lead from gasoline is clear from numerous studies in urban and remote settings in Europe and North America (Mahler et al. 2006). However, according to these authors, lead continues to contribute the largest amount of contamination among the most frequently investigated metals on the basis of comparison to predicted environmental concentrations, particularly in dense urban watersheds. For the marine environment, lead is one of the metals identified by OSPAR as chemicals for priority action. Concentrations in fish, shellfish, and sediments have generally fallen since 1990 (OSPAR Commission 2009). As much of the reduction in inputs of metals occurred before 2000, changes in environmental concentrations have been relatively small since 1998 as concentrations approach, but have not reached, background levels in large parts of the OSPAR area. Thus, it remains useful to monitor the presence and effects of lead in aquatic biota.

Lead causes a dose-dependent inhibition of δ-amino levulinic acid dehydratase (ALAD), which is an essential enzyme for the synthesis of hemoglobin in hemopoietic tissue. ALAD inhibition is recognized as a good indicator of lead exposure, which is quite specific and has been used in freshwater and marine fish, birds, and mammals. It is thus recommended in the JAMP Guidelines for Contaminant-Specific Biological Effects (OSPAR Agreement 2008–09). Lead's effects on ALAD have been also reported for amphibians (Arrieta et al. 2000) and reptiles (Overmann and Krajicek 1995). Although hemoglobin is not synthesized in most bivalves, ALAD activity is generally negatively correlated with Pb concentration in both freshwater (Company et al. 2008) and marine bivalve species (Kalman et al. 2008; Company et al. 2011).

4.3 Histocytological Biomarkers

Interest in histocytological biomarkers has been evident for at least two decades as shown by the early reviews published for invertebrates and fish (Hinton 1993a, 1993b; Yevich and Yevich 1994). The most commonly used of these biomarkers are externally visible fish disease, internal fish disease, histopathology of gills and digestive glands in mollusks, and cytopathology (Table 4.3). They have been widely used in order to reveal biological impacts of a variety of contaminants/stresses in field studies.

TABLE 4.3

Evaluation of Histocytological Biomarkers for Biomonitoring of Marine Pollution

Biomarkers	Ecological Relevance	Sensitivity	Specificity	Dose–Response Relationship	Confounding Factors	Technical Difficulties	Cost-Effectiveness
External fish disease							
Fin erosion	M	H	L	H	M	L	H
Skeletal malformation	M	H	H	H	H	L	H
Epidermal hyperplasia	M	H	L	H	?	L	H
Operculum abnormalities	M	?	H	?	M	M	M–H
Internal fish disease							
Liver histopathology	H	M–H	L	H	M	H	M
Gill histopathology	H	H	L	H	L–M	H	M
Kidney histopathology	H	L	L	?	M	H	L
Oocyte atresia	H	H	L	H	H	M	M–L
Embryonic effect	H	H	L	H	M	H	M
Macrophage aggregates	M	H	L	H	H	M	M
Histopathology of mollusks	H	H	L	?	?	H	M–L
Cytopathology							
Lysosome integrity	L–M	H	L	H	H	M	H
Lipopigment content	L–M	H	M	M–H	?	M–H	M–L
Peroxisome proliferation	L–M	M	M	M–L	H	M–H	M

Source: Au, D.W.T., *Mar. Pollut. Bull.*, 48, 817–834, 2004. With permission.

Note: L, low; M, medium; H, high; ?, unknown.

4.3.1 Responses to Organic Contaminants

Monitoring of oil spills may be based on both invertebrates and fish models. Histological and ultrastructural studies may provide useful information on the effects of pollutants on fauna, especially in acute exposures, as reported by Giari et al. (2012). Following the wreck of the *Amoco Cadiz* in Brittany, France, responsible for the spillage of 223,000 tons of petroleum, histological abnormalities were shown in oysters for many years (Berthou et al. 1987). Neoplasia was observed in soft-shell clams *Mya arenaria* collected from oil-impacted sites (Yevich and Barszcz 1977 cited by Yevich and Yevich 1994). Biomonitoring studies assessing the residual biological effects of pollution caused by the wreck of the *Haven* in 1991 on marine organisms in the Ligurian Sea (Italy) were carried out in 1997 (Viarengo et al. 2007) and 1999 (Pietrapiana et al. 2002). Fish with different habitats and feeding habits were collected from two differently impacted areas and a control site: *Lepidorhombus boscii, Mullus barbatus, Merluccius merluccius, Boops boops,* and *Uranoscupus scaber.* In addition to this, mussels (*Mytilus galloprovincialis*) were caged along the coast affected by the *Haven* disaster. Significant biological responses were observed in lysosomal membrane stability, neutral lipid and lipofuscin accumulation, and micronucleus frequency in mussels caged at two sites close to the *Haven* wreck (Viarengo et al. 2007). By using hepatic tissue damage such as the presence of necrotic and tumor-like aspects (Pietrapiana et al. 2002) as biomarkers of oxidative stress and genotoxicity (Viarengo et al. 2007), it was shown that benthic fish displayed a stress syndrome, whereas few biological effects were noted in species that had no direct contact with the bottom. However, the fact that *M. barbatus* (swimming near the sediment–water interface and eating benthic prey) remained unaffected suggested interspecific differential sensitivity (Pietrapiana et al. 2002). The determination of PAH content of mussel and fish tissues and/or the assessment of EROD activity were very useful in order to be able to point out the sources of biological effects, either the residual effects of the oil spill or chronic habitat pollution (Viarengo et al. 2007).

An important study of the chemical exposure of fish (*Pleuronectes vetulus, Platichthys stellatus, Genyonemus lineatus,* and *Pleuronectes americanus*) was carried out on Pacific and Atlantic North American coasts using liver toxicopathological lesions (Myers et al. 1998). Risk factors were noticeably enhanced for numerous pollutants (particularly PAHs, DDT, chlordanes). The hepatotoxicity threshold for PAHs was determined at 940 (680–1200) $\mu g\ kg^{-1}$ and, in the most polluted sites, prevalence reached 40% of the individuals. Histopathological examination of fish is relevant to document the presence or absence of toxicopathological liver lesions involved in hepatocarcinogenesis in English sole (*Parophrys vetulus*), and these have been causally related to exposure to PAHs (Myers et al. 2008 and literature quoted therein). Lesions that occur early in the histogenesis of liver neoplasia can occur in young of the year (<1 year of age) from PAH-contaminated sites, whereas preneoplastic and neoplastic hepatic lesions are rarely detected in English sole less than 3 years of age. Such lesions are therefore considered effective long-term biomarkers of PAH exposure.

Yuen et al. (2007) examined morphofunctional changes in the intestine of juvenile carnivorous fish (*Epinephelus coioides*) upon dietary exposure to environmentally realistic concentrations of the model PAH B[*a*]P for 4 weeks. Significant hyperplasia of basal enterocytes of mucosal folds was detected shortly after 3 days' exposure to 12.5 μg B[*a*]P g^{-1} body weight. These impairments were reversible in the fish upon the abatement of dietary B[*a*]P. Yuen et al. (2007) concluded that "realistic levels of food borne B[*a*]P could induce sublethal toxicity in *E. coioides.*" Recently, Xing et al. (2012) have reported that, in addition to oxidative stress, the commonly used pesticides atrazine and chlorpyrifos, separately or as a mixture, cause histopathological effects in the common carp *Cyprinus carpio*.

4.3.2 Responses to Metal Contamination

Various histocytological effects have been shown after exposure to metals either in the field or in the laboratory. In *in situ* exposure of mussels *Mytilus edulis* to an effluent of the titanium dioxide industry, impairments were observed mainly in the absorbing organs (gills and mantle) by comparison with controls (Ballan-Dufrançais et al. 1990). However, in all the organs observed, the regression of mitochondrial cristae evokes an alteration of cellular respiration. The authors hypothesized that this may be due to shell closure in response to the bad quality of water, thus generating hypoxia. All spermatozoa showed morphological alterations with swollen membrane systems and a lack of mitochondrial cristae, suggesting that these spermatozoa may be nonfunctional (Figure 4.2). Structural alterations of mitochondria (e.g., mitochondria with a swollen external membrane and deprived of cristae) were also observed in the gill epithelium of oysters *Crassostrea gigas* exposed to arsenic in seawater (Ettajani et al. 1996).

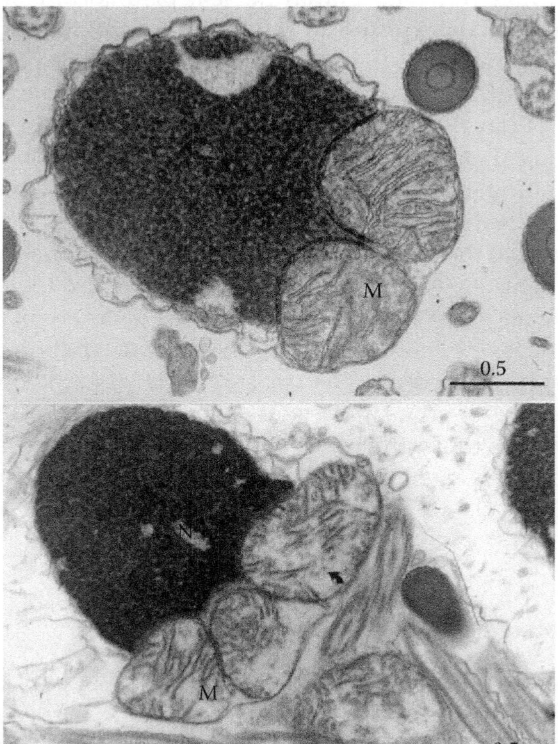

FIGURE 4.2
Spermatozoa of mussels (*Mytilus edulis*) from a reference population (top) compared to specimens exposed in the field to effluents of the titanium dioxide industry (bottom). Note the lack of mitochondrial cristae in exposed mussels. (After Ballan-Dufrançais, C. et al., *Ann. Inst. Océanogr.*, 66, 1–17, 1990. With permission.)

In situ exposure of the New Zealand mudsnail *Potamopyrgus antipodarum* along a gradient of Cd and Zn pollution induced histological lesions of the digestive gland, with hypertrophy of calcium cells and vacuolization of digestive cells (Gust et al. 2011). In Chinook salmon (*Oncorhynchus tshawytscha*) and rainbow trout (*Oncorhynchus mykiss*), exposure to copper (25 μg L^{-1} for 4 h) significantly reduced the number of olfactory receptors (Hansen et al. 1999). In zebrafish embryos, a copper dose of 68 μg L^{-1}, responsible for histological impairment (Johnson et al. 2007), is in the same order of magnitude of environmental concentrations corresponding to low densities of fish populations in Michigan lakes (34.0 μg L^{-1}) (Ellenberger et al. 1994). In both studies, histological damage parallels behavioral disturbances.

4.3.3 Responses to Nanoparticles

As nanotechnologies are rapidly expanding, histocytological studies have been recently reported in this particular field of ecotoxicology. Because many nanoparticles (NPs) are electron-dense and electron probe microanalysis allows the determination of the elemental constituents of metallic particles, microscopic techniques are useful tools with which to examine both the uptake and effects of NPs in organisms.

The liver of sticklebacks (*Gasterosteus aculeatus*) exposed to CdS NPs (5, 50, or 500 μg Cd L^{-1} for 21 days) exhibited hepatocellular nuclear pleomorphism, with the most severe cases recorded in individuals exposed to the highest dose; this liver pathology was not observed in the control treatment (Sanders et al. 2008). Toxicity tests were performed to investigate possible harmful effects on medaka (*Oryzias latipes*) exposed to nano-iron (0, 0.5, 5, 50 μg mL^{-1}) for 14 days. No significant change was found in the liver and the brain, whereas histopathological changes and morphological alterations were shown in the gill and intestine (cell swelling, hyperplasia, and granulomas, etc.), which suggest that deleterious effects occur as a result of direct contact with nano-iron. Direct exposure through the alimentary canal led to the accumulation of nano-iron in intestine tissues as confirmed by microanalysis (Li et al. 2009). In the same species, exposure to Ag NPs (100–1000 μg L^{-1} for 70 days at early life stages of development) induced a variety of morphological malformations such as edema, spinal abnormalities, finfold abnormalities, heart malformations, and eye defects in Japanese medaka (*Oryzias latipes*). Histopathological observations also confirmed the occurrence of abnormal eye development induced by Ag NPs (Wu et al. 2010). Ultrastructural changes in the midgut of the microcrustacean *Daphnia magna* upon exposure for 48 h to CuO NPs (at their 48 h EC_{50} level = 4.0 mg CuO L^{-1}) but not to bulk CuO (also at 48 h EC_{50} levels = 175 mg CuO L^{-1}) indicate nanosize-related adverse effects (Heinlaan et al. 2011).

4.3.4 Responses to Mixed Contamination

From the review by Au (2004) summarized in Table 4.3, one of the main limitations of the use of histocytological biomarkers is their low specificity, with the exception of certain disturbances linked to the presence of endocrine disrupting compounds in the medium such as intersex (Chapters 8 and 9). In sites impacted predominantly by a particular class of contaminants, as shown above for petroleum or metals, interpretation difficulties are limited, but the situation is more bothersome in the case of complex mixtures. However, histocytological biomarkers are highly appreciated for their sensitivity to chemical stress in field studies.

4.3.4.1 Marine and Brackish Environments

Histopathological examination of the black quahog *Artica islandica* (a bivalve) proved useful for the biomonitoring of dump sites, revealing tumors of the heart, fusion of gill filaments, and swelling of the interlamellar connective tissue of the gills (Yevich and Yevich 1994). The sea anemone *Cerianthopsis americanus* was successfully used to assess the spatiotemporal effects of dredged spoils: accumulation of cellular debris and necrosis in all areas of the body, vacuolated epidermis, and loss of mucous secretory cells close to the dumping site were attenuated after 1 year (Yevich and Yevich 1994). In mussels *Mytilus edulis*, exposure to heat effluents from a power plant was responsible for cilia loss from the gills and necrosis of the digestive diverticula (Yevich and Yevich 1994). In the framework of the U.S. Mussel Watch Program, different impairments have been detected, particularly in areas impacted by high population density and industrial activity; these impairments include extensive parasitism, interference of parasites with reproduction, and hematopoietic tumors (Yevich and Yevich 1994).

Associations between contaminant exposure and liver and skin tumor prevalence were evaluated in fish, brown bullheads *Ameiurus nebulosus*, from the watershed of the tidal Potomac River, USA (Pinkley et al. 2001). These authors found statistically significant differences in liver tumor prevalences in the Anacostia (50 to 60% depending on the season), an urban tributary designated as a region of concern; the Neabsco (17%), a tributary with petroleum inputs from runoff and marinas; the Quantico (7%), near a Superfund site that released OC contaminants; and the Tuckahoe (10%) as a reference. Skin tumor prevalences were significantly different: Anacostia (10–37%), Neabsco 3%, Quantico 3%, and Tuckahoe 0%. Evidence was found of higher PAH exposure in Anacostia fish, but a cause–effect linkage could not be established.

Another important field study dealing with liver histopathology in the Baltic flounder (*Platichthys flesus*) was carried out in 2001 and 2002 in four coastal sampling areas of the Baltic Sea: Kvädö fjärden (Swedish east coast, reference area), Klaipeda–Butinge (Lithuanian coast), Gulf of Gdansk (Polish coast), and Wismar Bay (German coast) (Lang et al. 2006). In total, 83.0% of the 436 female flounder examined were affected by liver lesions, out of which 74.3% were assigned to the category of nonspecific, 3.4% to the category of early toxicopathological nonneoplastic, 4.6% to the category of pre-neoplastic, and 0.7% to the category of neoplastic lesions.

The prevalence of toxicopathological liver lesions in demersal fish was studied to determine whether wastewater discharge could affect fish health (Basmadjian et al. 2008). Fish livers were sampled at different distances from the wastewater outfall on the San Pedro Shelf, California, for a 15-year period (1988–2003). The prevalence of toxicopathological lesion classes neoplasms (NEO), preneoplastic foci of cellular alteration (FCA), and hydropic vacuolation (HYDVAC), varied among species and locations. Prevalence of HYDVAC, NEO, and FCA in white croaker (*Genyonemus lineatus*) was 15.2%, 2.0%, and 0.7%, respectively. Bigmouth sole (*Hippoglossina stomata*) had a prevalence of FCA and NEO of 1.3% and 0.35%, respectively. In hornyhead turbot (*Pleuronichthys verticalis*), the prevalence of FCA and NEO was 3.4% and 0.37%, respectively. Consistent spatial differences for lesion prevalence were not demonstrated, and Basmadjian et al. (2008) underlined "the analytical difficulties of detecting a possible point source impact when the effect is rare, correlated with the size/age structure of the population, and may be caused by exposure to unknown multiple sources."

Fish diseases, including several variables associated with liver neoplasia, were investigated in the flatfish *Limanda limanda* from geographically distinct offshore marine sites in the Irish Sea and the North Sea (Stentiford et al. 2010). The authors concluded that age was an important factor when assessing fish population health status but did not explain all the differences established between sites. These differences may be related to anthropogenic contaminants, but other natural factors such as population genetics and migration must not be ignored.

4.3.4.2 Freshwater Environments

Brown trout (*Salmo trutta f. fario*) were exposed for weeks to water diverted (bypass systems) from two differently polluted streams, the Körsch and the Krähenbach (Triebskorn et al. 1997). Fish liver ultrastructural lesions were classified as belonging to one of the following three categories: (1) control; (2) slight deviations from the control with some visible pathologies; (3) major changes from the control with obvious pathologies. For six to eight specimens per fish group, each specimen was assigned to one of the three categories after examination of at least 50 cells and, subsequently, a mean assessment value (MAV) was calculated. The results are depicted in Figure 4.3. MAVs obtained for controls, Körsch-exposed trout, and Krähenbach-exposed specimens remained rather stable during different seasons, whereas clear intersite differences are shown between fish originating from the moderately polluted Krähenbach waters and the highly polluted Körsch waters.

Histopathological alterations of gill, liver, and spleen were studied in feral fish from three freshwater ecosystems that experience different types of contaminant stress (Teh et al. 1997). Certain organ and tissue lesions, detected microscopically, were common to

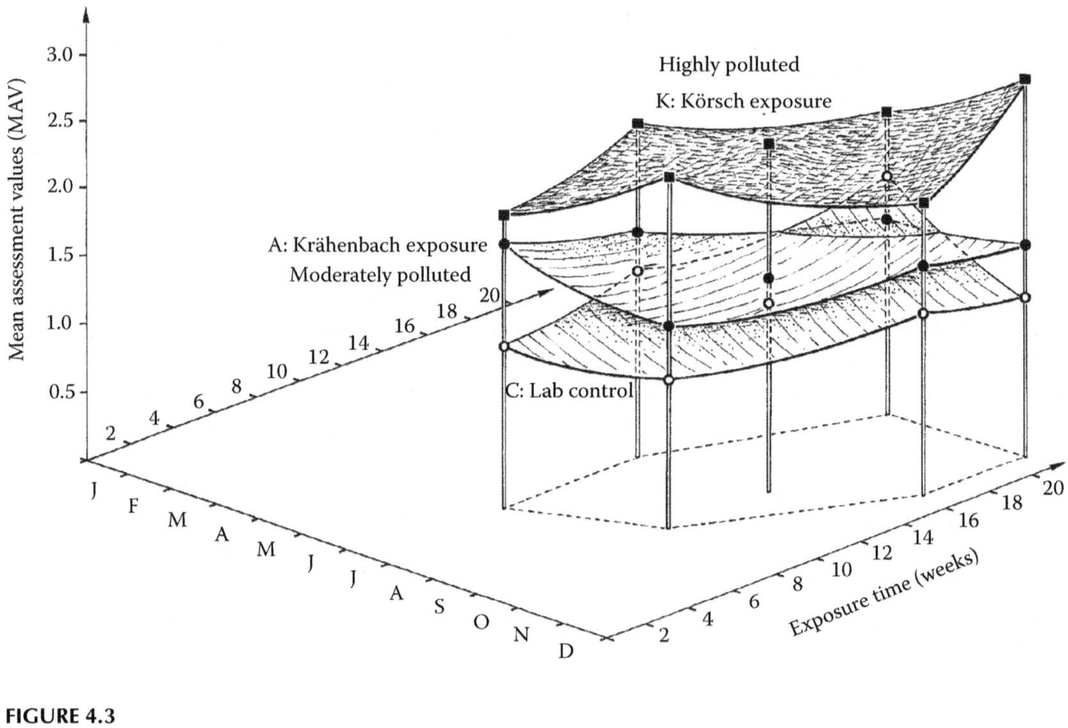

FIGURE 4.3
Intersite differences of fish liver ultrastructural lesions in specimens exposed to moderately polluted (Krähenbach) or highly polluted (Körsch) waters. (After Triebskorn, R. et al., *J. Aquat. Ecosyst. Stress Recov.*, 6, 57–73, 1997. With permission.)

fish from both reference and contaminated sites. On the other hand, the finding of specific lesions only in fish from contaminated sites suggested a contaminant etiology, particularly when they were similar to those observed in laboratory exposures to specific contaminants (Teh et al. 1997).

Assessment of liver tissues was carried out in the sharptooth catfish *Clarias gariepinus* from two dams in South Africa known to be multipolluted (metals, endocrine disrupting chemicals) despite being situated within a protected urban nature reserve (Marchand et al. 2008). Histopathological alterations included structural alterations in 27% of studied specimens, granular or fatty degeneration of hepatocytes (98% and 25%, respectively), hepatocyte nuclear alterations (90%), an increase in melanophage centers (32%), and necrosis of liver tissue (14%). By using a standardized quantification of histopathological alterations, the authors were able to distinguish between the degrees of impact at these two sites (Marchand et al. 2009). In the same species, Abdel-Moneim and Abdel-Mohsen (2010) examined the ultrastructural changes in hepatocytes of specimens from a polluted location and a relatively clean area in Lake Mariut, Egypt. Fish hepatocytes from the polluted area showed accumulation of heterochromatin, enlarged nucleoli, and an extremely folded nuclear envelope. The most frequent pathological modifications were the swelling of mitochondria and cristae regression.

4.4 Conclusions

The use of cortisol impairment as a biomarker, however conceptually attractive, presents considerable difficulties. Hontela (2000) stresses the need for "very specific sampling protocols since several factors influence cortisol secretion," first of all the stress of capture and handling. In addition to this problem of feasibility, conflicting results have been shown in the present review of the literature. In a review encompassing many more species than aquatic organisms, Busch and Hayward (2009) highlight a steep increase in the number of conservation-related field studies that measure glucocorticoid hormones (corticosterone or cortisol) as markers for stress. Since glucocorticoids have key roles in vital functions (animal performance including growth and metabolism, fetal development), it may be argued that, in addition to being able to reveal the presence of chemical stressors, cortisol is a biomarker with added ecological value as described for arctic fish and polar bears (Letcher et al. 2010). Despite a great potential for informing conservation, interpretation of the results of endocrine tools is often complicated.

AChE activity proved to be a responsive biomarker in different biological models, with decreased values at sites influenced by agricultural, urban, and industrial activities. This is well recognized for environmental assessment in monitoring programs (Burgeot et al. 2010). It is not as specific for OP and carbamate pesticides as was initially believed, but this inconvenience may be turned to advantage, since AChE activity can be used as a generalist biomarker, representative of the physiological status of organisms (Leiniö and Lehtonen 2005). Different forms of ChE exist in invertebrates and fish, and they can exhibit variable susceptibility to environmental contaminants. However, in many studies, it is unclear if the authors have truly characterized the enzyme that they call AChE. A better fundamental knowledge of this biomarker would help in correctly interpreting field data.

It is also important to be aware of all confounding factors capable of modulating the response of AChE activity in the presence of neurotoxicants. However, it seems that the problem may be more or less crucial, depending on the biological model used for AChE determination. For instance, in the black tiger shrimp *Penaeus monodon*, Tu et al. (2012) report that the effect of the pesticide deltamethrin was independent of temperature and salinity. Moreover, in two invertebrates widely used in estuarine studies, the clam *Scrobicularia plana* is more sensitive to confounding factors than the ragworm *Nereis diversicolor* (Solé et al. 2009; Kalman et al. 2010; Fossi Tankoua et al. 2011).

However, AChE activity has an interesting potential as a biomarker of ecological interest since it is clearly linked to effects at higher levels of biological integration, particularly behavior that is important for the normal functioning of individuals and populations (Chapter 10) and even for survival.

The inhibition of ALAD is a specific and sensitive biomarker of pollution by lead, recognized in official monitoring strategies but with no ecological relevance. For instance, in different areas of a river contaminated with Pb mine tailing in Missouri, USA, Overmann and Krajicek (1995) showed that enzyme activity of snapping turtles *Chelydra serpentine* was depressed by 94% in the most impacted site. Despite substantial reduction in enzyme activity associated with high Pb concentrations in tissues, the physiology of the snapping turtles was not seriously affected.

Cellular pathologies triggered by chemical exposure may be early warning signals of deleterious effects on flora and fauna. Such responses may be particularly useful biomarkers if they are precursors of diseases, directly related to potential risks (Bannasch et al. 1989; Hinton and Lauren 1990; Köhler et al. 1992; Moore et al. 1994; Regoli 2000; Galloway et al. 2002; Köhler et al. 2002; Moore 2002; Carajaville et al. 2003; Marigomez and Baybay-Villacorta 2003). The association of histocytopathological biomarkers with biomarkers representing different levels of biological organization may be of great interest in order to understand the complex machinery that can lead to serious impacts on reproduction, growth, and survival. In some cases, histocytopathological biomarkers reveal not only structural changes but also functional changes. Reviewing the application of histocytopathological biomarkers in marine pollution monitoring, Au (2004) has highlighted the symptoms that are the most highly ecologically relevant (Table 4.3).

Li et al. (2009), who have observed histopathological changes and morphological alterations in intestine and gills of medaka (*Oryzias latipes*) exposed to nano-iron, suggest that effects on the outside wall of the intestine might promote impairments of normal digestion function. A reduction of the contact surface at the level of the gill epithelium is likely to occur, thus affecting gas and ion exchange, whereas a breakage occurring on the surface of gill filament and the secondary gill lamellae might provide a direct invasion route for exogenous chemicals.

From 1979 to 1982, Yevich and Yevich (1994) observed abnormalities of the byssus organ in mussels *Mytilus edulis* collected from a limited area (Prudence Island, Narragansett Bay, USA) but were unable to find the causative factor of the disease. In March 1981, there was a massive kill of mussels. In the byssus-bearing scallop *Chlamys varia* exposed to sublethal doses of silver in the laboratory, histological examination showed that the secretion of byssus threads was inhibited. A large number of individuals lost their byssus and became unable to attach themselves to the substratum (Berthet et al. 1992).

Disruption of sensory system function (e.g., Hansen et al. 1999; Johnson et al. 2007) may be responsible for the absence of detection of environmental pollutants, thus leading to inappropriate behavioral responses since once detected, avoidance (escape, valve closure in bivalves) allows a reduction of exposure, limiting toxic effects and improving survival (Chapter 10).

In addition to alterations of liver ultrastructure, the brown trout (*Salmo trutta f. fario*) studied by Triebskorn et al. (1997) showed a significant decrease in swimming velocity in a highly polluted stream, perhaps as a consequence of less energy being available for locomotion since metabolic enzyme studies in parallel fish groups indicated that catabolic, energy-providing mechanisms were activated (Konradt et al. 1996 quoted in Triebskorn et al. 1997). These findings were in agreement with the liver ultrastructure showing an increase in the number of mitochondria and a reduction of glycogen storage. As shown by these authors, histocytological examinations can help to detect energy metabolism impairments. Structural alterations of mitochondria may be interpreted as a disturbance of cellular respiratory mechanisms (Ballan-Dufrançais et al. 1990; Ettajani et al. 1996). Histochemistry clearly provides added value to histological examination, giving access to changes in cellular or tissue energy reserves such as the depletion of glycogen discussed above after Triebskorn et al. (1997), or the depletion and recovery of glycogen in interstitial tissues of oysters *Crassostrea gigas* exposed to silver then allowed to depurate in clean water (Berthet et al. 1990), a pattern that may be explained by the energy cost of combating environmental contaminants (Chapter 3). Histological lesions of the digestive gland observed in the New Zealand mudsnail *Potamopyrgus antipodarum* exposed to multimetal pollution in the field, apparently explain at least partly the decrease in energy reserves (triglycerides and proteins), juvenile growth, and adult fecundity at the most contaminated site (Gust et al. 2011).

Early life stages of organisms are generally the most sensitive to stress. Thus, effects on germinal cells reported above as the consequence of exposure to effluents (Ballan-Dufrançais et al. 1990) or genotoxicants (Lewis and Galloway 2009; Lacaze et al. 2010; Devaux et al. 2011) may be suspected to affect the success of reproduction. Embryotoxicity (such as histological impairments and associated behavioral disturbances described by Johnson et al. 2007) or malformations associated with other impairments of embryo–larval development in zebrafish embryos exposed to ZnO NPs described by Bai et al. (2010) is also a source of concern from this point of view. We have not discussed in this chapter the histopathological impairments that occur as a consequence of pollution by endocrine disruptors, such as imposex in gastropods and intersex in fish and bivalves (Chapters 8 and 9), but they also have potential to predict ecological disturbances through altered success of reproduction.

In conclusion, biomarkers of damage reviewed in this chapter as well as lysosomal biomarkers (Chapter 5), biomarkers of immunotoxicity (Chapter 6), endocrine disruption (Chapters 8 and 9), and genotoxicity (Chapter 13) provide precise information on the health status of individuals. However, in most cases, it may be difficult to extrapolate to the population level. Between infra- and supra-individual effects, a number of mechanisms can interfere to mitigate and repair damage. In particular, the ecological significance of oxidative damage cannot be assessed independently of antioxidant defense and repair mechanisms (Metcalfe and Alonso-Alvarez 2010). In order to have an ecological significance and therefore value, any infra-individual biomarker must be linked to a key process in the functioning of organisms and their progeny. Among the approaches used to study pollutant responses in aquatic organisms, those associated with the success of reproduction are particularly able to provide relevant toxicological as well as ecological information. Reproduction is indeed at a crossroads of numerous processes, notably hormone levels, genetic changes, energy metabolism, and behavior. Even when it has not been established that damage will lead to population depletion or local extinction through cascading events, it is nevertheless important to know the effects of environmental toxicants before they affect higher levels of organization, and many biomarkers of damage make this possible. It has been proposed that the "precautionary principle" could be applied as soon as a biochemical biomarker shows a value deviating significantly from normal levels (Lopez-Barea 1994, in Flammarion 2000).

Acknowledgment

Thanks are due to Professor Inoue, University of Osaka, who has kindly provided us with Figure 4.1.

References

Abdel-Moneim, A.M., and H.A. Abdel-Mohsen. 2010. Ultrastructure changes in hepatocytes of catfish *Clarias gariepinus* from Lake Mariut, Egypt. *J. Environ. Biol.* 31:715–20.

Abele, D., J.P. Vázquez-Medina, and T. Zenteno-Savín. 2012. *Oxidative Stress in Aquatic Ecosystems.* Chichester, UK: Wiley-Blackwell.

Ahmad, I. et al. 2011. Lipid peroxidation vs. antioxidant modulation in the bivalve *Scrobicularia plana* in response to environmental mercury—Organ specificities and age effect. *Aquat. Toxicol.* 103:150–8.

Andersen, T.H. et al. 2006. Acute and chronic effects of pulse exposure of *Daphnia magna* to dimethoate and pirimicarb. *Environ. Toxicol. Chem.* 25:1187–95.

Arrieta, M.A. et al. 2000. Blood lead concentration and delta-aminolevulinic acid dehydratase activity in adult *Bufo arenarum*. *Arch. Physiol. Biochem.* 108:275–80.

Au, D.W.T. 2004. The application of histo-cytopathological biomarkers in marine pollution monitoring: A review. *Mar. Pollut. Bull.* 48:817–34.

Bai, W. et al. 2010. Toxicity of zinc oxide nanoparticles to zebrafish embryo: A physicochemical study of toxicity mechanism. *J. Nanopart. Res.* 12:1645–54.

Ballan-Dufrançais, C., A.Y. Jeantet, and J. Coulon. 1990. Cytological features of mussels (*Mytilus edulis*) *in situ* exposed to an effluent of the titanium dioxide industry. *Ann. Inst. Océanogr*. 66:1–17.

Bannasch, P. et al. 1989. Significance of sequential cellular changes inside and outside foci of altered hepatocytes during hepatocarcinogenesis. *Toxicol. Pathol.* 4:617–28.

Barata, C. et al. 2007. Combined use of biomarkers and in situ bioassays in *Daphnia magna* to monitor environmental hazards of pesticides in the field. *Environ. Toxicol. Chem.* 26:370–9.

Basmadjian, E. et al. 2008. Liver lesions in demersal fishes near a large ocean outfall on the San Pedro Shelf, California. *Environ. Monit. Assess.* 138:239–53.

Beauvais, S.L. et al. 2000. Physiological measures of neurotoxicity of diazinon and malathion to larval rainbow trout (*Oncorhynchus mykiss*) and their correlation with behavioral measures. *Environ. Toxicol. Chem.* 19:1875–80.

Beauvais, S.L. et al. 2001. Cholinergic and behavioral neurotoxicity of carbaryl and cadmium to larval rainbow trout (*Oncorhynchus mykiss*). *Ecotoxicol. Environ. Saf.* 49:84–90.

Bechshøft, T.Ø. et al. 2011. Distribution of vitamins A (retinol) and E (α-tocopherol) in polar bear kidney: Implications for biomarker studies. *Sci. Total Environ.* 409:3508–11.

Bechshøft, T.Ø. et al. 2012. Measuring environmental stress in East Greenland polar bears, 1892–1927 and 1988–2009: What does hair cortisol tell us? *Environ. Intern.* 45:15–21.

Berthet, B. et al. 1992. Bioaccumulation, toxicity and physico-chemical speciation of silver in bivalve molluscs: Ecotoxicological and health consequences. *Sci. Tot. Environ.* 125:97–122.

Berthet, B., C. Amiard-Triquet, and R. Martoja. 1990. Chemical and histological effects of depuration in *Crassostrea gigas* Thunberg exposed previously to silver. *Water Air Soil Pollut.* 50:355–69.

Berthou, F. et al. 1987. The occurrence of hydrocarbons and histopathological abnormalities in oysters for seven years following the wreck of the Amoco Cadiz in Brittany (France). *Mar. Environ. Res.* 23:103–33.

Blokhina, O., E. Virolainen, and K.V. Fagerstedt. 2003. Antioxidants, oxidative damage and oxygen deprivation stress: A review. *Ann. Bot.* 91:179–94.

Boily, M., J. Thibodeau, and M. Bisson. 2009. Retinoid metabolism (LRAT, REH) in the liver and plasma retinoids of bullfrog, *Rana catesbeiana*, in relation to agricultural contamination. *Aquat. Toxicol.* 91:118–25.

Braathen, M. et al. 2004. Relationships between PCBs and thyroid hormones and retinol in female and male polar bears. *Environ. Health Perspect.* 112:826–33.

Brewer, S.K. et al. 2001. Behavioral dysfunctions correlate to altered physiology in rainbow trout (*Oncorhynchus mykiss*) exposed to cholinesterase-inhibiting chemicals. *Arch. Environ. Contam. Toxicol.* 40:70–6.

Brown, R.J. et al. 2004. Differential sensitivity of three marine invertebrates to copper assessed using multiple biomarkers. *Aquat. Toxicol.* 66:267–78.

Burgeot, T. et al. 2010. Acetylcholinesterase: Methodology development of a biomarker and challenges of its application for biomonitoring. ICES CM Code: F-25. In: *Intern Council for the Exploration of the Sea*, Annual Science Conference, 20–24 September 2010, Nantes, France. e-paper: http://www.ices.dk/products/CMdocs/CM-2010/F/F2510.pdf.

Busch, D.S., and L. S. Hayward. 2009. Stress in a conservation context: A discussion of glucocorticoid actions and how levels change with conservation-relevant variables. *Biol. Conserv.* 142:2844–53.

Cailleaud, K. et al. 2007. Effects of salinity and temperature on the expression on enzymatic biomarkers in *Eurytemora affinis* (Calanoida, Copepoda). *Comp. Biochem. Physiol.* 147A:841–9.

Cailleaud, K. et al. 2009. Tidal influence on the distribution of hydrophobic organic contaminants in the Seine Estuary and biomarker responses on the copepod *Eurytemora affinis*. *Environ. Pollut.* 157:64–71.

Canesi, L. et al. 2011. Bivalve molluscs as a unique target group for nanoparticle toxicity. *Mar. Environ. Res.* 76:16–21.

Carajaville, M.P. et al. 2003. Peroxisome proliferation as a biomarker in environmental pollution assessment. *Microsc. Res. Technol.* 61:117–20.

Carballo, M. et al. 2005. A survey of potential stressor-induced physiological changes in carp (*Cyprinus carpio*) and barbel (*Barbus bocagei*) along the Tajo River. *Environ. Toxicol.* 20:119–25.

Champoux, L. et al. 2006. Contamination and biomarkers in the Great Blue Heron, an indicator of the state of the St. Lawrence river. *Ecotoxicology* 15:83–96.

Chen, L. et al. 2012. Alteration in retinoid status after long-term exposure to PBDEs in zebrafish (*Danio rerio*). *Aquat. Toxicol.* 120–121:11–8.

Choi, J.Y. et al. 2011. Acetylcholine (ATC)–cleaving cholinesterase (ChE) activity as a potential biomarker of pesticide exposure in the Manila clam, *Ruditapes philippinarium*, of Korea. *Mar. Environ. Res.* 71:162–8.

Company, R. et al. 2008. Using biochemical and isotope geochemistry to understand the environmental and public health implications of lead pollution in the lower Guadiana River, Iberia: A freshwater bivalve study. *Sci. Total Environ.* 405:109–19.

Company, R. et al. 2011. Source and impact of lead contamination on δ-aminolevulinic acid dehydratase activity in several marine bivalve species along the Gulf of Cadiz. *Aquat. Toxicol.* 101:146–54.

Cooper, N.L., and J.R. Bidwell. 2006. Cholinesterase inhibition and impacts on behavior of the Asian clam, *Corbicula fluminea*, after exposure to an organophosphate insecticide. *Aquat. Toxicol.* 76:258–67.

Damiens, G. et al. 2004. Evaluation of biomarkers in oyster larvae in natural and polluted conditions. *Comp. Biochem. Physiol.* 138C:121–8.

Damiens, G. et al. 2007. Integrated biomarker response index as a useful tool for environmental assessment evaluated using transplanted mussels. *Chemosphere* 66:574–83.

De Lafontaine, Y. et al. 2000. Biomarkers in zebra mussels (*Dreissena polymorpha*) for the assessment and monitoring of water quality of the St Lawrence River (Canada). *Aquatic Toxicol.* 50:51–71.

Devaux, A. et al. 2011. Reproduction impairment following paternal genotoxin exposure in brown trout (*Salmo trutta*) and Arctic charr (*Salvelinus alpinus*). *Aquat. Toxicol.* 101:405–11.

Devier, M.H. et al. 2005. One-year monitoring survey of organic compounds (PAHs, PCBs, TBT), heavy metals and biomarkers in blue mussels from the Arcachon Bay, France. *J. Environ. Monitor.* 7:224–40.

Domingues, I. et al. 2010. Cholinesterase and glutathione-*S*-transferase activities in freshwater invertebrates as biomarkers to assess pesticide contamination. *Environ. Toxicol. Chem.* 29:5–18.
Duquesne, S. 2006. Effects of an organophosphate on *Daphnia magna* at suborganismal and organismal levels: Implications for population dynamics. *Ecotoxicol. Environ. Saf.* 65:145–50.
Duquesne, S., and E. Küster. 2010. Biochemical, metabolic, and behavioural responses and recovery of *Daphnia magna* after exposure to an organophosphate. *Ecotoxicol. Environ. Saf.* 73:353–9.
Durou, C. et al. 2007. Biomonitoring in a clean and a multi-contaminated estuary based on biomarkers and chemical analyses in the endobenthic worm *Nereis diversicolor*. *Environ. Pollut.* 148:445–58.
Ellenberger, S.A., P.C. Baumann, and T.W. May. 1994. Evaluation of effects caused by high copper concentrations in Torch Lake, Michigan, on reproduction of yellow perch. *Jour. Great Lakes Res.* 20:531–6.
Ettajani, H. et al. 1996. Fate and effects of soluble or sediment-bound arsenic in oysters (*Crassostrea gigas* Thun). *Arch. Environ. Contam. Toxicol.* 31:38–46.
Flammarion, P. 2000. *Mesure d'un biomarqueur de pollution chez des poissons d'eau douce. Validation et optimisation*. Vol 15. Cemagref Editions.
Forget, J. et al. 1999. Joint action of pollutant combinations (pesticides and metals) on survival (LC_{50} values) and acetylcholinesterase activity of *Tigriopus brevicornis* (Copepoda, Harpacticoïda). *Environ. Toxicol. Chem.* 18:912–18.
Forget, J., B. Beliaeff, and G. Bocquené. 2003. Acetylcholinesterase activity in copepods (*Trigriopus brevicornis*) from the Vilaine river estuary, France, as a biomarker of neurotoxic contaminants. *Aquat. Toxicol.* 62:195–204.
Fossi Tankoua, O. et al. 2010. Towards a comprehensive methodology for assessing the health status of coastal and estuarine ecosystems. In: *Intern Council for the Exploration of the Sea*, Annual Science Conference, 20–24 September 2010, Nantes, France. e-paper: http://www.ices.dk/products/CMdocs/CM-2010/F/F1510.pdf.
Fossi Tankoua, O. et al. 2011. Potential influence of confounding factors (size, salinity) on biomarker tools in the sentinel species *Scrobicularia plana* used in monitoring programmes of estuarine quality. *Environ. Sci. Pollut. Res.* 18:1253–63.
Frasco, M.F. et al. 2005. Do metals inhibit acetylcholinesterase (AChE)? Implementation of assay conditions for use of AChE activity as a biomarker of metal toxicity. *Biomarkers* 10:360–75.
Fulton, M.H., and P.B. Key. 2001 Acetylcholinesterase inhibition in estuarine fish and invertebrates as an indicator of organophosphorus insecticide exposure and effects. *Environ. Toxicol. Chem.* 20:37–45.
Gagnaire, B. et al. 2008. Cholinesterase activities as potential biomarkers: Characterization in two freshwater snails, *Potamopyrgus antipodarum* (Mollusca, Hydrobiidae, Smith 1889) and *Valvata piscinalis* (Mollusca, Valvatidae, Müller 1774). *Chemosphere* 71:553–60.
Galgani, F., and G. Bocquéné. 2000. Molecular biomarkers of exposure of marine organisms to organophosphorus pesticides and carbamates. In: *Use of Biomarkers for Environmental Quality Assessment*, ed. L. Lagadic et al., 113–37. Enfield, NH: Science Publishers.
Galloway, T.S. et al. 2002. Rapid assessment of marine pollution using multiple biomarkers and chemical immunoassays. *Environ. Sci. Technol.* 36:2219–26.
Garcia-de la Parra, L.M. et al. 2006. Effects of methamidophos on acetylcholinesterase activity, behavior, and feeding rate of the white shrimp (*Litopenaeus vannamei*). *Ecotoxicol. Environ. Saf.* 65:372–80.
Gaworecki, K.M. et al. 2009. Biochemical and behavioral effects of diazinon exposure in hybrid striped bass. *Environ. Toxicol. Chem.* 28:105–12.
Giari, L. et al. 2012. The impact of an oilspill on organs of bream *Abramis brama* in the Po River. *Ecotoxicol. Environ. Saf.* 77:18–27.
Gopal, K. et al. 1985. Neurobehavioural changes in freshwater fish *Channa punctatus* exposed to endosulfan. *J. Adv. Zool.* 6:74–80.
Guilhermino, L. et al. 1998. Should the use of inhibition of cholinesterase as a specific biomarker for organophosphate and carbamate pesticides be questioned? *Biomarkers* 3:157–63.

Gust, M. et al. 2011. Comprehensive biological effects of a complex field poly-metallic pollution gradient on the New Zealand mudsnail *Potamopyrgus antipodarum* (Gray). *Aquat. Toxicol.* 101:100–8.

Halliwell, B., and J.M.C. Gutteridge. 2007. *Free Radicals in Biology and Medicine*. Oxford: Oxford University Press.

Hansen, J.A. et al. 1999. Chinook salmon (*Oncorhynchus tshawytschaya*) and rainbow trout (*Oncorhynchus mykiss*) exposed to copper: Neurophysiological and histological effects on the olfactory system. *Environ. Toxicol. Chem.* 18:1979–91.

Heath, A.G. et al. 1997. Physiological responses of fathead minnow larvae to rice pesticides. *Ecotoxicol. Environ. Saf.* 37:280–8.

Heinlaan, M. et al. 2011. Changes in the *Daphnia magna* midgut upon ingestion of copper oxide nanoparticles: A transmission electron microscopy study. *Water Res.* 45:179–90.

Hernandez-Moreno, D. et al. 2010. Brain acetylcholinesterase, malondialdehyde and reduced glutathione as biomarkers of continuous exposure of tench, *Tinca tinca*, to carbofuran or deltamethrin. *Sci. Total Environ.* 408:4976–83.

Hinton, D.E., and D.J. Lauren. 1990. Liver structural alterations accompanying chronic toxicity in fishes: Potential biomarkers of exposure. In: *Biomarkers of Environmental Contamination*, J.F. McCarthy and L.K. Shugart, 17–37. Boca Raton, FL: Lewis Publishers.

Hinton, D.E. 1993a. Toxicologic histopathology of fishes: A systemic approach and overview. In: *Pathobiology of Marine and Estuarine Organisms*, ed. J.A. Couch and J.W. Fournie, 177–215. Boca Raton, FL: CRC Press.

Hinton, D.E. 1993b. Cells, cellular responses, and their markers in chronic toxicity of fishes. In: *Aquatic Toxicology: Molecular, Biochemical and Cellular Perspectives*, ed. D.C. Malins and G.K. Ostrander, 207–239. Boca Raton, FL: Lewis Publishers.

Hoguet, J., and P.B. Key. 2007. Activities of biomarkers in multiple life stages of the model crustacean, *Palaemonetes pugio*. *J. Exp. Mar. Biol. Ecol.* 353:235–44.

Hontela, A. 2000. Endocrine biomarkers: Hormonal indicators of sublethal toxicity in fishes. In: *Use of Biomarkers for Environmental Quality Assessment*, ed. L. Lagadic, 187–204. Enfield, NH: Science Publishers.

Hopkins, W.A., and C.T. Winne. 2006. Influence of body size on swimming performance of four species of neonatal natricine snakes acutely exposed to a cholinesterase-inhibiting pesticide. *Environ. Toxicol. Chem.* 25:1208–13.

Inoue, D., K. Sei, and M. Ike. 2010. Disruption of retinoic acid receptor signaling by environment pollutants. *J. Health Sci.* 56:221–30.

Iwanowicz, L.R. et al. 2009. Aroclor 1248 exposure leads to immunomodulation, decreased disease resistance and endocrine disruption in the brown bullhead, *Ameiurus nebulosus*. *Aquat. Toxicol.* 93:70–82.

Jebali, J. et al. 2011. Characterization and evaluation of cholinesterase activity in the cockle *Cerastoderma glaucum*. *Aquat. Biol.* 13:243–50.

Jenssen, B.M. et al. 2001. PCBs, TEQs, and plasma retinol in grey heron (*Ardea cinera*) hatchlings from two rookeries in Norway. *Chemosphere* 44:483–9.

Johnson, A., E. Carew, and K.A. Sloman. 2007. The effects of copper on the morphological and functional development of zebrafish embryos. *Aquat. Toxicol.* 84:431–8.

Jørgensen, E.H. et al. 2001. Influence of *o'p*-DDD on the physiological response to stress in Arctic charr (*Salvelinus alpinus*). *Aquat. Toxicol.* 54:179–93.

Kalman, J. et al. 2008. Is δ-aminolevulinic acid dehydratase activity in bivalves from south-west Iberian Peninsula a good biomarker of lead exposure? *Mar. Environ. Res.* 66:38–40.

Kalman, J. et al. 2010. Assessment of the influence of confounding factors (weight, salinity) on the response of biomarkers in the estuarine polychaete *Nereis diversicolor*. *Biomarkers* 15:461–9.

Kirby, M.F. et al. 2000. The use of cholinesterase activity in flounder (*Platichthys flesus*) muscle tissue as a biomarker of neurotoxic contamination in UK estuaries. *Mar. Pollut. Bull.* 40:780–91.

Klaine, S.J. et al. 2008. Nanomaterials in the environment: Behavior, fate, bioavailability, and effects. *Environ. Toxicol. Chem.* 27:1825–51.

Köhler, A., E. Wahl, and K. Soffker. 2002. Functional and morphological changes of lysosomes as prognostic biomarkers of toxic liver injury in a marine flatfish (*Platichthys flesus* (L.)). *Environ. Toxicol. Chem.* 21:2434–44.

Köhler, A., H. Deisemann, and B. Lauritzen. 1992. Ultrastructural and cytochemical indices of toxic injury in dab liver. *Mar. Ecol. Prog. Ser.* 91:141–53.

Kopecka, J., and J. Pempkowiak. 2008. Temporal and spatial variations of selected biomarker activities in flounder (*Platichthys flesus*) collected in the Baltic proper. *Ecotoxicol. Environ. Saf.* 70:379–91.

Körner, O. et al. 2008. Water temperature and concomitant waterborne ethinylestradiol exposure affects the vitellogenin expression in juvenile brown trout (*Salmo trutta*). *Aquat. Toxicol.* 90:188–96.

Kristoff, G. et al. 2006. Inhibition of cholinesterase activity by azinphos-methyl in two freshwater invertebrates: *Biomphalaria glabrata* and *Lumbriculus variegates*. *Toxicology* 222:185–94.

Labrot, F. et al. 1996. *In vitro* and *in vivo* studies of potential biomarkers of lead and uranium contamination: Lipid peroxidation, acetylcholinesterase, catalase and glutathione peroxidase activities in three non-mammalian species. *Biomarkers* 1:21–28.

Lacaze, E. et al. 2010. Genotoxicity assessment in the amphipod *Gammarus fossarum* by use of the alkaline Comet assay. *Mutat. Res.* 700:32–8.

Lang, T. et al. 2006. Liver histopathology in Baltic flounder (*Platichthys flesus*) as indicator of biological effects of contaminants. *Mar. Pollut. Bull.* 53: 488–96.

Lehtonen, K.K., and S. Leiniö. 2003. Effects of exposure to copper and malathion on metallothionein levels and acetylcholinesterase activity of the mussel *Mytilus edulis* and the clam *Macoma balthica* from the Northern Baltic Sea. *Bull. Environ. Contam. Toxicol.* 71:489–96.

Lehtonen, K.K. et al. 2003. Accumulation of nodularin-like compounds from the cyanobacterium *Nodularia spumigena* and changes in acetylcholinesterase activity in the clam *Macoma balthica* during short-term laboratory exposure. *Aquat. Toxicol.* 64:461–76.

Leiniö, S., and K.K. Lehtonen. 2005. Seasonal variability in biomarkers in the bivalves *Mytilus edulis* and *Macoma balthica* from the northern Baltic Sea. *Comp. Biochem. Physiol.* 140C:408–21.

Lerner, D.T., B.T. Björnsson, and S.D. McCormick. 2007. Effects of aqueous exposure to polychlorinated biphenyls (Aroclor 1254) on physiology and behavior of smolt development of Atlantic salmon. *Aquat. Toxicol.* 81:329–36.

Letcher, R.J. et al. 2010. Exposure and effects assessment of persistent organohalogen contaminants in arctic wildlife and fish. *Sci. Total Environ.* 408: 2995–3043.

Lewis, C., and T. Galloway. 2009. Reproductive consequences of paternal genotoxin exposure in marine invertebrates. *Environ. Sci. Technol.* 43:928–33.

Li, H.C. et al. 2009. Effects of waterborne nano-iron on medaka (*Oryzias latipes*): Antioxidant enzymatic activity, lipid peroxidation and histopathology. *Ecotox. Environ. Saf.* 72:684–92.

Linde-Arias, A.R. et al. 2008a. Multibiomarker approach in fish to assess the impact of pollution in a large Brazilian river, Paraiba do Sul. *Environ. Pollut.* 156:974–9.

Linde-Arias, A.R. et al. 2008b. Biomarker in an invasive fish species, *Oreochromis niloticus*, to assess the effects of pollution in a highly degraded Brazilian river. *Sci. Total Environ.* 399:186–92.

Lushchak, V.I. 2011. Environmentally induced oxidative stress in aquatic animals. *Aquat. Toxicol.* 101:13–30.

Mahler, B.J. et al. 2006. Trends in metals in urban and reference lake sediments across the Unites States, 1970 to 2001. *Environ. Toxicol. Chem.* 25:1698–709.

Mann, R.M. et al. 2009. Amphibians and agricultural chemicals: Review of the risks in a complex environment. *Environ. Pollut.* 157:2903–27.

Marchand, M.J. et al. 2009. Histopathological alterations in the liver of the Sharptooth catfish *Clarias gariepinus* from polluted aquatic systems in South Africa. *Environ. Toxicol.* 24:133–47.

Marigomez, I., and L. Baybay-Villacorta. 2003. Pollutant-specific and general lysosomal responses in digestive cells of mussels exposed to model organic chemicals. *Aquat. Toxicol.* 64:235–57.

Metcalfe, N.B., and C. Alonso-Alvarez. 2010. Oxidative stress as a life-history constraint: The role of reactive oxygen species in shaping phenotypes from conception to death. *Funct. Ecol.* 24:984–96.

Milla, S. et al. 2009. Corticosteroids: Friends or foes of teleost fish reproduction? *Comp. Biochem. Physiol.* 153A:242–51.
Miller, G.G. et al. 2002. *In vitro* toxicity and interactions of environmental contaminants (Arochlor 1254 and mercury) and immunomodulatory agents (lipopolysaccharide and cortisol) on thymocytes from lake trout (*Salvelinus namaycush*). *Fish Shellfish Immunol.* 13:11–26.
Monteiro, M. et al. 2005. Characterization of the cholinesterases present in head tissues of the estuarine fish *Pomatoschistus microps*: Application to biomonitoring. *Ecotoxicol. Environ. Saf.* 62: 341–7.
Moore, M.N. 2002. Biocomplexity: The post-genome challenge in ecotoxicology. *Aquat. Toxicol.* 59:1–15.
Moore, M.N. 2006. Do nanoparticles present ecotoxicological risks for the health of the aquatic environment? *Environ. Intern.* 32:967–76.
Moore, M.N. et al. 1994. An integrated approach to cellular biomarkers in fish. In: *Non-Destructive Biomarkers in Vertebrates*, ed. M.C. Fossi and C. Leonzio, 171–97. Boca Raton, FL: Lewis Publishers.
Moreira-Santos, M. et al. 2005. Short-term sublethal (sediment and aquatic roots of floating macrophytes) assays with a tropical chironomid based on postexposure feeding and biomarkers. *Environ. Toxicol. Chem.* 24:2234–42.
Moreira, S.M. et al. 2006. Effects of estuarine sediment contamination on feeding and on key physiological functions of the polychaete *Hediste diversicolor*: Laboratory and *in situ* assays. *Aquatic Toxicol.* 78:186–201.
Mos, L. et al. 2007. Contaminant-associated disruption of vitamin A and its receptor (retinoic acid receptor α) in free-ranging harbour seals (*Phoca vitulina*). *Aquat. Toxicol.* 81:319–28.
Myers, M.S. et al. 1998. Toxicopathic hepatic lesions as biomarkers of chemical contaminant exposure and effects in marine bottomfish species from the Northeast and Pacific Coasts, USA. *Mar. Pollut. Bull.* 37:92–113.
Myers, M.S. et al. 2008. Improved flatfish health following remediation of a PAH-contaminated site in Eagle Harbor, Washington. *Aquat. Toxicol.* 88:277–88.
Nascimento, C.R.B., M.M. Souza, and C.B.R. Martinez. 2012. Copper and the herbicide atrazine impair the stress response of the freshwater fish *Prochilodus lineatus*. *Comp. Biochem. Physiol.* 155C:456–61
Novák, J., M. Benisek, and K. Hilscherova. 2008. Disruption of retinoid transport, metabolism and signaling by environment pollutants. *Environ. Int.* 34:898–913.
Nyman, M. et al. 2003. Contaminant exposure and effects in Baltic ringed and grey seals as assessed by biomarkers. *Mar. Environ. Res.* 55:73–99.
Oliveira, M.M. et al. 2007. Brain acetylcholinesterase as a marine pesticide biomarker using Brazilian fishes. *Mar. Environ. Res.* 63:303–12.
Oliveira, M., M. Pacheco, and M.A. Santos. 2011. Fish thyroidal and stress responses in contamination monitoring—An integrated biomarker approach. *Ecotoxicol. Environ. Saf.* 74:1265–70.
OSPAR Agreement 2008-09. JAMP Guidelines for Contaminant-Specific Biological Effects http://www.ospar.org/content/content.asp?menu=00900301400135_000000_000000.
OSPAR Commission. 2009. CEMP assessment report: 2008/2009 Assessment of trends and concentrations of selected hazardous substances in sediments and biota. http://www.ospar.org/content/content.asp?menu=00200304000116_000000_000000.
Overmann, S.R., and J.J. Krajicek. 1995. Snapping turtles (*Chelydra serpentine*) as biomonitors of lead contamination of the big river in Missouri old lead belt. *Environ. Toxicol. Chem.*14:689–95.
Payne, J.F. et al. 1996. Acetylcholinesterase, an old biomarker with a new future? Field trials in association with two urban rivers and a paper mill in Newfoundland. *Mar. Pollut. Bull.* 32:225–31.
Pessoa, P.C. et al. 2011. Cholinesterase inhibition and behavioral toxicity of carbofuran on *Oreochromis niloticus* early stages. *Aquat. Toxicol.* 105:312–20.
Peter, M.C.S. 2011. The role of thyroid hormones in stress response of fish. *Gen. Comp. Endocrinol.* 172:198–210.
Pietrapiana, D. et al. 2002. Evaluating the genotoxic damage and hepatic tissue alterations in demersal fish species: A case study in the Ligurian Sea (NW-Mediterranean). *Mar. Pollut. Bull.* 44:238–43.

Pinkley, A.E. et al. 2001. Tumor prevalence and biomarkers of exposure in brown bullheads (*Ameirus nebolosus*) from the tidal Potomo River, USA, watershed. *Environ. Toxicol. Chem.* 20:1196–205.
Printes, L.B., and A. Callaghan. 2004. A comparative study on the relationship between acetylcholinesterase activity and acute toxicity in *Daphnia magna* exposed to anticholinesterase insecticides. *Environ. Toxicol. Chem.* 23:1241–7.
Printes, L.B., M.D.E. Fellowes, and A. Callaghan. 2008. Clonal variation in acetylcholinesterase biomarkers and life history traits following OP exposure in *Daphnia magna*. *Ecotoxicol. Environ. Saf.* 71:519–26.
Regoli, F. 2000. Total oxyradical scavenging capacity (TOSC) in polluted and translocated mussels: A predictive biomarker of oxidative stress. *Aquat. Toxicol.* 50:351–61.
Ricciardi, F. et al. 2010. Biomarker responses and contamination levels in crabs (*Carcinus aestuarii*) from the Lagoon of Venice: An integrated approach in biomonitoring estuarine environments. *Water Res.* 44:1725–36.
Rolland, R.M. 2000. A review of chemically-induced alterations in thyroid and vitamin A status from field studies of wildlife and fish. *J. Wildlife Dis.* 36:615–35.
Romani, R. et al. 2006. Organophosphate-resistant forms of acetylcholinesterases in two scallops—The Antarctic *Adamussium colbecki* and the Mediterranean *Pecten jacobaeus*. *Comp. Biochem. Physiol.* 145B:188–96.
Roméo, M. et al. 2006. Responses of *Hexaplex* (*Murex*) *trunculus* to selected pollutants. *Sci. Total Environ.* 359:135–44.
Rosa, C. et al. 2007. Vitamin A and E tissue distribution with comparisons to organochlorine concentrations in the serum, blubber and liver of the bowhead whale (*Balaena mysticetus*). *Comp. Biochem. Physiol.* 148B:454–62.
Routti, H. et al. 2010. Hormone, vitamin and contaminant status during the moulting/fasting period in ringed seals (*Pusa [Phoca] hispida*) from Svalbard. *Comp. Biochem. Physiol.* 155A:70–6.
Sanchez, W. et al. 2008. Biomarker responses in wild three-spined stickleback (*Gasterosteus aculeatus* L.) as a useful tool for freshwater biomonitoring: A multiparametric approach. *Environ. Int.* 34:490–8.
Sánchez-Hernandez, J.C. 2001. Wildlife exposure to organophosphorus insecticides. *Rev. Environ. Contam. Toxicol.* 172:21–63.
Sandahl, J.F. et al. 2005. Comparative thresholds for acetylcholinesterase inhibition and behavioral impairment in Coho Salmon exposed to chlorpyrifos. *Environ. Toxicol. Chem.* 24:136–45.
Sanders, M.B. et al. 2008. Exposure of sticklebacks (*Gasterosteus aculeatus*) to cadmium sulfide nanoparticles: Biological effects and the importance of experimental design. *Mar. Environ. Res.* 66:161–3.
Scaps, P., and O. Borot. 2000. Acetylcholinesterase activity of the polychaete *Nereis diversicolor*: Effects of temperature and salinity. *Comp. Biochem. Physiol.* 125C:377–83.
Scaps, P. et al. 1996. Biochemical and enzymatic characterization of an acetylcholinesterase from *Nereis diversicolor* (Annelida, Polychaeta): Comparison with the cholinesterases of *Eisenia fetida* (Annelida, Oligochaeta). *Biol. Bull.* 190:396–402.
Scholz, N.L. et al. 2006. Dose-additive inhibition of Chinook salmon acetylcholinesterase activity by mixtures of organophosphate and carbamate insecticides. *Environ. Toxicol. Chem.* 25:1200–7.
Sies, H. 1991. Oxidative stress: Introduction. In: *Oxidative stress, Oxidants and Antioxidants*, ed. H. Sies, I–XV. Oxford: Academic Press.
Simms, W., and P.S. Ross. 2000. Vitamin A physiology and its application as a biomarker of contaminant-related toxicity in marine mammals: A review. *Toxicol. Ind. Health* 16:291–302.
Solé, M., J. Kopecka-Pilarczyk, and J. Blasco. 2009. Pollution biomarkers in two estuarine invertebrates, *Nereis diversicolor* and *Scrobicularia plana*, from a marsh ecosystem in SW Spain. *Environ. Int.* 35:523–31.
Solé, M. et al. 2010. Effects on feeding rate and biomarker responses of marine mussels experimentally exposed to propranolol and acetaminophen. *Anal. Bioanal. Chem.* 396:649–56.
Sparling, D.W., G.M. Fellers, and L.L. McConnell. 2001. Pesticides and amphibian population declines in California, USA. *Environ. Toxicol. Chem.* 20:1591–5.

Stentiford, G.D. et al. 2010. Effect of age on liver pathology and other diseases in flatfish: Implications for assessment of marine ecological health status. *Mar. Ecol. Prog. Ser.* 411:215–30.
Teh, S.J., S.M. Adams, and D.E. Hinton. 1997. Histopathologic biomarkers in feral freshwater fish populations exposed to different types of contaminant stress. *Aquat. Toxicol.* 37:51–70.
Tellis, M.S., D. Alsop, and C.M. Wood. 2012. Effects of copper on the acute cortisol response and associated physiology in rainbox trout. *Comp. Biochem. Physiol.* 155C:281–9.
Tessier, L. et al. 2000. Anomalies on capture nets of *Hydropsyche slossonae* larvae (Trichoptera; Hydropsychidae), a potential indicator of chronic toxicity of malathion (organophosphate insecticide). *Aquat. Toxicol.* 50:125–39.
Tintos, A. et al. 2008. β-Naphtoflavone and benzo(*a*)pyrene treatment affect liver intermediary metabolism and plasma cortisol levels in rainbow trout *Oncorhynchus mykiss*. *Ecotoxicol. Environ. Saf.* 63:180–6.
Tlili, S. et al. 2010. Linking changes at sub-individual and population levels in *Donax trunculus*: Assessment of marine stress. *Chemosphere* 81:692–700.
Triebskorn, R. et al. 1997. Induction of heat shock proteins, changes in liver ultrastructure, and alterations of fish behavior: Are these biomarkers related and are they useful to reflect the state of pollution in the field? *J. Aquat. Ecosyst. Stress Recov.* 6:57–73.
Tsangaris, C. et al. 2010. Multiple biomarkers of pollution effects in caged mussels on the Greek coastline. *Comp. Biochem. Physiol.* 151C:369–78.
Tu, H.T. et al. 2012. Combined effects of deltamethrin, temperature and salinity on oxidative stress biomarkers and acetylcholinesterase activity in the black tiger shrimp (*Penaeus monodon*). *Chemosphere* 86:83–91.
Viarengo, A. et al. 2007. A biomonitoring study assessing the residual biological effects of pollution caused by the HAVEN wreck on marine organisms in the Ligurian Sea (Italy). *Arch. Environ. Contam. Toxicol.* 53:607–16.
Vioque-Fernández, A. et al. 2009. Assessment of Doñana National Park contamination in *Procambarus clarkii*: Integration of conventional biomarkers and proteomic approaches. *Sci. Total Environ.* 407:1784–97.
Weis, J.S. et al. 2001. Effects of contaminants on behavior. Biochemical mechanisms and ecological consequences. *Biosciences* 51:209–17.
Widder, P.D., and J.R. Bidwell. 2006. Cholinesterase activity and behaviour in chlorpyrifos-exposed *Rana sphenocephala* tadpoles. *Environ. Toxicol. Chem.* 25:2446–54.
Wiseman, S., and M.M. Vijayan. 2011. Aroclor 1254 disrupts liver glycogen metabolism and enhances acute stressor-mediated glycogenolysis in rainbow trout. *Comp. Biochem. Physiol.* 154C:254–60.
Wogram, J. et al. 2001. Effects of parathion on acetylcholinesterase, butyrylcholinesterase, and carboxylesterase in three-spined stickleback (*Gasterosteus aculeatus*) following short-term exposure. *Environ. Toxicol. Chem.* 20:1528–31.
Wu, Y. et al. 2010. Effects of silver nanoparticles on the development and histopathology biomarkers of Japanese medaka (*Oryzias latipes*) using the partial-life test. *Aquat. Toxicol.* 100:160–7.
Xing, H.J. et al. 2012. Histopathological changes and antioxidant response in brain and kidney of common carp exposed to atrazine and chlorpyrifos. *Chemosphere* 88:377–83.
Xuereb, B. et al. 2009a. Acetylcholinesterase activity in *Gammarus fossarum* (Crustacea Amphipoda): Linking AChE inhibition and behavioural alteration. *Aquat. Toxicol.* 94:114–22.
Xuereb, B. et al. 2009b. Acetylcholinesterase activity in *Gammarus fossatum* (Crustacea Amphipoda). Intrinsic variability, reference levels, and a reliable tool for field surveys. *Aquat. Toxicol.* 93:225–33.
Yevich, P.P., and C.A. Yevich. 1994. Use of histopathology in biomonitoring marine invertebrates. In: *Biomonitoring of Coastal Waters and Estuaries*, ed. K.J.M. Kramer, 179–204. Boca Raton, FL: CRC Press.
Yuen, B.B.H. et al. 2007. Induction and recovery of morphofunctional changes in the intestine of juvenile carnivorous fish (*Epinephelus coioides*) upon exposure to foodborne benzo[*a*]pyrene. *Aquat. Toxicol.* 82:181–94.
Zinkl, J.G. et al. 1991. The effects of cholinesterase inhibiting insecticides on fish. In: *Cholinesterase-Inhibiting Insecticides—Their Impact on Wildlife and the Environment*, ed. P. Mineau, 233–54. Amsterdam: Elsevier.

5

Linking Lysosomal Biomarkers and Ecotoxicological Effects at Higher Biological Levels

Michael N. Moore, Aldo G. Viarengo, Paul J. Somerfield, and Susanna Sforzini

CONTENTS

5.1 Introduction

Distress signals from environmental sentinel animals such as marine mussels can be used as biomarkers of harmful effects and exposure to stressors including chemical contaminants. Such biomarkers include responses at the molecular, cellular, and physiological levels of biological organization; and some of these can be prognostic for pathological reactions to environmental stress (Depledge et al. 1993; Depledge 1994, 1999; Moore et al. 2004a, b). A major challenge facing ecotoxicologists is to integrate individual biomarker responses into a set of tools and indices capable of detecting and monitoring the degradation in health of a particular type of sentinel organism.

However, although we are currently lacking fully integrated explanatory frameworks for evaluating complex environmental information and predicting harmful biological effects and their subsequent consequences for environmental health, considerable progress has been achieved (Allen and Moore 2004; Moore et al. 2006a). And although it is clearly recognized that stress-induced changes at the population/assemblage/ecosystem/human health levels of biological organization are the ultimate concern, they are generally too complex and far removed from the causative events to be of much use in developing tools for the early detection and prediction of the consequences of environmental stress (Depledge et al. 1993; Moore et al. 2004a, b).

Consequently, it is only at the lower levels of biological organization that we will have the reasonable expectation of developing a basis of mechanistic understanding of how different environmental conditions can modulate organismal function, which in turn will ultimately help in linking causality with predictability of response (Livingstone et al. 2000; Marigómez and Baybay-Villacorta 2003). This is partly attributable to our ability to make certain generalizations about biological organization and function at the molecular and cellular levels, which rapidly disappears as we ascend through the organizational hierarchy. Consequently, distress signals at the molecular, cellular, and physiological levels of organisation should be capable of providing early warning biomarkers indicating reduced performance, some of which may be prognostic for impending pathology and severe damage to health of the animal (Depledge et al. 1993; Depledge 1994; Moore 2002; Galloway et al. 2002, 2004; Moore et al. 2006a, b, 2007).

In consideration of these factors, the focus of this chapter is relating stress-induced changes in processes at the subcellular level, specifically the intracellular lysosomal system, to adverse effects at higher cellular physiological responses and pathological reactions.

5.1.1 Can Lysosomal Function Be Used to Assess Health of Aquatic Environment?

Responses of the lysosomal–vacuolar system may provide a solution to the question of prognostic biomarkers, since injurious lysosomal reactions frequently precede cell and tissue pathology. Lysosomal perturbations have been widely used as early indicators of adverse effect to various factors, including pollutant exposure (Moore 2002; Galloway et al. 2004; Moore et al. 2004a, b). Consequently, lysosomal function can be used across a range of animals, including annelids, mollusks, crustaceans, and fish, to detect responses to environmental stress (Köhler et al. 1992, 2002; Lowe et al. 1992, 1995a, b; Svendsen and Weeks 1995; Wedderburn et al. 1998; Cajaraville et al. 2000; Lekube et al. 2000; Hwang et al. 2002; Galloway et al. 2004; Hankard et al. 2004).

Lysosomes are highly conserved multifunctional cellular organelles present in almost all cells of eukaryotic organisms from yeast to humans (Figure 5.1). Their function in the cellular economy includes the degradation of redundant or damaged organelles (e.g., mitochondria and endoplasmic reticulum) and longer-lived proteins as part of autophagic cellular turnover (Klionsky and Emr 2000). Lysosomes are also involved in the digestion of materials ingested by endocytosis and phagocytosis (i.e., intracellular digestion).

Lysosomes are involved in normal physiological responses as well as many cell injury and disease processes; these include augmented sequestration and autophagy of organelles and proteins (Figure 5.1; Moore 1990, 2002; Klionsky and Emr 2000; Cuervo 2004). Stress-induced macroautophagy, such as that triggered by nutrient deprivation, is regulated by the mTOR (mammalian target of rapamycin) kinase in eukaryotic cells from yeast to mammals (Klionsky and Emr 2000). Such reactions have been widely documented for many adaptive and developmental physiological and disease processes, and lysosomal

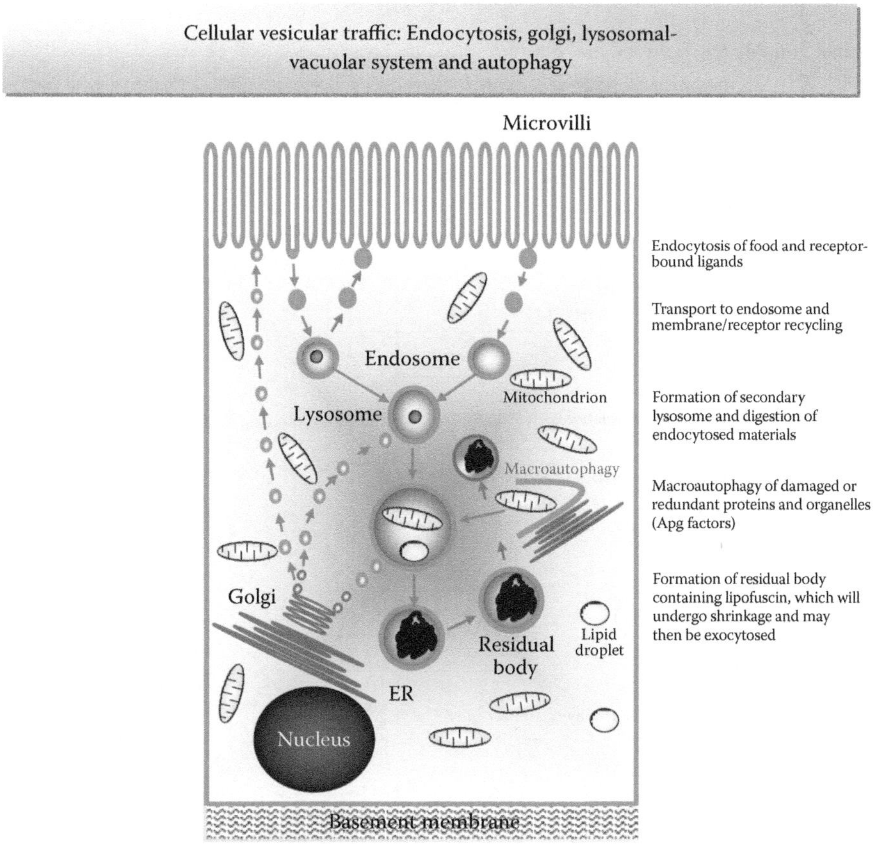

FIGURE 5.1
Diagrammatic representation of a columnar epithelial cell (e.g., molluscan digestive gland cell) showing role of lysosomal system in digestion of endocytosed food and autophagic breakdown of cellular constituents.

responses have been shown to be involved in generalized reactions to environmental stress (Moore 1985, 1990, 2002; Cajaraville et al. 1995a, b; Köhler et al. 2002).

The functional stability of the lysosomal membrane is a good indicator of lysosomal integrity and has been used widely to measure responses to environmental perturbation in fish and mollusks (Hwang et al. 2002; Köhler et al. 2002; Moore 2002; Moore and Allen 2002; Moore et al. 2004a, b). Increasing use of lysosomal stability methods has demonstrated correlations in blue mussels with oxyradical scavenging capacity, potential for protein synthesis (translational efficiency of ribosomes); genotoxicity (micronucleus formation—inverse correlation), physiological scope for growth, tumorigenesis in flounder liver (inverse correlation), and larval viability in oysters (Figure 5.2; Regoli 2000; Köhler et al. 2002; Allen and Moore 2004; Kalpaxis et al. 2004; Ringwood et al. 2004).

Lysosomes are also remarkable for the vast and diverse array of chemicals and pharmaceuticals that they can sequester and accumulate (De Duve et al. 1974; Moore 1990, 2002; Rashid et al. 1991; Moore et al. 2004a, b). These range from metal ions such as iron, copper and mercury, transuranics, asbestos, polycyclic aromatic hydrocarbons (PAHs), heterocyclics, antipsychotic drugs, to nanoparticles, to name but a few (De Duve et al. 1974;

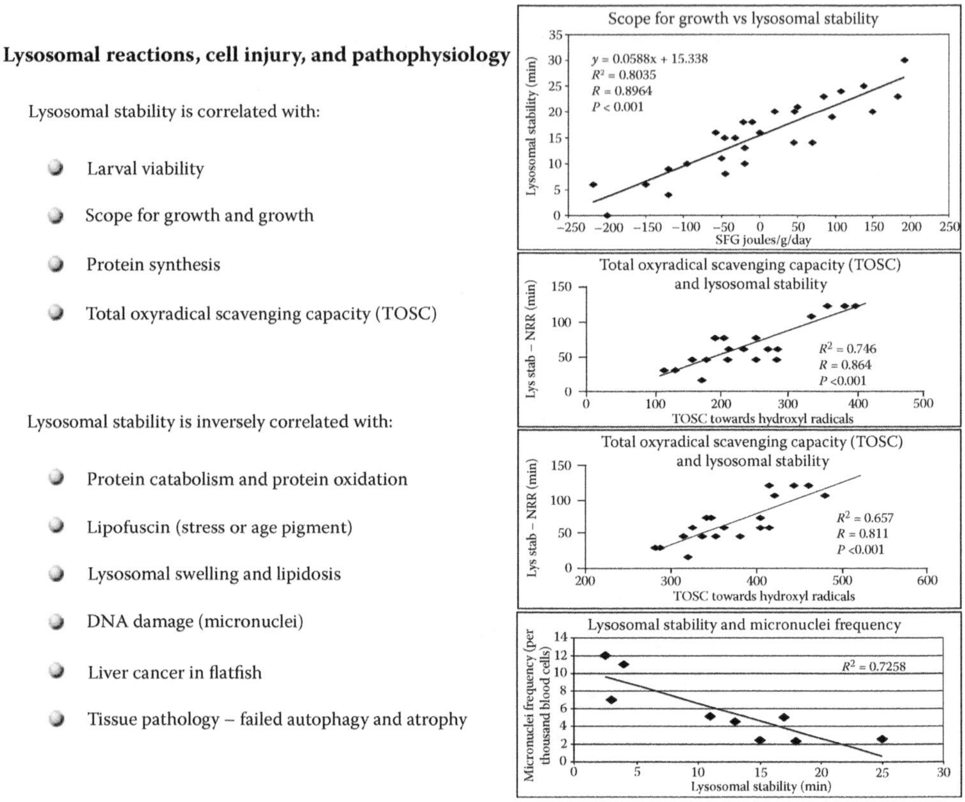

FIGURE 5.2
Correlations between lysosomal stability and other cellular indicators or biomarkers. Small boxed graphs are lysosomal stability versus scope for growth, total oxyradical scavenging capacity (TOSC), and genotoxicity (micronucleus formation).

Moore 1985; Moore et al. 1985, 1997; Nott and Moore 1987; Rashid et al. 1991; Panyam and Labhasetwar 2003; Gould 2004; Howard 2004).

The theme of lysosomal responses and pathological reactions as health-related indicators or biomarkers will be developed more fully in the following sections, as will the wider potential ecotoxicological significance of such changes induced by environmental stressors.

5.2 Background and Rationale

5.2.1 Background on Lysosomal and Autophagic Processes in Cellular Physiology and Disease

Pathological reactions of lysosomes to environmental stressors, including toxic chemical contaminants, frequently involve destabilizing changes in the lysosomal membrane and the induction of autophagy (Moore and Halton 1973, 1977; Moore 1976, 1980, 1985, 1990,

2002, 2004; Moore et al. 1978, 1979, 1984, 1985, 1986, 1987, 1996, 2006a, b, c; Moore and Clarke 1982; Axiak et al. 1988; Cajaraville et al. 1995a, b; Da Ros et al. 2000; Marigómez and Baybay-Villacorta 2003; Marigómez et al. 2005a, b). Autophagy, which is the degradation of cellular components in lysosomes, is implicated in many disease processes, cell death, and adaptive responses (Cuervo 2004). Regulation of this highly conserved group of cellular processes appears to be very similar in eukaryotic organisms ranging from yeasts to man (Klionsky and Emr 2000; Cuervo 2004; Levine 2005).

Other related lysosomal perturbations can also occur such as lysosomal swelling, accumulation of lipid (lipidosis), and age pigment or lipofuscin (lipofuscinosis)—all of these changes have been described in molluscan digestive gland cells (Moore 1976, 1988; Moore et al. 1978, 1979, 2006a; Lowe et al. 1981; Nott and Moore 1987; Krishnakumar et al. 1990, 1994; Regoli 1992; Cajaraville et al. 1995a, b, 2000; Etxeberria et al. 1995; Marigómez at al. 1996, 2005a, b; Domouhtsidou and Dimitriadis 2001; Marigómez and Baybay-Villacorta 2003; Kalpaxis et al. 2004).

There is a substantial body of literature documenting the harmful effects of copper and PAHs on both blue (*Mytilus* spp.) and green (*Perna* spp.) mussels; much of this involves responses of the lysosomal–vacuolar system in the cells of the digestive gland (Viarengo 1989; Krishnakumar et al. 1990, 1994; Moore 1990; Viarengo and Nott 1993). Autophagy of cellular components is associated with the lysosomal reactions and includes autophagic sequestration of copper-induced metallothionein (Viarengo and Nott 1993; Viarengo et al. 1985), microautophagy induced by phenanthrene, and phospholipidosis induced by anthracene and phenanthrene (Lüllmann-Rauch 1979; Nott and Moore 1987).

In eukaryotic cells, the first tier of defense against oxidative damage is provided by xenobiotic transporters, biotransformation enzymes, and antioxidant protection enzymes, such as superoxide dismutase and catalase (Livingstone 2001; Moore 2004; Chapter 3). Lysosomal autophagy provides a second line of defense, by removing oxidatively damaged proteins and impaired organelles and even portions of the nucleus and DNA (Bergamini et al. 2003; Cuervo 2004). Consequently, when the first line of defensive systems is overcome, autophagy protects the cell against the harmful effects of damaged and malfunctioning proteins, which can form aggregates that will accumulate irreversibly in cells (Grune et al. 2004; Kiffen et al. 2004; Moore 2004; Moore et al. 2006a, b, c). Autophagy may also serve as a third tier of defense: when autophagic capabilities are compromised, the autophagic system can trigger programmed cell death (Lockshin and Zakeri 2004) to remove irreversibly damaged cells and thereby maintain organ integrity.

5.2.2 Development and Use of Lysosomal Biomarkers in Assessing Effects of Environmental Stressors

Understandably, the concerns of environmental managers and regulators are largely focused on the ecosystem level and not individual animals (Xu et al. 2002; Rice 2003). Unfortunately, the necessary epidemiological (actually *epizootiological*) data for pollutant impact on sentinel animals that should permit a more comprehensive understanding of possible causal links between animal and ecosystem health are often limited or fragmentary, both spatially and temporally (Xu et al. 2002; Rice 2003). Consequently, alternative interdisciplinary approaches will be required to identify such links, and we are proposing that the health status of representative sentinel species (e.g., blue and green mussels, oysters, clams, periwinkles, crabs, flatfish) be used in evaluating health of the environment (Depledge et al. 1993; Depledge 1994, 1999; Allen and Moore 2004).

Lysosomal biomarkers including lysosomal membrane stability (lysosomal enzyme latency in frozen tissue sections and lysosomal dye retention in living cells), lysosomal swelling, lipid accumulation, and lipofuscin have already been used in many studies for assessing the health of marine environmental sentinel animals (Moore et al. 2006a; Moore 2008; Shaw et al. 2011). These methods have been described in detail in methodological manuals produced for International Council for the Exploration of the Sea (ICES) and the Mediterranean Pollution Programme (UNEP–MEDPOL) and elsewhere (Moore et al. 2004a, b, 2008). Lysosomal membrane stability is one of the biomarkers that were intercalibrated in the framework of the ICES (BELQUAM; www.ices.dk/committe/acom/comwork/report/2010/Special%20Requests/EC%20Northern%20hake%20FMSY.pdf) and of the UNEP-MAP activities (Viarengo et al. 2000). The aim of this Quality Assurance procedure was to obtain high-quality data from large biomonitoring studies that, as in the case of the Mediterranean biomonitoring program, involves the activity of 17 research institutes in the different countries.

In order to objectively interpret lysosomal and other biomarker data, an expert system has been developed (Dagnino et al. 2007). The expert system is able to integrate biomarker results into a five-level health status index. The expert system is based on a set of rules derived from data available on responses to natural and contaminant-induced stress of marine mussels. Parameters are integrated on the basis of their level of biological organization, biological significance, mutual interrelationship, and qualitative trends in a stress gradient. The expert system can therefore rank the biological effects of different levels of pollution. The system represents a simple tool for the risk assessment of the harmful impact of contaminants by providing a clear indication of the degree of stress syndrome induced by pollutants in mussels (Dagnino et al. 2007; Shaw et al. 2011).

5.2.3 Aim and Objectives

In this chapter, our aim is to explore the utility of lysosomal responses as prognostic biomarkers for putative pathophysiology that will permit prediction of animal health status, and the development of a realistic integrated conceptual model for lysosomal reactions in cell injury and pathology. The following questions will be addressed in respect of achieving this goal:

1. What is the generic significance of the strong evolutionary conservation of lysosomal processes?
2. How can the sensitivity of lysosomal processes be harnessed for evaluation of environmental impacts?
3. What properties of lysosomes determine their ability to sequester a very broad spectrum of chemical pollutants?
4. What properties of lysosomes enable their functional perturbation to be mechanistically linked to key processes in cellular physiology?
5. How can lysosomal biomarkers be related to health status?
6. Is there sufficient evidence for a correlation and perhaps functional relationships with ecological parameters?

These questions are strongly interlinked and consequently will be largely considered from an essentially systems perspective.

5.3 Lysosomal Responses and Pathological Reactions to Environmental Stressors

Pathological reactions involving the lysosomal system are often linked to augmented autophagic sequestration of cellular components; and numerous studies have shown that many environmental stressors including pollution are capable of inducing cellular autophagy in the cells of social amoebae (slime molds, myxomycetes), aquatic mollusks, and fish (Köhler et al. 2002; Moore et al. 2004a, 2006a, 2007; Dondero et al. 2006; Einsporn and Koehler 2008a, b). Protistans, earthworms, aquatic mollusks, and flatfish are now widely used in monitoring the health of the environment, while fish and mollusks also provide about 20% of global animal protein consumed by mankind, hence raising issues of food safety and security of supply (Deutsch et al. 2007).

Autophagy is often considered to be primarily a survival strategy in multicellular organisms, which is initiated by stressors (e.g., restricted nutrients, hyperthermia, hypoxia, and salinity increase; Klionsky and Emr 2000; Cuervo 2004; Moore et al. 2006a, b, c). However, recent evidence indicates that autophagy is much more than just a survival process and is, in fact, intimately involved in cell physiology (Moore et al. 1980, 2006a; Moore 1988, 2004; Cuervo 2004; Lockshin and Zakeri 2004). Augmented autophagy is apparently controlled by switching off the mTOR kinase: mTOR signaling is involved in many aspects of cell growth regulation and has been implicated in some cancers (Asnaghi et al. 2004; Proud 2002; Levine 2005). mTOR kinase is also coupled with a nutrient sensing pathway, and is switched off by lack of nutrients (see review by Proud 2002). This kinase is evolutionarily conserved in eukaryotes and has been variously described in yeast, nematodes, mollusks, insects, crustaceans, and mammals (Klionsky and Emr 2000; Beaumont et al. 2001; Cammalleri et al. 2003; Levine 2005; Klionsky et al. 2007). The mTOR signaling system is classically switched off by nutrient deprivation, with resultant up-regulation of autophagy in mammals, which has been described in mussels and marine and terrestrial snails as well (Moore and Halton 1973, 1977; Bayne et al. 1978, 1979; Moore et al. 1978, 1979, 1985, 1986; Proud 2002; Bergamini et al. 2003; Cuervo 2004). Hypoxia also switches off the mTOR protein and has previously been demonstrated to induce autophagy in eukaryotes including mussels (Moore et al. 1979; Proud 2002; Wouters et al. 2005; Hipkiss 2006), as have both increases and decreases in salinity (Moore and Halton 1973, 1977; Moore et al. 1980, 1985; Pipe and Moore 1985).

Lysosomal and autophagic responses and reactions to environmental stressors have been widely documented in the scientific literature. As well as being used in a variety of marine sentinel species, they have also been used in amoebic slime molds and earthworms for assessing the impacts of soil contamination.

Tests involving lysosomal biomarkers in marine bivalves have been extensively used in this context, including blue and green mussels and oysters (Moore et al. 2004, 2008). The flatfish, flounder, and dab, have also been used, particularly in the North Sea (Köhler et al. 1992; Moore et al. 2004a, b, 2008).

5.4 Linking Lysosomal Biomarkers with Other Ecotoxicological Effects

5.4.1 Physiological Relevance of Lysosomal and Autophagic Responses

Physiological responses and pathological reactions to environmental stressors in molluscan digestive gland cells (hepatocyte analogues), phagocytic hemocytes (blood cells), and fish hepatocytes frequently involve destabilizing changes in the lysosomal membrane and the induction of autophagy (Moore and Halton 1973, 1977; Moore 1976, 1980, 1985, 1990, 2002, 2004; Moore et al. 1978, 1979, 1984, 1985, 1986, 1987, 1996, 2006a, b, c, 2007; Moore and Clarke 1982; Axiak et al. 1988; Köhler et al. 1992, 2002; Cajaraville et al. 1995a, b; Da Ros et al. 2000; Marigómez and Baybay-Villacorta 2003; Marigómez et al. 2005a, b).

Damaged cellular constituents and redundant products are removed by lysosomal autophagy, but this is also critically involved in the continuous basal turnover of intracellular components (Hawkins and Day 1996; Ryazanov and Nefsky 2002; Tavernarakis and Driscoll 2002; Cuervo 2004; Levine 2005). Autophagy is up-regulated in times of stress or physiological change by breaking down longer-lived proteins and organelles, and recycling the products into protein synthesis and energy-production pathways: the process allows cells to be temporarily self-sustaining during periods when nutrients are restricted (Bergamini et al. 2003; Cuervo 2004; Levine 2005). New evidence also indicates that autophagy may have a protective role in the context of oxidative stress (Bergamini et al. 2003; Cuervo 2004; Moore et al. 2006b, c, 2007; Moore 2008). Nutrient deprivation–induced autophagy in mollusks appears to confer some resistance to the toxicity of both reactive oxygen species (ROS)–generating PAHs and copper (Viarengo et al. 1987; Moore 2004; Moore et al. 2006a, b, c). Normal tidal fluctuations in salinity, food and oxygen, however, do not induce a stress syndrome (Bayne et al. 1978, 1979; Moore et al. 1979, 1982, 1987; Moore 1980; Widdows et al. 1981, 1982).

5.4.2 Accumulation of Pollutant Chemicals in Lysosomes and Oxidative Stress

As mentioned previously, lysosomes are noted for their ability to accumulate a very wide range of noxious substances including many metals and organic xenobiotics (Figure 5.3). There is also a substantial body of literature documenting the harmful effects of copper and PAHs on both blue and green mussels, with much of this involving responses of the lysosomal–vacuolar system in the cells of the digestive gland (Viarengo 1989; Krishnakumar et al. 1990, 1994; Moore 1990; Viarengo and Nott 1993). The processes involved in the lysosomal sequestration of metal ions have been described by Viarengo and Nott (1993).

Autophagy of cellular components is associated with pollutant-induced lysosomal reactions and includes autophagic sequestration of copper-induced metallothionein (Viarengo et al. 1985; Viarengo and Nott 1993), microautophagy induced by phenanthrene, and phospholipidosis induced by anthracene and phenanthrene (Lüllmann-Rauch 1979; Nott and Moore 1987). The internal environment of secondary lysosomes is acidic, and this contributes to the ionic trapping of PAHs and heterocyclic compounds including neutral red and acridine orange (Figure 5.3; Allison and Young 1969; Lüllmann-Rauch 1979; Rashid et al. 1991; Lowe et al. 1992).

Intracellular lysosomes are also a site for generation of ROS, including oxyradicals, as demonstrated by Winston et al. (1991) in isolated digestive cells of mussels (Brunk and Terman 2002). Within the lysosomes of normal unstressed digestive gland cells, ROS are probably generated by transition metal ions, such as iron and copper, which accumulate in lysosomes from exogenous sources, such as algal and microbial food, and also by autophagic degradation of endogenous metalloproteins (Brunk and Terman 2002; Moore et al. 2006a, 2007). There is also increasing evidence that endocytosed nanoparticles may contribute to the generation of ROS (Howard 2004; Klaine et al. 2008; Bottero et al. 2011; Kroll et al. 2011). Although digestive cell lysosomes spontaneously generate oxyradicals such as superoxide, they also contain a superoxide dismutase, which may protect the lysosomal membrane from excessive oxidative damage (Winston et al. 1991; Livingstone et al. 1992).

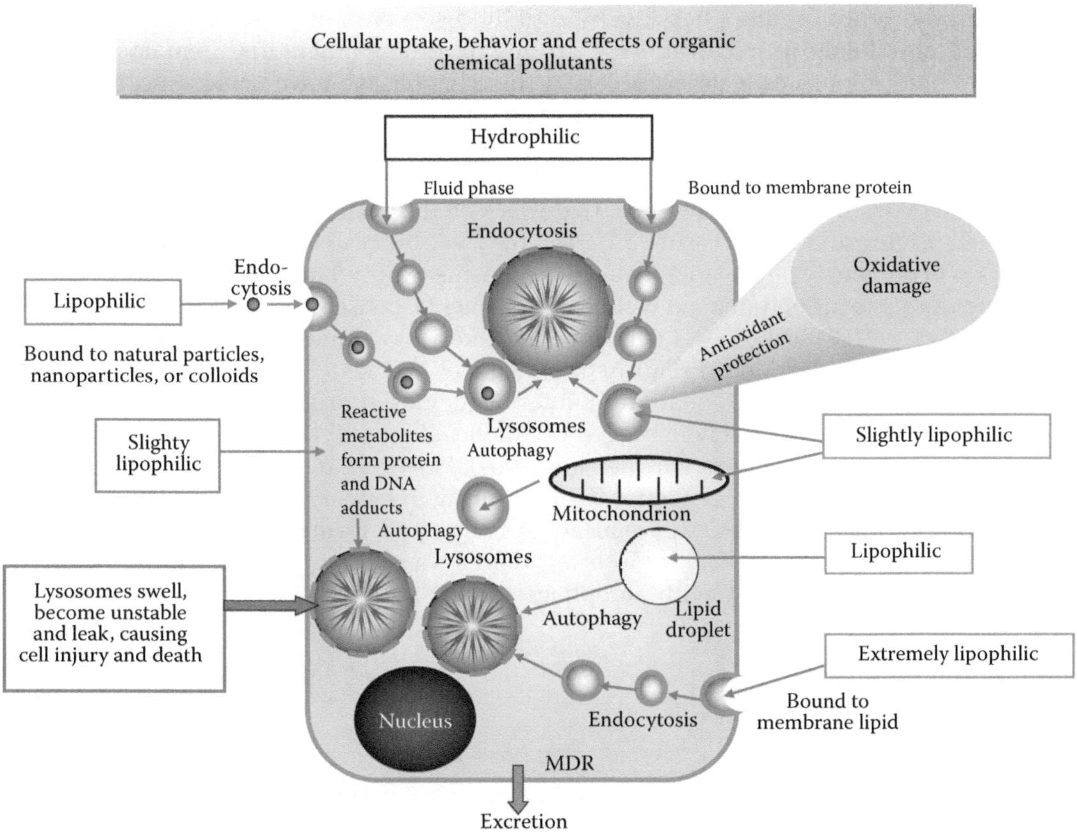

FIGURE 5.3
Diagrammatic representation of the various routes of uptake (both diffusion and endocytotic) of contaminant xenobiotic chemicals into the cell that can result in their accumulation in the lysosomal compartment, based on their physical chemical characteristics. (Modified from Moore, M.N. et al., *Mutat. Res.*, 552, 247–268, 2004a; Moore, M.N. et al., *ICES Techniques in Marine Environmental Sciences* (ICES, Copenhagen), 36, 1–31, 2004b; Rashid, F. et al., *Histochem. J.*, 23, 450–459, 1991.)

Exposure of mussels to copper and some PAHs also results in increased ROS, as also does reimmersion in seawater following a period of anoxia as described above (Regoli 2000; Livingstone 2001; Moore et al. 2007). Reimmersion following anoxia is probably analogous to reperfusion injury (Robin et al. 2007). Copper exposure also increases the concentration of protein carbonyls and lipofuscin (Kirchin et al. 1992; Moore et al. 2007). Lushchak (2011) provides a recent account of the many other pollutants that can enhance ROS levels in aquatic animals, including a variety of trace metals, pesticides, and oil products.

5.4.3 Environmental Stress and Failed Autophagy Leading to Pathology and Disease

Autophagy is implicated in many disease processes, cell death, and adaptive responses (Cuervo 2004; Levine 2005). This group of at least three related processes (i.e., macro- and microautophagy and chaperonin-mediated autophagy) is essential for maintenance of cellular homeostasis (Cuervo 2004), and morphological evidence has long indicated that lysosomal autophagy is a highly conserved mechanism evolutionarily (Cuervo 2004).

Although the genes for autophagy were initially discovered in yeast, analogues of those genes are now being found in diverse groups of animals including nematode worms, fruit flies, and mammals (Cuervo 2004). In this chapter, we have considered the role of lysosomal autophagy as a protective system against the consequences of oxidative stress, and will subsequently discuss its significance in the development of tolerance in aquatic ecosystems.

Pollutant exposure often causes oxidative attack on the protein machinery and organelles of the cell (Livingstone 2001; Figure 5.1). An effective ability to up-regulate the autophagic process will obviously be advantageous to organisms exposed to pollutants, since this will facilitate the removal of damaged cellular constituents, conserve cell function (Figure 5.4; Kirchin et al. 1992; Cuervo 2004; Lockshin and Zakeri 2004), and reduce the amount of age pigment (lipofuscin) produced (Figure 5.4; Grune et al. 2004). Lipofuscin accumulates in lysosomes as a result of peroxidation of autophagocytosed proteins associated with protein aggregates and oxidatively damaged organelles (Figure 5.4), and was previously considered to be just cellular junk (Grune et al. 2004). However, new evidence indicates that lipofuscin binds iron, which generates ROS, resulting in exacerbation of oxidative damage (Brunk and Terman 2002; Dailianis et al. 2003). Brunk and Terman (2002) also hypothesize that lipofuscin binds lysosomal hydrolases, thereby inhibiting protein degradation. The outcome is failed autophagy with autophagic accumulation of essentially undegradable

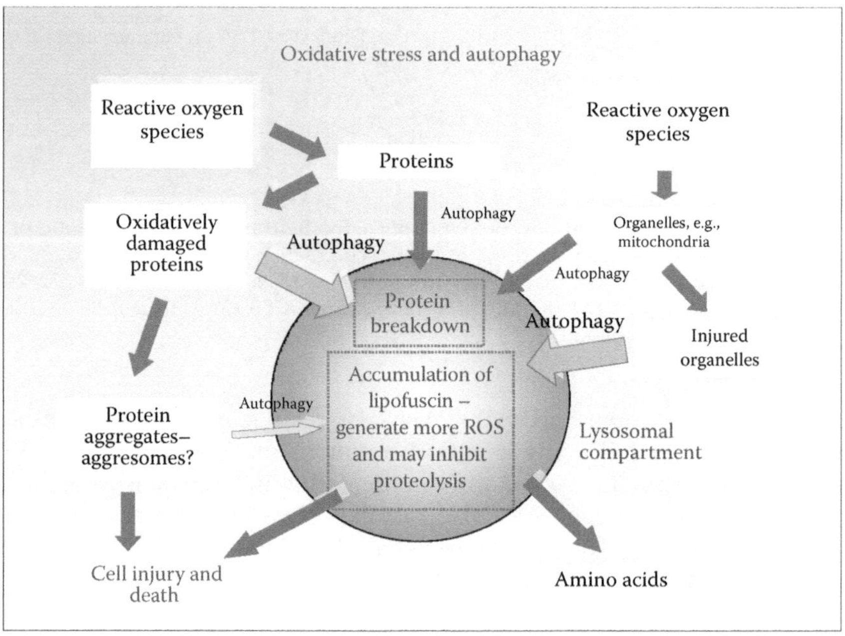

FIGURE 5.4
Conceptual mechanistic model for effects of autophagy and ROS on proteins and cellular organelles. The model shows normal autophagic turnover of proteins and organelles (arrows) with superimposed augmented autophagy (large arrows), which is postulated as having a protective role against oxidative stress. If augmented autophagy of damaged organelles and proteins is impaired, as occurs in cells with lysosomal damage and autophagic degradative dysfunction, then harmful products can accumulate contributing to cell injury (see Moore et al. 2006a, 2006b, 2007). (Modified from Moore, M.N. et al., *Aquat. Toxicol.*, 84:80–91, 2007.)

damaged organelles, proteins, phospholipids, and lipids contributing to further lipofuscin formation (Moore 1988; Krishnakumar et al. 1994; Brunk and Terman 2002; Moore et al. 2006a, b, c).

Failed autophagy has been observed in mollusks, fish, and mammals exposed to pollutants (Moore 1988; Allen and Moore 2004; Moore et al. 2004a, 2006a, b, c). Reports on other animals tend to be largely descriptive, so systematic observations of other phyletic groups would be beneficial to the development of a generic understanding of the role of autophagy in ameliorating the harmful effects of pollutants.

5.4.4 Linking Lysosomal and Autophagic Responses to Higher Level Effects

Since autophagy is probably one of the most highly evolutionarily conserved molecular processes in the eukaryotes (Klionsky and Emr 2000; Cuervo 2004), it is hypothesized that up-regulation of this system may play a significant role in adaptation to environments where oxidative stress is a major selection pressure, and where the facility to rapidly delete and recycle oxidatively damaged proteins and organelles will provide an evolutionary advantage (Figure 5.4).

Many mollusks have such a capability for augmented autophagy, including mussels and marine and terrestrial snails (Moore and Halton 1973; Bayne et al. 1978; Lowe et al. 1981; Moore et al. 1985, 2006a; Pipe and Moore 1986a, b; Moore 2004; Marigómez et al. 2005b). Augmented autophagy induced by short-term starvation has a protective effect against copper and phenanthrene-mediated oxidative stress in mussels (Moore 2004; Shaw et al. 2011). Further support comes from a study of blue mussels *Mytilus edulis* from a highly contaminated site in Norway that showed evidence of tolerance, in comparison with less contaminated mussels (Figure 5.5; Moore 1988). Samples from four sites along a pollution gradient showed that lysosomal stability in mussels from the most contaminated site was more similar to that in mussels from the cleanest site, than their counterparts from intermediate sites. Although these heavily contaminated mussels had relatively high lysosomal stability, they showed clear evidence of autophagy, placing them in Tier 2, whereas mussels from intermediate sites with low lysosomal stability were clearly in Tier 3 (Figure 5.6; Moore 1988). The inference here is that up-regulated autophagy may have been selected in this heavily contaminated population (Moore 1988). Lysosomal stability was also strongly correlated with macrobenthic diversity (H') in spatially adjacent sites (Figure 5.5; $R = 0.946$, $P < 0.01$), and although these are just correlations, they may be a reflection of the universality of autophagic reactions to stress.

In addition to studies of single populations and species, many studies have examined changes in the structure of communities along marine pollution gradients (Gray et al. 1988). Indeed, this is a standard component of most pollution monitoring schemes. One common finding is that responses are often manifest at taxonomic levels higher than species, commonly as high as phyla (Olsgard et al. 1998). Organisms within different phyla are considered to have different sensitivities to stress (Warwick and Clarke 1993), raising the interesting possibility that this reflects long-established evolutionary differences in the physiology of organisms within different phyla. Although autophagic mechanisms are highly conserved among eukaryotes, we hypothesize that ability to up-regulate this process varies among phyla, and this underlies differences in the ability of different animals to colonize parts of the coastal and estuarine environment and also contributes to their differential sensitivities to anthopogenic stress. We propose that representative organisms from a range of phyla making up functional ecological assemblages be assessed to

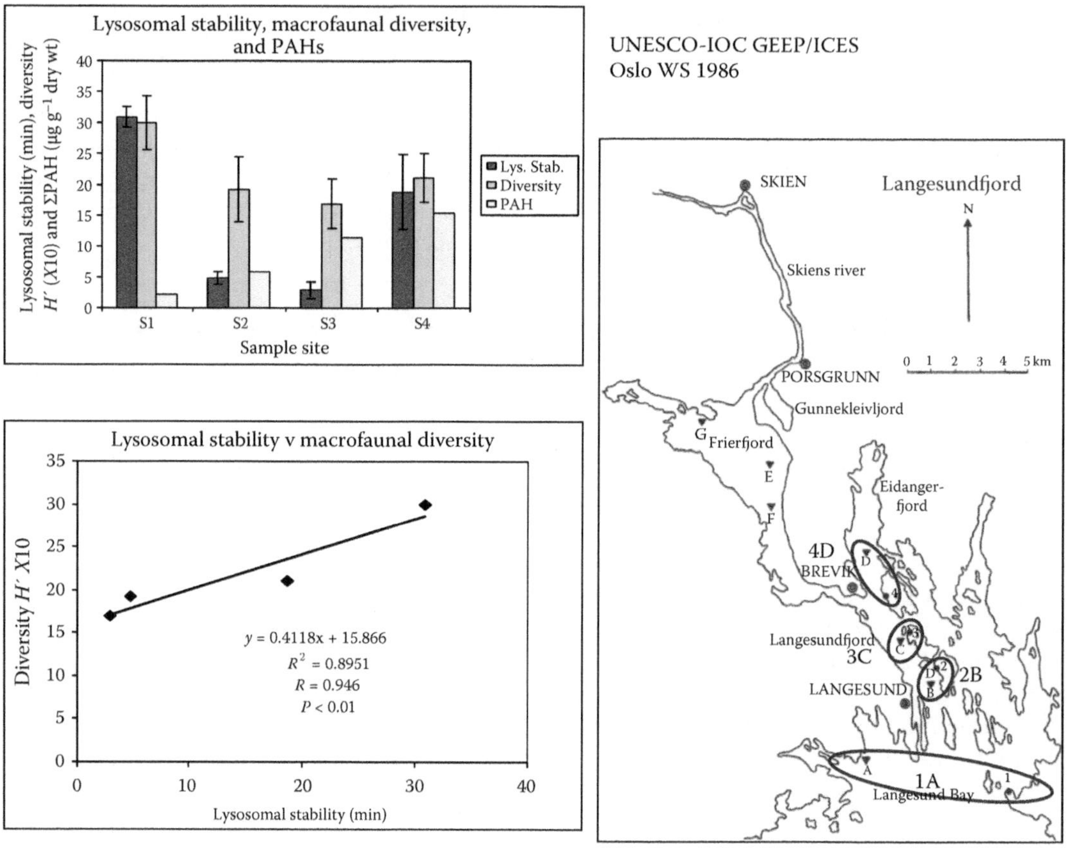

FIGURE 5.5
Macrofaunal diversity from four sites (A–D) and lysosomal stability (mean ± 95% CL, $n = 10$) in mussel *Mytilus edulis* digestive gland cells from four sites (1–4), sampled along a pollutant gradient in Norway (indicated by ΣPAHs in mussels), shows that mussels from the most contaminated site 4 are more similar to the cleanest mussels at site 1 than are mussels from sites 2 and 3. Macrofaunal diversity H' (mean ± 95% CL, $n = 4$) was measured at adjacent sites (A–D). (C) Lysosomal stability is directly correlated with macrofaunal diversity. Lysosomal stability is used here as an integrated indicator of cellular well-being, that can be used as a generic yardstick for interpreting the responses of many biomarkers (Allen and Moore 2004; Moore et al. 2004a, b; 2006a). (Data are derived from Gray, J.S. et al., *Mar. Ecol. Prog. Ser.*, 46, 151–165, 1988; and Moore, M.N., *Mar. Ecol. Prog. Ser.*, 46, 81–89, 1988. With permission.)

determine the extent to which their relative capabilities for up-regulation of autophagy and the degradative (retrograde) component of protein turnover contribute to their ability to survive natural and anthopogenic stress.

Consequently, it is probably a reasonable assumption that lysosomal autophagy does in fact appear to provide a second line of defense against oxidative stress, and the capability to effectively up-regulate this process is probably a significant factor contributing to the ability of some organisms to tolerate stressful and polluted environments. However, further studies are required to substantiate this claim.

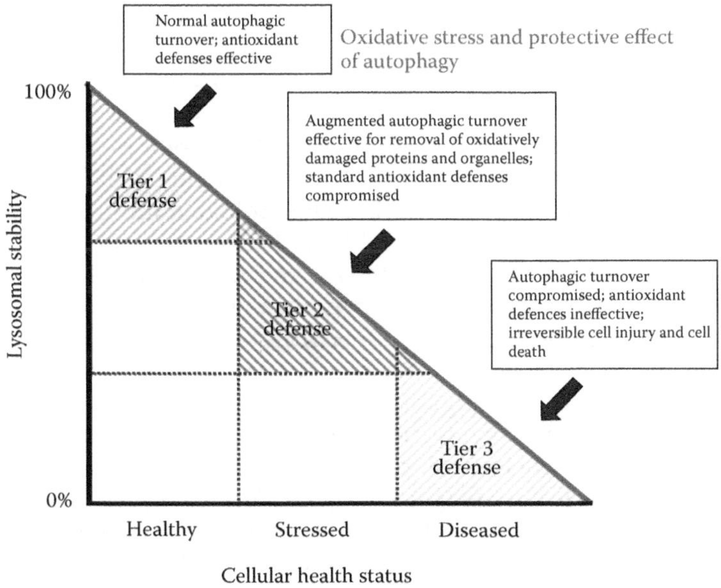

FIGURE 5.6
Relationship between lysosomal stability and cellular health status, showing how autophagic degradation acts as a second tier of defense against oxidative stress, and possibly as a third tier when autophagic capabilities are compromised and the autophagic system may trigger cell death (i.e., apoptotic—type I, autophagic—type II, or necrotic—type III) for removal of irreversibly damaged cells. Lysosomal stability is used here as an integrated indicator of cellular well-being, that can be used as a generic yard-stick for interpreting the responses of many biomarkers. (From Allen, J.I., Moore M.N., *Mar. Environ. Res.*, 58, 227–232, 2004; Moore, M.N. et al. *Mutat. Res.*, 552, 247–268, 2004a; Moore, M.N. et al., *ICES Techniques in Marine Environmental Sciences* (ICES, Copenhagen), 36, 1–31, 2004b; Moore, M.N. et al., *Mar. Environ. Res.*, 61, 278–304, 2006a. With permission.)

5.5 Can Lysosomal Biomarkers Be Developed for New Applications?

5.5.1 Understanding the Broader Significance of Autophagy in Evolution of Tolerance to Fluctuating Environments and Pollutant Stress

Given that autophagic turnover is strongly evolutionarily conserved in the eukaryotes, it is conjectured that up-regulation of autophagy may play a significant role in adaptation to environments where induced oxidative stress is a major physiological problem, and where the facility to rapidly sequester and recycle oxidatively damaged proteins and organelles will provide a selective advantage (Cuervo 2004; Moore et al. 2006c, 2007). Estuarine and coastal environments, where fluctuating salinity, nutrient, temperature, and oxygen regimes are all important selective drivers and also inducers of autophagy, are likely to be areas where it would be an adaptive advantage to be able to minimize the effects of oxidative stress (Moore et al. 2006c, 2007). Furthermore, because near-shore and estuarine environments are those most likely to be subjected to adverse human impacts, we have a scenario where effective antioxidant protection may provide an evolutionary advantage (Moore et al. 2006c). Consequently, we would expect to observe functional ecological groupings of organisms in coastal and estuarine environments that have generic and highly effective first-tier standard protective enzymes and antioxidants, as well as second-tier autophagic antioxidant defensive systems for dealing with oxidative stress (Figure 5.6; Moore et al. 2006c; Moore 2008). The corollary of this is that an ability for rapid up-regulation of autophagy is a probable necessity for acquisition of tolerance in high stress environmental scenarios (Moore 2008). Support for this hypothesis comes from the previously described study of blue mussels from a heavily polluted site in Norway that show evidence of tolerance, in comparison with less contaminated mussels (Moore et al. 2006c).

Changes in lysosomes have been used as biomarkers of aging in a wide range of organisms including nematodes, fruit flies, mollusks, and mammals (Hole et al. 1992; Cuervo and Dice 2000; Lockshin and Zakeri 2004). In general, there is a trend for decreasing proteolytic capability with increased age that has been linked with a gradual decline in the efficiency of the autophagic process (Cuervo and Dice 2000; Ryazonov and Nefsky 2002; Bergamini et al. 2003). However, Bergamini et al. (2003) have proposed that repeated triggering of the autophagic system by nutrient deprivation or caloric restriction will prevent the decline in proteolytic capacity and, hence, contribute to increased life span, probably through the maintenance of more efficient cellular housekeeping. This may parallel the situation with animals such as mussels that live in an environment where autophagy is repeatedly switched on and off as discussed above, thus maintaining an effective capacity for the removal of altered proteins, membranes, and organelles that are damaged by ROS and hypoxia-induced methylglyoxal (Tavernarakis and Driscoll 2002; Lockshin and Zakeri 2004; Moore et al. 2006c, 2007). Further investigation of the significance of autophagy in conferring resistance to stress is required, but the possibility raises provocative questions about the role of ongoing and fluctuating low levels of stress in the evolution of stress tolerance.

Animals such as marine mussels and periwinkles (e.g., *Littorina* species) are robust animals that frequently live in fluctuating environments such as estuaries where they are subjected to variable nutritional, temperature, and salinity regimes, as well as repeated air exposure (which results in hypoxia) and reimmersion in seawater (Moore et al. 2006c, 2007). Consequently, this essentially stressful fluctuating environment will tend to trigger repeated autophagic events, which by effectively removing inappropriately altered proteins and damaged or redundant cellular constituents will result in more efficient cellular housekeeping and help to minimize the formation of harmful lipofuscin and other aggregates (Kirchin et al. 1992; Klionsky and Emr 2000; Ryazanov and Nefsky 2002; Bergamini et al. 2003). This more efficient cellular functionality may underpin the ability of mussels to survive, and often thrive, in environments that are subject to man-made stresses such as chemical pollution (Moore et al. 2006c, 2007).

5.5.2 Extending Lysosomal Biomarkers to Nonmarine Environments

Given the increasing demands for environmentally sustainable food production, it is essential to have effective indicators of the health and growth of the prospective food animals, whether these are farmed (aquaculture) or from the wild, as well as the quality of their environment. Indicators of animal well-being and potential for growth are provided by cellular biomarkers of lysosomal and autophagic processes in both mollusks and fish (Moore 1985; Köhler et al. 2002; Moore et al. 2006a, 2007). Other studies have shown that lysosomal biomarkers can also be used in protistans, coelenterates, annelid worms, and crustaceans (Moore and Stebbing 1976; Svendsen and Weeks 1995; Brown et al. 2004; Svendsen et al. 2004; Dondero et al. 2006; Galloway et al. 2006; Nolde et al. 2006; Sforzini et al. 2008).

Since Giambérini and Cajaraville (2005) showed that the digestive gland lysosomal response in freshwater zebra mussels *Dreissena polymorpha* was related to exposure time and to cadmium concentration, the potential of this biomarker in freshwater biomonitoring has been widely explored. After *in vitro* exposure to the antimicrobial agent triclosan and the antibiotic trimethoprim, the lysosomal stability assay applied to zebra mussel hemocytes did not show any significant differences as compared to baseline levels (Binelli et al. 2009a), whereas *in vivo* exposure to triclosan induced a clear destabilization of lysosomal membranes (Binelli et al. 2009b), with the *in vivo* effect of trimethoprim remaining moderate (Binelli et al. 2009c). The same team also revealed destabilization of the lysosomal membrane in zebra mussels after exposure to the pharmaceuticals paracetamol (Parolini et al. 2010) and ibuprofen (Parolini et al. 2011) and the illicit drug cocaine (Binelli et al. 2012). A multibiomarker approach carried out to investigate the state of contamination of the River Lambro/River Po confluence in Italy revealed a destabilization of zebra mussel lysosomal membranes most likely induced by polychlorobiphenyls and dichlorodiphenyl-trichloroethanes (Binelli et al. 2010). In another freshwater bivalve (*Unio pictorum*) transplanted into a metal polluted river basin, the loss of lysosomal membrane stability clearly discriminated among sites (Guidi et al. 2010). The sensitivity of the lysosomal membrane to the modulation of cell redox equilibrium was well illustrated since a multiple regression analysis showed that more than 80% of the neutral red retention test variance was explained by the sum of antioxidant responses (Guidi et al. 2010).

The fact that lysosome stability represents a key component of cell responses to chemical stress was also shown in freshwater snails *Lymnaea stagnalis* exposed to the herbicides fomesafen or atrazine (Russo et al. 2007, 2009) and land snails *Helix aspersa* (exposed to the fungicide copper oxychloride in the laboratory or in the field; Snyman et al. 2000, 2002) and *Eobania vermiculata* collected from polluted areas in the field or treated with trace metals in the laboratory (Itziou and Dimitriadis 2011).

In vivo and *in vitro* exposures of the neotropical freshwater fish *Prochilodus lineatus* to lead have also indicated that this metal induces a destabilization of the lysosomal membrane (Monteiro et al. 2011).

The social amoeba *Dictyostelium discoideum* has been utilized recently as an experimental model to study the biological effects of adverse environmental conditions. In particular, the cellular changes induced by inorganic as well as organic pollutants on the activity of the lysosomal vacuolar system have been studied (Dondero et al. 2006). The results showed that nanomicromolar concentrations of mercury or benzo[*a*]pyrene are able to destabilize the lysosomal membranes in the amoebae, thus confirming the high sensitivity of the neutral red retention assay (Lowe et al. 1992; Lowe and Pipe 1994; Dondero et al. 2006; Moore et al. 2006a; Sanzi Cortez et al. 2012) used in these experiments.

The relationship between the destabilization of lysosomal membranes and the autophagic rate has been established using d2EGFP recombinant amoebae (Dondero et al. 2006). A d2EGFP strain characterized by a sufficiently rapid turnover has been utilized to verify the possible effects induced by pollutant concentrations able to destabilize lysosomal membranes. In the amoebae, the results confirmed that destabilization of lysosomal membranes is associated with a parallel increase in the catabolic rate of d2EGFP. This clearly indicates that the activation of the lysosomal vacuolar system in amoebae maintained in a standard medium suitable to support cell growth and division, represents the initial event of an increased autophagic activity itself representing an important aspect of the physiological response of the protozoan to environmental stressors (Dondero et al. 2006).

In the field of terrestrial ecotoxicology, earthworms (e.g., *Eisenia* sp., *Lumbricus* sp.) are widely used as model organisms to assess soil quality (Spurgeon et al. 2003; Römbke et al. 2005). In addition to mortality and reproduction, as typical high-level endpoints, considerable attention has been focused on evaluating more sensitive physiological changes induced by bioavailable toxic chemicals (Scott-Fordsmand and Weeks 2000). Lysosomal membrane stability is one of the most commonly used biomarkers on earthworms. This parameter has shown to be highly sensitive and able to detect the effects of minimal amounts of toxic chemicals, in both laboratory and field conditions (Svendsen et al. 1996; Weeks and Svendsen 1996; Rocco et al. 2011; Sforzini et al. 2011a). A reduction of lysosomal membrane stability is often associated with an increase in the lysosome/cytoplasm volume ratio, a parameter indicating an unphysiological level of cell catabolism (Moore and Viarengo 1987; Moore et al. 2008).

Recently, a battery of biomarkers suitable for proving pollutant-induced physiological changes at different levels of functional complexity has been developed on *Eisenia andrei* (Caselli et al. 2006; Gastaldi et al. 2007; Sforzini et al. 2011a, b), one of the most commonly used earthworm species in ecotoxicology (OECD 1984, 2004; van Gestel et al. 1993; Saint-Denis et al. 1998; Robidoux et al. 2000). The set of physiological parameters includes, among others, different lysosomal parameters at cellular level (i.e., lysosomal membrane stability, lipofuscin, and neutral lipid lysosomal content) as well as tissue damage (i.e., lysosome/cytoplasm volume ratio). Following an approach similar to that used for the mussels (Dagnino et al. 2007), an expert system for biomarker data management to evaluate the health status of the earthworms has been developed (Sforzini et al. 2011a). In the expert system procedures, lysosomal membrane stability and lysosome/cytoplasm ratio are of particular relevance since they help to provide a reliable interpretation of the physiological meaning in the changes observed in bell-shaped biomarkers responses/reactions (such as enzymatic activities) that may give false negative results. In particular, only when relevant alterations of both parameters are observed, does the environmental expert system interpret the data of bell-shaped biomarkers as lying in the decreasing part of the response curve. In these conditions, the results indicate an augmented autophagic sequestration of cellular components, a process that may lead to severe damage of cells and of tissue functions.

The successful transfer and application of lysosomal biomarker methodology to protistans and earthworms underline the ubiquity of lysosomal responses and reactions to environmental stressors. It further indicates that these biomarkers will probably have utility in marine and estuarine protistans and annelids.

In the framework of OSPAR (2011), lysosomal stability has been proposed as a global health status indicator in biomonitoring. This recommendation is based on the fact that lysosomal membrane stability (LMS) appears to be largely independent of taxon, so "the absolute values for measurement of lysosomal stability (NRR and cytochemical method) are directly comparable. Furthermore, measurements of this biomarker in animals from climatically and physically diverse terrestrial and aquatic ecosystems also indicate that it is potentially a universal indicator of health status."

5.6 Conclusions

There is overwhelming evidence that the lysosomal–autophagic system is ancient in evolutionary terms and intimately tied in to the overall economy of eukaryotic cells. This system can, therefore, be considered in a relatively generic manner across the phyletic range.

Investigations of responses of the lysosomal–vacuolar system to environmental stressors indicate that the system is highly sensitive and very strongly linked in a mechanistic manner to higher-level physiological processes of protection, protein and cellular turnover, and growth. The lysosomal system is probably a particular target for chemical pollution as a result of its ability to accumulate a broad spectrum of environmental pollutants. The possibility of lysosomal biomarker reactions having an interpretative capacity in terms of ecosystem level effects is still tentative, but the current results are exciting.

References

Allen, J.I., and M.N. Moore. 2004. Environmental prognostics: Is the current use of biomarkers appropriate for environmental risk evaluation. *Mar. Environ. Res.* 58:227–32.

Allison, A.C., and M.R. Young. 1969. Vital staining and fluorescence microscopy of lysosomes. In: *Lysosomes in Biology and Pathology*, ed. J.T. Dingle and H.B. Fell, 600–28. Amsterdam: Elsevier.

Asnaghi, L. et al. 2004. mTOR: A protein kinase switching between life and death. *Pharmacol. Res.* 50:545–9.

Axiak, V., J.J. George, and M.N. Moore. 1988. Petroleum hydrocarbons in the marine bivalve *Venus verrucosa:* Accumulation and cellular responses. *Mar. Biol.* 97:225–30.

Bayne, B.L. et al. 1978. Further studies on the effects of stress in the adult on the eggs of *Mytilus edulis. J. Mar. Biol. Assoc. U.K.* 58:825–41.

Bayne, B.L. et al. 1979. Measurement of the responses of individuals to environmental stress and pollution: Studies with bivalve molluscs. *Philos. Trans. R. Soc. Lond.* 286B:563–81.

Bergamini, E. et al. 2003. The anti-ageing effects of caloric restriction may involve stimulation of macroautophagy and lysosomal degradation, and can be intensified pharmacologically. *Biomed. Pharmacother.* 57:203–8.

Beaumont, V. et al. 2001. Phosphorylation and local presynaptic protein synthesis in calcium- and calcineurin-dependent induction of crayfish long-term facilitation. *Neuron* 32:489–501.

Binelli, A. et al. 2009a. Cytotoxic and genotoxic effects of *in vitro* exposure to triclosan and trimethoprim on zebra mussel (*Dreissena polymorpha*) hemocytes. *Comp. Biochem. Physiol.* 150C: 50–6.

Binelli, A. et al. 2009b. *In vivo* experiments for the evaluation of genotoxic and cytotoxic effects of triclosan in zebra mussel hemocytes. *Aquat. Toxicol.* 91:238–44.

Binelli, A. et al. 2009c. A multi-biomarker assessment of the impact of the antibacterial trimethoprim on the non-target organism zebra mussel (*Dreissena polymorpha*). *Comp. Biochem. Physiol.* 150C:329–36.

Binelli, A. et al. 2010. Multi-biomarker approach to investigate the state of contamination of the R. Lambro/R. Po confluence (Italy) by zebra mussel (*Dreissena polymorpha*). *Chemosphere* 79:518–28.

Binelli, A. et al. 2012. Illicit drugs as new environmental pollutants: Cyto-genotoxic effects of cocaine on the biological model *Dreissena polymorpha*. *Chemosphere* 86:906–11.

Bottero, J.V. et al. 2011. Manufactured metal and metal-oxide nanoparticles: Properties and perturbing mechanisms of their biological activity in ecosystems. *C. R. Geosci.* 343:168–76.

Brown, R.J. et al. 2004. Differential sensitivity of three marine invertebrates to copper assessed using multiple biomarkers. *Aquat. Toxicol.* 66:267–78.

Brunk, U.T., and A. Terman. 2002. Lipofuscin: Mechanisms of age-related accumulation and influence on cell function. *Free Radic. Biol. Med.* 33:611–9.

Cajaraville, M.P. et al. 1995a. Lysosomes as cellular markers of environmental pollution: Time- and dose-dependant responses of the digestive lysosomal system of mussels after petroleum hydrocarbon exposure. *Environ. Toxicol. Water Qual.* 10:1–8.

Cajaraville, M.P. et al. 2000. The use of biomarkers to assess the impact of pollution in coastal environments of the Iberian peninsula: A practical approach. *Sci. Total Environ.* 247:295–311.
Cajaraville, M.P et al. 1995b. Cellular biomarkers as useful tools in the biological monitoring of environmental pollution: Molluscan digestive lysosomes. In: *Cell Biology in Environmental Toxicology*, ed. M.P. Cajaraville, 29–55. Bilbao: University of the Basque Country Press Service.
Cammalleri, M. et al. 2003. Time-restricted role for dendritic activation of the mTOR–p70S6K pathway in the induction of late-phase long-term potentiation in the CA1. *Proc. Natl. Acad. Sci. U. S. A.* 100:14368–73.
Caselli, F. et al. 2006. *In vitro* characterization of cholinesterases in the earthworm *Eisenia andrei*. *Comp. Biochem. Physiol.* 43C:416–21.
Cuervo, A.M. 2004. Autophagy: In sickness and in health. *Trends Cell Biol.* 14:70–7.
Cuervo, A.M., and J.F. Dice. 2000. When lysosomes get old. *Exp. Gerontol* 35:119–131.
Dagnino, A. et al. 2007. Integration of biomarker data into an organism health index: Development of an expert system and its validation with field and laboratory data in mussels. *Biomarkers* 12:155–72.
Dailianis, S. et al. 2003. Evaluation of neutral red retention assay, micronucleus test, acetylcholinesterase activity and a signal transduction molecule (cAMP) in tissues of *Mytilus galloprovincialis* (L.), in pollution monitoring. *Mar. Environ. Res.* 56:443–70.
Da Ros, L., F. Meneghetti, and C. Nasci. 2000. Field application of lysosomal destabilisation indices in the mussel *Mytilus galloprovincialis*: Biomonitoring and transplantation in the Lagoon of Venice (North-East Italy). *Mar. Environ. Res.* 54:817–22.
De Duve, C. et al. 1974. Commentary. Lysosomotropic agents. *Biochem. Pharmacol.* 23:2495–531.
Depledge, M.H. 1994. Genotype toxicity: Implications for individuals and populations. *Environ. Health Perspect.* 102:101–4.
Depledge, M.H. 1999. Recovery of ecosystems and their components following exposure to pollution. *J. Aquat. Ecosyst. Health* 6:199–206.
Depledge, M.H. et al. 1993. The conceptual basis of the biomarker approach. In: *Biomarkers — Research and Application in the Assessment of Environmental Health*, ed. D.G. Peakall and L.R. Shugart, 15–29. Berlin: Springer.
Deutsch, L. et al. 2007. Feeding aquaculture growth through globalization: Exploitation of marine ecosystems for fishmeal. *Global Environ. Change* 17:238–49.
Domouhtsidou, G.P., and V.K. Dimitriadis. 2001. Lysosomal and lipid alterations in the digestive gland of mussels, *Mytilus galloprovincialis* (L.) as biomarkers of environmental stress. *Environ. Pollut.* 115:123–37.
Dondero, F. et al. 2006. Cellular responses to environmental contaminants in amoebic cells of the slime mould *Dictyostelium discoideum*. *Comp. Biochem. Physiol.* 143C:150–57.
Einsporn, S., and A. Köhler. 2008a. Immuno-localisations (GSSP) of subcellular accumulation sites of phenanthrene, Aroclor 1254 and lead (Pb) in relation to cytopathologies in the gills and digestive gland of the mussel *Mytilus edulis*. *Mar. Environ. Res.* 66:185–6.
Einsporn, S., and A. Köhler. 2008b. Electron-microscopic localization of lipophilic chemicals by an antibody-based detection system using the blue mussel *Mytilus edulis* as a model system. *Environ. Toxicol. Chem.* 27:554–60.
Etxeberria, M., M.P. Cajaraville, and I. Marigómez. 1995. Changes in digestive cell lysosomal structure as biomarkers of environmental stress in the Urdaibai Estuary (Biscay Coast, Iberian Peninsula). *Mar. Pollut. Bull.* 30:599–603.
Galloway, T.S. et al. 2002. Rapid assessment of marine pollution using multiple biomarkers and chemical immunoassays. *Environ. Sci. Technol.* 36:2219–26.
Galloway, T.S. et al. 2004. A multibiomarker approach to environmental assessment. *Environ. Sci. Technol.* 38:1723–31.
Galloway, T.S. et al. 2006. The ECOMAN project: A novel approach to defining sustainable ecosystem function. *Mar. Pollut. Bull.* 53:186–94.
Gastaldi, L. et al. 2007. Application of a biomarker battery for the evaluation of the sublethal effects of pollutants in the earthworm *Eisenia andrei*. *Comp. Biochem. Physiol.* 146C:398–405.

Giambérini, L., and M.P. Cajaraville. 2005. Lysosomal responses in the digestive gland of the freshwater mussel, *Dreissena polymorpha*, experimentally exposed to cadmium. *Environ. Res.* 98:210–4.

Gould, P. 2004. Nanoparticles probe biosystems. *Mater. Today* 7:36–43.

Gray, J.S. et al. 1988. Analysis of community attributes of the benthic macrofauna of Frierfjord/Langesundfjord and in a mesocosm experiment. *Mar. Ecol. Prog. Ser.* 46:151–65.

Grune, T. et al. 2004. Decreased proteolysis caused by protein aggregates, inclusion bodies, plaques, lipofuscin, ceroid, and "aggresomes" during oxidative stress, ageing and disease. *Int. J. Biochem. Cell Biol.* 36:2519–30.

Guidi, P. et al. 2010. Antioxidant, genotoxic and lysosomal biomarkers in the freshwater bivalve (*Unio pictorum*) transplanted in a metal polluted river basin. *Aquat. Toxicol.* 100:75–83.

Hankard, P.K. et al. 2004. Biological assessment of contaminated land using earthworm biomarkers in support of chemical analysis. *Sci. Total Environ.* 330:9–20.

Hawkins, A.J.S., and A.J. Day. 1996. The metabolic basis of genetic differences in growth efficiency among marine animals. *J. Exp. Mar. Biol. Ecol.* 203:93–115.

Hipkiss, A. 2006. Accumulation of altered proteins and ageing: Causes and effects. *Exp. Gerontol.* 41:464–73.

Hole, L.M., M.N. Moore, and D. Bellamy. 1992. Age-related differences in the recovery of lysosomes from stress-induced pathological reactions in marine mussels. *Mar. Environ. Res.* 34:75–80.

Howard, V. 2004. Small particles – big problems. *Int. Lab. News* 34:28–9.

Hwang, H.M., T.L. Wade, and J.L. Sericano. 2002. Relationship between lysosomal membrane destabilization and chemical body burden in eastern oysters (*Crassostrea virginica*) from Galveston Bay, Texas, USA. *Environ. Toxicol. Chem.* 21:1268–71.

Itziou, A., and V.K. Dimitriadis. 2011. Introduction of the land snail *Eobania vermiculata* as a bioindicator organism of terrestrial pollution using a battery of biomarkers. *Sci. Total Environ.* 409:1181–92.

Kalpaxis, D.L. et al. 2004. Biomonitoring of Gulf of Patras, N. Peloponnesus, Greece. Application of a biomarker suite including evaluation of translation efficiency in *Mytilus galloprovincialis* cells. *Environ. Res.* 94:211–20.

Kiffen, R. et al. 2004. Activation of chaperone-mediated autophagy during oxidative stress. *Mol. Biol. Cell* 15:4829–40.

Kirchin, M.A. et al. 1992. The role of oxyradicals in intracellular proteolysis and toxicity in mussels. *Mar. Environ. Res.* 34:315–20.

Klaine, S.J. et al. 2008. Nanomaterials in the environment: Behaviour, fate, bioavailability, and effects. *Environ. Toxicol. Chem.* 27:1825–51.

Klionsky, D.J., and S.D. Emr. 2000. Autophagy as a regulated pathway of cellular degradation. *Science* 290:1717–21.

Klionsky, D.J., A.M. Cuervo, and P.O. Seglen. 2007. Methods for monitoring autophagy from yeasts to human. *Autophagy* 3:181–206.

Köhler, A., H. Deisemann, and B. Lauritzen. 1992. Ultrastructural and cytochemical indices of toxic injury in dab liver. *Mar. Ecol. Prog. Ser.* 91:141–53.

Köhler, A., E. Wahl, and K. Söffker. 2002. Functional and morphological changes of lysosomes as prognostic biomarkers of toxic liver injury in a marine flatfish (*Platichthys flesus* (L)). *Environ. Toxicol. Chem.* 21:2434–44.

Krishnakumar, P.K., P.K. Asokan, and V.K. Pillai. 1990. Physiological and cellular responses to copper and mercury in the green mussel *Perna viridis* (Linnaeus). *Aquat. Toxicol.* 18:163–74.

Krishnakumar, P.K., E. Casillas, and U. Varanasi. 1994. Effect of environmental contaminants on the health of *Mytilus edulis* from Puget Sound, Washington, USA: I. Cytochemical measures of lysosomal responses in the digestive cells using automatic image analysis. *Mar. Ecol. Prog. Ser.* 106:249–61.

Kroll, A. et al. 2011. Cytotoxicity screening of 23 engineered nanomaterials using a test matrix of ten cell lines and three different assays. *Part. Fibre Toxicol.* 8:9. http://www.particleandfibretoxicology.com/content/8/1/9.

Lekube, X., M.P. Cajaraville, and I. Marigomez. 2000. Use of polyclonal antibodies for the detection of changes induced by cadmium in lysosomes of aquatic organisms. *Sci. Total Environ.* 247:201–12.

Levine, B. 2005. Eating oneself and uninvited guests: Autophagy-related pathways in cellular defense. *Cell* 120:159–62.

Livingstone, D.R. 2001. Contaminant-stimulated reactive oxygen species production and oxidative damage in aquatic organisms. *Mar. Pollut. Bull.* 42:656–66.

Livingstone, D.R. et al. 1992. Antioxidant enzymes in the digestive gland of the common mussel, *Mytilus edulis. Mar. Biol.* 112:265–76.

Livingstone, D. R. et al. 2000. Development of biomarkers to detect the effects of organic pollution on aquatic invertebrates: Recent molecular, genotoxic, cellular and immunological studies on the common mussel (*Mytilus edulis* L.) and other mytilids. *Int. J. Environ. Pollut.* 13:56–91.

Lockshin, R.A., and Z. Zakeri. 2004. Apoptosis, autophagy, and more. *Int. J. Biochem. Cell Biol.* 36:2405–19.

Lowe, D.M., and R.K. Pipe. 1994. Contaminant induced lysosomal membrane damage in marine mussel digestive cells: An *in vitro* study. *Aquat. Toxicol.* 30:357–65.

Lowe, D.M., M.N. Moore, and K.R. Clarke. 1981. Effects of oil on digestive cells in mussels: Quantitative alterations in cellular and lysosomal structure. *Aquat. Toxicol.* 1:213–26.

Lowe, D.M., M.N. Moore, and B.M. Evans. 1992. Contaminant impact on interactions of molecular probes with lysosomes in living hepatocytes from dab *Limanda limanda. Mar. Ecol. Prog. Ser.* 91:135–40.

Lowe, D.M., V.U. Fossato, and M.H. Depledge. 1995a. Contaminant induced lysosomal membrane damage in blood cells of mussels *M. galloprovincialis* from the Venice Lagoon: An *in vitro* study. *Mar. Ecol. Prog. Ser.* 129:189–96.

Lowe, D.M., C. Soverchia, and M.N. Moore. 1995b. Lysosomal membrane responses in mussels to experimental contaminant exposure. *Aquat. Toxicol.* 33:105–12.

Lushchak, V.I. 2011. Environmentally induced oxidative stress in aquatic animals. *Aquat. Toxicol.* 101:13–30.

Lüllmann-Rauch, R. 1979. Drug-induced lysosomal storage disorders. In: *Lysosomes in Applied Biology and Therapeutics*, vol. 6, ed. J.T. Dingle, P.J. Jacques, and I.H. Shaw, 49–130. Amsterdam: Elsevier.

Marigómez, I., and L. Baybay-Villacorta. 2003. Pollutant-specific and general lysosomal responses in digestive cells of mussels exposed to model organic chemicals. *Aquat. Toxicol.* 64:235–57.

Marigómez, I., U. Izagirre, and X. Lekube. 2005a. Lysosomal enlargement in digestive cells of mussels exposed to cadmium, benzo[*a*]pyrene and their combination. *Comp. Biochem. Physiol.* 141C:188–93.

Marigómez, I. et al. 1996. Structural changes in the digestive lysosomal system of sentinel mussels as biomarkers of environmental stress in Mussel-Watch programmes. *Comp. Biochem. Physiol.* 113 C:291–97.

Marigómez, I. et al. 2005b. Comparison of cytochemical procedures to estimate lysosomal biomarkers in mussel digestive cells. *Aquat. Toxicol.* 75:86–95.

Monteiro, V. et al. 2011. *In vivo* and *in vitro* exposures for the evaluation of the genotoxic effects of lead on the neotropical freshwater fish *Prochilodus lineatus. Aquat. Toxicol.* 104:291–8.

Moore, M.N. 1976. Cytochemical demonstration of latency of lysosomal hydrolases in digestive cells of the common mussel, *Mytilus edulis* and changes induced by thermal stress. *Cell Tissue Res.* 175:279–87.

Moore, M.N. 1980. Cytochemical determination of cellular responses to environmental stressors in marine organisms. *Rapp. P-V. Reun. Cons. Int. Explor. Mer.* 170:7–15.

Moore, M.N. 1985. Cellular responses to pollutants. *Mar. Pollut. Bull.* 16:134–9.

Moore, M.N. 1988. Cytochemical responses of the lysosomal system and NADPH-ferrihemoprotein reductase in molluscs to environmental and experimental exposure to xenobiotics. *Mar. Ecol. Prog. Ser.* 46:81–9.

Moore, M.N. 1990. Lysosomal cytochemistry in marine environmental monitoring. *Histochem. J.* 22:187–91.

Moore, M.N. 2002. Biocomplexity: The post-genome challenge in ecotoxicology. *Aquat. Toxicol.* 59:1–15.

Moore, M.N. 2004. Diet restriction induced autophagy: A protective system against oxidative- and pollutant-stress and cell injury. *Mar. Environ. Res.* 58:603–7.

Moore, M.N. 2008. Autophagy as a second level protective process in conferring resistance to environmentally-induced oxidative stress. *Autophagy* 4:254–6.

Moore, M.N., and J.I. Allen. 2002. A computational model of the digestive gland epithelial cell of the marine mussel and its simulated responses to aromatic hydrocarbons. *Mar. Environ. Res.* 54:579–84.

Moore, M.N., A. Bubel, and D.M. Lowe. 1980. Cytology and cytochemistry of the pericardial gland cells of *Mytilus edulis* and their lysosomal responses to injected horseradish peroxidase and anthracene. *J. Mar. Biol. Assoc. U.K.* 60:135–49.

Moore, M.N., and K.R. Clarke. 1982. Use of microstereology and cytochemical staining to determine the effects of crude oil-derived aromatic hydrocarbons on lysosomal structure and function in a marine bivalve mollusc, *Mytilus edulis. Histochem. J.* 14:713–8.

Moore, M.N., and D.W. Halton. 1973. Histochemical changes in the digestive gland of *Lymnaea truncatula* infected with *Fasciola hepatica. Z. Parasitenk.* 43:1–16.

Moore, M.N., and D.W. Halton. 1977. Cytochemical localization of lysosomal hydrolases in the digestive cells of littorinids and changes induced by larval trematode infection. *Z. Parasitenk.* 53:115–22.

Moore, M.N., and A.R.D. Stebbing. 1976. The quantitative cytochemical effects of three metal ions on a lysosomal hydrolase of a hydroid. *J. Mar. Biol. Assoc. U.K.* 56:995–1005.

Moore, M.N., and A. Viarengo. 1987. Lysosomal membrane fragility and catabolism of cytosolic proteins: Evidence for a direct relationship. *Experientia* 43:320–3.

Moore, M.N., D.M. Lowe, and P.E.M. Fieth. 1978. Lysosomal responses to experimentally injected anthracene in the digestive cells of *Mytilus edulis. Mar. Biol.* 48:297–302.

Moore, M.N., D.M. Lowe, and S.L. Moore. 1979. Induction of lysosomal destabilisation in marine bivalve molluscs exposed to air. *Mar. Biol. Lett.* 1:47–57.

Moore, M.N., R.K. Pipe and S.V. Farrar. 1982. Lysosomal and microsomal responses to environmental factors in *Littorina littorea* from Sullom Voe. *Mar. Pollut. Bull.* 13:340–5.

Moore, M.N. et al. 1984. Responses of the mussel *Mytilus edulis* to copper and phenanthrene: Interactive effects. *Mar. Environ. Res.* 14:167–83.

Moore, M.N., J.A. Mayernick, and C.S. Giam. 1985. Lysosomal responses to a polynuclear aromatic hydrocarbon in a marine snail: Effects of exposure to phenanthrene and recovery. *Mar. Environ. Res.* 17:230–3.

Moore, M.N. et al. 1986. Lysosomal and microsomal responses to oil-derived hydrocarbons in *Littorina littorea.* In: *Oceanic Processes in Marine Pollution—Biological Processes and Waste in the Ocean*, Vol. 1, ed. J.M. Capuzzo and D.R. Kester, 89–96. Melbourne, FL: Krieger Publishing.

Moore, M.N. et al. 1987. Molecular, cellular and physiological effects of oil-derived hydrocarbons on molluscs and their use in impact assessment. *Phil. Trans. R. Soc. (Lond.)* 316B:603–23.

Moore, M.N., C. Soverchia, and M. Thomas. 1996. Enhanced lysosomal autophagy of intracellular proteins by xenobiotics in living molluscan blood cells. *Acta Histochem. Cytochem.* 29(Suppl.):947–8.

Moore, M.N. et al. 1997. Uptake of a non-calorific, edible sucrose polyester oil and olive oil by marine mussels and their influence on uptake and effects of anthracene. *Aquat. Toxicol.* 39:307–20.

Moore, M.N. et al. 2004a. An integrated biomarker-based strategy for ecotoxicological evaluation of risk in environmental management. *Mutat. Res.* 552:247–68.

Moore, M.N., D.M. Lowe, and A. Köhler. 2004b. Measuring lysosomal membrane stability. *ICES Techniques in Marine Environmental Sciences* (ICES, Copenhagen) 36:1–31.

Moore, M.N., J.I. Allen, and A. McVeigh. 2006a. Environmental prognostics: An integrated model supporting lysosomal stress responses as predictive biomarkers of animal health status. *Mar. Environ. Res.* 61:278–304.

Moore, M.N. et al. 2006b. Lysosomal and autophagic reactions as diagnostic and predictive indicators of environmental pollutant toxicity in aquatic animals. *Autophagy* 2:217–20.

Moore, M.N. et al. 2006c. Autophagy: Role in surviving environmental stress. *Mar. Environ. Res.* 62:S420–5.

Moore, M.N. et al. 2007. Autophagic and lysosomal reactions to stress in the hepatopancreas of blue mussels. *Aquat. Toxicol.* 84:80–91.

Moore, M.N. et al. 2008. Lysosomes and autophagy in aquatic animals. In: *Methods in Enzymology*, Vol. 451, ed. D. Klionsky, 582–620. Burlington: Academic Press/Elsevier.

Nolde, N. et al. 2006. Lysosomal membrane stability in laboratory- and field-exposed terrestrial isopods *Porcellio scaber* (Isopoda, Crustacea). *Environ. Toxicol. Chem.* 25:2114–22.

Nott, J.A., and M.N. Moore. 1987. Effects of polycylic aromatic hydrocarbons on molluscan lysosomes and endoplasmic reticulum. *Histochem. J.* 19:357–68.

OECD Guideline for testing of chemicals, no 207. 1984. *Earthworm Acute Toxicity Tests*. Paris, France: Organization for Economic Cooperation and Development.

OECD Guideline for testing of chemicals, no 222. 2004. *Earthworm Reproduction Test (Eisenia fetida/andrei)*. Paris, France: Organization for Economic Cooperation and Development.

Olsgard, F., P.J. Somerfield, and M.R. Carr. 1998. Relationships between taxonomic resolution, macrobenthic community patterns and disturbance. *Mar. Ecol. Prog. Ser.* 172:25–36.

OSPAR. 2011. Further development of guidance on integrated monitoring and assessment of chemicals and biological effects. Annex 2: Lysosomal stability as a global health status indicator in biomonitoring.

Panyam, J., and V. Labhasetwar. 2003. Biodegradable nanoparticles for drug and gene delivery to cells and tissues. *Adv. Drug Deliv. Rev.* 55:329–47.

Parolini, M. et al. 2010. Multi-biomarker approach for the evaluation of the cyto-genotoxicity of paracetamol on the zebra mussel (*Dreissena polymorpha*). *Chemosphere* 79:489–98.

Parolini, M., A. Binelli, and A. Provini. 2011. Chronic effects induced by ibuprofen on the freshwater bivalve *Dreissena polymorpha*. *Ecotoxicol. Environ. Saf.* 74:1586–94.

Pipe, R.K., and M.N. Moore. 1985. The ultrastructural localization of acid hydrolases in developing oocytes of *Mytilus edulis*. *Histochem. J.* 17:939–49.

Pipe, R.K., and M.N. Moore. 1986a. Arylsulphatase activity associated with phenanthrene induced digestive cell deletion in the marine mussel *Mytilus edulis*. *Histochem. J.* 18:557–64.

Pipe, R.K., and M.N. Moore. 1986b. An ultrastructural study on the effects of phenanthrene on lysosomal membranes and distribution of the lysosomal membranes and distribution of the lysosomal enzyme β-glucuronidase in digestive cells of the periwinkle *Littorina littorea*. *Aquat. Toxicol.* 8:65–76.

Proud, C.G. 2002. Regulation of mammalian translation factors by nutrients. *FEBS J.* 269:5338–49.

Rashid, F., R.W. Horobin, and M.A. Williams. 1991. Predicting the behaviour and selectivity of fluorescent probes for lysosomes and related structures by means of structure–activity models. *Histochem. J.* 23:450–9.

Regoli, F. 1992. Lysosomal responses as a sensitive index in biomonitoring heavy metal pollution. *Mar. Ecol. Prog. Ser.* 84:63–9.

Regoli, F. 2000. Total oxyradical scavenging capacity (TOSC) in polluted and translocated mussels: A predictive biomarker of oxidative stress. *Aquat. Toxicol.* 50:351–61.

Rice, J. 2003. Environmental health indicators. *Ocean Coastal Man.* 46:235–59.

Ringwood, A.H. et al. 2004. Linkages between cellular biomarker responses and reproductive success in oysters, *Crassostrea virginica*. *Mar Environ. Res.* 58:151–5.

Robidoux, Y.P. et al. 2000. Chronic toxicity of energetic compounds in soil determined using earthworm (*Eisenia andrei*) reproduction test. *Environ. Toxicol. Chem.* 19:1764–73.

Robin, E. et al. 2007. Oxidant stress during simulated ischemia primes cardiomyocytes for cell death during reperfusion. *J. Biol. Chem.* 282:19133–43.

Rocco, A. et al. 2011. Suitability of lysosomal membrane stability in *Eisenia fetida* as biomarker of soil copper contamination. *Ecotoxicol. Environ. Saf.* 74:984–8.

Römbke, J., S. Jänsch, and W. Didden. 2005. The use of earthworms in ecological soil classification and assessment concepts. *Ecotoxicol. Environ. Saf.* 62:249–65.

Russo, J., L. Lefeuvre-Orfila, and L. Lagadic. 2007. Hemocyte-specific responses to the peroxidizing herbicide fomesafen in the pond snail *Lymnaea stagnalis* (Gastropoda, Pulmonata). *Environ. Pollut.* 146:420–7.

Russo, J., L. Madec, and M. Brehélin. 2009. Haemocyte lysosomal fragility facing an environmental reality: A toxicological perspective with atrazine and *Lymnaea stagnalis* (Gastropoda,Pulmonata) as a test case. *Ecotoxicol. Environ. Saf.* 72:1719–26.

Ryazanov, A.G., and B.S. Nefsky. 2002. Protein turnover plays a key role in ageing. *Mech. Ageing Dev.* 123:207–13.

Saint-Denis, M. et al. 1998. Glutathione, glutathione-related enzymes, and catalase activities in the earthworm *Eisenia fetida andrei*. *Arch. Environ. Contam. Toxicol.* 35:602–14.

Sanzi Cortez, F. et al. 2012. Biological effects of environmentally relevant concentrations of the pharmaceutical triclosan in the marine mussel Perna perna (Linnaeus, 1758). *Environ. Pollut.* 168:145–50.

Scott-Fordsmand, J.J., and J.M. Weeks. 2000. Biomarkers in earthworms. *Rev. Environ. Contam. Toxicol.* 165:117–59.

Sforzini, S. et al. 2008. Use of highly sensitive sublethal stress responses in the social amoeba *Dictyostelium discoideum* for an assessment of freshwater quality. *Sci. Tot. Environ.* 395:101–8.

Sforzini, S. et al. 2011a. Effects of dioxin exposure in *Eisenia andrei*: Integration of biomarker data by an Expert System to rank the development of pollutant-induced stress syndrome in earthworms. *Chemosphere* 85:934–42.

Sforzini, S. et al. 2011b. Genotoxicity assessment in *Eisenia andrei* coelomocytes: A study of the micronuclei induction and DNA damage in earthworms exposed to B[*a*]P and TCDD-spiked soils. *Mutat. Res.* 746:35–45.

Shaw, P.E. et al. 2011. Integration of biochemical, histochemical and toxicogenomic indices for the assessment of health status of mussels from the Tamar Estuary, UK. *Mar. Environ. Res.* 72:13–24.

Snyman, R.G., S.A. Reinecke, and A.J. Reinecke. 2000. Hemocytic lysosome response in the snail *Helix aspersa* after exposure to the fungicide copper oxychloride. *Arch. Environ. Contam. Toxicol.* 39:480–5.

Snyman, R.G., A.J. Reinecke, and S.A. Reinecke. 2002. Field application of a lysosomal assay as biomarker of copper oxychloride exposure, in the snail *Helix aspersa*. *Bull. Environ. Contam. Toxicol.* 69:117–22.

Spurgeon, D.J., J.M. Weeks, and C.A.M. van Gestel. 2003. A summary of eleven years progress in earthworm ecotoxicology. *Pedobiology* 47:588–606.

Svendsen, C., and J.M. Weeks. 1995. The use of a lysosome assay for the rapid assessment of cellular stress from copper to the freshwater snail *Viviparus contectus* (Millet). *Mar. Pollut. Bull.* 31:139–42.

Svendsen, C. et al. 1996. Use of an earthworm lysosomal biomarker for the ecological assessment of pollution from an industrial plastics fire. *Appl. Soil Ecol.* 3:99–107.

Svendsen, C. et al. 2004. A review of lysosomal membrane stability measured by neutral red retention: Is it a workable earthworm biomarker? *Ecotoxicol. Environ. Saf.* 57:20–9.

Tavernarakis, N., and M. Driscoll. 2002. Caloric restriction and lifespan: A role for protein turnover? *Mech. Ageing Dev.* 123:215–29.

van Gestel, C.A.M., E.M. Dirven-van Breemen, and R. Baerselman. 1993. Accumulation and elimination of cadmium, chromium and zinc and effects on growth and reproduction in *Eisenia andrei* (Oligochaeta, Annelida). *Sci. Total Environ.* 134:585–97.

Viarengo, A. 1989. Heavy metals in marine invertebrates: Mechanisms of the regulation and toxicity at the cellular level. *CRC Rev. Aquat. Sci.* 1:295–317.

Viarengo, A., and J.A. Nott. 1993. Mini-review. Mechanisms of heavy metal cation homeostasis in marine invertebrates. *Comp. Biochem. Physiol.* 104C:355–72.

Viarengo, A. et al. 1985. Detoxification of copper in the cells of the digestive gland of mussel: The role of lysosomes and thioneins. *Sci. Tot. Environ.* 44:135–45.

Viarengo, A. et al. 1987. Metallothioneins and lysosomes in metal toxicity and accumulation in marine mussels: The effect of cadmium in the presence and absence of phenanthrene. *Mar. Biol.* 94:251–7.

Viarengo, A. et al. 2000. Networking and expert–system analysis: Next frontier in biomonitoring. *Mar. Environ. Res.* 49:483–6.

Warwick, R.M., and K.R. Clarke. 1993. Comparing the severity of disturbance: A meta-analysis of marine macrobenthic community data. *Mar. Ecol. Prog. Ser.* 92:221–31.

Wedderburn, J. et al. 1998. Biomarkers of histochemical and cellular stress in *Carcinus maenas*: An *in situ* field study. *Mar. Environ. Res.* 46:321–4.

Weeks, J.M., and C. Svendsen. 1996. Neutral red retention by lysosomes from earthworm (*Lumbricus rubellus*) coelomocytes: A simple biomarker of exposure to soil copper. *Environ. Toxicol. Chem.* 15:1801–5.

Widdows, J. et al. 1981. Measurements of the responses of mussels to environmental stress and pollution in Sullom Voe: A baseline study. *Proc. R. Soc. Edin.* 80B:323–38.

Widdows, J. et al. 1982. Responses of *Mytilus edulis* L. on exposure to the water accommodated fraction of North Sea oil. *Mar. Biol.* 67:15–31.

Winston, G.W. et al. 1991. Lysosomal stability in *Mytilus edulis* L.: Potential as a biomarker of oxidative stress related to environmental contamination. *Arch. Environ. Contam. Toxicol.* 21:401–8.

Wouters, B.G. et al. 2005. Control of the response through regulation of mRNA translation. *Semin. Cell Dev. Biol.* 16:487–501.

Xu, F.-L. et al. 2002. System-level responses of lake ecosystems to chemical stresses using exergy and structural exergy as ecological indicators. *Chemosphere* 46:173–85.

6

Linking Immunotoxicity and Ecotoxicological Effects at Higher Biological Levels

Pauline Brousseau, Stéphane Pillet, Héloïse Frouin, Michel Auffret, François Gagné, and Michel Fournier

CONTENTS

6.1 General Overview

The function of the immune system is to fend off invasions by foreign pathogens and to prevent the growth of malignant cells by a variety of effector cells, molecules, and chemical messengers that together compose a highly sophisticated system. Consequently, complex interactions between the various components (cells, receptors, soluble factors) must occur in order to provide well-orchestrated immune defenses. Protection against foreign materials occurs through two different components of the immune system: the first, described as immediate, produces a nonspecific response identified as innate immunity. The second, described as long term, produces a specific response as well as an immunological memory identified as acquired immunity. This immunological memory is a key element, which favors induction of a rapid specific response in the case of repeated assaults by the same foreign material. In regard to aquatic species, the immune system of teleosts has a high degree of similarity with that of mammals. Based on evolutionary advances and specificity, the complexity of the immune system is less in lower vertebrates and in invertebrates. The immunological defenses of invertebrate aquatic species such as mollusks, crustaceans, sponges, and oligochaetes rely mainly on nonspecific innate immunity accompanied by antimicrobial peptides. The nonspecific immune response is the first line of defense against potential pathogenic organisms and includes phagocytosis, oxidative burst, cytotoxic activity, inflammatory reactions, and soluble mediators such as enzymes, antimicrobial peptides, and cytokine-like molecules.

Immunotoxicology, albeit a young science, is a rapidly expanding discipline that studies the deleterious effects of xenobiotics on the immune system. Substantial efforts have now been made in terms of the harmonization and validation of experimental approaches and methodologies to perform immunotoxicity studies (Brousseau et al. 2005) and to identify immunological responses that are sensitive to a large spectrum of xenobiotics in order to define immunological biomarkers. These efforts will serve as benchmarks for comparison with emerging high-throughput assays such as gene expression, proteomics, and metabolomics. In addition, harmonization of these techniques will also promote recognition for environmental risk assessments by both the national and international regulatory community.

Studies establishing the links between the effects of environmental contaminants on the immune system and the reduction of the host's resistance to infections or the development of cancer are the keystone of immunotoxicology, defined as the study of direct or indirect alterations of the immune system occurring as a result of exposure to chemicals such as pharmaceuticals, recombinant biological or environmental and occupational pollutants (Dean et al. 1979; Luster et al. 1988). A good correlation exists between changes in immunity tests and altered host resistance, since there have been no instances where host resistance was altered without affecting these tests. However, in some cases, immune changes can occur without corresponding alterations in host resistance that are perceived as early adaptative responses to stress (Luster et al. 1992a, 1992b, 1993). A "three-tier" approach was therefore recommended in which Tier 1 includes screening assays that would likely detect immunotoxic xenobiotics; Tier 2 allows the definition of the immune component(s) affected as well as the establishment of effects on host resistance; Tier 3 provides, in very general terms, approaches that could be used to identify the mode of action of the chemicals on the immune system (Dean et al. 1979; Luster et al. 1988). Subsequently, the sensitivity of various immune tests and the degree to which they agree with the commonly used host susceptibility tests has been addressed in order to define the immune correlate of adverse effects on host's resistance to infections and cancers (Luster et al. 1992a, 1992b, 1993; Luster and Gerberick 2010). These studies lead to the concept of immunological biomarkers or "immunomarkers" (Pillet and Nicolas 2005). From the ecological risk assessment perspective, biomarkers that cannot only predict impacts on the health status of the individual but also at the population levels are perceived as having added value. In a field study with wild clam (*Mya arenaria*) populations, specimens collected at sites with lower clam density and growth index had significantly higher levels of lipid peroxidation (a marker of inflammation), vitellogenin-like proteins, and metallothioneins (Blaise et al. 2003). Moreover, clams collected at polluted sites had fewer young, suggesting important changes in population structure. In another case, an extensive multibiomarker study revealed that the population metrics of *M. arenaria* living in polluted sites were more strongly related with phagocytosis in addition to energy status and sensitivity to temperature changes (Gagné et al. 2008a).

The development of immunotoxicology has revealed that contaminants can impact the immune system in the environment. Trace metals and organic pollutants, including polychlorinated biphenyls (PCBs), polychlorinated dibenzo-*p*-dioxins (PCDDs), a number of pesticides and more recently polybrominated diphenyl ethers (PBDEs), have been shown to alter normal immune functions in practically all animal models as well as in many wildlife species (Krzystyniak et al. 1995; Luebke et al. 1997; Voccia et al. 1999; Galloway and Handy 2003).

The immune system is quite well conserved through animal taxa and is sensitive to environmental stress. During evolution, the immune system has developed its effector mechanisms, through host–infectious agent interactions, thereby protecting the host against potential pathogens.

By definition, immunological biomarkers or immunomarkers are biomarkers that result from alterations to the immune system caused by exposure to one or more xenobiotics. Immunotoxic effects may occur either (1) by immunosuppression, which is a decline in host resistance toward a large spectrum of infectious agents, as well as the increased incidence of cancers, or (2) by immunostimulation, whose consequences are more complex and usually characterized by a qualitatively abnormal response such as hypersensitivity reactions like allergies or asthma, flulike reactions, and the occurrence of autoimmune diseases for the acquired immune component at the innate level, increased phagocytosis and mediators of inflammation could be observed as well (Luster et al. 1992a, 1992b, 1993; Revillard 1994; Kouassi et al. 2003). The outstanding value of the immune system relies on its high sensitivity, intimately linked to its sophisticated structure and integration of its function.

6.2 Case of Invertebrates—Bivalves

6.2.1 Introduction

A growing body of data on immunological biomarkers comes from experiments using bivalves as animal models to study the fundamental effects of toxins on the innate immune level. Bivalves, being filter-feeding organisms, are able to accumulate in their tissues xenobiotics at concentrations considerably higher than those found in the ambient waters (Phillips 1995). They also represent an experimental model of choice for both *in vitro* and *in vivo* approaches with organisms exposed in aquaria, caged animals, and indigenous specimens.

Invertebrates, like all other organisms, are permanently exposed to antigenic stimulation. Whereas in vertebrates the immune system includes acquired or adaptive immunity and innate or natural immunity, in invertebrates protection of the organism is assured solely by innate or natural immunity (Mydlarz et al. 2006). However, a primitive immunoglobulin (Ig) system seems to exist in invertebrates as indicated by the presence of a molluskan defense molecule (MDM) (Hoek et al. 1996). MDM is preferentially expressed in hemocytes and in granular cells. It is released in the hemolymph as a cytokine for nonself recognition and is involved in the agglutination and opsonization of foreign microparticles in the snail. However, the role of Ig in immune systems is not well understood at present, and more research will be needed to understand these fundamental aspects of immunological functions in invertebrates. Allograft inflammatory factor-1 has also been found in sponges and abalones; this is a cytokine involved in the recognition and in the rejection of allografts (Müller and Müller 2003; De Zoysa et al. 2010). In bivalve mollusks, natural immunity is composed of cellular components (mostly hemocytes) and humoral components consisting of molecules circulating in the hemolymph. Hemocytes, distributed in all tissues, represent the main immune cell population. No site of hemopoiesis has been described yet in this group, but observation of some cytological diversity suggests that cell maturation occurs among circulating hemocytes. These cells are capable of performing several immune functions such as phagocytosis (Lopez et al. 1997; Beaven and Paynter 1999; Canesi et al. 2002a, 2002b) and cytotoxicity through the production of reactive oxygen intermediates (Anderson et al. 1992; Pipe 1992). They can also secrete high molecular weight antibacterial substances as well as short peptides also presenting antibacterial activity (mytilins, myticins, defensins, and mytomycins). Hemocytes can also produce antibacterial enzymes (proteinases, glycosidases, sulfatases, lysozymes, etc.) and bacterial enzyme inhibitors involved in blocking enzymes produced by bacteria to help the infectious process. Finally, hemocytes have been shown to interact with several proteins circulating in hemolymph and be responsible for the opsonization and aggregation of bacteria (gigalins, galectins, integrins, and ficolins) (reviewed by Auffret 2005).

6.2.2 Hemocytes, Front Line of Immunocompetence

The involvement of circulating hemocytes as the first line of defense against invading microorganisms is well established (Cheng et al. 1977). There are several types of hemocytes based on their morphology and size. They represent a heterogeneous population consisting of at least two subpopulations as determined by enzymatic and morphological characterization in most species of bivalves, except in scallops (Auffret 1988; Brousseau et al. 2000; Fournier et al. 2001; Pampanin et al. 2002). Granulocytes are granular cells with a small nucleus. As for hyalinocytes, they are agranular, with a central nucleus surrounded by little cytoplasm (Pampanin et al. 2002). Indeed, hyalinocytes are relatively heterogeneous in size with diameters ranging from 4 to 17 μm (Bachère et al. 1988). The granulocyte/hyalinocyte ratio was found to be dependent on the species and flow cytometric measurements have revealed great interindividual variations. A parallel evaluation of leukocyte count and phenotyping was considered a good indicator for immunotoxic potential in other species such as humans, monkeys, and frogs (Luster et al. 1992a, 1992b, 1993). Indeed, an increased number of hemocytes was observed after *in vivo* exposure of bivalves to heat stress (Renwrantz 1990), pathogens (Oubella et al. 1993; Parisi et al. 2008), and xenobiotics (Renwrantz 1990; Anderson et al. 1992; Coles et al. 1994). Sami et al. (1992) also showed that the percentage and the relative size of small hemocytes of the oyster *Crassostrea virginica* could be modulated *in situ* by exposure to polycyclic aromatic hydrocarbons (PAHs). We believe that the profile of hemocytes by flow cytometry has good potential for use as an effective biomarker of exposure.

6.2.3 Phagocytosis, Marker of Effect

The sensitivity of phagocytosis to chemicals has been the object of numerous studies. It was observed that exposure of the bivalve *Mercenaria mercenaria* to phenol reduced the phagocytic activity of hemocytes (Fries and Tripp 1980). Long-term exposures to benzo(*a*)pyrene, pentachlorophenol, and hexachlorobenzene (HCB) weakly stimulate phagocytosis (Anderson et al. 1981; Frouin et al. 2007). Bouchard et al. (1999) showed that phagocytosis decreased with increasing doses of tributyltin (TBT) and dibutyltin (DBT). The toxicity of butyltin on the hemocytes of the four species of bivalves (*Mytilus edulis, Mya arenaria, Spisula polynima, Mactromeris polynyma*) was: MBT < TBT < DBT. The comparison of the relative sensitivity of species has shown that the blue mussel (*M. edulis*) was more tolerant of butyltin compounds than the other species studied.

Research studies on phagocytosis in several species of freshwater and marine bivalves have demonstrated a classic curve for the *in vitro* toxicity of mercury: (1) lack of cytotoxicity and stimulation of phagocytosis; (2) absence of cytotoxicity and normal phagocytosis; (3) low cytotoxicity and dramatic inhibition of phagocytosis; and (4) relatively high cytotoxicity accompanied by a drastic inhibition of phagocytosis (Brousseau et al. 2000; Sauvé et al. 2002). It seems that cadmium and zinc, like organic and inorganic mercury, have the potential to induce a stimulation of phagocytosis in some bivalve species (Brousseau et al. 2000; Sauvé et al. 2002). Stimulation of phagocytosis at low concentrations of metals has also been observed in rodents, reptiles, birds, fish, cattles, whales, and primates (Dieter et al. 1983; Voccia et al. 1994; De Guise et al. 1996b). However, no stimulation of phagocytosis was observed in the clam *Mya arenaria* exposed *in vivo* to low concentrations of both forms of mercury (Fournier et al. 2001).

It is noteworthy that the *in vitro* toxicity of CH_3HgCl and $HgCl_2$ followed similar patterns in some species of bivalves (*Cyrtodaria siliqua, Mya truncata, Elliptio complanata, Dreissena polymorpha*) (Sauvé et al. 2002). Organic mercury is more toxic than the inorganic form for several wildlife species (Fournier et al. 2000b), including bivalves (Brousseau et al. 2000). For example, a study with *Mya arenaria* exposed *in vivo* at different concentrations of mercury showed that doses of 10^{-6} M of $HgCl_2$ did not induce any difference in cell viability by the 28th day of exposure, whereas CH_3HgCl was found to be cytotoxic at 10^{-6} M from the 14th day of exposure (Fournier et al. 2001). In addition, phagocytosis was significantly decreased after 28 days of exposure to 10^{-6} M of $HgCl_2$, whereas a suppression of phagocytosis was induced by methyl mercury at 10^{-6} M after 7 days of exposure. Results clearly show a preponderance of the bioaccumulation of CH_3HgCl versus $HgCl_2$ (Watras and Bloom 1992; Inza et al. 1997). Comparison of *in vitro* toxicity thresholds of organic and inorganic mercury suggests that some bivalve species may metabolize or absorb mercury by different processes and therefore do not demonstrate the same response or relative sensitivity for each form of mercury (Sauvé et al. 2002).

The lipid and protein contents of the digestive gland were measured in clams exposed *in vivo* to different concentrations of organic and inorganic mercury. No differences in the contents of lipids and proteins were observed for the different treatments except for dead animals exposed to higher concentrations (Fournier et al. 2001). However, phagocytosis was decreased. This reinforced the fact that phagocytosis is a sensitive biomarker in the context of *in vivo* exposures to different mercury compounds.

6.2.4 Oxidative Burst and Lysozymes

Production of oxygen derivatives (mainly superoxide radicals (ROS)) by hemocytes plays an important role in killing foreign agents in invertebrates (Adema et al. 1991). Exposures of invertebrates to high concentrations of trace metals inhibit their production (Anderson et al. 1992). However, at concentrations measured in the environment there would be no effect on this parameter (Anderson et al. 1992; Coles et al. 1995). It is interesting to note that exposures to high concentrations of fluoranthene stimulate the production of superoxide radicals in the blue mussel *Mytilus edulis* (Coles et al. 1994). These opposite results may be due to a specific interaction between the contaminant involved (either metal or organic compound) and the target such as loss of membrane permeability and the production of oxygen derivatives or defense mechanisms against oxidative stress induced (Pipe and Coles 1995).

Several contaminant chemicals affect the release and activity of enzymes such as lysozyme. Results vary depending on the type of enzyme and the level of exposure considered. Toxic stress induced after exposure can result in many cases in an increased release and activity of certain enzymes and/or reduce the activities of others (Pickwell and Steinert

1984; Chu and Lapeyre 1989). More recently, Gagnaire et al. (2007) measured the reduced gene expression of several enzymes in hemocytes from oysters *Crassostrea gigas* exposed to pesticides.

6.2.5 Comparison between Markers and Adverse Effects

Pipe and Coles (1995) have attempted to correlate the response of different immunological biomarkers and the presence of a variety of contaminants in *Mytilus galloprovincialis* living in different sections of the Venice lagoon. The biomarkers consisted of hemocyte number and viability, phagocytosis, the release of lysosomal enzymes, and the production of oxygen derivatives. Contaminants measured were metals and organochlorine (OC) compounds including PCBs. All measured parameters showed seasonal fluctuations. In many cases, a correlation was found between the immunological response and contaminant concentration measured in mussel tissues. In addition, hemocytes appear more abundant at sites with low metal concentrations compared to the total number of circulating hemocytes in mussels at more contaminated sites. A tendency to inhibition of phagocytosis was observed in mussels exposed to high concentrations of metals, a phenomenon also encountered after exposure to PAHs. In mussels exposed to low concentrations of contaminants, an increase in the production of oxygen species was observed. A negative correlation between the production of oxygen derivatives and decreasing concentrations of organic contaminants was also noted.

Moreover, the use of multiple immunomarkers could be of value in understanding the mode of action of chemicals, especially for complex mixtures. Municipal effluents are well known to be major sources of contamination in aquatic ecosystems. In addition to trace metals and polyaromatic hydrocarbons, they contain pharmaceutical and personal care products (Kummerer 2001), as well as well-known endocrine disruptors such as 17α-,17β-ethynylestradiol (active ingredient in birth control pills), nonylphenol (breakdown product of alkylphenol polyethoxylate surfactants), bisphenol A, and the natural estrogen 17β-estradiol (Sumpter and Jobling 1995; Sabik et al. 2003; Vethaak et al. 2005). Effluent discharges also contain potentially pathogenic microorganisms (Farcy et al. 2011). Municipal effluents have been the object of many studies using freshwater bivalves.

Caged *E. complanata*, a freshwater mussel, placed downstream of an effluent discharge point revealed that hemocyte viability was significantly affected between sites, and phagocytosis was significantly reduced (Bouchard et al. 2009). In another study, phagocytic activity, nitrite oxide (NO), and cyclooxygenase (COX) activity were induced in hemocytes of *E. complanata* exposed to aeration lagoon effluents containing several pharmaceuticals as well as high bacterial loads (Gagné et al. 2006a, 2006b, 2012). Discriminant function analysis has enabled the separation of the effects of microparticles and microorganisms on the immune parameters of *E. complanata*. Although the effects on intracellular ROS and thiol levels were more pronounced with exposure to filtered effluent, phagocytosis, NO levels, COX, and lysozyme activities were higher with unfiltered effluent. This is in agreement with a previous study in rainbow trout *Oncorhynchus mykiss* exposed *in vivo* to the same municipal effluent (Hébert et al. 2008). Phagocytosis and natural cytotoxic cell activities were more affected by the insoluble (particles and microorganisms) fraction of the municipal effluent. However, the presence of estrogenic compounds could also influence the immune system in mollusks (Canesi et al. 2007a, 2007b, 2007c). Exposures to estrogenic chemicals found in municipal effluents (17α-ethynyl estradiol, mestranol, nonylphenol, nonylphenol monoethoxylate carboxylate, bisphenol A, benzophenone) were shown to decrease phagocytosis, lysosome membrane stability, and lysozyme release in mussels.

The production of NO in hemocytes was also shown to be enhanced by opiate receptors involved in the silencing of the immune response (Stefano et al. 1993). The presence of opiates (pain killers, anti-diarrheal and antitussive drugs) in municipal effluents could also impact the immune system in mollusks. The presence of microorganisms however, could stimulate phagocytosis. This was also revealed in freshwater fish, spottail shiners *Notropis hudsonius*, collected upstream and downstream of the municipal effluent dispersion plume from a large city on the Saint Lawrence River, Canada (Ménard et al. 2010). This study showed that although phagocytosis was stimulated by the presence of bacteria in the water column, cells from fish located downstream of the plume of a municipal discharge had relatively lower activity.

In field exposures of *E. complanata*, lysozyme activity was not significantly increased in downstream sites where hemocyte viability was unaffected (Bouchard et al. 2009; Farcy et al. 2011). In field exposures to municipal effluents, COX activity in hemocytes was not impaired (Bouchard et al. 2009). COX is involved in the formation of proinflammatory prostaglandins in tissues. Interestingly, COX activity was significantly decreased by exposure to the filtered effluent, indicating an anti-inflammatory effect consistent with the presence of COX inhibitors, which are more active after the removal of microorganisms (Gagné et al. 2008b). COX expression could be increased by NO in vertebrates (Dokmeci 2004), and this seems to be corroborated by the significant induction of NO production and COX activity in the bivalve hemocytes exposed to the effluent. This further suggests that NO production is mainly linked to bacterial load to support oxidative burst after phagocytosis rather than to exposure to xenobiotics. This agrees with a previous study with these freshwater mussels exposed to urban waste water (Bouchard et al. 2009), and studies in vertebrates reporting that bacterial infection causes up-regulation of NO production (Wheeler et al. 1997). Recent studies have revealed that vitellogenin, an estrogen-dependent energy-rich protein reserve in eggs, can also act at the immune system level by killing bacteria through an interaction with lipopolysaccharide (LPS) and lipoteichoic acid in fish (Zhang et al. 2005; Shi et al. 2006; Li et al. 2009). This was corroborated in freshwater mussels exposed to municipal effluents where phagocytosis activity better explained the observed changes in vitellogenin-like proteins in the freshwater mussels (Bouchard et al. 2009).

6.2.6 Conclusions

The immune system in mussels is essentially based on innate processes dominated by phagocytosis and antimicrobial peptides. The cellular component consists of granulocytes and hyalinocytes, the former accounting for most of the phagocytic activity. Bivalves also release a number of cell mediators involved in inflammation, which contributes to oxidative stress. For example, mussels exhibit NO synthase activity, which is involved in the oxidative burst after phagocytosis (Banakou and Dailianis 2010). The activity of arachidonate COX, which is blocked by nonsteroidal anti-inflammatory drugs, was shown to be induced by exposure to urban effluents, suggesting the involvement of this enzyme during oxidative stress events (Gagné et al. 2005). Besides this information and several immune functions reported in other groups of invertebrates such as natural killer (NK) activity, anticancer responses have never or rarely been signaled in bivalve mollusks. There is also a gap in knowledge regarding the role of each subpopulation of hemocytes and how the immune system is regulated. Furthermore, despite evidence of endocrine and nervous systems' interaction with hemocyte functions (Homo-Delarche and Dardenne 1993), mechanisms of action are still the object of research. Moreover, several results clearly indicate that factors other than the level of contamination, such as gender difference sensitivity, seasonal variation, and temperature, modulate immune system functions in natural populations (Duchemin et al. 2007; Gagné et al. 2008a, 2009; Pichaud et al. 2009; Greco et al. 2011).

Immunotoxicological assessment in ecotoxicology may therefore require an integrated approach based on a comprehensive battery of biomarkers. Bivalve studies point out the need to base this approach on metabolic pathways, allowing the monitoring of molecular events to be correlated with cellular responses in order to increase the robustness of the assessment. Genomic tools may therefore represent the ideal tools to fulfill this need (Douville et al. 2010).

6.3 Case of Marine Mammals

Marine mammals are valuable indicators of the health of the aquatic ecosystem, because of their "top predator" position in the trophic food chain, their long life span, and their long biological half-time of pollutant elimination. Although it is difficult to directly assess the risks associated with the exposure of wildlife to complex environmental mixtures of chemicals, marine mammals can served as useful indicators of ecosystem contamination because they can integrate contaminant signals from aquatic food chains and generate a "state of the environment" report.

In the aquatic environment, persistent toxic substances (PTSs) are accumulated by invertebrates and fish initially. Food ingestion (coupled with metabolism and excretion) is the dominant pathway for bioaccumulation of chemicals in marine mammals. Transfer within the food web can lead to biomagnification (the successive increase in concentration of a chemical with increasing trophic levels). Marine mammals, as homeotherms, maintain their body temperature and therefore have greater energy requirements for the same body weight than poikilotherms (e.g., fish) or terrestrial homeotherms. Thus, their metabolic rate and caloric requirements are higher. As a result, there is a compensatory increase in consumption of prey by marine mammals, and the largest biomagnification factors for PTSs are usually observed between fish and marine mammals (Muir et al. 1999).

6.3.1 Exposure of Marine Mammals to PTSs

Marine mammals accumulate high levels of trace metals (Das et al. 2003) and have extensive fat stores that can serve as depots to accumulate high levels of lipophilic compounds such as PAHs, OCs including DDT (dichloro-diphenyl-trichloroethane) and PCBs, and organobromines (OBs) such as PBDEs and other flame retardants (Tanabe et al. 1994; Rayne et al. 2004; Hall and Thomas 2007; Kajiwara et al. 2008a; Frouin et al. 2011). Proximity to contaminant sources represents the most evident factor influencing exposure to PTSs, near-shore feeders inhabiting industrialized coastal areas having higher contaminant levels than pelagic feeders in deeper waters (Meng et al. 2009). However, levels of pollutants in marine mammals also depend on other biological, physiological, spatial, and temporal confounding factors. The position in the food chain, the food web structure, and the nature of the prey also directly influence the levels of exposure to PTSs in marine mammals (Hansen et al. 1990; Ross et al. 2000; Rayne et al. 2004).

The pattern of distribution of contaminants within the organism is tissue and pollutant specific. For example, trace metals are mostly concentrated in the liver and kidney of marine mammals (Das et al. 2003). On the contrary, the highest lipophilic contaminant concentrations are usually encountered in the blubber (Severinsen et al. 2000). General biological traits of the individual, particularly age and gender, also affect the tissue levels of pollutants (Krahn et al. 2009). Contaminant burdens may differ between age class, partly as a result of historical changes in environmental contamination and the long life span of most marine mammals (Honda and Tatsukawa 1983). Whereas persistent pollutant concentrations increase in males during their lifetime, females transfer these substances to their offspring during each pregnancy and lactation (Wagemann et al. 1988; Tanabe et al. 1994; Borrell et al. 1995). The intensive offloading of contaminants via gestation and lactation constitutes a major elimination and detoxification mechanism of lipophilic contaminants stored in the fatty tissue of females and is also a significant route of exposure for suckling pups or calves at a moment in their development when they could be more susceptible to the immunotoxic effects of these substances (Beckmen et al. 2003; Pillet et al. 2005).

Metabolism or biotransformation of pollutants within marine mammals can also affect pollutant concentration patterns. Interestingly, the by-products are often highly reactive intermediates, and have been shown to elicit toxic effects in marine mammals. For example, DBT, a metabolite of TBT, exhibits much stronger immunological toxicity to harbor seals *Phoca vitulina* than the parent compound (Frouin et al. 2008).

6.3.2 Immunotoxicity of PTSs in Marine Mammals

The demonstrated exposure of marine mammals to PTSs leads to the question of the potential immunotoxic effects and the consequences on the health of marine mammal populations. The possible contribution of exposure to immunotoxic environmental pollutants became a major concern in late 1980s and early 1990s when a severe epizootic resulted in massive lethality in harbor (*Phoca vitulina*) and gray seal (*Halichoerus grypus*) populations inhabiting a relatively polluted area in northern Europe. In Canada, the high prevalence of cancers in the beluga whale *Delphinapterus leucas* population of the St. Lawrence Estuary (Quebec, Canada) concurs with the impact of anthropogenic chemicals on the immune system of this heavily contaminated population.

A great deal of attention was therefore accorded to assess the risk of immunotoxic effects resulting from exposure to environmental contaminants in marine mammals, especially on populations heavily exposed to anthropogenic chemicals. Consequently, the effort to apply the recommended tier approach promoted the development of immunological tools needed to address the immune response of marine mammals. It turns out that the immune system of marine mammals appears very similar to that of terrestrial mammals. NK cell cytotoxic activity, phagocytosis, and inflammatory cytokines have been demonstrated in pinnipeds and cetaceans, revealing the existence of an innate immune response (De Guise et al. 1995a; Ross et al. 1996; De Guise et al. 1997b; Pillet et al. 2000; Noda et al. 2003). The innate immune response represents the host's first line of defense against pathogens and is essential to the initiation of the adaptive immune response. In contrast to the innate immune response, the adaptive response is specific to the pathogen and consequently more efficient. The adaptive immune response involves humoral and cell-mediated immunity, and both have been identified in several species of marine mammals. The B lymphocytes are the main effectors of the humoral immune response and their antigen-recognition molecules are Ig. Membrane-bound Ig on the B-cell surface serves as the cell's receptor for antigen. The Ig of the same antigen specificity is secreted as an antibody (Ab) by terminally differentiated B cells after the binding of the antigen to the Ig at the cell's surface. Although gross similarities have been shown to exist in the structures of the different Ig classes, species-specific differences in binding affinities require the development of new reagents for each species (King et al. 1994; Lundqvist et al. 2002; Mancia et al. 2006, 2007; Nollens et al. 2007; Ruiz et al. 2009). The capacity of lymphocytes to proliferate can be unspecifically triggered by mitogens and has been demonstrated to be a valuable immunomarker (Luster et al. 1993). Mitogen and antigen-induced lymphocyte proliferation assays have been developed for pinnipeds (De Swart et al. 1993; DiMolfetto-Landon et al. 1995; Lalancette et al. 2003) and cetaceans (Beineke et al. 2004). As observed in terrestrial mammals, including laboratory animal models such as rodents, concanavalin A (ConA), and phytohemagglutinin (PHA) mainly induce the proliferation of T cells, whereas LPS primarily induces the proliferation of B cells. Meanwhile, the pokeweed mitogen (PWM) appears to have an impact on the proliferation of both B and T lymphocytes.

The relative proportion of the different immune cell subpopulations in immune tissues is also a highly relevant immunomarker (Luster and Gerberick 2010). In mammals, these populations are characterized by the presence of specific cell-surface proteins: the cluster of differentiation (CD). Unfortunately, the low interspecies homology between CD leads to poor interspecies cross-reactivity, strongly limiting the use of commercially available antibodies to identify immune cell types in marine mammals. However, several studies report successful detection of leukocyte subpopulations in marine mammal tissues using either homemade specific or cross-reactive antibodies (De Guise et al. 1997a, 1998a; Romano et al. 1999; De Guise et al. 2002; Beineke et al. 2007; Seibel et al. 2010). Expression of key signaling cell-surface molecules such as the major histocompatibility complex (MHC) has also been successfully demonstrated in marine mammal species (De Guise et al. 1997a; Zabka and Romano 2003; Seibel et al. 2010).

Cytokines are the main signaling proteins responsible for the communications between immune cells, as well as between immune cells and other cell types in the host's body. Several studies have succeeded in the identification and characterization of these essential messengers and their receptors in marine mammals (King et al. 1993, 1995; St-Laurent et al. 1999; St-Laurent and Archambault 2000). The cytokines are regulated at the transcription level, and, given the lack of species-specific antibodies recognizing marine mammal cytokines, reverse transcription-polymerase chain reaction has been widely used for their detection in pinnipeds and cetaceans (for review, see Beineke et al. 2010). Interestingly, beluga whale as well as harbor seal T lymphocytes and NK cells have been shown to be responsive to human recombinant interleukin IL-2, indicative of the presence of an IL-2–like receptor on these cells (De Swart et al. 1993; DiMolfetto-Landon et al. 1995; De Guise et al. 1996a; Ross et al. 1996).

The cytokines play an essential role in the regulation of the immune response and the resistance to pathogens. However, the consequence of an effect on a particular cytokine is difficult, and somewhat irrelevant to address, considering the complexity of the interactions in the highly intricate cytokine network. This complexity represents the major drawback for the use of cytokines as immunomarkers. Nevertheless, it is essential to pursue efforts to develop the characterization of cytokines in marine mammals considering their role in the immune response.

The characterization of the immune system of marine mammals remains limited compared to our knowledge of humans and rodents, but all the tools developed over past decades on the drivers of immunotoxicology research have significantly contributed to our understanding of the marine mammal immune response (Sitt et al. 2010).

6.3.3 Immunotoxic Effects of Environmental Pollutants in Marine Mammals

As discussed previously, marine mammals are exposed to a worrying cocktail of potentially immunotoxic compounds. However, establishing the link between environmental exposures and immunotoxic effects is complicated by logistic, technical, and ethical considerations.

To our knowledge, only one study has addressed the question of the immunotoxicity of environmental pollutants in a semicontrolled *in vivo* exposure. In that study, two groups of 11 captive harbor seals *Phoca vitulina* were fed herrings (*Clupea harengus*) from either the relatively uncontaminated Atlantic Ocean or from the contaminated Baltic Sea over the course of 2.5 years. The concentrations of persistent organic pollutants in the Baltic Sea herrings were 4–10 times higher than in the Atlantic Ocean herrings. Within 1 year, results of *in vitro* tests of immune function using peripheral blood mononuclear cells isolated from whole blood of the harbor seals revealed significantly lower T-cell function and NK cell response in the seals fed with Baltic Sea herrings. The specific immune response of the seals fed Baltic herrings was also affected, showing lower *in vitro* proliferative response to antigens after immunization, and lower *in vivo* delayed-type hypersensitivity and antibody responses to ovalbumin (Vos et al. 2003). These effects indicated a general alteration of the immune response.

Considering the logistic as well as technical and ethical constraints associated with that semicontrolled *in vivo* study, *in vitro* exposures represent an alternative and complementary approach. Moreover, they enable discrimination between direct and indirect effects on immune cells, and investigation of the mechanisms of toxicity at cellular and molecular levels. Most of the immune functions characterized in marine mammals have been demonstrated to be sensitive to *in vitro* exposures at environmentally relevant doses of the major environmental contaminants measured in marine mammal tissues. These include trace metals (De Guise et al. 1996b; Pillet et al. 2000; Camara Pellisso et al. 2008; Das et al. 2008; Kakuschke et al. 2008; Dufresne et al. 2010), OCs, and butyltins (De Guise et al. 1998b; Nakata et al. 2002; Frouin et al. 2008), as well as PBDEs (Frouin et al. 2010b). The use of a harbor seal cell line also proved to be useful in addressing the mechanism of toxicity in marine mammal cells (Frouin et al. 2010a). Additional studies are therefore desirable to better characterize this cell line in order to better comprehend the relevance of this tool.

Few studies have addressed the question of immunotoxic effects in the field not only because of the logistic and technical constraints but also considering the difficulty of dealing with confounding factors such as age, gender, physiological status, or historical exposure.

The proliferative response of peripheral blood lymphocytes to mitogens has been demonstrated to be impacted by exposure to OCs in different marine mammal populations. The responses to PHA, ConA, and PWM were significantly lower in the more contaminated Baltic gray seal *Halichoerus grypus* pups as compared to the less contaminated pups from the Atlantic (Sormo et al. 2009). Negative correlations between lymphocyte response to mitogens with increasing PCB and other OC plasma levels were also reported in free-ranging bottlenose dolphins (*Tursiops truncatus*) and polar bears (*Ursus maritimus*) (Lahvis et al. 1995; Lie et al. 2005). Interestingly, although the responses were negatively correlated to the more persistent dioxin-like compounds in the more contaminated pups from the Baltic Sea, a positive correlation with the non-dioxin, AhR-independent PCBs was observed in the pups from the Atlantic (Sormo et al. 2009). A similar positive correlation between PCB mitogen-induced lymphocyte proliferation was also shown in free-ranging harbor seal pups from southern British Columbia (Levin et al. 2005). Bernhoft et al. (2000) reported a negative correlation between IgG and the sum of PCB levels in the plasma of polar bears as well as levels of PCB congeners 99, 194, and 206 and HCB. These studies reflect the impact of PTSs on both cellular and humoral immune responses of marine mammals.

Confounding factors such as age, gender, and hormonal and nutritional status not only modulate the intake and bioaccumulation of PTSs as detailed above, they also strongly affect the immune response of marine mammals (Ross et al. 1993, 1994; Hall et al. 2003a; Lalancette et al. 2003). The endocrine system exerts a strong control on the immune response in mammals and any physiological or chemically induced modulation of the hormonal status could have profound impacts on the immune response, therefore affecting sensitivity to immunotoxic compounds (Pillet et al. 2006). Numerous environmental contaminants including PCBs and PBDEs have a direct impact on the endocrine system of marine mammals (Hall et al. 2003b; Sormo et al. 2005; Tabuchi et al. 2006). Most of these "endocrine disruptor" compounds also have demonstrated immunotoxic effects that could therefore result from either a direct effect on immune cells or from the chemically induced alterations of the hormonal balance. An analogous situation is observed with vitamin status. PCBs or hexachlorocyclohexanes appeared to modulate vitamin A and E homeostasis in pinnipeds and polar bear (Jenssen et al. 2003; Debier et al. 2004; Mos et al. 2007) with potential consequences on the immune response (Mora et al. 2008). Moreover, effects of retinol on morbillivirus infections have recently been partially elucidated (Trottier et al. 2009). These indirect impacts of vitamin A on pinniped resistance to phocine distemper virus (PDV) certainly deserve more attention.

6.3.4 Environmental Contaminants and Health Risks in Marine Mammals

Several studies from different geographical areas have addressed the impact of environmental pollutants on marine mammal species, investigating morphological changes of certain organs or an increased susceptibility to infectious diseases of animals inhabiting contaminated regions or with higher contaminant body burdens. Higher levels of PCBs and mercury in tissues were associated with prevalence of infections in harbor porpoises (*Phocoena phocoena*) and bottlenose dolphins (*Tursiops truncatus*) stranded in the United Kingdom and Germany (Siebert et al. 1999; Bennett et al. 2001; Jepson et al. 2005; Davison et al. 2011). The pathologies reported in these studies included severe parasitic pneumonia often associated with secondary bacterial infections, as well as bacterial septicemia, which is usually associated with an immunosuppressive status. Thymic atrophy and splenic depletion were associated with increasing levels of PCBs and PBDEs in the subcutaneous blubber of harbor porpoises from northern European waters (Beineke et al. 2005). This study also pointed out a positive correlation between high PBDE levels, poor health status, and lymphoid depletion of the thymus in *Phocoena phocoena*. In general, lymphoid depletion was associated with bacterial infections, such as severe dermatitis, pneumonia, peritonitis, or myocarditis. The most common causes of pathological lesions were not primary pathogenic agents, such as highly pathogenic viruses, but rather nematodes and opportunistic bacterial infections, again suggesting a deficient immune system (Beineke et al. 2005). Recently, Nakayama et al. (2009) have demonstrated a significant association between butyltin levels and the parasitic infection status of lung nematodes in finless porpoises (*Neophocaena phocaenoides*).

In the Saint-Lawrence Estuary beluga whale (*Delphinapterus leucas*) population, cancer was the second leading cause of death overall, and tumors have been identified in 27% of adult animals, representing the highest rate of cancer in a cetacean population (Martineau et al. 2002). This high incidence of neoplastic diseases has been potentially attributed to carcinogenic effects and/or reduced antitumoral immune responses due to environmental factors, including high environmental levels of PAH (De Guise et al. 1995b; Martineau et al. 2002). The high occurrence of opportunistic infections and parasitic infestations reported in these beluga whales strongly supports the hypothesis of a chemically induced immunodeficiency as a possible cause of the above-mentioned reduced antitumoral immune response (Martineau et al. 2002).

The consequences and the virulence of epidemics on populations are multifactorial, and so is the etiology of cancer (Thompson and Hall 1993). One of these factors is undoubtedly the fitness of the immune system, which in turn, as discussed above, can be affected by many factors, including exposure to anthropogenic compounds. The difficulty of linking exposure to potentially immunotoxic chemicals with a higher occurrence of a disease or an increase of clinical signs lies in the plasticity of each of the organizational levels of the system (i.e., molecular, cellular, tissue, organ, individual, and population). Fortunately, every single perturbation at the molecular or cellular level will not translate into clinical signs or pathology. However, this plasticity is limited and can also be influenced by different factors. Therefore, for a specific group within a population, an individual can be basically categorized as "resistant," "normal," or "sensitive" to a particular pathogen (Figure 6.1).

Morbilliviruses are known to be highly virulent and to cause serious mortality events in a sero-negative population of mammals (Barrett 1999). However, exposure to immunosuppressive compounds could contribute to the general stress affecting the immune system, amplifying the mass mortalities associated with epizootic episodes or tumor occurrence. This has been partly supported by several studies reporting higher concentrations of PTSs in animals that have died from morbillivirus infections, as compared with free-ranging animals or animals that died from physical trauma (Kuehl et al. 1991; Hall et al. 1992; Aguilar and Borrell 1994; Kajiwara et al. 2008b).

Moreover, serological evidence has established that harbor and gray seals from the less contaminated Canadian East coast have been infected with a virus that was similar or identical to PDV before 1988 with no signs of unusual mortalities (Henderson et al. 1992; Ross et al. 1992). Differences in species sensitivity to immunotoxic compounds, and particularly to PCBs, could also explain the higher sensitivity of harbor seals as compared to gray seals (Hammond et al. 2005).

In conclusion, in marine mammals and humans, the establishment of a relationship linking exposure to PTSs, the impacts on the immune system, and the host's reduced resistance to infection and cancer is complicated by numerous confounding factors. However, there is a very convincing body of evidence indicating that environmental pollution can significantly contribute to stress on the immune response resulting in major consequences on the health of marine mammal populations.

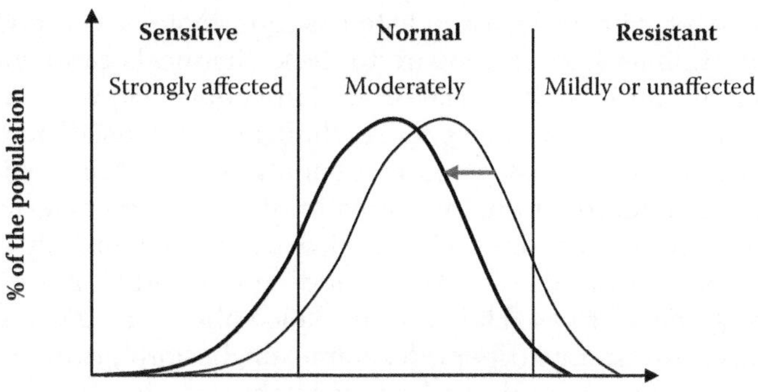

FIGURE 6.1
Resistance to a pathogen. Shift from normal resistance to a pathogen in a population after exposure to immunotoxic stress. The exposed population (heavy line) will be at risk of developing clinical signs and disease compared to the unexposed population (fine line).

6.4 Meaningfulness of Immune Biomarkers in Relation to Individual and Population Health

As aforementioned, several biomarkers of the immune response (immunomarkers) are shared among vertebrates and, to a lesser extent, with invertebrates (Fournier et al. 2000a; Kouassi et al. 2003). These immunomarkers can be affected by xenobiotics such as drugs and pollutant chemicals (Brousseau et al. 1997a, 1997b; Fournier et al. 2000a, 2000b). However, effects on immune functions do not necessarily imply an impact on host resistance to pathogens or an increased susceptibility to cancer. Environmental doses and ways of exposure are important factors to take into account to address the relevance of the effect on immunomarkers in terms of risk assessment analysis. As with other systems, the immune system has a great "plasticity," and effects at lower levels of biological organization (molecular and cellular) have to overcome detoxification and compensation processes before they affect higher levels of organization (tissue and organism).

Immunotoxic effects of xenobiotics have been reported in every major immune tissue in vertebrates, including bone marrow, thymus, spleen, and lymph nodes. Effects at cellular level have also been characterized. Xenobiotics can strongly impair the functions of macrophages, monocytes, or B and T cells with impacts on the nonspecific immune response as well as on the humoral and cell-mediated immune response (Kouassi et al. 2003; Fournier et al. 2000a; Brousseau et al. 1997a). These effects can result in immunodepression or immunostimulation, leading to hypersensitivity and autoimmunity disorders.

Immunodepression can result in a decrease of host resistance to parasite or viral and bacterial infections, and as previously described, environmental exposure to immunotoxic chemicals has been pinpointed to explain the high mortality during several epizootics that affected marine mammal populations inhabiting highly polluted areas. Severe immunodepression can be easily identified by histological alterations of lymphoid tissues, immunophenotyping, and *ex vivo* functional assays. In contrast, moderate long-term immunodepression is much more difficult to detect, and can lead to an increase of the incidence of cancer.

Polybrominated biphenyls, PCBs, dioxins, furans, and organochlorinated insecticides have been reported to induce immunodepression (Voccia et al. 1999). Trace metals (mercury, cadmium, lead, etc.) have also been listed as immunotoxic compounds (Omara et al. 1998; Zelikoff and Thomas 1998). Exposure to these chemicals can generally result in an increase of host susceptibility to bacterial or viral infections. However, given the complexity and the specificity of the immune response, challenges against different types of pathogens have to be included in the risk assessment analysis of immunotoxic compounds.

In case of chemical contamination, immunostimulation is characterized by an aberrant stimulation of the immune response. Different drugs and chemicals have been used to qualitatively modify the immune response including interferon, interleukins, levamisole, and dithiocarbamate derivatives (Haller 2001; Holcombe et al. 2001). Other compounds including isoprinosine, muramyl dipeptide, azimexon, bestatin, tuftsin, and pyrimidinols are strong immunostimulants and are known to shorten the disease period of several infections (Hadden 1987). However, the beneficial effects of these compounds are generally counterbalanced by strong side effects such as induction of autoimmune disorders (Krzystyniak et al. 1995; Kimber 2002). Although some trace metals have been demonstrated to have potentiator effects at lower levels of exposure (Bernier et al. 1995), the consequences of an immunostimulation resulting from exposure to environmental chemicals are yet not well defined and are poorly understood.

In ecotoxicology, a tier approach is used to evaluate the immunotoxicity of environmental pollutants. Tier 1 addresses the *in vitro* effects of chemicals on immune cells from the investigated species. *In vitro* exposures to a single compound or to cocktails can be considered, and effects on phagocytic activity, lymphocyte proliferation, and oxidative burst, in parallel with a cytotoxicity assay, are generally tested as immunomarkers. This first tier not only allows the identification of potential immunotoxic compounds in the targeted species, it also provides a range of concentrations within which compound(s) are immunotoxic but not cytotoxic. This represents essential information as the second tier aims to investigate *in vivo* laboratory exposure to doses inducing immunotoxic effects without systemic toxicity. In this second tier of the approach, efforts have to be made to design experiment settings that reproduce environmentally relevant doses and ways of exposure. As in the first tier, exposures to a single compound or cocktails can be considered. Three or more doses have to be tested in order to establish dose–response curves for the different immunomarkers. These provide essential information to adequately select the panel of markers used in the third tier. In most cases, NK cells cytotoxic activity, phenotyping, specific antibody production, and phagocytosis are selected (Luster et al. 1992a, 1992b; FDA 1993; Fournier et al. 2000a). Additional immunomarkers such as the synthesis of proinflammatory cytokines can also be investigated, depending on whether the xenobiotic(s) present in the environment is suspected to act on inflammation processes. The third tier addresses the detection of immunotoxic effects in the field in relation to exposure to environmental pollutants and the potential consequences on health status.

The integration of immunotoxicology as part of a risk assessment analysis is relatively new. The few studies that address the link between the effects on immunomarkers and the consequences on host susceptibility to pathogens and cancers provide large databases underpinning the identification of highly predictive immune assays (Luster et al. 1993). Ongoing studies are presently addressing the dose–effect and dose–response curves of different immunomarkers and xenobiotics in order to determine the "no observable adverse effect level" (NOAEL). The establishment of this NOAEL is complicated by the existence of confounding factors and the different sensitivity of the immunological assays (Luster et al. 1993). When these values have been determined for a large number of xenobiotics in different species, the next challenge will be the use of these models in field studies to support the risk assessment analysis. Therefore, the development of models of exposure that clearly identify the dose, the method of exposure, the metabolism, and the bioavailability of chemicals in the environment for key species is strongly needed.

Effects of xenobiotics on the immune response can vary from immunodepression to immunostimulation. Moreover, the same chemical can actually decrease specific immunomarkers while stimulating others, and a single compound can stimulate an immune function at lower doses while acting as a strong inhibitor at higher doses. The method and duration of exposure, gender, age, stress, nutritional status, and the relative complexity of the immune system of the investigated organism have been highlighted as important confounding factors that can modulate the immunotoxic effects of toxicants. In this regard, stress induction during sampling requires particular attention since it has been demonstrated to strongly affect immune functions. Accordingly, noninvasive sampling methods have to be emphasized in order to minimize as much as possible any stress during sampling. Additionally, efforts have to be made to standardize protocols including interlaboratory exercises in order to establish reference values with standard toxics in specific animal models.

Regarding the extreme complexity of the immune response, immunological risk assessment should include several immunomarkers, ideally covering the different arms of the immune response. Differential cell count and phenotyping, histology, cytotoxic activity of NK cells, and phagocytosis allow a relatively rapid detection of an impact on the immune system. Other immunomarkers such as delayed-type hypersensitivity or T cell activation address the mechanisms of immunotoxicity in more detail. In the future, efforts should also be made to integrate challenges with pathogens in the immunological risk analysis. Again, this should include a standardized process with reference pathogens.

In conclusion, since the emergence of this new discipline, immunotoxicology studies have revealed that the immune system is highly sensitive to xenobiotics. Therefore, the integration of immunomarkers into risk assessment analysis appears very promising. Although immunomarkers cannot be used to causally identify the source of an immunotoxic effect, they do indicate an impairment of the immune response, enabling us to identify populations at risk. It is also important to recognize that impairment of the immune system can be latent. Without an appropriate immunological survey, a weakened immune competence will remain undetected until the population is challenged by a pathogen, actually contributing to the severity of the epidemic. As illustrated by the still open question of the contribution of potential immunotoxic effects induced by chemicals on the severity of morbillivirus epizootics in marine mammals inhabiting heavily polluted areas, a clear link between environmental exposure to pollutants and an increased sensitivity to pathogens and cancer will remain very difficult to establish. However, immunotoxicology significantly contributes to the gathering of a body of evidence leading to a more precise definition of the risk associated with exposure to xenobiotics.

References

Adema, C.M. et al. 1991. Molluscan hemocyte-mediated cytotoxicity: The role of reactive oxygen intermediates. *Rev. Aquat. Sci.* 4:201–23.

Aguilar, A., and A. Borrell. 1994. Abnormally high polychlorinated biphenyl levels in striped dolphins (*Stenella coeruleoalba*) affected by the 1990–1992 Mediterranean epizootic. *Sci. Total. Environ.*, 154:237–47.

Anderson, R.S. et al. 1981. Effects of environmental pollutants on immunological competency of the clam *Mercenaria mercenaria*: Impaired bacterial clearance. *Aquat. Toxicol.* 1:187–95.

Anderson, R.S. et al. 1992. Superoxide anion generation by *Crassostrea virginica* hemocytes as measured by nitroblue tetrazolium reduction. *J. Invert. Pathol.* 59:303–7.

Auffret, M. 1988. Bivalve hemocyte morphology. *Am. Fish. Soc. Spec. Publ.* 18:169–77.

Auffret, M. 2005. Bivalves as models for marine immunotoxicology. In: *Investigative Immunotoxicology*, ed. H. Tryphonas, M. Fournier, B. Blakley, J.E.G. Smits, P. Brousseau, 29–48. Boca Raton, FL: Taylor & Francis.

Bachère, E. et al. 1988. Separation of *Crassostrea gigas* hemocytes by density gradient centrifugation and counterflow centrifugal elutriation. *Dev. Comp. Immunol.* 12:549–59.

Banakou, E., and S. Dailianis. 2010. Involvement of Na^+/H^+ exchanger and respiratory burst enzymes NADPH oxidase and NO synthase, in CD-induced lipid peroxidation and DNA damages in hemocytes of mussels. *Comp. Biochem. Physiol.* 152C:346–52.

Barrett, T. 1999. Morbillivirus infections, with special emphasis on morbilliviruses of carnivores. *Vet. Microbiol.* 69:3–13.

Beaven, A.E., and K.T. Paynter. 1999. Acidification of the phagosome in *Crassostrea virginica* hemocytes following engulfment of zymosan. *Biol. Bull.* 196:26–33.

Beckmen, K.B. et al. 2003. Organochlorine contaminant exposure and associations with hematological and humoral immune functional assays with dam age as a factor in free-ranging northern fur seal pups (*Callorhinus ursinus*). *Mar. Pollut. Bull.* 46:594–606.

Beineke, A. et al. 2004. Development of a lymphocyte-transformation-assay for peripheral blood lymphocytes of the harbor porpoise and detection of cytokines using the reverse-transcription polymerase chain reaction. *Vet. Immunol. Immunopathol.* 98:59–68.

Beineke, A. et al. 2005. Investigations of the potential influence of environmental contaminants on the thymus and spleen of harbor porpoises (*Phocoena phocoena*). *Environ. Sci. Technol.* 39:3933–8.

Beineke, A. et al. 2007. Phenotypical characterization of changes in thymus and spleen associated with lymphoid depletion in free-ranging harbor porpoises (*Phocoena phocoena*). *Vet. Immunol. Immunopathol.* 117:254–65.

Beineke, A. et al. 2010. Immunology of whales and dolphins. *Vet. Immunol. Immunopathol.* 133:81–94.

Bennett, P.M. et al. 2001. Exposure to heavy metals and infectious disease mortality in harbour porpoises from England and Wales. *Environ. Pollut.* 112:33–40.

Bernhoft, A. et al. 2000. Possible immunotoxic effects of organochlorines in polar bears (*Ursus maritimus*) at Svalbard. *J. Toxicol. Environ. Health* 59A:561–74.

Bernier, J. et al. 1995. Immunotoxicity of selected heavy metals in relation to Great Lakes. *Environ. Health Perspect.* 103:23–34.

Blaise, C. et al. 2003. Bivalve population status and biomarker responses in *Mya arenaria* clams (Saguenay Fjord, Quebec, Canada). *Fresenius Environ. Bull.* 12:956–60.

Borrell, A. et al. 1995. Age trends and reproductive transfer of organochlorine compounds in long-finned pilot whales from the Faroe Islands. *Environ. Pollut.* 88:283–92.

Bouchard, B. et al. 2009. An *in-situ* study of the impacts of urban wastewater on the immune and reproductive systems of the freshwater mussel *Elliptio complanata*. *Comp. Biochem. Physiol.* 150C:132–40.

Bouchard, N. et al. 1999. Effects of butyltin compounds on hemocytes phagocytosis activity of three marine bivalves. *J. Toxicol. Environ. Chem.* 18:519–22.

Brousseau, P. et al. 1997a. Marqueurs immunologiques. In: *Utilisation des biomarqueurs en écotoxicologie*, ed. L. Lagadic et al., 288–314. Paris, France: Masson.

Brousseau, P. et al. 1997b. Evaluation of earthworm exposure to contaminated soil by cytometric assay of coelomocyte phagocytosis in *Lumbricus terrestris* (Oligochaeta). *Soil Biol. Biochem.* 29:681–4.

Brousseau, P. et al. 2000. Flow cytometry as a tool to monitor the disturbance of phagocytosis in the clam *Mya arenaria* hemocytes following *in vitro* exposure to heavy metals. *Toxicology* 142:145–56.

Brousseau, P. et al. 2005. Approaches to immunotoxicity testing: The need for national and international harmonization/standardization of immunotoxicity methods. In: *Investigative Immunotoxicology*, ed. H. Tryphonas, M. Fournier, B. Blakley, J.E.G. Smits, and P. Brousseau, 421–429. Boca Raton, FL: Taylor & Francis.

Camara Pellisso, S. et al. 2008. Determination of the immunotoxic potential of heavy metals on the functional activity of bottlenose dolphin leukocytes *in vitro*. *Vet. Immunol. Immunopathol.* 121:189–98.

Canesi, L. et al. 2002a. Bacteria–hemocyte interactions and phagocytosis in marine bivalves. *Microsc. Res. Technol.* 57:469–76.

Canesi, L. et al. 2002b. Bacterial killing by *Mytilus* hemocyte monolayers as a model for investigating the signaling pathways involved in mussel immune defence. *Mar. Environ. Res.* 54:547–51.

Canesi, L. et al. 2007a. Effects of 17β-estradiol on mussel digestive gland. *Gen. Comp. Endocrinol.* 153:40–6.

Canesi, L. et al. 2007b. Effects of Triclosan on *Mytilus galloprovincialis* hemocyte function and digestive gland enzyme activities: Possible modes of action on non target organisms. *Comp. Biochem. Physiol.* 145C:464–72.

Canesi, L. et al. 2007c. Immunomodulation of *Mytilus* hemocytes by individual estrogenic chemicals and environmentally relevant mixtures of estrogens: *In vitro* and *in vivo* studies. *Aquat. Toxicol.* 81:36–44.

Cheng, T.C., and K.R. Auld. 1977. Hemocytes of the pulmonate gastropod *Biomphalaria glabrata*. *J. Invert. Pathol.* 30:119–22.

Chu, F.L.E., and J.F. Lapeyre. 1989. Effects of environmental factors and parasitism on hemolymph lysozyme and protein of American oyster (*Crassostrea virginica*). *J. Invert. Pathol.* 54:224–32.
Coles, J.A. et al. 1994. Effects of fluoranthene on the immunocompetence of the common marine mussel, *Mytilus edulis*. *Aquat. Toxicol.* 30:367–79.
Coles, J.A. et al. 1995. Alteration of the immune response of the common marine mussel, *Mytilus edulis*, resulting from exposure to cadmium. *Dis. Aquat. Org.* 22:59–65.
Das, K. et al. 2003. Heavy metals in marine mammals. In: *Toxicology of Marine Mammals*, ed. J.G. Vos, G.D. Bossart, M. Fournier, and T. O'Shea, 135–67. London: Taylor & Francis.
Das, K. et al. 2008. Mercury immune toxicity in harbour seals: Links to *in vitro* toxicity. *Environ. Health* 7:52–69.
Davison, N.J. et al. 2011. Infection with *Brucella ceti* and high levels of polychlorinated biphenyls in bottlenose dolphins (*Tursiops truncatus*) stranded in south-west England. *Vet. Rec.* 169:14.
De Guise, S. et al. 1995a. Immune functions in beluga whales (*Delphinapterus leucas*): Evaluation of phagocytosis and respiratory burst with peripheral blood leukocytes using flow cytometry. *Vet. Immunol. Immunopathol.* 47:351–62.
De Guise, S. et al. 1995b. Possible mechanisms of action of environmental contaminants on St. Lawrence beluga whales (*Delphinapterus leucas*). *Environ. Health Perspect.* 103(Suppl 4):73–7.
De Guise, S. et al. 1996a. Immune functions in beluga whales (*Delphinapterus leucas*): Evaluation of mitogen-induced blastic transformation of lymphocytes from peripheral blood, spleen and thymus. *Vet. Immunol. Immunopathol.* 50:117–26.
De Guise, S. et al. 1996b. Effects of *in vitro* exposure of beluga whale splenocytes and thymocytes to heavy metals. *Environ. Toxicol. Chem.* 15:1357–64.
De Guise, S. et al. 1997a. Phenotyping of beluga whale blood lymphocytes using monoclonal antibodies. *Dev. Comp. Immunol.* 21:425–33.
De Guise, S. et al. 1997b. Immune functions in beluga whales (*Delphinapterus leucas*): Evaluation of natural killer cell activity. *Vet. Immunol. Immunopathol.* 58:345–54.
De Guise, S. et al. 1998a. Characterization of a monoclonal antibody that recognizes a lymphocyte surface antigen for the cetacean homologue to CD45R. *Immunology* 94:207–12.
De Guise, S. et al. 1998b. Effects of *in vitro* exposure of beluga whale leukocytes to selected organochlorines. *J. Toxicol. Environ. Health* 55A:479–93.
De Guise, S. et al. 2002. Monoclonal antibodies to lymphocyte surface antigens for cetacean homologues to CD2, CD19 and CD21. *Vet. Immunol. Immunopathol.* 84:209–21.
De Swart, R.L. et al. 1993. Mitogen and antigen induced B and T cell responses of peripheral blood mononuclear cells from the harbour seal (*Phoca vitulina*). *Vet. Immunol. Immunopathol.* 37:217–30.
De Zoysa, M. et al. 2010. Allograft inflammatory factor-1 in disk abalone (*Haliotis discus discus*): Molecular cloning, transcriptional regulation against immune challenge and tissue injury. *Fish Shellfish Immunol.* 29:319–26.
Dean, J.H. et al. 1979. Assessment of immunobiological effects induced by chemicals, drugs or food additives: I. Tier testing and screening approach. *Drug Chem. Toxicol.* 2:5–17.
Debier, C. et al. 2004. An unexpected parallelism between Vitamin A and PCBs in seal milk. *Aquat. Toxicol.* 68:179–83.
Dieter, M.P. et al. 1983. Immunological and biochemical responses in mice treated with mercuric chloride. *Toxicol. Appl. Pharmacol.* 68:218–28.
DiMolfetto-Landon, L. et al. 1995. Blastogenesis and interleukin-2 receptor expression assays in the harbor seal (*Phoca vitulina*). *J. Wildl. Dis.* 31:150–8.
Dokmeci, D. 2004. Ibuprofen and Alzheimer's disease. *Folia Med.* 46:5–10.
Douville, M. et al. 2010. Characterization of commercial microbial products by polymorphic DNA markers and enzymatic activity diversity: Occurrence and potential effects on freshwater mussels exposed to municipal effluents. *Res. J. Biotechnol.* 5:31–48.
Duchemin, M.B., M. Fournier, and M. Auffret. 2007. Seasonal variations of immune parameters in diploid and triploid pacific oysters, *Crassostrea gigas*. *Aquaculture* 264:73–81.
Dufresne, M.M. et al. 2010. Comparative sensitivity of harbour and grey seals to several environmental contaminants using *in vitro* exposure. *Mar. Pollut. Bull.* 60:344–9.

Farcy, E. et al. 2011. Short-term physiological effects of a xenobiotic mixture on the freshwater mussel *Elliptio complanata* exposed to municipal effluents. *Environ. Res.* 111:1096–106.
FDA 1993. Draft Redbook II.
Fournier, M. et al. 2000a. Biomarkers in immunotoxicology: Evolutionary perspective. In: *Environmental Endocrine Disruptors*, ed. L. Guillette and D. Crain. New York: Taylor & Francis, 335 pp.
Fournier, M. et al. 2000b. Phagocytosis as a biomarker of immunotoxicity in wildlife species exposed to environmental xenobiotics. *Am. Zool.* 40:412–20.
Fournier, M. et al. 2001. Effects of *in vivo* exposure of *Mya arenaria* to organic and inorganic mercury on phagocytic activity of hemocytes. *Toxicology* 161:201–11.
Fries, C.R., and M.R. Tripp. 1980. Depression of phagocytosis in *Mercenaria* following chemical stress. *Dev. Comp. Immunol.* 4:233–44.
Frouin, H. et al. 2007. Physiological effects of polycyclic aromatic hydrocarbon soft-shell clam *Mya arenaria*. *Aquat. Toxicol.* 82:120–34.
Frouin, H. et al. 2008. Toxic effects of tributyltin and its metabolites on harbour seal (*Phoca vitulina*) immune cells *in vitro*. *Aquat. Toxicol.* 90:243–51.
Frouin, H. et al. 2010a. Toxic effects of various pollutants in 11B7501 lymphoma B cell line from harbour seal (*Phoca vitulina*). *Toxicology* 270:66–76.
Frouin, H. et al. 2010b. Effects of individual polybrominated diphenyl ether (PBDE) congeners on harbour seal immune cells *in vitro*. *Mar. Pollut. Bull.* 60:291–8.
Frouin, H. et al. 2011. PBDEs in serum and blubber of harbor, grey and harp seal pups from Eastern Canada. *Chemosphere* 82:663–9.
Gagnaire, B. et al. 2007. Combination of a pesticide exposure and a bacterial challenge: In vivo effects on immune response of Pacific oyster, *Crassostrea gigas* (Thunberg). *Aquat. Toxicol.* 84: 92–102.
Gagné, F. et al. 2005. Inflammatory properties of municipal effluents to *Elliptio complanata* mussels—Lack of effects from anti-inflammatory drugs. *Comp. Biochem. Physiol.* 141C:332–7.
Gagné, F. et al. 2006a. Occurrence of pharmaceutical products in a municipal effluent and toxicity to rainbow trout (*Oncorhynchus mykiss*) hepatocytes. *Ecotoxicol. Environ. Saf.* 64:329–36.
Gagné, F. et al. 2006b. Effects of selected pharmaceutical products on phagocytic activity in *Elliptio complanata* mussels. *Comp. Biochem. Physiol.* 143C:179–86.
Gagné, F. et al. 2008a. Relationships between intertidal clam population and health status of the soft-shell clam *Mya arenaria* in the St. Lawrence Estuary and Saguenay Fjord (Quebec, Canada). *Environ. Int.* 34:30–43.
Gagné, F. et al. 2008b. Immunotoxic effects on freshwater mussels of a primary-treated wastewater before and after ozonation: A pilot plant study. *Ecotoxicol. Environ. Saf.* 69:366–73.
Gagné, F. et al. 2009. Impacts of pollution in feral *Mya arenaria* populations: The effects of clam bed distance from the shore. *Sci. Total Environ.* 407:5844–54.
Gagné, F. et al. 2012. Immunotoxic potential of aeration lagoon effluents for the treatment of domestic wastewaters in the freshwater mussel *Elliptio complanata*. *J. Environ. Sci.* 24: DOI: 10.1016/S1001-0724(1011)60862-60860.
Galloway, T., and R. Handy. 2003. Immunotoxicity of organophosphorous pesticides. *Ecotoxicology* 12:345–63.
Greco, L. et al. 2011. Physiological effects of temperature and herbicide water exposure on the soft-shell clam *Mya arenaria* (Mollusca: Bivalvia). *Environ. Toxicol. Chem.* 30:132–41.
Hadden, J.W. 1987. Immunorestoration in secondary immunodeficiency. In: *Immunotoxicology*, ed. A. Berlin, J. Dean, M.H. Draper, E.M.B. Smith, F. Spreafico, 104–24. Dordrecht: Martinus Nijhoff Publishers.
Hall, A.J., and G.O. Thomas. 2007. Polychlorinated biphenyls, DDT, polybrominated diphenyl ethers, and organic pesticides in United Kingdom harbor seals (*Phoca vitulina*)—Mixed exposures and thyroid homeostasis. *Environ. Toxicol. Chem.* 26:851–61.
Hall, A.J. et al. 1992. Organochlorine levels in common seals (*Phoca vitulina*) which were victims and survivors of the 1988 phocine distemper epizootic. *Sci. Total Environ.* 115:145–62.

Hall, A.J. et al. 2003a. The immunocompetence handicap hypothesis in two sexually dimorphic pinniped species—Is there a sex difference in immunity during early development? *Dev. Comp. Immunol.* 27:629–37.
Hall, A.J. et al. 2003b. Polybrominated diphenyl ethers (PBDEs) in grey seals during their first year of life—Are they thyroid hormone endocrine disrupters? *Environ. Pollut.* 126:29–37.
Haller, D.G. 2001. Adjuvant chemotherapy for colon cancer: The US experience. *Tumori* 87:S86–8.
Hammond, J.A. et al. 2005. Comparison of polychlorinated biphenyl (PCB) induced effects on innate immune functions in harbour and grey seals. *Aquat. Toxicol.* 74:126–38.
Hansen, C.T. et al. 1990. Zinc, cadmium, mercury and selenium in minke whales, belugas, and narwhals from west Greeland. *Polar. Biol.* 10:529–39.
Hébert, N. et al. 2008. The effects of a primary-treated municipal effluent on the immune system of rainbow trout (*Oncorhynchus mykiss*): Exposure duration and contribution of suspended particles. *Comp. Biochem. Physiol.* 148C:258–64.
Henderson, G. et al. 1992. Demonstration of antibodies in archival sera from Canadian seals reactive with a European isolate of phocine distemper virus. *Sci. Total Environ.* 115:93–8.
Hoek, R.M. et al. 1996. A new Ig-superfamily member, molluscan defence molecule (MDM) from *Lymnaea stagnalis*, is down-regulated during parasitosis. *Eur. J. Immunol.* 26:939–44.
Holcombe, R.F. et al. 2001. Investigating the role of immunomodulation for colon cancer prevention: Results of an *in vivo* dose escalation trial of levamisole with immunologic endpoints. *Cancer Detect. Prev.* 25:183–91.
Homo-Delarche, F., and M. Dardenne. 1993. The neuro-immune axis. *Springer Semin. Immunopathol.* 14:221–38.
Honda, K., and R. Tatsukawa. 1983. Distribution of cadmium and zinc in tissues and organs, and their age-related changes in striped dolphins, *Stenella coeruleoalba*. *Arch. Environ. Contam. Toxicol.* 12:543–50.
Inza, B. et al. 1997. Tissue distribution of inorganic mercury, methylmercury and cadmium in the Asiatic clam (*Corbicula fluminea*) in relation to the contamination levels of the water column and sediment. *Chemosphere* 35:2817–36.
Jenssen, B.M. et al. 2003. Negative relationship between PCBs and plasma retinol in low-contaminated free-ranging gray seal pups (*Halichoerus grypus*). *Environ. Res.* 93:79–87.
Jepson, P.D. et al. 2005. Relationships between polychlorinated biphenyls and health status in harbor porpoises (*Phocoena phocoena*) stranded in the United Kingdom. *Environ. Toxicol. Chem.* 24:238–48.
Kajiwara, N. et al. 2008a. Polybrominated diphenyl ethers (PBDEs) and organochlorines in melon-headed whales, *Peponocephala electra*, mass stranded along the Japanese coasts: Maternal transfer and temporal trend. *Environ. Pollut.* 156:106–14.
Kajiwara, N. et al. 2008b. Persistent organic pollutants (POPs) in Caspian seals of unusual mortality event during 2000 and 2001. *Environ. Pollut.* 152:431–42.
Kakuschke, A.S. et al. 2008. Metal-induced impairment of the cellular immunity of newborn harbor seals (*Phoca vitulina*). *Arch. Environ. Contam. Toxicol.* 55:129–36.
Kimber, I. 2002. Reduction, refinement and replacement: Putting the immune system to work—The FRAME annual lecture. *Altern. Lab. Anim.* 30:569–79.
King, D.P. et al. 1993. Identification and partial characterization of common seal (*Phoca vitulina*) and grey seal (*Haliochoerus grypus*) interleukin-6-like activities. *Dev. Comp. Immunol.* 17:449–58.
King, D.P. et al. 1994. Identification, characterisation, and measurement of immunoglobulin concentrations in grey (*Haliocherus grypus*) and common (*Phoca vitulina*) seals. *Dev. Comp. Immunol.* 18:433–42.
King, D.P. et al. 1995. Leucocyte interleukin-1-like activity in the common seal (*Phoca vitulina*) and grey seal (*Halichoerus grypus*). *J. Comp. Pathol.* 113:253–61.
Kouassi, E. et al. 2003. Système immunitaire. In: *Environnement et santé publique: Fondements et pratiques*, ed. M. Gérin, P. Gosselin, S. ordier, C. Viau, P. Quénel, E. Dewailly. Edisem, 1098 pp. Acton Vale: Paris.
Krahn, M.M. et al. 2009. Effects of age, sex and reproductive status on persistent organic pollutant concentrations in "Southern Resident" killer whales. *Mar. Pollut. Bull.* 58:1522–9.

Krzystyniak, K. et al. 1995. Approaches to the evaluation of chemical-induced immunotoxicity. *Environ. Health Perspect.* 103:17–22.

Kuehl, D.W. et al. 1991. Chemical residues in dolphins from the US Atlantic coast including Atlantic bottlenose obtained during the 1987/88 mass mortality. *Chemosphere* 22:1071–84.

Kummerer, K. 2001. Drugs in the environment: Emission of drugs, diagnostic aids and desinfectants into wastewater by hospitals in relation to other sources—A review. *Chemosphere* 45:957–69.

Lahvis, G.P. et al. 1995. Decreased lymphocyte responses in free-ranging bottlenose dolphins (*Tursiops truncatus*) are associated with increased concentrations of PCBs and DDT in peripheral blood. *Environ. Health Perspect.* 103:67–72.

Lalancette, A. et al. 2003. Contrasting changes of sensitivity by lymphocytes and neutrophils to mercury in developing grey seals. *Dev. Comp. Immunol.* 27:735–47.

Levin, M. et al. 2005. Association between lymphocyte proliferation and polychlorinated biphenyls in free-ranging harbor seal (*Phoca vitulina*) pups from British Columbia, Canada. *Environ. Toxicol. Chem.* 24:1247–52.

Li, Z. et al. 2009. Vitellogenin is a cidal factor capable of killing bacteria via interaction with lipopolysaccharide and lipoteichoic acid. *Mol. Immunol.* 46:3232–9.

Lie, E. et al. 2005. Does high organochlorine (OC) exposure impair the resistance to infection in polar bears (*Ursus maritimus*)?: Part II. Possible effect of OCs on mitogen- and antigen-induced lymphocyte proliferation. *J. Toxicol. Environ. Health* 68A:457–84.

Lopez, C. et al. 1997. Differential phagocytic ability of the circulating haemocytes types of the carpet shell clam *Ruditapes decussatus* (mollusca: Bivalva). *Dis. Aquat. Org.* 30:209–15.

Luebke, R.W. et al. 1997. Aquatic pollution-induced immunotoxicity in wildlife species. *Fundam. Appl. Toxicol.* 37:1–15.

Lundqvist, M.L. et al. 2002. Cloning of the IgM heavy chain of the bottlenose dolphin (*Tursiops truncatus*), and initial analysis of VH gene usage. *Dev. Comp. Immunol.* 26:551–62.

Luster, M.I., and G.F. Gerberick. 2010 Immunotoxicology testing: Past and future. *Methods Mol. Biol.* 598:3–13.

Luster, M.I. et al. 1988. Development of a testing battery to assess chemical-induced immunotoxicity: National Toxicology Program's guidelines for immunotoxicity evaluation in mice. *Fundam. Appl. Toxicol.* 10:2–19.

Luster, M.I. et al. 1992a. Risk assessment in immunotoxicology: I. Sensitivity and predictability of immune tests. *Fundam. Appl. Toxicol.* 18:200–10.

Luster, M.I. et al. 1992b. Qualitative and quantitative experimental models to aid in risk assessment for immunotoxicology. *Toxicol. Lett.* 64–65:71–8.

Luster, M.I. et al. 1993. Risk assessment in immunotoxicology: II. Relationships between immune and host resistance tests. *Fundam. Appl. Toxicol.* 21:71–82.

Mancia, A. et al. 2006. The immunoglobulin G heavy chain (IGHG) genes of the Atlantic bottlenose dolphin, *Tursiops truncatus. Comp. Biochem. Physiol.* 144B:38–46.

Mancia, A. et al. 2007. Characterization of the immunoglobulin A heavy chain gene of the Atlantic bottlenose dolphin (*Tursiops truncatus*). *Vet. Immunol. Immunopathol.* 118:304–9.

Martineau, D. et al. 2002. Cancer in wildlife, a case study: Beluga from the St. Lawrence Estuary, Quebec, Canada. *Environ. Health Perspect.* 110:285–92.

Ménard, L. et al. 2010. The impact of urban pollution on the immune system of spottail shiners (*Notropis hudsonius*) in the St. Lawrence River. *Fresenius Environ. Bull.* 19:1369–74.

Meng, X.Z. et al. 2009. Polybrominated diphenyl ethers in pinnipeds stranded along the southern California coast. *Environ. Pollut.* 157:2731–6.

Mora, J.R. et al. 2008. Vitamin effects on the immune system: Vitamins A and D take centre stage. *Nat. Rev. Immunol.* 8:685–98.

Mos, L. et al. 2007. Contaminant-associated disruption of vitamin A and its receptor (retinoic acid receptor alpha) in free-ranging harbour seals (*Phoca vitulina*). *Aquat. Toxicol.* 81:319–28.

Muir, D. et al. 1999. Spatial and temporal trends and effects of contaminants in the Canadian Arctic marine ecosystem: A review. *Sci. Total Environ.* 230:83–144.

Müller, W.E.G., and I.M. Müller. 2003. Origin of the metazoan immune system: Identification of the molecules and their functions in sponges. *Integr. Comp. Biol.* 43:281–92.
Mydlarz, L.D. et al. 2006. Innate immunity environmental drivers and disease ecology of marine and freshwater invertebrates. *Ann. Rev. Ecol. Evol. System.* 37:251–88.
Nakata, H. et al. 2002. Evaluation of mitogen-induced responses in marine mammal and human lymphocytes by *in-vitro* exposure of butyltins and non-ortho coplanar PCBs. *Environ. Pollut.* 120:245–53.
Nakayama, K. et al. 2009. Temporal and spatial trends of organotin contamination in the livers of finless porpoises (*Neophocaena phocaenoides*) and their association with parasitic infection status. *Sci. Total Environ.* 407:6173–8.
Noda, K. et al. 2003. Evaluation of the polymorphonuclear cell functions of bottlenose dolphins. *J. Vet. Med. Sci.* 65:727–9.
Nollens, H.H. et al. 2007. Development and validation of monoclonal and polyclonal antibodies for the detection of immunoglobulin G of bottlenose dolphins (*Tursiops truncatus*). *J. Vet. Diagn. Invest.* 19:465–70.
Omara, F. et al. 1998. Immunotoxicity of essential elements. In: *Immunotoxicology of Environmental and Occupational Metals*, ed. J. Zelikoff and P.T. Thomas, 231–62. Padstow, UK: Taylor & Francis.
Oubella, R. et al. 1993. Experimentally induced variation in hemocyte density for *Ruditapes philippinarum* and *R. decussatus* (*Mollusca bivalvia*). *Dis. Aquat. Org.* 15:193–7.
Pampanin, D.M. et al. 2002. Air exposure and functionality of *Chamelea gallina* haemocytes: Effects on haematocrit, adhesion, phagocytosis and enzyme contents. *Comp. Biochem. Physiol.* 131A:605–14.
Parisi, M.G. et al. 2008. Differential involvement of mussel hemocytes sub-populations in the clearance of bacteria. *Fish. Shellfish Immunol.* 25:834–40.
Phillips, D.J.H. 1995. The chemistries and environmental fates of trace metals and organochlorines in aquatic ecosystem. *Mar. Pollut. Bull.* 31:193–200.
Pichaud, N. et al. 2009. Metabolic capacities and immunocompetence in sea scallops (*Placopecten magellanicus*) at different ages and life stages. *J. Shellfish Res.* 28:865–76.
Pickwell, G.V., and S.A. Steinart. 1984. Serum biochemical and cellular responses to experimental cupric ion challenge in mussels. *Mar. Environ. Res.* 14:245–365.
Pillet, S., and J.M. Nicolas. 2005. Biomarkers of immunotoxic effects in the context of environmental risk assessment. In: *Investigative Immunology*, ed. H. Tryphonas, M. Fournier, B. R. Blakley, J.E.G. Smits, P. Brousseau. Boca Raton, FL: Taylor & Francis.
Pillet, S. et al. 2000. *In vitro* exposure of seal peripheral blood leukocytes to different metals reveal a sex-dependent effects of zinc on phagocytic activity. *Mar. Pollut. Bull.* 40:921–7.
Pillet, S. et al. 2005. Sex-specific effects of neonatal exposures to low levels of cadmium through maternal milk on development and immune functions of juvenile and adult rats. *Toxicology* 209:289–301.
Pillet, S. et al. 2006. Immunomodulatory effects of estradiol and cadmium in adult female rats. *Toxicol. Sci.* 92:423–32.
Pipe, R.K. 1992. Generation of reactive oxygen metabolites by the hemocytes of the mussel *Mytilus edulis*. *Dev. Comp. Immunol.* 16:111–22.
Pipe, R.K., and J.A. Coles. 1995. Environmental contaminants influencing immune function in marine bivalve molluscs. *Fish. Shellfish Immunol.* 5:581–95.
Rayne, S. et al. 2004. PBDEs, PBBs and PCNs in three communities of free-ranging killer whales (*Orcinus orca*) from the Northeastern Pacific Ocean. *Environ. Sci. Technol.* 38:4293–9.
Renwrantz, L. 1990. Internal defence system of *Mytilus edulis*. In: *Studies in Neuroscience, Neurobiology of Mytilus edulis*, ed. G.B. Stefano, 256–75. Manchester: Manchester University Press.
Revillard, J.P. 1994. Therapeutic prospects in autoimmunity. *Rev. Praticien* 44:93–7.
Romano, T.A. et al. 1999. Molecular cloning and characterization of CD4 in an aquatic mammal, the white whale *Delphinapterus leucas*. *Immunogenetics* 49:376–83.
Ross, P.S. et al. 1992. Antibodies to phocine distemper virus in Canadian seals. *Vet. Rec.* 130:514–6.
Ross, P.S. et al. 1993. Immune function in free-ranging harbor seal (*Phoca vitulina*) mothers and their pups during lactation. *J. Wildl. Dis.* 29:21–9.

Ross, P.S. et al. 1994. Relative immunocompetence of the newborn harbour seal, *Phoca vitulina*. *Vet. Immunol. Immunopathol.* 42:331–48.

Ross, P.S. et al. 1996. Suppression of natural killer cell activity in harbour seals (*Phoca vitulina*) fed Baltic Sea herring. *Aquat. Toxicol.* 34:71–84.

Ross, P.S. et al. 2000. High PCB concentrations in free-ranging Pacific killer whales, *Orcinus orca*: Effects of age, sex and dietary preference. *Mar. Pollut. Bull.* 40:504–15.

Ruiz, C.L. et al. 2009. Baseline circulating immunoglobulin G levels in managed collection and free-ranging bottlenose dolphins (*Tursiops truncatus*). *Dev. Comp. Immunol.* 33:449–55.

Sabik, H. et al. 2003. Occurrence of alkylphenol polyethoxylates in the St. Lawrence River and their bioconcentration by mussels (*Elliptio complanata*). *Chemosphere* 51:349–56.

Sami, S. et al. 1992. Alterations in cytometry characteristics of hemocytes from the American oyster *Crassostrea virginica* exposed to a polycyclic aromatic hydrocarbon (PAH) contaminated environment. *Mar. Biol.* 113:247–52.

Sauvé, S. et al. 2002. Phagocytic activity of marine and freshwater bivalves: *In vitro* exposure of hemocytes to metals (Ag, Cd, Hg and Zn). *Aquat. Toxicol.* 58:189–200.

Seibel, H. et al. 2010. Immunohistochemical characterization of selected cell markers for the detection of hematopoietic cells in formalin-fixed, paraffin wax-embedded lymphoid tissues of harbor seals (*Phoca vitulina*) and walruses (*Odobenus rosmarus rosmarus*). *Vet. Immunol. Immunopathol.* 137:305–9.

Severinsen, T. et al. 2000. Spatial distribution of persistent organochlorines in ringed seal (*Phoca hispida*) blubber. *Mar. Environ. Res.* 49:291–302.

Shi, X., S. Zhang, and Q. Pang. 2006. Vitellogenin is a novel player in defense reactions. *Fish Shellfish Immunol.* 20:769–72.

Siebert, U. et al. 1999. Potential relation between mercury concentrations and necropsy findings in cetaceans from German waters of the North and Baltic seas. *Mar. Pollut. Bull.* 38:285–95.

Sitt, T. et al. 2010. Cellular immune responses in cetaceans immunized with a porcine erysipelas vaccine. *Vet. Immunol. Immunopathol.* 137:181–9.

Sormo, E.G. et al. 2005. Thyroid hormone status in gray seal (*Halichoerus grypus*) pups from the Baltic Sea and the Atlantic Ocean in relation to organochlorine pollutants. *Environ. Toxicol. Chem.* 24:610–6.

Sormo, E.G. et al. 2009. Immunotoxicity of polychlorinated biphenyls (PCB) in free-ranging gray seal pups with special emphasis on dioxin-like congeners. *J. Toxicol. Environ. Health* 72A:266–76.

St-Laurent, G., and D. Archambault. 2000. Molecular cloning, phylogenetic analysis and expression of beluga whale (*Delphinapterus leucas*) interleukin 6. *Vet. Immunol. Immunopathol.* 73:31–44.

St-Laurent, G. et al. 1999. Molecular cloning and phylogenetic analysis of beluga whale (*Delphinapterus leucas*) and grey seal (*Halichoerus grypus*) interleukin 2. *Vet. Immunol. Immunopathol.* 67:385–94.

Stefano, G.B. et al. 1993. Opiate-like substances in an invertebrate, an opiate receptor on invertebrate and human immunocytes, and a role in immunosuppression. *Proc. Natl. Acad. Sci. U. S. A.* 90:11099–103.

Sumpter, J.P., and S. Jobling. 1995. Vitellogenesis as a biomarker for estrogenic contamination of the aquatic environment. *Environ. Health Perspect.* 103:173–8.

Tabuchi, M. et al. 2006. PCB-related alteration of thyroid hormones and thyroid hormone receptor gene expression in free-ranging harbor seals (*Phoca vitulina*). *Environ. Health Perspect.* 114:1024–31.

Tanabe, S. et al. 1994. Global contamination by persistent organochlorines and their ecotoxicological impact on marine mammals. *Sci. Total Environ.* 154:163–77.

Thompson, P.M., and A.J. Hall. 1993. Seals and epizootics—What factors might affect the severity of mass mortalities? *Mammal Rev.* 23:149–54.

Trottier, C. et al. 2009. Retinoids inhibit measles virus through a type I IFN-dependent bystander effect. *FASEB* 23:3203–12.

Vethaak, A.D. et al. 2005. An integrated assessment of estrogenic contamination and biological effects in the aquatic environment of The Netherlands. *Chemosphere* 59:511–24.

Voccia, I. et al. 1994. *In vitro* mercury-related cytotoxicity and functional assay of immune cells of rainbow trout (*Oncorhynchus mykiss*). *Aquat. Toxicol.* 29:37–48.

Voccia, I. et al. 1999. Immunotoxicity of pesticides: A review. *Toxicol. Ind. Health* 15:119–32.
Vos, J.G. et al. 2003. The effects of chemicals contaminants on immune functions in harbour seals: Results of a semi-field study. In: *Toxicology of Marine Mammals*, ed. J.G. Vos, G.D. Bossart, M. Fournier, T.J. O'Shea, 558–70. New York: Taylor & Francis.
Wagemann, R. et al. 1988. Trace metals and methyl mercury: Associations and transfer in harp seal (*Phoca groenlandica*) mothers and their pups. *Mar. Mammal. Sci.* 4:339–55.
Watras, C.J., and N.S. Bloom. 1992. Mercury and methylmercury species: The role of aqueous-phase chemical speciation. *Environ. Toxicol. Chem.* 11:1373–6.
Wheeler, M.A. et al. 1997. Bacterial infection induces nitric oxide synthase in human neutrophils. *J. Clin. Invest.* 99:110–16.
Zabka, T.S., and T.A. Romano. 2003. Distribution of MHC II (+) cells in skin of the Atlantic bottlenose dolphin (*Tursiops truncatus*): An initial investigation of dolphin dendritic cells. *Anat. Rec.* 273A:636–47.
Zelikoff, J., and P.T. Thomas. 1998. *Immunotoxicology of Environmental and Occupational Metals*, 231–62. Padstow, UK: Taylor & Francis.
Zhang, S.C. et al. 2005. Hemagglutinating and antibacterial activities of vitellogenin. *Fish Shellfish Immunol.* 19:93–5.

7

Sentinel Species

Brigitte Berthet

CONTENTS

The increasingly precise techniques of analytical chemistry allow us to identify and to quantify, in all biological or physical matrices, an increasing number of xenobiotics and their metabolites, but these measures supply no information about the toxicological or ecotoxicological risks of such contaminants. For that purpose, toxicity tests, carried out in the laboratory on a limited number of species, were established, allowing definition of the nature of the risk as well as any associated dose–effect relationship. However, these tests do not take into account the influence of the multiple parameters present in the environment that intervene under natural conditions to affect toxicity: abiotic ecological factors (such as temperature, salinity, conditions of oxygenation), interspecific variability and the interactions between species, the heterogeneity of populations in their interactions with pollutants, or between interactions between pollutants and abiotic factors of the environment (Smolders et al. 2004). The biological monitoring of the environment, or "biomonitoring," has the objective of integrating these various aspects, particularly by using sentinel species as models.

7.1 Concept of Sentinel Species

Programs involving sentinel species in which data on organisms exposed to contaminants in the environment are regularly and systematically collected and analyzed, represent a mechanism used to estimate the risks of contaminants for the health of mankind and/or that of the environment (NRC 1991).

When a xenobiotic is introduced into an environment (Figure 7.1), its first effects are observed at the level of the organism, then the population, the community, and ultimately at the level of the whole ecosystem. The earlier the detection of these disturbances, the better are the possibilities of remediation and the opportunity to define the contaminant source, explaining, therefore, the interest shown in sentinel species and in the study of their biomarkers. Sentinel organisms are indeed particular bioindicators used to reveal precociously the presence and the toxicity of one or several pollutants in the environment. Such substances are often bioaccumulated by these organisms, for example, filter-feeding mollusks, but in particular they provoke changes in biological parameters at various levels of biological organization (molecular, cellular, physiological, individual, etc.) leading to dysfunction of variable ecotoxicological importance. Thus, although they remain present in the environment, particular organisms can reveal changes in their health status or modifications of their behavior that might have an influence on their capacity to reproduce and their long-term survival in this environment, potentially leading to disturbance of the structure and the functioning of the ecosystem.

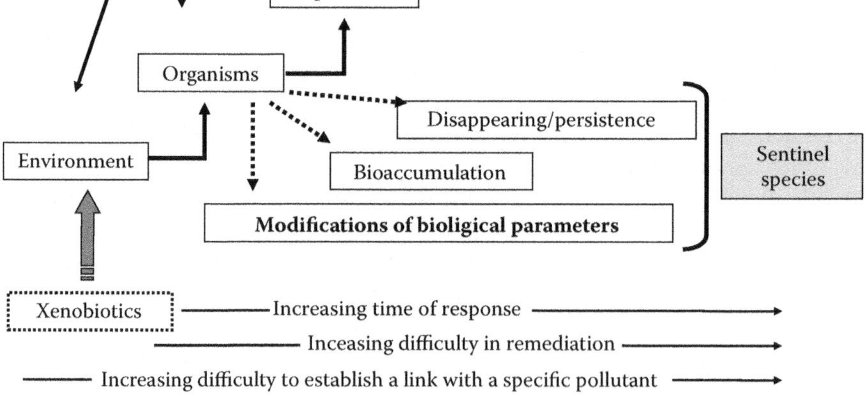

FIGURE 7.1
Positions of sentinel species in the various levels of biological organization.

7.1.1 Development of Concept and the Species Used

The notion of sentinel species has existed for a long time: in the nineteenth century, miners took a canary with them down into a mine. The appearance of symptoms of asphyxiation in the bird warned them of the presence in the subterranean galleries of dangerous concentrations of toxic gas. The use and the definition of the term of "sentinel species" evolved over time, and in different contexts, explaining why today there are disagreements and differences in the use of the concept of sentinel species between authors. Phillips (1994) considered that species used for biomonitoring are referred to under the several generic names of "bioindicators," "bioaccumulative indicators," "biomonitors," and "sentinel organisms," and that these names are generally badly used. As often defined today, sentinel species may be considered as any species providing a warning of a dysfunction or an imbalance of the environment, or, more restrictively, a warning of the dangers of substances to human health. In the widest acceptance of the term, the sentinel species can be (Figure 7.1):

- A bioindicator species, providing information by its absence (or presence) and/or of abundance of individuals
- A bioaccumulative species, the capacity of which to accumulate contaminants at levels much higher than those in the environment, designates it as the material of choice for the monitoring of the degree of contamination of the environment
- A species for which modifications of biological parameters at one or various levels of organization (molecular, cellular, physiological, organism, or behavioral) are used to estimate the risks associated with the presence of contaminants

So, the terms of bioindicators and bioaccumulators are often used as synonyms for sentinel species. However, although the two terms are indeed associated with the same objective of using living organisms to inform us about the quality of the environment and its consequences, there is a conceptual difference between them. To avoid problems resulting from different understandings of their separate meanings, and to maintain the integrity of their use, Dauvin et al. (2010) recommended a clear definition of the meaning of the terms used. It is therefore the definition of a sentinel species as a species showing modifications of biological parameters (indeed biomarkers) that is retained in this chapter.

The first sentinel species used were chosen on the criteria of convenience of easy collection, availability, ease of breeding in the laboratory, and good genetic stability, but without ecological relevance. An example is the freshwater crustacean *Daphnia pulex*, widely used to test the toxicity of effluents in the 1970s. Chapman et al. (1994) showed that the tests carried out on the reproduction of this species showed it to be more sensitive to pollutants than other species of daphnids. Nevertheless, *D. pulex* has been replaced little by little by *Daphnia magna* or *Ceriodaphnia dubia*, both of which are easier to breed in the laboratory.

A second criterion often advanced to justify the choice of a species as a sentinel species is its ease of collection in the field and subsequent manipulation in the laboratory. Mussels (bivalve mollusks of the family Mytilidae) are typical examples, with moreover a large power of bioaccumulation of pollutants that have caused them to be widely used in biomonitoring programs of coastal zones such as "Mussel Watch" in the United States and in the Netherlands or the "Réseau d'Observation de la Contamination Chimique" of the marine environment (ROCCH) in France. Furthermore, the biology of mussels, as well as their mechanisms of biotransformation, and the mechanisms of toxicity of xenobiotics, are well known, allowing a good understanding of the fluctuations observed in the biomarkers measured in them (Michel et al. 1998).

However, several authors (Chapman 2002; McCarty et al. 2002; Galloway et al. 2004; Tabor and Aguirre 2004) have expressed regret that the criteria of selection of these species do not integrate enough of an ecological dimension. More recently, De Lange et al. (2010) proposed the assessment of ecosystem vulnerability by using biological and ecological information at different hierarchical levels, in order to bring more ecology into environmental risk assessment, indeed into ecotoxicology, a view supported by Luoma and Rainbow (2010).

In 2001, Chapman showed the utility and relevance of aquatic oligochaetes in ecological risk assessment, but he noted that, because their taxonomy is difficult and uncertain, they have largely been neglected or mistakenly considered as representing only pollution tolerant taxa. Nevertheless, more recently, Tixier et al. (2011) indicated that now their taxonomy and autecology are sufficiently known to allow identification to species level and that, when considered at the species level, the oligochaete community represents a highly diverse group with a wide range of ecological roles and a wide range of sensitivity to pollution. Thus, they proposed an approach of sediment quality assessment based on oligochaete community analysis to evaluate the ecological status of storm water ponds.

Furthermore, as pointed out by Lam and Gray (2001), it is not uncommon that in the laboratory, the choice of the species to be tested falls on a particular genotype of the species, whereas in the natural environment populations are made up from a mixture of genotypes. These same authors also expressed doubts on the choice of species that are easy to breed and manipulate in the laboratory, considering that such species are the very species that are often intrinsically strongly tolerant to any type of stress, natural or resulting from contamination. Athrey et al. (2007) tested the effect of such reduced genetic variation in the least killifish *Heterandria formosa*. They used laboratory populations from an eight-generation experiment designed to study selection for resistance to Cd, compared to a control laboratory population and to fish from the original source population. The authors drew two conclusions from this study. The first one is that the loss of genetic variation in genes not directly under selection means that populations that have managed to adapt genetically to an environmental pollutant might be less able to adapt to other environmental stressors. Athrey et al. (2007) underlined that this could be exacerbated by negative effects on fitness-associated characteristics that are sometimes associated with contaminant adaptation. The second conclusion is that the loss of genetic variation in their laboratory lines appeared to be due to effective population sizes being much lower than actual population sizes, pointing out the necessity of maintaining large laboratory population sizes of organisms used in environmental toxicology studies.

7.1.2 Criteria for Selection of Sentinel Species

In order to reflect the intensity of contaminant exposures, to integrate the effects of the various parameters acting in the environment, and to be able to determine the effects of exposures that may be of weak intensity but endure over a long period, specimens of sentinel species collected in the field must be representative of the local environment and meet several conditions. However, in certain cases, specimens may also be transplanted into the habitat of interest, and maintained at the site in enclosures [active biomonitoring (ABM)] in order to follow any change (increase or decrease) in contaminant exposure, and the associated impacts on the individuals, limiting any effects of interspecific biotic factors that might affect interpretation.

Sentinel species have been sought in all environments. Numerous authors (e.g., Rand 1985; NRC 1991; White 2004) have drawn up lists of the characteristics considered to be of greatest importance in these species, but there have been different interpretations, in particular, according to the definition used for a sentinel species and the kind of exposure (in laboratory or in field). The most usually quoted desirable characteristics are discussed below:

- Their sedentarity in the site studied, so that the condition of the individuals can be directly correlated to the level of pollution at that site. Animals moving over long distances or visiting the studied sites only occasionally cannot specifically reflect the level of contamination at the site of capture. In this respect, sessile or weakly mobile animals are thus preferred.
- Their ease of identification, capture, or collection all year round, requiring a good knowledge of the ecology of the species and its adaptations to the spatiotemporal variations of the environmental factors in estuarine and coastal zones (temperature, salinity) or the nature of any reproductive migrations.
- A sufficient population size so that any impact of regular collections on population structure and density is negligible. Furthermore, the size of the population available for collection conditions the speed of collecting, thereby affecting the feasibility and the cost of the monitoring program.
- A wide and known distribution range that not only allows comparisons between various sites, but also increases the probability of finding sites that are as weakly impacted as possible by the pollution. However, since it is nearly impossible to find totally cosmopolitan species, closely related species, whether of local or other origin, can be used. So, in the historic example of the Mussel Watch, various species of mussels (genera *Mytilus* and *Perna*) or oysters (genera *Crassostrea* and *Saccostrea*) have been preferred. Other closely related taxa have also been used to supplement mussels and oysters, such as various species of zebra mussels (genus *Dreissena*) in the region of the Great Lakes, or another bivalve, *Chama sinuata*, in the Florida Keys (NOAA 2005). Similarly, oysters (*Crassostrea gigas*) and mussels (*Mytilus* spp.) have been used since 1979 in the framework of the ROCCH for the monitoring of French coasts, whereas similar programs have been set up in numerous other coastal areas (Chapter 15).
- A longevity of several years and a resistance to pollutants in the environment, allowing individual specimens to be in contact with contaminants over long enough periods for any changes in biomarkers to be measurable. This persistence of individuals in the environment also allows changes in parameters to be studied over an extended period, and possibly to reveal any reversibility of ecotoxicological damage if measures have been taken to reduce the level of contaminant exposure. However, the majority of the relevant species have a much lower longevity than humans, and the disorders caused by pollutants appear more quickly, without the latency periods observed in longer-lived animals.
- Dose–effect and cause–effect relationships must be consistently established. Primarily, the species has to present a clear response (effect) to a measurable increase in the level of contaminant (dose). The sentinel species, although resistant to contaminants, does nevertheless have to show sufficient sensitivity to these substances that dysfunctions are apparent, and that these are measurable by

techniques in current use. This relationship is often difficult to establish because of distance to the contaminant source, and of interactions with other stress factors. Nevertheless, it has been well identified in some examples, such as that of the imposex in the dog whelk, *Nucella lapillus*, living in areas contaminated by tributyltin (Chapter 9). Moreover, responses of the organisms to the pollutants must be early enough to reveal any potential degradation as soon as it occurs, so that substantial ecological damage can be avoided and before any remediation becomes extremely problematic.

- The biology of the species has to be sufficiently well known to differentiate the signal from the background noise (Figure 1.6). Indeed, intraspecific variability of the biological responses used as biomarkers has been often observed, making the interpretation of data difficult. Furthermore, it is important to be able to make use of control organisms held in parallel, in order to differentiate the changes caused by the contaminants from the natural fluctuations (seasonal, interannual) of the parameter studied. Finally, the characterization of the individuals must be perfectly mastered if the biomarkers to be studied show variation according to sex, age, etc.

The interest of the general public in particular species is also a criterion sometimes necessarily taken into account for the selection of sentinel species. This interest raises new criteria such as recreational value (fishing of salmonids, sports fishing of swordfish), economic value (commercial species), protection status of the species, and any role as flagship species as in the cases of charismatic megafauna such as dugongs, tortoises, seals and sea lions, whales and dolphins, or polar bears.

7.1.3 Active Biomonitoring

7.1.3.1 Context

Numerous biotic factors (age, size and weight, sex, stage of sexual maturity) are at the origin of any increase in the intraspecific variability of the responses of sentinel species to contaminant exposure. In order to work on statistically homogeneous populations and to try to be free from this intraspecific variability encountered with individuals taken in their natural environment, ABM by transplantation can be used (NRC 1991; De Kock and Kramer 1994; Olivier et al. 2002; Wepener et al. 2005; Ji et al. 2010; Lu et al. 2010). ABM uses a set of organisms of known age, and thus at the same stage of sexual maturity, homogeneous in size and weight. These organisms maintained in enclosures are then placed in sites of particular interest. Besides promoting a reduced variability in the responses of the organisms, this methodology also allows control of the period of exposure to contaminants and, at least in theory, to have individuals available for easy sampling at designated times (nevertheless, in the field, incidents such as loss of enclosures during storms or acts of malevolence are still to be feared!). ABM operations particularly allow the comparison of various sites, and the distribution of a single species in contaminated sites and/or in control ones. ABM has been extensively used in environmental toxicology, and has been shown to give clearer results than passive biomonitoring since organisms already present *in situ* may have adapted to the local pollutants. In aquatic ecosystems, transplanting mollusks and fish from a reference site to a polluted area is a feasible strategy for biomonitoring the effects of environmental changes. For example, this technique has been used since 1996 in the Biological Integrators Network on the Mediterranean coast (RINBIO

2002). Based on the bioaccumulative capacities of the mussels, this network required the implementation of transplants because of the relatively low availability of local mussels to provide cover along all the French Mediterranean Coast, and because no local closely related species fulfilled enough of the necessary criteria to be chosen as a replacement.

The use of transplants can only be envisaged if the conditions of the environment allow the species to acclimatize. Conversely, this technique can be also used to try to explain the causes of the absence of a species in an environment, the ecological conditions of which seem nevertheless to be adequate to meet its requirements, or to eliminate some hypotheses on such absences. An example is that of the mussel *Mytilus edulis*, absent from the Gironde estuary (with the exception of its mouth), whereas oysters (*Crassostrea gigas*) are widely present, although abiotic factors are nevertheless apparently favorable. By transplantations, Geffard et al. (2004) showed that the metal contamination of the estuary could not be the cause of this absence, the transplanted mussels developing there as well as oysters, with effective mechanisms of metal detoxification and without deleterious effects on their condition indices.

7.1.3.2 Implementation of Biomonitoring

7.1.3.2.1 Risk of Invasive Species

For commercial (furs, food products, etc.) and decorative purposes (stretch of water), in leisure activities (ornamental fish keeping, hunts, fishing) or as subjects of study, man has voluntarily introduced into his environment species that do not usually live there. Few of those succeed in prospering in this new environment, but those that do succeed can be the cause of real ecological disasters. By definition, transplanted sentinel species are able to acclimatize in their new environment, and thus to colonize it. Before deciding to introduce nonnative species in any ABM program, it is thus important to think about the consequences. So, in the example presented above (Geffard et al. 2004), the transplantation of mussels into the Gironde estuary appears to have been without consequences, because the mussels naturally have the opportunity of settling there, but do not so.

7.1.3.2.2 Life History

An important factor to be taken into account in the use of sentinel species is the exposure history of the transplanted individuals. In this respect, it is important to pay attention to the history of exposure to pollutants, but also to natural ecological parameters such as bathymetric level on the subtidal zone.

It has been well established for several decades that, within the same species, populations previously exposed to anthropogenic compounds are able to cope with pollutants more efficiently than "naive" individuals. Takamura et al. (1989) compared the responses of 118 strains of benthic cyanophyceae and eukaryotic algae, from rivers that are unpolluted or polluted by mining activities (Cu and Zn), on exposure to Cu by measuring their photosynthetic activity (Table 7.1). They showed that the exposure concentration of Cu that caused 50% inhibition of photosynthesis (IC_{50}) was positively correlated with the Cu concentrations in the river from which the algae had been isolated.

Berthet et al. (1992) showed that for identical contaminations with silver, carried out at the same season but with animals of different origins, oysters *Crassostrea gigas* of the same size and at the same stage of sexual maturity, could accumulate the metal in quantities varying by one order of magnitude. Furthermore, bioaccumulation at different levels was associated with different fractionations of silver between the cytosolic and particulate

TABLE 7.1

Parameters of Linear Regression Showing Relationship between Cu Exposure Concentration Giving an Inhibition of 50% of Photosynthesis and Cu Concentration in River from Which Algae Were Isolated

Taxa	*n*	*a*	*b*	*r*
Bacillariophyceae	47	1.2	0.79	0.68*
Charophyceae	10	1.8	0.65	0.76*
Cyanophyceae	19	1.2	0.36	0.60*
Chlorophyceae (Chlorococcales)	24	2.6	0.32	0.42*
Chlorophyceae (Chaetophorales)	9	1.3	0.77	0.64
Achnanthes minutissima (bacillariophyceae)	9	1.9	0.74	0.78*
Achnanthes linearis (bacillariophyceae)	4	1.9	0.58	0.75
Nitzschia palea (bacillariophyceae)	3	1.3	0.50	0.99
Synedra rumpens (bacillariophyceae)	3	0.7	2.4	1.00*
Chlorella vulgaris (chlorococcale)	4	2.5	0.64	0.59
Scenedesmus acutus (chlorococcale)	5	3.1	0.13	0.18
Klebsormidium klebsii (charophyceae)	3	1.7	0.64	0.98

Source: Takamura, N. et al., *J. Appl. Phycol.*, 1, 39–52, 1989.
* Significant correlation.

forms of storage in the oysters. These differences were attributed to different genetic characteristics having led to different responses of the organisms in the presence of contaminants. Furthermore, Wallner-Kersanach et al. (2000) showed that the rate of elimination of copper in oysters is influenced by their environmental history: the metal is more easily eliminated (30% after 30 days) in animals temporarily (60 days) transplanted into a copper-rich site, than in those which had permanently resided in the same contaminated site (elimination limited to 9% after 30 days). The importance of exposure history was also revealed by Nigro et al. (2006) and by Regoli and Principato (1995) studying, respectively, the differences in the sensitivity of genotoxic and cellular responses, and in the contents of glutathione and of some enzymes involved in the antioxidative defenses, of mussels native to a polluted zone or transplanted from a control zone. For early and very sensitive biomarkers (e.g., lysosomal membrane integrity and DNA strand breaks), the origin of the mussels had no influence, whereas for biomarkers of chronic contamination (the appearance of micronuclei or lipofuscin accumulation) a higher sensitivity was observed in specimens originating from the contaminated site (Nigro et al. 2006). Regoli and Principato (1995) did not observe differences between mussels subjected to a chronic exposure (natives) and those transplanted (acute exposure) in glutathione levels or in glyoxalase I activity, whereas different activities were observed for catalase and alkaline phosphatase between the two mussel groups. Bolognesi et al. (2004) noted that a parallel analysis of caged and wild mussels may improve the characterization of the studied area: wild mussels express higher genetic damage (micronuclei frequency) as a result of cumulative exposure, whereas, in contrast, caged mussels show higher DNA strand breaks as a result of recent exposure.

Chronic exposure to contaminants can lead to the acquisition of a certain tolerance in individuals (Amiard-Triquet et al. 2011). Natural selection tends to favor these individuals to the detriment of the ones that are more sensitive to the pollutants, and leads to an increase of the representation of resistance genes over successive generations (see Chapter 14). So, the environmental history of the parents must also be taken into account, and not only that of the animals intended for transplantation. This aspect can be illustrated by the work of Damiens et al. (2006), who studied the effects of copper contamination on some biomarkers [concentrations of metallothioneins (MT) and thiobarbituric acid reactive substances, catalase (CAT) activity] in oyster larvae (larvae D of *Crassostrea gigas*) obtained from parents that lived in areas differentially contaminated by this metal. They concluded that larvae of oysters from a strongly copper-contaminated estuary (Bidassoa) are better protected against the toxicity of the metal than the larvae of oysters living in an environment with low copper exposure (Bassin d'Arcachon). Collin et al. (2010) exposed to pesticides (lindane, metolachlor, carbofuran), at environmentally relevant concentrations, two populations of oysters from sites differing in pollution levels, particularly pesticide exposure: Arcouest (low pollution) and the Vilaine estuary (high pollution). They observed significant differential gene and protein expression between the two populations, and concluded that—as gene expression is often heritable within populations and supposing that the Arcouest and Vilaine oyster populations have different genetic backgrounds—these differences could be explained by a combination of environmental and genetic components.

Consequently, whenever it is possible, it is recommended, for bivalves in particular, to use farmed (mariculture) animals for transplantation. For species for which such breeding sources are unavailable, particular attention must be paid to the absence of pollution in the site of origin of the transplants.

If exposure history in terms of exposure to pollutants influences future responses of the organisms, it is also important to pay attention to history of exposure to other ecological parameters that may act as confounding factors. Camus et al. (2004) showed, for example, that according to their bathymetric position intertidally or subtidally, mussels are more or less accustomed to periods of emersion and immersion, and then react differently to the oxidizing stress generated by the metabolization of pollutants, or simply resulting from collection and transport during the transplantation process. More generally, Honkoop et al. (2003) have tested the impact on their physiology (scope for growth) of the stress of handling animals (i.e., mussels), moving them out of their habitat to a new habitat, and caging them. They have used combinations of treatment of mussels including reciprocal transplants, but also false ones consisting of moving mussels but putting them back into their place of origin or in zones of the same contaminant exposure. They concluded that to allow for the confounding effects of moving and disturbance on the physiology of the animals, correct control treatments should be included in the study.

7.1.3.2.3 Period of Latency

The use of transplanted sentinel organisms involves a period of latency between the arrival of the individuals in the environment and the beginning of sampling. According to Couillard et al. (1995a) or Geffard et al. (2002), several months are necessary for the organisms (bivalves in both cases) to reach a state of equilibrium with their environment in terms of their total accumulated metal concentration, and changes over time are also observed in the physicochemical forms of storage of the accumulated metals (Evtushenko et al. 1986; Couillard et al. 1995b; Mouneyrac et al. 1999). In the same way, the condition indices of the transplanted animals tend to get closer to those of the resident individuals, and to move away from those of individuals remaining at the site of origin, as, for example, for oysters transplanted from the Bay of Bourgneuf to the Gironde estuary, and conversely (Geffard, personal communication). Using double transplantations of mussels (*Mytilus galloprovincialis*) between a polluted harbor area and in a weakly impacted site, Regoli et al. (2004) were able to associate several biomarkers of oxidative stress [antioxidants, total oxyradical scavenging capacity (TOSC)] and changes in cellular structures (integrity of the DNA, lysosomal stability), and observed that not all have the same periods of change, and that the transplantation itself causes an antioxidant response in the initial phase of the experiment. Camus et al. (2004), carrying out the same type of experiment to estimate the impacts of the anthropological activities in the Venice urban zone and in various points of Venice lagoon, also demonstrated the influence of transplantation on biomarkers of oxidative stress. Camus et al. (2004) indicated that 2 weeks are necessary to observe significant differences in TOSC between mussels transplanted into the urban zone and those at the control site, whereas the return to the initial values in the control group is realized only after 4 weeks.

7.2 Concept of Ecosystem Approach

Any selection of a sentinel species has to be justified by a recognized link to the structure or the functioning of the ecosystem under surveillance, so that its use brings "a part of the complete puzzle," integrating the interrelations between the various constituents of the studied system. In recent years, several authors (Chapman 2002; McCarty et al. 2002; Galloway et al. 2004; Tabor and Aguirre 2004; Hagger et al. 2009) have expressed regret that the selection of sentinel species has been based more on anthropomorphic criteria than ecological ones. For Chapman (2002), the species used have too often been selected for their ease of collection and manipulation or breeding in laboratory, whereas their roles in the ecosystem (place in the food chain, the structure of the community, and their functions in the ecosystem) have been neglected. In order to introduce more ECO into ECOtoxicological programs, in natural environments as in the laboratory, it is advisable to study biomarkers in sentinel species that are more representative of the ecosystem. In a project entitled "The ECOMAN project," Galloway et al. (2004) recommend a multispecific approach where the species are selected on the variety of their habitats (rocky, sandy, muddy) in estuarine or coastal environments and of their food strategies (filter- or deposit-feeders, grazers or predators). Table 7.2 shows the various representative organisms of the main phyla of invertebrates that are easy to identify and are widely represented in the coastal zones of Northern Europe, selected for their biological and ecological complementarity.

However, if this ecological approach is indispensable, it is not sufficient as pointed out by Chapman (2002), who recommends focus on keystone species.

7.2.1 Keystone Species

Species interact with numerous constituents of the ecosystem. Where there are no ecological equivalents present, the disturbance of a species, such as by contaminant exposure, can cause deleterious modifications of the quantitative structure of the community, perturb

TABLE 7.2

Common Coastal Organisms from United Kingdom and Northern Europe Coasts Selected for Study in ECOMAN Program

Phylum	Species	Habitat	Feeding Strategy
Echinodermata	*Asterina gibbosa*	Rocky shores, rock pools	Omnivorous scavenger of microorganisms, seaweed and invertebrates
	Asterias rubens	Rocky mid-shore to sublittoral	Carnivore
Urochordata	*Ciona intestinalis*	Sporadic distribution around the lower shoreline	Sessile suspension feeder
Annelida	*Nereis diversicolor*	Burrows in estuarine mud, rocky, muddy sand	Omnivore
	Arenicola marina	Mid to lower shore sand	Deposit feeder
Mollusca	*Patella vulgata*	Intertidal rocky shore	Herbivorous grazer of microorganisms and algae
	Littorina littorea	Rocky coasts, estuaries, mud flats, mid shore to sublittoral	Grazer of microorganisms, detritus and algae
	Nucella lapillus	Mid shore to shallow rocky sublittoral	Carnivore
	Mytilus edulis	Below mid-tide level, rocky shores and estuaries	Sessile filter feeder
	Cerastoderma edule	Middle to lower shore, sand and muddy sand	Filter feeder
	Scrobicularia plana	High shore intertidal soft sediment, burrows up to 20 cm below the surface	Filter and deposit feeder
Arthropoda	*Neomysis integer*	Euryhaline inshore waters	Omnivore
	Carcinus maenas	Rocky coasts to soft sediment, estuaries	Omnivore

Source: Modified from Galloway, T.S. et al., *Mar. Pollut. Bull.*, 53, 186–194, 2006.

the spatial organization of the community or temporal successions therein. Such species are therefore keystone species. The examples that follow illustrate some levels of actions of these keystone species.

7.2.1.1 Roles in Food Webs

Some species have a particular importance in an ecosystem through their activities of predation or by the biomass that they themselves offer as prey.

- Among the approximately 3000 species of fishes present in coral reefs, which are among the most biologically productive and diverse ecosystems in the world and among the world's most fragile and endangered ecosystems, the humphead parrotfish *Bolbometopon muricatum* is the only predator of the coral whose action can be considered significant. Figure 7.2 shows its impact on the ecosystem: by its bioerosion,

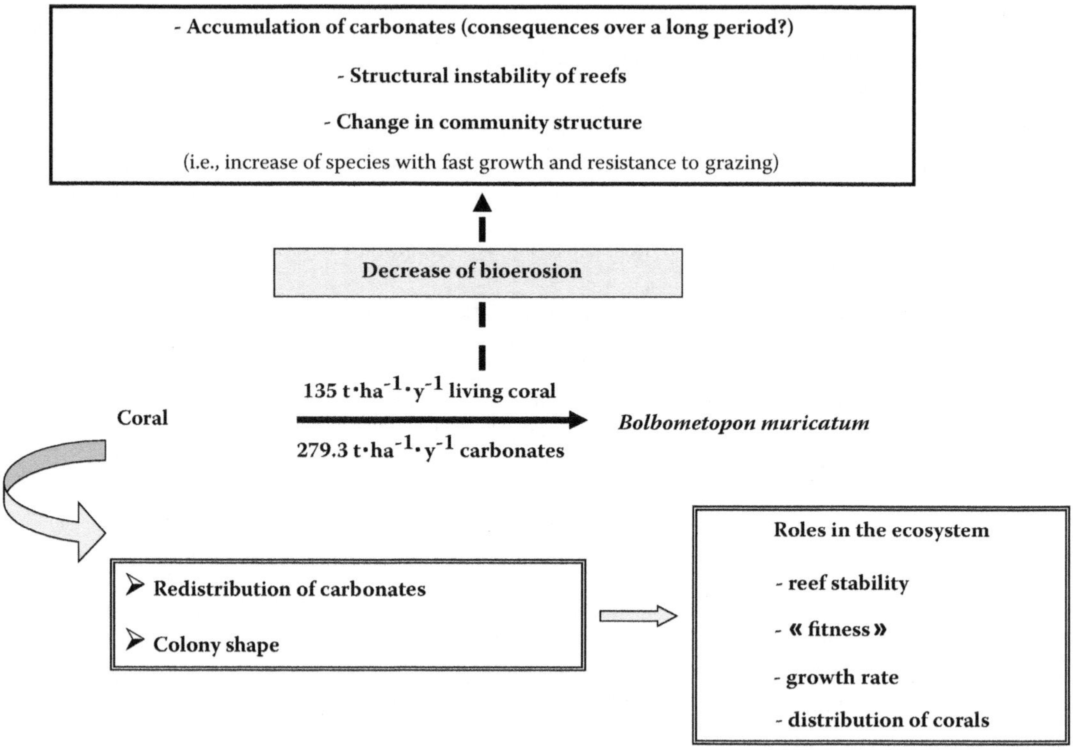

FIGURE 7.2
Impacts of the bioerosion by *Bolbometopon muricatum* on the coral ecosystem and the consequences of its depletion. (After Bellwood, D.R. et al., *Ecol. Lett.*, 6, 281–285, 2003.)

it intervenes in the growth of reefs and their spatial distribution, on the shape of coral colonies and their fitness, and on their stability in case of storms or of cyclones (Bellwood et al. 2003). The humphead parrotfish is also a major factor in the redistribution of carbonates in the ecosystem. According to these authors, its depletion because of pollution and overexploitation is one of the reasons that can explain the regression of coral reefs and the observed changes in the structure of coral communities. These same authors are also concerned, in the longer term, about consequences, which could be dramatic, of the accumulation of carbonates in the reef system.

- In northwest European estuarine areas, the endobenthic annelid *Nereis diversicolor* (Figure 7.3a) represents an important biomass; according to the environmental conditions, its biomass can vary from 39 to less than 1 g m^{-2} (Scaps 2002). In the same manner, the infaunal tellinid bivalve *Scrobicularia plana* (Figure 7.3b) plays a key role in the structure and the functioning of estuarine and coastal mudflats because of its density of up to 2000 ind m^{-2} (Orvain 2005). Both species are common prey for numerous predators (crabs, fishes, birds), and so they are particularly important food reserves, the rarefaction of which can have important consequences for all the species for which they constitute the basic food source.
- The work of Weis et al. (2001, 2011) has shown the cascading effects of imbalance that can result from the decrease in the density of a population of predators and/or of prey species, and from any decrease in their capacity, respectively, to carry out predation or to escape from predators.

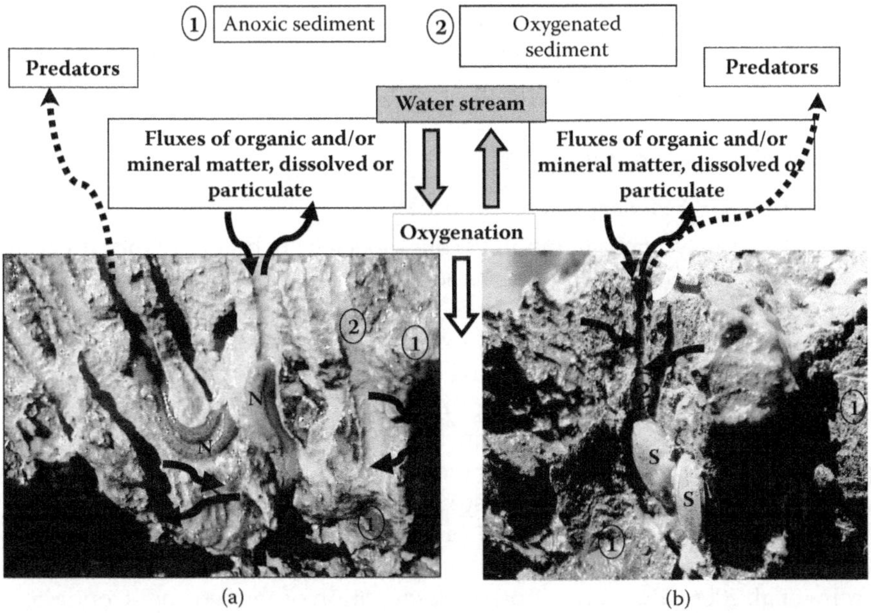

FIGURE 7.3
(a) *Nereis diversicolor* and (b) *Scrobicularia plana*: keystone species of the sedimentary estuarine compartment through their biomass and their activities of bioturbation.

These authors, studying the ecological consequences of the effects of contaminants on the behavior of the mummichog *Fundulus heteroclitus* (a fish) and its main prey, the grass shrimp *Palaemonetes pugio*, showed that the capacity of mummichog for predation was decreased by contaminant exposure, such that they then consume more detritus that are of low food value. This lower quality nutrition supply, associated with hormonal and/or neurological disturbances, in turn, makes the mummichog more vulnerable to predation by other fish or by the blue crab *Callinectes sapidus*. The changes in the behavior of *F. heteroclitus*, a keystone species of the estuarine ecosystem, and modifications of the structure of its populations, thus generate in turn changes in the whole ecosystem, in particular by leading to an increase in the population density and the size of the local grass shrimp, but also a modification of the quantity of available prey for higher levels in local food chains (Weis et al. 2001).

However, the effects on community structure go beyond only the trophic relations between some species: Kneib and Stiven (1982), studying the effects of various classes of size of killifish, *F. heteroclitus*, on the abundance of diverse benthic invertebrates, showed that the density of the benthos was bigger in cages containing large adult individual fish. Kneib (1988) highlighted that, by controlling populations of the grass shrimp *P. pugio*, the large killifish had an indirect positive effect on the dwarf mud anemone *Nematosella vectensis*.

Recently, Weis et al. (2011) completed their behavioral and ecological observations on the effects of pollutant exposure on other common species living in the same ecosystem (common killifish; grass shrimp; blue crabs; fiddler crabs, *Uca pugnax*; and bluefish, *Pomatomus saltatrix*). They showed for each species of the ecosystem, alterations of feeding and trophic interactions, but with different consequences for each. They pose questions for future studies: (1) Are organisms at higher trophic levels more likely to be affected by pollutants? (2) Are certain taxonomic groups more likely to be successful under contaminant stress? (3) Are certain traits associated with being more successful in contaminated areas? (4) Are the particular contaminants at a given site important in determining which organisms are more susceptible than others? Although they provide elements of answers to some of these questions, Weis et al. (2011) call for further work to confirm or generalize their results.

7.2.1.2 Roles in Biogeochemical Cycles

7.2.1.2.1 Bioturbation of Sediment and Transformation of Organic Matter

Within the sedimentary compartment, certain species such as the annelid *N. diversicolor* (Figure 3.7a), by their activities of bioturbation, play a key role in the transfer of pollutants, or nutrients, between the sediment and the overlying aqueous phase (Mermillod-Blondin et al. 2005; Banta and Andersen 2003). By excavating burrows in the sediment, the worms increase the surface area of exchange between sediment and the water; they move organic and mineral material including various pollutants, modifying their distribution and promoting their homogenous distribution in the sediment column; and they contribute to the oxygenation of the sediment close to the burrows. Furthermore, because they usually feed on detrital organic matter on the surface of sediments, *N. diversicolor* significantly influence the speed of decomposition of organic matter in the estuarine ecosystem, and any disturbances of their activities will have a direct repercussion on this decomposition rate (Moreira et al. 2006). By their action of oxygenation of the sediment, endobenthic polychaetes such as *N. diversicolor* or *Arenicola marina* also stimulate the local aerobic microflora and so strengthen their impact on the processes of organic matter degradation and biotransformation of xenobiotics (Banta and Andersen 2003). Timmermann et al. (2011) studied the effects of bioturbation by the common lugworm *A. marina* on the fate of oil hydrocarbons (alkanes and polycyclic aromatic hydrocarbons) *in situ* during a simulated oil spill in a shallow coastal area in a fjord in Denmark. They observed stimulation of both physical removal and microbial degradation of oil compounds in deeper sediments layers by *A. marina*, whereas the fate of oil compounds deposited in surface layers was affected by other bioturbating infauna such as *N. diversicolor*, *Corophium volutator*, and *Hydrobia* spp. Studying the release of a polychlorobiphenyl (PCB) ([^{14}C]PCB 32) from Baltic Sea sediment to water, Hedman et al. (2009) indicate that biological reworking of the sediment by the amphipod *Monoporeia affinis* (a particle biodiffuser) or the polychaete *Marenzelleria* sp. (a bioirrigator) can be just as, or even more, important than physical resuspension for the remobilization of sediment-bound contaminants.

In freshwater ecosystems, various species contribute to the decomposition of leaves including the crustaceans *Gammarus pulex* and *Asellus aquaticus*, and larvae of insects. Their feeding is a major rate-limiting step in the processing of stream detritus. Thus, these detritivores have a crucial role in the biogeochemical cycle of organic matter, and if they are impacted by pollutants in their environment, this function is no longer performed or at least is reduced (Maltby et al. 2002; Kunz et al. 2010 and literature cited).

7.2.1.2.2 Role of Primary Producers in Biogeochemical Cycle of Xenobiotics

Pigeot (2001) studied the role of various benthic communities (88 species or communities) on the fate of certain metals (Cd, Cu, Pb, and Zn) in a coastal ecosystem, the Marennes-Oléron basin, Fr. He observed that the benthos stores, on average, 8 kg cadmium, 65 kg lead, 450 kg copper, and 6 tons zinc. Although oysters store the majority of cadmium (47.9%), copper (89.3%), and zinc (94.5%), it is the primary producers (microphytobenthos + dwarf eelgrass, *Zostera noltii* + sea purslane, *Atriplex portulacoides*) which immobilize the great majority of lead (90.3%), while playing a significant role in the storage of the three other elements (Figure 7.4a–d). These primary producers, in particular the microphytobenthos, thus play a major role in the transfer of these elements to microphagous, suspension-feeding, deposit-feeding (including sediment-ingesting) animals, and consequently to all aquatic food webs. However, when integrating in a temporal dimension (rate of renewal of the various organisms), the fluxes are much higher than the stored quantities (200 kg $year^{-1}$, 4 tons $year^{-1}$, 4 tons $year^{-1}$, and 12 tons $year^{-1}$ for Cd, Pb, Cu, and Zn, respectively). Furthermore, because of their very high productivity, this mobilization is essentially attributable to the primary producers, in particular, the microphytobenthos (Figure 7.4e–h).

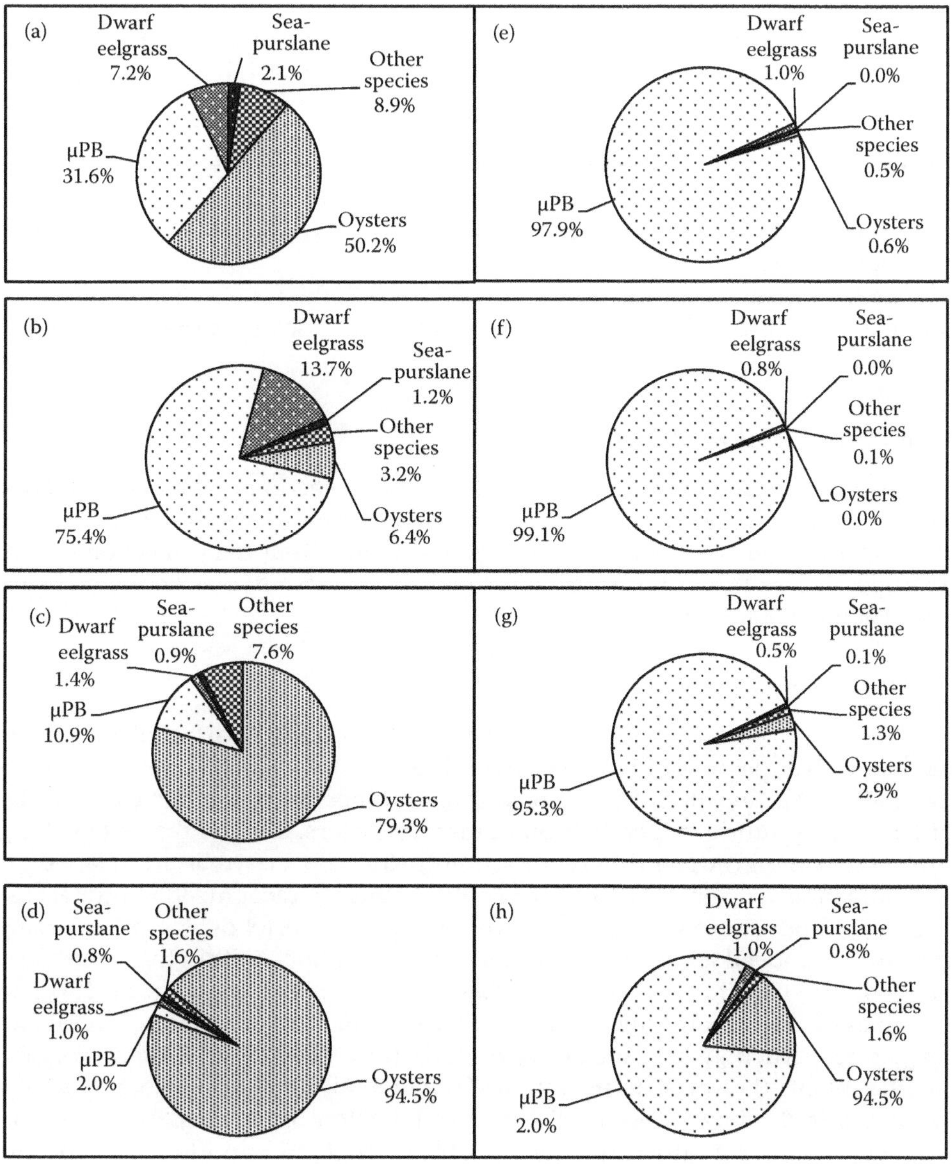

FIGURE 7.4
Quantities (%) of (a) Cd, (b) Pb, (c) Cu, and (d) Zn fixed by the major benthic organisms and fluxes (%) of (e) Cd, (f) Pb, (g) Cu, and (h) Zn associated with these organisms in the Marennes-Oléron basin (according to Pigeot 2001). μPB, microphytobenthos.

7.2.1.2.3 Filtering Activities

Filter-feeding species, in particular bivalve mollusks or sponges, are well known for their action on the cycling of organic matter. Indeed, by their activity of filtration of the seston, they accelerate the sedimentation of particularly organic matter into biodeposits via feces and/or pseudofeces, as well as the pollutants associated with the organic material (Dobson and Mackie 1998; Björk et al. 2000). Thus, filter-feeders are responsible for an important part of the transfer of pollutants between the water column and the sediment, with, as a consequence, a modification of their retention times in coastal zones (Björk et al. 2000), a decontamination of the aqueous phase and in parallel an enrichment of the sediment, as well as an increase in the bioavailability of the pollutants to deposit feeders including sediment ingestors (Turner and Millward 2002). Cho et al. (2004) have investigated, under laboratory conditions, the extent of biodeposition and bioconcentration of PCBs by zebra mussels (*Dreissena polymorpha*) collected from the Hudson River. They concluded that, since zebra mussels filter a wide range of particles, they can increase not only the residence time of PCBs in the Hudson River through redeposition to sediments, but also the transmission of PCBs to animals up the food chain, through exposure to benthic organisms at a lower trophic level.

By taking into account these various keystone species, the functioning of the ecosystem will doubtless be protected, but this will not resolve the problem of the conservation of its biodiversity.

7.2.2 Conservation of Biodiversity

In this respect, two questions in particular arise: the protection of the most sensitive species, and the extrapolation of the data acquired on keystone species to all the species constituting the ecosystem. A commonly held proposition is that the protection of the most sensitive species, or of their most sensitive biological stages, in an ecosystem ensures the protection of the whole community.

7.2.2.1 Interspecific Variability

The validation of a sentinel species is based on the possibility of extrapolating the results obtained on such a species to all the species of the ecosystem. However, even between very closely related taxa, sensitivity to xenobiotics and other contaminants can vary by several orders of magnitude, depending on numerous factors, such as the taxonomic group or, within the same group, a different natural history. Species that are relatively sensitive or, contrastingly, insensitive to a toxic contaminant are recognized in numerous taxonomic groups, from microorganisms to invertebrates and vertebrates, from microalgae to higher plants, as indeed are contaminant-tolerant populations within the same species. In a recent book, *Tolerance to Environmental Contaminants* (Amiard-Triquet et al. 2011), numerous examples of interspecific and intraspecific variability are described. Several chapters are devoted to particular taxa including bacteria (Imfeld et al. in Amiard-Triquet et al. 2011), higher plants (Frérot et al. in Amiard-Triquet et al. 2011), and insects (Ghanim and Ishaaya in Amiard-Triquet et al. 2011), whereas Johnston (in Amiard-Triquet et al. 2011) and Berthet et al. (in Amiard-Triquet et al. 2011) highlighted and illustrated the fact that species belonging to most of the major taxonomic groups, and organisms living in different media (terrestrial, freshwater, estuarine, and marine organisms), show examples of the development of tolerance to chemicals. Some examples chosen from different taxa further illustrate these points.

7.2.2.1.1 Microorganisms and Invertebrates

- The effects of Cd, Cu, and Zn were examined in numerous species of freshwater microorganisms by taking as a reference parameter the dissolved exposure concentration of metals associated with IC_{50} for photosynthesis (Takamura et al. 1989). In Figure 7.5, these data are shown for three taxa, selecting only strains from environments where the level of metal contamination was low, in order to eliminate any interference by the presence of any acquired tolerance. Cyanophyceae (blue green bacteria) are sensitive to the three metals, whereas Chlorophyceae (green algae) tend to show less sensitivity. Within each of these taxa, the interspecific differences of IC_{50} can reach 1 (Cd, Zn) to 2 (Cu) orders of magnitude.
- Bielmyer et al. (2010) exposed three species of coral to dissolved copper (ranging from 2 to 20 μg L^{-1}) for 5 weeks: the western Atlantic species *Acropora cervicornis* (a threatened species of staghorn coral), *Montastraea faveolata* (star coral), and the Indo-Pacific species *Pocillopora damicornis* (cauliflower coral), each containing different algal symbionts (*Symbiodinium* A3, C1, or D1a). The three coral species exhibited significantly different sensitivities to copper, with toxic effects occurring in *A. cervicornis* and *P. damicornis* at copper concentrations as low as 4 μg L^{-1}. Copper exposure affected zooxanthellar photosynthesis in *A. cervicornis* and *P.*

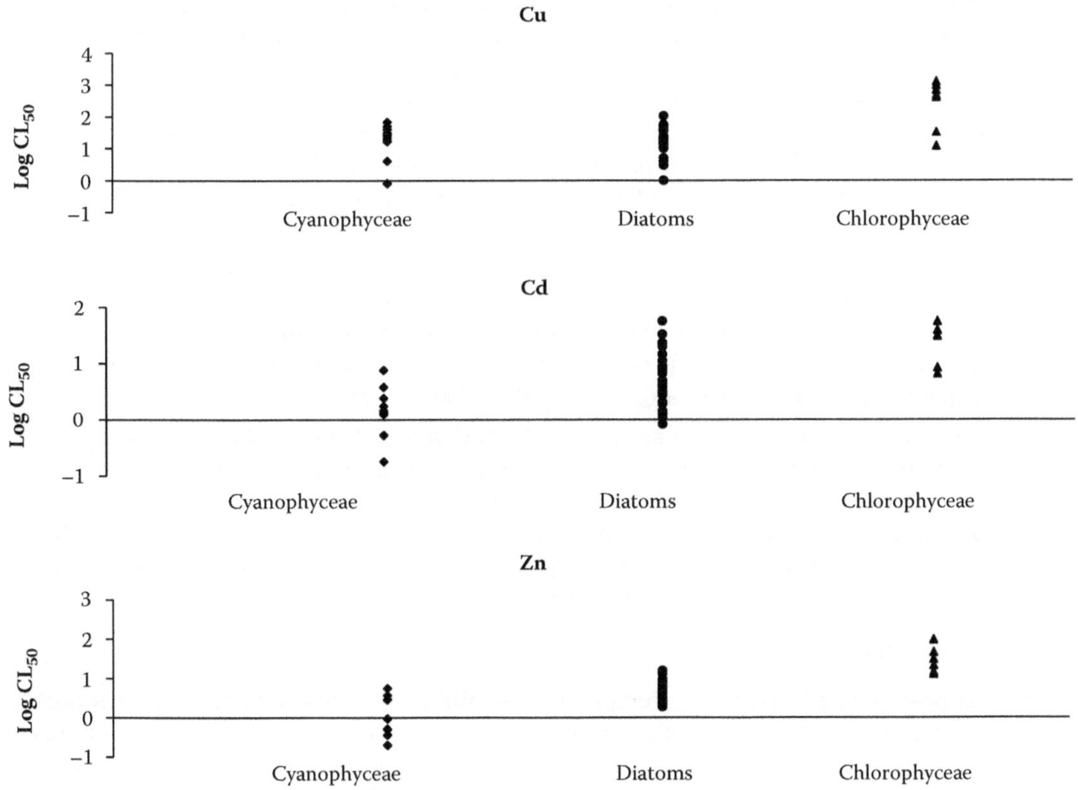

FIGURE 7.5
Comparison of the inhibition of photosynthesis by metals in different taxa of microorganisms. Log IC_{50} in μg L^{-1} (Cu) or mg L^{-1} (Cd, Zn). (After Takamura, N. et al., *J. Appl. Phycol.*, 1, 39–52, 1989.)

damicornis, and carbonic anhydrase was significantly decreased in *A. cervicornis* and *M. faveolata*. Likewise, significant decreases in skeletal growth were observed in *A. cervicornis* and *P. damicornis* after copper exposure.

- De Pirro and Marshall (2005) compared the physiological and mortality responses to copper (Cu^{2+}) of limpets from two different gastropod mollusk lineages: the Prosobranchia (Patellogastropoda: *Helcion concolor* and *Cellana capensis*) and the Pulmonata (*Siphonaria serrata* and *Siphonaria capensis*). The *Siphonaria* limpets were nearly one order of magnitude more insensitive to copper than the patellogastropod limpets. Moreover, copper exposure (0.25 ppm Cu^{2+} for 2 h) induced heart rate depression in *Siphonaria* spp. (to around 50% of the baseline rate), whereas their tissue copper concentrations remained at relatively high control levels. Copper exposure (0.25 ppm Cu^{2+} for 2 h) had no effect on heart rate of the patellogastropod limpets, but led to a significant increase in tissue copper. These authors suggest that the better regulation of internal copper levels by *Siphonaria* may be related to their exclusive possession of the Cu-requiring respiratory protein hemocyanin in the blood, and that a dependence on relatively high aerobic metabolism by the patellogastropod limpets would limit their capacity for isolation and pollutant avoidance.
- Responses to exposure to dissolved silver were compared between four filter-feeding bivalve mollusks, closely related phylogenetically and sharing a common mode of feeding. Sensitive species (a scallop *Chlamys varia* and a mussel *Mytilus galloprovincialis*) and more insensitive ones (an oyster *Crassostrea gigas* and a zebra mussel *Dreissena polymorpha*) were discerned. Toxic impairments were very variable according to the species: respiration in the mussel, byssus secretion in the scallop, transitory energy impairment in the oyster. Relative sensitivity to silver exposure was not based on an interspecific difference in physiological patterns of silver bioaccumulation because the (sensitive) scallop and the (insensitive) oyster are strong bioaccumulators, whereas both mussels (one is sensitive, the other one tolerant) are poor bioaccumulators of silver (Berthet et al. 1992).
- In two species of crustaceans, the isopod *Platynympha longicaudata* and the caridean decapod *Palaemon (as Leander) intermedius*, exposed *in situ* to the impact of a lead smelter discharge, it was shown that this preexposure had generated an increased tolerance to metals only in the case of the first of these two species during later exposures in laboratory (Ross et al. 2002).
- Particular interest has been shown to the tolerance of the nontarget species that are exposed during the agricultural use of pesticides. Strains of daphnids (freshwater crustaceans) can develop a genetic resistance to ethyl-parathion, allowing them to survive doses 10 times higher than the sensitive strains. However, this remains very modest compared to insects where the range reaches a factor of 100 (Barata et al. 2001 and quoted literature).

7.2.2.1.2 Vertebrates

- It is often considered that interspecific variability of sensitivity to contaminants is connected to the very high variability of the physiological adaptive strategies of invertebrates. In the vertebrates with more developed homeostatic processes, a higher homogeneity of responses might therefore be expected. In fact, for four species of marine teleost fish, interspecific differences in behavior were shown in response to the same sublethal dose of copper. These differences were attributed to a differential sensitivity of the neurophysiological systems between the species, or to differences in the rate of uptake of the pollutant in connection with the relative surface of the body and the gills, knowing that copper and other metals can affect the coat of mucus present on the skin and gills of fish (Scarfe et al. 1982). Kalman et al. (2010) compared the toxicity of Cd (intraperitoneal injection of 2.5 mg $CdCl_2$ kg^{-1} body weight) in two commercial fish species, *Sparus aurata* and *Solea senegalensis*. These authors showed that MTLP synthesis in *S. aurata* probably led to an enhanced ability to cope with the toxicant, whereas MTLP induction in *S. senegalensis* was inhibited by the accumulated cadmium, as reflected by the mortality rate of this species; the authors suggest that MTLPs did not play a main role in metal detoxification by metal sequestration in this latter species. Eyckmans et al. (2011) have investigated differences in a Cu-induced oxidative stress response between three freshwater fish species [rainbow trout (*Oncorhynchus mykiss*), common carp (*Cyprinus carpio*), and gibel carp (*Carassius auratus gibelio*)]. They exposed fishes to sublethal Cu concentrations and determined antioxidative enzyme (superoxide dismutase, glutathione reductase and CAT) activities and antioxidant (reduced glutathione and reduced ascorbate) concentrations in gill samples collected from 1 h to 1 month of Cu exposure. They concluded that rainbow trout might use their high reduced GSH concentrations as a first defense against metal exposure, whereas gibel carp and common carp rely more on MT in combination with antioxidant enzymes.

In fishes, interspecific differences of toxicity of organophosphate pesticides can exceed one to two orders of magnitude. Differences of toxicity of diazinon, for example, may be explained by differences in the features of the target enzyme [acetylcholinesterase (AChE)] and in the metabolization of the compound in the liver (Keizer et al. 1995). Studying the response of brain AChE to methyl paraoxon in more than 20 species of marine fishes, Oliveira et al. (2007) suggest a possible linkage between brain AChE activity levels and the phylogenetic origin of the fish. To use AChE as biomarker for pesticides in the marine environment, these authors recommend as choices as sentinel species those fish that are among the most pesticide-sensitive for they have the highest levels of brain AChE.

7.2.2.1.3 Consequences at the Community Level

As shown by the previous various examples, biological communities are composed of different species, and the inherent sensitivity of each is highly variable. In a contaminated environment, the most sensitive organisms are lost, whereas relatively insensitive or tolerant ones survive and are able to reproduce, leading to a new community that is—as a whole—more tolerant to the contaminant concerned.

In situ, numerous studies carried out on oil spills have shown that crustaceans, in particular amphipods, are very sensitive to hydrocarbons, whereas their impact on polychaetes is much more limited. In a similar manner, copepods seem to be more sensitive to hydrocarbons (and also to anoxia) than nematodes (Gómez Gesteira and Dauvin 2000). However, within these groups, an increase in the abundance of opportunist organisms (opportunist polychaetes, oligochaetes, and nematodes) has been observed, in addition to a diminution and a low abundance of the species considered sensitive to the pollution (Gómez Gesteira and Dauvin 2000 and quoted literature). Similar observations have been reported by Wentsel et al. (1977), who showed that the populations of oligochaetes increase in lacustrine sediments of contaminated areas (Cd, Zn, and Cr), whereas on the contrary those of chironomids disappear in these same sites. In contrast, the populations of oligochaetes are minimal in sediments in this same lake that are slightly or not contaminated, whereas chironomid populations here are at their greatest.

In response to the European Water Framework Directive frame (EC 2000) on the use of indices and benthic bioindicators to estimate the quality of coastal and estuarine waters, various indices based on differences in sensitivity between species and species succession along a gradient of organic pollution have been developed. Dauvin and Ruellet (2007) proposed an index based on the ratio of the opportunist polychaetes to amphipods (benthic opportunistic polychaetes/amphipods). Of use in regions of organic pollution, this index can also be used in the case of pollution by hydrocarbons. Recently, this ratio has been tested successfully by Andrade and Renaud (2011) along the Norwegian continental shelf for offshore oil and gas impacts.

This differential sensitivity of organisms is also the basis of the "pollution-induced community tolerance" concept proposed by Blanck et al. (1988), developed and exemplified by Tlili and Montuelle (in Amiard-Triquet et al. 2011) and Imfeld et al. (in Amiard-Triquet et al. 2011) for microorganisms, but also demonstrated in several studies on periphyton (Blank et al. 2003; Rotter et al. 2011) or nematodes (Millward and Grant 2000; Beyrem et al. 2010). Neira et al. (2011) examined Cu contamination effects on macrobenthic communities

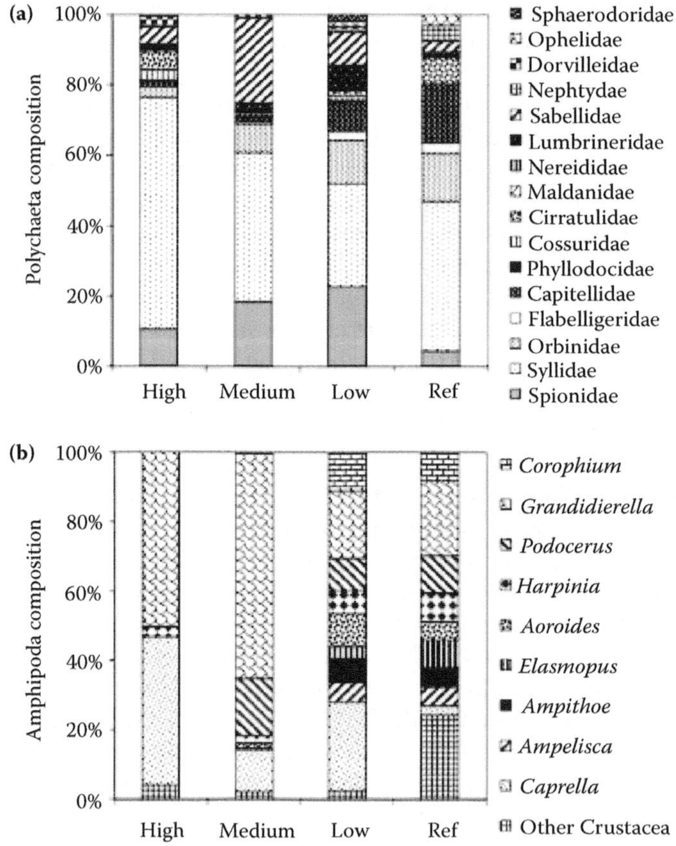

FIGURE 7.6
Percentage composition of polychaete families (a) and peracarid crustaceans (b) at different study sites. (Reprinted from Marine Pollution Bulletin, 62, Macrobenthic community response to copper in Shelter Island Yacht Basin, San Diego Bay, California, 701–717, Copyright 2011, with permission from Elsevier.)

within Shelter Island Yacht Basin (California). They identified 48 taxa with annelid polychaetes (16 families) and peracarid crustaceans (one class and three orders) as the dominant groups (overall >77% of total). Among the polychaetes, there was a higher richness at "low" Cu and reference sites, than at higher Cu sites (Figure 7.6a). Syllid polychaetes were the dominant group overall, and syllids had an enhanced percentage contribution in the "high" contaminated sites, whereas the percentage contributions of spionid, orbiniid, and capitellid polychaetes were lower at "high" Cu sites (Figure 7.6a). Among the crustaceans, peracarids were the most affected when sediment Cu was high (Figure 7.6b); the number of amphipods was reduced to only four species (with a dominance of two of them, representing ≈93% of the total crustaceans) in these contaminated sites, whereas at "low" Cu and reference sites, 10 species were present, in relatively equitable proportions (Figure 7.6b).

7.2.2.2 Intraspecific Variability

7.2.2.2.1 Sensitive Stages of Organisms

The reproductive and young growth phases of organisms attract the attention of ecotoxicologists who consider that they are the most sensitive stages, and therefore key to the preservation of the species in the ecosystem. Pessoa et al. (2011) exposed early life stages of tilapia *Oreochromis niloticus* at 9 days post-hatch in semistatic experiments to sublethal concentrations of the carbamate insecticide carbofuran (8.3–397 μg L^{-1}). Their results showed effects on cholinesterase activity and a suite of behavioral endpoints (vision, growth and predator avoidance skills at environmentally significant concentrations, and effects on swimming and prey capture only at relatively high concentrations) that are directly related to important ecological mechanisms relevant for their survival, growth, and recruitment to the adult population. In a synthesis on the use of embryos and larvae of mollusks in toxicity tests published in 1999, His et al. confirmed the hypothesis of the relative sensitivity of early life stages, indicating, however, some considerable exceptions such as resistance to cadmium or insecticides. In order to test the total toxicity of dredging materials, Alzieu et al. (2003) considered bioassays of embryotoxicity in the oyster *Crassostrea gigas* or tests on copepodites (juvenile stages) of the copepod crustacean *Tigriopus brevicornis* to be sensitive and discriminating, but reported young sea bass or the nauplius larvae of brine shrimp (*Artemia*) to show low sensitivity. Other authors have also demonstrated that larvae or young do not systematically always show the greatest sensitivities to pollutants. For example, Lotufo and Fleeger (1997) exposed two species of meiobenthic copepods (*Schizopera knabeni* and *Nitocra lacustris*) to phenanthrene, and observed that, whereas for the first species the nauplius stage is effectively the most sensitive, for the second, it is the mature females which present a higher sensitivity than all other stages. Stubblefield et al. (1999) showed that for rainbow trout, *Oncorhynchus mykiss*, the young were approximately three times less sensitive to the toxic effects of zinc and cadmium than the adults. Furthermore, these authors indicated that the acclimatization of young rainbow trout to sublethal concentrations of cadmium increases their tolerance by a factor of 25, whereas this tolerance is increased only by 4.8 for the adults. Moreover, as demonstrated by Kammenga and Riksen (1996), the intensity of the response of a population to xenobiotics is not always determined by the most sensitive biological stage. Indeed, exposing two species of nematodes to copper, they showed that the population growth rate of the two species is reduced in a similar manner, whereas one of these species shows a higher sensitivity for other life history traits (juvenile survival, duration of the juvenile period and of the period of reproduction, and the daily rate of reproduction).

7.2.2.2.2 Origins of Sentinel Species

Berthet et al. (in Amiard-Triquet et al. 2011) have reviewed the sources of intraspecific variability of tolerance in sentinel species exposed to contaminants, which may lead to differences in survival of organisms. They discussed the use of wild populations versus "laboratory rats," of populations taken in stable environments versus fluctuating ones and of naive populations versus exposed ones. Concerning the first point, it is often considered that the genetic impoverishment of laboratory populations is an important bias in ecotoxicological tests. According to the origin of the strain (a relatively clean or polluted site), the clones used in ecotoxicological bioassays may vary in their tolerance to xenobiotics and other contaminants, and the results obtained may thus lead to under- or overestimation of the risk for a natural population. This can be taken into account by the application of safety factors. On the second point, Berthet et al. (in Amiard-Triquet et al. 2011) indicate that in laboratory studies, test organisms are almost always kept in optimal conditions of temperature, oxygen, salinity, food, etc., in order to optimize performance in the control, and to isolate the effects of the tested chemical. On the contrary, in their natural environment, organisms are rarely in optimal conditions, and have to cope with environmental stresses. Thus, for a long time, euryoecious species, such as estuarine organisms, have been considered to be more tolerant to xenobiotics than stenoecious ones, but this assertion has still not been verified (Berthet et al. and Hummel et al. in Amiard-Triquet et al. 2011), and that a substantial part of uncertainty in toxicity tests could be explained by natural factors.

The mechanisms that allow organisms to cope with the presence of various contaminants in their environment have been recently reviewed by Johnston (in Amiard-Triquet et al. 2011) and Rainbow and Luoma (in Amiard-Triquet et al. 2011) for metal tolerance, Regoli et al. (in Amiard-Triquet et al. 2011) for antioxidant defenses, Roméo and Wirgin (in Amiard-Triquet et al. 2011) for biotransformation of organic contaminants, and Mouneyrac and Roméo (in Amiard-Triquet et al. 2011) for stress proteins. They may be individual physiological acclimations after exposure to a contaminant (mechanisms of defense); in this case, the tolerance may be lost if exposures are not maintained, or genetic adaptations (heritable) when a selective force applied by a contaminant is maintained across multiple generations.

In all cases, both inter- and intraspecific variabilities of sensitivity of organisms to pollutants have to be taken into account in order to correctly interpret the results of toxicity tests, and to avoid, as much as possible, any under- or overestimation of risk. Because the sensitivity of a given species depends on the pollutant involved, the ecotoxicological endpoint measured, and the type of test used, several species and endpoints have to be investigated in risk assessment, in order to reach a sound extrapolation at the level of the whole ecosystem.

7.2.3 Choice of Physical Compartment

In the aquatic environments, ecotoxicological research has, over a long period, mainly concerned species that are representative of the water column. On the other hand, although sediment quantitatively represents the most contaminated environment, both by organic and inorganic pollutants (Gagnon and Fisher 1997; Bernes and Naylor 1998), the endofauna and endoflora were relatively little taken into account during this period. Moreover, Chapman and Wang (2001) underline that most methods of evaluating the toxicities of sediment are especially applicable to freshwater or marine sediments, but few take into account the specific characteristics of estuaries. These authors attempted to produce an inventory of the species used in tests of the overall toxicity of estuarine sediments. They noted a relative phylogenetic poverty, with a wide dominance of crustaceans, in particular amphipods, whereas they wondered at the absence of some taxa such as the aquatic oligochaetes or the chironomids (midge larvae), and at the low representation of fish. In the same work, they regret that too many tested species are more marine than estuarine, and that allowances for salinity levels are insufficient. Authors too often try to fit the salinity of a sediment to the available model species, thereby modifying the physicochemical forms of pollutants and therefore their bioavailability and potential toxicity, rather than testing the relevant sediment while maintaining its initial physicochemical characteristics, and selecting the species to be tested on these criteria. Chapman and Wang (2001) thus end on the necessity to develop studies of this physical compartment in estuaries, regions that are particularly subjected to anthropogenic contributions, often in complex mixtures, and to pay more attention to the physicochemical conditions of the laboratory tests in order to be able to compare the laboratory results to those *in situ* with more justification. In this perspective, few endobenthic estuarine species have been selected as biological models in ecotoxicological programs, typically the ragworm *Nereis diversicolor* and the bivalve *Scrobicularia plana*, which are indeed keystone species of the estuarine ecosystem. With the aim of a general appreciation of the quality of the estuarine environments, these two deposit feeding estuarine invertebrates appear as an interesting complement to filter feeders representative of the quality of the water column, allowing a wider appreciation of the contamination of sediments *in situ*, and over relatively long periods (Moreira et al. 2006; Poirier et al. 2006; Coelho et al. 2008; Sole et al. 2009; Fossi Tankoua et al. 2010, 2012). In Quebec, it is a suspension feeding endobenthic bivalve mollusk, *Mya arenaria*, which has been retained as sentinel species (Siah et al. 2003; Frouin et al. 2007; Gagné et al. 2009). Indeed, ubiquitous in the northern hemisphere (through anthropogenic introduction), this clam, besides its local economic interest, plays a key role in estuarine food webs in Quebec.

Among fishes, the European eel, *Anguilla anguilla*, a bottom-dwelling species has been chosen by Piva et al. (2011) and Benedetti et al. (2012) as a sentinel species for assessing sediment quality because of its frequency in European estuaries, brackish and freshwater basins, and its tolerance to a wide range of biotic and abiotic factors. Moreover, ecotoxicological studies have shown their capacity to accumulate elevated concentrations of chemicals due to their high lipid content (Maes et al. 2005; Belpaire et al. 2011), and several biomarkers have been characterized that respond in this fish to the main classes of environmental pollutants (Van der Oost et al. 1996; Valbonesi et al. 2011).

7.3 Species Used as Sentinels

Many species, animal or plant, can be used to provide indications on the quality of the environment, but their performance and hence suitability in this role may well turn out to differ according to the biomarker chosen. This factor, added to the list of criteria already evoked in the choice of a sentinel species, may further quickly limit the choice for a particular objective (Depledge and Hopkin 1995; UNEP/RAMOGE 1999).

In a review on the use of MT as a biomarker, Amiard et al. (2006) show that for this biomarker, three taxa of invertebrates are mainly used in coastal and estuarine regions where they play key roles: mollusks, crustaceans, and annelid worms. Figure 7.7 shows that among the mollusks, it is the bivalves that are mostly used in this role (65%), whereas decapods dominate the crustaceans (70%), and polychaetes are the annelids most often quoted (81%).

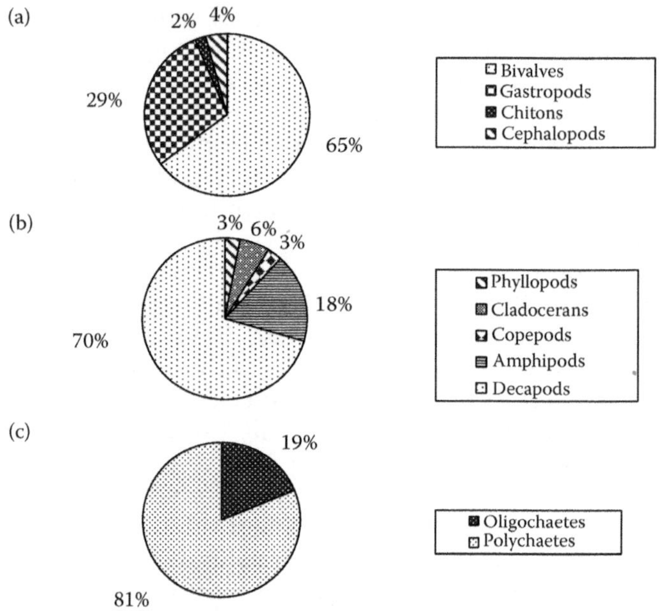

FIGURE 7.7
Percentages of species of mollusks (a), crustaceans (b), and annelids (c) in which metallothionein induction by metals has been shown. (After Amiard, J.C. et al., *Aquat. Toxicol.*, 76, 160–202, 2006.)

Although mollusks are often recommended for use to estimate lysosomal membrane integrity, DNA strand breaks, or MT concentrations, they are, on the other hand, of less interest in investigations, for example, of the catalytic activity of P450. Other taxa are then recommended such as particular teleost fishes: red mullet (*Mullus barbatus*) or bass (*Serranus* spp.) (UNEP/RAMOGE 1999). In reverse again, although only few data for glutathione-*S*-transferase (GST) are supplied by fish (Roméo et al. 1994; Castro et al. 2004), GST is successfully used as biomarker of exposure in mollusks (Roméo et al. 2003; Buffet et al. 2011) and annelids (Durou et al. 2007; Buffet et al. 2011). In a recent review, Amiard-Triquet et al. (2012) have highlighted the importance of polychaetes in ecotoxicological studies, because of the robustness of many polychaete worms, and because many biomarkers measured in these animals have proven to be generally more sensitive to pollution factors than to natural factors (e.g., salinity of the studied area, size of the worms), thereby avoiding many of the problems caused by so-called "confounding factors." Amiard-Triquet et al. (2012) conclude that ragworms of the family Nereidae appear to represent the most commonly used taxon of polychaetes for the determination of biomarkers.

On the other hand, mixed populations may supply different data to those obtained on isolated organisms. Chapman et al. (1982), using mixed populations of freshwater oligochaetes (*Limnodrilus hoffmeisteri* and *Tubifex tubifex*), demonstrated that mixed populations of these two species showed higher tolerance to cadmium, mercury, and sodium pentachlorophenol, but lower tolerance to anoxia, than isolated populations. Similar conclusions were reached by Keilty et al. (1988), who showed that, during contamination of sediment by endrin, *L. hoffmeisteri* benefits from the presence of the oligochaete *Stylodrilus heringianus*, with a decrease of its mortality, an increase of its mass, and a stimulation of its bioturbatory activity.

7.4 Conclusions

From the preceding discussion, it can be concluded that the selection of a single species as a unique sentinel organism is unrealistic. Thus, the choice has to turn toward a set of species that can represent at best the various functional aspects of an ecosystem. According to the specific objectives under consideration, various approaches (described by Figure 7.8) are possible:

- In an analysis of the environmental risks associated with the discharge of a xenobiotic contaminant, bioassays can be carried out on various species, according to a methodology approved by national and international authorities such as the European Commission (2000, 2003). After the application of a safety factor, guideline values can be established, fixing quantitative limits not to be exceeded in order to protect the quality of the environment. Nevertheless, the evaluation of risk could benefit from the use of biomarkers in controlled exposures.
- When the objective is the protection of the biodiversity in an area of particular interest, the use of groups of bioindicator species allows a long-term evaluation of contaminant effects on the resident populations and communities. The criteria of absence or presence of sensitive species can be investigated and core biomarkers of stress can be measured in these organisms, but it will remain difficult to establish relationships between cause and effect.
- The objectives of the present work are to develop a biomarker tools using sentinel species recognized as successful in the manifestation of biomarkers revealing contamination of the environment, and the resulting damage to the organisms that will reverberate at the level of the population, the community, and ultimately, the ecosystem. Their use would be in keeping with the continuity of programs monitoring the quality of the environment such as "Mussel Watch," ROCCH or MED POL, the action plan for the Mediterranean Sea within the framework of the environmental program of the United Nations.

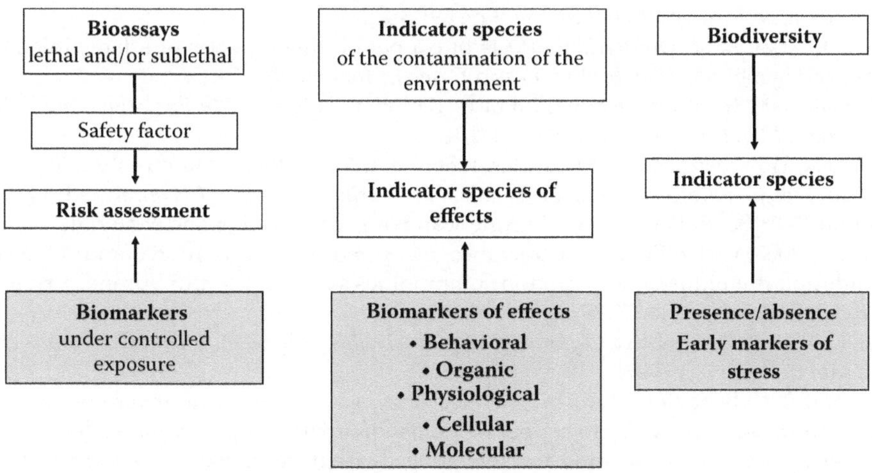

FIGURE 7.8
Different uses of sentinel species according to the chosen objectives.

References

Alzieu, C., F. Quiniou, and R. Delesmont. 2003. L'approche globale: Scores de risque, tests de toxicité, indices biotiques. In: *Bioévaluation de la qualité environnementale des sédiments portuaires et des zones d'immersion*, coord. C. Alzieu, 175–90. Ifremer.

Amiard, J.C. et al. 2006. Metallothioneins in aquatic invertebrates: Their role in metal detoxification and their use as biomarkers. *Aquat. Toxicol.* 76:160–202.

Amiard-Triquet, C., P.S. Rainbow, and M. Roméo. 2011. *Tolerance to Environmental Contaminants*. Boca Raton, FL: CRC Press.

Amiard-Triquet, C., C. Mouneyrac, and B. Berthet. 2012. Polychaetes in ecotoxicology. In: *Encyclopedia of aquatic ecotoxicology*, ed. C. Blaise and J.F. Ferard. Berlin: Springer (in press).

Andrade, H., and P.E. Renaud. 2011. Polychaete/amphipod ratio as an indicator of environmental impact related to offshore oil and gas production along the Norwegian continental shelf. *Mar. Pollut. Bull.* 62:2836–44.

Athrey, N.R.G., P.L. Leberg, and P.L. Klerks. 2007. Laboratory culturing and selection for increased resistance to cadmium reduce genetic variation in the lesat killifish, *Heterandria formosa*. *Environ. Toxicol. Chem.* 9:1916–21.

Banta, G.T., and O. Andersen. 2003. Bioturbation and the fate of sediment pollutants. Experimental case studies of selected infauna species. *Vie Milieu* 53:233–48.

Barata, C. et al. 2001. Biochemical factors contributing to response variation among resistant and sensitive clones of *Daphnia magna* Straus exposed to ethyl parathion. *Ecotoxicol. Environ. Saf.* 49:155–63.

Bellwood, D.R., A.Q.S. Hoey, and J.H. Choat. 2003. Limited functional redundancy in high diversity systems: Resilience and ecosystem function on coral reefs. *Ecol. Lett.* 6:281–5.

Belpaire, C. et al. 2011. What can we learn from monitoring PCBs in the European eel? A Belgian experience. *Environ. Int.* 37:354–64.

Benedetti, M. et al. 2012. A multidisciplinary weight of evidence approach for classifying polluted sediments: Integrating sediment chemistry, bioavailability, biomarkers responses and bioassays. *Environ. Int.* 38:17–28.

Bernes, C., and M. Naylor. 1998. *Persistent Organic Pollutants. A Swedish View of an International Problem*. Stockholm: Swedish Environmental Protection Agency.

Berthet, B. et al. 1992. Bioaccumulation, toxicity and physico-chemical speciation of silver in bivalves molluscs: Ecotoxicological and health consequences. *Sci. Total Environ.* 125:97–122.

Beyrem, H. et al. 2010. Effects of two lubricant oils on marine nematode assemblages in a laboratory microcosm experiment. *Mar. Environ. Res.* 69:248–53.

Bielmyer, G.K. et al. 2010. Differential effects of copper on three species of scleractinian corals and their algal symbionts (*Symbiodinium* spp.). *Aquat. Toxicol.* 97:125–33.

Björk, M. et al. 2000. In situ determination of PCB biodeposition by *Mytilus edulis* in a Baltic coastal ecosystem. *Mar. Ecol. Prog. Ser.* 194:193–201.

Blanck, H., S.A. Wängberg, and S. Molander. 1988. Pollution-induced community tolerance—A new tool. In: *Function Testing of Aquatic Biota for Estimating Hazards of Chemicals*, ed. J.J. Cairns and J.R. Pratt, 219–30. Philadelphia, PA: American Society for Testing and Materials.

Blank, H. et al. 2003. Variability in zinc tolerance, measured as incorporation of radio-labeled carbon dioxide and thymidine, in periphyton communities sampled from 15 European river stretches. *Arch. Environ. Toxicol.* 44:17–29.

Bolognesi, C. et al. 2004. Genotoxicity biomarkers in *Mytilus galloprovincialis*: Wild versus caged mussels. *Mutat. Res.* 552:153–62.

Buffet, P.E. et al. 2011. Behavioural and biochemical responses of two marine invertebrates *Scrobicularia plana* and *Hediste diversicolor* to copper oxide nanoparticles. *Chemosphere* 84:166–74.

Camus, L. et al. 2004. Total oxyradical scavenging capacity responses in *Mytilus galloprovincialis* transplanted into the Venice lagoon (Italy) to measure the biological impact of anthropogenic activities. *Mar. Pollut. Bull.* 49:801–8.

Castro, B.B. et al. 2004. An in situ bioassay integrating individual and biochemical responses using small fish species. *Ecotoxicology* 13:667–81.
Chapman, P.M. 2001. Utility and relevance of aquatic oligochaetes in ecological risk assessment. *Hydrobiologia* 463:149–69.
Chapman, P.M. 2002. Integrating toxicology and ecology: Putting the "eco" into ecotoxicology. *Mar. Pollut. Bull.* 44:7–15.
Chapman, P.M., and F. Wang. 2001. Assessing sediment contamination in estuaries. *Environ. Toxicol. Chem.* 20:3–22.
Chapman, P.M., M.A. Farrel, and R.O. Brinkhurst. 1982. Effects of species interactions on the survival and respiration of *Limnodrilus hoffmeisteri* and *Tubifex tubifex* (Oligochaeta, tubificidae) exposed to various pollutants and environmental factors. *Water Res.* 16:1405–8.
Chapman, P.M. et al. 1994. Refinery water (intake and effluent) quality: Update of 1970s with 1990s toxicity testing. *Environ. Toxicol. Chem.* 13:897–909.
Cho, Y.-C., R.C. Frohnhoefer, and G.-Y. Rhee. 2004. Bioconcentration and redeposition of polychlorinated biphenyls by zebra mussels (*Dreissena polymorpha*) in the Hudson River. *Water Res.* 38:769–77.
Coelho, J.P. et al. 2008. The role of two sediment-dwelling invertebrates on the mercury transfer from sediments to estuarine trophic web. *Estuar. Coast. Shelf Sci.* 78:505–12.
Collin, H. et al. 2010. Response of the Pacific oyster *Crassostrea gigas*, Thunberg 1793, to pesticide exposure under experimental conditions. *J. Exp. Biol.* 213:4010–7.
Couillard, Y. et al. 1995a. Field transplantation of freshwater bivalve, *Pyganodon grandis*, across a metal contamination gradient: I. Temporal changes in metallothionein and metal (Cd, Cu and Zn) concentrations in soft tissues. *Can. J. Fish Aquat. Sci.* 52:690–702.
Couillard, Y. et al. 1995b. Field transplantation of freshwater bivalve, *Pyganodon grandis*, across a metal contamination gradient: II. Metallothionein response to Cd and Zn exposure, evidence for cytotoxicity, and links to effects at higher levels of biological organization. *Can. J. Fish Aquat. Sci.* 52:703–15.
Damiens, G. et al. 2006. Metal bioaccumulation and metallothionein concentrations in larvae of *Crassostrea gigas*. *Environ. Pollut.* 140:492–9.
Dauvin, J.C., and T. Ruellet. 2007. Polychaete/amphipod ratio revisited. *Mar. Pollut. Bull.* 55:215–24.
Dauvin, J.C., D. Bellan, and D. Bellan-Santini. 2010. Benthic indicators: From subjectivity to objectivity—Where is the line? *Mar. Pollut. Bull.* 60:947–53.
De Kock, W.C., and K.J.M. Kramer. 1994. Active biomonitoring (ABM) by translocation of bivalve Molluscs. In: *Biomonitoring of Coastal Waters and Estuaries*, ed. K.J.M. Kramer, 51–84. Boca Raton, FL: CRC Press.
De Lange, H.J., S. Sala, and J.H. Faber. 2010. Ecological vulnerability in risk assessment—A review and perspectives. *Sci. Total Environ.* 408:3871–9.
De Pirro, M., and D.J. Marshall. 2005. Phylogenetic differences in cardiac activity, metal accumulation and mortality of limpets exposed to copper: A prosobranch–pulmonate comparison. *J. Exp. Mar. Biol. Ecol.* 322:29–37.
Depledge, M.H., and S.P. Hopkin. 1995. Methods to assess effects on brackish, estuarine, and near-coastal water organisms. In: SCOPE 53. *Methods to Assess the Effects of Chemicals on Ecosystems*, ed. R.A. Lithurst, P. Bourdeau, and R.J. Tardiff, 125–49.
Dobson, E.P., and G.L. Mackie. 1998. Increased deposition of organic matter, polychlorinated biphenyls, and cadmium by zebra mussels (*Dreissena polymorpha*) in western Lake Erie. *Can. J. Fish Aquat. Sci.* 55:1131–9.
Durou, C. et al. 2007. Biomonitoring in a clean and a multi-contaminated estuary based on biomarkers and chemical analyses in the endobenthic worm *Nereis diversicolor*. *Environ. Pollut.* 148:445–58.
EC. 2000. Directive 2000/60/Ce du parlement européen et du conseil du 23 octobre 2000 établissant un cadre pour une politique communautaire dans le domaine de l'eau (JO L 327 du 22 décembre 2000). European Commission, Bruxelles.

EC. 2003. Technical guidance document in support of Commission Directive 93/67/EEC on risk assessment for new notified substances and Commission Regulation (EC) N° 1488/94 on risk assessment for existing substances and Commission Directive (EC) 98/8 on biocides. European Commission, Bruxelles.

Evtushenko, Z.S., N.N. Belcheva, and O.N. Lukyanova. 1986. Cadmium accumulation in organs of the scallop *Mizuhopecten yessoensis*: 2. Subcellular distribution of metals and metal-binding proteins. *Comp. Biochem. Physiol.* 83C:377–83.

Eyckmans, M. et al. 2011. Exposure to waterborne copper reveals differences in oxidative stress response in three freshwater fish species. *Aquat. Toxicol.* 103:112–20.

Fossi Tankoua, O. et al. 2010. Towards a comprehensive methodology for assessing the health status of coastal and estuarine ecosystems. ICES CM 2010/F:15, e-paper http://www.ices.dk/products/CMdocs/CM-2010/CM2010.pdf.

Fossi Tankoua, O. et al. 2012. Intersite variations of a battery of biomarkers at different levels of biological organisation in the estuarine endobenthic worm *Nereis diversicolor* (Polychaeta, Nereididae). *Aquat. Toxicol.* http://dx.doi.org/10.1016/j.aquatox.2012.02.016.

Gagnon, C., and N.S. Fisher. 1997. The bioavailability of sediment-bound Cd, Co and Ag to the mussel *Mytilus edulis*. *Can. J. Fish Aquat. Sci.* 54:147–56.

Galloway, T.S. et al. 2004. Ecosystem management bioindicators: The ECOMAN project—A multi-biomarker approach to ecosystem management. *Mar. Environ. Res.* 58:233–7.

Galloway, T.S. et al. 2006. The ECOMAN project: A novel approach to defining sustainable ecosystem function. *Mar. Pollut. Bull.* 53:186–94.

Geffard, A., J.C. Amiard, and C. Amiard-Triquet. 2002. Kinetics of metal elimination in oysters from a contaminated estuary. *Comp. Biochem. Physiol.* 131C:281–93.

Geffard, A. et al. 2004. Comparative study of metal handling strategies in bivalves *Mytilus edulis* and *Crassostrea gigas*: A multidisciplinary approach. *J. Mar. Biol. Assoc. U.K.* 84:1–10.

Gomez Gesteira, J.L., and J.C. Dauvin. 2000. Amphipods are good bioindicators of the impact of oil spills on soft-bottom macrobenthic communities. *Mar. Pollut. Bull.* 40:1017–27.

Hagger, J.A. et al. 2009. Application of biomarkers to assess the condition of European Marine Sites. *Environ. Pollut.* 157:2003–10.

Hedman, J.E., J. Stempa Tocca, and J.S. Gunnarsson. 2009. Remobilization of polychlorinated biphenyl from Baltic Sea sediment: Comparing the roles of bioturbation and physical resuspension. *Environ. Toxicol. Chem.* 28:2241–9.

His, E., R. Beiras, and M.N.L. Seaman. 1999. The assessment of marine pollution—Bioassays with bivalve embryos and larvae. *Adv. Mar Biol.* 37:1–178.

Honkoop, P.J.C. et al. 2003. Appropriate experimental design for transplanting mussels (*Mytilus* sp.) in analyses of environmental stress. *J. Exp. Mar. Biol. Ecol.* 297:253–68.

Ji, Y. et al. 2010. Fish transplantation and stress-related biomarkers as useful tools for assessing water quality. *J. Environ. Sci.* 22:1826–32.

Kalman, J. et al. 2010. Comparative toxicity of cadmium in the commercial fish species *Sparus aurata* and *Solea senegalensis*. *Ecotoxicol. Environ. Saf.* 73:306–11.

Kammenga, J.E., and J.A.G. Riksen. 1996. Comparing differences in species sensitivity to toxicants: Phenotypic plasticity versus concentration–response relationships. *Environ. Toxicol. Chem.* 15:1649–53.

Keilty, T.J., D.S. White, and P.F. Landrum. 1988. Sublethal responses to endrin in sediment by *Limnodrilus hoffmeisteri* (Tubificidae), and in mixed-culture with *Stylodrilus heringianus* (Lumbriculidae). *Aquat. Toxicol.* 13:227–49.

Keizer, J. et al. 1995. Enzymological differences of AChE and diazinon hepatic metabolism: Correlation of in vitro data with the selective toxicity of diazinon to fish species. *Sci. Total Environ.* 171:213–20.

Kneib, R.T. 1988. Testing for indirect effects of predation in an intertidal soft-bottom community. *Ecology* 69:1795–1805.

Kneib, R.T., and A.E. Stiven. 1982. Benthic invertebrate responses to size and density manipulations of the common mummichog, *Fundulus heteroclitus*, in an intertidal salt marsh. *Ecology* 63:1518–32.

Kunz, P.Y., C. Kienle, and A. Gerhardt. 2010. *Gammarus* spp. In aquatic ecotoxicology and water quality assessment: Toward integrated multilevel tests. *Rev. Environ. Contam. Toxicol.* 205:1–76.

Lam, P.K., and J.S. Gray. 2001. Predicting effects of toxic chemicals in the marine environment. *Mar. Pollut. Bull.* 42:169–73.

Lotufo, G.R., and J.W. Fleeger. 1997. Effects of sediment-associated phenanthrene on survival, development and reproduction of two species of meiobenthic copepods. *Mar. Ecol. Prog. Ser.* 15:91–102.

Lu, G.H. et al. 2010. Active biomonitoring of complex pollution in Taihu Lake with *Carassius auratus*. *Chemosphere* 79:588–94.

Luoma, S.N., and P.S. Rainbow. 2010. Linking new science frontiers in metal ecotoxicology to better risk management: Lateral thinking. In: *Surface Chemistry, Bioavailability and Metal Homeostasis in Aquatic Organisms: An Integrated Approach*, ed. N.R. Bury and R.D. Handy, 1–28. Essential reviews in Experimental Biology, vol. 2. London, UK: Society for Experimental Biology (SEB).

Maes, G.E. et al. 2005. The catadromous European eel *Anguilla anguilla* (L.) as a model for freshwater evolutionary ecotoxicology: Relationship between heavy metal bioaccumulation, condition and genetic variability. *Aquat. Toxicol.* 73:99–114.

Maltby, L. et al. 2002. Evaluation of the *Gammarus pulex* in situ feeding assay as a biomonitor of water quality: Robustness, responsiveness, and relevance. *Environ. Toxicol. Chem.* 21:361–8.

McCarty, L.S., M. Power, and K.R. Munkittrick. 2002. Bioindicators versus biomarkers. In: *Ecological Risk Assessment. Hum. Ecol. Risk. Assess.* 8:159–64.

Mermillod-Blondin, F., F. François-Carcaillet, and R. Rosenberg. 2005. Biodiversity of benthic invertebrates and organic matter processing in shallow marine sediments: An experimental study. *J. Exp. Mar. Biol. Ecol.* 315:187–209.

Michel, X. et al. 1998. Indicateurs biochimiques de pollution des ecosystems côtiers: Experience du groupe interface chimie-biologie des ecosystems marins (GICBEM). In: *Utilisation de biomarqueurs pour la surveillance de la qualité de l'environnement*, ed. L. Lagadic, T. Caquet, J.C. Amiard, and F. Ramade, 9–32. Paris: Lavoisier Tec & Doc.

Millward, R.N., and A. Grant. 2000. Pollution-induced tolerance to copper of nematode communities in the severely contaminated Restronguet Creek and adjacent estuaries, Cornwall, United Kingdom. *Environ. Toxicol. Chem.* 19:454–61.

Moreira, S.M. et al. 2006. Effects of estuarine sediment contamination on feeding and on key physiological functions of the polychaete *Hediste diversicolor*. Laboratory and in situ assays. *Aquat. Toxicol.* 78:186–201.

Mouneyrac, C., B. Berthet, and J.C. Amiard. 1999. Cd distribution in the tissues of oysters (*Crassostrea gigas*) exposed chronically *in situ*. *Water Air Soil Pollut.* 112:187–96.

Neira, C. et al. 2011. Macrobenthic community response to copper in Shelter Island Yacht Basin, San Diego Bay, California. *Mar. Pollut. Bull.* 62:701–17.

Nigro, M. et al. 2006. Cellular biomarkers for monitoring estuarine environments: Transplanted versus native mussels. *Aquat. Toxicol.* 77:339–47.

NOAA. 2005. ccma.nos.noaa.gov/stressors/pollution/nsandt/mw_details.html. Updated 18 August 2006.

NRC, Committee on Animals as Monitors of Environmental Hazards. 1991. *Animals as Sentinels of Environmental Health Hazards*. Washington, D.C.: National Research Council, National Academy Press.

Oliveira, M.M. et al. 2007. Brain acetylcholinesterase as a marine pesticide biomarker using Brazilian fishes. *Mar. Environ. Res.* 63:303–12.

Olivier, F., M. Ridd, and D. Klumpp. 2002. The use of transplanted cultured tropical oysters (*Saccostrea commercialis*) to monitor Cd levels in North Queensland coastal waters (Australia). *Mar. Pollut. Bull.* 44:1051–62.

Orvain, F. 2005. A model of sediment transport under the influence of surface bioturbation: Generalisation to the facultative suspension-feeder *Scobicularia plana*. *Mar. Ecol. Prog. Ser.* 286:43–56.

Pessoa, P.C. et al. 2011. Cholinesterase inhibition and behavioral toxicity of carbofuran on *Oreochromis niloticus* early life stages. *Aquat. Toxicol.* 105:312–20.
Phillips, D.J.H. 1994. Macrophytes as biomonitors of trace metals. In: *Biomonitoring of Coastal Waters and Estuaries*, ed. K.J.M. Kramer, 85–106. Boca Raton, FL: CRC Press.
Pigeot, J. 2001. Approche écosystèmique de la contamination métallique du compartiment biologique benthique des littoraux charentais: Exemple du bassin de Marennes-Oléron. Thèse de Doctorat, Université de La Rochelle.
Piva, F. et al. 2011. Assessing sediment hazard through a weight evidence approach with bioindicator organisms: A practical model to elaborate data from sediment chemistry, bioavailability, biomarkers and ecotoxicological bioassays. *Chemosphere* 83:475–85.
Poirier, L. et al. 2006. A suitable model for the biomonitoring of trace metal bioavailabilities in estuarine sediments: The annelid polychaete *Nereis diversicolor*. *J. Mar. Biol. Assoc. U.K.* 86:71–82.
Rand, G.M. 1985. Behavior. In: *Fundamentals of Aquatic Toxicology: Methods and Applications*, ed. G.M. Rand and S.R. Petrocelli, 221–63. Washington, D.C.: Hemisphere Publishing.
Regoli, F., and G. Principato. 1995. Glutathione, glutathione-dependent and antioxidant enzymes in mussel, *Mytilus galloprovincialis*, exposed to metals under field and laboratory conditions: Implications for the use of biochemical biomarkers. *Aquat. Toxicol.* 31:143–64.
Regoli, F. et al. 2004. Time-course variations of oxyradical metabolism, DNA integrity and lysosomal stability in mussels, *Mytilus galloprovincialis*, during a field translocation experiment. *Aquat. Toxicol.* 68:167–78.
RINBIO. 2002. www.corse.eaufrance.fr/rlm/reseaux-resultats-surveillance/rinbio.htm.9k; Updated 25/01/2002.
Roméo, M. et al. 1994. Heavy metal content and biotransformation enzymes in two fish species from the NW Mediterranean. *Mar. Ecol. Prog. Ser.* 107:15–22.
Roméo, M. et al. 2003. Mussel transplantation and biomarkers as useful tools for assessing water quality in the NW Mediterranean. *Environ. Pollut.* 122:369–78.
Ross, K. et al. 2002. Genetic diversity and metal tolerance of two marine species: A comparison between populations from contaminated and reference sites. *Mar. Pollut. Bull.* 44:671–9.
Rotter, S. et al. 2011. Active bio-monitoring of contamination in aquatic systems—An in situ translocation experiment applying the PICT concept. *Aquat. Toxicol.* 101:228–36.
Scaps, P. 2002. A review of the biology, ecology and potential use of the ragworm *Hediste diversicolor* (O.F. Müller) (Annelida: Polychaeta). *Hydrobiologia* 470:203–18.
Scarfe, A.D. et al. 1982. Locomotor behavior of four marine teleosts in response to sublethal copper exposure. *Aquat. Toxicol.* 2:335–53.
Smolders, R., L. Bervoets, and R. Blust. 2004. In situ and laboratory bioassays to evaluate the impact of effluent discharges on receiving aquatic ecosystems. *Environ. Pollut.* 132:231–43.
Solé, M., J. Kopecka-Pilarczyk, and J. Blasco. 2009. Pollution biomarkers in two estuarine invertebrates, *Nereis diversicolor* and *Scobicularia plana*, from a Marsh ecosystem in SW Spain. *Environ. Int.* 35:523–31.
Stubblefield, W.A. et al. 1999. Acclimation-induced changes in the toxicity of zinc and cadmium to rainbow trout. *Environ. Toxicol. Chem.* 18:2875–81.
Tabor, G.M., and A.A. Aguirre. 2004. Ecosystem health and sentinel species: Adding an ecological element to the proverbial "canary in the mineshaft." *EcoHealth* 1:226–8.
Takamura, N., F. Kasai, and M.M. Watanabe. 1989. Effects of Cu, Cd and Zn on photosynthesis of freshwater benthic algae. *J. Appl. Phycol.* 1:39–52.
Timmermann, K. et al. 2011. Effects of bioturbation on the fate of oil in coastal sandy sediments—An in situ experiment. *Chemosphere* 82:1358–66.
Tixier, G. et al. 2011. Ecological risk assessment of urban stormwater ponds: Literature review and proposal of a new conceptual approach providing ecological quality goals and the associated bioassessment tools. *Ecol. Indic.* 11:1497–506.
Turner, A., and G.E. Millward. 2002. Suspended particles: Their role in estuarine Biochemical cycles. *Estuar. Coastal Shelf Sci.* 55:857–3.

UNEP/RAMOGE. 1999. *Manuel sur les biomarqueurs recommandés pour le programme de biosurveillance MED POL*. Athènes: PNUE, 41–92.

Valbonesi, P. et al. 2011. Cholinesterase activities and sensitivity to pesticides in different tissues of silver European eel, *Anguilla anguilla*. *Comp. Biochem. Physiol.* 154C:353–9.

Van der Oost, R. et al. 1996. Biomonitoring of aquatic pollution with feral eel (*Anguilla anguilla*): II. Biomarkers: Pollution-induced biochemical responses. *Aquat. Toxicol.* 36:189–222.

Wallner-Kersanach, M. et al. 2000. Accumulation and elimination of trace metals in a transplantation experiment with *Crassostrea rhizophorae*. *Arch. Environ. Contam. Toxicol.* 38:40–5.

Weis, J.S. et al. 2001. Effects of contaminants on behavior: Biochemical mechanisms and ecological consequences. *Bioscience* 51:209–17.

Weis, J.S. et al. 2011. Living in a contaminated estuary: Behavioral changes and ecological consequences for five species. *Bioscience* 61:375–85.

Wentsel, R., A. McIntosh, and V. Anderson. 1977. Sediment contamination and benthic macroinvertebrate distribution in a metal-impacted lake. *Environ. Pollut.* 14:187–93.

Wepener, V. et al. 2005. Active biomonitoring in freshwater environments: Early warning signals from biomarkers in assessing biological effects of diffuse sources of pollutants. *Phys. Chem. Earth* 30:751–61.

White, G.J. 2004. Selection of ecological indicators for monitoring terrestrial systems. In: *Environmental Monitoring*, ed. G.B. Wiersma, 263–82. Boca Raton, FL: CRC Press.

8

Impairments of Endocrine Functions: Causes and Consequences

Jean-Claude Amiard, Arnaud Chaumot, Mickaël Couderc, Jeanne Garric, Olivier Geffard, and Benoît Xuereb

CONTENTS

8.1 Introduction

The problem of endocrine disruptor compounds (EDCs) became apparent with observations in the field of major noxious effects such as the thinning of egg shells of certain birds (due to dichlorodiphenyltrichloroethane, DDT), the appearance of a penis (imposex) in female marine gastropods [due to tributyltin (TBT); cf. Chapter 9], and reports of reproductive disorders in seals from the Baltic [exposed to polychlorobiphenyl (PCBs), to DDT and their metabolites], these various disturbances ending in a drastic reduction of the populations concerned. Historically, reports of abnormal sexual development in reptiles (Guillette et al. 1994, 1995) or birds (Fry 1995), as well as feminization responses in male fish (Jobling and Sumpter 1993; White et al. 1994; Jobling et al. 1995), have suggested an endocrine disruption effect with environmental chemicals functioning as estrogens. Similar hypotheses have been advanced in relation to an increased risk for breast cancer in women (Falk et al. 1992; Wolff et al. 1993) and an observed decrease in human semen quality (Carlsen et al. 1993; Auger et al. 1995). Endocrine disturbances have also been revealed in various species of fish, including premature maturity of females, intersexuality (Chapter 9), and the induction of vitellogenin (VTG). The first observations of abnormalities of thyroid function resulting from exposure to chemical compounds were made on wild organisms, mainly birds and fish of the Great Lakes region (Black et al. 1974; Moccia et al. 1986).

Approximately 550 substances are now suspected to have an action on hormones (CEC 2001). More than 100 naturally occurring and synthetic substances have been reported to have effects on thyroid function or thyroid hormone (TH) metabolism (Pearce and Braverman 2009).

Wastewater treatment plant (WWTP) effluents are a major source of potential or known EDCs in the aquatic environment (Gust et al. 2010a; Tetreault et al. 2011; Vajda et al. 2011 and references quoted by these authors). The steroidal natural estrogens 17β-estradiol (E2), estrone (E1), estriol, and synthetic 17α-ethynylestradiol (EE2) (contraceptive pills), as well as estrogenic alkylphenols and bisphenol A, have been identified in such effluents. Pharmaceuticals and personal care products are also wastewater contaminants, the endocrine disrupting potential of which is poorly documented since few compounds have been inventoried or regulated worldwide (Caliman and Gavrilescu 2009). In the Pearl River system, south China, the equivalent concentrations of hormonal activities in some sites were greater than the lowest-observed-effect concentrations reported in the literature, suggesting the potential for adverse effects on local aquatic organisms (Zhao et al. 2011).

Owing to the importance of sediments on the transport, fate, and bioavailability of EDCs, sediment contamination by EDCs has been investigated in more than 30 countries (Gong et al. 2011 and literature quoted therein). In the Pearl River system, these authors have shown that the spatial distributions of 4-*tert*-octylphenol, 4-nonylphenol, and bisphenol A (BPA), and the steroid estrogens estrone and 17β-estradiol in the sediments were related to the discharge of domestic and industrial wastewater along the river system. In 54 coastal sediment samples collected from nine locations off central-southern Chile, Bertin et al. (2011) have shown that total estrogenic loads (estrone, 17β-estradiol, estriol, 17α-ethinylestradiol) were correlated with the size of human populations served by local sewage plants. Steroid estrogens in sediments can be directly bioaccumulated by benthic species of invertebrates and fish but are also transferred through the food web (Bertin et al. 2011 and literature cited therein). An abundant literature exists on the sources and environmental fates of numerous contaminants that were already of interest to ecotoxicologists before their endocrine disrupting potential was discovered (pesticides, PCBs, fire retardants, cadmium) (Eisler 2007).

It is thus necessary to have good estimates of EDC exposures, and for this purpose we have chemical techniques and state-of-the-art biological tests to be detailed below. Because many molecules are suspected to behave as EDCs, it is impossible, for both technical and cost-effective reasons, to determine the concentrations of all such compounds. Moreover, chemical analyses do not provide information about the toxicity of complex mixtures of these compounds, and it is in mixtures that they nearly always occur in the environment. Another possible approach is to reveal the presence of EDCs with *in vitro* and *in vivo* tests designed to identify estrogenic, antiestrogenic, androgenic, antiandrogenic, and thyroid effects (e.g., Jugan et al. 2007; Kinani et al. 2010; Terrien et al. 2011). Toxicity identification evaluation (TIE) methods have been recently adapted for the detection of EDCs in environmental samples. When the presence of EDCs is revealed in a sample (water, effluent, sediment, or biological tissue) by biotests, chemical analyses are then used to detect and quantify the specific contaminants responsible for any hormonal disturbance (Amiard et al. 2009; Kinani et al. 2010).

In contrast to aquatic vertebrates for which the impact of endocrine disruptors (ED) on reproductive function has been widely studied and documented, leading to the development and validation of several specific biomarkers to predict alterations in reproductive functions (concentrations in steroid hormones and induction of VTG synthesis in male and juvenile organisms) (Matthiessen 2003; Langston et al. 2005), few studies have been conducted on invertebrates, in particular, mollusks and crustaceans, despite their obvious ecological importance (deFur et al. 1999; deFur 2004; Oetken et al. 2004). This is largely attributed to the lack of knowledge on the endocrine regulation of the reproductive process in these species, which are routinely used in ecotoxicology (Porte et al. 2006). In crustaceans, hormonal regulation is mainly based on ecdysteroid and terpenoid hormones in females and androgenic gland hormone (AGH) in males (Charniaux-Cotton 1954). Because of this specificity of endocrine systems, ED biomarkers available in fish cannot be easily used and/or should not be directly applied to protostome invertebrates (i.e., ecdysozoans and lophotrochozoans). Indeed, as illustrated for the steroid hormone signaling pathway (Thornton 2003; Markov et al. 2009), modern endocrine systems in metazoans result from a complex evolution of molecular players such as hormone receptors or enzymes implicated in hormone synthesis. This diversifying biological evolution has given rise to divergent endocrine systems between the main metazoan lineages with specific modes of hormonal regulation. The impact of endocrine disruptors on reproductive function has been widely studied since the 1990s. Currently, other types of disruption are studied in order to investigate the general reproduction and development of exposed species. This is specially the case for thyroid function disruption as THs play a major role in growth, reproduction, and development, in particular, neurological development.

The ultimate purpose of the evaluation of ecological risk is to protect communities and ecosystems from the noxious effects caused by chemical pollutants. Disturbance of sex hormones and other hormones by miscellaneous chemical contaminants can have cascading effects on the growth and reproduction of individuals, leading to changes at higher levels of biological organization, populations, and communities (Kidd et al. 2007). Ecoepidemiological studies show changes at these levels, but any causal link of these changes to EDCs is usually tenuous.

8.2 Generalities

8.2.1 Normal Mechanisms of Hormonal Action

In nearly all complex multicellular animals, there are two main systems—the nervous and the endocrine systems—controlling and coordinating processes within the body. The endocrine system is based on chemical messengers, the hormones, which are secreted into the blood (or other extracellular fluids), and can reach all parts of the body.

The endocrine system has three main components: the endocrine glands, the hormones, and the receptors. The endocrine glands are situated at various sites in the body, and the cells in these glands secrete specific chemicals called hormones. The hormones circulate in the body via the bloodstream and modulate cellular or organ functions by binding with receptors in the target cells. The receptors in the target cells, once activated by binding of the hormone, regulate the functions and processes in the tissue through interactions with the cell's DNA or other complex intracellular signaling processes.

In vertebrates, the main hormones are those released in the hypothalamus that stimulate pituitary activity, and then stimulate thyroid, adrenal, gonadal, and pancreatic activity. The THs contribute to the regulation of metabolism, growth and development, behavior, and puberty. The corticosteroid hormones and catecholamines secreted in adrenal glands also intervene in metabolism and behavior. Insulin and glucagon secreted in the pancreas regulate blood sugar levels. The sex steroid hormones (androgens and estrogens) produced in the gonads regulate development and growth, reproduction, immunity, onset of puberty, and behavior. The production and circulating levels of hormones are controlled by means of negative feedback processes.

The endocrine system is important for the control and regulation of all major functions and processes of the body: energy control, reproduction, immunity, behavior, growth, and development.

8.2.2 Mechanisms of Endocrine Disruption: Principle of Lock and Key

8.2.2.1 Mechanisms of Disruption

Potential sites of toxic action of pollutants on the endocrine system in fish and other organisms concern (1) the perception of environmental stimuli by the nervous system, the synthesis and release of hypothalamic hormones (RH/IH); (2) the synthesis and release of hypophyseal hormones [GTH, ACTH, thyroid stimulating hormone (TSH)]; (3) the response of endocrine glands under the control of the hypophysis (gonads, interrenal tissue, thyroid gland) and other glands (e.g., endocrine pancreas); (4) the synthesis and release of hormones by other endocrine glands; (5) the response of target organs to hormones; and (6) the metabolism and elimination of hormones by excretory organs (Hontela 2000).

Some chemicals can act on the endocrine system to disturb the homeostatic mechanisms of the body or to initiate processes at abnormal times in the life cycle. First, they may mimic the biological activity of a hormone by binding to a cellular receptor, leading to an unwarranted response by initiating the cell's normal response to the naturally occurring hormone at the wrong time or to an excessive extent (agonistic effect). Second, they may bind to the receptor but not activate it. Instead, the presence of the chemical on the receptor will prevent binding of the natural hormone (antagonistic effect). Third, they may bind to transport proteins in the blood, thus altering the amounts of natural hormones present in the circulation. Fourth, they may interfere with the metabolic processes in the body, affecting the synthesis or breakdown rates of the natural hormones.

The normal mechanism of the action of hormones, for example, steroid hormones, includes three stages. After hormones bind to the receptor, this couple is internalized in the nucleus, where it can interact with other proteins to bind to specific promoter regions of DNA (Figure 8.1). The genes under hormonal regulation possess in their promoter region an element of response in the hormone/receiver (RRE) complex. The binding of the complex on this element of response is going to provoke the expression of the gene until then silent (Figure 8.1). The endocrine disrupter can exert its toxic action either by preventing or modifying binding to the receptor, or by settling on the promoter region of the gene, for example, by methylation.

Endocrine disruptors can bind to the hormonal receptor, and the hormonal signal may be enhanced or decreased (Figure 8.2). This is an example of lock and key. As a rule, a single key can match a given lock. But in reality, several keys can enter the lock, making it turn or blocking it. This is all the more true as the number of locks (hormonal receptors) and keys (EDCs) is multiplied.

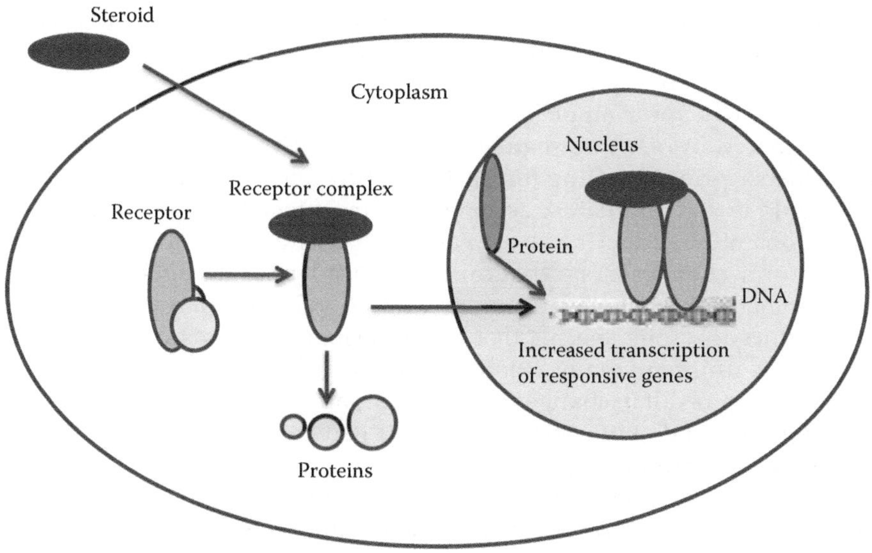

FIGURE 8.1
Regulatory role of steroidal hormones. (After Landis, W.G., Yu, M.H., *Introduction to Environmental Toxicology. Impacts of Chemicals upon Ecological Systems*, Lewis Publishers, Boca Raton, FL, 2004. With permission.)

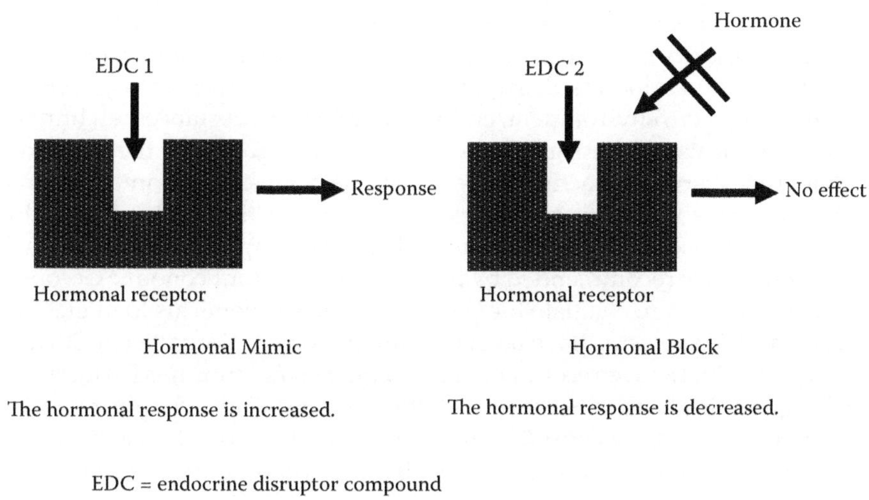

FIGURE 8.2
Mechanisms of action of endocrine disrupters; principle of lock and key.

8.2.2.2 What Are Endocrine Disruptors?

For the EU, "An endocrine disruptor is an exogenous substance or mixture that alters function(s) of the endocrine system and consequently causes adverse health effects in an intact organism, or its progeny, or (sub)populations" (EU 1998). The US Environmental Protection Agency (1997) has adopted the definition proposed by Kavlock et al. (1996): "An exogenous agent that interferes with the synthesis, secretion, transport, binding, action, or elimination of natural hormones in the body which are responsible for the maintenance or homeostasis, reproduction, development, and/or behavior." Consistent with this definition, the agency has also stated that it "...does not consider endocrine disruption to be an adverse effect *per se*, but rather to be a mode or mechanism of action potentially leading to other outcomes, for example carcinogenic, reproductive, or developmental effects, routinely considered in reaching regulatory decisions. Evidence of endocrine disruption alone can influence priority setting for further testing and the assessment of the results of this testing could lead to regulatory action if adverse effect are shown to occur."

Classically, potential endocrine disrupters are classified into four categories: natural hormones, natural chemicals, synthetically produced pharmaceuticals, and man-made chemicals. The natural hormones from any animal, released into the environment, and chemicals produced by one species that exert hormonal actions on other animals (e.g., human hormones unintentionally reactivated during the processing of human waste in sewage effluent) may result in changes to aquatic organisms. The natural chemicals include plant toxins (the so-called phytoestrogens, such as genistein or coumestrol) and fungal toxins. The synthetically produced pharmaceuticals that are intended to be highly hormonally active (contraceptive pills and treatments for hormone-responsive cancers) may also be detected in sewage effluent. The man-made chemicals and by-products released into the environment include some pesticides (e.g., DDT and other chlorinated compounds), plasticizers (bisphenol A, phthalates), degradation products of some alkylphenols, and a number of industrial chemicals (e.g., PCBs, dioxin, cadmium).

Priority lists of substances have been published in the United States and Europe. In 2009, the USEPA has established a list of initial pesticide active ingredients and pesticide inert ingredients to be screened under the Federal Food, Drug, and Cosmetic Act. Toward the establishment of a priority list of substances, the EU General Directorate for the Environment has funded studies on endocrine disruptors, the results of which are compiled in a database (http://ec.europa.eu/environment/endocrine/strategy/substances_en.htm#top). Using expert advice, chemicals were assigned to one of three categories: (1) the substance has evidence of endocrine disrupting activity in at least one species using intact animals; (2) the substance has at least some *in vitro* evidence of biological activity related to endocrine disruption; (3) the substance has no evidence of endocrine disrupting activity, or no data available. The methodologies recommended by the Organisation for Economic Co-operation and Development and USEPA to evaluate the potential of these chemicals to interact with estrogen, androgen, and TH systems have been recently reviewed (Borgert et al. 2011; Dang et al. 2011). Borgert et al. (2011) have recommended a sequence for running Endocrine Disruptor Screening Program/Endocrine Screening Battery assays: (1) screening for estrogen and androgen activity *in vitro* and *in vivo*; (2) screening *in vitro* for aromatase and steroidogenesis; (3) multimodal *in vivo* mammalian screening; (4) *in vivo* nonmammalian screening.

8.3 Effects of EDCs on Reproductive System

8.3.1 Estrogenic Activities

Numerous organic compounds released in the aquatic environment as a result of industrial, agricultural, and domestic activities can mimic estrogens and have demonstrated estrogenic activities both *in vitro* and *in vivo*. Among them are substances no longer used in developed countries (polychlorinated biphenyls, organochlorinated pesticides), but also numerous surfactants, plasticizers, pesticides, and pharmaceuticals still widely used at present.

Estrogenic activities are elicited through several mechanisms of action. Xenoestrogens can bind to mammalian intracellular estrogen receptors (ERs). After binding and nuclear translocation, they can directly act as ligand-inducible transcription factors, and specifically regulate the expression of target genes. Such a mechanism is generally referred to as the genomic or "classical" pathway of estrogen action (Porte et al. 2006). Some can produce effects similar to those of estrogens without being mediated by the ER. They bind other receptors able to interact with estrogen responsive elements, or can go through other receptor and/or signal transduction pathways, including interactions with a variety of binding globulins, growth factors, different receptor systems, and/or steroidogenic enzymes (Gillesby and Zacharewski 1998). One example of such a pathway is the binding of xenoestrogens (as PCBs or dioxins) with the aryl hydrocarbon receptor (AhR), which is involved in Cyp P450 activity, and 17β-estradiol metabolism (Gillesby and Zacharewski 1998; Navas and Segner 1998).

The effect of xenoestrogens is dependent on both the dose and the developmental stage at which exposure occurred. Numerous studies have demonstrated that exposure to estrogens at critical times during fetal development may irreversibly influence responsive tissues as well as subsequent growth. These effects may be manifested as structural, functional, or long-term pathologic changes that are not limited to one sex (McLachlan and Arnold 1996).

Hormonal-mimic activities (estrogenic, androgenic) of pure chemicals and of environmental mixtures such as wastewater, river water, or sediment pore water can be searched for by using *in vitro* cellular tests. Several tools exist and are implemented to detect multiple hormonal activities in environmental samples (Van der Linden et al. 2008; Swart et al. 2011). They are based on steroid-responsive reporter cell lines developed for human cells [E-Screen, CALUX (Chemically Activated Luciferase EXpression)], or yeast. Briefly, these tests are based on tumoral cells proliferation (MCF-7, CEF) (White et al. 1994) or by making use of the fact that steroid receptors are transcription factors that induce transcription of target genes after binding to specific DNA sequences in their promoter. These DNA sequences are linked to the gene of a readily measurable protein, such as β-galactosidase in yeast assay (McLachlan and Arnold 1996; Routledge and Sumpter 1996) or firefly luciferase in ER and AR CALUX assays (Sonneveld et al. 2005).

One of the most commonly used *in vivo* biomarkers of steroidogenesis impairment is the quantification of VTG. VTG is the major precursor of egg yolk proteins, which provide energy reserves for embryonic development in oviparous organisms. In mature females, VTG is generally synthesized in response to endogenous estrogens, whereas in males, the VTG gene, although present, is normally silent. However, it may be activated by (xeno) estrogens (Matozzo et al. 2008). EDC-induced changes in VTG may be examined either at the level of the gene by using polymerase chain reaction (PCR) (Biales et al. 2007; Nadzialek et al. 2010) or at the level of the protein by using the enzyme-linked immunosorbent assay (ELISA) test or by liquid chromatography tandem mass spectrometry (LC-MS/MS) as recently proposed by Simon et al. (2010).

8.3.2 Antiestrogenic Activities

Xenobiotics can exert antiestrogenic activities on aquatic vertebrates, as has been reported in several fish species exposed to AhR agonists. AhR is a ligand activating transcription factor that controls the activities of several genes encoding phase I and II biotransformation enzymes such as cytochrome P450 1A1, and mediates sex steroid hormone-related actions.

Besides eliciting several toxic effects (immunotoxicity, hepatotoxicity, carcinogenicity), compounds, such as PCB 77, PCB 126, Aroclor 1254 (polychlorobiphenyls), or benzo[*a*]pyrene (BaP), can lead to reproductive process impairment in inhibiting fish VTG secretion or reducing gonadal development (Smeets et al. 1999; Vaccaro et al. 2005).

Some of the most potent AhR agonists, PCBs, polychlorinated dibenzo-*p*-dioxins, or polychlorinated dibenzofurans, can inhibit some E2 responses in rodents, suggesting a relationship between AhR activation and antiestrogenicity. 2,3,7,8-Tetrachlorodibenzo-*p*-dioxin can induce estrogenic action or inhibit estrogen-induced effects in various tissues because of AhR–ER cross-talk. Inhibitory AhR and ER cross-talk have been demonstrated in mammalian cancers (Safe et al. 1991).

Several mechanisms have been elucidated that partially explain the AhR-mediated antiestrogenic activities of dioxins and related compounds. These include increased estrogen metabolism, down-regulation of ER protein levels and binding activity, and decreased ER-mediated gene expression (Gillesby and Zacharewski 1998). The ER nuclear receptor and AhR are both critical for respective VTG and Cyp 1A expression. Arukwe et al. (2001) stressed that interaction between the AhR and ER in sexually maturing fish can cause adverse reproductive effect through disruption of estrogen-regulated reproductive processes.

The physiological functions of the AhR during development appear to be ancestral to adaptive functions. Consequently, differences in sensitivity to the developmental toxicity of dioxins and related chemicals may have had their origin in the evolution of the dioxin-binding capacity of AhR in the vertebrate lineage (Hahn 2002).

8.3.3 Effects of Complex Mixtures Containing Substances with Endocrine Potential

As underlined by Tyler and Jobling (2008), the phenomenon of estrogenic-containing effluents has been established widely across Europe, United States, Japan, and China. Tiny quantities of several estrogenic substances in urban effluents lead to measurable estrogenic effects in freshwater (Jobling et al. 2002, 2003; Silva et al. 2002), and brackish environments as shown in wild populations of flounders *Platichthys flesus* from four estuaries in Great Britain (Tyne, Crouch, Thames, and Mersey) (Allen et al. 1999a, 1999b). The estrogenic effects on these fish are revealed by a 4- to 6-fold increase in VTG concentrations by comparison to controls and by the presence of intersex (oocytes imbricated in testicles). Such effects were also shown in several other fish species such as *Notropis hudsonius* (Aravindakshan et al. 2004), *Rutilus rutilus* (Bjerregaard et al. 2006), *Oncorhynchus mykiss* (Gagné et al. 2006), and other salmonids (Lahnsteiner et al. 2006). Recently, Vajda et al. (2011) linked serious perturbations, revealed as intersex, gonad deformities, and feminization, measured in the wild white suckers (*Castotomus commersoni*) sampled in wastewater effluent–dominated Colorado streams, to endocrine active chemicals (steroidal hormones, estrogenic alkylphenols, and bisphenol A) present in the WWTP effluents (Woodling et al. 2006).

Chemicals acting similarly as mimetic estrogens can act in combination according to the principle of concentration addition. Consequently, when present as a mixture in the receiving waters, they are likely to pose a significant environmental risk, even when each substance is present at below the threshold of detectable effects (Brian et al. 2007). Such additive effects are a topic of concern for the long-term future of exposed ichthyological populations. Estrogenic effects have also been observed in other aquatic vertebrates such as the frog *Xenopus laevis* (Coady et al. 2005).

Despite the coexistence of several classes of EDCs (estrogenic, androgenic) in most aquatic ecosystems, there is still limited information regarding their combined effects. Bugel et al. (2011) demonstrated that contaminants within Newark Bay, USA, exert both estrogenic and antiestrogenic responses that result in an overtly antiestrogenic phenotype in the fish *Fundulus heteroclitus*, with reduced egg production due to inhibition of vitellogenesis. In many European estuaries, from the UK to France and Portugal, intersex has been revealed in the endobenthic bivalve *Scrobicularia plana* (Langston et al. 2007; Gomes et al. 2009; Fossi Tankoua et al. 2012). These disturbances are most probably attributable to the presence of EDCs but, at least in the French sites, no direct correlation was found between percentage of intersex and proportion of cellular responses in bioassays or organic compound concentrations detected by chemical analyses (Fossi Tankoua, Budzinski, and Aït Aïssa, personal communication).

In a recent study, Sárria et al. (2011) have exposed juvenile black-striped pipefish (*Syngnathus abaster*) to the estrogenic chemical ethinylestradiol (EE2) and the androgenic chemical TBT, single and in binary mixtures. The obtained results indicated that EE2, TBT, or their combined exposure, influence morphological and behavioral development during the early life of the pipefish. The authors stressed that mixtures significantly modulate newborn responses in distinct ways when compared to individual chemical exposures and the combined effects of the selected contaminants are far from being predictable. This unpredictability underlines the importance of considering mixture effects in environmental risk assessment and particularly the issue of mixtures of EDCs acting through dissimilar modes of action (MOAs).

The presence of steroidogenesis disruptors in the environment may be also revealed through the assessment of different end-points linked to reproduction such as the Daphnia 21d reproduction test (Dang et al. 2012) and the Fish Sexual Development Test (FSDT) (Holbech et al. 2012) included in the OECD working program. Compiling 142 data sets for 21-day fish assays and 38 data sets for a fish sexual development test, encompassing 62 chemicals with different MOAs, Dang et al. (2011) concluded that fecundity and gonad histology are the most sensitive endpoints for fathead minnow *Pimephales promelas*, medaka *Oryzias latipes*, and zebrafish *Danio rerio* in 21-day fish assays, whereas secondary sex characteristics are less sensitive endpoints and are likely to be inadequate to detect all known MOAs. Intersex in fish and bivalves and imposex in gastropods are well documented (Chapter 9). In certain cases, even sex reversal has been observed in fish (Iwamatsu et al. 2006) and frogs (Pettersson et al. 2006). The reproductive status of fish and invertebrates is also commonly used by determining gonadosomatic indices (e.g., Vijayavel and Balasubramanian 2008 in crustaceans; Nadzialek et al. 2010 in fish; Pellerin et al. 2009 in bivalves) or sexual maturation stages (Siah et al. 2003; Durou and Mouneyrac 2007). Indeed, temporal changes in the maturation of sexual products may be influenced by environmental stress, but this tool suffers a lack of specificity because natural factors (temperature, food availability) can interact with contamination factors.

All these different bioassays *in vitro* and *in vivo* can generally be applied both in the laboratory and in the field. An intermediate step consists of field experiments such as caging, which is discussed in Chapter 2. Field experiments have already been used in ecotoxicological studies with EDCs, providing results of great value because of their environmental realism (Gust et al. 2010b; Vajda et al. 2011).

8.4 Thyroid Endocrine Disruptors

Several classes of chemicals are now suspected or even proved to be disruptive to thyroid function at several levels. However, evidence for thyroid disrupting activity is often not clear, because the thyroid is a relatively robust gland in adults. It is able to compensate for moderate or mild disruption by a thyroid hyperplasia known as goiter. Thus, the impact of thyroid disruption during the development of individuals is different from that in an adult. Indeed, THs have an essential role in fetal development. Prenatal exposure to thyroid disruptors will have consequences that can be very damaging, particularly in the developing brain.

The regulation of thyroid function is a finely tuned negative feedback mechanism of circulating THs at the hypothalamic and pituitary levels, that maintain relatively stable serum levels of THs with each individual having it set at a specific point (Feldt-Rasmussen et al. 1980).

The mechanisms involved in thyroid homeostasis are numerous and complex, and environmental chemicals may interfere at all levels (Figure 8.3). Thyroid function is affected by binding proteins, peripheral metabolism, alterations in the thyroid gland, and clearance (Boas et al. 2011). In a review, Brown et al. (2004) describe the thyroid cascade and the effects of contaminants on the teleost fish thyroid.

Unlike mammals and amphibians that have an organized pair of thyroid glands, the majority of teleost fishes have multiple thyroid follicles, individually or in clusters, scattered among the afferent branchial arterioles in the ventral region of the pharynx. Nevertheless, the basic structure and function of the thyroid follicle is fairly conserved across vertebrates (Paris et al. 2008a). The thyroid follicles are formed by a basement membrane surrounding epithelial cells and enclose a lumen. Iodine is obtained by follicular cells from the blood via a Na^+/I symport (NIS). It is then transported to the epithelial surface where it will form thyroxin [3,5,3,5-tetraiodo-L-thyronine (T4)] by incorporating I into thyroglobulin. The resulting protein complex is stored in the follicle. Activation of the neuroendocrine hypothalamo–pituitary–thyroid (HPT) axis induces the production of THs (Eales 2006; Blanton and Specker 2007; Zoeller et al. 2007). More precisely, under hypothalamic control, the pituitary secretes TSH, which proceeds to the thyroid gland to activate synthesis of THs.

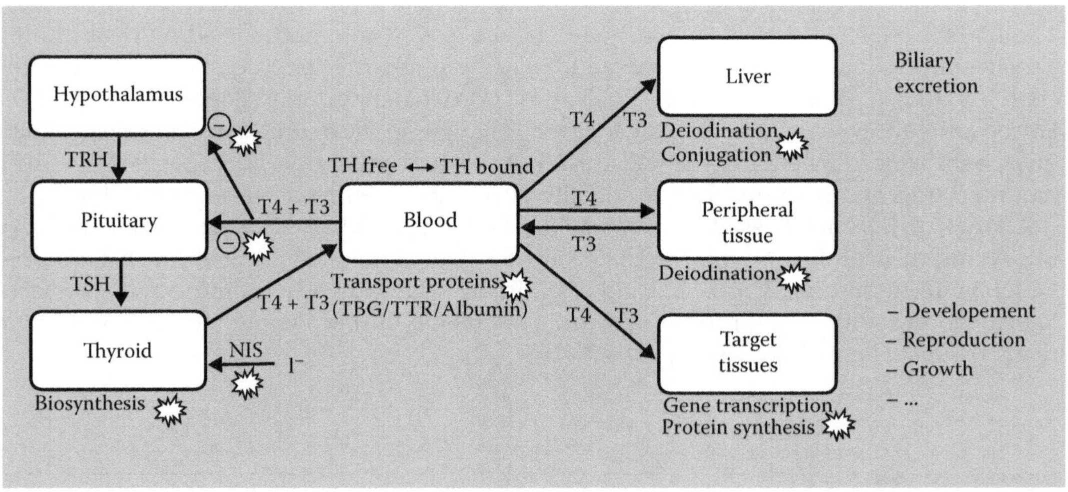

FIGURE 8.3
Basic elements of thyroid function. TRH, thyrotropin-releasing hormone; TSH, thyroid stimulating hormone; TH, thyroid hormone; NIS, sodium iodide symporter; T3, triiodothyronine; T4, thyroxine; TBG, thyroxine-binding globulin; TTR, transthyretin; (✸) main thyroid disruption endpoints.

In mammals (De Vito et al. 1999; Ahmed et al. 2008), amphibians (Fort et al. 2007), and teleosts (Brown et al. 2004; Blanton and Specker 2007), T4 is the main product released by thyroid follicles into the circulation. It is then bound to TH-binding proteins (transport proteins) and carried to target tissues. In peripheral tissues, T4 is converted into triiodothyronine [3,5,3-triiodo-L-thyronine (T3)] by action of a complex of deiodinase enzymes that removes one of the two outer ring iodines (Blanton and Specker 2007; Fort et al. 2007; Eales 2006; Zoeller et al. 2007). T3 has higher affinity than T4 to the thyroid receptors (TRα or TRβ); it is the most active hormone and is responsible for most biological effects. Thyroxin and T3 are rendered inactive by the loss of an inner ring iodine catalyzed by another group of deiodinase enzymes and eliminated by biliary excretion. The thyroid condition is usually evaluated by measuring T4 and T3; moreover, it is these two hormones that are generally associated with the term "thyroid hormone." THs are evolutionarily conserved molecules present in all extant vertebrates and some invertebrates (Heyland et al. 2005). The action mechanisms of THs through nuclear receptors that influence gene expression is also strongly conserved across the vertebrate taxa (Whitfield et al. 1999; Bertrand et al. 2004; Buchholz et al. 2006). In addition, because of this highly conserved nature of TH chemistry, synthesis, signaling, and regulation, environmental factors that affect thyroid function or TH signaling in one species may well affect thyroid function or TH signaling in others including humans (Miller et al. 2009).

Molecular signaling pathways regulated by THs affect development, energy balance, and metabolism for all taxonomic groups of vertebrates. It is understood that THs regulate, at least in part, the classical larva-to-juvenile metamorphosis of teleosts and amphibians (Paris et al. 2008a) and the embryo-to-larva transformation of teleosts (Liu et al. 2002). Concerning amphibian metamorphosis, many of the morphological changes are TH-dependent, among them the resorption of the tail, development of limbs, and restructuring of the head (Shi 2000; Buchholz et al. 2005; Carr and Norris 2006). Among teleosts, THs are essential for early development, particularly for larval–juvenile transitions and induction of metamorphosis (craniofacial rearrangement and loss of bilateral symmetry) in flatfish (Yamano et al. 1994; Inui et al. 1995; Power et al. 2001; Yamano 2005; Schreiber et al. 2006; Shiao and Wang 2006; Blanton and Specker 2007; Klaren et al. 2008). Many studies have correlated the necessity of THs for gonadal development and reproduction in most teleosts (Weber et al. 1992; Cyr and Eales 1996; Van der Ven et al. 2006; Mukhi and Patiño 2007). In mammals (Zoeller and Rovet 2004), THs are also essential for development. The crucial role of THs in brain development for mammalian and nonmammalian species has been further described (Ahmed et al. 2008; Koibuchi 2008; Leonard 2008; Ahmed et al. 2010; Sigrun and Heike 2010).

Given the fundamental role of TH for physiologic function and normal development in all vertebrates and some invertebrates, the identification of environmental factors that may adversely affect thyroid function and/or TH signaling is very important as is an evaluation of their ability to adversely affect ecosystems. Main endpoints used for the assessment of HPT disruption in most mammals, teleosts, and amphibians are as follows: central effects–thyroid histopathology; plasma binding proteins and plasma TH levels; TH deiodination and metabolism; target tissue and receptor level endpoints; whole animal endpoints (Brucker-Davis 1998; Pearce and Braverman 2009; Carr and Patiño 2010).

Some EDCs have a structure relatively similar to that of thyroxin (T4) and triiodothyronine (T3). Thus, these compounds are mainly characterized by their ability to have direct actions on TH receptors, inhibit thyroidal iodine uptake, inhibit TH synthesis (thyroperoxidase activity), alter deiodinase functions in peripheral tissues, decrease T4 half-life by activating hepatic enzymes, and displace TH from TH-binding proteins (transport protein). Among these compounds, PCBs, brominated flame retardants [polybrominated diphenyl ether (PBDEs) and tetrabromobisphenol (TBBPA) in particular], perchlorates, or alkylphenols are frequently found in the environment.

In rodents, PCB exposure is associated with a decreased T4 level. In a study on wild fish in San Francisco Bay, Brar et al. (2010) have shown that T3 and T3/T4 ratio are correlated with PCB exposures. Biological effects of PBDEs in animals seem to be similar to those of PCBs. In rodents, amphibians (*Xenopus laevis*), and fish (*Pimephales promelas*), decreased T4 levels have been reported (Kudo et al. 2006; Kuriyama et al. 2007; Lema et al. 2008). It has been reported that PBDEs during prenatal exposure of rats cause long-term changes in motor activity (hyperactivity) and disrupt performance in learning and memory tests (Kuriyama et al. 2005). A recent study (Park et al. 2011) on tetrabromoethylcyclohexane, a potential substitute for PBDEs hexabromocyclodecane has shown a significant reduction in total plasma thyroxin and a significant increase in mean thyroid epithelial cell height in the juvenile brown trout *Salmo trutta*.

The brominated flame retardant TBBPA has biological effects observed across many species (Darnerud 2003; Saegusa et al. 2009). In amphibians, marked molecular effects were observed *in vivo* on TH signaling (Fini et al. 2007). TBBPA could affect TH binding to thyroid receptors and the transport of TH by the TH-binding protein, transthyretin (Kitamura et al. 2005). In rodents, TBBPA induces neurobehavioral effects (Nakajima et al. 2009). Long-term exposure of European flounder (*Platichthys flesus*) to TBBPA increased levels of the TH thyroxin (T4) in correlation with internal concentrations of the test compound. The authors indicated a possible competition of TBBPA for plasma protein binding. Triiodothyronin (T3) levels were not affected, and histology showed no signs of altered thyroid gland activity (Kuiper et al. 2007).

Perchlorates are chemicals with well-known antithyroidal effects (Bernhardt et al. 2011; Li et al. 2011; Boas et al. 2012). They inhibit thyroidal iodine uptake in all vertebrates, because they have a greater affinity for NIS than iodine, and the structure of the symporter is largely conserved in all vertebrates (Carr and Patiño 2010). In teleosts and amphibians, it has been widely demonstrated that perchlorates decrease T4 concentration in blood and tissue, cause thyroid histopathological alterations, and induce trouble in growth and metamorphosis failure (Carr and Patiño 2010). In fish, more precisely zebrafish, perchlorate induced hypothyroidism, which causes a halt to spawning (Mukhi and Patiño 2007). Another class of chemicals, nonylphenols, induced a significant decrease of thyroxin levels in fish, whereas no effect on triiodothyronine concentrations was detected, thus inhibiting the TH balance (Zaccaroni et al. 2009). No histopathological changes were detected in the thyroid. Other organic chemicals and also metals have been described or suspected to cause thyroid disruption (Brucker-Davis 1998; Brown et al. 2004; Miller et al. 2009; Pearce and Braverman 2009; Carr and Patiño 2010).

Thyroid disruption properties can be detected with *in vitro* cellular tests. Jugan et al. (2007) have developed a cellular assay (PC-DR-LUC) that can be used to identify compounds that have thyromimetic or TH-antagonistic effects on thyroid receptors. Kitamura et al. (2005) showed a thyromimetic effect of TBBPA on *in vitro* cellular test using a GH3 pituitary cell line. *In vitro* bioassays indicate that some of polycyclic aromatic hydrocarbon derivatives have agonist or potentiation activity with thyroid receptors and have a strong TTR-binding activity (Bekki et al. 2009). Fini et al. (2007) have developed an *in vivo* test for the detection of thyroid receptor agonists and for the determination of effects of TH production inhibitors, including inhibitors of TH synthesis, antagonists acting at the receptor, and deiodinase inhibitors.

Real-life exposure of humans and wildlife to a wide spectrum of synthetic chemicals makes it difficult to prove the causative thyroid effect of one single agent. Therefore, it is more prudent to talk of association rather than causation, and the role of potential cocontaminants should always be considered (Brucker-Davis 1998).

8.5 Impairment of Endocrine Functions in Crustaceans

8.5.1 Endocrine Regulation of Reproduction in Crustaceans

Crustaceans are major representatives of ecdysozoans in aquatic ecosystems. The endocrine system of crustaceans involves neuropeptides secreted by a neurohemal complex, the X organ–sinus gland complex. These neurohormones regulate the secretion of ecdysteroid, sesquiterpenoid, and glycopeptide hormones via three main endocrine structures: the Y organ, the mandibular organ, and the androgenic glands in males (Figure 8.4).

The crustacean hyperglycemic hormone (CHH) family, which includes about 150 members, is the archetype of crustacean neurohormones (review by Montagné et al. 2010). CHHs are neuropeptides secreted by the X organ–sinus gland complex. They are subdivided into two subfamilies. Type I peptides, the CHHs *sensu stricto*, are pleiotropic hormones involved in the regulation of energy and ion metabolism. Type II peptides—the molt-inhibiting hormones, the gonad- or vitellogenesis-inhibiting hormones, and the mandibular organ–inhibiting hormones—are functionally specialized in hormonal regulation of the molting and reproductive processes. Several studies have suggested that biogenic amines, secreted in the pericardial organ, act as neurohormones in crustaceans (Cooke and Sullivan 1982). Biogenic amines could be implicated in reproductive process regulation. For example,

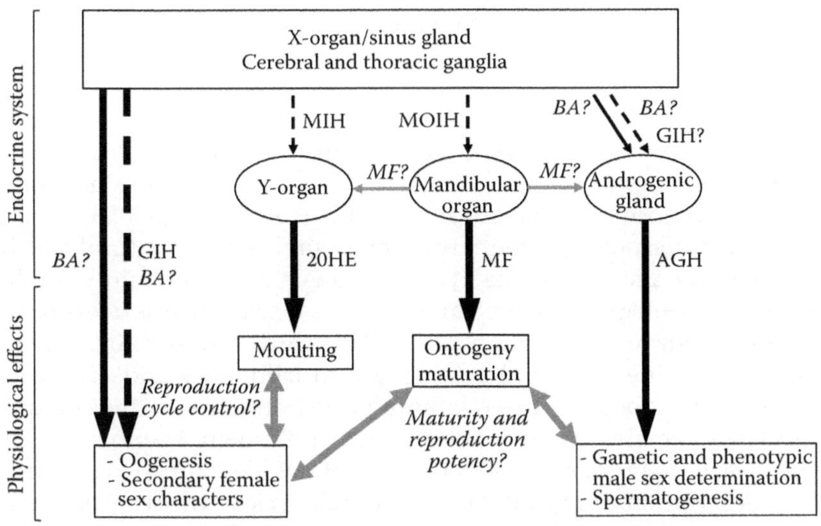

FIGURE 8.4
Synthetic representation of major endocrine organs, their secreted hormones, and their target tissues involved in male and female reproduction in crustaceans. Stimulatory effects, inhibitory effects, and physiological event interactions are indicated with solid arrows, dashed arrows, and gray double arrows, respectively. AGH, androgenic gland hormone; BA, biogenic amines (e.g., serotonin and dopamine); GIH, gonad-inhibiting hormone; MF, methyl farnesoate; MIH, molt-inhibiting hormone; MOIH, mandibular organ-inhibitory hormone; 20HE, 20-hydroxyecdysone.

serotonin (i.e., 5-hydroxytryptamine) tends to enhance gametogenesis in decapod species, whereas dopamine tends to inhibit it (reviewed by Mazurová et al. 2008).

20-Hydroecdysone is known to be the most active form of ecdysteroid. Commonly called molting hormone, it enhances the molting process such as apolysis and new cuticle secretion, which are essential for the growth and development of crustaceans (Lachaise et al. 1993). Because of the close relationship between molt and reproduction cycles in female organisms, this hormone certainly plays a key role in vitellogenesis regulation (Charniaux-Cotton 1973; Subramoniam 2000). Ecdysteroids trigger their regulatory activity through interaction with the ecdysteroid receptor (EcR). To regulate transcription, EcR needs to dimerize with the orphan retinoid X receptor (RXR). EcR and RXR belong to the superfamily of nuclear receptors. However, although EcR structure and functioning in insects is well documented, our knowledge of EcR in crustaceans remains limited (Verhaegen et al. 2010). Similarly, the natural ligand(s) for invertebrate RXRs is still under debate.

The sesquiterpenoid methyl farnesoate (MF) is the equivalent of the juvenile hormone of insects. MF is involved as a morphogen in the postembryonic larval stage of development, in the positive regulation of the molt cycle, as well as in stimulation of reproductive maturation in female decapods (Laufer and Biggers 2001; Nagaraju et al. 2003). Experimental evidence also suggests that MF is a male sex determinant in cladocerans in which cyclical parthenogenesis occurs (Olmstead and LeBlanc 2001). Insect RXRs are capable of binding terpenoid hormones (reviewed by LeBlanc 2007). However, the regulatory activity of MF by binding to RXR or the EcR–RXR complex has not yet been demonstrated in crustaceans.

In malacostracans, the wider taxon that includes decapods, AGH positively regulates sexual differentiation and the maintenance of the male reproductive system, spermatogenetic activity, and differentiation of secondary sexual characteristics (Charniaux-Cotton 1965; Charniaux-Cotton and Payen 1988). In females, the androgenic glands do not develop, and the sexual differentiation of ovaries is spontaneously induced and oogenesis is directly regulated by a gonad-inhibiting neurohormone.

8.5.2 Impact of EDCs on Crustaceans

By analogy to the *in vivo* approaches used for aquatic vertebrates, in particular in fish, the measurements of hormone levels, associated receptor quantities and/or activities of enzymes involved in hormone metabolism should also constitute relevant approaches to assess the endocrine-disrupting potential of a compound and/or a natural matrix. However, studies reporting such measurements in crustaceans have been mainly conducted in the framework of ecophysiological research and on large species such as decapods with potential applications in aquaculture (LeBlanc 2007). As regards ecotoxicologically relevant species (i.e., with a short life cycle and easy to use in laboratory and *in situ* experiments) such as branchiopods, copepods, isopods, amphipods, and mysids, the characterization of hormone regulation remains limited or even largely unknown. The study of ecotoxicologically relevant species is burdened with (1) a lack of detailed knowledge concerning their life cycle (e.g., description of reproductive and molt cycles), which is essential to understand and to interpret the evolution of hormone levels, and (2) analytical limitations owing to small amounts of biological material available.

Impact assessment of EDCs on invertebrates has received particular attention over the past 10 years (Verslycke et al. 2007; Mazurová et al. 2008). Crustacean studies have been structured around two main axes. The first concerns the toxicity assessment of compounds known or suspected to be ED in vertebrate species, such as 17α-ethinylestradiol, bisphenol A, nonylphenols, diethylstilbestrol, and polychlorobiphenyls in a context of risk assessment. The second focuses on the study of the impact of compounds designed to modulate the regulation of juvenile hormones found in insects (pesticides), in order to evaluate their effect on nontarget species, because of the similarity of endocrine systems within arthropods, and to improve knowledge on the endocrine regulation of crustaceans.

For known vertebrate endocrine disruptors, most studies have focused on the impact assessment of these compounds on growth, reproduction, and molt in many species of crustaceans, many of which have standardized or chronic bioassays available. In a context of risk assessment, although these studies have been shown to be relevant by taking into account the toxicity of these compounds to a large number of living species in aquatic systems, they do not provide (1) the potential of these compounds as EDCs in crustaceans, (2) knowledge on modulation of hormone regulation in relation to the effects observed, and (3) the role played by steroid hormones in endocrine regulation in crustaceans. In the copepod *Eurytemora affinis*, Forget-Leray et al. (2005) showed the adverse effects on the development and sex ratio of several vertebrate EDCs. Mortality and reduced fertility, reproduction, and growth may result from a disturbance of hormone regulation, but these responses may also result from other mechanisms such as modulation of energy allocation, disturbance of the nervous system, and/or transfer of genetic abnormalities after exposure to genotoxic compounds. Indeed, several studies on marine crustaceans (*Tisbe battagliai, Nicotra spinipes*, and *Palaemon elegans*) have shown that compounds such as 17α-ethinyl estradiol, 17β-ethinyl estradiol, estrone, and 4-nonylphenol have few adverse effects on survival, development, reproduction, and sex ratio (Lye et al. 2008). Similarly, Hannas et al. (2011) clearly show that the measurement of VTG mRNA levels in *Daphnia magna* can be used in biomonitoring exposure of daphnids to some environmental chemicals but is not useful as an indicator of exposure to estrogenic chemicals.

To a lesser extent, a more specific exploration of endocrine disruption has been conducted for the major insect and crustacean hormones (ecdysteroids and terpenoids) and analog chemicals such as insect growth regulators (IGRs), which are pesticides specifically developed to mimic, interfere with, and/or block the action pathway of ecdysteroids and juvenile hormone in insects. IGRs are by definition arthropod-specific EDCs. Thus, pesticides designed as juvenile hormone mimics such as fenoxycarb, pyriproxyfen, or methoprene, showed deleterious effects on the embryo–larval development and molt cycle in decapods (e.g., McKenney et al. 2004; Tuberty and McKenney 2005) as well as in mysidaceans (e.g., McKenney and Matthews 1990; Ghekiere et al. 2006a) and cladocerans (Olmstead and LeBlanc 2001), for exposure concentrations much lower than lethal concentrations. These insecticides also caused alterations of reproductive functions such as inhibition of oogenesis and vitellogenesis (e.g., Tokishita et al. 2006; Ghekiere et al. 2006b), delays in sexual maturity, and/or decreased fertility (e.g., Olmstead and LeBlanc 2001; McKenney 2005). More recently, several studies have shown that juvenile hormone and analog chemicals induced male neonate production in parthenogenetic cladoceran species (for a review, see Tatarazako and Oda 2007).

8.5.3 Tools for Monitoring Crustacean Exposure to EDCs

The *in vitro* measurement of the (ant)agonist potential of chemicals toward a specific hormonal receptor is a relevant approach for screening the endocrine-disrupting potential of a compound and/or extracts of a natural matrix. Clément et al. (1993) proposed a rapid *in vitro* bioassay for the detection of ecdysteroid agonistic and antagonistic activities based on the ecdysteroid receptor–specific responses of a tumorous blood cell line of *Drosophila melanogaster*. To our knowledge, this is the only *in vitro* bioassay available using a specific nuclear receptor present in crustaceans and involved in reproduction. Among more than 80 environmental contaminants tested using this bioassay, including industrial chemicals, pesticides, pharmaceuticals, phytoestrogens, as well as vertebrate steroids, few compounds were defined as ecdysteroid receptor (ant)agonists (Dinan et al. 2001).

In the field, several observations have led to the suspicion of the occurrence of EDCs within wild crustacean populations. The incidence of intersex individuals among a wide range of taxa is certainly the most convincing and the most thoroughly documented of these observations (for a review, see LeBlanc 2007). For example, Jungmann et al. (2004) reported intersex occurrence up to 24% within a population of *Gammarus fossarum* (a freshwater amphipod crustacean). However, and for the reasons discussed above in relation to the available knowledge on the endocrine regulation of reproductive processes in crustaceans, very few specific molecular biomarkers are available in crustaceans for assessment of exposure to EDCs. By analogy to the method applied in fish, the measurement of VTG-like proteins has been proposed as a biomarker of AGH pathway disruption (LeBlanc 2007; Matozzo et al. 2008). In crustaceans, the use of specific methods for VTG assessment in an ecotoxicological context has been limited to the measurement of gene expression encoding this protein in only a few species, such as *Daphnia magna* (Tokishita et al. 2006), the copepod *Tigriopus japonicus* (Lee et al. 2008), and *Gammarus fossarum* (Xuereb et al. 2011). VTG production was more commonly assessed at the protein level, using indirect methods such as an ELISA (Sagi et al. 1999; Oberdörster et al. 2000; Volz and Chandler 2004; Ghekiere et al. 2006b) or alkali-labile phosphate (ALP) measurement (Gagné and Blaise 2004; Huang and Chen 2004; Gagné et al. 2005; Huang et al. 2006). Recently, a method based on LC-MS/MS has been developed allowing absolute quantification of this protein in the freshwater amphipod *Gammarus fossarum* (Simon et al. 2010). Among the studies using VTG as a specific ED biomarker, most were conducted on females (Gagné and Blaise 2004; Martín-Díaz et al. 2004; Volz and Chandler 2004; Gagné et al. 2005; Ghekiere et al. 2006a, 2006b; Huang et al. 2006; Poynton et al. 2007), showing that VTG synthesis is susceptible to different contaminants (e.g., xenoestrogens, pesticides, metals, or complex mixtures such as urban and industrial wastes). However, females display natural fluctuations in VTG levels during their reproductive cycle (Meusy and Junera 1974; Jasmani et al. 2000; Okumura and Aida 2000; Jayasankar et al. 2006). Therefore, accurate use and reliable interpretation of data using VTG measurements in females requires detailed knowledge of their reproductive cycle to discriminate between natural fluctuations and chemical-induced variations (Hannas et al. 2011; Xuereb et al. 2011, Jubeaux et al. 2012); however, this is rarely the case in the literature (Lee and Noone 1995; Gagné and Blaise 2004; Ghekiere et al. 2006b). To our knowledge, only the studies of Huang and Chen (2004) on the decapod *Neocaridina denticulata*, Sanders et al. (2005) on *Palaemon elegans*, and Xuereb et al. (2011) and Jubeaux et al. (2012) on *Gammarus fossarum* have developed and used VTG measurement (i.e., protein or transcript) as an ED biomarker in male crustaceans. In addition, here again, the vertebrate point of view leading to the interpretation of the induction of VTG in males as an ED could be debated in invertebrates because of pleiotropic functions of VTG proteins (e.g., oxidative stress). Indeed, duplications and functional shifts in VTG-like proteins have occurred during metazoan evolution (Smolenaars et al. 2007). This has been exemplified in decapods, in which the protein ortholog of VTG in most metazoans plays a role in clotting, whereas egg yolk protein is produced by a VTG paralog gene belonging to the subfamily of apolipoproteins (Avarre et al. 2007). Therefore, a functional validation of such a classical biomarker routinely used in vertebrates has to be undertaken with caution before application in crustaceans (Xuereb et al. 2011; Jubeaux et al. 2012).

In parallel with development carried out to propose specific molecular biomarkers and as recommended by Verslycke et al. (2007), several approaches based on the simultaneous use of biological variables directly under endocrine regulation, such as oocyte growth, molt, and fertility, have been proposed as tools based on the weight of evidence to diagnose and identify compounds suspected of being ED in crustaceans. For example, Tatarazako and Oda (2007) proposed using the production of males in *Daphnia magna* as a specific indicator of exposure to compounds having a similar effect to that of MF. Indeed, MF antagonist activity can be measured by exposing crustaceans to MF to stimulate male offspring production and assessing the ability of chemicals to block this action (Wang et al. 2005). Thus, dieldrin appeared antagonistic at low, endogenous levels of MF, but agonistic in the absence of MF (LeBlanc 2007). However, this approach cannot be applied to crustaceans with sexual reproduction. Similarly, Mu and LeBlanc (2002) proposed using the intermolt cycle duration in *D. magna* as a specific response to exposure to ED antagonists to ecdysteroids, by comparing the effects obtained in organisms after an exposure to a compound alone and in combination with the 20-hydroxyecdysone, known as the active form of molting hormone in crustaceans. Finally, Geffard et al. (2010) developed a biotest that specifically tests the impact of pollutants on the molt cycle, embryonic development, and oocyte growth (secondary vitellogenesis) and production in *G. fossarum*. As previously demonstrated, all of these biological variables are perfectly synchronized during the reproductive cycle of this species under uncontaminated conditions; consequently, for each molt stage there is a specific oocyte surface and embryonic stage. Any disruption of the synchronism of these endpoints by a chemical contaminant would highlight a possible endocrine disruption mechanism.

8.6 Endocrine Control and Disruption in Mollusks

In mollusks, neuropeptides produced by neurosecretory ganglia have been described to play a role in several physiological processes such as reproduction, energy, and ion regulation. Among the best-known neuropeptides—gonadotropic, egg-laying hormones, insulin-like and FMRFamide peptides—the gonadotropic and egg laying hormones are the main neuropeptides identified in gastropods to play roles in sexual differentiation, gonad maturation, egg laying, and behavior (Joose 1988; Nassel 1996; LeBlanc et al. 1999; LaFont 2000; de Jong-Brink et al. 2001; López-Vera et al. 2008).

Although their role in invertebrates is still under discussion, vertebrate-like steroids, including 17β-estradiol (E2) and testosterone (T) have been described in bivalves and gastropods as well. Janer and Porte (2007) reviewed studies in which sexual steroids were detected in invertebrates, and key steps of steroidogenesis leading to androgens or estrogens described in species from different phyla. Similar to vertebrates, a combination of cytochrome P450 and steroid dehydrogenase catalyzes the conversion of cholesterol to sex steroids in molluscan species (Porte et al. 2006). Nevertheless, the transformation rates of some metabolites could be low (Gottfried and Dorfman 1970; De Longcamp et al. 1974). Some temporal variations in their levels have been linked to possible roles in the reproductive cycle in bivalves (reviewed by Lafont and Mathieu 2007) and gastropods (Sternberg et al. 2008; Gust et al. 2011).

Steroid conjugation and esterification have also been described in mollusks. Only a small quantity of testosterone and estradiol is free in *Marisa cornuarietis*, whereas the major part is esterified (Janer et al. 2006). Other steroids have been found in the form of esterified products in bivalves (e.g., 17β-estradiol in *Mytilus galloprovinciallis*, Janer et al. 2005; and *Mytilus edulis*, Labadie et al. 2007). The physiological role of the esterification of steroids is not elucidated; nevertheless, it seems to be a factor in the regulation of homeostasis of the free active forms, likely to be a target of pollutants. For instance, some studies have shown the ability of endocrine disruptors to alter steroid metabolism in invertebrates and particularly in mollusks. *In vitro* studies have shown that TBT inhibited aromatase activity in the bivalves *Ruditapes decussata* and *Crassostrea gigas* (Morcillo et al. 1998; Le Curieux-Belfond et al. 2001). Exposure to TBT also decreased testosterone sulfation and testosterone esterification in *Littorina littorea* (Ronis and Mason 1996) and the mudsnail *Ilyanassa obsoleta* (Gooding et al. 2003). Steroid levels were also impaired in the mudsnail *Potamopyrgus antipodarum* exposed to the specific vertebrate aromatase inhibitor, fadrozole (Gust et al. 2010b).

In vertebrates, the effects of steroids are mediated by intracellular or cell surface receptors. In recent studies, ER orthologs were identified in several mollusks, e.g., *Aplysia californica* (Thornton et al. 2003), *Thais clavigera* (Kajiwara et al. 2006), *Octopus vulgaris* (Keay et al. 2006), *C. gigas* (Matsumoto et al. 2007), *Marisa cornuarietis* (Bannister et al. 2007), *Nucella lapillus* (Castro et al. 2007), and *P. antipodarum* (Stange et al. 2012), but these receptors seem to be constitutively active without the ability of estrogen binding.

An alternative candidate hormone signaling pathway is retinoic acid acting via the RXR. Several studies provide substantial evidence that the retinoid signaling pathway is involved in reproductive development and is the target through which TBT causes imposex (Sternberg et al. 2008). Other hypotheses could be that the active molluscan sex hormone is not E2 itself, but a derivate (Baker 2004; Paris et al. 2008b), or that a receptor, capable of E2 binding, has not still been identified.

Moreover, besides the "classical" genomic pathway, recent studies have shown that E2 can act through nongenomic mechanisms (reviewed by Janer and Porte 2007) in mollusks causing rapid effects. These alternative pathways of EDCs can involve membrane ERs or receptor-independent mechanisms, with possible both direct local effect (such as modulation of ion fluxes) and regulation of gene transcription secondary to modulation of kinase cascades, and are likely to impair functions other than reproduction, such as, in particular, immune function (Stefano et al. 2003; Canesi et al. 2004).

Results suggest that synthesis of VTG-like protein occurs in bivalve gonads (Li et al. 1998; Matsumoto et al. 2003) under control of both estrogen and neuropeptides (Osada et al. 2003). Similar to fish, VTG-like protein measurements have been used to assess the effect of potential xenoestrogens in bivalves, by using mainly indirect methods such as use of ALPs both in laboratory and field studies (review of Matozzo et al. 2007; Porte et al. 2006). Gagné et al. (2002) showed modifications in VTG-like proteins of bivalves exposed to polluted environments, prompting interest in the study of VTG-like proteins in bivalves.

Nevertheless, knowledge on the functioning of the molluscan endocrine system and its role in controlling reproduction remains insufficient, and, except for the presence or degree of imposex in gastropods, no biomarker is yet available in mollusks to clearly identify endocrine disruption likely to have been caused by environmental exposure to xenoestrogens.

8.7 Ecological Consequences of Endocrine Disruption

Field and experimental studies at environmentally relevant concentrations (e.g., Li et al. 2011; Sarria et al. 2011; Hatef et al. 2012) have established that EDCs represent a serious threat for the future of our environment. In 2005, Mills and Chichester considered that linking endocrine disruption and reproductive impairment with an ecologically relevant impact on the sustainability of real fish populations remained an open challenge. Weis et al. (2011)

conclude that thyroid impairments could contribute to behavioral changes with ecological consequences for estuarine fish and crustaceans (Table 10.2). Several studies with the fish *Pimephales promelas* show an effect of EDCs on several parameters important for population maintenance. Lange et al. (2011) have observed that a concentration as low as 0.2 ng EE2 L^{-1} was sufficient to induce a decrease in the number of eggs per female, concentration of VTG in plasma, and the weight and size of larvae, whereas at 4.0 ng EE2 L^{-1}, the sex ratio was female biased. In mature females exposed to 29–60 ng EE2 L^{-1}, both an increase and a decrease in VTG concentrations were associated with a reduced reproductive capacity (a reduction in the frequency of spawning that led to an overall reduction in the number of eggs spawned) (Thorpe et al. 2007). Grist et al. (2003) have calculated that the intrinsic growth rate of the population would be reduced to 0 if the environmental concentration of EE2 was between 3.1 and 3.4 ng EE2 L^{-1}. Miller et al. (2007) have shown a highly significant correlation between relative fecundity of *P. promelas* and VTG concentration in plasma. Yet, several molecules, such as 17α-trembolone and 17β-trembolone, induce a consistent decrease of VTG concentrations for experimental overloads of a few ng L^{-1}. These authors have modeled changes of fish population size depending on the concentration of VTG in the plasma. A reduction of VTG concentration by 50% would be responsible for population extinction after 20 years, a reduction by 75%, after 10 years. In killifish *Fundulus heteroclitus* from Newark Bay, reduced egg production due to inhibition of vitellogenesis has been reported (Bugel et al. 2011). However, the Newark Bay killifish population continues to sustain itself, as is evidenced by young-of-the-year. A small percentage of the population appears to perform the majority of the reproductive duties and may be why this population persists. In another fish species, *Poecilia reticulata*, exposure to doses of 112 ng EE2 L^{-1} induced a dramatic change in the sex ratio, which was highly female-biased, and significant reductions in male sperm count with a subsequent drop in fertility in the exposed population (Kristensen et al. 2005). Daouk et al. (2011) performed a life-cycle exposure using the zebrafish model and mixtures representative of some environmental situations in terms of dietary doses and composition and containing mainly non-dioxin-like congeners. Under these realistic conditions, the fertilization rate of fish spawns was decreased. Several mechanisms that could underlie the reduction of fertility are reviewed, and additional examples of reduction are quoted.

In the fish *Rutilus rutilus* from wild populations living in effluent-contaminated rivers in the United Kingdom, the intersex condition reduced reproductive performance by up to 76% for the most feminized individuals, demonstrating a significant adverse effect of intersex on reproductive performance. Feminization of male fish is likely to be an important determinant of reproductive performance in rivers where there is a high prevalence of moderately to severely feminized males (Harris et al. 2011), a situation frequently observed in the field (Chapter 9).

The most convincing study showing a collapse of a fish population after exposure to a synthetic estrogen has been carried out by Kidd et al. (2007). In a lake amended with 5–6 ng EE2 L^{-1} from summer 2001 to fall 2003, females of the fish *Pimephales promelas* had higher VTG concentrations after exposure to EE2 when compared with reference samples from another lake (reference lake fish had 2.5% of the VTG concentrations observed in treated fish). More strikingly, in males, VTG concentrations were 3 orders of magnitude greater than reference samples. Fish density was highly variable at the reference site, but dramatically decreased in the presence of EE2. As a consequence, in the amended lake, no recruitment occurred from 2002 to 2004.

Linking endocrine disruption and reproductive impairment with an ecologically relevant impact has also been demonstrated in invertebrates (Chapter 9). In New Zealand, the mudsnail *Potamopyrgus antipodarum* exposed in the field to WWTP effluents, cascading effects from infra- to supraindividual effects were observed. An increase in testosterone was reported downstream of the discharges, and increases and decreases in 17β-estradiol levels according to site. Reproduction was significantly impacted downstream from the discharges with 60–70% decrease of embryos without shells after 3–4 weeks' exposure (Gust et al. 2010a). Intersex is commonly observed in bivalves *Scrobicularia plana* living in European estuaries. However, at this stage, no clear link may be established between intersex and populational effects (Langston et al. 2007; Gomes et al. 2009; Fossi Tankoua et al. 2012).

8.8 Conclusions

Caliman and Gavrilescu (2009) and Vandenberg et al. (2012) have highlighted a number of discrepancies between classic toxicology assumptions and the specificity of EDCs. Contrary to dose–response curves showing that the greater the exposure to a chemical, the greater the health effect, some EDCs exhibit effects only at dilute or large concentrations yielding U-shaped or J-shaped dose–response curves. Generally, a chemical cannot have opposite effects, whereas EDCs such as dioxin exhibit an estrogen-like effect in fetal organisms but an antiestrogen effect in adults. As also admitted for carcinogenic compounds, most hormones in the bloodstream occur in dilute concentrations, and even very dilute levels of endocrine-disrupting chemicals can prove dangerous, suggesting the absence of any threshold effect.

Understanding the gap between the endocrine potencies of chemical mixtures such as released in effluents and their real impact in the wild is an outcome that is highly desirable. Even if *in vitro* assays present some limitations in terms of estimation of the activities of estrogens or androgens released into the aquatic environment, they are able to give a good overview of the impact of effluent in certain cases, as demonstrated for a wild population of gudgeon *Gobio gobio* (Nadzialek et al. 2010). According to these authors, it could be interesting to include *in vitro* assays more in laboratory studies to establish direct correlations, so the actual significance of estrogenicity screening for predicting *in vivo* effects can be assessed. They conclude that an integrated picture of the estrogenicity of a given compartment can only be achieved if chemical analysis, estrogenic activity screening, and *in vivo* research in the field are combined.

Ankley et al. (2010) have defined an adverse outcome pathway (AOP) as "a conceptual construct that portrays existing knowledge concerning the linkage between a direct molecular initiating event (e.g., a molecular interaction between a xenobiotic and a specific biomolecule) and an adverse outcome at a biological level of organization relevant to risk assessment." This concept is very relevant to describe the case of EDC effects (Figure 8.5). An AOP involves cascading effects observed at different levels of biological organization. If the links between these different levels are well established from a mechanistic point of view, or even if they are based on strong correlations, population models may be used to formalize these links, with an interesting potential for predictive risk assessment of chemicals in a regulatory context (Ankley et al. 2010; Galic et al. 2010). Despite effects of EDCs on sexual behavior (courtship, territorial behavior) being well documented (Chapter 10), no clear links have been established with population effects. Thus, it is necessary to study in depth the relationships between reproductive behavior, reproductive success, and mechanisms controlling population dynamics. In the case of EDCs, models have already been used to predict the fate of fish populations (Miller and Ankley 2004; Brown et al. 2005; Miller et al. 2007).

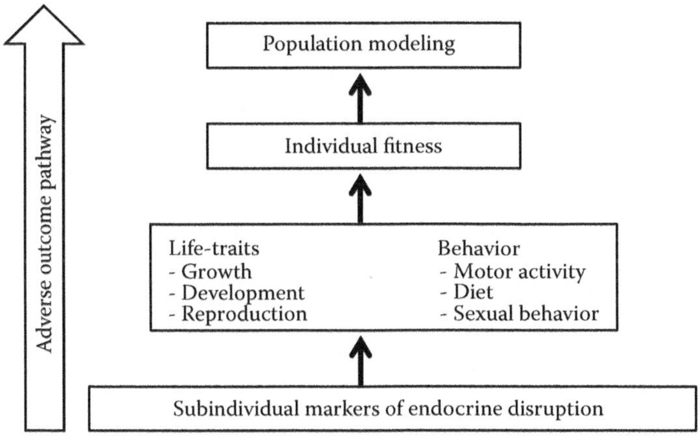

FIGURE 8.5
Simplified representation of concept of an adverse outcome pathway applied to effects of EDCs on steroid, thyroid, and molt hormones. (After Coulaud, R. 2012. Modélisation et changement d'échelles pour l'évaluation écotoxicologique: application à deux macroinvertébrés aquatiques, *Gammarus fossarum* (Crustacé amphipode) et *Potamopyrgus antipodarum* (Mollusque Gastéropode) Thesis Lyon University. With permission.)

The status of methods for estrogen-specific effects monitoring has been recently reviewed (OSPAR agreement 2008–2009). VTG and gonadal intersex measurements have been intercalibrated in the framework of BEQUALM (Biological Effects Quality Assurance in Monitoring Programmes in the European Community) and are recommended for use by the International Council for the Exploration of the Sea.

Endocrine disruptors affect biological processes linked to reproduction and development. So, it is evident that, in order to distinguish between natural stress and stress induced by EDCs, it is indispensable to have a good knowledge of the normal functioning of endocrine metabolism depending on phylogeny, sex, and development status. For wildlife, and particularly for invertebrates, only relatively sparse knowledge of endocrine mechanisms and relevant biomarkers is yet available to predict and to survey the impact of complex mixtures of endocrine disrupters in the field.

Moreover, in addition to biological factors, extrinsic factors can also influence reproduction and development, such as temperature and food availability. Seasonality not only affects temperature (e.g., Sieratowicz et al. 2011) but also the production of estrogens and their conjugates (Kumar et al. 2011), and hypoxia has been recently recognized as acting like an endocrine disruptor (Murphy et al. 2009; Wu 2009; Thomas and Rahman 2012). With the aim of using biomarkers to investigate the presence and effect of EDCs in the environment, it is necessary to evaluate the influence of all such confounding factors.

References

Ahmed, O.M. et al. 2008. Thyroid hormones states and brain development interactions. *Int. J. Dev. Neurosci.* 26:147–209.

Ahmed, O.M. et al. 2010. Effects of experimentally induced maternal hypothyroidism and hyperthyroidism on the development of rat offspring: I. The development of the thyroid hormones–neurotransmitters and adenosinergic system interactions. *Int. J. Dev. Neurosci.* 28:437–54.

Allen, Y. et al. 1999a. Survey of estrogenic activity in United Kingdom estuarine and coastal waters and its effects on gonadal development of the flounder *Platichthys flesus*. *Environ. Toxicol. Chem.* 18:1791–800.

Allen, Y. et al. 1999b. The extent of oestrogenic contamination in the UK estuarine and marine environments—Further surveys of flounder. *Sci. Total Environ.* 233:5–20.

Amiard, J.C. et al. 2009. Quantification of contaminants. In: *Environmental Assessment of Estuarine Ecosystems*, ed. C. Amiard-Triquet and P.S. Rainbow, 31–57. Boca Raton, FL: CRC Press.

Ankley, G.T. et al. 2010. Adverse outcome pathways: A conceptual framework to support ecotoxicology research and risk assessment. *Ecotoxicol. Environ. Saf.* 29:730–41.

Aravindakshan, J. et al. 2004. Consequences of xenoestrogen exposure on male reproductive function in spottail shiners (*Nortropis hudsonius*). *Toxicol. Sci.* 78:156–65.

Arukwe, A. et al. 2001. *In vivo* modulation of nonylphenol-induced zonagenesis and vitellogenesis by antiestrogen, 3,3′4,4′-tetrachlorobiphenyl (PCN-77) in juvenile fish. *Environ. Toxicol. Pharmacol.* 10:5–15.

Auger, J. et al. 1995. Decline in semen quality among fertile men in Paris during the past 20 years. *N. Engl. J. Med.* 332, 281–5.

Avarre, J.C. et al. 2007. Apolipocrustacein, formerly vitellogenin, is the major egg yolk precursor protein in decapod crustaceans and is homologous to insect apolipophorin II/I and vertebrate apolipoprotein B. *BMC Evol. Biol.* 7:3.

Baker, M.E. 2004. Co-evolution of steroidogenic and steroid-inactivating enzymes and adrenal and sex steroid receptors. *Mol. Cell. Endocrinol.* 215:55–62.

Bannister, R. et al. 2007. Novel estrogen receptor-related transcripts in *Marisa cornuarietis*; a freshwater snail with reported sensitivity to estrogenic chemicals. *Environ. Sci. Technol.* 41: 2643–50.

Bekki, K. et al. 2009. Evaluation of toxic activities of polycyclic aromatic hydrocarbon derivatives using *in vitro* bioassays. *J. Health Sci.* 55:601–10.

Bertin, A., P.A. Inostroza, and R.A. Quinones. 2011. Estrogen pollution in a highly productive eco system off central-south Chile. *Mar. Pollut. Bull.* 62:1530–7.

Bertrand, S. et al. 2004. Evolutionary genomics of nuclear receptors: From twenty-five ancestral genes to derived endocrine systems. *Mol. Biol. Evol.* 21:1923–37.

Biales, A.D. et al. 2007. Quantification and associated variability of induced vitellogenine gene transcripts in fathead minnow (*Pimephales promelas*) by quantitative real-time polymerase chain reaction assay. *Environ. Toxicol. Chem.* 26:287–96.

Bjerregaard, L.B., B. Korsgaard, and P. Bjerregaard. 2006. Intersex in wild roach (*Rutilus rutilus*) from Danish sewage effluent-receiving streams. *Ecotoxicol. Environ. Saf.* 64:321–8.

Black, J.J. et al. 1974. Thyroid enlargement in Lake Erie Coho salmon. *J. Natl. Cancer Inst.* 53:725–9.

Blanton, M.L., and J.L. Specker. 2007. The hypothalamic–pituitary–thyroid (HPT) axis in fish and its role in fish development and reproduction. *Crit. Rev. Toxicol.* 37:97–115.

Boas, M. et al. 2012. Thyroid effects of endocrine disrupting chemicals. *Mol. Cell. Endocrinol.* 355:240–8.

Borgert, C.J. et al. 2011. Evaluation of EPA's Tier 1 Endocrine Screening Battery and recommendations for improving the interpretation of screening results. *Reg. Toxicol. Pharmacol.* 59: 397–411.

Brar, N.K. et al. 2010. Evidence for thyroid endocrine disruption in wild fish in San Francisco Bay, California, USA. Relationships to contaminant exposures. *Aquat. Toxicol.* 96:203–15.

Brian, J.V. et al. 2007. Evidence of estrogenic mixture effects on the reproductive performance of fish. *Environ. Sci. Technol.* 41:337–44.

Brown, A.R. et al. 2005. Predicting the effects of endocrine disrupting chemicals on healthy and disease impacted populations of perch (*Perca fluviatilis*). *Ecol. Model.* 189:377–95.

Brown, S.B. et al. 2004. Contaminant effects on the teleost fish thyroid. *Environ. Toxicol. Chem.* 23:1680–701.

Brucker-Davis, F. 1998. Effects of environmental synthetic chemicals on thyroid function. *Thyroid* 8:827–56.

Buchholz, D.R. et al. 2005. Gene-specific changes in promoter occupancy by thyroid hormone receptor during frog metamorphosis. Implications for developmental gene regulation. *J. Biol. Chem.* 280:41222–8.

Buchholz, D.R. et al. 2006. Molecular and developmental analyses of thyroid hormone receptor function in *Xenopus laevis*, the African clawed frog. *Gen. Comp. Endocrinol.* 145:1–19.

Bugel, S.M., L.A. White, and K.R. Cooper. 2011. Decreased vitellogenin inducibility and 17β-estradiol levels correlated with reduced egg production in killifish (*Fundulus heteroclitus*) from Newark Bay, NJ. *Aquat. Toxicol.* 105:1–12.

Caliman, F.A., and M. Gavrilescu. 2009. Pharmaceuticals, personal care products and endocrine disrupting agents in the environment—A review. *Clean* 37:277–303.

Canesi, L. et al. 2004. Rapid effects of 17 beta-estradiol on cell signaling and function of *Mytilus* hemocytes. *Gen. Comp. Endocrinol.* 136:58–71.

Carlsen, E. et al. 1993. Decreasing quality of semen. *Br. Med. J.* 306:461.

Carr, J.A., and D.O. Norris. 2006. The thyroid gland. In: *Endocrine Disruption: Biological Bases for Health Effects in Wildlife and Humans*, ed. D.O. Norris and J.A. Carr, 87–110. New York: Oxford University Press.

Carr, J.A., and R. Patiño. 2010. The hypothalamus–pituitary–thyroid axis in teleosts and amphibians: Endocrine disruption and its consequences to natural populations. *Gen. Comp. Endocrinol.* 290:299–312.

Castro, L. et al. 2007. The estrogen receptor of the gastropod *Nucella lapillus*: Modulation following exposure to an estrogenic effluent? *Aquat. Toxicol.* 84:465–8.

CEC. 2001. Communication from the commission to the Council and the European Parliament on the implementation of the Community Strategy for Endocrine Disrupters—a range of substances suspected of intefering with the hormone systems of human and wildlife. COM (2001) 262 final.

Charniaux-Cotton, H. 1954. Découverte chez un crustacé amphipode (*Orchestia gammarella*) d'une glande endocrine responsable de la différenciation de caractères sexuels primaires et secondaires mâles. [Discovery in an amphipod crustacean (*Orchestia gammarella*) of an endocrine gland responsible for the differentiation of primary and secondary male sex characteristics.] *C. R. Acad. Sci.* 239:780–2.

Charniaux-Cotton, H. 1965. Hormonal control of sex differentiation in invertebrates. In: *Organogenesis*, ed. R.L. Dehaan and H. Ursprung, 701–40. New York: Holt, Rinehart, and Winston.

Charniaux-Cotton, H. 1973. Description et contrôle de l'ovogénèse chez les Crustacés supérieurs. *Ann. Biol. Anim. Biochem. Biophysiol.* 13:21–30.

Charniaux-Cotton, H., and G. Payen. 1988. Crustacean reproduction. In: *Endocrinology of Selected Invertebrate Types*, ed. H. Laufer and R.G.H. Downer, 279–303. New York: A.R. Liss.

Clément, C.Y. et al. 1993. Assessment of a microplate-based bioassay for the detection of ecdysteroid-like or antiecdysteroid activities. *Insect Biochem. Mol. Biol.* 23:187–93.

Coady, K.K. et al. 2005. Effects of atrazine on metamorphosis, growth, laryngeal and gonadal development, aromatase activity, and sex steroid concentrations in *Xenopus laevis*. *Ecotoxicol. Environ. Saf.* 62:160–73.

Cooke, I.M., and S.E. Sullivan. 1982. Hormones and neurosecretion. In: *The Biology of Crustacea*, ed. H. Atwood and D. Sandeman, 206–91. New York: Academic Press.

Cyr, D.G., and J.G. Eales. 1996. Interrelationships between thyroidal and reproductive endocrine systems in fish. *Rev. Fish Biol. Fish.* 6:165–200.

Dang, Z.C. et al. 2011. Endpoint sensitivity in fish endocrine disruption assays: Regulatory implications. *Toxicol. Lett.* 202:36–46.

Dang, Z. et al. 2012. Evaluation of the Daphnia magna reproduction test for detecting endocrine disruptors. *Chemosphere* 88:514–23.

Daouk, T. et al. 2011. Long-term food-exposure of zebrafish to PCB mixtures mimicking some environmental situations induces ovary pathology and impairs reproduction ability. *Aquat. Toxicol.* 105:270–8.

Darnerud, P.O. 2003. Toxic effects of brominated flame retardants in man and in wildlife. *Environ. Int.* 29:841–53.

deFur, P.L. 2004. Use of invertebrates in testing for endocrine disruptors. *ILAR J.* 45:484–93.

deFur, P.L. et al. 1999. Endocrine Disruption in Invertebrates: *Endocrinology, Testing and Assessment.* Pensacola, FL: SETAC Press.

de Jong-Brink, M. Maat, and C.P. Tensen. 2001. NPY in invertebrates: Molecular answers to altered functions during evolution. *Peptides* 22:309–15.

De Longcamp, D. et al. 1974. The *in vitro* biosynthesis of steroids by the gonad of the mussel (*Mytilus edulis*). *Gen. Comp. Endocrinol.* 22:116–27.

De Vito, M. et al. 1999. Screening methods for thyroid hormone disruptors. *Environ. Health Perspect.* 107:407–15.

Dinan, L. et al. 2001. Screening of environmental contaminants for ecdysteroid agonist and antagonist activity using the *Drosophila melanogaster* B-II cell in vitro assay. *Environ. Toxicol. Chem.* 20:2038–46.

Durou, C., and C. Mouneyrac. 2007. Linking steroid hormone levels to sexual maturity index and energy reserves in *Nereis diversicolor* from clean and polluted estuary. *Gen. Comp. Endocrinol.* 150:106–13.

Eales, J.G. 2006. Modes of action and physiological effects of thyroid hormones in fish. In: *Fish Endocrinology*, vol. 2, ed. M. Reinecke, G. Zaccone, and B.G. Kapoor, 767–808. Plymouth, UK: Science Publishers.

Eisler, R. 2007. *Eisler's Encyclopedia of Environmentally Hazardous Priority Chemicals.* Amsterdam: Elsevier.

EU. 1998. http://ec.europa.eu/environment/endocrine/definitions/endodis_en.htm.

European Commission 1997. European Workshop on the Impact of Endocrine Disrupters on Human Health and the Environment. Environment and Climate Research Programme, DG XII, European Commission, Report EUR 17549.

Falk, F.J. et al. 1992. Pesticides and polychlorinated biphenyl residues in human breast lipids and their relation to breast cancer. *Arch. Environ. Health* 47:143–6.

Feldt-Rasmussen, U. et al. 1980. Long-term variability in serum thyroglobulin and thyroid related hormones in healthy subjects. *Acta Endocrinol.* 95:328–34.

Fini, J.B. et al. 2007. An *in vivo* multiwell-based fluorescent screen for monitoring vertebrate thyroid hormone disruption. *Environ. Sci. Technol.* 41:5908–14.

Forget-Leray, J. et al. 2005. Impact of endocrine toxicants on survival, development, and reproduction of the estuarine copepod *Eurytemora affinis* (Poppe). *Ecotoxicol. Environ. Saf.* 60:288–94.

Fort, D.J. et al. 2007. The hypothalamic–pituitary–thyroid (HPT) axis in frogs and its role in frog development and reproduction. *Crit. Rev. Toxicol.* 37:117–61.

Fossi Tankoua, O. et al. 2012. Physiological status and intersex in the endobenthic bivalve *Scrobicularia plana* from thirteen estuaries in Northwest France. *Environ. Pollut.* 167:70–7.

Fry, D.M. 1995. Reproductive effects in birds exposed to pesticides and industrial chemicals. *Environ. Health Perspect* 103(Suppl. 7):165–71.

Gagné, F., and C. Blaise. 2004. Shell protein characteristics and vitellogenin-like proteins in brine shrimp *Artemia franciscana* exposed to municipal effluent and 20-hydroxyecdysone. *Comp. Biochem. Phys.* 138C:515–22.

Gagné F. et al. 2002. Alterations of the biochemical properties of female gonads and vitellins in the clam *Mya arenaria* at contaminated sites in the Saguenay fjord. *Mar. Environ. Res.* 53:295–310.

Gagné, F. et al. 2005. Altered exoskeleton composition and vitellogenesis in the crustacean *Gammarus* sp. collected at polluted sites in the Saguenay Fjord, Quebec, Canada. *Environ. Res.* 98:89–99.

Gagné, F., C. Blaise, and C. André. 2006. Occurrence of pharmaceutical products in a municipal effluent and toxicity to rainbow trout (*Oncorhynchus mykiss*) hepatocytes. *Ecotoxicol. Environ. Saf.* 64:329–36.

Galic, N. et al. 2010. Potential application of population models in the European ecological risk assessment of chemicals: II. Review of models and their potential to address environmental protection aims. *Integr. Environ. Assess. Manag.* 6:338–60.

Geffard, O. et al. 2010. Ovarian cycle and embryonic development in *Gammarus fossarum*: Application for reproductive toxicity assessment. *Environ. Toxicol. Chem.* 29:2249–59.

Ghekiere, A. et al. 2006a. Non-target effects of the insecticide methoprene on molting in the estuarine crustacean *Neomysis integer* (Crustacea: Mysidacea). *J. Exp. Mar. Biol. Ecol.* 332:226–34.
Ghekiere, A. et al. 2006b. Effects of methoprene, nonylphenol, and estrone on the vitellogenesis of the mysid *Neomysis integer*. *Gen. Comp. Endocrinol.* 147:190–5.
Gillesby, B., and T. Zacharewski. 1998. Exoestrogens: Mechanisms of action and strategies for identification and assessment. *Environ. Toxicol. Chem.* 17:3–14.
Gomes, T., M. Gonzalez-Rey, and M.J. Bebianno. 2009. Incidence of intersex in male clams *Scrobicularia plana* in the Guadiana Estuary (Portugal). *Ecotoxicology* 18:1104–9.
Gong, J. et al. 2011. Occurrence of endocrine-disrupting chemicals in riverine sediments from the Pearl River Delta, China. *Mar. Pollut. Bull.* 63:556–63.
Gooding, M.P. et al. 2003. The biocide tributyltin reduces the accumulation of testosterone as fatty acid esters in the mud snail (*Ilyanassa obsoleta*). *Environ. Health Perspect.* 111:426–30.
Gottfried, H., and R.I. Dorfman. 1970. Steroids of invertebrates. V. The *in vitro* biosynthesis of steroids by the male-phase ovotestis of the slug (*Ariolimax californicus*). *Gen. Comp. Endocrinol.* 15:120–38.
Grist, E.P.M. et al. 2003. Estimating the effects of 17α-ethynyestradiol on populations of the fathead minnow *Pimephales promelas*: Are conventional toxicological endpoints adequate? *Environ. Sci. Technol.* 37:1609–16.
Guillette, L.J. Jr. et al. 1994. Developmental abnormalities of the gonad and abnormal sex hormone concentrations in juvenile alligators from contaminated and control lakes in Florida. *Environ. Health Perspect.* 102:680–8.
Guillette, L.J. Jr. et al. 1995. Gonadal steroidogenesis *in vitro* from juvenile alligators obtained from contaminated or control lakes. *Environ. Health Perspect.* 103(suppl. 4):31–6.
Gust, M. et al. 2010a. *In situ* biomonitoring of freshwater quality using the New Zealand mudsnail *Potamopyrgus antipodarum* (Gray) exposed to waste water treatment plant (WWTP) effluent discharges. *Water Res.* 44:4517–28.
Gust, M. et al. 2010b. Sensitivity of New Zealand mudsnail *Potamopyrgus antipodarum* (Gray) to a specific aromatase inhibitor. *Chemosphere* 79:47–53.
Gust, M. et al. 2011. Natural variability and response interpretation of fecundity, vertebrate-like sex-steroid levels and energy status in the New Zealand mudsnail *Potamopyrgus antipodarum* (Gray). *J. Gen. Comp. Endocrinol.* 172:243–50.
Hahn, M.E. 2002. Aryl hydrocarbon receptors: Diversity and evolution. *Chem. Biol. Interact.* 141:131–60.
Hannas, B.R. et al. 2011. Regulation and dysregulation of vitellogenin mRNA accumulation in daphnids (*Daphnia magna*). *Aquat. Toxicol.* 101:351–7.
Harris, C.A. et al. 2011. The consequences of feminization in breeding groups of wild fish. *Environ. Health Perspect.* 119:306–11.
Hatef, A. et al. 2012. Adverse effects of bisphenol A on reproductive physiology in male goldfish at environmentally relevant concentrations. *Ecotox. Environ. Saf.* 76:56–62.
Heyland, A. et al. 2005. Cross-kingdom hormonal signaling: An insight from thyroid hormone functions in marine larvae. *J. Exp. Biol.* 208:4355–61.
Holbech, H. et al. 2012. Comparison of zebrafish (*Danio rerio*) and fathead minnow (*Pimephales promelas*) as test species in the Fish Sexual Development Test (FSDT). *Comp. Biochem. Physiol.* 155C:407–15.
Hontela, A. 2000. Endocrine biomarkers: Hormonal indicators of sublethal toxicity in fishes. In: *Use of Biomarkers for Environmental Quality Assessment*, ed. L. Lagadic, T. Caquet, J.-C. Amiard, and F. Ramade, 187–204, Chapter 9. Enfield, NH: Science Publishers.
Huang, D.J., and H.C. Chen. 2004. Effects of chlordane and lindane on testosterone and vitellogenin levels in green neon shrimp (*Neocaridina denticulata*). *Int. J. Toxicol.* 23:91–5.
Huang, D.-J. et al. 2006. Reproduction obstacles for the female green neon shrimp (*Neocaridina denticulata*) after exposure to chlordane and lindane. *Chemosphere* 64:11–6.
Inui, Y. et al. 1995. The role of thyroid hormone in tissue development in metamorphosing flounder. *Aquaculture* 135:87–98.
Iwamatsu, T. et al. 2006. Testosterone content of developing eggs and sex reversal in the medaka (*Oryzias latipes*). *Gen. Comp. Endocrinol.* 145:67–74.

Janer, G., and C. Porte. 2007. Sex steroids and potential mechanisms of non-genomic endocrine disruption in invertebrates. *Ecotoxicology* 16:145–60.
Janer, G. et al. 2005. Effects of 17 beta-estradiol exposure in the mussel *Mytilus galloprovincialis*: A possible regulating role for steroid acyltransferases. *Aquat. Toxicol.* 75:32–42.
Janer, G. et al. 2006. The effect of organotin compounds on gender specific androstenedione metabolism in the freshwater ramshorn snail *Marisa cornuarietis*. *J. Steroid Biochem. Mol. Biol.* 99:147–56.
Jasmani, S. et al. 2000. Hemolymph vitellogenin levels during ovarian development in the kuruma prawn *Penaeus japonicus*. *Fish. Sci.* 66:535–9.
Jayasankar, V. et al. 2006. Dynamics of vitellogenin synthesis in juvenile giant freshwater prawn *Macrobrachium rosenbergii*. *J. Exp. Zool.* 305A:440–8.
Jobling, S., and J.P. Sumpter. 1993. Detergent components in sewage effluent are weakly oestrogenic to fish: An *in vitro* study using rainbow trout (*Oncorhynchus mykiss*) hepatocytes. *Aquat. Toxicol.* 27:361–72.
Jobling, S. et al. 1995. A variety of environmentally persistent chemicals, including some phthalate plasticizers, are weakly estrogenic. *Environ. Health Perspect.* 103:582–7.
Jobling, S. et al. 2002. Wild intersex roach (*Rutilis rutilis*) have reduced fertility. *Biol. Reprod.* 67:515–24.
Jobling, S. et al. 2003. Comparative responses of molluscs and fish to environmental estrogens and an estrogenic effluent. *Aquat.Toxicol.* 65:205–20.
Joose, J. 1988. The hormones of molluscs. In: *Endocrinology of Selected Invertebrate Types*, ed. Haufer, H. and R.G.H. Downer, 89–140. New York: Alan R. Liss, Inc.
Jubeaux, G. et al. 2012. Vitellogenin-like proteins in the freshwater amphipod *Gammarus fossarum* (Koch, 1835): Functional characterization throughout reproductive process, potential for use as an indicator of oocyte quality and endocrine disruption biomarker in males. *Aquat. Toxicol.* 112–113:72–82.
Jugan, M.L. et al. 2007. A new bioluminescent cellular assay to measure the transcriptional effects of chemicals that modulate the alpha-1 thyroid hormone receptor. *Toxicol. In Vitro.* 21:1197–205.
Jungmann, D. et al. 2004. Intersexuality in *Gammarus fossarum* Koch—A common inducible phenomenon? *Arch. Hydrobiol.* 159:511–29.
Kajiwara, M. et al. 2006. Tissue preferential expression of estrogen receptor gene in the marine snail, *Thais clavigera*. *Gen. Comp. Endocrinol.* 148:315–26.
Kavlock, R.J. et al. 1996. Research needs for risk assessment of health and environmental effects of endocrine disruptors: A review of the U.S. EPA-sponsored workshop. *Environ. Health Perspect.* 104:715–40.
Keay, J. et al. 2006. The *Octopus vulgaris* estrogen receptor is a constitutive transcriptional activator: Evolutionary and functional implications. *Endocrinology* 147:3861–9.
Kidd, K.A. et al. 2007. Collapse of a fish population after exposure to a synthetic estrogen. *Proc. Natl. Acad. Sci. U.S.A.* 104:8897–901.
Kinani, S. et al. 2010. Bioanalytical characterisation of multiple endocrine- and dioxin-like activities in sediments from reference and impacted small rivers. *Environ. Pollut.* 158:74–83.
Kitamura, S. et al. 2005. Comparative study of the endocrine-disrupting activity of bisphenol A and 19 related compounds. *Toxicol. Sci.* 84:249–59.
Klaren, P.H.M. et al. 2008. The thyroid gland and thyroid hormones in Senegalese sole (*Solea senegalensis*) during early development and metamorphosis. *Gen. Comp. Endocrinol.* 155:686–94.
Koibuchi, N. 2008. The role of thyroid hormone on cerebellar development. *Cerebellum.* 7:530–3.
Kristensen, T., E. Baatrup, and M. Bayley. 2005. 17α-Ethinylestradiol reduces the competitive reproductive fitness of the male guppy (*Poecilia reticulata*). *Biol. Reprod.* 72:150–6.
Kudo, Y. et al. 2006. *In vitro* and *in vivo* analysis of the thyroid system-disrupting activities of brominated phenolic and phenol compounds in *Xenopus laevis*. *Toxicol. Sci.* 92:87–95.
Kuiper, R.V. et al. 2007. Long-term exposure of European flounder (*Platichthys flesus*) to the flame-retardants tetrabromobisphenol A (TBBPA) and hexabromocyclododecane (HBCD). *Ecotoxicol. Environ. Saf.* 67:349–60.
Kumar, V. et al. 2011. How seasonality affects the flow of estrogens and their conjugates in one of Japan's most populous catchments. *Environ. Pollut.* 159:2906–12.

Kuriyama, S.N. et al. 2005. Developmental exposure to low-dose PBDE-99: Effects on male fertility and neurobehavior in rat offspring. *Environ. Health Perspect.* 113:149–54.

Kuriyama, S.N. et al. 2007. Developmental exposure to low-dose PBDE-99: Tissue distribution and thyroid hormone levels. *Toxicology* 242:80–90.

Labadie, P. et al. 2007. Identification of the steroid fatty acid ester conjugates formed *in vivo* in *Mytilus edulis* as a result of exposure to estrogens. *Steroids* 72:41–9.

Lachaise, F. et al. 1993. The molting activity of crustaceans: Localization, activity, and endocrine control. *J. Crustacean Biol.* 13:198–234.

LaFont, R. 2000. The endocrinology of invertebrates. *Ecotoxicology* 9:41–57.

Lafont, R., and M. Mathieu. 2007. Steroids in aquatic invertebrates. *Ecotoxicology* 16:109–30.

Lahnsteiner, F. et al. 2006. Effect of 17β-estradiol on gamete quality and maturation in two salmonid species. *Aquat. Toxicol.* 79:124–31.

Landis, W.G., and M.H. Yu. 2004. *Introduction to Environmental Toxicology. Impacts of Chemicals Upon Ecological Systems.* 3rd edition. Boca Raton, FL: Lewis Publishers, 484 pp.

Lange A. et al. 2011. Implications of persistent exposure to treated wastewater effluent for breeding in wild roach (*Rutilus rutilus*) populations. *Environ. Sci. Technol.* 45:1673–9.

Langston, W.J. et al. 2005. Partioning, bioavailability and effects of oestrogens and xeno-oestrogens in aquatic environment. *J. Mar. Biol. Assoc. U.K.* 85:1–31.

Langston, W.J., G.R. Burt, and B.S. Chesman. 2007. Feminisation of male clams *Scrobicularia plana* from estuaries in Southwest UK and its induction by endocrine-disrupting chemicals. *Mar. Ecol. Progr. Ser.* 333:173–84.

Laufer, H., and W.J. Biggers. 2001. Unifying concepts learned from methyl farnesoate for invertebrate reproduction and post-embryonic development. *Am. Zool.* 41:442–57.

LeBlanc, G.A. 2007. Crustacean endocrine toxicology: A review. *Ecotoxicology* 16:61–81.

LeBlanc, G.A. et al. 1999. The endocrinology of invertebrates. In: *Endocrine Disruption in Invertebrates: Endocrinology, Testing, and Assessment*, ed. P.L. deFur et al. Pensacola, FL: SETAC Presse.

Le Curieux-Belfond, O. et al. 2001. Androgen metabolism in oyster *Crassostrea gigas*: Evidence for 17 beta-HSD activities and characterization of an aromatase-like activity inhibited by pharmacological compounds and a marine pollutant. *J. Steroid Biochem. Mol. Biol.* 78:359–66.

Lee, K.-W. et al. 2008. Molecular cloning, phylogenetic analysis and developmental expression of a vitellogenin (Vg) gene fron the intertidal copepod, *Tigriopus japonicus*. *Comp. Biochem. Phys. B* 150B:395–402.

Lee, R.F., and T. Noone. 1995. Effect of reproductive toxicants on lipovitellin in female blue crabs, *Callinectes sapidus*. *Mar. Environ. Res.* 39:151–4.

Lema, S.C. et al. 2008. Dietary exposure to 2,2′,4,4′-tetrabromodiphenyl ether (PBDE-47) alters thyroid status and thyroid hormone-regulated gene transcription in the pituitary and brain. *Environ. Health Perspect.* 116:1694–9.

Leonard, J.L. 2008. Non-genomic actions of thyroid hormone in brain development. *Steroids* 73:1008–12.

Li, Q. et al. 1998. Changes in vitellin during oogenesis and effect of estradiol-17 beta on vitellogenesis in the Pacific oyster *Crassostrea gigas*. *Invert. Reprod. Dev.* 33:87–93.

Li, W. et al. 2011. Regulation of iododinases and sodium iodide symporter mRNA expression by perchlorate in larvae and adult Chinese rare minnow (*Gobiocypris rarus*). *Mar. Pollut. Bull.* 63:350–5.

Liu, Y.W. et al. 2002. Thyroid hormones are important for embryonic to larval transitory phase in zebrafish. *Differentiation* 70:36–45.

López-Vera, E., M.B. Aguilar, and E.P. Heimer de la Cotera. 2008. FMRFamide and related peptides in the phylum Mollusca. *Peptides* 29:310–7.

Lye, C.M. et al. 2008. Effects of 4-nonylphenol on the endocrine system of shore crab, *Carcinus maenas*. *Environ. Toxicol.* 23:309–18.

Markov, G.V. et al. 2009. Independent elaboration of steroid hormone signaling pathways in metazoans. *Proc. Natl. Acad. Sci. U. S. A.* 106: 11913–8.

Martin-Diaz, M. et al. 2004. Influence of salinity in hemolymph vitellogenin of the shore crab *Carcinus maenas*, to be use as a biomarker of contamination. *Bull. Environ. Contam. Toxicol.* 73:870–7.

Matozzo, V. et al. 2008. Vitellogenin as a biomarker of exposure to estrogenic compounds in aquatic invertebrates: A review. *Environ. Int.* 34:531–45.

Matsumoto, T. et al. 2003. Molecular characterization of a cDNA encoding putative vitellogenin from the Pacific oyster *Crassostrea gigas*. *Zool. Sci.* 20:37–42.

Matsumoto, T. et al. 2007. Oyster estrogen receptor: cDNA cloning and immunolocalization. *Gen. Comp. Endocrinol.* 151:195–201.

Matthiessen, P. 2003. Endocrine disruption in marine fish. *Pure Appl. Chem.* 75:2249–61.

Matthiessen, P., and P.E. Gibbs. 1998. Critical appraisal of the evidence for tributyltin-mediated endocrine disruption in mollusks. *Environ. Toxicol. Chem.* 17:37–43.

Mazurová, E. et al. 2008. Endocrine regulation of the reproduction in crustaceans: Identification of potential targets for toxicants and environmental contaminants. *Biologia* 63:139–50.

McKenney, C.L. 2005. The influence of insect juvenile hormone agonists on metamorphosis and reproduction in estuarine Crustaceans. *Integr. Comp. Biol.* 45:97–105.

McKenney, J.C.L., and E. Matthews. 1990. Influence of an insect growth regulator on the larval development of an estuarine shrimp. *Environ. Pollut.* 64:169–78.

McKenney, C.L. et al. 2004. Comparative embryonic and larval developmental responses of estuarine shrimp (*Palaemonetes pugio*) to the juvenile hormone agonist fenoxycarb. *Arch. Environ. Contam. Toxicol.* 47:463–70.

McLachlan, J.A., and S.F. Arnold. 1996. Environmental estrogens: Found internally, certain compounds are important signals; found in the environment, they can become just so much noise. *Am. Sci.* 84:452–61.

Meusy, J.-J., and H.Y.C. Junera. 1974. Data concerning the female protein fraction in *Orchestia gammarella* Palas (Crustacea Amphipoda), during inter-moult stage and sexual rest. [Données sur la synthèse de la fraction protéique femelle chez *Orchestia gammarella* Palas (Crustacé Amphipode), au cours de l'intermue et chez les femelles en repos sexuel.] *C. R. Acad. Sci.* 279:587–90.

Miller, D.H., and G.T. Ankley. 2004. Modeling impacts on populations: Fathead minnow (*Pimephales promelas*) exposure to the endocrine disruptor 17[beta]-trenbolone as a case study. *Ecotoxicol. Environ. Saf.* 59:1–9.

Miller, D.H. et al. 2007. Linkage of biochemical responses to population-level effects: A case study with vitellogenine in the fathead minnow (*Pimephales promelas*). *Environ. Toxicol. Chem.* 26:521–7.

Miller, M.D. et al. 2009. Thyroid-disrupting chemicals: Interpreting upstream biomarkers of adverse outcomes. *Environ. Health Perspects.* 117:1033–41.

Mills, L.J., and C. Chichester. 2005. Review of evidence: Are endocrine-disrupting chemicals in the aquatic environment impacting fish populations? *Sci. Total Environ.* 343:1–34.

Moccia, R.D. et al. 1986. A quantitative assessment of thyroid histopathology of herring gulls (*Larus argentus*) from the Great Lakes and a hypothesis on the causal role of environmental contaminants. *J. Wildl. Dis.* 22:60–70.

Montagné, N. et al. 2010. Molecular evolution of the crustacean hyperglycemic hormone family in ecdysozoans. *BMC Evol. Biol.* 10:62.

Morcillo, Y. et al. 1998. Effects of tributyltin on the Phase I testosterone metabolism and steroid titres of the clam *Ruditapes decussata*. *Aquat. Toxicol.* 42:1–13.

Mu, X.Y., and G.A. LeBlanc. 2002. Environmental anti-ecdysteroids alter embryo development in the crustacean *Daphnia magna*. *J. Exp. Zool.* 292:287–92.

Mukhi, S., and R. Patiño. 2007. Effects of prolonged exposure to perchlorate on thyroid and reproductive function in zebrafish. *Toxicol. Sci.* 96:246–54.

Murphy, C.A. et al. 2009. Testing and applying a fish vitellogenesis model to evaluate laboratory and field biomarkers of endocrine disruption in Atlantic croaker (*Micropogonias undulatus*) exposed to hypoxia. *Environ. Toxicol. Chem.* 28:1288–303.

Nadzialek, S. et al. 2010. Understanding the gap between the estrogenicity of an effluent and its real impact into the wild. *Sci. Total Environ.* 408:812–21.

Nagaraju, G.P.C. et al. 2003. Methyl farnesoate stimulates gonad development in *Macrobrachium malcolmsonii* (H. Milne Edwards) (Decapoda, Palaemonidae). *Crustaceana* 76:1171–8.

Nakajima, A. et al. 2009. Neurobehavioral effects of tetrabromobisphenol A, a brominated flame retardant, in mice. *Toxicol. Lett.* 189:78–83.
Nassel, D.R. 1996. Peptidergic neurohormonal control systems in invertebrates. *Curr. Opin. Neurobiol.* 6:842–50.
Navas, J.M., and H. Segner. 1998. Antiestrogenic activity of anthropogenic and natural chemicals. *Environ. Sci. Pollut. Res.* 5:75–82.
Oberdörster, E. et al. 2000. Purification of vitellin from grass shrimp *Palaemonetes pugio*, generation of monoclonal antibodies, and validation for the detection of lipovitellin in Crustacea. *Comp. Biochem. Physiol.* 127C:199–207.
Oetken, M. et al. 2004. Evidence for endocrine disruption in invertebrates. *Intern. Rev. Cytol.* 236:1–44.
Okumura, T., and K. Aida. 2000. Hemolymph vitellogenin levels and ovarian development during the reproductive and non reproductive molt cycles in the giant freshwater prawn *Macrobrachium rosenbergii*. *Fish. Sci.* 66:678–85.
Olmstead, A.W., and G.A. LeBlanc. 2001. Temporal and quantitative changes in sexual reproductive cycling of the cladoceran *Daphnia magna* by a juvenile hormone analog. *J. Exp. Zool.* 290:148–55.
Osada, M. et al. 2003. Vitellogenin synthesis in the ovary of scallop, *Patinopecten yessoensis*: Control by estradiol-17 beta and the central nervous system. *J. Exp. Zool.* 299A:172–9.
OSPAR Agreement 2008–09. JAMP Guidelines for Contaminant-Specific Biological Effects. OSPAR Commission, Monitoring guidelines N°2008–9.
Paris, M. et al. 2008a. The history of a developmental stage: Metamorphosis in chordates. *Genesis* 46:657–72.
Paris, M. et al. 2008b. An amphioxus orthologue of the estrogen receptor that does not bind estradiol: Insights into estrogen receptor evolution. *BMC Evol. Biol.* 8:219.
Park, B.J. et al. 2011. Thyroid axis disruption in juvenile brown trout (*Salmo trutta*) exposed to the flame retardant β-Tetrabromoethylcyclohexane (β-TBECH) via the diet. *Environ. Sci. Technol.* 45:7923–7.
Pearce, E.N., and L.E. Braverman. 2009. Environmental pollutants and the thyroid. *Best Pract. Res. Clin. Endocrinol. Metab.* 23:801–13.
Pellerin, J. et al. 2009. Identification des facteurs explicatifs des variations à long terme de la condition physiologique de *Mya arenaria* dans le parc marin Saint-Laurent-Saguenay. *Rev. Sci. Eau* 22:271–89.
Pettersson, I. et al. 2006. Persistent sex-reversal and oviducal agenesis in adult *Xenopus* (*Silurana*) *tropicalis* frogs following larval exposure to the environmental pollutant ethynylestradiol. *Aquat. Toxicol.* 79:356–65.
Porte, C. et al. 2006. Endocrine disruptors in marine organisms: Approaches and perspectives. *Comp. Biochem. Physiol.* 143C:303–15.
Power, D.M. et al. 2001. Thyroid hormones in growth and development of fish. *Comp. Biochem. Physiol.* 130C:447–59.
Poynton, H.C. et al. 2007. *Daphnia magna* ecotoxicogenomics provides mechanistic insights into metal toxicity. *Environ. Sci. Technol.* 41:1044–50.
Ronis, M.J.J., and A.Z. Mason. 1996. The metabolism of testosterone by the periwinkle (*Littorina littorea*) *in vitro* and *in vivo*: Effects of tributyl tin. *Mar. Environ. Res.* 42:161–6.
Routledge, E.J., and J.P. Sumpter. 1996. Estrogenic activity of surfactant and some of their degradation products assessed using a recombinant yeast screen. *Environ. Toxicol. Chem.* 15:241–8.
Saegusa, Y. et al. 2009. Developmental toxicity of brominated flame retardants, tetrabromobisphenol A and 1,2,5,6,9,10-hexabromocyclododecane, in rat offspring after maternal exposure from mid-gestation through lactation. *Reprod. Toxicol.* 28:456–67.
Safe, S. et al. 1991. 2,3,7,8-Tetrachlorodibenzo-*p*-dioxin (TCDD) and related compounds as antiestrogens: Characterization and mechanism of action. *Pharmacol. Toxicol.* 69:400–9.
Sagi, A. et al. 1999. A newly established ELISA showing the effect of androgenic gland on secondary-vitellogenic–specific protein in the hemolymph of the crayfish *Cherax quadricarinatus*. *Gen. Comp. Endocrinol.* 115:37–45.

Sanders, M.B. et al. 2005. Larval development and vitellin-like protein expression in *Palaemon elegans* larvae following xeno-oestrogen exposure. *Integr. Comp. Biol.* 45:51–60.
Sárria, M.P. et al. 2011. The unpredictable effects of mixtures of androgenic and estrogenic chemicals on fish early life. *Environ. Int.* 37:418–24.
Schreiber, A.M. et al. 2006. Asymmetric craniofacial remodeling and lateralized behavior in larval flatfish. *J. Exp. Biol.* 209:610–21.
Shi, Y.B. 2000. *Amphibian Metamorphosis: From Morphology to Molecular Biology*. New York: Wiley-Liss.
Shiao, J.C., and P.P. Wang. 2006. Thyroid hormones are necessary for the metamorphosis of tarpon, *Megalops cyprinoides leptocephali*. *J. Exp. Mar. Biol. Ecol.* 331:121–32.
Siah, A. et al. 2003. Delayed gametogenesis and progesterone levels in soft-shell clams (*Mya arenaria*) in relation to *in situ* contamination to organotins and heavy metals in the St Lawrence river (Canada). *Comp. Biochem. Physiol.* 135C:145–56.
Sieratowicz, A. et al. 2011. Reproductive toxicity of bisphenol A and cadmium in *Potamopyrgus antipodarum* and modulation of bisphenol A effects by different test temperature. *Environ. Pollut.* 159:2766–74.
Sigrun, H., and H. Heike. 2010. Review: Thyroid hormone action during brain development: More questions than answers. *Mol. Cell. Endocrinol.* 315:19–26.
Silva, E., N. Rajapakse, and A. Kortenkamp. 2002. Something from "nothing"—Eight weak estrogenic chemicals combined at concentrations below NOECs produce significant mixture effects. *Environ. Sci. Technol.* 36:1751–6.
Simon, R. et al. 2010. Mass spectrometry assay as an alternative to the enzyme-linked immunosorbent assay test for biomarker quantitation in ecotoxicology: Application to vitellogenin in Crustacea (*Gammarus fossarum*). *J. Chromatogr.* 1217A:5109–15.
Smeets, J.M.W. et al. 1999. The anti-estrogenicity of Ah receptor agonists in carp *Cyprinus carpio* hepatocytes. *Toxicol. Sci.* 52:178–88.
Smolenaars, M.M.W. et al. 2007. Molecular diversity and evolution of the large lipid transfer protein superfamily. *J. Lipid Res.* 48:489–502.
Sonneveld, E. et al. 2005. Development of androgen- and estrogen responsive bioassay, members of a panel of human cell line-based highly selective steroids responsive bioassays. *Toxicol. Sci.* 83:136–48.
Stange, D. et al. 2012. Freshwater mudsnail (*Potamopyrgus antipodarum*) estrogen receptor: Identification and expression analysis under exposure to xeno-hormones. *Environ. Ecotoxicol. Saf.* 75:94–101.
Stefano, G.B. et al. 2003. Estrogen signaling at the cell surface coupled to nitric oxide release in *Mytilus edulis* nervous system. *Endocrinology* 144:1234–40.
Sternberg, R.M. et al. 2008. The contribution of steroidal androgens and estrogens to reproductive maturation of the eastern mud snail *Ilyanassa obsoleta*. *Gen. Comp. Endocrinol.* 156:15–26.
Subramoniam, T. 2000. Crustacean ecdysteriods in reproduction and embryogenesis. *Comp. Biochem. Phys.* 125C:135–56.
Swart, J.C. et al. 2011. The implementation of a battery of *in vivo* and *in vitro* bioassays to assess river water for estrogenic endocrine disrupting chemicals. *Ecotoxicol. Environ. Saf.* 74:138–43.
Tatarazako, N., and S. Oda. 2007. The water flea *Daphnia magna* (Crustacea, Cladocera) as a test species for screening and evaluation of chemicals with endocrine disrupting effects on crustaceans. *Ecotoxicology* 16:197–203.
Terrien, X. et al. 2011. Generation of fluorescent zebrafish to study endocrine disruption and potential crosstalk between thyroid hormone and corticosteroids. *Aquat. Toxicol.* 105:13–20.
Tetreault, G.R. et al. 2011. Intersex and reproductive impairment of wild fish exposed to multiple municipal wastewater discharges. *Aquat. Toxicol.* 104:278–90.
Thomas, P., and M.S. Rahman. 2012. Extensive reproductive disruption, ovarian masculinization and aromatase suppression in Atlantic croaker in the northern Gulf of Mexico hypoxic zone. *Proc. Roy. Soc.* 279B:28–38.
Thornton, J.W. 2003. Nonmammalian nuclear receptors: Evolution and endocrine disruption. *Pure Appl. Chem.* 75:1827–39.

Thornton, J.W. et al. 2003. Resurrecting the ancestral steroid receptor: Ancient origin of estrogen signaling. *Science* 301:1714–7.
Thorpe, K.L. et al. 2007. Associations between altered vitellogenin concentrations and adverse health effects in fathead minnow (*Pimephales promelas*). *Aquat. Toxicol.* 85:176–83.
Tokishita, S. et al. 2006. Organization and repression by juvenile hormone of a vitellogenin gene cluster in the crustacean, *Daphnia magna*. *Biochem. Biophys. Res. Commun.* 345:362–70.
Tuberty, S.R., and C.L. McKenney. 2005. Ecdysteroid responses of estuarine crustaceans exposed through complete larval development to juvenile hormone agonist insecticides. *Integr. Comp. Biol.* 45:106–17.
Tyler, C.R., and S. Jobling. 2008. Roach, sex and gender-bending chemicals: The feminization of wild fish in English Rivers. *BioSciences* 58:1051–9.
USEPA. 1997. Special report on Environmental Endocrine Disruption: An Effects Assessment and Analysis. Office of Research and Development, EPA/630/R-96/012, Washington, D.C.
USEPA. 2009. [EPA–HQ–OPPT–2004–0109; FRL–8399–7]. Final List of Initial Pesticide Active Ingredients and Pesticide Inert Ingredients to be Screened Under the Federal Food, Drug, and Cosmetic Act. Federal Register, Vol 74, N° 71, April 15, 2009, 17579–17585.
Vaccaro, E. et al. 2005. Effects of 17β-estradiol, 4-nonylphenol and PCB 126 on the estrogenic activity and phase 1 and 2 biotransformation enzymes in male sea bass (*Dicentrarchus labrax*). *Aquat. Toxicol.* 75:293–305.
Vajda, A.M. et al. 2011. Demasculinization of male fish by wastewater treatment plant effluent. *Aquat. Toxicol.* 103:213–21.
Vandenberg, L.N. et al. 2012. Hormones and endocrine-disrupting chemicals: Low-dose effects and nonmonotonic dose responses. *Endocr. Rev.* 33:378–455.
Van der Linden, S.C. et al. 2008. Detection of multiple hormonal activities in waste water effluents and surface water, using a panel of steroid receptor CALUX bioassays. *Environ. Sci. Technol.* 42:5814–20.
Van der Ven, L.T. et al. 2006. Effects of the antithyroid agent propylthiouracil in a partial life cycle assay with zebrafish. *Environ. Sci. Technol.* 40:74–81.
Verhaegen, Y. et al. 2010. The brown shrimp (*Crangon crangon* L.) ecdysteroid receptor complex: Cloning, structural modeling of the ligand-binding domain and functional expression in an EcR-deficient Drosophila cell line. *Gen. Comp. Endocrinol.* 168:415–23.
Verslycke, T. et al. 2007. Mysid crustaceans as standard models for the screening and testing of endocrine-disrupting chemicals. *Ecotoxicology* 16:205–19.
Vijayavel, K., and M.P. Balasubramanian. 2008. Reproductive dysfunction induced by naphthalene in an estuarine crab *Scylla serrata* with reference to vitellogenesis. *Ecotoxicol. Environ. Saf.* 69:89–94.
Volz, D.C., and G.T. Chandler. 2004. An enzyme-linked immunosorbent assay for lipovitellin quantification in copepods: A screening tool for endocrine toxicity. *Environ. Toxicol. Chem.* 23:298–305.
Wang, H.Y. et al. 2005. The screening of chemicals for juvenoid-related endocrine activity using the water flea *Daphnia magna*. *Aquat. Toxicol.* 74:193–204.
Weber, G.M. et al. 1992. Patterns of thyroxine and triiodothyronine in serum and follicle bound oocytes of the tilapia, *Oreochromis mossambicus*, during oogenesis. *Gen. Comp. Endocrinol.* 85:392–404.
Weis, J.S. et al. 2011. Living in a contaminated estuary: Behavioral changes and ecological consequences for five species. *Biosciences* 61:375–85.
White, R. et al. 1994. Environmentally persistent alkylphenolic compounds are estrogenic *Endocrinology* 135:175–82.
Whitfield, G.K. et al. 1999. Steroid hormone receptors: Evolution, ligands, and molecular basis of biologic function. *J. Cell. Biochem.* 32–33:110–22.
Wolff, M.S. et al. 1993. Blood levels of organochlorine residues and risk of breast cancer. *J. Natl. Cancer Inst.* 85, 648–52.
Woodling, J.D. et al. 2006. Intersex and other reproductive disruption of fish in wastewater effluent dominated Colorado streams. *Comp. Biochem. Physiol.* 144C:10–5.
Wu, R.S.S. 2009. Effects of hypoxia on fish reproduction and development. In: *Fish Physiology*, ed. J.G. Richards, A.P. Farrell, and C.J. Brauner, 79–141, Volume 27. London: Academic Press.

Xuereb, B. et al. 2011. Vitellogenin-like gene expression in freshwater amphipod *Gammarus fossarum* (Koch, 1835): Functional characterization in females and potential for use as endocrine disruption biomarker in males. *Ecotoxicology* 20:1286–99.
Yamano, K. 2005. The role of thyroid hormone in fish development with reference to aquaculture. *JARQ* 39:161–8.
Yamano, K. et al. 1994. Cloning of thyroid hormone receptor genes expressed in metamorphosing flounder. *Dev. Genet.* 15:378–82.
Zaccaroni, A. et al. 2009. Thyroid hormones as a potential early biomarker of exposure to 4-nonylphenol in adult male shubunkins (*Carassius auratus*). *Sci. Total Environ.* 407:3301–6.
Zhao, J.L. et al. 2011. Screening of multiple hormonal activities in surface water and sediment from the Pearl River system, South China, using effect-directed *in vitro* bioassays. *Environ. Toxicol. Chem.* 30:2208–15.
Zoeller, R.T., and J. Rovet. 2004. Timing of thyroid hormone action in the developing brain: Clinical observations and experimental findings. *J. Neuroendocrinol.* 16:809–18.
Zoeller, R.T. et al. 2007. General background on the hypothalamo–pituitary–thryoid (HPT) axis. *Crit. Rev. Toxicol.* 37:11–53.

9

Impairments of Endocrine Functions: Case Studies

Matthew J. Gubbins, Martial Huet, Reinier M. Mann, and Christophe Minier

CONTENTS

9.1 Imposex

9.1.1 Inputs of Tributyltin into Marine Waters

Tributyltin (TBT), an organometallic molecule, is one of the most potent pesticides ever introduced into the environment. Its introduction into marine ecosystems is primarily attributable to its use as the active compound in antifouling paints. Fouling is caused by the settlement of plants and animals on submerged surfaces, with deleterious consequences. Friction forces on the hulls of marine vessels are enhanced, thus decreasing speed and causing much higher fuel costs. Fouling also accelerates surface erosion and can block subsurface ducts by colonization of their walls.

Many toxic substances, such as copper or mercury, have been used in marine antifouling paints (Alzieu et al. 1987). The search for more active compounds led in the late 1960s to the use of TBT. It proved highly effective, and as a result was used for this purpose worldwide.

9.1.2 Imposex: A Bioindicator Specific for TBT Contamination

9.1.2.1 TBT Is Responsible for Masculinization of Female Gastropods

Female masculinization in marine gastropods was called "imposex" by Smith (1971), derived from the words "impose" and "sex." In the Plymouth area, Blaber (1970) had observed a penis-like structure in females of the dogwhelk *Nucella lapillus*. Subsequently, Féral (1974) encountered the same phenomenon in another snail species, *Ocenebra erinacea*. She showed that imposex was not linked to sequential hermaphroditism: all species affected by imposex are gonochoristic, meaning that they have separate sexes. She also observed imposex-free populations (Féral 1976a, b), and by means of transplants she showed that the phenomenon has an environmental origin (Féral 1980b). The implication of TBT was established by Smith in *Ilyanassa* (*Nassarius*) *obsoleta*. He first linked the presence of affected animals to the vicinity of harbor areas (Smith 1981a, b). Subsequently, he showed that the causative agent originated from the boats themselves (Smith 1981c). Among all compounds tested, only TBT was able to induce imposex (Smith 1981d).

9.1.2.2 Relation between Imposex Development and TBT Contamination Levels

Imposex consists of the superimposition of male sexual organs on female gastropods; a penis, a prostate, and in between them, a spermatogenic duct called the *vas deferens* (Gibbs et al. 1991). The development of these male organs in a female is directly proportional to TBT concentrations in the seawater and the tissues (Gibbs et al. 1987; Oehlmann et al. 1992). In instances of high contamination, female sterilization is caused either by loss of integrity of the "capsule gland" (Gibbs et al. 1990) or by obstruction of the genital opening through uncontrolled proliferation of the *vas deferens* (Gibbs and Bryan 1986).

An imposex development scale has been proposed by Gibbs et al. (1987) and is presented in Figure 9.1. Stage 0 represents the uncontaminated condition: no signs of masculinization. The first masculinization stage shows an outline of a penis behind the right eyestalk. At stage 2, a *vas deferens* starts developing near the genital papilla. The penis has already a similar form to the male, although it is reduced in size. Stage 3 is characterized by a longer penis, which prolongs into the proximal portion of the *vas deferens*. The distal portion of this duct is also longer than in stage 2. Stage 4 shows the fusion of the two portions of the *vas deferens*, and the penis has become even longer. In highly contaminated conditions, the *vas deferens* proliferates around the genital papilla, which can lead to obstruction of this its opening. The female is unable to copulate and cannot eject her eggs anymore; she has become sterile (stage 5). When aborted egg capsules are present in the capsule gland, stage 6 is reached. At stages 5 and 6, the female penis is similar in size and form to that of normal males. In extreme cases, the ovary is reversed into testis (Gibbs et al. 1988). This progress of masculinization led to the definition of two bioindices of TBT contamination.

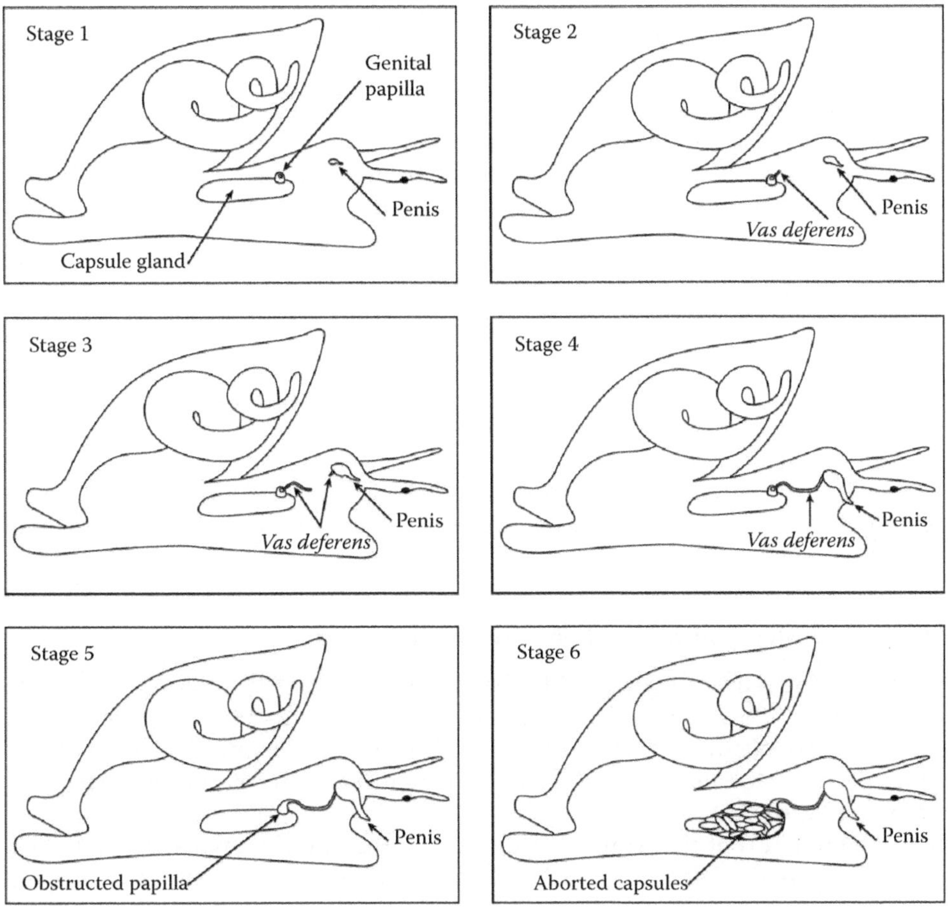

FIGURE 9.1
Imposex development stages.

The Relative Penis Size Index (RPSI), established by Bryan et al. (1986), is based on penis length of females and defined as:

$$\text{RPSI} = \frac{(\text{mean female penis length})^3}{(\text{mean male penis length})^3} \times 100.$$

The mean of encountered imposex development stages (Figure 9.1) is the second index. This Vas Deferens Sequence Index (VDSI) was proposed by Gibbs et al. (1987).

9.1.2.3 Do Any Other Factors Influence Imposex Development?

The RPS and VDS indices permit a specific estimation of TBT pollution only if no other factors intervene with imposex development.

Analyses of museum samples revealed that imposex did not occur before TBT use (Bright and Ellis 1990). The first evidence for imposex was reported by Griffith and Castagna (1962), soon after introduction of TBT in antifouling paints. They described a "vestigial penis" in some *Urosalpinx cinerea* females from the Bay of Chincoteague, USA.

To ascertain that TBT is the sole masculinizating compound, Bryan et al. (1988) tested a collection of organotins on *N. lapillus*. Although TBT was shown not to be the only compound capable of inducing imposex, all other organotins tested were far less potent. This, together with the observation that TBT is the major organotin present in the environment, forced the conclusion that imposex reflects principally TBT pollution.

The question remains whether factors could diminish female masculinization. An inhibition of imposex development was obtained by Bettin et al. (1996) in animals exposed to both TBT and a mix of estrone and 17β-estradiol (1 μg L^{-1}). However, such high concentrations of xenoestrogens are never observed in the environment. One biological factor is known to inhibit penis growth: parasitism by trematode flatworms. This infection is visible to the naked eye and measurements of penis length in parasitized individuals have to be discarded. It should be noted that, in this case, *vas deferens* development is not affected.

Therefore, imposex measurement does seem to be a valid and specific biological indicator of TBT pollution, even if it is not possible to exclude whether other factors could also influence female masculinization.

9.1.3 Biological Mechanisms for Imposex Development

A summary of the studies published by Féral (1977, 1978, 1979, 1980a), Féral and Le Gall (1982), Bettin et al. (1996), Oberdörster and McClellan-Green (2000), Heidrich et al. (2001), Saitoh et al. (2001), Nishikawa et al. (2004), Nishikawa (2006), and Sternberg et al. (2010), led to a hypothesis for the molecular mechanisms that could be implicated in imposex development (Figure 9.2).

It should be noted that this simple diagram represents only a working hypothesis distilled from experiments where biases may be caused by nonphysiological doses, injection

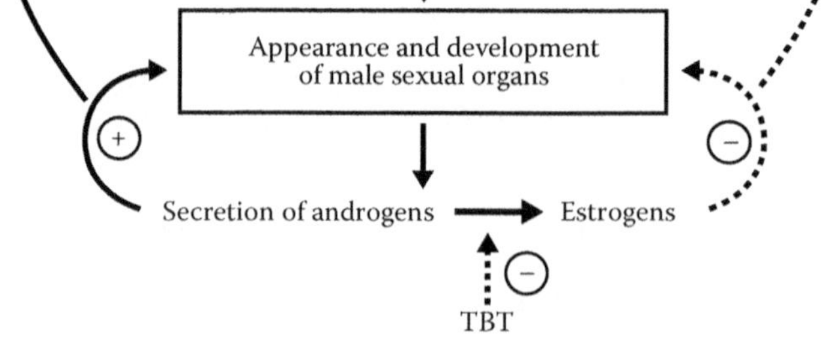

FIGURE 9.2
Hypothetical molecular pathways implicated in imposex development.

routes, or the absence of replicate experiments throughout the physiological (reproductive) seasonal states of the animals. In the hypothesis, TBT is fixed onto the retinoid X receptor in the cerebral ganglia. This leads to signaling to the pedal ganglia, which in turn start secreting "morphogenetic penis factor" (MPF). This molecule activates the establishment of male sexual organs. Subsequently, these organs will secrete testosterone and via positive retro-control, MPF will continue stimulating their own growth. Moreover, this retrocontrol is reinforced because TBT inhibits conversion of testosterone into estradiol, and thus the negative estrogenic retrocontrol network.

Moreover, several other factors, not shown in Figure 9.2, take part in the establishment and development of male sexual organs in gastropods. Among them is the "de-differentiation penis factor" (DPF) produced by the pleural ganglia. This factor causes penis regression in the absence of its antagonist MPF. Both males and females, independent of the presence of a penis, produce DPF. Development, growth, or regression of the penis depend on the interplay between these two antagonistic hormones. And this balance is influenced by environmental factors such as pollution by TBT.

Fluctuation in MPF and DPF levels in correlation with age or season may explain why females are refractory or, on the contrary, responsive to the masculinizing effects of TBT. These seasonal variations probably explain the penis loss observed in males of *Littorina littorea* by Streiff and Le Breton (1970). Oberdörster et al. (2005) reported that TBT pollution caused a MPF/DPF equilibrium change leading to the absence of seasonal penis regression in males of *Ilyanassa obsoleta*.

Steroid levels are not only controlled by an aromatase, a member of the large cytochrome P450 superfamily, which transforms testosterone to estradiol. A second control mechanism implicates steroid conjugation. Conjugated steroids are less hydrophobic and therefore easier to excrete. In *L. littorea*, TBT was shown to inhibit this steroid conjugation, which leads to higher free testosterone levels (Ronis and Mason 1996).

9.1.4 Temporal Evolution of TBT Contamination

Dissemination of TBT into the environment started in the late 1960s. At the beginning of the 1980s, a link was established between the observed decrease in oyster spat-fall in Arcachon Bay, France, and the presence of TBT. His and Robert (1980) demonstrated the mortality of larvae of the pacific oyster, *Crassostrea gigas*, with environmentally relevant concentrations of TBT. TBT was also suggested as the agent responsible for the "chambering" of the oyster's shell (Alzieu et al. 1981). This aberration of calcification diminishes the commercial value of the oysters. This, together with the reduced spat-fall, endangered the oyster industry, and a French law was quickly passed in 1982 that limited the use of TBT. Subsequently, several other countries reacted by establishing similar regulations. The International Maritime Organisation (IMO) banned its use on small vessels (<25 m) in 1987. In 2003, new applications of the substance to ships hulls were banned, and finally, in September 2008, a total ban of TBT use was brought into force by the IMO.

9.1.4.1 Case Study: Brest Harbor

As legislations differ from one country to another, temporal changes in the levels of imposex vary locally. Since 1992, a survey of *N. lapillus* imposex has been undertaken on a yearly basis in Brest, France, and the surrounding area. Follow-up results from 34 field stations under the heavy influence of the Brest harbor area (Huet et al. 2004) are shown in Figure 9.3. The temporal trends of three groups of imposex intensities are shown: stage

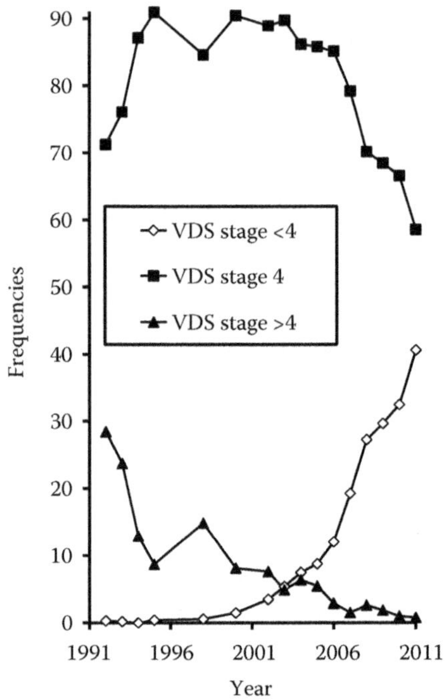

FIGURE 9.3
Temporal change of imposex intensity in the area directly influenced by Brest harbor (cf. Huet et al. 2004) from 1992 to 2011.

4, stages below, and stages above (i.e., sterility). In 1992, female sterilization was about 30% and diminished to less than 1% in 2011, except between 1995 and 1998. The decrease is continuous, except for the brief increase in 1998. The launching of the French aircraft carrier *Charles de Gaulle* in 1995, and its subsequent departure to Toulon (Mediterranean Sea), seems to represent the only valid explanation for this observed increase in TBT pollution effects. Due to the absence of sampling in 1996 and 1997, it is not clear to what extent female sterilization occurred between 1995 and 1998. VDS stage 4 is the most frequent. Whereas the frequency of stage 4 individuals increased continually between 1992 and 1995, its abrupt decrease in 1998 corresponds to the abundance peak of the higher stages. In 2000, VDS levels were observed to return to 1995 values, as was also the case for the incidence of female sterility. An inverse change between stage 4 levels and higher stages is detected between 1992 and 2000. Subsequently, the proportions of these high imposex development stages diminish in parallel. This decrease is concomitant with the increase in VDS stages below 4. This effects-based study clearly suggests that a decrease in TBT pollution occurred in the Brest area between 1992 and 2011. This occurred between 10 and 30 years after the establishment of laws limiting TBT usage. Although in this period, shipping traffic remained almost stable in Brest, the improvement of environmental conditions observed is probably due to a shift to free association antifouling paints over the same period with lower rates of TBT loss (Birchenough et al. 2003).

The vicinity of Brest harbor is a highly polluted area where TBT levels normally prevent any reproduction of a TBT-sensitive species such as *N. lapillus*. As a result of the improvement of environmental conditions over time, the extent of this zone of effect has become more and more reduced. In spite of the very high levels of TBT in the water, a few individuals were always observed in this zone of high impact. How then can the presence in 1992 of small *N. lapillus* populations near the civil and military harbors of Brest (Huet et al. 1996) be explained? Was there a progressive migration from external areas to the heavily affected zone, or did in some way resistance develop to the devastating TBT effects? A similar event was encountered at Dumpton Gap (UK) by Gibbs (1993) and in Galicia (Spain) by Barreiro et al. (1999). It was called Dumpton syndrome and proven to be a genetically based imposex resistance (Gibbs 2005). The possible presence of this phenomenon must be appreciated when determining imposex intensity measures (resistant individuals are not suitable bioindicators and must be omitted from the calculation of imposex indices).

9.1.4.2 Case Study: Sullom Voe, Shetland

The oil terminal based at Sullom Voe in Shetland has received large volumes of liquid petroleum gas and crude oil tanker traffic since 1978. Sullom Voe Association, through the Shetland Oil Terminal Environmental Advisory Group (SOTEAG), has funded long-term monitoring of the state of the environment surrounding the terminal, including both surveys of rocky shore community structure and (after the discovery of imposex in dogwhelks in the area in the early 1980s) effects of TBT (Gubbins 2009). This now represents one of the longest time series of TBT effects recording with data from 1987 to 2011, and clearly demonstrates the recovery (post-TBT inputs) both of the imposex condition in discrete dogwhelk populations as well as the recruitment of this species back to areas where populations had previously expired due to female sterility caused by TBT exposure (Gubbins et al. 2010).

TBT inputs to Sullom Voe have been declining since the 1980s through reduced tanker shipping traffic (Figure 9.4), changes to free association antifouling paints with reduced TBT leaching rates, and IMO restrictions on use on large vessels in 2003 and 2008, which put a stop on the use of TBT-based paints.

Monitoring of dogwhelk populations has taken place since 1987 at a wide range of sites (20) both close to the oil terminal jetties where tankers and TBT sources are based and further afield in the open waters of Yell Sound, removed from the sources of TBT (Figure 9.5).

Thus, it has been possible to compare the temporal trend of imposex intensity at both heavily impacted and reference sites in the area. Typically, imposex is highest at sites within the confined waters of Sullom Voe and close to the terminal. Sites further away in Yell Sound show much lower levels of imposex, typical for background levels of contamination, and sites on the "boundary" between Sullom Voe and Yell Sound show intermediate levels of imposex (Figure 9.5). Large pelagic fishing vessels are likely responsible for the high level of imposex shown on the West shore of Yell Sound in the figure (this has since recovered to much lower values) (Gubbins 2009).

Historically, imposex has resulted in the sterility of female dogwhelks throughout the voe, but only resulted in the complete expiry of populations at sites very close to the terminal itself. Metrics of imposex (VDSI, RPSI) have declined substantially at monitoring sites within the voe. RPSI values have declined substantially since the early 1990s and are now close to zero (Figure 9.6a). VDSI has declined very slowly over the same period, but until 2007 was still >4 at most sites within the voe with many sterile females present in populations. After 2007 there was a very substantial decline in VDSI across all voe populations (Figure 9.6b), and in 2009 for the first time no sterile females were observed anywhere in the study area, a situation that persists in 2011 (Figure 9.6c).

Concurrently with these observations of declining TBT inputs and effects on dogwhelk reproductive capacity, rocky shore community monitoring since 1981 in the same area has observed both a recovery of adult and juvenile dogwhelk populations to heavily affected

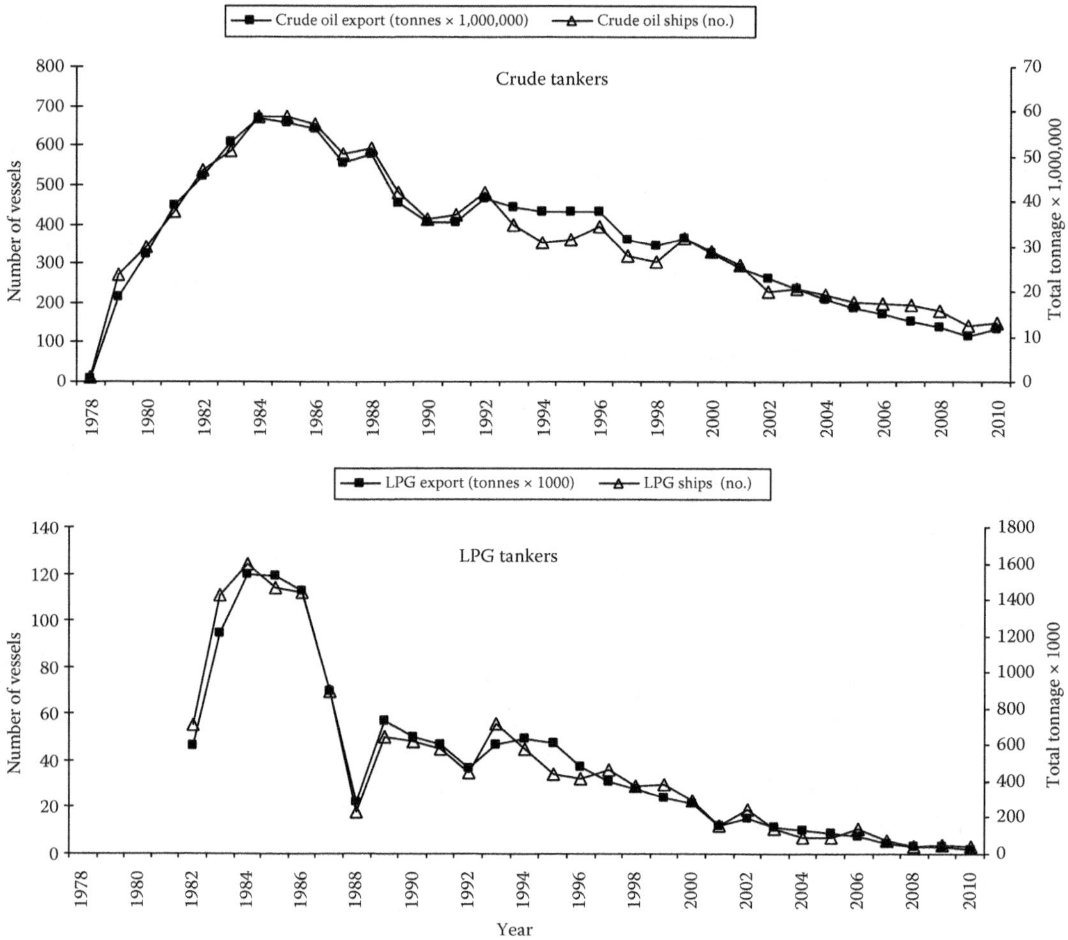

FIGURE 9.4
Number and total tonnage of crude and LPG tankers visiting Sullom Voe.

sites such as The Kames (site 12) (Figure 9.7) and the reestablishment of dogwhelk populations to some sites where they had been previously absent close to the jetties of the terminal (Figure 9.8) (Moore and Gubbins 2012).

Figure 9.8 shows how, at some heavily impacted sites close to the oil terminal, dogwhelk populations vanished by the 1990s and did not start recovering until after 2000, when the incidences of female sterility were lower across the voe as a whole (Figure 9.6c). At one site (Jetty 3), dogwhelks were found in 2011 for the first time since 1987. These temporal trend data, when interpreted together with imposex, suggest that the effects of TBT can be evident at many levels of biological organization, with impacts on community structure through expiration (and subsequent recovery) of populations at some sites.

9.1.5 Ecological Consequences

From a holistic ecological point of view, antifouling paints are in a certain way advantageous as they reduce fuel consumption, protect from erosion, and prevent dispersion

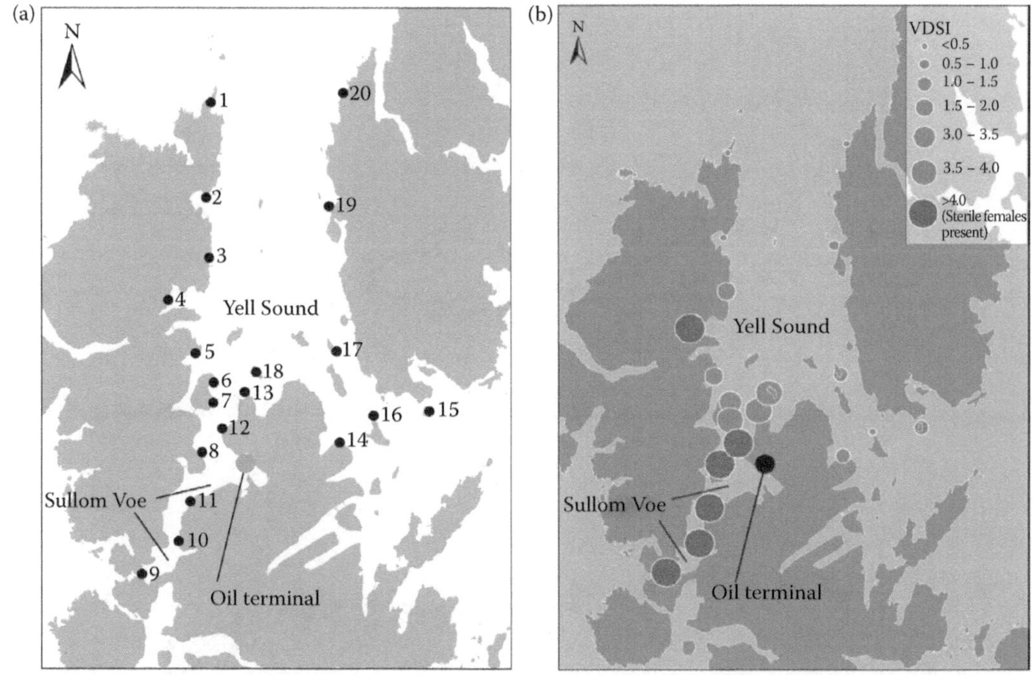

FIGURE 9.5
Monitoring sites in Sullom Voe and Yell Sound, showing (a) the position of the voe, Yell Sound, and the oil terminal, and (b) the intensity of imposex development in dogwhelks (*N. lapillus*) measured as Vas Deferens Sequence Index (VDSI). Data from 2004 shown for example purposes.

of invasive species (Champ 2000). This environmental benefit is, however, clearly undermined by the negative consequences caused by the toxicity of the active compounds. Unfortunately, in the case of TBT contamination, the deleterious effects were only realized after a very long period of contamination at very low concentrations. Almost uniquely, these consequences have been demonstrated at the molecular (endocrine disruption), histological (imposex), individual (sterility), population (expiry), and community (intertidal community structure) levels. It is worth noting that these biological responses have been measured at levels far below the detection limits of the chemical analyses due to accumulation of the compound in the biological tissues and the high sensitivity of the indicator species chosen to monitor the morphological effects (imposex) of TBT. A clear correlation between TBT concentration and imposex intensity is established. However, it is thus far unknown how the animals accumulate the compound over time, especially with regard to the duration and intensity of contamination events. A biological response that assimilates pollution fluctuations is much more informative, reliable, and sensitive for estimating environmental conditions than chemical analysis of a discrete environmental sample. The partial irreversibility of the phenomenon (Féral 1980c) demonstrates that on cessation of the pollution, a return to the initial state of the ecosystem takes time, is not immediate, and will probably never be attained. Indeed, ecosystems are in an evolutionary equilibrium and, as no "control ecosystem" is available, no one can assert that an ecosystem will eventually attain the same state if pollution had not occurred. Because of the occurrence of genetic resistance (Dumpton Syndrome), a profound modification of the genetic structure

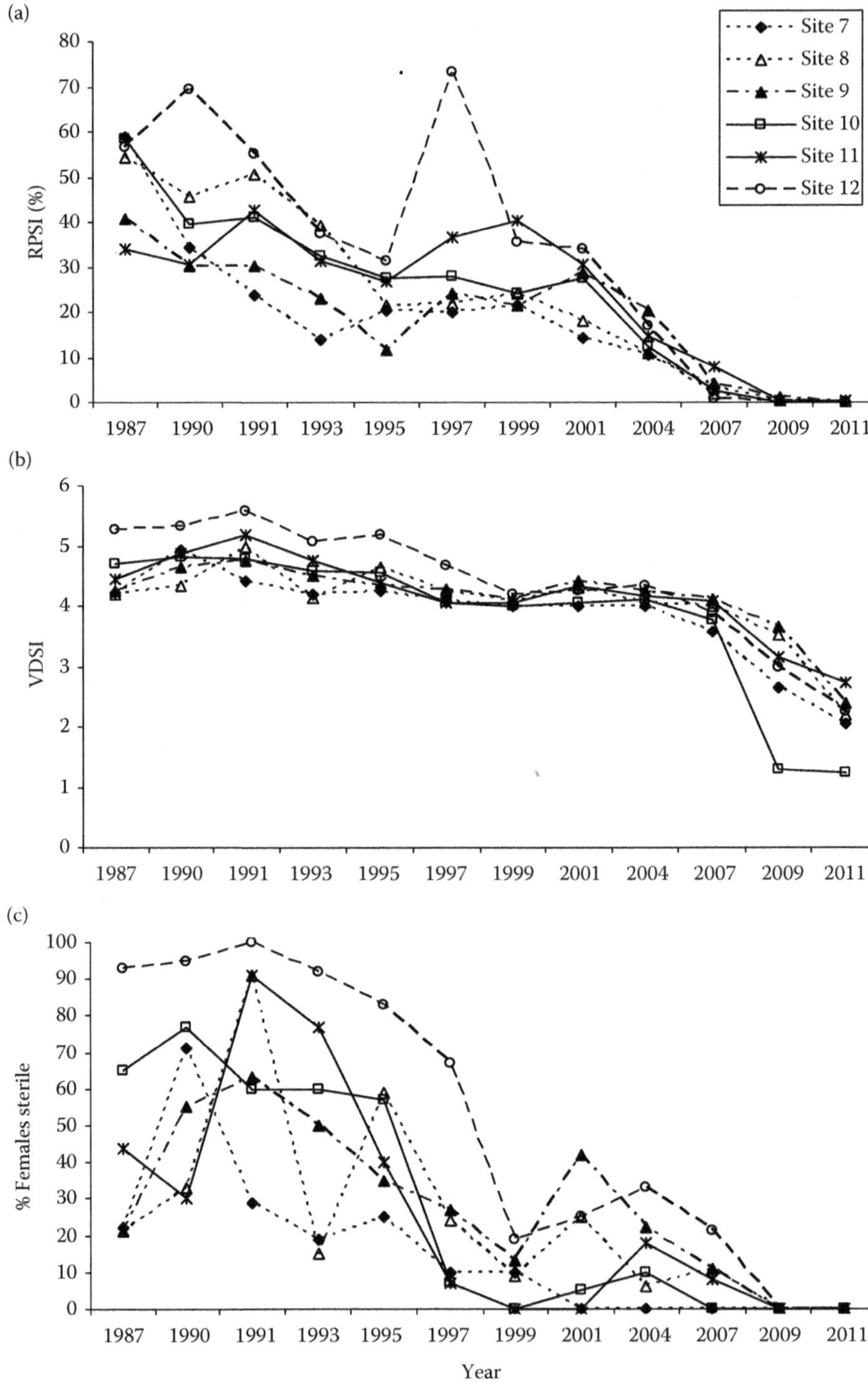

FIGURE 9.6
Temporal trend in dogwhelk imposex metrics RPSI, (a) VDSI (b) and incidence of sterile females in populations (c) at six monitoring sites inside Sullom Voe where populations have been most heavily affected by exposure to TBT. See Figure 9.5 for site location details.

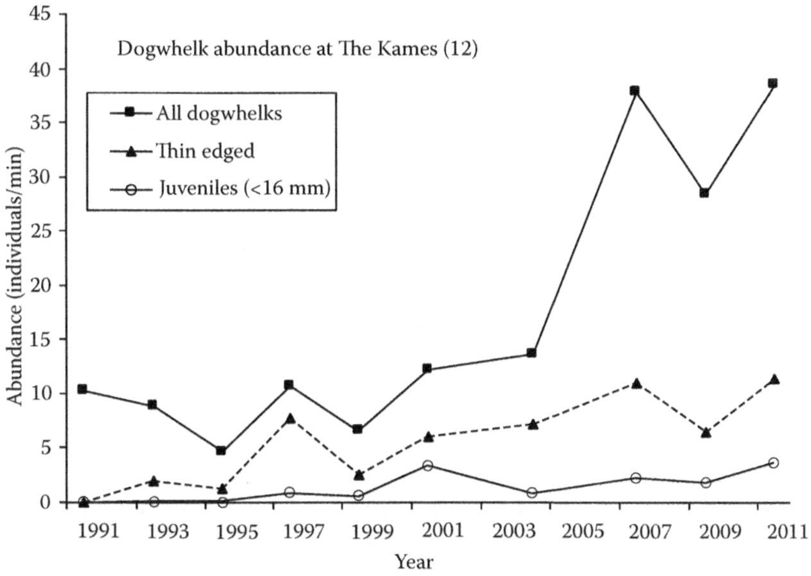

FIGURE 9.7
Increasing abundance of adult, subadult (thin edged shells), and juvenile (<16 mm shell length) dogwhelks at monitoring site close to The Kames (site 6). Abundances are semiquantitative, expressed as a sampling rate (time taken to collect individuals by the same field sampler).

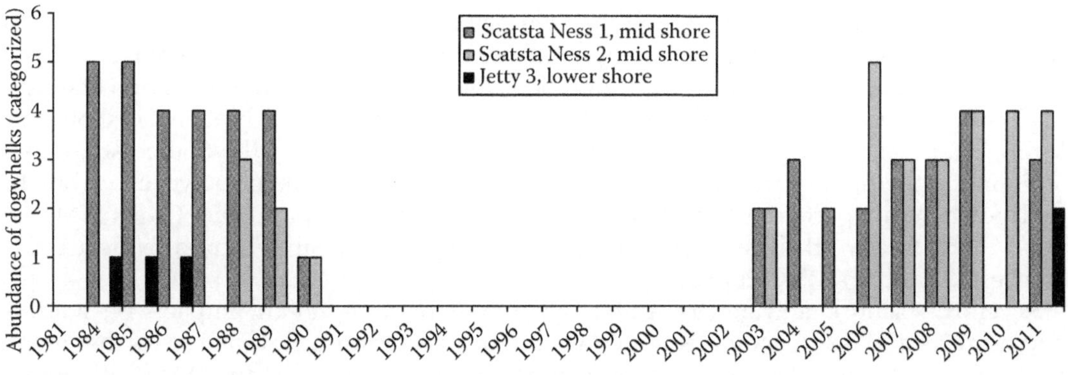

FIGURE 9.8
Temporal changes in relative abundance of dogwhelks (all size classes) at three monitoring sites close to the oil terminal in Sullom Voe. Abundances are based on the following classification: 0 = absent, 1 (rare) <0.25/m^2, 2 (occasional) 0.25–0.5/m^2, 3 (frequent) 0.5–2.5/m^2, 4 (common) 2.5–7.5/m^2, 5 (abundant) 7.5–25/m^2.

of the populations, leading to even more profound and lasting effects on ecosystem structure, may occur.

Finally, it is also worth noting that imposex is a unique bioindicator due to its "sexual orientation." In contrast to all other known sex change phenomena that feminize or demasculinize (see below), it causes masculinization.

9.2 Occurrence and Effects of Estrogenic and Antiandrogenic Compounds in Seine Estuary

9.2.1 Seine Estuary: A Heavily Modified Environment

Before its industrial management in the mid-nineteenth century, the Seine was a shallow river surrounded by large wetlands that regulated its hydrological flow and allowed the reproduction and development of many species. Significant changes have been made since then to ensure the navigability of the river and to promote the industrial and urban development of the area. Major drainage works and embankments along the Seine estuary enabled the port of Le Havre and the adjacent industrial area to increase significantly in size. The estuary was framed with artificial banks (dikes) at the cost of regular dredging to maintain a shipping channel. The facilities reduced the volume of the estuary, altered its hydrosedimentary regulation, and significantly reduced the intertidal area, which decreased from 130 km^2 in 1834 to only 31 km^2 in 1978 (Avoine 1981). Since then, deposition of sediments in the downstream part of the estuary has led to a very rapid increase of the salt marsh at the expense of mudflats that can no longer fulfill their ecological functions (water treatment, habitat essential for protection and reproduction of living species).

The Seine estuary now suffers from significant anthropogenic pressure. It is the natural receptacle of a watershed of nearly 80,000 km^2, which houses 18 million people (more than a quarter of the French population), 40% of French industries, 50% of the national river traffic, and approximately 50,000 km^2 of agricultural land dedicated to intensive production (Guézennec 1999). This human pressure results in high levels of contaminants released into the air and water (Minier et al. 2006a; Motelay-Massei et al. 2007; Amiard et al. 2009). Indeed, the national monitoring network of the French coast, which conducts regular chemical measurements, clearly indicates the Seine estuary to be a hotspot of pollution at national level (Claisse 1989; RNO 2006). Identification and quantification of industrial compounds banned since the 1980s, such as polychlorinated biphenyls, show that levels in tissues of zebra mussels (*Dreissena polymorpha*) living in the Seine estuary are still among the highest in the world. Similarly, the concentrations of polycyclic aromatic hydrocarbons are particularly high, resulting in elevated toxicity and genotoxicity (Minier et al. 2006a). Many "classical" contaminants such as mercury, silver, and dioxins are also at levels of concern (RNO 2006).

Recently, scientific and general public awareness of contaminant impacts on wildlife and humans and the development of new analytical techniques, have led to further monitoring programs and the discovery of new contaminants in the environment. It appears that pharmaceutical compounds, such as analgesics, antidepressants, synthetic hormones, antibiotics, and lipid-lowering drugs are regularly discharged and found in sediment and water of the Seine (Budzinski et al. 2005). Yet, the toxicity of these mixtures of contaminants has never been evaluated and needs to be addressed.

9.2.2 Estrogenic and Antiandrogenic Contamination of Seine Estuary

Endocrine disrupters are compounds that exert a deleterious effect on organisms, not through a classical mechanism of toxicity but by altering an existing control system in the body, the endocrine system. By interfering with this system, endocrine disrupting compounds (EDCs) affect the essential coordination and organization of living organisms, thus having major consequences on reproduction or development (see Chapter 8).

Unlike many other compounds, this new class of contaminants is defined by its biological activity and not by physical or chemical characteristics. Indeed, many molecules with different chemical structures show evidence or potential evidence for endocrine disrupting effects, and a list of 320 compounds has been compiled (European Commission 2007). As a consequence, it is advisable to search for EDCs not by using chemical or physical detection methods but by using screening methods that may highlight their biological activities.

One possible way to detect EDCs is the use of genetically engineered cells in which key receptors are expressed. For instance, the yeast estrogen screen (YES) consists of yeast cells expressing the human receptor to estradiol and possessing a genetic construct used to reveal the activation of the receptor. It involves the estrogen response element and a reporter gene (a bacterial β-galactosidase enzyme) that is synthesized upon activation (Figure 9.9). This system involving hormones (such as estradiol) and their receptors is indeed an essential component of the endocrine system of vertebrates. The use of the YES test can reveal the presence of estrogenic compounds able to interact with and activate the receptor and the subsequent transcriptional activity in exposed cells. This bioassay has been validated for the detection of a wide range of receptor agonists such as estradiol in 17β-estradiol, estrone, estradiol, and xenoestrogens (alkylphenols and bisphenol A) (Routledge and Sumpter 1996, 1997). Beside the estrogen receptor, many other nuclear receptors can be used for screening purposes.

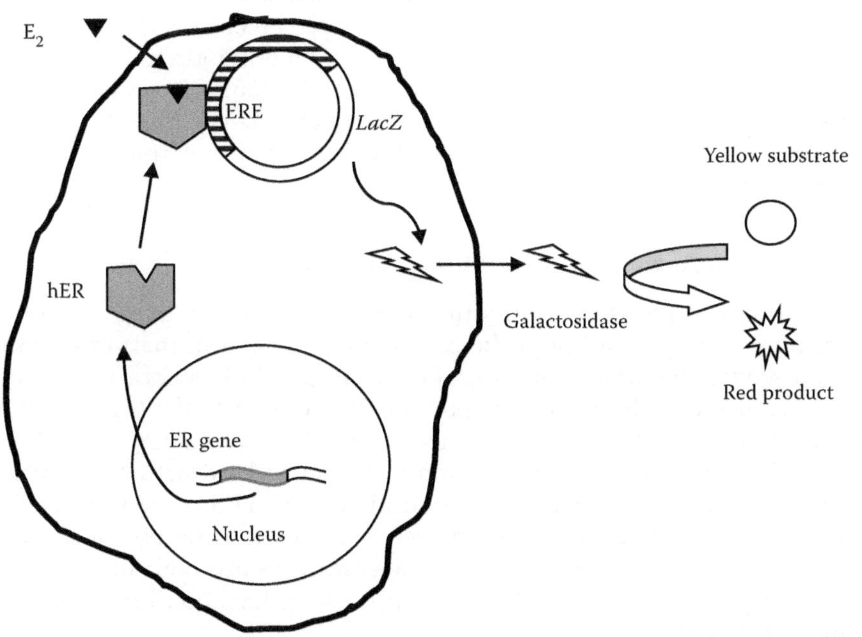

FIGURE 9.9
Simplified diagram of gene construct used to measure estrogenic activity in YES. The gene encoding human α receptor to estradiol (ERH) is inserted into nuclear DNA of yeast and allows receptor expression in cytoplasm. Upon binding to receptor, estradiol (E_2) activates the nuclear complex that binds to response element in plasmid (ERE) and then leads to expression of galactosidase gene (*LacZ* gene). Enzyme secreted into culture medium catalyzes conversion of a chromogenic yellow substrate into an easily quantifiable red product by measurement of optical density. (From Routledge, E.J., Sumpter, J.P., *Environ. Toxicol. Chem.*, 15, 241–248, 1996. With permission.)

Monitoring of EDCs using YES or YAS (yeast androgen screen) tests has been conducted for water and sediment samples in the estuary of the Seine. Potency to elicit a response is easily compared to the response engendered by the natural hormone, estradiol (or testosterone), and expressed as estradiol (or testosterone) equivalents present in the sample. Waters near the discharge of sewage from the Seine estuary contain concentrations of several tens of nanogram per liter estradiol equivalents, whereas their concentrations in the waters of the estuary after dilution by sea water fall generally below detection limits. Most biologically active compounds are associated with suspended matter and possibly sediment (Peck et al. 2004). Sediment analysis revealed high estrogenic activity in many samples along the Seine River, highlighting the widespread contamination throughout the river and estuary. The concentrations vary according to sampling site but suggest an urban origin for most of the detected activity (Minier et al. 2007).

The biologically active samples were subjected to chemical analysis by high pressure liquid chromatography and GC-tandem MS (gas chromatography coupled with dual mass spectroscopy) to identify the main compounds responsible for the measured response. The active compounds present consist primarily of natural steroids (estradiol, estrone), synthetic steroids originating from oral contraceptives (ethinylestradiol), and industrial detergents used in various cleaning products, such as dispersing agents in formulation pesticides and as an additive in some plastic compounds (alkylphenols) (Gallien-Landriau 2003).

Measurements using the YAS assays also revealed androgenic and antiandrogenic activities of water and sediment extracts. The latter are measured by using their ability to act as antagonists of testosterone for the human androgen receptor expressed in yeast cells. Their potency is compared to flutamide, a well-known antiandrogen. Results show that very high amounts of antiandrogen compounds can be found in sediments collected in the Seine estuary. Values range from 0.1 to 10 μg flutamide equivalents/g sediment. The identities of those compounds are yet to be elucidated, but they might contribute to the effects on the aquatic fauna.

9.2.3 Xenoestrogen Accumulation in Mussel Tissues

Mussels are aquatic organisms commonly used in monitoring programs because they accumulate large amounts of lipophilic compounds present in their environment, thus reflecting the environmental contamination (Phillips 1978). Zebra mussels (*Dreissena polymorpha*) are used in the upstream part of the Seine estuary (Minier et al. 2006a), whereas marine mussels (*Mytilus edulis*) are sampled in the marine part of the estuary (Minier et al. 2006b). After homogenization of tissue and extraction of lipophilic compounds, quantification of estrogenic compounds by the YES assay has highlighted the occurrence of these compounds across the estuary at concentrations of about 1 to 25 ng estradiol equivalents per gram of tissue. Concentrations are high in the Seine estuary (and in many other French rivers), but they fall below detection limits in coastal areas away from the estuary.

Laboratory experiments have demonstrated that the ability of mussels to bioconcentrate 17β-estradiol in their tissues is particularly high; the corresponding bioconcentration factors are about 900 for zebra mussels (Peck et al. 2007) and more than 2000 for marine mussels (Labadie et al. 2007). Figure 9.10 shows the results of an exposure experiment using the marine mussel *M. edulis*. Mussels accumulate the natural hormone rapidly in large amounts but, when placed in an environment free of this hormone, only eliminate approximately 50% of the accumulated amount. The E_2 efflux is extremely low after a few

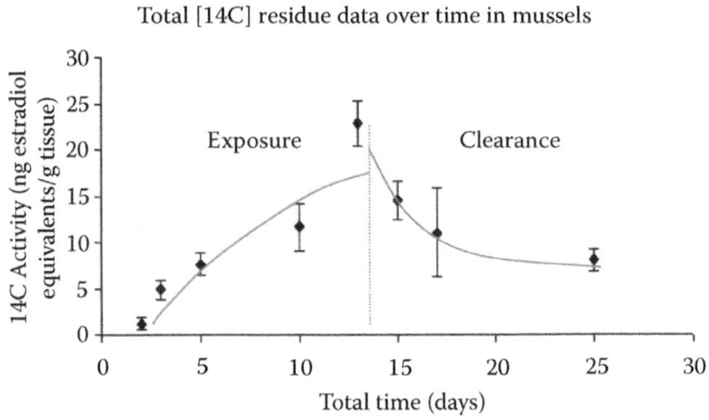

FIGURE 9.10
Concentration of radiolabeled estradiol in tissues of mussels, *Mytilus edulis*. Mussels (both sexes) were exposed to 10 ng L^{-1} E_2 ^{14}C for 13 days (exposure period) and then placed in clean water for 12 days (depuration period). Concentration of radiolabeled E_2 in tissues of mussels was determined by liquid scintillation.

days and the tissue concentration then remains stable, suggesting that these mussels, in the environment, would accumulate and hold large amounts of hormones in their tissues; they thus would allow the transfer of contaminants to potential consumers.

These experiments have also revealed that mollusks are able to rapidly metabolize estradiol into various estradiol esters, which are biologically inactive on the hormone receptor. This esterification of steroids appears to be common among various mollusk species (De Souza and De Oliveira 1976; Janer et al. 2005a,b, 2006) and is catalyzed from C16 fatty acids in *M. edulis* (Labadie et al. 2007). Because the activity is lost, quantification performed using the YES assay does not reflect the total amount of steroid hormones accumulated in mussels. To be more accurate, saponification of the extracts that allows the steroids to be released should be performed. Such measurements made using the YES assay after saponification of mussel tissue extracts can multiply by 5 to 10 the amounts of steroids obtained without saponification.

It should be noted that the esterification of estradiol is unlikely to be able to protect a potential consumer of mussels since saponification takes place in the digestive tract, and the steroids can probably be easily absorbed. The total concentrations measured in marine mussels *M. edulis* sampled in the port of Le Havre (near the outlet of the sewage treatment plant) can reach 100 ng estradiol equivalents per gram of tissue. This concentration is among the highest concentrations currently reported in the scientific literature.

Steroid hormones are active at very low concentrations in vertebrates where their role is well documented. In contrast, the role of sex steroids in bivalves is still a subject of debate. The cloning of genes encoding the receptor for estradiol and vitellogenin (VTG) in *M. edulis* and the study of their expression in the presence of steroid hormones did not show a significant action of these compounds on the expression level of the cloned genes (Puinean et al. 2006). Measurements of the levels of expression of these genes in mussels collected at different sites more or less exposed to estrogenic contaminants (including Le Havre) showed no clear relationship between these variables. This suggested that the genes studied may not be (at least directly) under the control of sex steroids, in contrast to what is known in vertebrates. Alternatively, mussel could only be sensitive to E_2 action during a short period in their reproductive cycle. Indeed, under laboratory exposure, a significant increase in VTG and estradiol receptor mRNA expression was observed in mussels exposed to estrogens at the early stage of gametogenesis. In contrast, mature mussels displayed no statistically significant change in both gene mRNA expressions (Ciocan et al. 2010). Other reproduction-specific genes, including testis-specific kinases, vitelline lysin, and envelope protein, are estrogen-regulated during development (Ciocan et al. 2011). This suggests that the reproductive physiology of mollusks may be susceptible to damage by environmental estrogens at certain points in their gametogenesis process. This is especially true for serotonin and cyclooxygenases (involved in prostaglandin synthesis), which are both important in the endocrinology of mollusks (Cubero-Leon et al. 2010).

9.2.4 Accumulation and Molecular Effects of Estrogens or Antiandrogens in Fish

Endocrine disrupters are accumulated in fish from the Seine estuary. Detection of these compounds is particularly evident in fish bile, where many xenobiotics are excreted and found in high concentrations (Gibson et al. 2005; Smith and Hill 2006). The same compounds as those identified in zebra mussels (mainly steroids) are found in bile of male roach *Rutilus rutilus* from the Seine River. However, a larger diversity of compounds has sometimes been observed in male flounder (*Platichthys flesus*) living in the estuary and in the Baie de Seine. Indeed, some compounds whose chemical nature has not been elucidated so far are present in flounder bile, in addition to natural and synthetic steroids and alkylphenols and bisphenol A (Denier et al. 2008). This increased diversity may be associated with the habitat of the benthic fish that could lead flounder to be exposed to a larger number of compounds present in sediments. A lower diversity of compounds is observed in bile from roach living in rivers and whose lifestyle is not benthic. Nevertheless, as the most downstream part of the estuary is host to numerous industries and to the large city of Le Havre, the possibility that flounder are exposed to compounds present in significant amounts originating from these specific sources cannot be ruled out.

Unlike in mussels, molecular effects of xenoestrogens can easily be demonstrated in fish by measuring VTG expression in male fish (see Chapter 2). Indeed, VTG is a phospholipoglycoprotein used as reserve substance in eggs in order to allow the development of fish larvae. It is naturally synthesized by female fish in response to estradiol secreted by the ovaries during the reproductive cycle. However, the occurrence of VTG in male fish reveals an exogenous exposure to xenoestrogens and is a biomarker of endocrine disruption (Van der Ven et al. 2003).

This protein can be measured in plasma or liver of fish and, in the Seine estuary, it has been demonstrated in male roach and flounder. VTG concentration can be very high in some male roach sampled in the Seine, with values reaching 10,000 ng mL^{-1} in plasma. Eighty percent of the roach analyzed show very significant plasma concentrations, exceeding 200 ng mL^{-1} (Minier et al. 2005; Geraudie et al. 2010a). The average concentration observed in flounder plasma collected in 2001 was 125,000 ng mL^{-1}. Forty-nine percent of the samples were above 1000 ng mL^{-1}, a threshold value signifying a clear disruption of the endocrine system. Compared to plasma VTG measured in flounder from other large European estuaries, this result shows that the Seine estuary is one of the most contaminated estuaries in Europe along with the Mersey and the Tyne (England) but ahead of other major estuaries such as the Thames, the Scheldt, and the Elbe, still considered to be contaminated (Kleinkauf et al. 2004; Vethaak et al. 2005).

The presence of large amounts of VTG has been linked to problems of excretion, leading to kidney dysfunction, since male fish cannot eliminate these large plasma proteins.

Moreover, fish with high plasma VTG would have a significantly reduced fertility (Nash et al. 2004).

Occurrence of EDCs may affect hormone synthesis. P450 aromatase catalyzes the conversion of C19 androgens to C18 estrogens, which is thought to be essential for the regulation of the reproductive function. Measurements of aromatiase activity (AA) in polluted sites showed that activity is significantly up-regulated in sites with fish showing high levels of plasma VTG concentration, thus suggesting the occurrence of estrogenic compounds and their involvement in AA modulation (Geraudie et al. 2011). Furthermore, a significant decrease in plasma estradiol (50–60%) and 11-ketotestosterone (76–84%) was measured, indicating that EDCs may have interfered with the normal sex steroid synthesis of male roaches (Geraudie et al. 2010b).

9.2.5 Fish Feminization and Other Consequences

The most obvious endocrine disrupting effect in fish is the occurrence of intersex fish living in the Seine estuary (Minier et al. 2000a, b). These intersex fish have testes, generating not only spermatozoa but also oocytes. Histological sections thus show, among a generally normal male tissue, some typically female cells including primary and secondary oocytes (Figure 9.11). However, histological examinations of gonads through a sexual cycle show that the intersex fish generally cannot ensure a full and normal development of secondary oocytes. They do not become eggs and degenerate prematurely. The cells must be eliminated and thus represent a waste of reproductive potential and a significant energy expenditure that might be harmful. It is important to note that such intersex fish can be generated experimentally by exposing fish to EDCs during a crucial window of development, thus showing the involvement of these compounds, including steroids, in the phenomenon (Baroiller and D'Cotta 2001). Some treatments may even completely reverse the sex of the organism. What is particularly revealing about the extent of the effects in the Seine estuary is that many species are affected. Intersex individuals were observed in

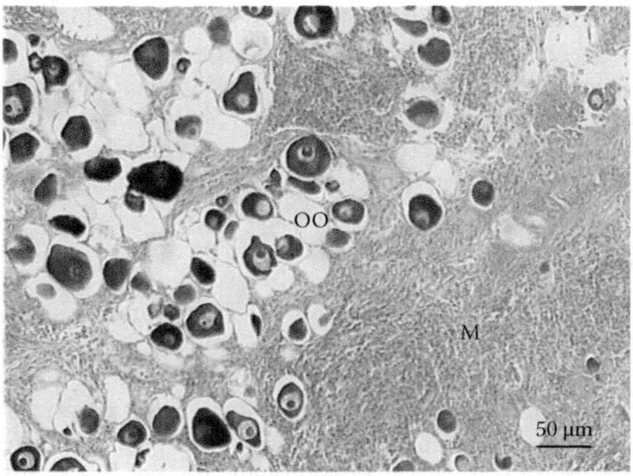

FIGURE 9.11
Histological section of testis of flounder (*Platichthys flesus*) showing occurrence of degenerating primary oocytes (OO) in male tissue (M).

almost all species studied so far, including roach (*R. rutilus*), gudgeon (*Gobio gobio*), chub (*Squalius cephalus*), rudd (*Scardinius erythrophthalmus*), and bream (*Abramis brama*). Species such as flounder (*P. flesus*) and smelt (*Osmerus eperlanus*), which are euryhaline fish spending only part of their life in the estuary and therefore may be less exposed, are nevertheless affected. It was suggested that these effects could not be found in the sea because the dilution of the compounds would be too large, yet intersex individuals of swordfish (*Xiphias gladius*) and red mullet *Mullus barbatus* have been found particularly in the Mediterranean Sea (De Metrio et al. 2003; Martín-Skilton et al. 2006).

The most noticeable effects observed in fish populations studied seem to indicate that EDCs lead to a feminization of individuals. Yet laboratory studies show that it is possible to completely reverse the phenotypic sex (Nimrod and Benson 1998). Such effects could possibly occur in the field. Unfortunately, there is no genetic sex marker for roach and flounder. Despite this lack of genetic tool, and although a causal relationship between the occurrence of intersex fish and specific EDCs in the field cannot be firmly established, it was found that the number of males is significantly lower than those of females in the Seine estuary. Male roach, flounder, and gobies (*Pomatoschitus minutus*) account only for less than 40% of the fish population in the Seine estuary. These figures may be indicative of an effect on the population level on sexual development.

Other cytological changes accompany these histological changes. The frequency of necrosis observed in the gonads of intersex fish is twice that of normal individuals. These alterations indicate that molecular mechanisms are disrupted in the reproductive organs and may further reduce the reproductive capacity of the fish. Another molecular alteration is often observed: numerous oocytes in intersex fish from the Seine estuary undergo abnormal cell divisions leading to the appearance of several nuclei within a cell. Such an effect has been reported only twice in the literature. The first time was from an experiment conducted on rats with very high doses of a very potent EDC, diethylstilbestrol, whose effects were dramatic in men (Igushi et al. 1990). The second time was after a major spill of several pesticides in Lake Apopka in Florida that led to major effects in populations of alligators (*Alligator mississippiensis*) (Guillette et al. 1994). Steroid hormone plasma concentrations were significantly altered, and male alligators developed only a small penis. A decrease in the number of individuals was observed in the 10 years after the spill. The cytological changes observed in fish gonads are the result of alterations in crucial molecular mechanisms and may be indicative of other potential effects. Indeed, EDCs could be responsible for long-term effects by affecting the DNA. It should be noted that some EDCs are carcinogenic; steroids, in particular, are tumor promoters and could be involved in tumor development in vertebrates (Hankinson and Eliassen 2007).

9.2.6 Long-Term Consequences

The Seine estuary appears to be the recipient of many contaminants, some of which clearly affect the endocrine system of fish. The most obvious effect is feminization of male fish that produce female proteins and the testes of which show developing oocytes. The recurring and important question is "What are the effects on populations in the long term?" It may seem obvious, since reproduction is the basis of population turnover, that reductions in population number will occur. Such effects are, however, still largely speculative in natural environments. Nevertheless, it is important to note that the observed effects on fish in the Seine estuary are comparable to those that led to the disappearance of the population of minnows (*Pimephales promelas*) after 7 years of experimentation in a lake in Ontario (Kidd et al. 2007).

9.3 Biomarkers of Endocrine Disruption in Amphibians

9.3.1 Introduction

For more than 20 years, the phenomenon of amphibian population decline has provided the impetus for research into the effects of agricultural chemicals in amphibians. Despite the fact that not a single population decline can be unreservedly ascribed to the effects of exposure to agricultural chemicals, suspicions remain strong—possibly a legacy of concerns left in the wake of Rachel Carson's *Silent Spring* (1962), which still stands today as an important milestone in our understanding of the dangers posed by pesticides in wildlife, and a publication that spawned five decades of ecotoxicological research.

When it comes to amphibians, there is an irony that needs to be recognized—that agriculture waters (dams, weirs, irrigation channels, rice paddies, drainage channels, etc.) provide breeding habitat for amphibians (Knutson et al. 2004). Whether indeed these artificially maintained and regulated waters are suitable for the full range of amphibian species in a geographical region is debatable (Shulse et al. 2010), but frogs and other amphibians do successfully colonize and persist in wetlands associated with agriculture (Herzon and Helenius 2008; Choung et al. 2011), despite the often high concentrations of various agricultural pesticides (Harris et al. 1998; Bishop et al. 1999; Bérubé et al. 2005; McDaniel et al. 2008; Bishop et al. 2010a; Spolyarich et al. 2011).

One of the difficulties with trying to discover the underlying causes of extinctions and population declines, is that the primary witnesses are unavailable for questioning, and the smoking gun can seldom be identified. Invariably, we need to look for clues among remnant species and individuals. The persistence of amphibian populations in agricultural landscapes provides a paradigm for the examination of pesticide effects in wildlife in general. A recently perceived threat to wildlife is that of endocrine disruption, which is epitomized by the highly publicized feminization of fish downstream of sewage treatment plants (Jobling et al. 1996; Tarrant et al. 2008) and alligators exposed to pesticides in Lake Apopka, Florida (Milnes and Guillette 2008).

Since the 1990s, amphibians living in agricultural landscapes have provided a laboratory for the assessment of the potential of biomarkers that reflect the risks associated with pesticide exposure. One particular pesticide that has drawn a great deal of attention is atrazine. Indeed, there are few more controversial issues in the discipline of ecotoxicology than that surrounding atrazine and the consequences of exposure in amphibians, and the story is signposted by numerous biomarkers that have been used to build the case for pesticide induced endocrine disruption. The amphibian–atrazine story provides a valuable case study for which several biomarkers of exposure and effect have played a central role, particularly as biomarkers of the risk to recruitment, because almost all of the research has specifically examined the effects of atrazine on the reproductive potential of amphibians. Much has already been written on this issue and numerous reviews are available—some of them partisan (Solomon et al. 2008; Hayes et al. 2011), some of them less so (Mann et al. 2009; Bishop et al. 2010b; Rohr and McCoy 2010), but the picture is far from complete.

9.3.2 Ambiguous Gender and Gonadal Histology

In 2002, two studies came to light regarding apparent endocrine effects of atrazine in frogs. The first was a study by Tavera-Mendoza (2001), some of which was later published as two

TABLE 9.1

Terminology for Gonadal Abnormalities in Amphibians

Term	Description
Ambisexuality	Immature state of gonads during larval development. The gonad has potential for development into ovarian or testicular tissue
Testicular oocytes	Oocytes present in the testis regardless of state of maturity
Rudimentary testicular oocytes	Oocytes present as a normal part of early development, and which disappear with maturation
Testicular oogenesis	Genesis of oocytes in testis tissue
Testicular dysgenesis	Abnormal or testicular development or degeneration of testicular tissue
Segmented gonads	Gonads segmented as discrete subunits and separated by connective tissue
Intersex	Phenotypic sex different from genotypic sex
Mixed gonadal tissue	Testicular and ovarian tissue occurs within the same individual, either unilaterally appearing as distinct ovaries and testes
Ovotestis	Occurrence of ovarian and testicular tissue in the same gonad, typically with >30% of one or other tissue type. Not necessarily indicative of functional hermaphroditism
Functional hermaphroditism	Both male and female reproductive structures, and fully functional as a male or female
Developmental hermaphroditism	Ovotestis present during early sexual development. Further development of males concomitant with degeneration of oocytes. Residual oocytes are not functionally relevant
Rudimentary hermaphroditism	Nonfunctional form of hermaphroditism in which a few germ cells of opposite sex are present in gonad

Source: Modified from Hecker, M. et al., *Rev. Environ. Contam. Toxicol.*, 187, 103–131, 2006.

articles (Tavera-Mendoza et al. 2002a, b). The authors presented histological evidence of gonadal abnormalities (see Table 9.1 for descriptions) in African clawed frogs (*Xenopus laevis*) after exposure to 21 μg atrazine L^{-1} during sexual differentiation of the gonads. Specifically, they described degeneration of testicular tissue, manifested as testicular resorption and a decline in spermatogonial cell nests (Tavera-Mendoza et al. 2002a), and resorption of oogonia in ovaries (Tavera-Mendoza et al. 2002b). In the same year, Hayes et al. (2002a) also described gonadal abnormalities following exposure of *X. laevis* tadpoles throughout their larval development to atrazine at 25 μg L^{-1}, but also at the much lower concentration of 0.1 μg L^{-1}. Those abnormalities were predominantly gross morphological abnormalities of the gonads, including multiple gonads or a mixture of testes and gonads (hermaphroditism). The identity of tissue types (i.e., testis or ovary) was validated with histology. Both groups of authors implicated increased aromatase activity as the mechanism for the effect of atrazine (see below); however, apart from involving the gonads, the actual morphologies described by Tavera-Mendoza et al. (2002a) and Hayes et al. (2002a) were quite different. Tavera-Mendoza et al. (2002a) described gonadal dysgenesis, whereas Hayes et al. were describing gonadal ambiguities and emasculation of male frogs.

Since these two studies were published, the use of histopathological markers as biomarkers of exposure and effects of atrazine have been central to amphibian studies focusing on endocrine disruption. Following up on their studies in *X. laevis*, Hayes et al. (2002b, 2003) published data describing testicular dysgenesis among American native frogs (*Rana pipiens*) and the occurrence of mixed sex gonads characterized by the presence of testicular oocytes (TOs; see Figure 9.12), again at 25 μg L^{-1} and the much lower concentration of 0.1 μg L^{-1}. Exposed males developed TOs (29% and 8% at 0.1 and 25 ppb, respectively) or

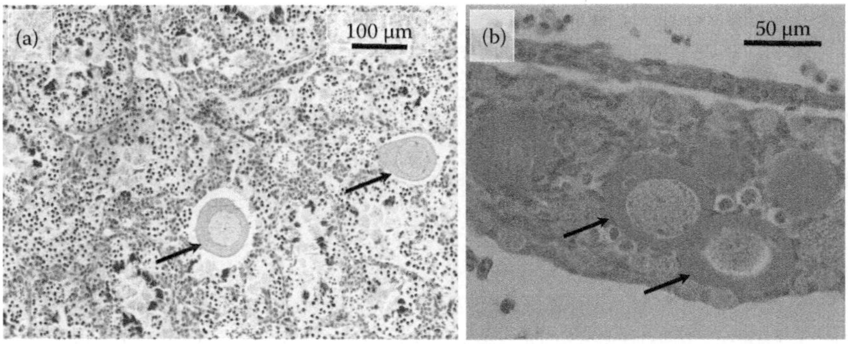

FIGURE 9.12
Testis tissue from (a) an adult male green frog (*Rana clamitans*) collected from an agricultural landscape displaying testicular oocytes (arrows) (photo donated by Susan Bolden; for more information, see Skelly et al. 2010), and (b) a juvenile leopard frog (*Rana sphenocephala*) displaying ovotestes (>30% area with testicular oocytes) following laboratory exposure to 1 μg L^{-1} of atrazine (photo donated by Sara Storrs-Méndez; for more information, see Storrs-Méndez and Semlitsch 2010).

gonadal dysgenesis (36% and 12% at 0.1 and 25 ppb, respectively). The more pronounced response at the lower concentration was emphasized by the authors, who advocated an "inverted U" response as one that could be expected within the realm of endocrine disruption (Hayes 2005). Several subsequent studies have consistently failed to demonstrate similar gonadal abnormalities with incidences significantly greater than control animals among several species, including *X. laevis* (Coady et al. 2005; Oka et al. 2008; Kloas et al. 2009a) *R. pipiens* (Orton et al. 2006; Langlois et al. 2010a), *Hyla versicolor* (LaFiandra et al. 2008), or *Limnodynastes tasmaniensis* (Spolyarich et al. 2010), although exposure to estradiol as a positive control will produce TOs (see below).

Concurrent with their laboratory studies with *R. pipiens*, Hayes et al. (2003) described incidences of these same gonadal morphologies among leopard frogs collected from across the midwest of the United States in regions of low or high atrazine use. Male frogs examined at these sites were shown to exhibit morphologies that were similar to those induced by atrazine in the laboratory. Gonadal dysgenesis was observed in 28% of male frogs at one site where atrazine was widely used, a corn field in Nebraska where the atrazine was reported at a concentration of 0.8 ppb. Moreover, varying incidences of TOs were observed among male leopard frogs at sites where atrazine was detected. The site with the highest incidence of TOs (92%), a "wildlife area" in Wyoming, had relatively low concentrations of atrazine (0.2 ppb; the detection limit for the study), and although Hayes (2005) has suggested that low atrazine concentrations present a greater risk because of an "inverted U" response, a U.S. Environmental Protection Agency (EPA) data evaluation report (Steeger et al. 2003) observed that a clear dose relationship was not evident in the data of Hayes et al. (2003).

One of the difficulties in interpreting the occurrence of ambiguous morphologies such as TOs and ovotestis (OVT; >30% infiltration of testis with TOs) is the general plasticity of amphibians with regard to sexual phenotype, because:

1. Environmental conditions during development can alter sexual phenotype;
2. Ambisexuality and developmental hermaphroditism are natural phenomena among many amphibians.

Although gender in amphibians is genetically determined, temperature and exogenously applied sex hormones (or hormone mimics) can influence sexual phenotype. Temperature-induced sex determination does not appear to have any ecological relevance (Hayes 1998) in the same way as it does in some reptiles, because the low temperatures required to promote female phenotype is outside the range of temperatures normally encountered by amphibians during development. Despite this assertion, low temperatures were proposed as the reason for the occurrence of TOs among *X. laevis* that otherwise had not been exposed to a hormone or hormone mimic. In a mesocosm study, Jooste et al. (2005) described a high incidence of TOs among Nieuwkoop and Faber stage 66 *X. laevis* metamorphs after exposure to atrazine (1.0 to 25 μg L^{-1}) as well as control animals during larval development. However, this was not consistent with previous laboratory studies that notably failed to observe the same (see above), and may have been as a consequence of the low temperatures (10°C) experienced during the early stages of the study (Hayes 2005; Hayes et al. 2006). Another explanation for the occurrence of TOs among *X. laevis* in some studies (Jooste et al. 2005), but not in others (e.g., Kloas et al. 2009a), is that the test animals were derived from phylogenitically distinct populations (Du Preez et al. 2009), and that TOs were a normal phenomenon in some populations.

The effects of hormones and hormone mimics are, of course, the subject of this chapter. The potential for an environmental contaminant to act as an estrogen mimic is measured by the extent to which the effects simulate the effects of exogenously applied estrogen. Generally speaking, exposure to estradiol (usually between 1 and 100 nM) is used as a positive control, and will result in significantly higher proportions of females and ambiguous males that display TOs or other gonadal abnormalities (Coady et al. 2005; Oka et al. 2008; Kloas et al. 2009a; Langlois et al. 2010a; Storrs-Méndez and Semlitsch 2010).

The main obstacle to the interpretation of gonad morphology as a biomarker of endocrine disruption in amphibians is the natural occurrence of hermaphroditism during development. Developmental hermaphroditism appears to be a common phenomenon among many amphibians. Storrs-Méndez and Semlitsch (2010) demonstrated that TOs occurred naturally during gonad development, and frequently persisted beyond metamorphosis in the testes of several species of native American frogs, including American toads (*Bufo americanus*), southern leopard frogs (*Rana sphenocephala*), and gray tree frogs (*Rana versicolor*). Storrs-Méndez and Semlitsch (2010) also demonstrated that among southern leopard frog larvae (*R. sphenocephala*), ambiguity during development, including a skewed prevalence of female phenotypes and the occurrence of TOs and OVT, was normal while they transitioned through various stages of gonadal development, with no such morphologies evident in juvenile frogs 120 days after metamorphosis. It is notable, therefore, that the leopard frogs collected by Hayes et al. (2002b, 2003) from various locations in the USA, were "small individuals in an attempt to sample newly metamorphosed animals," and raises the possibility that field sampling of young animals may promote the observation of TOs and OVT in field-collected animals. As an aside, this may also be the explanation for the high prevalence of gonadal TOs observed among newly metamorphosed leopard frogs that had been exposed during larval development to glyphosate based herbicides containing polyoxyethylene tallowamine (Howe et al. 2004).

Persistence of hermaphroditic morphologies into adulthood can be expected (Storrs-Méndez and Semlitsch 2010), and this is supported by both contemporary and historical literature indicating the occurrence of TOs in a small to medium proportion (0–16%) of adult, field-collected frogs and represented by numerous species (Witschi 1921, 1929; Reeder et al. 1998, 2005; Smith et al. 2005; Murphy et al. 2006b; McDaniel et al. 2008; Hyne et al. 2009; Skelly et al. 2010; Spolyarich et al. 2011). Developmental hermaphroditism helps explain why TOs were frequently observed in control animals that had not otherwise been exposed to atrazine (Coady et al. 2004, 2005; Orton et al. 2006; Du Preez et al. 2009; Storrs-Méndez and Semlitsch 2010). Notwithstanding the importance of phylogenetic variation among test animals described above, the complete absence of hermaphroditic morphologies among control animals in several high-profile studies (Oka et al. 2008; Kloas et al. 2009a; Langlois et al. 2010a) perhaps suggests that under optimal conditions for growth and development, hermaphroditic morphologies are less likely to persist in postmetamorphic subadults.

The studies by Kloas et al. (2009a) provide a case in point. These authors described the results of two parallel studies in two separate laboratories that examined the effects of atrazine (0.01 to 100 μg L^{-1}) on the larval development of *X. laevis*. As indicated above, they were unable to demonstrate any hermaphroditic morphologies in either control or exposed animals. Their studies came about because of the methodological shortcomings indentified in prior studies (US EPA 2007). The criticisms placed particular importance on factors relating to animal husbandry and biological loading in tests. Implicit within these criticisms is an acknowledgment that other environmental co-stressors can affect the outcome of toxicity bioassays (for a review of the effects of co-stressors, see Mann et al. 2009). It is important to note that trade-offs between somatic and gonadal development have been documented among *X. laevis* reared under conditions of environmental stress (McCoy et al. 2007). In order to address these shortcomings, Kloas and colleagues adopted a stringent feeding regime, and a flow-through system to ensure high water quality (Lutz et al. 2008). In so doing, these authors provided a level of control over environmental parameters not described previously in studies with amphibians, and perhaps provided conditions for growth and development that were not available in studies where environmental controls were lacking (e.g., the mesocosm studies of Jooste et al. 2005), or even available to animals in the wild, which frequently breed in isolated ephemeral ponds where environmental conditions can be extreme. In the wake of these landmark studies, the challenge for amphibian ecotoxicologists was, and remains, to reintroduce environmental variability in a manner that allows quantification of biotic and abiotic stressors and the effects they have on the toxicity of environmental contaminants (Mann et al. 2009).

Despite the caution required when interpreting gonadal abnormalities in amphibians, a few field studies continue to report disproportionately high rates of gonadal abnormalities among frogs collected from agricultural zones (McCoy et al. 2008; McDaniel et al. 2008) and urban or industrial zones (Skelly et al. 2010). McDaniel et al. (2008) described a high incidence (42%) of TOs among "adult [leopard] frogs and large juveniles (greater than 10 g)" collected from regions of intensive row agriculture in southern Ontario, Canada, which was not observed in this same species (*R. pipiens*) collected from reference sites (7%). Again, the emphasis was on leopard frogs, as the same phenomenon was not observed among green frogs (*Rana clamitans*) that displayed TOs among 10% and 11% of male frogs sampled from nonagricultural and agricultural sites, respectively.

McCoy et al. (2008) described an increased prevalence of gonadal abnormalities in giant toads (*Bufo marinus*; known in Australia as cane toads) along a gradient characterized by increasing agricultural intensity. All abnormalities described for these toads were based on gross anatomical features that were not confirmed by histology. Abnormalities included mixed gonadal morphologies such as presence of both male and female reproductive structures (i.e., testes as well as vitellogenic or nonvitellogenic ovaries and/or oviduct) or abnormal development of the Bidder's organ (i.e., multiple or vitellogenic Bidder's organs without ovaries or oviducts). Histological examination of these tissues would likely have added an extra layer of valuable data, and no doubt confirmed the coexistence of both male and female reproductive structures, and possibly infiltration of testicular tissues with oocytes.

It is interesting to note at this point, that although much of the data published in recent years have sought to compare frogs from agricultural zones with those from nonagricultural zones, a recent study by Skelly et al. (2010) documented a much higher prevalence of TOs among male frogs collected from urban and suburban environments in the Connecticut River Valley. The authors sampled green frogs (*R. clamitans*), the same species for which there had been no difference in TO prevalence when comparing agricultural and nonagricultural zones in Ontario (McDaniel et al. 2008), and reported 0%, 6%, 16%, and 21% incidence rates of TOs among green frogs from undeveloped landscapes, agricultural landscapes, urban landscapes, and suburban landscapes, respectively.

The occurrence of gonadal abnormalities is not necessarily indicative of an absence of function, and histology offers other biomarkers of testicular functionality. The various phases of spermatozoa production can be quantified in histological sections, although only a few studies have exploited this biomarker. Hayes et al. (2010) described reduced sperm production in *X. laevis* exposed to 2.5 $\mu g\ L^{-1}$ of atrazine and reduced fertility in subsequent mating trials. Orton et al. (2006) also described significantly fewer spermatids and significantly more spermatocytes among *R. pipiens* metamorphs after larval exposure to 10 $\mu g\ L^{-1}$ of atrazine, and Hecker et al. (2005a) described a slight but nonsignificant reduction in spermatozoa among adult male *X. laevis* exposed for 49 days to 10 and 100 $\mu g\ L^{-1}$ of atrazine. Similar diminution of testis function was not found in leopard frogs collected from agricultural areas where atrazine was in use (McDaniel et al. 2008).

9.3.3 Looking beyond the Gonad

The specter that has overshadowed the various atrazine studies in the past 10 years is that of the unexplained toxicity reported by Hayes et al. (2002a, 2003). The demonstrable occurrence of TOs among field-collected frogs and in laboratory studies—inconsistent though they may be—has kept the focus on gonad morphology as the primary biomarker of endocrine disruption in amphibians. Gonad histology provides a direct biomarker of reproductive status. However, as with any organism or chemical exposure scenario, a suite of different biomarkers is required in order to establish cause and effect relationships. Hayes et al. (2002a, 2003) also adopted other biomarkers of endocrine disruption, which fall into two categories: those that are indicative of effects on secondary sex characteristics and those that probe the mechanisms of endocrine disruption.

9.3.4 Biomarkers of Secondary Sex Characteristics

Among most sexually dimorphic species, male frogs advertise to female frogs using distinct calls. Anatomically, this ability is conferred upon male frogs by a developmental elaboration of the larynx. Hayes et al. (2002a) reported a decrease in the size of larynges of postmetamorphic *X. laevis* after exposure of larvae to >1 $\mu g\ L^{-1}$ of atrazine. This effect on this secondary sex characteristic was confirmed in a later study (Hayes et al. 2010) following larval exposure to 2.5 $\mu g\ L^{-1}$ atrazine. The latter study differed from the earlier studies in that all larvae used in the exposure were produced from the mating of sex-reversed adult females (i.e., ZZ females; see Hayes 1998; Wallace et al. 1999), which on a genetic basis are expected to be 100% male. Other studies have not been able to demonstrate similar effects in laryngeal dimensions (Carr et al. 2003; Coady et al. 2005; Smith et al. 2005). Hayes et al. (2010) also described diminution of other secondary sex characteristic following larval exposure to 2.5 μg L^{-1} atrazine, including nuptial pad and breeding gland development, measured as cross-sectional areas in histological sections, and a reduced propensity to enter amplexus and breed.

9.3.5 Biomarkers of a Mechanism for Emasculation/Feminization

Gender in amphibians is genetically predetermined, although as we have observed above, sexual phenotype, or at least several subphenotypic characteristics, can be influenced by environmental factors. However, the manifestation of phenotype is determined by physiological processes that can be followed through the expression of various biomarkers.

Gonadal development is controlled by the sex hormones, estrogen and testosterone. Disruption of sexual differentiation by environmental contaminants is commonly associated with mimicry of estrogen. One of the most common biomarkers for estrogenic effects, particularly in fish, is an increase in circulating VTG (Tarrant et al. 2008). However, increased titers of VTG have not been observed among amphibians exposed to atrazine (Mosconi et al. 2005; McDaniel et al. 2008; Oka et al. 2008). Therefore, the mechanism by which atrazine exerts an endocrine effect appears not be via increased induction of estrogenic pathways leading to the production of VTG (Sanderson 2006).

Both estrogen and testosterone are steroids with a common metabolic genesis. Estrogen is metabolized from testosterone by a single metabolic step catalyzed by the enzyme aromatase (Figure 9.13). Both Tavera-Mendoza (2002a) and Hayes (2005) speculated that the alleged endocrine disruptive effects of atrazine were mediated via an induction of aromatase. In high doses, atrazine has been demonstrated to increase aromatase activity in mammalian cell lines (e.g., Holloway et al. 2008), and conversely, inhibition of aromatase, thereby preventing the conversion of testosterone to estrogen, effectively resulting in male phenotype in otherwise genetically female newts (*Pleurodeles waltl*) (Chardard and Dournon 1999). Therefore, it was reasonable to hypothesize that atrazine-induced increases in aromatase expression or activity could reduce the concentrations of circulating testosterone and promote the development (or perhaps the retention) of female characteristics. In support of this hypothesis, exposure of adult male *X. laevis* to 2.5–250 μg L^{-1} of atrazine for a period of 36–46 days resulted in significantly lower titers of testosterone in the plasma of *X. laevis* (Hayes et al. 2002a; Hecker et al. 2005b; Hayes et al. 2010).

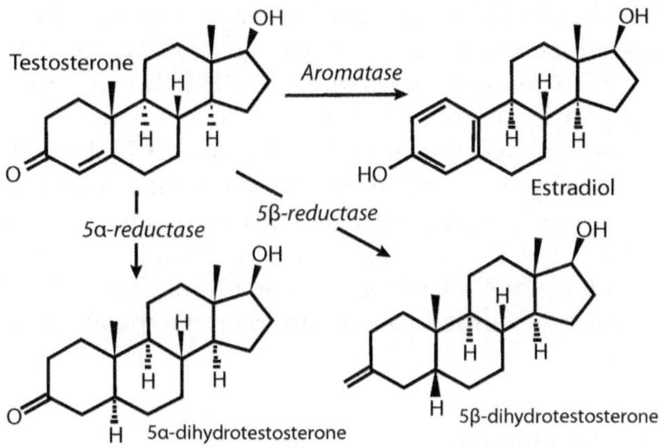

FIGURE 9.13
Pathways for testosterone catabolism.

Furthermore, adult male frogs (*Rana esculenta* and *Rana lessonae*) sampled from agricultural areas presented with lower plasma titers of testosterone (and higher plasma titers of estrogen) than frogs sampled from "pristine" areas (Mosconi et al. 2005). McDaniel et al. (2008) also described an inverse relationship between environmental atrazine concentrations and plasma testosterone and ketotestosterone. Murphy et al. (2006a) and Hecker et al. (2004) were not similarly able to correlate reduced testosterone with atrazine exposure in field exposed frogs, but overall, the evidence for reduced androgens following exposure to atrazine in adult male frogs is persuasive. However, it remains unclear if the reduced testosterone titers described in adult male frogs bears any relevance to the likely levels of physiologically active testosterone during gonadal development, and needs to be examined.

Despite the persuasive link between atrazine exposure and reduced testosterone in adult frogs in laboratory and field studies, nobody has yet demonstrated increased aromatase gene expression (Hecker et al. 2005b; Oka et al. 2008; Kloas et al. 2009b; Langlois et al. 2010a) or increased aromatase activity (Hecker et al. 2004; Coady et al. 2005; Hecker et al. 2005a, b; Murphy et al. 2006a; Langlois et al. 2010a) in an amphibian during sexual differentiation of the gonads as a consequence of exposure to atrazine. A recent study by Hayes et al. (2010) demonstrated complete sex reversal of four (10% of exposed animals) African clawed frogs (*X. laevis*) that had been exposed to 2.5 μg L^{-1} of atrazine throughout their larval and postmetamorphic development. Two of these phenotypically female frogs were examined for aromatase gene expression, and indeed, were demonstrated to express aromatase in their gonads (Hayes et al. 2010). However, this is to be expected in functionally female frogs, and is not necessarily indicative of the mechanism of emasculation during sexual differentiation. Taken as a whole, the evidence for induction of aromatase as the mechanism of atrazine-induced endocrine disruption in amphibians is not strong. So, what else could produce an estrogen-like response without involving aromatase?

Catabolism of testosterone also occurs via two other pathways: one leading to the formation of 5α-dihydrotestosterone (potent androgen), and a second leading to the formation of 5β-dihydrotestosterone (5β-DHT; nonpotent androgen) (Figure 9.13; Langlois et al. 2010b). Although the two DHT isomers are chemically similar, they are stereochemically distinct, and their formation catalyzed by evolutionarily distinct enzymes (Langlois et al. 2010b). In *X. laevis*, the activity of 5α-reductase appears to be unaffected by atrazine at a concentration of 25 μg L^{-1} (Kloas et al. 2009b). In contrast, Langlois et al. (2010a) reported increased activity of hepatic 5β-reductase among Gosner stage 42 *R. pipiens* tadpoles after exposure to 1.8 μg L^{-1} of atrazine. Although the role of 5β-DHT during amphibian development remains unknown, any up-regulation of 5β-reductase as a consequence of exposure to an environmental contaminant may result in lower levels of biologically active androgens such as testosterone, thereby resulting in increased prevalence of female and ambiguous phenotypes. In the same study (Langlois et al. 2010a), exposure to 1.8 μg L^{-1} of atrazine also resulted in significantly skewed sex ratios (1:1.4 male/female) and increased expression of estrogen receptor α (*eralpha*) mRNA in the tadpole brain. *Eralpha* is activated upon estrogen binding, and has been recognized as a biomarker of exposure to estrogenic chemicals (Lutz et al. 2005).

9.3.6 Conclusions on Amphibians

Two points of view have developed regarding the environmental risk posed by atrazine for amphibians, and both camps present their cases eloquently (Solomon et al. 2008; Hayes et al. 2011). As a consequence, the reader cannot help but be confused by the complexities of the issue. Nevertheless, a recent meta-analysis of all studies indicate that atrazine consistently affects amphibian development (Rohr and McCoy 2010), which likely occurs as a consequence of altered steroid balance. The most pronounced and dramatic outcome of these imbalances is complete sex reversal (Oka et al. 2008; Hayes et al. 2010), but more commonly results in retarded or abnormal gonadal development (e.g., Storrs-Méndez and Semlitsch 2010) and reduced reproductive success (Hayes et al. 2010).

All evidence for an endocrine effect in amphibians is reliant on the correct interpretation of biomarker data. Knowledge of various potential morphological, behavioral, and biochemical biomarkers is slowly building and will eventually lead to a better understanding of the complex nature of endocrine disruption in amphibians and other fauna. Along the way, several promising biomarkers have already emerged into prominence and subsequently receded to become but one part of the greater picture. The occurrence of TOs in testicular tissue was one of the first biomarkers to come into our view, and is likely to be a strong biomarker of developmental stress and possibly endocrine disruption after exposure to agricultural pesticides and other anthropogenic chemicals. However, in the case of atrazine exposure, the occurrence of TOs may well have been a "red herring" because of the inconsistencies between studies, and the occurrence of TOs as a natural phase in the sexual development of many species. Similarly, the hypothesis that the endocrine disruptive action of atrazine is mediated by the induction of aromatase in amphibians is not strongly supported by the literature, and other enzymatic pathways may provide better mechanistic explanations for the effects of atrazine.

Biomarkers can only be determined in animals that continue to live in an environment, and as indicated at the beginning of this discussion, amphibian populations in agricultural landscapes provide a convenient laboratory for examining these issues. However, we must be careful about interpreting data about endocrine disruption in amphibians as a cause for amphibian population decline. Low-land amphibians that colonize agricultural and other anthropogenic landscapes are generally r-strategists with high fecundity and rapid reproduction. The occurrence of developmental abnormalities in even a large proportion of individuals is an evolutionary trade-off offset by rapid adaptation to otherwise hostile environments. The reader of the handful of studies that have documented histological abnormalities among field-collected frogs, cannot help but observe that the authors were able to collect a large number of frogs (e.g., McCoy et al. 2008; McDaniel et al. 2008; Skelly et al. 2010), and it is unlikely that the species being examined in these studies are actually those species that we need to be concerned about.

References

Alzieu, C. et al. 1981. Influence des peintures antisalissures à base d'organostanniques sur la calcification de la coquille de l'huître *Crassostrea gigas*. *Rev. Trav. Pêches Marit.* 45:101–16.

Alzieu, C., G. Barbier, and J. Sanjuan. 1987. Evolution des teneurs en cuivre des huîtres du bassin d'Arcachon: Influence de la législation sur les peintures antisalissures. *Océanol. Acta* 10:463–8.

Amiard, J.C. et al. 2009. Quantification of contaminants. In: *Environmental Assessment of Estuarine Ecosystems. A Case Study*. Amiard-Triquet C. and P.S. Rainbow (eds), pp. 31–57. Boca Raton: CRC Press.

Avoine, J. 1981. L'estuaire de la Seine: Sédiments et dynamique sédimentaire. Thèse de 3e cycle, Université de Caen.

Baroiller, J.F., and H. D'Cotta. 2001. Environment and sex determination in farmed fish. *Comp. Biochem. Physiol.* 130C:399–409.

Barreiro, R., M. Quintela, and J.M. Ruiz. 1999. Aphally and imposex in *Nucella lapillus* from Galicia (NW Spain): Incidence, geographical distribution and consequences for the biomonitoring of TBT contamination. *Mar. Ecol. Prog. Ser.* 185:229–38.

Bérubé, V.E. et al. 2005. Plasma retinoid profile in bullfrogs, *Rana catesbeiana*, in relation to agricultural intensity of sub-watersheds in the Yamaska River drainage basin, Québec, Canada. *Aquat. Toxicol.* 71:109–20.

Bettin, C., J. Oehlmann, and E. Stroben. 1996. TBT-induced imposex in marine neogastropods is mediated by an increasing androgen level. *Helg. Meeres.* 50:299–317.

Birchenough, A.C. et al. 2003. Re-colonisation and recovery of populations of dogwhelks *Nucella lapillus* (L.) on shores formerly subject to severe TBT contamination. *Mar. Pollut. Bull.* 44:652–9.

Bishop, C.A. et al. 1999. Anuran development, density and diversity in relation to agricultural activity in the Holland River watershed, Ontario, Canada (1990–1992). *Environ. Monit. Assess.* 57:21–43.

Bishop, C.A. et al. 2010a. Hatching success and pesticide exposures in amphibians living in agricultural habitats of the South Okanagan Valley, British Columbia, Canada (2004–2006). *Environ. Toxicol. Chem.* 29:1593–603.

Bishop, C.A., T. McDaniel, and S.R. de Solla. 2010b. Atrazine in the environment and its implications for amphibians and reptiles. In: *Ecotoxicology of Amphibians and Reptiles*, 2nd Edition, ed. D.W. Sparling, G. Linder, C.A. Bishop, and S.K. Krest, 225–59. Pensacola, FL: Society of Environmental Toxicology and Chemistry (SETAC).

Blaber, S.J.M. 1970. The occurrence of a penis-like outgrowth behind the right tentacule in spent females of *Nucella lapillus* (L.). *Proc. Malac. Soc. Lond.* 39:231–3.

Bright, D.A., and D.V. Ellis. 1990. A comparative survey of imposex in northeast Pacific neogastropods (prosobranchia) related to tributyltin contamination, and choice of a suitable bioindicator. *Can. J. Zool.* 68:1915–24.

Bryan, G.W. et al. 1986. The decline of the gastropod *Nucella lapillus* around south-west England: Evidence for the effect of tributyltin from antifouling paints. *J. Mar. Biol. Assoc. U.K.* 66:611–40.

Bryan, G.W., P.E. Gibbs, and G.R. Burt. 1988. A comparison of the effectiveness of tri-*n*-butyltin chloride and five other organotin compounds in promoting the development of imposex in dogwhelk, *Nucella lapillus*. *J. Mar. Biol. Assoc. U.K.* 68:733–44.

Budzinski, H. et al. 2005. Nouvelles classes de contaminants de l'estuaire de la Seine: Alkylphénols, hormones stéroïdiennes, antibiotiques, analgésiques et hypolipémiants. Rapport Seine-Aval. Editions Seine-Aval.

Carr, J.A. et al. 2003. Response of larval *Xenopus laevis* to atrazine: Assessment of growth, metamorphosis, and gonadal and laryngeal morphology. *Environ. Toxicol. Chem.* 22:396–405.

Carson, R. 1962. *Silent Spring*. Boston: Houghton Mifflin.

Champ, M.A. 2000. A review of organotin regulatory strategies, pending actions, related costs and benefits. *Sci. Total Environ.* 258:21–71.

Chardard, D., and C. Dournon. 1999. Sex reversal by aromatase inhibitor treatment in the newt *Pleurodeles waltl*. *J. Exp. Zool.* 283:43–50.

Choung, C.B. et al. 2011. Developmental toxicity of two common corn pesticides to the endangered southern bell frog (*Litoria raniformis*). *Environ. Pollut.* 159:2648–55.

Ciocan, C.M. et al. 2010. Effects of estrogen exposure in mussels, *Mytilus edulis,* at different stages of gametogenesis. *Environ. Pollut.* 158:2977–84.

Ciocan, C.M. et al. 2011. Identification of reproduction-specific genes associated with maturation and estrogen exposure in a marine bivalve *Mytilus edulis*. *PLoS ONE* 6(7):e22326. doi:10.1371/journal.pone.0022326.

Claisse, D. 1989. Chemical contamination of French coasts: The results of ten years mussel watch. *Mar. Pollut. Bull.* 20:523–8.

Coady, K.K. et al. 2004. Effects of atrazine on metamorphosis, growth, and gonadal development in the green frog (*Rana clamitans*). *J. Toxicol. Environ. Health 67A*:941–57.

Coady, K.K. et al. 2005. Effects of atrazine on metamorphosis, growth, laryngeal and gonadal development, aromatase activity, and sex steroid concentrations in *Xenopus laevis*. *Ecotoxicol. Environ. Saf.* 62:160–73.
Commission of the European Communities. 2007. Commission staff working document on the implementation of the "Community Strategy for Endocrine Disrupters"—A range of substances suspected of interfering with the hormone systems of humans and wildlife. SEC 1635.
Cubero-Leon, E. et al. 2010. Estrogens disrupt serotonin receptor and cyclooxygenase gene expression in the gonads of mussels (*Mytilus edulis*). *Aquat. Toxicol.* 98:178–87.
De Metrio, G. et al. 2003. Evidence of a high percentage of intersex in the Mediterranean swordfish (*Xiphias gladius* L.). *Mar. Pollut. Bull.* 46:358–61.
De Souza, A.M., and D.N. De Oliveira. 1976. Esterification of cholesterol and cholesterol ester hydrolysis by the hemolymph of the mollusc *Biomphalaria glabrata*. *Comp. Biochem. Physiol.* 53B:345–7.
Denier, X. et al. 2012. Investigation of estrogenic contamination in the Seine estuary: Plasma vitellogenin and estrogenicity identification evaluation of flounder (*Platichthys flesus*) bile content. *Aquat. Toxicol.* (submited).
Du Preez, L.H. et al. 2009. Population-specific incidence of testicular ovarian follicles in *Xenopus laevis* from South Africa: A potential issue in endocrine testing. *Aquat. Toxicol.* 95:10–6.
Féral, C. 1974. Etude comparée des populations d'*Ocenebra erinacea* (L.) de Granville et d'Arcachon. *Haliotis* 4:129–34.
Féral, C. 1976a. Etude statistique de la présence d'un tractus génital mâle externe chez les femelles d'un mollusque gastéropode gonochorique: *Ocenebra erinacea* (L.). *Cah. Biol. Mar.* 17:61–76.
Féral, C. 1976b. Répartition géographique des femelles à tractus génital mâle externe chez *Ocenebra erinacea* (L.), espèce gonochorique. *Haliotis* 7:29–30.
Féral, C. 1977. Analyse expérimentale de la morphogenèse et du cycle du pénis chez les femelles d'*Ocenebra erinacea*, espèce gonochorique. *Haliotis* 6: 267–71.
Féral, C. 1978. Présence des facteurs morphogénétique et dédifférenciateur du pénis chez un mollusque prosobranche gonochorique *Ocenebra erinacea* (L.). *C. R. Hebd. Séances Acad. Sci. Paris, Sér. D* 287:1235–7.
Féral, C. 1979. Etude des facteurs régissant l'apparition d'un pénis chez les femelles d'*Ocenebra erinacea* L. (mollusque gastéropode gonochorique) de la station d'Arcachon. *C. R. Hebd. Séances Acad. Sci. Paris, Sér. D* 289:331–4.
Féral, C. 1980a. Apparition de femelles à tractus génital mâle externe chez *Ocenebra erinacea* (L.), mollusque gastéropode gonochorique de la station de Granville: Recherche des facteurs côntrolant l'apparition de cette anomalie. *C. R. Hebd. Séances Acad. Sci. Paris, Sér. D* 290:1003–6.
Féral, C. 1980b. Influence de la qualité de l'eau de mer sur la différenciation d'un tractus génital mâle externe chez les femelles d'un mollusque gastéropode gonochorique: *Ocenebra erinacea* L. *C. R. Hebd. Séances Acad. Sci. Paris, Sér. D* 291:775–8.
Féral, C. 1980c. Variations dans l'évolution du tractus génital mâle externe des femelles de trois gastéropodes prosobranches gonochoriques de stations atlantiques. *Cah. Biol. Mar.* 21:479–91.
Féral, C., and S. Le Gall. 1982. Induction expérimentale par un polluant marin (le tributylétain), de l'activité neuroendocrine contrôlant la morphogenèse du pénis chez les femelles *d'Ocenebra erinacea* (mollusque, prosobranche gonochorique). *C. R. Hebd. Séances Acad. Sci. Paris, Sér. III* 295:627–30.
Gallien-Landriau, I. 2003. Etude de l'altération fonctionnelle du système reproducteur par les perturbateurs endocriniens. Caractérisation des effets, identification des xéno-estrogènes impliqués et conséquences sur les populations de poissons en estuaire et Baie de Seine. Thèse d'Université, Université du Havre.
Geraudie, P. et al. 2010a. Roach (*Rutilus rutilus*) reproduction: A study of biochemical and histological parameters in a low contaminated site. *Fish. Physiol. Biochem.* 36:767–77.
Geraudie, P., M. Gerbron, and C. Minier. 2010b. Seasonal variations and alteration of sex steroid levels during the reproductive cycle of the roach (*Rutilus rutilus*). *Mar. Environ. Res.* 69:S53–5.
Geraudie, P. et al. 2011. Brain cytochrome P450 aromatase activity in roach (*Rutilus rutilus*): Seasonal variations and impact of environmental contaminants. *Aquat. Toxicol.* 105:378–84.

Gibbs, P.E. 1993. A male genital defect in the dog-whelk, *Nucella lapillus* (Neogastropoda), favouring the survival of a population in a TBT-polluted area. *J. Mar. Biol. Assoc. U.K.* 73:667–78.
Gibbs, P.E. 2005. Male genital defect (Dumpton Syndrome) in the dog-whelk *Nucella lapillus* (Neogastropoda): Mendelian inheritance inferred, based on laboratory breeding experiments. *J. Mar. Biol. Assoc. U.K.* 85:143–50.
Gibbs, P.E., and G.W. Bryan. 1986. Reproductive failure in populations of the dog-whelk, *Nucella lapillus*, caused by imposex induced by tributyltin from antifouling paints. *J. Mar. Biol. Assoc. U.K.* 66:767–77.
Gibbs, P.E. et al. 1987. The use of the dog-whelk, *Nucella lapillus*, as an indicator of tributyltin (TBT) contamination. *J. Mar. Biol. Assoc. U.K.* 67:507–23.
Gibbs, P.E., P.L. Pascoe, and G.R. Burt. 1988. Sex change in the female dog-whelk, *Nucella lapillus*, induced by tributyltin from antifouling paints. *J. Mar. Biol. Assoc. U.K.*, 68:715–31.
Gibbs, P.E. et al. 1990. Reproductive abnormalities in female *Ocenebra erinacea* (gastropoda) resulting from tributyltin-induced imposex. *J. Mar. Biol. Assoc. U.K.* 70:639–56.
Gibbs, P.E., P.L. Pascoe, and G.W. Bryan. 1991. Tributyltin-induced imposex in stenoglossan gastropods: Pathological effects on the female reproductive system. *Comp. Biochem. Physiol.* 100:231–5.
Gibson, R.W., C. Tyler, and E.M. Hill. 2005. Mixtures of estrogenic contaminants in bile of fish exposed to wastewater treatment works effluents. *Environ. Sci. Technol.* 39:2461–71.
Griffith, G.W., and M. Castagna. 1962. Sexual dimorphism in oyster drills. *Chesapeake Sci.* 3:215–7.
Gubbins, M.J. 2009. SOTEAG rocky shore monitoring programme. TBT contamination in Sullom Voe, Shetland. 2009 dogwhelk survey. *Scot. Mar. Fresh. Sci.* 1.
Gubbins, M.J. et al. 2010. Long time series data showing recent recovery of gastropod populations from effects of tri-butyl tin at the Shetland Oil Terminal. ICES Annual Science Conference, Nantes, France, September 2010. Paper: ICES CM 2010/F:09.
Guézennec, L. 1999. Seine-Aval: Un estuaire et ses problèmes. Programme scientifique Seine Aval. Plouzané (France): Editions Ifremer.
Guillette, L.J. et al. 1994. Developmental abnormalities of the gonad and abnormal sex hormone concentrations in juvenile alligators from contaminated and control lakes in Florida. *Environ. Health Perspect.* 102:680–8.
Hankinson, S.E., and A.H. Eliassen. 2007. Endogenous estrogen, testosterone and progesterone levels in relation to breast cancer risk. *J. Steroid Biochem. Mol. Biol.* 106:24–30.
Harris, M.L. et al. 1998. The functional integrity of northern leopard frog (*Rana pipiens*) and green frog (*Rana clamitans*) populations in orchard wetland: I. Genetics, physiology and biochemistry of breeding adults and young-of-the year. *Environ. Toxicol. Chem.* 17:1338–50.
Hayes, T.B. 1998. Sex determination and primary sex differentiation in amphibians: Genetic and developmental mechanisms. *J. Exp. Zool.* 281:378–99.
Hayes, T.B. 2005. Welcome to the revolution: Integrative biology and assessing the impact of endocrine disruptors on environmental and public health. *Integr. Comp. Biol.* 45:321–29.
Hayes, T.B. et al. 2002a. Hermaphroditic, demasculinized frogs after exposure to the herbicide atrazine at low ecologically relevant doses. *Proc. Natl. Acad. Sci. U.S.A.* 99:5476–80.
Hayes, T.B. et al. 2002b. Feminization of male frogs in the wild. *Nature* 419:895–6.
Hayes, T. et al. 2003. Atrazine-induced hermaphroditism at 0.1 ppb in American leopard frogs (*Rana pipiens*): Laboratory and field evidence. *Environ. Health Perspect.* 111:568–75.
Hayes, T.B. et al. 2006. Characterization of atrazine-induced gonadal malformations in African clawed frogs (*Xenopus laevis*) and comparisons with effects of an androgen antagonist (cyproterone acetate) and exogenous estrogen (17β-estradiol): Support for the demasculinization/feminization hypothesis. *Environ. Health Perspect.* 114:134–41.
Hayes, T.B. et al. 2010. Atrazine induces complete feminization and chemical castration in male African clawed frogs (*Xenopus laevis*). *Proc. Natl. Acad. Sci. U.S.A.* 107:4612–17.
Hayes, T.B. et al. 2011. Demasculinization and feminization of male gonads by atrazine: Consistent effects across vertebrate classes. *J. Steroid Biochem. Mol. Biol.*, 127:64–73.
Hecker, M. et al. 2004. Plasma sex steroid concentrations and gonadal aromatase activities in African clawed frogs (*Xenopus laevis*) from South Africa. *Environ. Toxicol. Chem.* 23:1996–2007.

Hecker, M. et al. 2005a. Plasma concentrations of estradiol and testosterone, gonadal aromatase activity and ultrastructure of the testis in *Xenopus laevis* exposed to estradiol or atrazine. *Aquat. Toxicol.* 72:383–96.

Hecker, M. et al. 2005b. Effects of atrazine on CYP19 gene expression and aromatase activity in testes and on plasma sex steroid concentrations of male African clawed frogs (*Xenopus laevis*). *Toxicol. Sci.* 86:273–80.

Hecker, M. et al. 2006. Terminology of gonadal anomolies in fish and amphibians resulting from chemical exposures. *Rev. Environ. Contam. Toxicol.* 187:103–31.

Heidrich, D.D., S. Steckelbroecke, and D. Klingmuller. 2001. Inhibition of human P450 aromatase activity by butyltins. *Steroids* 66:763–9.

Herzon, I., and J. Helenius. 2008. Agricultural drainage ditches, their biological importance and functioning. *Biol. Conserv.* 141:1171–83.

His, E., and R. Robert. 1980. Action d'un sel organo-métallique, l'acétate de tributyle-étain, sur les œufs et les larves D de *Crassostrea gigas* (Thunberg). *Int. Counc. Explor. Sea Comm. Meet. (Mariculture Comm)* 27:1–10.

Holloway, A.C. et al. 2008. Atrazine-induced changes in aromatase activity in estrogen sensitive target tissues. *J. Appl. Toxicol.* 28:260–70.

Howe, C. et al. 2004. Toxicity of glyphosate-based pesticides to four North American frog species. *Environ. Toxicol. Chem.* 23:1928–38.

Huet, M., Y.M. Paulet, and M. Le Pennec. 1996. Survival of *Nucella lapillus* in a tributyltin-polluted area in west Brittany: A further example of a male genital defect (Dumpton Syndrome) favouring survival. *Mar. Biol.* 125:543–9.

Huet, M., Y.M. Paulet, and J. Clavier. 2004. Imposex in *Nucella lapillus*: A ten year survey in NW Brittany. *Mar. Ecol. Prog. Ser.* 270:153–61.

Hyne, R.V. et al. 2009. Distribution of frogs in rice bays within an irrigated agricultural area: Links to pesticide usage and farm practice. *Environ. Toxicol. Chem.* 28:1255–65.

Janer, G. et al. 2005a. Testosterone conjugating activities in invertebrates: Are they targets for endocrine disruptors? *Aquat. Toxicol.* 71:273–82.

Janer, G., G.A. LeBlanc, and C. Porte. 2005b. Androgen metabolism in invertebrates and its modulation by xenoandrogens: A comparative study. *Ann. N. Y. Acad. Sci.* 1040:354–6.

Janer, G. et al. 2006. Sexual dimorphism in esterified steroid levels in the gastropod *Marisa cornuarietis*: The effect of xenoandrogenic compounds. *Steroids* 71:435–44.

Jobling, S. et al. 1996. Inhibition of testicular growth in rainbow trout (*Oncorhynchus mykiss*) exposed to estrogenic alkylphenolic chemicals. *Environ. Toxicol. Chem.* 15:194–202.

Jooste, A.M. et al. 2005. Gonadal development of larval male *Xenopus laevis* exposed to atrazine in outdoor microcosms. *Environ. Sci. Technol.* 39:5255–61.

Kidd, K.A. et al. 2007. Collapse of a fish population after exposure to a synthetic estrogen. *Proc. Natl. Acad. Sci.* 104:8897–901.

Kleinkauf, A. et al. 2004. Abnormally elevated VTG concentrations in flounder (*Platichthys flesus*) from the Mersey Estuary (UK)—A continuing problem. *Ecotoxicol. Environ. Saf.* 58:324–34.

Kloas, W. et al. 2009a. Does atrazine influence larval development and sexual differentiation in *Xenopus laevis*? *Toxicol. Sci.* 107:376–84.

Kloas, W. et al. 2009b. Does atrazine affect larval development and sexual differentiation of South African clawed frogs? In: *Trends in Comparative Endocrinology and Neurobiology*, ed. H. Vaudry, E.W. Roubos, G.M. Coast, and M. Vallarino, 437–40. Oxford: Blackwell Publishing.

Knutson, M.G. et al. 2004. Agricultural ponds support amphibian populations. *Ecol. Appl.* 14:669–84.

Labadie, P. et al. 2007. Identification of the steroid fatty acid ester conjugates formed *in vivo* in *Mytilus edulis* as a result of exposure to estrogens. *Steroids* 72:41–9.

LaFiandra, E.M., K.J. Babbitt, and S.A. Sower. 2008. Effects of atrazine on anuran development are altered by the presence of a nonlethal predator. *J. Toxicol. Environ. Health* 71A:505–11.

Langlois, V.S. et al. 2010a. Low levels of the herbicide atrazine alter sex ratios and reduce metamorphic success in *Rana pipiens* tadpoles raised in outdoor mesocosms. *Environ. Health Perspect.* 118:552–7.

Langlois, V.S. et al. 2010b. Evolution of steroid-5α-reductases and comparison of their function with 5β-reductase. *Gen. Comp. Endocrinol.* 166:489–97.

Lutz, I., S. Blodt, and W. Kloas. 2005. Regulation of estrogen receptors in primary cultured hepatocytes of the amphibian *Xenopus laevis* as estrogenic biomarker and its application in environmental monitoring. *Comp. Biochem. Physiol.* 141C:384–92.

Lutz, I. et al. 2008. Development, standardization and refinement of procedures for evaluating effects of endocrine active compounds on development and sexual differentiation of *Xenopus laevis. Anal. Bioanal. Chem.* 390:2031–48.

Mann, R.M. et al. 2009. Amphibians and agricultural chemicals: Review of the risks in a complex environment. *Environ. Pollut.* 157:2903–27.

Martín-Skilton, R. et al. 2006. First evidence of endocrine disruption in red mullets—*Mullus barbatus*—from the NW Mediterranean sea. *Environ. Pollut.* 141:60–8.

McCoy, K.A. et al. 2007. Tradeoffs between somatic and gonadal investments during development in the African clawed frog (*Xenopus laevis*). *J. Exp. Zool.* 307A:637–46.

McCoy, K.A. et al. 2008. Agriculture alters gonadal form and function in the toad *Bufo marinus. Environ. Health Perspect.* 116:1526–32.

McDaniel, T.V. et al. 2008. Potential endocrine disruption of sexual development in free ranging male northern leopard frogs (*Rana pipiens*) and green frogs (*Rana clamitans*) from areas of intensive row crop agriculture. *Aquat. Toxicol.* 88:230–42.

Milnes, M.R., and Guillette, L.J. 2008. Alligator tales: New lessons about environmental contaminants from a sentinel species. *BioScience* 58:1027–36.

Minier, C. et al. 2000a. Flounder health status in the Seine Bay. A multibiomarker study. *Mar. Environ. Res.* 50:69–73.

Minier, C. et al. 2000b. An investigation of the incidence of intersex fish in Seine-Maritime and Sussex regions. *Analusis* 28:801–6.

Minier, C. et al. 2005. Détermination de l'origine et des conséquences pour la faune aquatique de certaines perturbations endocriniennes dans les régions de Manche Orientale. Rapport Interreg IIIa.

Minier, C. et al. 2006a. A pollution monitoring pilot study involving chemical analysis and biomarker measurements in the Seine estuary using zebra mussels (*Dreissena polymorpha*). *Environ. Toxicol. Chem.* 25:112–9.

Minier, C. et al. 2006b. Multixenobiotic resistance protein expression in *Mytilus edulis, Mytilus galloprovincialis* and *Crassostrea gigas* from the French coasts. *Mar. Ecol. Prog. Ser.* 322:155–68.

Minier, C. et al. 2007. Analyse des risques associés aux perturbateurs endocriniens en Manche Orientale Rapport Interreg IIIa. Project number EEC/156. 90 pp.

Moore, J., and M.J. Gubbins. 2012. Surveys of dogwhelks *Nucella lapillus* in the vicinity of Sullom Voe, Shetland, August 2011. A report to SOTEAG from Aquatic Survey & Monitoring Ltd., Cosheston, Pembrokeshire and Marine Scotland Science, Aberdeen.

Mosconi, G. et al. 2005. Plasma sex steroid and thyroid hormones profile in male water frogs of the *Rana esculenta* complex from agricultural and pristine areas. *Gen. Comp. Endocrinol.* 142:318–24.

Motelay-Massei, A. et al. 2007. PAHs in the bulk atmospheric deposition of the Seine river basin: Source identification and apportionment by ratios, multivariate statistical techniques and scanning electron microscopy. *Chemosphere* 67:312–21.

Murphy, M.B. et al. 2006a. Plasma steroid hormone concentrations, aromatase activities and GSI in ranid frogs collected from agricultural and non-agricultural sites in Michigan (USA). *Aquat. Toxicol.* 77:153–66.

Murphy, M.B. et al. 2006b. Atrazine concentrations, gonadal gross morphology and histology in ranid frogs collected in Michigan agricultural areas. *Aquat. Toxicol.* 76:230–45.

Nash, J.P. et al. 2004. Long-term exposure to environmental concentrations of the pharmaceutical ethynylestradiol causes reproductive failure in fish. *Environ. Health Perspect.* 112:1725–33.

Nimrod, A.C., and W.H. Benson. 1998. Reproduction and development of Japanese medaka following an early life stage exposure to xenoestrogens. *Aquat. Toxicol.* 44:141–56.

Nishikawa, J.-I. 2006. Imposex in marine gastropods may be caused by binding of organotins to retinoid X receptor. *Mar. Biol.* 149:117–24.

Nishikawa, J.-I. et al. 2004. Involvement of the retinoid X receptor in the development of imposex caused by organotins in gastropods. *Environ. Sci. Technol.* 38:6271–6.

Oberdorster, E., and P. McClellan-Green. 2000. The neuropeptide APGWamide induces imposex in the mud snail, *Ilyanassa obsoleta*. *Peptides* 21:1323–30.

Oberdörster, E., J. Romano, and P. McClellan-Green. 2005. The neuropeptide APGWamide as a penis morphogenic factor (PMF) in gastropod mollusks. *Int. Comp. Biol.* 45:28–32.

Oehlmann, J., E. Stroben, and P. Fioroni. 1992. The rough tingle *Ocenebra erinacea* (Neogastropoda: Muricidae): An exhibitor of imposex in comparison to *Nucella lapillus*. *Helg. Meeres.* 46:311–28.

Oka, T. et al. 2008. Effect of atrazine on metamorphosis and sexual differentiation in *Xenopus laevis*. *Aquat. Toxicol.* 87:215–26.

Orton, F., J.A. Carr, and R.D. Handy. 2006. Effects of nitrate and atrazine on larval development and sexual differentiation in the northern leopard frog *Rana pipiens*. *Environ. Toxicol. Chem.* 25:65–71.

Peck, M.R. et al. 2004. Sediments are major sinks of steroidal estrogens in two United Kingdom rivers. *Environ. Toxicol. Chem.* 23:945–52.

Peck, M.R. et al. 2007. Profiles of environmental and endogenous estrogens in the zebra mussel *Dreissena polymorpha*. *Chemosphere* 69:1–8.

Phillips, D.J.H. 1978. Use of biological indicator organisms to quantitate organochlorine pollutants in aquatic environments: A review. *Environ. Pollut.* 16:167–229.

Puinean, A.M. et al. 2006. Laboratory exposure to 17β-estradiol fails to induce vitellogenin and estrogen receptor gene expression in the marine invertebrate *Mytilus edulis*. *Aquat. Toxicol.* 79:376–83.

Reeder, A.L. et al. 1998. Forms and prevalence of intersexuality and effects of environmental contaminants on sexuality in cricket frogs (*Acris crepitans*). *Environ. Health Perspect.* 106:261–6.

Reeder, A.L. et al. 2005. Intersexuality and the cricket frog decline: Historic and geographic trends. *Environ. Health Perspect.* 113:261–5.

RNO. 2006. Surveillance du Milieu Marin. Travaux du RNO. Edition 2006. Ifremer et Ministère de l'Ecologie et du Développement Durable ISSN 1620-1124.

Rohr, J.R., and K.A. McCoy. 2010. A qualitative meta-analysis reveals consistent effects of atrazine on freshwater fish and amphibians. *Environ. Health Perspect.* 118:20–32.

Ronis, M.J.J., and A.Z. Mason. 1996. The metabolism of testosterone by the periwinkle (*Littorina littorea) in vitro* and *in vivo*: Effects of tributyl tin. *Mar. Environ. Res.* 42:161–6.

Routledge, E.J., and J.P. Sumpter. 1996. Estrogenic activity of surfactants and some of their degradation products assessed using a recombinant yeast screen. *Environ. Toxicol. Chem.* 15:241–8.

Routledge, E.J., and J.P. Sumpter. 1997. Structural features of alkylphenolic chemicals associated with estrogenic activity. *J. Biol. Chem.* 272:3280–8.

Saitoh, M. et al. 2001. Tributyltin or triphenyltin inhibits aromatase activity in the human granulosa-like tumor cell line KGN. *Biochem. Biophys. Res. Commun.* 289:198–204.

Sanderson, J.T. 2006. The steroid hormone biosynthesis pathway as a target for endocrine-disrupting chemicals. *Toxicol. Sci.* 94:3–21.

Shulse, C.D. et al. 2010. Influences of design and landscape placement parameters on amphibian abundance in constructed wetlands. *Wetlands* 30:915–28.

Skelly, D., S. Bolden, and K. Dion. 2010. Intersex frogs concentrated in suburban and urban landscapes. *EcoHealth* 7:374–9.

Smith, B.S. 1971. Sexuality in the American mud snail, *Nassarius obsoletus* Say. *Proc. Malac. Soc. Lond.* 39:377.

Smith, B.S. 1981a. Male characteristics in female *Nassarius obsoletus*: Variations related to locality, season and year. *Veliger* 23:212–6.

Smith, B.S. 1981b. Reproductive anomalies in stenoglossan snails related to pollution from marinas. *J. Appl. Toxicol.* 1:15–21.

Smith, B.S. 1981c. Male characteristics on female mud snails caused by antifouling bottom paints. *J. Appl. Toxicol.* 1:22–5.

Smith, B.S. 1981d. Tributyltin compounds induce male characteristics on female mud snails *Nassarius obsoletus* = *Ilyanassa obsoleta*. *J. Appl. Toxicol.* 1:141–4.

Smith, E.E. et al. 2005. Assessment of laryngeal muscle and testicular cell types in *Xenopus laevis* (Anura Pipidae) inhabiting maize and non-maize growing areas of South Africa. *Afr. J. Herpetol.* 54:69–76.

Smith, M.D., and E.M. Hill. 2006. Profiles of short chain oligomers in roach (*Rutilus rutilus*) exposed to waterborne polyethoxylated nonylphenols. *Sci. Total Environ.* 356:100–11.

Solomon, K.R. et al. 2008. Effects of atrazine on fish, amphibians, and aquatic reptiles: A critical review. *Crit. Rev. Toxicol.* 38:721–72.

Spolyarich, N. et al. 2010. Growth, development and sex ratios of Spotted Marsh Frog (*Limnodynastes tasmaniensis*) larvae exposed to atrazine and a herbicide mixture. *Chemosphere* 78:807–13.

Spolyarich, N. et al. 2011. Morphological abnormalities in frogs from a rice-growing region in NSW, Australia, with investigations into pesticide exposure. *Environ. Monit. Assess.* 173:397–407.

Steeger, T.M. et al. 2003. Data evaluation report on atrazine-Induced hermaphroditism at 0.1 ppb in American leopard frogs (*Rana pipiens*): Laboratory and field evidence. United States Environmental Protection Agency.

Sternberg, R.M. et al. 2010. Environmental-endocrine control of reproductive maturation in gastropods: Implications for the mechanism of tributyltin-induced imposex in prosobranchs. *Ecotoxicology* 19:4–23.

Storrs-Méndez, S.I., and R.D. Semlitsch. 2010. Intersex gonads in frogs: Understanding the time course of natural development and role of endocrine disruptors. *J. Exp. Zool. B Mol. Dev. Evol.* 314:57–66.

Streiff, W., and J. Le Breton. 1970. Etude endocrinologique des facteurs régissant la morphogenèse et la régression du pénis chez un mollusque prosobranche gonochorique *Littorina littorea* L. *C. R. Hebd. Séances Acad. Sci. Paris* 270:547–9.

Tarrant, H. et al. 2008. Raised plasma vitellogenin in male wild brown trout (*Salmo trutta*) near a wastewater treatment plant in Ireland. *Environ. Toxicol. Chem.* 27:1773–9.

Tavera-Mendoza, L.E. 2001. Influence of atrazine on gonadal differentiation in *Xenopus laevis* tadpoles during metamorphosis. MSC thesis, Concordia University, Montreal, QC, Canada.

Tavera-Mendoza, L.E. et al. 2002a. Response of the amphibian tadpole (*Xenopus laevis*) to atrazine during sexual differentiation of the testis. *Environ. Toxicol. Chem.* 21:527–31.

Tavera-Mendoza, L.E. et al. 2002b. Response of the amphibian tadpole (*Xenopus laevis*) to atrazine during sexual differentiation of the ovary. *Environ. Toxicol. Chem.* 21:1264–7.

US-EPA. 2007. White paper on the potential for atrazine to affect amphibian gonadal development. United States Environment Protection Agency; Office of Prevention, Pesticides, and Toxic Substances; Office of Pesticide Programs; Environmental Fate and Effects Division, Washington, DC, USA.

Van der Ven, L.T.M. et al. 2003. Vitellogenin expression in zebrafish *Danio rerio*: Evaluation by histochemistry, immunohistochemistry, and *in situ* mRNA hybridization. *Aquat. Toxicol.* 65:1–11.

Vethaak, A.D. et al. 2005. An integrated assessment of estrogenic contamination and biological effects in the aquatic environment of The Netherlands. *Chemosphere* 59:511–24.

Wallace, H., G.M. Badawy, and B.M. Wallace. 1999. Amphibian sex determination and sex reversal. *Cell. Mol. Life Sci.* 55:901–9.

Witschi, E. 1921. Development of gonads and transformation of sex in the frog. *Am. Nat.* 55:529–38.

Witschi, E. 1929. Studies on sex differentiation and sex determination in the frog: III. Rudimentry hermaphroditism and Y chromosome in *Rana temporaria*. *J. Exp. Zool. A* 54:157–223.

10

Behavioral Ecotoxicology

Claude Amiard-Triquet and Jean-Claude Amiard

CONTENTS

10.1 Introduction

In the present state of knowledge, most core biomarkers of defense or damage do not allow the prediction of effects at higher levels of biological integration (Forbes et al. 2006), thus hampering their use for the ecological risk assessment of chemically contaminated ecosystems. However, for environmental toxicologists, it is intuitively evident that behavioral changes can induce effects at population and community levels (Grue et al. in Dell'Omo 2002). For instance, a reduced swimming capability of fish may reduce their ability to capture prey and avoid predators, and may interfere with social and reproductive behavior with a potential decrease in the fitness of wild populations (Weis et al. 2001; Vieira et al. 2009). On the other hand, behavioral disturbances have their origins in various biochemical and physiological impairments such as sensory system dysfunction, neurotoxicity, endocrine disruption, or energy disturbances. Thus, behavior may be considered a key link among cascading effects from infra- to supraorganismal responses. The relevance of behavioral ecotoxicology is reinforced by numerous studies showing that behavioral impairments occur at realistic doses in laboratory experiments, even after test organisms have been exposed for their whole life in a contaminated environment (Amiard-Triquet 2009; Hellou et al. 2011). Owing to the large number of studies dealing with pesticides, the relationships between acetylcholinesterase (AChE) inhibition (used as a biomarker of exposure to organophosphate and carbamate pesticides) and behavioral impairments are well documented for aquatic biota (see Table 4.1). In addition, recent studies dealing with the ecological risk of emerging contaminants have shown that behavioral biomarkers are responsive to contamination due for instance to nanoparticles (Bilberg et al. 2011; Buffet et al. 2011; Chen et al. 2011; Barros Amorim and Scott-Fordsmand 2012; Truong et al. 2012), phytoestrogens (Clotfelter et al. 2006), antibiotics (Yu et al. 2011), brominated flame retardants (BDE-47, Chen et al. 2012) or nonylphenols (Cailleaud et al. 2011).

10.2 Physiological Mechanisms Inducing Behavioral Impairments

Sensory systems are key for normal behavior since detection of environmental toxicants—but also natural compounds emitted by prey, congeners, and predators—govern avoidance ability and inter- and intraspecific relationships, particularly reproduction. Many studies have shown important links between hormones and behavior. Thus, it is highly probable that endocrine disruptors (Chapter 8) can have behavioral effects. Behavior is also influenced by the levels of neurotransmitters including the core damage biomarker AChE (Chapter 4). Energy-consuming locomotion activities are needed for the implementation of most behaviors; thus, it is expected that contaminants that impair energy metabolism (Chapter 11) probably influence behavioral responses.

10.2.1 Disruption of Sensory System Function

When exposed to an environmental contaminant, an organism will react differently depending on its ability (or lack thereof) to detect the presence of this compound (Figure 10.1). Once detected, avoidance (escape, valve closure in bivalves) allows a reduction of exposure, limiting noxious effects and improving survival. In the absence of detection and appropriate behavioral responses, organisms endure maximal exposure that can lead to lethal effects and reduced longevity.

Communication by chemosignals is often based on pheromones that can be chemically similar to internal chemical messengers such as steroids or neuropeptides. Production

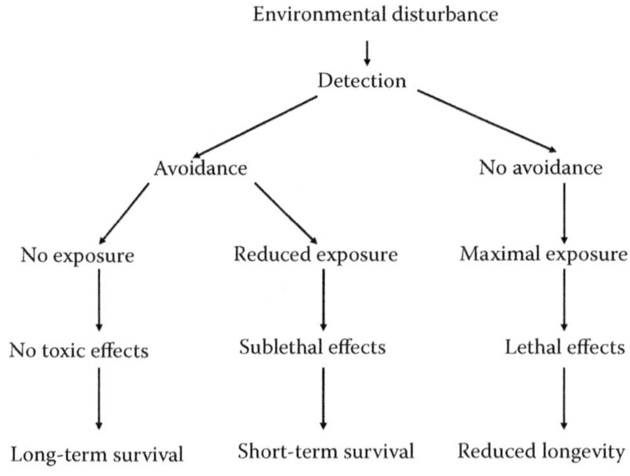

FIGURE 10.1
Different types of responses of organisms exposed to contaminants in their medium and consequences for their health and survival.

of, dissemination of, and reaction to chemosignals are often hormone dependent, especially for those that have links with reproductive functions (for more details, see review by Adkins-Regan and Weber in Dell'Omo 2002). In male parr of brown trout *Salmo trutta*, the pyrethroid pesticide cypermethrin had effects on endocrine responses to female odors and reproductive behavior (Jaensson et al. 2007).

Different pollutants can influence behavior through effects at the interface between an organism and the ambient medium. In fish and invertebrates, chemoreceptors—but also light receptors and mechanoreceptors—may be affected by chemicals. Toxicants can act directly at the surface or indirectly by diffusion through teguments, thus reducing or modifying the information that reaches the brain (Scarfe et al. 1982; Blaxter and Hallers-Tjabbes 1992; Scott and Sloman 2004). Subsequent changes in the response of effectors have potential to influence behavioral performance. Thus, in the Norway lobster *Nephrops norvegicus*, several mechanisms may be involved in the effect of manganese exposure on foraging behavior (Krång and Rosenqvist 2006). This metal may compete with natural odorants for binding sites on chemosensory receptors, thereby masking chemical signals from food, as has been shown with mercury in coho salmon *Oncorhynchus kisutch*. In hypoxic areas, precipitation of manganese dioxide onto the surface of the antennules could decrease the permeability of cuticle and thereby disturb odor stimuli diffusion and perception. Manganese could also affect chemoreception by blocking neural signals involved in odor transduction, as shown in another crustacean, *Panulirus argus*. In *N. norvegicus*, the effect of manganese on antennular chemosensory neurons is unknown, but a blocking effect on calcium-activated steps in the neuromuscular system has been reported (Krång and Rosenqvist 2006 and literature cited therein).

Impact of metals on the neuromasts of the lateral line induces an inability in fish to perform appropriate responses to water currents. However, the cadmium dose needed to induce such damage in the sea bass *Dicentrarchus labrax* was 10 times higher than the concentration occurring in highly polluted estuaries (Faucher et al. 2006). In zebrafish embryos, the lowest copper dose (68 $\mu g\ L^{-1}$) responsible for both histological and behavioral impairment (Johnson et al. 2007) is in the same order of magnitude of environmental concentrations corresponding to low densities of fish populations in Michigan lakes (34.0 $\mu g\ L^{-1}$) (Ellenberg et al. 1994). For comparison again, in Norway, freshwater fish are present only when the dissolved copper concentration is less than 60.0 $\mu g\ L^{-1}$ and some humic acids are present (Eisler 2007).

Many studies have demonstrated the ability of copper to result in olfactory impairments in fish, particularly salmonids (Hansen et al. 1999a; Sandahl et al. 2007; McIntyre et al. 2008; Green et al. 2010; Baldwin et al. 2011). Carreau and Pyle (2005) showed that exposure to copper at the embryonic stage of fish *Pimephales promelas* affects their ability to answer chemical alarm stimuli later in their life. Because many populations of Pacific salmon (*Oncorhynchus* spp.) are listed as threatened or endangered species under the Endangered Species Act, protectiveness of water quality criteria for copper in United States waters is focusing on avoidance and olfactory responses to copper by fish (Meyer and Adams 2010; Deforest et al. 2011).

Many studies have been devoted to the effects of metals, but impairments may also be induced by a surfactant (linear alkylbenzene sulfonate) in fish (Olsén and Höglund 1985), by pesticides in zooplankton (Hanazato 2001) and fish (Moore and Waring 1996, 1998, cited by Scott and Sloman 2004; Moore et al. 2007), polychlorobiphenyls (PCBs) in fish larvae (McCarthy et al. 2003), and silver nanoparticles in fish (Bilberg et al. 2011). In fish larvae contaminated by the organochlorine pesticide DDT through parental transfer via the egg yolk, chemical and visual signals and locomotion were impaired (Faulk et al. 1999).

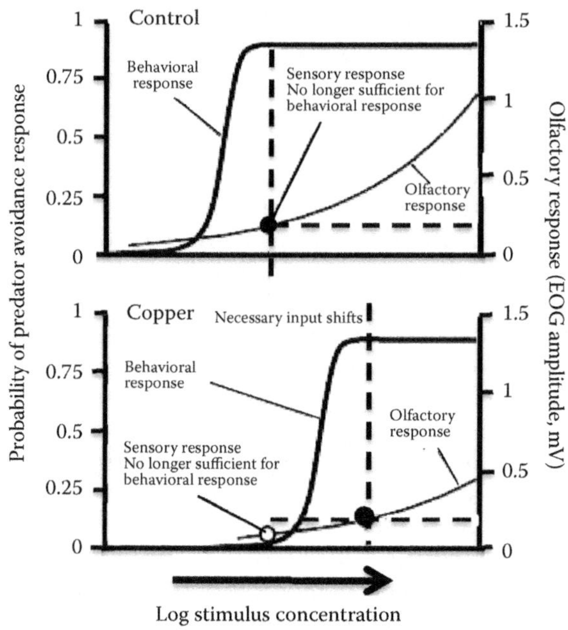

FIGURE 10.2
Conceptual model illustrating how shifts in olfactory sensitivity can result in corresponding shifts in predator avoidance behavior. In unexposed fish, a threshold concentration of alarm pheromone (left vertical dashed line) is required to generate an olfactory response (horizontal dashed line) that will be sufficient to trigger an alarm response. Following exposure to copper, a shift in olfactory sensitivity increases strength of stimulus needed to reach this physiological and behavioral threshold (right vertical dashed line), and the previous stimulus now effectively fails to elicit alarm behavior. (After Sandahl, J.F. et al., *Environ. Sci. Technol.*, 41, 2998–3004, 2007. With permission.)

Such detrimental effects can occur at low doses. Chronic exposure to (relatively) low concentrations of waterborne cadmium (0.2, 2.0, and 20 μg L^{-1}) during embryonic and larval development was claimed to result in the long-term hindrance of antipredator behavior in zebrafish *Danio rerio* (Kusch et al. 2008). Olfactory toxicity has been observed, for instance, in coho salmon *O. kisutch* exposed to commonly used pesticides and certain soluble compounds (glyphosate, iodocarb) at doses lower than guideline concentrations for water quality in Canada (Tierney et al. 2006). Similarly, the sensory physiology of juvenile coho salmon is significantly impaired by copper at concentrations as low as 2 μg L^{-1}, a concentration commonly detected in stormwater runoff from urban landscapes (Sandahl et al. 2007). This copper-induced chemosensory deprivation has the potential to cause increased predation mortality in exposed coho salmon via corresponding shifts in predator avoidance (Figure 10.2).

10.2.2 Neural and Hormonal Mechanisms of Behavior

Behavioral mechanisms have been mainly studied in mammals and birds. However, they include general characteristics that are highly conserved throughout evolution, and others that are more diverse and comparatively specific to certain taxa. In their review, Adkins-Regan and Weber (in Dell'Omo 2002) underlined that a number of traits of vertebrate nervous system organization were remarkably similar from lamprey to primates. Numerous characteristics of individual neuron physiology are highly conserved in both vertebrates and invertebrates. Synaptic transmission occurs in similar manner in different taxa, whereas intraspecific variations are recorded depending on the type of neuron, its localization in the nervous system, and the location of the synapse on the sending and receiving neurons. Processes involved in synaptic transmission, particularly the role of neurotransmitters such as acetylcholine, are potential targets for pharmacological and toxicological impairments. In addition to modifying AChE activity (Chapter 4), exposure to environmental neurotoxicants can also affect brain levels of other neurotransmitters such as serotonin (5-HT), dopamine, or gamma-aminobutyric acid.

In the freshwater fish *Channa punctata*, exposure to endosulfan induces a decrease in 5-HT and AChE activity with changes in surfacing behavior (Gopal et al. 1985). Dopamine concentration was also affected but differently according to the level and duration of endosulfan exposure. In the fish *Pimephales promelas* exposed to mercury (1.69 to 13.57 μg L^{-1} as mercuric chloride $HgCl_2$ for 10 days), foraging and capture speed were disturbed but neither differences in brain levels of norepinephrine, serotonin, and dopamine or its metabolite L-DOPA were shown nor any inhibition of AChE activity (Grippo and Heath 2003). On the contrary, in the fish *Mugil cephalus*, exposure to mercury was linked to a progressive loss of locomotion capability and serotonin level was decreasing (Thomas et al. 1981). Grippo and Heath (2003) suggested that such discrepancies could be due to environmentally realistic versus acute exposure procedures. However, in the estuarine fish *Fundulus heteroclitus* collected from a contaminated area and showing high brain mercury concentrations, a reduction of general activity was correlated to low brain concentrations of serotonin (Weis et al. 2001). In the fish *Pimephales promelas*, exposure to lead induces various behavioral disturbances associated with changes in neurotransmitter concentrations (serotonin, norepinephrine and its metabolite, vanillylmandelic acid) (Weber et al. 1991, 1997). Similarly, in rainbow trout (*Oncorhynchus mykiss*) exposed to lead and cadmium, physiological profiles were affected with a trend toward high activity of serotonin in dominant individuals, but the social hierarchy established before exposure was not affected (Sloman et al. 2005). These discrepancies could, again, have their origin in exposure conditions (0.5 or 1.0 mg Pb L^{-1} for 4 weeks in the study of Weber et al. 1991; 0.046 or 0.325 mg Pb L^{-1} for 48 h in the study of Sloman et al. 2005). In the coral reef fish *Thalassoma bifasciatum*, the link between serotonin and aggressive behavior has been shown after exposure to fluoxetine (a drug used as inhibitor of serotonin resorption), which is still present in effluents after the treatment of wastewater (Perreault et al. 2003). In hybrid striped bass (*Morone saxatilis* × *M. chrysops*), sublethal exposure to fluoxetine decreases the ability to capture prey in association with decreases in brain serotonin activity (Gaworecki and Klaine 2008). In the goldfish *Carassius auratus*, DDT induces interferences with behavior, either innate or acquired, and neurotransmitters (high concentration of brain serotonin and depletion of dopamine concentrations) (Weis and Weis 1974; McDonald 1979). In *Fundulus grandis*, PCBs affect both locomotion and dopamine and norepinephrine concentrations (Fingerman and Russell 1980).

Adkins-Regan and Weber (in Dell'Omo 2002) have listed molecules (neurotransmitters, neuromodulators, neurohormones, and hormones) considered as involved in behaviors important for an ecology's organisms as a consequence of their actions on the nervous system. Among these regulators, many are neuropeptides, some of which are widely distributed such as vasopressin–oxytocin-like molecules (nonapeptides), GnRH (gonadotropin releasing hormone, decapeptides), or P substance (family of tachykinins) that are present in both vertebrates and invertebrates. Steroid hormones are also numerous and widely distributed. The relationship between steroid hormones and behavior is well established in different fish species (Bjerselius et al. 2001; Sisneros et al. 2004; Knapp and Neff 2006). However, the mechanisms behind these observations are not clearly understood and generally are extrapolated from studies with birds and mammals.

Thyroid hormones are essential for the normal development of brain, and mammalian studies have shown that alterations of concentrations of these hormones have consequences on neurobehavioral development and neurotransmitter levels. A study dealing with European eel *Anguilla anguilla* upstream migration suggested that the thyrotropic axis was activated during the colonization of river habitats by elvers (Imbert et al. 2008). Juvenile specimens of Atlantic salmon *Salmo salar* exposed to PCB (Aroclor 1254 at 1 or 10 $\mu g\ L^{-1}$) during smolting exhibit a dose-dependent reduction in preference for seawater. In parallel, plasma triiodothyronine was reduced 35–50%, whereas plasma thyroxine was unaffected (Lerner et al. 2007). Impairment of thyroid function could contribute to behavioral disturbances observed in the estuarine fish *F. heteroclitus* from a contaminated estuary (Weis et al. 2001). In the Atlantic cod *Gadus morhua*, altered thyroid function has been shown to influence spontaneous activity and feeding (Castonguay and Cyr 1998, quoted by Weis et al. 2001). In zebrafish (*Danio rerio*) embryos waterborne exposed to BDE-47 at 1.25, 5, and 20 µM starting from 6 h post-fertilization (hpf), Chen et al. 2012 showed an inhibition of the axonal growth of primary and secondary motor neurons during the early developmental stages, suggesting the functional relevance of structural changes, in agreement with impaired motor behavior.

10.2.3 Energy Metabolism Impairments

Many studies have shown a link between the ability of animals to perform normal behavior in polluted media and several aspects of energy metabolism. In fish, a review by Scott and Sloman (2004) indicates that different metals and numerous pesticides are responsible for alteration in metabolic substrate levels, a pattern that could result from enzymatic system dysfunction (enzymes involved in carbohydrate and/or protein metabolism). These metabolic impairments result in alterations of maintenance (changes in oxygen consumption and ventilation rate) and reduced swimming performance.

In trout (*Salmo trutta fario*) exposed for several weeks in water diverted from two polluted streams (bypass systems) in the field, biomarkers were determined at different levels of biological organization (swimming behavior, stress protein concentrations, liver ultrastructure). Swimming velocity significantly dropped in trout in the most polluted stream, perhaps as a consequence of less energy being available for locomotion. In agreement with an increasing number of mitochondria and a depletion of glycogen storage in the liver, catabolic, energy-providing mechanisms were activated (Triebskorn et al. 1997). In the ragworm *Hediste (Nereis) diversicolor* exposed in the field or in the laboratory to contaminated sediment, Moreira et al. (2006a) have shown a depletion of the post-exposure feeding rate associated with increased lactate dehydrogenase activity that reveals an increased contribution of anaerobic metabolism. This metabolic change suggests an urgent need for additional energy to ameliorate chemical stress.

Coping with toxicants has an associated metabolic cost (Mouneyrac et al. in Amiard-Triquet et al. 2011). The relationship between behavioral and metabolic disruption is depicted in Figure 10.3 in the case of invertebrates exposed to increasing levels of metals. At low exposure, crustaceans are able to detoxify absorbed metals, but behaviors with a high energy cost are significantly impaired. With increasing exposure, the detoxification capacity of metallothioneins is overwhelmed, and this results in increasing metal

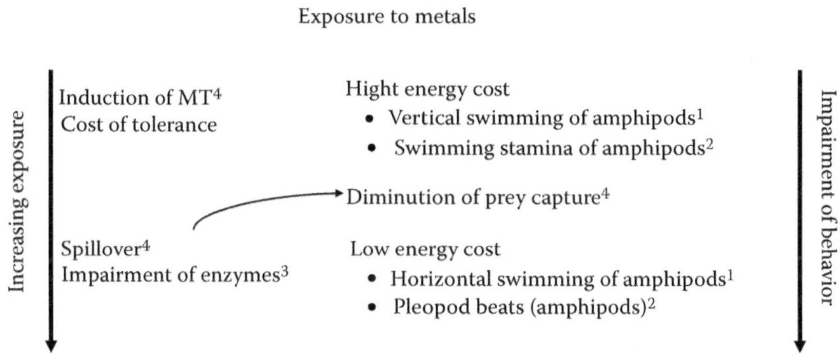

FIGURE 10.3
Relationship between behavioral and metabolic disruption: case of aquatic invertebrates exposed to increasing levels of metals. ([1]Wallace and Estephan 2004; [2]Lawrence and Poulter 1998; [3]Louie and Meade 1999; [4]Wallace et al. 2000). (From Amiard-Triquet, C. et al., eds, *Tolerance to Environmental Contaminants*, Boca Raton, FL, CRC Press.)

concentrations bound to high molecular weight proteins, that is to say, enzymes. With the impairment of enzymes, toxic effects may appear such as the diminution of prey capture observed in the grass shrimp *Palaemonetes pugio*. At this stage, even behaviors with lower energy cost may be disturbed.

10.3 Behavioral Effects of Aquatic Pollutants

In Table 10.1, we have listed studies showing the effect of different classes of chemicals on different categories of behavior in different zoological taxa.

Many organisms are able to detect chemicals in their environment and to avoid them. Avoidance of contaminated food results in the reduction of toxicant uptake but also, of food intake as a source of energy (see Chapter 11). Organisms can also detect and avoid contaminants present in water and sediments. Fish are able to react at low metal concentrations, of the order of magnitude of those encountered in the field in contaminated areas (Scherer and McNicol 1998; Hansen et al. 1999a and literature quoted therein). Fish can also react to the presence of organic compounds such as polycyclic aromatic hydrocarbons (PAHs). Invertebrates are also responsive to the presence of different classes of contaminants in their medium. On the other hand, Tierney et al. (2011) provide evidence for a behavioral preference toward environmental concentrations of urban-use herbicides in the fish *Oncorhynchus kisutch*. Consequently, these juvenile Coho salmon place themselves in conditions of exposure that alter their attraction response to L-alanine, an amino acid prevalent in food that therefore represents a cue necessary for survival.

The ability of fish to avoid their predators often falls off as a consequence of sublethal exposure to toxicants. As discussed in Section 10.2.1, this may be due to chemosensory deprivation delaying the appropriate response, but also to disturbances in schooling behavior of prey species (Nakayama et al. 2005). The literature provides numerous examples of reduced survival of contaminated fish in the presence of their predators (see review by Scott and Sloman 2004).

Locomotion is a key factor in performing an appropriate behavior since it has a direct impact on interactions between prey and predators, relationships of dominance/

TABLE 10.1

Impairment of Different Categories of Behavior in Different Zoological Taxa Exposed to Different Classes of Chemicals

Zoological Taxa	Species	Chemical Classes	Contaminants	References
Avoidance				
Fish[a]	*Oncorhynchus kisutch*	Metal	Cu	Sandahl et al. 2007
Fish[a]	*Oncorhynchus mykiss*	Metals	Cd + Cu + Pb + Zn	Hansen et al. 1999a
Fish[b]	*Coregonus clupeaformis*	Metals	Cu, Pb, Zn	Scherer and McNicol 1998
Fish[b]	*Pimephales promelas*	PAH	Fluoranthene	Farr et al. 1995
Fish[a]	Larval *Pimephales promelas*	Pyrethroid insecticide	Esfenvalerate	Floyd et al. 2008
Fish[a]	Larval *Pimephales promelas*	Pharmaceuticals	Antidepressants	Painter et al. 2009
Fish[a]	Larval *Danio rerio*	Metal	Cd	Kusch et al. 2008
Crustacean[b]	*Gammarus pulex*	PAH		De Lange et al. 2006b
Crustacean[b]	*Corophium volutator*	Sediments (harbors)	HAP	Hellou 2011
Crustacean[b]	*Eohaustorius estuarius*	Sediments	Creosote	Kravitz et al. 1999
Crustacean[b]	*Monoporeia affinis*	Sediments	Zinc pyrithione	Eriksson Wiklund et al. 2006
Insect[a]	Mosquito larva	Insecticide pyrethroid	Fenvalearate	Reynaldi et al. 2011
Bivalve[a]	*Corbicula fluminea*	Metals	Cd Cu U Hg	Tran et al. 2003 Tran et al. 2004 Fournier et al. 2004 Tran et al. 2007
Gastropod[b]	*Physella columbiana*	Metals	Cd, Pb, Zn	Lefcort et al. 2004
Gastropod[b]	*Ilyanassa obsoleta*	Sediments (harbors)	HAP	Hellou 2011
Annelid[b]	*Hirudo medicinalis*	River sediments Nuclear plant cooling water Metals	Mixture	Petrauskiene 2003
Locomotion				
Reptiles	*Seminatrix pygaea* *Nerodia rhombifer*	Pesticide carbamate	Carbaryl	Hopkins et al. 2005
Amphibian	*Rana pipiens*	PCB		Glennemeier and Denver 2001
Fish	*Micropogonias undulatus*	Organochlorine pesticide	*o,p*′-DDT	Faulk et al. 1999
Fish	*Clupea pallas*	Petroleum hydrocarbons	Hydrosoluble fraction	Kennedy and Farrell 2006
Fish	*Salmo trutta f. fario*	Polluted water in the field	Pesticides, HAP, PCB, metals	Triebskorn et al. 1997
Fish	*Oncorhynchus mykiss*	Metals	Cu	McGeer et al. 2000
Fish	*Oncorhynchus mykiss*	Pesticide carbamate	Carbaryl	Beauvais et al. 2001
Fish	*Oncorhynchus mykiss*	Pesticides OP	Dianizon, malathion	Beauvais et al. 2000; Brewer et al. 2001
Fish	*Oncorhynchus kisutch*	Insecticide OP	Chlorpyrifos	Sandahl et al. 2005

TABLE 10.1 (Continued)

Impairment of Different Categories of Behavior in Different Zoological Taxa Exposed to Different Classes of Chemicals

Zoological Taxa	Species	Chemical Classes	Contaminants	References
Fish	*Fundulus heteroclitus*	Insecticide OP	Azinphosmethyl	Van Dolah et al. 1997
	Sciaenops ocellatus			Van Dolah et al. 1997
Fish	Larval *Danio rerio*	Nanoparticles	TiO_2	Chen et al. 2011
Fish	*Danio rerio*	Contaminated food	Selenomethionine	Thomas and Janz 2011
Fish	*Potamoschistus microps*	Metals	Cu, Hg	Vieira et al. 2009
Fish	*Pimephales promelas*	Pyrethroid insecticide	Esfenvalerate	Floyd et al. 2008
Crustacean	*Litopenaeus vannamei*	Insecticide OP	Methamidophos	García-de la Parra et al. 2006
Crustacean	*Neomysis integer*	Metals	Cd	Roast et al. 2001
Crustacean	*N. integer*	Insecticide OP	Chlorpyrifos	Roast et al. 2000
Crustacean	*Gammarus lawrencianus*	Metals	Cd	Wallace and Estephan 2004
Crustacean	*Gammarus duebeni*	Different molecules	Cu, benzo(*a*)pyrene pentachlorophenol	Lawrence and Poulter 1998
Crustacean	*Gammarus pulex*	Polluted waters in the field	Metals, organic xenobiotics	Gerhardt 1996
Crustacean	*G. pulex*	Pharmaceutical Cationic surfactant		De Lange et al. 2006a
Crustacean	*Corophium volutator*	Crude oil	Water accommodated fraction	Kienle and Gerhardt 2008
Crustacean	*Daphnia magna*	PCB		Schmidt et al. 2006
Crustacean	*Eurytemora affinis*	Surfactants	Nonylphenol	Cailleaud et al. 2011
Annelid	*Lumbriculus variegatus*	Metals	Cu	O'Gara et al. 2004
Burrowing				
Crustacean	*Sadura entomon*	Metals	Sediment + Cu, Cd, Fe	Pynnonen 1996
Bivalve	*Protothaca staminea*	Petroleum hydrocarbons	Oiled sediments	Pearson et al. 1981
Bivalve	*Protothaca staminea*	Metals	Sediment + Cu	Phelps et al. 1983
Bivalves	*Macoma balthica* *Cerastoderma edule* *Abra alba*	Pesticides	Sediment + Parathion, Malathion, Methylparathion	Møhlenberg and Kiøboe 1983
Bivalve	*Macomona liliana*	Metals, pesticide	Sediment + Cu + chlordane	Roper and Hickey 1994
Bivalve	*M. liliana*	Metals	Sediment + Cu, + Zn	Roper et al. 1995
Bivalve	*Tapes philippinarum*	4-Nonylphenol		Matozzo et al. 2004
Bivalve	*Tapes semidecussatus*	Sediments	Contamination gradient	Byrne and O'Halloran 2001
Bivalve	*Scrobicularia plana*	Sediments	Contamination gradient	Byrne and O'Halloran 2001 Boldina-Cosqueric et al. 2010

(*continued*)

TABLE 10.1 (Continued)

Impairment of Different Categories of Behavior in Different Zoological Taxa Exposed to Different Classes of Chemicals

Zoological Taxa	Species	Chemical Classes	Contaminants	References
Bivalve	*S. plana*	Metals	Cu	Bonnard et al. 2009
Bivalve	*S. plana*	Metal, nanoparticles	Soluble Cu, CuO NP	Buffet et al. 2011
Bivalve	*Ruditapes philippinarum*	Metals	Sediments + Cd	Shin et al. 2002
Bivalve	*Corbicula fluminea*	Insecticide OP	Chlorpyrifos	Cooper and Bidwell 2006
Annelid	*N. diversicolor*	Metals	Cu	Bonnard et al. 2009
Annelid	*N. diversicolor*	Sediments	Contamination gradient	Kalman et al. 2009 Mouneyrac et al. 2010
Annelids	*N. diversicolor* *Scoloplos armiger*	Pesticides	Sediment + Parathion, Malathion, Methylparathion	Møhlenberg and Kiøboe 1983
Feeding Behavior				
Amphibian	*Rana pipiens*	PCB		Glennemeier and Denver 2001
Fish	*Oncorhynchus kisutch*	Insecticide OP	Chlorpyrifos	Sandahl et al. 2005
Fish	*Salvelinus namaycush*	Metals	Cd	Kislalioglu et al. 1996
Fish	*Salvelinus fontinalis*	Metals	Cd	Riddell et al. 2005
Fish	*Morone* sp.	Pharmaceuticals	Antidepressant fluoxetine	Gaworecki and Klaine 2008
Fish	*Morone* sp.	Organophosphate pesticide	Diazinon	Gaworecki et al. 2009
Fish	*Pimephales promelas*	Metals	Hg	Grippo and Heath 2003
Fish	*P. promelas*	Pyrethroid insecticide	Esfenvalerate	Floyd et al. 2008
Fish	*Fundulus heteroclitus*	Contamination in the field	Mainly Hg	Weis et al. 2001
Fish	*Cyprinus carpio*	Metals	Cu, Zn	Kuz'mina 2011
Insect	*Hydropsyche slossonae*	Insecticide OP	Malathion	Tessier et al. 2000
Insect (larvae)	*Chironomus xanthus*	Pyrethroid insecticide	Deltamethrin	Moreira-Santos et al. 2005
Crustacean	*Carcinus maenas*	Metals Sediments (estuaries)	Cd	Moreira et al. 2006b
Crustacean	*C. maenas*	PAHs in the field and in the lab		Dissanayake et al. 2010
Crustacean	*Nephrops norvegicus*	Metals	Mn	Krång and Rosenqvist 2006
Crustacean	*Palaemonetes pugio*	Metals	Cu	Wallace et al. 2000
Crustacean	*P. pugio*	Contamination in the field		Perez and Wallace 2004
Crustacean	*Gammarus pulex*	Effluents		Maltby et al. 2000, 2002

TABLE 10.1 (Continued)

Impairment of Different Categories of Behavior in Different Zoological Taxa Exposed to Different Classes of Chemicals

Zoological Taxa	Species	Chemical Classes	Contaminants	References
Crustacean	*G. pulex*	Metals	Zn	Wilding and Maltby 2006
Crustacean	*Gammarus fossarum*	Contamination in the field		Coulaud et al. 2011
Crustacean	*Daphnia magna*	Metals Pesticides	Cd, Cu, Zn Pyrethroids, lindane	McWilliam and Baird 2002b
Crustacean	*D. magna*	Contamination in the field	Pesticides	Barata et al. 2007
Bivalve	*S. plana*	Nanoparticles	CuO	Buffet et al. 2011
Gastropod	*Hydrobia ulvae*	Contamination in the field		Krell et al. 2011
Gastropod	*Potamopyrgus antipodarum*	Nanoparticles	Nano-CuO	Pang et al. 2012
Annelid	*Hediste diversicolor*	Metals Sediments (estuaries)	Cu	Moreira et al. 2006a; Moreira et al. 2005
Annelid	*Nereis (Hediste) diversicolor*	Sediments	Contamination gradient	Kalman et al. 2009
Annelid	*Neanthes arenaceodentata*	Metal PAH	Cu Phenanthrene	Rosen and Miller 2011
Reproductive Behavior				
Fish	*Gasterosteus aculeatus*	Hormone	Ethinyl estradiol	Bell 2001
Fish	*Carassius auratus*	Hormone	17β-Estradiol	Bjerselius et al. 2001
Fish	*Salmo trutta*	Pesticide	Cypermethrin	Jaensson et al. 2007
Fish	*Pimephales promelas*	Metal	Methylmercury	Sandheinrich and Miller 2006
Fish	*P. promelas*	Effluent waters		Garcia-Reyero et al. 2011

[a] Avoidance of predators.
[b] Avoidance of contaminants.

hierarchy, and foraging success. Literature data show that locomotion is affected in both fish and invertebrates (annelids, crustaceans) by different classes of contaminants (metals, pesticides, PAHs, pharmaceuticals, surfactants) or polluted waters in the field (Table 10.1).

In many endobenthic invertebrates, exposure to different classes of chemicals (petroleum hydrocarbons, pesticides, and metals) can disturb burrowing behavior. A chain of reflexes is needed to perform this behavior, and mechanisms inducing impairments would include histological damage to epithelium, neurotoxic effects, and avoidance (Pynnonen 1996).

Effects of chemicals on foraging behavior of fish has been excellently reviewed by Kasumyan (2001). In invertebrates (Table 10.1), disturbances of feeding behavior were demonstrated in many species using the method of "post-exposure feeding depression" initially designed by McWilliam and Baird (2002a). Impairments of prey capture and feeding rate have been observed in polychaetes, crustaceans, and fish (Table 10.1), including animals chronically exposed in the field to different anthropogenic contaminants (Weis et al. 2001; Perez and Wallace 2004; Kalman et al. 2009).

In fish, reproductive success depends on various complex behaviors. Underlying neural and hormonal mechanisms are relatively well known (Adkins-Regan and Weber in Dell'Omo 2002), but little information is available concerning dysfunctions resulting from the presence of potentially toxic chemicals in the environment (Table 10.1). Scott and Sloman (2004) have reviewed available data about nest building, courtship behaviors, spawning site selection, and nonreproductive social behaviors such as aggression, and formation and maintenance of hierarchies. Again, there is a long and diverse list of metals and organic contaminants that are able to impair such complex behavioral patterns. Migratory activity, which has a key role in the biology of salmonids, may be impaired by exposure of Atlantic salmon (*Salmo salar*) smolts at environmental doses of the pesticide atrazine (Moore et al. 2007).

10.4 Ecological Relevance of Behavioral Biomarkers

10.4.1 Environmental Realism of Behavioral Impairments

At sublethal doses, defense mechanisms are functional, and under these conditions, ecotoxicity tests seem more representative of effects due to chronic exposure in the field. Similarly, the duration of experiments can have a crucial influence on behavioral responses, depending on whether detoxification mechanisms have had enough time to be induced or not. Thus, several authors have reported that food consumption was initially reduced in the presence of various metals but partial or total recovery occurred within 3 weeks, suggesting the induction of compensatory mechanisms such as metallothionein induction (Bryan et al. 1995 and literature cited therein). On the other hand, in polychaetes *Hediste (Nereis) diversicolor* exposed for 96 h to reference sediments and sediments from contaminated sites, a depletion of feeding rates has been shown despite compensatory mechanisms functioning, although apparently insufficiently to eliminate all the contaminants' effects, particularly oxidative stress (Moreira et al. 2006a).

With the aim of increasing environmental realism, Riddell et al. (2005) have examined behavioral responses to sublethal cadmium doses in an experimental food web. In juvenile fish (*Salvelinus fontinalis*), the efficiency of prey capture was impaired at doses (0.5 μg L^{-1}) nearing the Canadian standard for the protection of aquatic life (0.2 μg L^{-1}). This comparison was based on nominal concentrations of cadmium added in experimental units. An additional step toward more realism involves considering the chemical speciation of dissolved metals that governs their bioavailability and hence toxicity (Fournier et al. 2004; Green et al. 2010; Meyer and Adams 2010). This is particularly important for estuarine species since free metal ion concentrations often depend on salinity, particularly in the case of cadmium. Sublethal concentrations of cadmium as $Cd^{2+}_{(aq)}$ (0.5 and 1.0 μg L^{-1}) were tested on the estuarine mysid crustacean *Neomysis integer* (Figure 10.4). Even at the lowest concentration, swimming behavior was impaired with fewer specimens being able to move forward into the current at stream velocities (3–9 cm s^{-1}) typical in their natural habitat. Hyperbenthic behavior was also modified since contaminated individuals were more numerous in the water column than their nonexposed conspecifics. The effective concentration of 0.5 μg L^{-1} was within the order of magnitude of concentrations found in

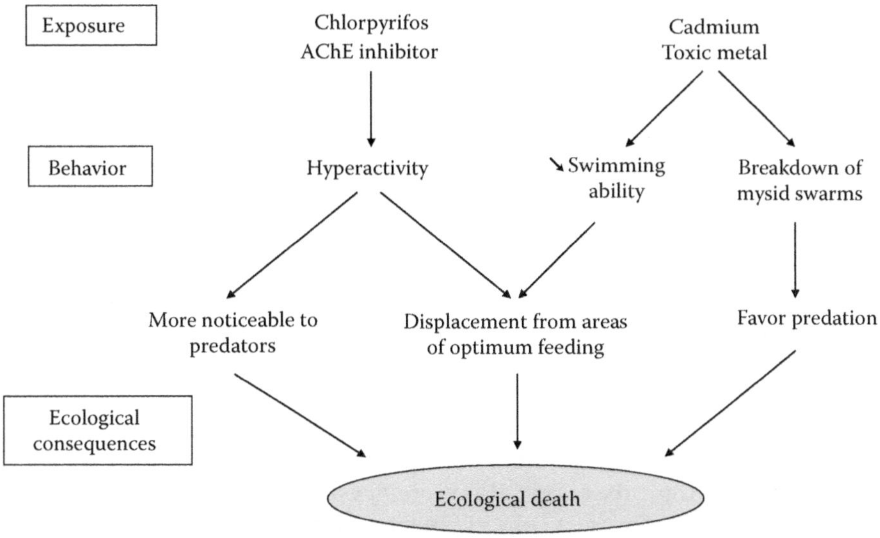

FIGURE 10.4
Ecological consequences for estuarine mysid crustacean *Neomysis integer* from exposure to realistic concentrations of contaminants. (After Roast, S.D. et al., *Aquat. Toxicol.*, 47, 227–241, 2000; Roast, S.D. et al., *Aquat. Toxicol.*, 52, 217–227, 2001.)

the Thames (0.53 μg $Cd^{2+}_{(aq)}$ L^{-1}), the Schelde (0.3 μg $Cd^{2+}_{(aq)}$ L^{-1}), and the Rhine (0.18 μg $Cd^{2+}_{(aq)}$ L^{-1}) (Asthaus 1992, quoted in Roast et al. 2001). Swimming was also disturbed in *N. integer* exposed to the organophosphate pesticide chlorpyrifos (Roast et al. 2000). After 7 days of exposure to 0.038 μg L^{-1}, hyperactivity was observed, and individuals swimming forward into a slow current (3 cm s^{-1}) were more numerous than in control conditions. The authors concluded that swimming behavior of mysids would be disturbed at concentrations corresponding to freshwater quality standards (in the absence of water quality criteria for the protection of aquatic life in estuaries). Using both laboratory and stream microcosm studies, Nørum et al. (2010) showed effects of the pyrethroid lambda-cyhalothrin on drift behavior in three species of freshwater macroinvertebrates. Low observed effect concentrations (LOECs) determined for *Leuctra nigra* (1 ng L^{-1}), *Gammarus pulex* (10 ng L^{-1}) and *Heptagenia sulphurea* (100 ng L^{-1}) were all within expected environmental concentrations. In the nematode *Caenorhabditis elegans*, exposure to sulfamethoxazole, one of the most common existing antibiotics in the environment, at environmentally relevant concentrations, Yu et al. (2011) have shown that behavior indicators exhibited a better sensitivity than growth indicator. They also demonstrated that behavior toxicity was transferable to the progeny, at least to the next generation.

Carrying out experiments over a duration significant compared to the life span of the species of interest, and/or tests that take place at a key stage of the life cycle, can also contribute to improving environmental realism. For instance, the possible sublethal alterations in smolt physiology and behavior due to PCB (Aroclor 1254) exposure of Atlantic salmon (*Salmo salar*) have been examined by Lerner et al. (2007) in fish exposed as yolk-sac larvae or as juveniles just before the parr–smolt transformation in April. After exposure, yolk-sac larvae were reared at ambient conditions for 1 year, until the peak of smolting the following May. These authors indicate that the effects of exposure may vary according to developmental stage. Exposure in the freshwater environment can inhibit preparatory adaptations that occur during smolting, thereby contributing to impairments of migratory behavior as shown in another species *Salvelinus alpinus*. This is in agreement with the hypothesis that PCBs could be one of the factors contributing to the decline of wild populations of Atlantic salmon as well as failures in salmon restoration (Lerner et al. 2007 and literature quoted therein). Denoël et al. (2012a) have used sets of behavioral biomarkers (lying, swirling, and feeding) to assess short-term effects of pesticides on frog tadpoles (*Rana temporaria*). Their results suggest that endosulfan is toxic to amphibians at environmental concentrations.

Field experiments offer an additional step with a view to increasing environmental realism. We have already discussed (Section 10.3) the method of postexposure feeding depletion. Similar approaches were carried out to examine escape behavior in crabs *Uca pugnax* exposed in the field to an oil spill (Krebs and Burns 1977) or social behavior in fish (*Gambusia* sp.) exposed to pulp and paper mill effluents including compounds with androgenic properties (Toft et al. 2004). Another possibility is the deployment of *in situ* tests to show the effects of local contamination. This strategy was able to reveal the effects of oiled sediments on the predation rate of crabs *Cancer magister* on bivalves *Protothaca staminea* (Pearson et al. 1981); effects of effluents on locomotion, activity phases, and feeding rate of amphipod crustaceans *Gammarus pulex* (Gerhardt 1996; Maltby et al. 2000, 2002); effects of contaminated sediments on feeding behavior of worms *Hediste diversicolor* and crabs *Carcinus maenas* (Moreira et al. 2006a, b; Kalman et al. 2009; Fossi Tankoua et al. 2012); effects of water from streams differing by their levels of chemical contaminants on swimming velocity of *Salmo trutta f. fario* (Triebskorn et al. 1997).

10.4.2 From Individuals to Communities: Cascading Effects of Behavioral Impairments

Avoidance of contaminated water, sediment, or food is frequently observed (Table 10.1) and can have a beneficial effect for wild populations (Figure 10.1). Even if the presence of a toxicant is detected, avoidance is not automatic as a consequence of interferences with other environmental factors. Many species are reactive to lighting levels; for instance, a given species can prefer a habitat highly contaminated but corresponding to its light preferendum to another clean habitat that does not meet this preferendum (e.g., Scherer and McNicol 1998). Beneficial at the individual level, avoidance may be responsible for local extinction of certain species at the ecosystem level. For instance, in North America, salmonids highly susceptible to metals have probably abandoned habitats in which the water quality was inadequate. According to Hansen et al. (1999b), the greater sensitivity of rainbow trout (*Oncorhynchus mykiss*) to metal mixtures may explain, in part, why rainbow trout populations appear to be more severely affected, compared to brown trout (*Salmo trutta*) populations in the Clark Fork River (Montana) subjected to historical contamination by mining and smelting operations.

Feeding impairment is a general stress response exhibited by a variety of organisms in response to various classes of chemical contaminants (Table 10.1). The amphipod crustacean *Gammarus pulex* is an important shredding detritivore in many streams. Inhibition of the *in situ* feeding rate of *G. pulex* deployed downstream of a variety of point-source discharges and experimental evidence of a correlation with leaf decomposition rate suggest that reductions in shredder feeding activity will also translate into reductions in detritus processing rates in riverine ecosystems (Maltby et al. 2002).

Several behavioral alterations have been observed in mysid crustaceans *Neomysis integer* exposed to an inhibitor of AChE activity or to a toxic metal (Figure 10.4). Potential ecological consequences have been examined (Roast et al. 2000, 2001). They include an increased predation pressure and a displacement from areas of optimum feeding that could lead indirectly to mysid mortalities in the natural environment, called "ecological death." Indeed, exposure of *N. integer* to 1.0 μg $Cd^{2+}_{(aq)}$ L^{-1} in the natural environment may lead to mysids being completely swept from the estuary, since in experiments by Roast et al. (2001) many cadmium-exposed mysids were unable to swim against a current as low as 3 cm s^{-1}.

Prey–predator interactions have a key role in controlling the composition and productivity of communities (Pearson et al. 1981; Mesa et al. 1994; Bryan et al. 1995). Table 10.1 provides examples of impairments of behavior of predators exposed to toxicants even under field conditions in polluted areas. For more examples, the reader can refer to reviews by Atchinson et al. (1987) and Grue et al. (in Dell'Omo 2002). Decreased food intakes have consequences in terms of energy budget (Chapter 11), growth, and reproduction. When the impacted species is an important link in the food chain, this can have repercussions at higher trophic levels. Considering prey species, fish subjected to stress may be more vulnerable to their predators than their nonexposed conspecifics (Mesa et al. 1994; Sandahl et al. 2007; Kusch et al. 2008). Impairments of burrowing behavior that facilitate predation are well documented in invertebrates exposed to different classes of chemical contaminants (Table 10.1).

Only few studies are available to make a judgment about the potential ecological consequences of reduced predation efficiency and increased prey vulnerability on populations and communities. Seven years after an oil spill, populations of the salt marsh crab *Uca pugnax* were still reduced as a consequence of the toxic effects of fuel oil on crab physiology and behavior (Krebs and Burns 1977). These authors underline that reduction of the crab population was ecologically significant since this species is important for the ecosystem as a consequence of its role in energy flow through the system. As predators, crabs *Cancer magister* can influence the species composition and diversity of soft-bottom communities: when a prey species (*Protothaca staminea*) becomes more vulnerable as a consequence of living in oiled sediment, predation by crabs is facilitated, thereby potentially reducing the presence of the prey species among the benthic macrofauna (Pearson et al. 1981). Janssens and Stoks (2012) have studied the vulnerability to predation in larvae of the damselfly *Enallagma cyathigerum* exposed to a pesticide (endosulfan) or a herbicide (Roundup). Although both contaminants induced maladaptive changes in overall activity, only for endosulfan this resulted in an increased mortality by predation as a consequence of different combined effects of these compounds on behavioral antipredator traits and escape swimming.

In a comprehensive study with two fish and three crustacean species chronically exposed in contaminated estuaries compared to conspecifics from a cleaner estuary, cascading effects have been shown, linking infra-individual effects to effects at the community level (Table 10.2 after Weis et al. 2011). In most cases, behavioral disturbances included a lower level of general activity, impaired prey capture, and increased vulnerability to predation. Altered thyroid status and altered levels of neurotransmitters may be responsible for these impairments in fish. As a consequence of feeding impairments, the diet of contaminated fish species and blue crabs was impoverished compared to their noncontaminated conspecifics. This poor diet can account in part for the reduced size observed in contaminated fish. In agreement with energy budget models (Maltby et al. 2001 and Chapter 12), population density of killifish *Fundulus heteroclitus* was reduced in the impacted estuary. Surprisingly, in crustaceans, size was greater for individuals from contaminated sites. In the cases of the grass shrimp *Palaemonetes pugio* and the blue crab

TABLE 10.2

Contamination-Induced Behavioral Changes and Ecological Consequences for Estuarine Fish and Crustaceans

Studied Effect	**Killifish (*Fundulus heteroclitus*)**	**Bluefish (*Pomatomus saltatrix*)**	**Grass Shrimp (*Palaemonetes pugio*)**	**Fiddler Crab (*Uca pugnax*)**	**Blue Crab (*Callinectes sapidus*)**
Feeding behavior	⇘	⇘	⇘	⇘	⇘
Diet	Detritus	Empty guts	NS	NS	Detritus
Predator avoidance	⇘	NS	=	⇗	⇗
General activity	⇘	⇘	⇘	⇘	NS
Size	⇘	⇘	⇗	⇗	⇗
Population density	⇘	NS	⇗	⇘	NS
Reproduction	⇗	NS	⇗	⇗	NS
Thyroid	√	√			
Neurotransmitters	√	√			

Source: Weis, J.S. et al., *Biosciences*, 61, 375–385, 2011. With permission.
Note: =, no effect observed; ⇗ and ⇘, positive and negative effects; √ (checkmarks), effects were observed; NS, not studied.

Callinectes sapidus, this has been attributed to a top–down effect, involving respectively the decreased predation by a depleted population of killifish in polluted sites, or a fishing ban on the crab, an edible species with chemical residues in their tissues reaching or exceeding regulation limits for health safety. In the case of the fiddler crab *U. pugnax*, the larger size at the contaminated site may be explained by less competition in a population with a lower density (as a result of impairment of larvae by contaminants and decreased recruitment) and decreased predation by the blue crab. The disturbance of feeding behavior and the consequences on predator–prey interactions in all five species are well established, but Weis et al. (2011) also highlighted a number of key questions that are important for the ecological effects of chemicals on communities: the biomagnification of some contaminants leading to higher exposure in predators than in prey; certain biological traits favoring tolerance (e.g., elimination of metal by molting in crustaceans) and more generally, the differential sensitivity and ability of different taxa to deal with and adapt to different types of contaminants.

Several studies have shown a link between exposure of fish to sublethal doses of various contaminants and reproductive behavior, but few authors have seriously examined potential consequences for reproduction success and population effects. In their review on fish, Jones and Reynolds (1997) mentioned two studies that directly measured a negative effect on reproductive success, whereas a possible decrease in the reproductive success of *Tilapia rendalli* was inferred in areas treated with pesticides in the Okavango Delta (Botswana). The behavior of fish larvae can also have a crucial role in reproductive success. DDT and PCBs transferred to the progeny through egg yolk are able to alter responses of larvae to warning signals. Even if these symptoms become less pronounced during development (exogenous food replacing the yolk as the energy source, biological dilution of contaminants in growing tissues), they are a clear handicap at a key phase of recruitment success. This point seems particularly critical since organochlorine concentrations in gonads and eggs of experimental fish were of the same order of magnitude as concentrations determined in fish contaminated in the wild (Faulk et al. 1999; McCarthy et al. 2003).

10.5 Conclusions

Older papers dealing with behavioral effects of chemicals were mainly based on acute exposures responsible for short-term effects (Eisler 2000; Grue et al. in Dell'Omo 2002). Behavioral responses are much more sensitive to chemical stress than lethal responses. On the other hand, they are not more sensitive than biochemical or physiological biomarkers classically proposed for environmental biomonitoring. This seems logical since a number of these biomarkers have a causal relationship with behavioral alterations. From the present review, it appears that these alterations may be induced at experimental concentrations that are realistic in comparison to those encountered in contaminated environments, including chronic exposures in the wild. In addition, good consistency has been shown between experimental and field studies, suggesting that extrapolation from the laboratory to the wild is not inappropriate, even if a cautious approach remains necessary. It is important to consider that, under experimental conditions, organisms are subjected to an unavoidable exposure, whereas in the natural environment, they may be able to avoid contaminated water masses, sediment, or food. It would be unwise, however, to consider that all populations will be able to flee polluted environments. It is also indispensable to take into account indirect effects that can occur in a polluted environment such as prey–predator interactions and competitive interactions between species (Lefcort et al. 1999 literature cited therein; Fleeger et al. 2003). The influence of abiotic factors (light, temperature, season) is also important (Triebskorn et al. 1997; Scherer and McNicol 1998). Moreover, natural environments are rarely exposed to only one kind of contaminant, and taking into account interactions between chemicals in mixtures is one of the major challenges for the future of ecotoxicology.

The use of behavioral biomarkers in the assessment of environmental risk is also limited because discrepancies exist between the effects of different contaminants at different levels of biological organization in different species. In many cases, discrepancies may result from acute versus chronic conditions of exposure. There is agreement in the literature to consider that organisms can adapt to chronic exposure since they have time to put into place efficient mechanisms of detoxification (Chapter 3), whereas in acute tests these mechanisms are immediately overwhelmed. This explanation is not always satisfactory, and behavioral ecotoxicology—as with other fields of ecotoxicology—suffers from a lack of fundamental research, most of the known mechanistic bases being derived from those elucidated in birds and mammals.

More precisely, the extrapolation of observations from one species in the laboratory to multiple species in the wild requires an understanding of the factors governing inherent sensitivity (e.g., Scarfe et al. 1982; Keizer et al. 1995). Because it is impossible to obtain a complete set of data for most species, behavioral ecotoxicologists must focus on a certain number of sentinel species, chosen according to criteria that are at least partly relevant in other fields of ecotoxicology (Chapter 7).

Despite these limitations, we can agree with Roast et al. (2000) in their claim that behavioral studies provide a better understanding of chemical impacts than lethal effects, particularly if behavior is altered at realistic doses. Behavior is an individual response that has clear links with biochemical and physiological responses (providing early warning biomarkers) on the one hand, and population effects resulting from reduced longevity and reduced reproductive success (ecologically relevant biomarkers) on the other hand. Many studies dealing with relationships between infra-individual impairments and behavioral changes are available, but few investigations have been devoted to the consequences of behavioral disturbances at supra-individual levels. However, Weis et al. (2001) have published a striking example showing the ecological relevance of behavioral biomarkers, and such comprehensive research, using key species for the structure and functioning of ecosystems as models, would permit a significant improvement of environmental risk assessment.

Behavioral biomarkers are taken into account in a number of environmental regulations, particularly in the United States (reviewed by Little in Dell'Omo 2002), but they are mainly ignored in European regulations. In order to convince the authorities that behavioral biomarkers should be integrated into environmental assessment, comprehensive field studies are needed to confirm what is intuitively evident to environmental toxicologists—that behavioral changes can induce effects at population and community levels (Grue et al. in Dell'Omo 2002). The research by Weis et al. (2011), summarized in Table 10.2, is an excellent example of what may be done, and this chapter also provides a number of examples supporting this point of view.

Acquisition and treatment of data including automated image capture or analysis of records of various animals' movements provide the environmentally justified implementation of behavioral tests *in situ* and in the laboratory. Bivalve closure responses upon detection of contaminants have been proposed as suitable aquatic pollution biosensors. Noninvasive methods and low-stress conditions are available using the Musselmonitor® and the *Dreissena* monitor® (Baldwin and Kramer 1994), and there are other techniques convenient for endobenthic species (Tran et al. 2003, 2004, 2007; Fournier et al. 2004). The Multispecies Freshwater Biomonitor, a commercial system designed by Gerhardt et al. (2002), has been successfully applied to study behavioral responses of diverse organisms to toxicants. More recently, Nørum et al. (2010) have proposed a fast and cost-effective video tracking methodology that may be applied for studying the potential effects of pesticides and other stressors on the locomotory behavior of freshwater invertebrates. For fish, different systems allow the description of simple (commercial system Ethovision, Noldus, Wageningen, The Netherlands) or much more complex behaviors such as sexual behavior with implications for the spatial orientation of individuals, their shape, and their interactions with their partners (Baatrup and Junge 2001; Denoël et al. 2012b). Moreover, behavioral analyses are nondestructive, allowing follow-up biochemical, physiological, or genetic analyses with the same specimens in multibiomarker approaches (Chapter 2).

References

Amiard-Triquet, C. 2009. Behavioral disturbances: The missing link between sub-organismal and supra-organismal responses to stress? Prospects based on aquatic research. *Hum. Ecol. Risk Assess. Behav. Ecotoxicol.* (Special issue) 15:87–110.

Amiard-Triquet, C., P.S. Rainbow, and M. Roméo, eds. 2011. *Tolerance to Environmental Contaminants.* Boca Raton, FL: CRC Press.

Atchinon, G.J., M.G. Henry, and M.B. Sandheinrich. 1987. Effects of metals on fish behavior: A review. *Environ. Biol. Fish.* 18:11–25.

Baatrup, E., and M. Junge. 2001. Antiandrogenic pesticides disrupt sexual characteristics in the adult male guppy (*Poecilia reticulata*). *Environ. Health Perspect.* 109:1063–70.

Baldwin, D.H., C.P. Tatara, and N.L. Scholz. 2011. Copper-induced olfactory toxicity in salmon and steelhead: Extrapolation across species and rearing environments. *Aquat. Toxicol.* 101:295–7.

Baldwin, I.G., and Kramer. 1994. Biological early warning systems. In: *Biomonitoring of Coastal Waters and Estuaries*, ed. K.J.M. Kramer, 1–23. Boca Raton, FL: CRC Press.

Barata, C. et al. 2007. Combined use of biomarkers and *in situ* bioassays in *Daphnia magna* to monitor environmental hazards of pesticides in the field. *Environ. Toxicol. Chem.* 26:370–9.

Barros Amorim, M.J., and J.J. Scott-Fordsmand. 2012. Toxicity of copper nanoparticles and $CuCl_2$ salt to *Euchytraeus albidus* worms: Survival, reproduction and avoidance responses. *Environ. Pollut.* 164:164–8.

Beauvais, S.L. et al. 2000. Physiological measures of neurotoxicity of diazinon and malathion to larval rainbow trout (*Oncorhynchus mykiss*) and their correlation with behavioral measures. *Environ. Toxicol. Chem.* 19:1875–80.

Beauvais, S.L. et al. 2001. Cholinergic and behavioral neurotoxicity of carbaryl and cadmium to larval rainbow trout (*Oncorhynchus mykiss*). *Ecotoxicol. Environ. Saf.* 49:84–90.

Bell, A.M. 2001. Effects of an endocrine disrupter on courtship and aggressive behaviour of male three-spined stickleback, *Gasterosteus aculeatus*. *Anim. Behav.* 62:775–80.

Bilberg, K. et al. 2011. Silver nanoparticles and silver nitrate cause respiratory stress in Eurasian perch (*Perca fluviatilis*). *Aquat. Toxicol.* 104:145–52.

Bjerselius, R. et al. 2001. Male goldfish reproductive behavior and physiology are severely affected by exogenous exposure to 17β-estradiol. *Aquat. Toxicol.* 53:139–52.

Blaxter, J.H.S., and C.C.T. Hallers-Tjabbes. 1992. The effect of pollutant on sensory systems and behaviour of aquatic animals. *Neth. J. Aquat. Ecol.* 26:43–58.

Boldina-Cosqueric, I. et al. 2010. Biochemical, physiological and behavioral markers in the endobenthic bivalve *Scrobicularia plana* as tools for the assessment of estuarine sediment quality. *Ecotoxicol. Environ. Saf.* 73:1733–41.

Bonnard, M., M. Roméo, and C. Amiard-Triquet. 2009. Effects of copper on the burrowing behaviour of estuarine and coastal invertebrates, the polychaete *Nereis diversicolor* and the bivalve *Scrobicularia plana*. *Health Environ. Risk Assess.* 15:11–26.

Brewer, S.K. et al. 2001. Behavioral dysfunctions correlate to altered physiology in rainbow trout (*Oncorhynchus mykiss*) exposed to cholinesterase-inhibiting chemicals. *Arch. Environ. Contam. Toxicol.* 40:70–6.

Bryan, M.D., G.J. Atchison, and M.B. Sandheinrich. 1995. Effects of cadmium on the foraging behavior and growth of juvenile bluegill, *Lepomis macrochirus*. *Can. J. Fish. Aquat. Sci.* 52:1630–8.

Buffet, P.E. et al. 2011. Behavioral and biochemical responses of two marine invertebrates *Scrobicularia plana* and *Hediste diversicolor* to copper oxide nanoparticles. *Chemosphere* 84:166–74.

Byrne, P.A., and J. O'Halloran. 2001. The role of bivalve molluscs as tools in estuarine sediment toxicity testing: A review. *Hydrobiologia* 465:209–17.

Cailleaud, K. et al. 2011. Changes in the swimming behavior of *Eurytemora affinis* (Copepoda, Calanoida) in response to a sub-lethal exposure to nonylphenols. *Aquat. Toxicol.* 102:228–31.

Carreau, N.D., and G.G. Pyle. 2005. Effect of copper exposure during embryonic development on chemosensory function of juvenile fathead minnows (*Pimephales promelas*). *Ecotoxicol. Environ. Saf.* 61:1–6.

Chen, T.H., C.Y. Lin, and M.C. Tseng. 2011. Behavioral effects of titanium dioxide nanoparticles on larval zebrafish (*Danio rerio*). *Mar. Pollut. Bull.* 63:303–8.

Chen, X. et al. 2012. BDE-47 disrupts axonal growth and motor behavior in developing zebrafish. *Aquat. Toxicol.* 120–121:35–44.

Clotfelter, E.D., and A.C. Rodriguez. 2006. Behavioral changes in fish exposed to phytoestrogens. *Environ. Pollut.* 144:833–9.

Cooper, N.L., and J.R. Bidwell. 2006. Cholinesterase inhibition and impacts on behavior of the Asian clam, *Corbicula fluminea*, after exposure to an organophosphate insecticide. *Aquat. Toxicol.* 76:258–67.

Coulaud, R. et al. 2011. *In situ* feeding assay with *Gammarus fossarum* (Crustacea): Modelling the influence of confounding factors to improve water quality biomonitoring. *Water Res.* 45:6417–29.

DeForest, D.K. et al. 2011. Are ambient water quality criteria for copper protective of olfactory impairment in fish? *Integr. Environ. Assess. Manag.* 7:145–6.

De Lange, H.J. et al. 2006a. Behavioural responses of *Gammarus pulex* (Crustacea, Amphipoda) to low concentrations of pharmaceuticals. *Aquat. Toxicol.* 78:209–216.

De Lange, H.J., V. Sperber, and E.T.H.M. Peeters. 2006b. Avoidance of polycyclic aromatic hydrocarbon-contaminated sediments by the freshwater invertebrates *Gammarus pulex* and *Asellus aquaticus*. *Environ. Toxicol. Chem.* 25:452–7.

Dell'Omo, G., ed. 2002. *Behavioral Ecotoxicology*. Chichester, West Sussex: John Wiley & Sons.

Denoël, M. et al. 2012a. Using sets of behavioral biomarkers to assess short-term effects of pesticide: A study case with endosulfan on frog tadpoles. *Ecotox*. 21:1240–50.

Denoël, M. et al. 2012b. Effects of a sublethal pesticide exposure on locomotor behavior: A video-tracking analysis in larval amphibians. *Chemosphere* (in press).

Eisler, R. 2000. *Handbook of Chemical Risk Assessment—Health Hazards to Humans, Plants and Animals*. Vols. 1–3. Boca Raton, FL: Lewis Publishers.

Eisler, R. 2007. *Eisler's Encyclopedia of Environmentally Hazardous Priority Chemicals*. Amsterdam: Elsevier.

Ellenberger, S.A., P.C. Baumann, and T.W. May. 1994. Evaluation of effects caused by high copper concentrations in Torch Lake, Michigan, on reproduction of yellow perch. *Jour. Great Lakes Res.* 20:531–6.

Eriksson Wiklund, A.K., T. Borjesson, and S.J. Wiklund. 2006. Avoidance response of sediment living amphipods to zinc pyrithione as a measure of sediment toxicity. *Mar. Pollut. Bull.* 52:96–9.

Farr, A.J., C.C. Chabot, and D.H. Taylor. 1995. Behavioral avoidance of fluoranthene by fathead minnows (*Pimephales promelas*). *Neurotoxicol. Teratol.* 17:265–71.

Faucher, K. et al. 2006. Impact of acute cadmium exposure on the trunk lateral line neuromasts and consequences on the "C-start" response behaviour of the sea bass (*Dicentrarchus labrax* L.; Teleostei, Moronidae). *Aquat. Toxicol.* 76:278–94.

Faulk, C.K., L.A. Fuiman, and P. Thomas. 1999. Parental exposure to *ortho*, *para*-dichlorodiphenyltrichloroethane impairs survival skills of Atlantic croaker (*Micropogonias undulatus*) larvae. *Environ. Toxicol. Chem.* 18:254–62.

Fingerman, S., and L. Russell. 1980. Effects of the polychlorinated biphenyl aroclor 1242 on locomotor activity and on the neurotransmitters dopamine and norepinephrine in the brain of the gulf killifish, *Fundulus grandis*. *Bull. Environ. Contam. Toxicol.* 25:682–7.

Fleeger, J.W., K.R. Carman, and R.M. Nisbet. 2003. Indirect effects of contaminants in aquatic ecosystems. *Sci. Total Environ.* 317:207–33.

Floyd, E.Y., J. Geist, and I. Wermer. 2008. Acute, sublethal exposure to a pyrethroid insecticide alters behavior, growth, and predation risk in larvae of the fathead minnow (*Pimephales promelas*). *Environ. Toxicol. Chem.* 27:1780–7.

Forbes, V.E., A. Palmqvist, and L. Bach. 2006. The use and misuse of biomarkers in ecotoxicology. *Environ. Toxicol. Chem.* 25:272–80.

Fossi Tankoua, O. et al. 2012. Intersite variations of a battery of biomarkers at different levels of biological organisation in the estuarine endobenthic worm *Nereis diversicolor* (Polychaeta, Nereididae). *Aquat. Toxicol.* 114–115:96–103.

Fournier, E. et al. 2004. Valve closure response to uranium exposure for a freshwater bivalve (*Corbicula fluminea*): Quantification of the influence of pH. *Environ. Toxicol. Chem.* 23:1108–14.

Garcia-de la Parra, L.M. et al. 2006. Effects of methamidophos on acetylcholinesterase activity, behavior, and feeding rate of the white shrimp (*Litopenaeus vannamei*). *Ecotoxicol. Environ. Saf.* 65:372–80.

Garcia-Reyero, N. et al. 2011. Behavioral and genomic impacts of a wastewater effluent on the fathead minnow. *Aquat. Toxicol.* 101:38–48.

Gaworecki, K.M., and S.J. Klaine. 2008. Behavioral and biochemical responses of hybrid striped bass during and after fluoxetine exposure. *Aquat. Toxicol.* 88:207–13.

Gaworecki, K.M. et al. 2009. Biochemical and behavioral effects of diazinon exposure in hybrid striped bass. *Environ. Toxicol. Chem.* 28:105–12.

Gerhardt, A. 1996. Behavioural early warning responses to polluted water. Performance of *Gammarus pulex* L. (Crustacea) and *Hydropsyche angustipennis* (Curtis) (Insecta) to a complex industrial effluent. *Environ. Sci. Pollut. Res.* 3:63–70.

Gerhardt, A., S. Schmidt, and S. Höss. 2002. Measurement of movement patterns of *Caenorhabditis elegans* (Nematoda) with the Multispecies Freshwater Biomonitor (MFB)—A potential new method to study a behavioral toxicity parameter of nematodes in sediments. *Environ. Pollut.* 120:513–6.

Glennemeier, K.A., and R.J. Denver. 2001. Sublethal effects of chronic exposure to an organochlorine compound on Northern Leopard Frog (*Rana pipiens*) tadpoles. *Environ. Toxicol.* 16:287–97.

Gopal, K. et al. 1985. Neurobehavioural changes in fresh water fish *Channa punctatus* exposed to endosulfan. *J. Adv. Zool.* 6:74–80.

Green, W.W. et al. 2010. Copper binding dynamics and olfactory impairment in fathead minnows (*Pimephales promelas*). *Environ. Sci. Technol.* 44:1431–7.

Grippo, M.A., and A.G. Heath. 2003. The effect of mercury on the feeding behavior of fathead minnows (*Pimephales promelas*). *Ecotoxicol. Environ. Saf.* 55:187–98.

Hanazato, T. 2001. Pesticide effects on freshwater zooplankton: An ecological perspective. *Environ. Pollut.* 112:1–10.

Hansen, J.A. et al. 1999a. Behavioral avoidance: Possible mechanism for explaining abundance and distribution of trout species in a metal-impacted river. *Environ. Toxicol. Chem.* 18:313–7.

Hansen, J.A. et al. 1999b. Chinook salmon (*Oncorhynchus tshawytscha*) and rainbow trout (*Oncorhynchus mykiss*) exposed to copper: Neurophysiological and histological effects on the olfactory system. *Environ. Toxicol. Chem.* 18:1979–91.

Hellou, J. 2011. Behavioral ecotoxicology, an "early warning" signal to assess environmental quality. *Environ. Sci. Pollut. Res.* 18:1–11.

Hopkins, W.A., C.T. Winne, and S.E. DuRant. 2005. Differential swimming performance of two natricine snakes exposed to a cholinesterase-inhibiting pesticide. *Environ. Pollut.* 133:531–40.

Imbert, H. et al. 2008. Relationships between locomotor behavior, morphometric characters and thyroid hormone levels give evidence of stage-dependent mechanisms in European eel upstream migration. *Horm. Behav.* 53:69–81.

Jaensson, A. et al. 2007. Effects of a pyrethroid pesticide on endocrine responses to female odours and reproductive behavior in male parr of brown trout (*Salmo trutta* L.). *Aquat. Toxicol.* 81:1–9.

Janssens, L., and R. Stoks. 2012. How does a pesticide pulse increase vulnerability to predation? Combined effects on behavioral antipredator traits and escape swimming. *Aquat. Toxicol.* 110–111:91–8.

Johnson, A., E. Carew, and K.A. Sloman. 2007. The effects of copper on the morphological and functional development of zebrafish embryos. *Aquat. Toxicol.* 84:431–8.

Jones, J.C., and J.D. Reynolds. 1997. Effects of pollution on reproductive behaviour of fishes. *Rev. Fish. Biol. Fish.* 7:463–91.

Kalman, J. et al. 2009. Assessment of the health status of populations of the ragworm *Nereis diversicolor* using biomarkers at different levels of biological organisation. *Mar. Ecol. Pogr. Ser.* 393: 55–67.

Kasumyan, A.O. 2001. Effects of chemicals pollutants on foraging behavior and sensitivity of fish to food stimuli. *J. Ichthyol.* 41:76–87.

Keizer, J. et al. 1995. Enzymological differences of AChE and diazinon hepatic metabolism: Correlation of *in vitro* data with the selective toxicity of diazinon to fish species. *Sci. Total Environ.* 171:213–20.

Kennedy, C.J., and A.P. Farrell. 2006. Effects of exposure to the water-soluble fraction of crude oil on the swimming performance and the metabolic and ionic recovery postexercise in Pacific herring (*Clupea pallasi*). *Environ. Toxicol. Chem.* 25:2715–24.

Kienle, C., and A., Gerhardt. 2008. Behavior of *Corophium volutator* (crustacean, Amphipoda) exposed to the water-accommodated fraction of oil in water and sediment. *Environ. Toxicol. Chem.* 27:599–604.

Kislalioglu, M., E. Scherer, and R.E. McNicol. 1996. Effects of cadmium on foraging behaviour of lake charr, *Salvelinus namaycush*. *Environ. Biol. Fish.* 46:75–82.

Knapp, R., and B.D. Neff. 2006. Relationships among steroid hormones, body size, paternity, and parental care behaviors in male bluegill sunfish. *Front Neuroendocrinol.* 27:25.

Krång, A.S., and G. Rosenqvist. 2006. Effects of manganese on chemically induced food search behaviour of the Norway lobster, *Nephrops norvegicus* (L.). *Aquat. Toxicol.* 78:284–91.

Kravitz, M.J. et al. 1999. Avoidance response of the estuarine amphipod *Eohaustorius estuarius* to polycyclic aromatic hydrocarbon-contaminated, field-collected sediments. *Environ. Toxicol. Chem.* 18:1232–5.

Krebs, C.T., and K.A. Burns. 1977. Long-term effects of an oil spill on populations of the salt-marsh crab *Uca pugnax*. *Science* 197:484–7.

Kusch, R.C., P.H. Krone, and D.P. Chivers. 2008. Chronic exposure to low concentrations of waterborne cadmium during embryonic and larval development results in the long-term hindrance of antipredator behavior in zebrafish. *Environ. Toxicol. Chem.* 27:705–10.

Kuz'mina, V.V. 2011. The influence of zinc and copper on the latency period for feeding and the food uptake in common carp, *Cyprinus carpio* L. *Aquat. Toxicol.* 102:73–8.

Lawrence, A.J., and C. Poulter. 1998. Development of a sub-lethal pollution bioassay using the estuarine amphipod *Gammarus duebeni*. *Water Res.* 32:569–78.

Lefcort, H. et al. 1999. Ramifications of predator avoidance: Predator and heavy-metal–mediated competition between tadpoles and snails. *Ecol. Appl.* 9:1477–89.

Lefcort, H. et al. 2004. Aquatic snails from mining sites have evolved to detect and avoid heavy metals. *Arch. Environ. Contam. Toxicol.* 46:478–84.

Lerner, D.T., B.T. Bjornsson, and S.D. McCormick. 2007. Effects of aqueous exposure to polychlorinated biphenyls (Aroclor 1254) on physiology and behavior of smolt development of Atlantic salmon. *Aquat. Toxicol.* 81:329–36.

Louie, A.Y., and T.J. Meade. 1999. Metal complexes as enzyme inhibitors. *Chem. Rev.* 99:2711–34.

Maltby, L. et al. 2000. Using single-species toxicity tests, community-level responses, and toxicity identification evaluations to investigate effluent impacts. *Environ. Toxicol. Chem.* 19:151–7.

Maltby, L. et al. 2001. Linking individual-level responses and population-level consequences. In: *Ecological Variability: Separating Natural from Anthropogenic Causes of Ecosystem Impairment*, ed. D.J. Baird and G.A.J. Burton, 27–82. Society of Environmental Toxicology and Chemistry (SETAC): Pensacola, FL.

Maltby, L. et al. 2002. Evaluation of the *Gammarus pulex in situ* feeding assay as a biomonitor of water quality: Robustness, responsiveness, and relevance. *Environ. Toxicol. Chem.* 21:361–8.

Matozzo, V., L. Ballarin, and M.G. Marin. 2004. Exposure of the clam *Tapes philippinarum* to 4-nonylphenol: Changes in anti-oxidant enzyme activities and re-burrowing capability. *Mar. Pollut. Bull.* 48:563–71.

McCarthy, I.D., L.A. Fuiman, and M.C. Alvarez. 2003. Aroclor 1254 affects growth and survival skills of Atlantic croaker *Micropogonias undulatus* larvae. *Mar. Ecol. Prog. Ser.* 252:295–301.

McDonald, T.O. 1979. Comparison of the effects of parathion and DDT on the concentrations of serotonin, norepinephrine and dopamine in the brain and retine/pigment epithelium/chloroid of the goldfish. PhD thesis, Tulane University.

McGeer, J.C. et al. 2000. Effects of chronic sublethal exposure to waterborne Cu, Cd or Zn in rainbow trout: 1. Iono-regulatory disturbance and metabolic costs. *Aquat. Toxicol.* 50:231–43.

McIntyre, J.K. et al. 2008. Chemosensory deprivation in juvenile Coho salmon exposed to dissolved copper under varying water chemistry conditions. *Environ. Sci. Technol.* 42:1352–8.

McWilliam, R.A., and D.J. Baird. 2002a. Application of postexposure feeding depression bioassays with *Daphnia magna* for assessment of toxic effluents in rivers. *Environ. Toxicol. Chem.* 21:1462–8.

McWilliam, R.A., and D.J. Baird. 2002b. Postexposure feeding depression: A new toxicity endpoint for use in laboratory studies with *Daphnia magna*. *Environ. Toxicol. Chem.* 21:1198–205.

Mesa, M.G. et al. 1994. Are all prey created equal? A review and synthesis of differential predation on prey in substandard condition. *J. Fish. Biol.* 45:81–96.

Meyer, J.S., and W.J. Adams. 2010. Relationship between biotic ligand model-based water quality criteria and avoidance and olfactory responses to copper by fish. *Environ. Toxicol. Chem.* 29:2096–103.

Møhlenberg, F., and T. Kiørboe. 1983. Burrowing and avoidance behaviour in marine organisms exposed to pesticide-contaminated sediment. *Mar. Pollut. Bull.* 14:57–60.

Moore, A. et al. 2007. The impact of a pesticide on migratory activity and olfactory function in Atlantic salmon (*Salmo salar* L.) smolts. *Aquaculture* 273:350–9.

Moreira-Santos, M. et al. 2005. Short-term sublethal (sediment and aquatic roots of floating macrophytes) assays with a tropical chironomid based on postexposure feeding and biomarkers. *Environ. Toxicol. Chem.* 24:2234–42.

Moreira, S.M. et al. 2005. A short-term sublethal *in situ* toxicity assay with *Hediste diversicolor* (polychaeta) for estuarine sediments based on postexposure feeding. *Environ. Toxicol. Chem.* 24:2010–8.

Moreira, S.M. et al. 2006a. Effects of estuarine sediment contamination on feeding and on key physiological functions of the polychaete *Hediste diversicolor*: Laboratory and *in situ* assays. *Aquat. Toxicol.* 78:186–201.

Moreira, S.M. et al. 2006b. An *in situ* postexposure feeding assay with *Carcinus maenas* for estuarine sediment-overlying water toxicity evaluations. *Environ. Pollut.* 139:318–29.

Mouneyrac, C., H. Perrein-Ettajani, and C. Amiard-Triquet. 2010. Influence of anthropogenic stress on fitness and behavior of a key-species of estuarine ecosystems, the ragworm *Nereis diversicolor*. *Environ. Pollut.* 158:121–8.

Nakayama, K. et al. 2005. Effects of polychlorinated biphenyls on the schooling behavior of Japanese medaka (*Oryzias latipes*). *Environ. Toxicol. Chem.* 24:2588–93.

Nørum, U. et al. 2010. Behavioural changes in three species of freshwater macroinvertebrates exposed to the pyrethroid lamda-cyhalothrin: Laboratory and stream microcosm studies. *Aquat. Toxicol.* 98:328–35.

O'Gara, B.A. et al. 2004. Copper-induced changes in locomotor behaviors and neuronal physiology of the freshwater oligochaete, *Lumbriculus variegatus*. *Aquat. Toxicol.* 69:51–66.

Olsén, K.H., and L.B. Höglund. 1985. Reduction by a surfactant of olfactory mediated attraction between juveniles of arctic charr, *Salvelinus alpinus* (L.). *Aquat. Toxicol.* 6:57–69.

Painter, M.M. et al. 2009. Antidepressants at environmentally relevant concentrations affect predator avoidance behavior of larval fathead minnows (*Pimephales promelas*). *Environ. Toxicol. Chem.* 28:2677–84.

Pang, C.F. et al. 2012. Effects of sediment-associated copper to the deposit-feeding snail, *Potamopyrgus antipodarum*: A comparison of Cu added in aqueous form or as nano- and micro-CuO particles. *Aquat. Toxicol.* 106–7:114–22.

Pearson, W.H. et al. 1981. Effects of oiled sediment on predation on the Littleneck Clam, *Protothaca staminea*, by the Dungeness Crab, *Cancer magister*. *Estuar. Coast. Shelf Sci.* 13:445–54.

Perez, M.H., and W.G. Wallace. 2004. Differences in prey capture in grass shrimp, *Palaemonetes pugio*, collected along an environment impact gradient. *Arch. Environ. Contam. Toxicol.* 46:81–9.

Perreault, H.A.N., K. Semsar, and J. Godwin. 2003. Fluoxetine treatment decreases territorial aggression in a coral reef fish. *Physiol. Behav.* 79:719–24.

Petrauskiene, L. 2003. Water and sediment toxicity assessment by use of behavioural responses of medicinal leeches. *Environ. Int.* 28:729–36.

Phelps, H.L. et al. 1983. Clam burrowing behavior: Inhibition by copper-enriched sediment. *Mar. Pollut. Bull.* 14:452–5.

Pynnonen, K. 1996. Heavy metal-induced changes in the feeding and burrowing behaviour of a Baltic isopod, *Saduria* (*Mesidotea*) *entomon* L. *Mar. Environ. Res.* 41:145–56.

Reynaldi, S., M. Meiser, and M. Liess. 2011. Effects of the pyrethroid fenvalerate on the alarm response and on the vulnerability of the mosquito larva *Culex pipiens* to the predator *Notonecta glauca*. *Aquat. Toxicol.* 104:56–60.

Riddell, D.J., J.M. Culp, and D.J. Baird. 2005. Behavioral responses to sublethal cadmium exposure within an experimental aquatic food web. *Environ. Toxicol. Chem.* 24:431–41.

Roast, S.D., J. Widdows, and M.B. Jones. 2000. Disruption of swimming in the hyperbenthic mysid *Neomysis integer* (Peracarida: Mysidacea) by the organophosphate pesticide chlorpyrifos. *Aquat. Toxicol.* 47:227–41.

Roast, S.D., J. Widdows, and M.B. Jones. 2001. Impairment of mysid (*Neomysis integer*) swimming ability: An environmentally realistic assessment of the impact of cadmium exposure. *Aquat. Toxicol.* 52:217–27.

Roper, D.S., and C.W. Hickey. 1994. Behavioural responses of the marine bivalve *Macomona liliana* exposed to copper- and chlordane-dosed sediments. *Mar. Biol.* 118:673–80.
Roper, D.S. et al. 1995. Burial, crawling and drifting behaviour of the bivalve *Macomona liliana* in response to common sediment contaminants. *Mar. Pollut. Bull.* 31:471–8.
Rosen, G., and K. Miller. 2011. A postexposure feeding assay using the marine polychaete *Neanthes arenaceodentata* suitable for laboratory and *in situ* exposures. *Environ. Toxicol. Chem.* 30:730–7.
Sandahl, J.F. et al. 2005. Comparative thresholds for acetylcholinesterase inhibition and behavioral impairment in Coho Salmon exposed to chlorpyrifos. *Environ. Toxicol. Chem.* 24:136–45.
Sandahl, J.F. et al. 2007. A sensory system at the interface between urban stormwater runoff and salmon survival. *Environ. Sci. Technol.* 41:2998–3004.
Sandheinrich, M.B., and K.M. Miller. 2006. Effects of dietary methylmercury on reproductive behavior of fathead minnows (*Pimephales promelas*). *Environ. Toxicol. Chem.* 25:3053–7.
Scarfe, A.D. et al. 1982. Locomotor behavior of four marine teleosts in response to sublethal copper exposure. *Aquat. Toxicol.* 2:335–53.
Scherer, E., and R.E. McNicol. 1998. Preference–avoidance responses of lake whitefish (*Coregonus clupeaformis*) to competing gradients of light and copper, lead, and zinc. *Water Res.* 32:924–9.
Schmidt, K., S. Pfugmacher, and G.B.O. Staaks. 2006. The influence of tributyltin chloride and polychlorinated biphenyls on swimming behavior, body growth, reproduction, and activity of biotransformation enzymes in *Daphnia magna*. *J. Freshw. Ecol.* 21:109–20.
Scott, G.R., and K.A. Sloman. 2004. The effects of environmental pollutants on complex fish behaviour: Integrating behavioural and physiological indicators of toxicity. *Aquat. Toxicol.* 68:369–92.
Shin, P.K.S., A.W.M. Ng, and R.Y.H. Cheung. 2002. Burrowing responses of the short-neck clam *Ruditapes philippinarum* to sediment contaminants. *Mar. Pollut. Bull.* 45:133–9.
Sisneros, J.A. et al. 2004. Seasonal variation of steroid hormone levels in an intertidal-nesting fish, the vocal plainfin midshipman. *Gen. Comp. Endocrinol.* 136:101–16.
Sloman, K.A. et al. 2005. Socially-mediated differences in brain monoamines in rainbow trout: Effects of trace metal contaminants. *Aquat. Toxicol.* 71:237–47.
Tessier, L. et al. 2000. Anomalies on capture nets of *Hydropsyche slossonae* larvae (Trichoptera; Hydropsychidae), a potential indicator of chronic toxicity of malathion (organophosphate insecticide). *Aquat. Toxicol.* 50:125–39.
Thomas, J.K., and D.M. Janz. 2011. Dietary selenomethionine exposure in adult zebrafish alters swimming performance, energetic and the physiological stress response. *Aquat. Toxicol.* 102:79–86.
Thomas, P., H.W. Wofford, and J.M. Neff. 1981. Biochemical stress responses of striped mullet (*Mugil cephalus* L.) to fluorene analogs. *Aquat. Toxicol.* 1:329–42.
Tierney, K.B. et al. 2006. Changes in juvenile Coho salmon electro-olfactogram during and after short-term exposure to current-use pesticides. *Environ. Toxicol. Chem.* 25:2809–17.
Tierney, K.B. et al. 2011. Evidence for behavioral preference toward environmental concentrations of urban-use herbicides in a model adult fish. *Environ. Toxicol. Chem.* 30:2046–54.
Toft, G., E. Baatrup, and J.L.J. Guillette. 2004. Altered social behavior and sexual characteristics in mosquitofish (*Gambusia holbrooki*) living downstream of a paper mill. *Aquat. Toxicol.* 70:213–22.
Tran, D. et al. 2003. Estimation of potential and limits of bivalve closure response to detect contaminants: Application to cadmium. *Environ. Toxicol. Chem.* 22:914–20.
Tran, D. et al. 2004. Copper detection in the Asiatic clam *Corbicula fluminea*: Optimum valve closure response. *Aquat. Toxicol.* 66:333–43.
Tran, D. et al. 2007. Inorganic mercury detection by valve closure response in the freshwater clam *Corbicula fluminea*: Integration of time and water metal concentration changes. *Environ. Toxicol. Chem.* 26:1545–51.
Triebskorn, R. et al. 1997. Induction of heat shock proteins, changes in liver ultrastructure, and alterations of fish behavior: Are these biomarkers related and are they useful to reflect the state of pollution in the field? *J. Aquat. Ecosyst. Stress Recov.* 6:57–73.

Truong, L. et al. 2012. Persistent adult zebrafish behavioral deficits results from acute embryonic exposure to gold nanoparticles. *Comp. Biochem. Physiol.* 155C:269–74.

Van Dolah, R.F. et al. 1997. Comparison of azinphosmethyl toxicity to juvenile red drum (*Sciaenops ocellatus*) and the mummichog (*Fundulus heteroclitus*). *Environ. Toxicol. Chem.* 16:1488–93.

Vieira, L.R. et al. 2009. Acute effects of copper and mercury on the estuarine fish *Pomatoschistus microps*: Linking biomarkers to behavior. *Chemosphere* 76:1416–27.

Wallace, W.G., and A. Estephan. 2004. Differential susceptibility of horizontal and vertical swimming activity to cadmium exposure in a gammaridean amphipod (*Gammarus lawrencianus*). *Aquat. Toxicol.* 69:289–97.

Wallace, W.G. et al. 2000. Alterations in prey capture and induction of metallothioneins in grass shrimp fed cadmium-contaminated prey. *Environ. Toxicol. Chem.* 19:962–71.

Weber, D.N. et al. 1997. Alterations in neurobehavioral responses in fishes exposed to lead and lead-chelating agents. *Am. Zool.* 37:354–62.

Weber, D.N. et al. 1991. Waterborne lead affects feeding neurotransmitter levels of juvenile fathead minnows (*Pimephales promelas*). *Aquat. Toxicol.* 21:71–80.

Weis, J.S. et al. 2001. Effects of contaminants on behavior. Biochemical mechanisms and ecological consequences. *Biosciences* 51:209–17.

Weis, J.S. et al. 2011. Living in a contaminated estuary: Behavioral changes and ecological consequences for five species. *Biosciences* 61:375–85.

Weis, P., and J.S. Weis. 1974. DDT causes changes in activity and schooling behavior in goldfish. *Environ. Res.* 7:68–74.

Wilding, J., and L. Maltby. 2006. Relative toxicological importance of aqueous and dietary metal exposure to a freshwater crustacean: Implications for risk assessment. *Environ. Toxicol. Chem.* 25:1795–801.

Yu, Z., L. Jiang, and D. Yin. 2011. Behavior toxicity to *Caenorhabditis elegans* transferred to the progeny after exposure to sulfamethoxazole at environmentally relevant concentrations. *J. Environ. Sci.* 23:294–300.

11

Origin of Energy Metabolism Impairments

Odile Dedourge-Geffard, Frédéric Palais, Alain Geffard, and Claude Amiard-Triquet

CONTENTS

11.1 Introduction

All biochemical and physiological processes in the life of organisms depend strictly on their energy metabolism. Organisms obtain the energy necessary for life from an external source, in the form of light energy in the case of autotrophic organisms or in the form of chemical energy contained in food for heterotrophic organisms. In the case of heterotrophic organisms, the obtaining of energy is controlled by feeding and the subsequent breakdown of food to release the energy contained (assimilation) (Figure 11.1). Feeding is controlled by two categories of factors: on one hand the availability of food species, and on the other the capacity of the feeder to capture and to consume this food. Decrease in feeding can result either from an avoidance behavior toward a food recognized as contaminated or from physiological disturbance limiting, for example, locomotory activity.

Once ingested, food is subjected to the action of enzymes, by the process of digestion, which represents all the processes by which raw food (containing carbohydrates, proteins, lipids) is transformed into smaller molecules available for absorption.

The liver and the pancreas of vertebrates as well as the digestive gland or hepatopancreas of mollusks and crustaceans play important roles in the digestion of food, by synthesizing and by secreting digestive enzymes. However, these organs may also be the target of the

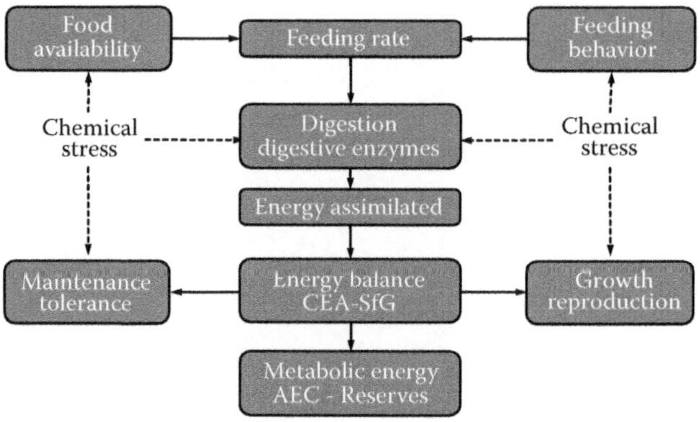

FIGURE 11.1
Potential effects of chemical stresses on energy metabolism. (Modified from De Coen, W.M., Janssen, C.R., *Hydrobiologia*, 367, 199–209, 1998; Dedourge-Geffard, O. et al., *Chemosphere*, 77, 1569–1576, 2009.)

chemical contaminants often transported there for detoxification. Therefore, organs such as the vertebrate liver or the invertebrate digestive gland or hepatopancreas constitute the main sites of accumulation of contaminants, and in particular contaminant trace metals, in aquatic animals. Given that the invertebrate digestive gland or hepatopancreas has a major role in digestion as well as in the detoxification and subsequent storage of toxic elements or compounds, it is interesting and relevant to improve our knowledge of the possible effect of pollutants on the activity of the digestive enzymes secreted by these organs.

Furthermore, within the framework of an evaluation of the quality of an environment, variations in enzymatic activities in organisms are often one of the first responses to chemical stress; they thus represent useful points in the confirmation of any toxic effects (Lagadic et al. 2000; Van der Oost et al. 2003) before later ones are perceptible at higher biological levels of organization (e.g., cellular, tissue, physiological, whole individual). Biomarkers related to energy metabolism are thus potentially very interesting prognostic tools, bearing in mind the close relationship existing between the energy balance and the health of individuals (Depledge and Fossi 1994; De Coen and Janssen 2003).

11.2 Food Availability and Quality in Polluted Media

Living organisms exhibit highly different sensitivities to chemical stress, particularly when they belong to different taxa but also, in some cases, when they are relatively closely related phylogenetically (Chapter 7). The presence of contaminants in the medium can exert direct toxic effects on the most sensitive species, inducing sublethal or even lethal effects, particularly in cases of acute accidental pollution events. In addition, indirect effects occur in communities exposed to chemical stress, with interactions between resources and consumers (Fleeger et al. 2003).

At the base of food webs, increased productivity or enhanced microorganism biomass has frequently been observed as a consequence of enrichment with organic matter. Despite the toxicity of some of their compounds, petroleum hydrocarbons represent a significant source of organic matter that can lead to a stimulation of bacterial microflora, favoring bacteria-consuming species. Similarly, pulp mill effluents can stimulate the growth of diatoms. On the other hand, herbicides are responsible for a selective decrease of primary producers, thus leading to a reduction in the abundance of their consumers as a consequence of limited food availability (Fleeger et al. 2003 and literature quoted therein). In this way, feeding would have an important role in the survival and physiological condition of *Daphnia* (freshwater cladoceran crustaceans), and, in the presence of the herbicide molinate, changes in the quantity/quality of phytoplankton on which they feed could be responsible for effects observed on their reproduction (Sánchez et al. 2004 and literature quoted therein).

At higher levels in the food web, interactions grow increasingly complex (Figure 3.2); nevertheless, it is clear that the tolerance of some populations/species to chemicals does not guarantee their durability if their major prey species are sensitive, thus reducing food and therefore energy available from their medium. For instance, a decline in recruitment of several littoral bivalve species in British estuaries strongly impacted by the use of tributyltin, would have influenced the distribution of certain wading bird species (Bryan and Gibbs 1991). In a lake impacted by mining activities, growth, fecundity, and lipid concentrations were impaired in the fish *Catostomus commersoni*, despite Cu and Zn tissue concentrations remaining within maintained limits of metal homeostasis. On the other hand, it was evident that—as a consequence of decreased diversity and density of macroinvertebrates—fish from this metal-rich lake invested much more energy in foraging (Munkittrick et al. 1991).

Considering not only two successive links of the trophic chain but the whole system, a review by Hanazato (2001) dealing with freshwater pelagic organisms, has shown that low pesticide concentrations directly damaging only relatively large taxa (e.g., *Daphnia*) may affect the population dynamics of smaller zooplankton species, particularly rotifers, indirectly through competitive relationships. The final consequence was an elongation of the food chain and a reduction of energy transfer efficiency from primary producers to top predators since each transfer from one level to the next corresponds to an energy loss.

In natural media subjected to pollutant inputs, decreased food availability or increased competition can limit the ability of a given species to maximize its energy uptake, thus leading to the consumption of unusual food and to reduced energy conversion (Morrow et al. 2004 and literature quoted therein). In mammals and fish, it has been reported that not only food quantity but also its quality could influence detoxification ability and stress. Fasting has no clear effects on general stress parameters (cortisol or glucose concentrations), whereas effects on biotransformation are more obvious, indicating generally a decrease of different enzyme activities in the liver (Morrow et al. 2004 and literature quoted therein).

In addition, food quality can interfere with energy metabolism through its degree of contamination. An impairment of lipid metabolism and storage was observed in the aquatic bird *Cairina moschata*, the Muscovy duck, exposed to dietary Cd (Lucia et al. 2010). When rainbow trout *Oncorhynchus mykiss* were fed oligochaete worms *Lumbriculus variegatus*, previously exposed to metal-rich sediments, the instantaneous growth rate became more and more reduced with increasing contamination of food (Figure 11.2, experimental groups A to E). According to Hansen et al. (2004), growth inhibition would be mainly a result of the reduction of food conversion efficiency in biomass rather than of a decrease in food uptake, which was limited in this study contrary to other observations.

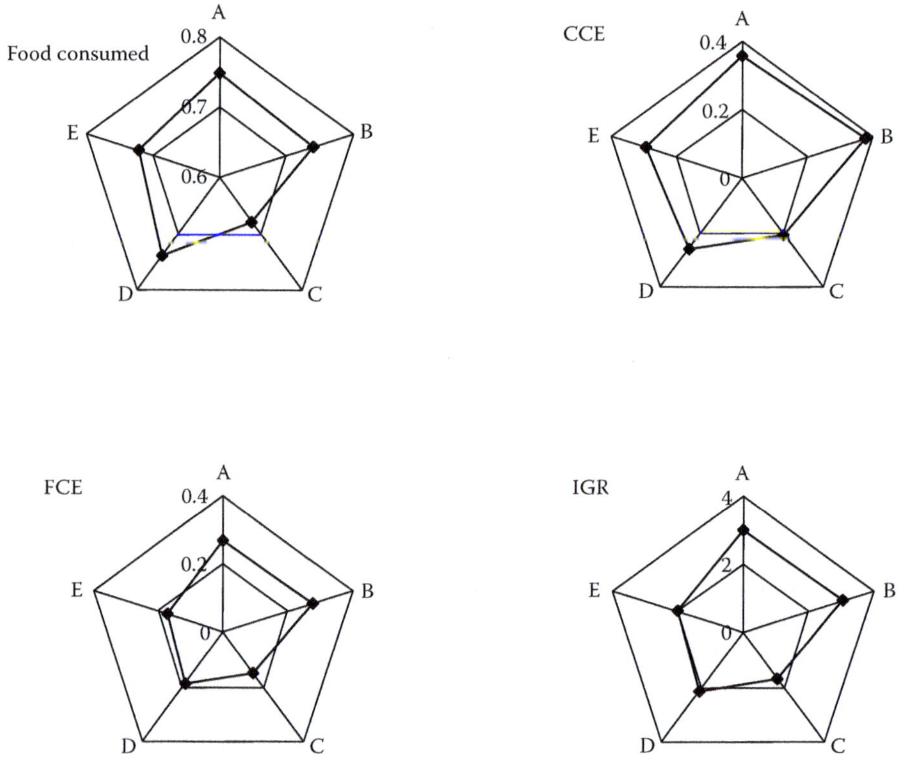

FIGURE 11.2
Feeding of rainbow trout (*Oncorhynchus mykiss*) with oligochaetes (*Lumbriculus variegatus*) contaminated with metals according to an increasing gradient from A to E. CCE, caloric conversion efficiency (g growth $kcal^{-1}$ food consumed); FCE, food conversion efficiency (g growth g^{-1} food consumed); IGR, instantaneous growth rate (% day^{-1}). (After Hansen, J.A. et al., *Environ. Toxicol. Chem.*, 23, 1902–1911, 2004.)

11.3 Impairments of Food Uptake

As reviewed previously (Chapter 10), food uptake/capture needs the achievement of complex behaviors that may be affected by impairments resulting from pollutant effects on nervous system functioning and locomotory behavior. Decreased feeding rate can reflect either physiological disturbances or avoidance behavior when chemosensors can detect the presence of contaminants in food. Experimental strategies have been designed to distinguish these two types of events. For instance, in the freshwater amphipod crustacean *Gammarus pulex*, a decreased feeding rate was observed as a consequence of dietary, waterborne, or mixed exposure to Zn for 6 days at concentrations of the order of magnitude encountered in the most contaminated sites in the United Kingdom (Wilding and Maltby 2006). As soon as exposure ceased, the normal feeding rate was immediately restored in specimens exposed to dietary Zn, showing that impairment was only due to avoidance. On the contrary, prolonged inhibition was observed in specimens exposed to waterborne Zn or to a mixed dietary dissolved exposure, indicating physiological impairment. In the terrestrial isopod crustacean *Oniscus asellus*, Köhler et al. (2000) have shown that avoidance of contaminated food and a subsequent reduction in food consumption were adaptive processes that helped these isopods to cope with chronic pollution in a terrestrial environment contaminated by mining activities. In contrast, isopods from a reference site were unable to distinguish between normal food and metal-rich food, and then consumption was similar for both. Food selection allows the reduction of toxic metal uptake and limits the induction of defense processes and associated energy costs. In the gastropod mollusk *Helix engaddensis*, the presence of Cd or Cu in the food induced decreases in feeding and growth rates. Inhibition due to Cd was irreversible, indicating a toxic effect, whereas fasting due to Cu was reversible, indicating that the gastropods were able to identify and avoid contaminated food (Swaileh and Ezzughayyar 2000).

Neurotoxicants such as insecticides can cause a reduction of filtration rates as a consequence of a loss of coordination as studied in zooplankton, particularly, *Daphnia* (Reynaldi et al. 2006 and literature quoted therein; Barata et al. 2007). In the presence of metals (Cd, Zn), feeding impairments in *Daphnia* may result from digestive poisoning resulting from the release of metals initially associated with the ingested microalgal food at the low pH level encountered in the digestive tract (Barata et al. 2002 and literature quoted therein). Whatever the classes of contaminants and the mechanisms involved in their effects on feeding, *Daphnia* suffer delayed growth and maturation, which could be at the origin of effects at the population level. In the decapod crustacean *Nephrops norvegicus*, food detection and uptake were disturbed at environmentally realistic concentrations of manganese. In the field, such Mn concentrations can last as long as weeks in relation to the hypoxia that commonly occurs in the habitat of this species. (Reducing conditions promote the presence of the relatively more soluble Mn^{2+} ion.) In addition, low levels of oxygen can directly affect *N. norvegicus*, obliging them to come out of their burrows and disturbing their feeding. For benthic organisms such as *N. norvegicus*, toxic effects of Mn can worsen the severe problems that hypoxia causes on a worldwide scale (Krång and Rosenqvist 2006 and literature cited therein).

In a freshwater predator, the leech *Nephelopsis obscura*, exposed to cadmium (5, 10, or 50 $\mu g\ L^{-1}$), only the highest dose tested induced a decrease in food ingestion, whereas the assimilation efficiency was unaffected by any of the tested doses. Despite reduced general activity and respiration rate, the growth rate was lower in specimens originating from the treatment with Cd at 50 $\mu g\ L^{-1}$, since the energy allocation was focused on mucus production and the sustainability of reproduction (Wicklum and Davies 1996).

In nematodes, modeling based on the Dynamic Energy Budget theory (Chapter 12) has been used to analyze the physiological modes of action of different toxicants (Alvarez et al. 2005, 2006). In *Acrobeloides nanus*, exposure to pentachlorophenol and carbendazim can induce a reduced energy uptake in the organism, and the same process can occur in *Caenorhabditis elegans* exposed to Cd and carbendazim.

To assess sediment toxicity, the gastropod *Hydrobia ulvae* was exposed to a gradient of estuarine sediments contaminated by mining activities (Shipp and Grant 2006). The feeding rate was determined by counting fecal pellets, which provide a relevant indicator of the quantity ingested during the hours preceding the assay. The highest production of fecal pellets was observed in individuals collected from the control site, and it decreased for individuals exposed to increasingly contaminated sediments. The production of fecal pellets was positively correlated with growth (incremental size increase) of the gastropods (Figure 11.3).

In fish, metal exposure can affect food uptake, but no general trend can be established. For instance, exposure of the rainbow trout *Oncorhynchus mykiss* to Cu induced an increase of appetite, suggesting a metabolic cost needing higher energy inputs, in agreement with increased oxygen consumption during phases of intense swimming activity. A similar

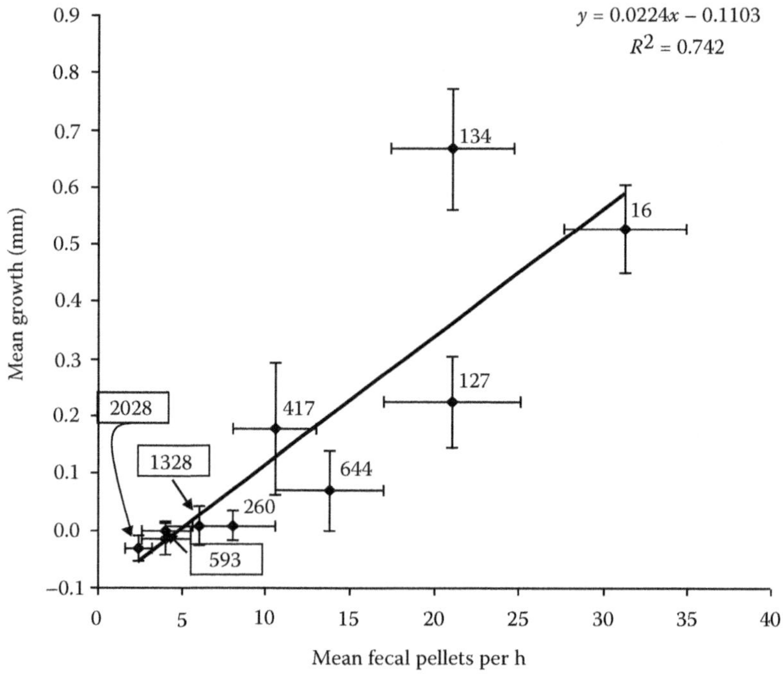

FIGURE 11.3
Correlation of feeding rate (assessed by using production of fecal pellets) with incremental growth (mm) over 28 days in marine gastropods *Hydrobia ulvae* exposed to metal-rich sediments. Value indicated nearby each experimental point is Cu concentration in sediment determined after extraction using HNO_3 in μg g^{-1}, dry weight. (After Shipp, E., Grant, A., *Environ. Toxicol. Chem.*, 25, 3246–3252, 2006.)

pattern was shown for the carp *Cyprinus carpio*. On the contrary, exposure of rainbow trout to Cd induced a decrease of appetite, perhaps related to hypoactivity likewise described in rainbow trout exposed to Al (McGeer et al. 2000 and literature quoted therein). On the other hand, assimilation efficiency was not affected, and under experimental conditions of *ad libitum* feeding (a situation contrasting with natural conditions at least at certain periods of the year when food is scarce), the growth rate was the same for all treatments. In embryos of another freshwater fish, the grayling *Thymallus thymallus*, exposure to methylmercury for the first 10 days of development, at concentrations that may be detected in eggs of carnivorous fish from lakes receiving atmospheric depositions of mercury, induced a permanent impairment of feeding behavior, still observable 3 years later (Fjeld et al. 1998). In young stages of *Cyprinus carpio* exposed to metals (Cd, Cu), the malformations and disturbances of swim bladder functions result in altered behavior, reduced locomotory and feeding activity, and in consequence—in reduced growth rate and viability of fish (Jezierska et al. 2009). Baldwin et al. (2009) have developed a model that explicitly links sublethal reductions in acetylcholinesterase activity to reductions in feeding behavior, food ration, growth, and size at migration in wild salmon populations (*Oncorhynchus tshawytscha*). Studying the impact of phenanthrene on gilthead seabream (*Sparus aurata*), Correia et al. (2007) concluded that the overall changes in behavior might not necessarily be an indication of reduced fitness, at least when the animals are monitored only over a short time. However, in the long term, lethargic juveniles have a less optimal feeding efficiency and thus reduced growth rate parameters (Correia et al. 2007 and literature quoted therein).

We also discuss in Chapter 10 that certain contaminants provoke hyperactivity of organisms. This may be another cause of the impairment of energy metabolism as clearly shown in larvae of the red drum (*Sciaenops ocellatus*) exposed to environmentally realistic doses of atrazine (Alvarez and Fuiman 2005). Despite high food rations, a decrease in growth rate paralleled behavioral changes, suggesting impairment of prey capture, ingestion, or protein metabolism. It appears possible that increased metabolic rates cannot be counterbalanced by a greater probability of encountering prey, expected from increased activity. In juvenile brook trout (*Salvelinus fontinalis*), exposed for 30 days to 5 µg Cd L^{-1}, activity was enhanced, clearly inducing a risk of cost to energy metabolism. Even a concentration 10 times lower (0.5 µg L^{-1}) was sufficient to provoke a decrease in prey capture rate and a subsequent decrease in energy uptake provided by food uptake. This may be compared to the threshold (0.2 µg Cd L^{-1}) considered relevant for the protection of aquatic life by Canadian authorities. Conversely, the foraging activity of longnose dace (*Rhinichthys cataractae*) was not affected by Cd, possibly as a result of species-specific relative Cd insensitivity (Riddell et al. 2005).

Weis et al. (2011) have examined the causes and consequences of behavioral disturbances in estuarine organisms, chronically exposed to contamination in the field (Table 10.2). As a consequence of feeding impairments, the diet of contaminated fish species and blue crabs *Callinectes sapidus* was impoverished compared to their noncontaminated conspecifics. This poor diet can account in part for the reduced size observed in contaminated fish.

11.4 Responses of Digestive Enzyme Activities to Chemical Stressors

Absorption of chemical contaminants by aquatic animals may follow two main routes—a direct aqueous route (by cutaneous and/or gill absorption) and an indirect dietary route (by ingestion followed by digestive absorption), the relative importance of these two routes depending on the biology of the organisms and the bioavailability of the contaminants to either route (Björk 1995; Wang and Fisher 1999; Rainbow 2002). According to exposure route and the subsequent fate of pollutants within the organism (absorption, distribution, metabolization/detoxification, accumulation, or elimination—all of which are species- and pollutant-specific), a toxic effect is more or less likely to occur on digestive enzyme activities. However, to date, there are very few studies that deal with the effects of chemical stressors on the digestive capacity of aquatic organisms and, in the specific case of invertebrates, they are almost exclusively concerned with exposure to trace metals (Table 11.1).

The activities of digestive enzymes depend on different parameters (intrinsic hydrolytic properties, rates of synthesis, rates of secretion for enzymes for extracellular activity) (Yan et al. 1996) on which a pollutant is likely to have an effect by interacting directly or indirectly with the protein. Metal ions are able to interact with enzymes by replacing (other) essential metals present in the active site (e.g., Ca^{2+}) or by combining with hydroxyl or sulfhydryl groups carried by the proteins (Li et al. 2008). This may result in modification of the three-dimensional conformation of the enzyme, eventually leading to an improvement in its lytic properties (induced activity) or, more generally, to an impairment in its properties (repressed activity).

This type of direct interaction between enzymes and trace metals has been highlighted in invertebrate species during *in vitro* exposures carried out on extracts of isolated

TABLE 11.1

Influence of Metals or Organic Contaminants on Digestive Enzyme Activities in Animals

Contaminant	Exposure Conditions	Species	Effects on Digestive Enzymes	Reference
Arsenic	*In vivo* 0.025 and 0.05 mg L^{-1}	*Cyprinus carpio*	Protease (–), Amylase (–), Lipase (–)	Sarita and Jain 2009
Cadmium	*In vitro* 50 mg L^{-1}	*Abramis brama* *Perca fluviatilis* *Stizostedion lucioperca*	Protease (–), amylase (–)	Kuz'mina et al. 2002
	In vitro 0.05 to 50 mg L^{-1}	11 freshwater fish species	Amylase (–) Protease (–)	Golovanova et al. 1999
	In vitro 0.05 to 50 mg L^{-1}	*Oreochromis mossambicus*	Carbohydrases (–)	Golovanova et al. 1994
	In vitro 0.05 to 50 mg L^{-1}	*Abramis brama*	Carbohydrases (–)	
	In vivo 5 mg L^{-1} (15 to 60 days)	*Oreochromis mossambicus*	Carbohydrases (–)	
	In vivo 3 to 25 µg L^{-1} (1 to 4 weeks)	*Corbicula fluminea*	Cellulase (+), Cellulase (–)	Barfield et al. 2001
	In vivo 0.28 to 2.56 mg L^{-1} (50 min to 6 h)	*Daphnia magna*	Cellulase (=), Amylase (=); Esterase (–), β-Galactosidase (–), Trypsin (–)	De Coen and Janssen 1998
	In vivo 0.8 to 570 µg L^{-1} (48 h)	*Daphnia magna*	β-Galactosidase (–), Trypsin (–); Cellulase (=), Amylase (=)	De Coen and Janssen 1997
	In vivo 0.8 to 570 µg L^{-1} (96 h)	*Daphnia magna*	Amylase (+), β-Galactosidase (+), Trypsin (+), Esterase (+), Cellulase (=)	
	In vivo 0.05 to 10 mg L^{-1} (24 h)	*Donax trunculus*	α-Amylase (–)	Mizrahi and Achituv 1989
	In vivo 27.81 mM	*Perna viridis*	Amylase (–)	Sabapathy and Teo 1992
Copper	*In vivo* 2 mg L^{-1} (11 days)	*Artemia salina*	Amylase (–), Trypsin (–)	Alayse-Danet et al. 1979
	In vitro 10^{-8} to 10^{-2} M	*Arenicola marina*	Protease (–), Esterase (–), Lipase (–), α and β-Glucosidase (–)	Chen et al. 2002
	In vitro 10^{-8} to 10^{-2} M	Echinoderm	Proteases (–)	

TABLE 11.1 (Continued)

Influence of Metals or Organic Contaminants on Digestive Enzyme Activities in Animals

Contaminant	Exposure Conditions	Species	Effects on Digestive Enzymes	Reference
	In vitro 10^{-8} to 10^{-2} M	Echiuran	Proteases (–)	
	In vitro 10^{-5} to 10^{-1} M	*Arenicola marina*	Protease (+ or –)	Chen and Mayer 1998
	In vivo 1.9 mg L^{-1} and 1.0 mg L^{-1}	*Lamellidens marginalis*	Amylase (–)	Sonawane 2004
	In vivo 80 µg L^{-1} (30 days)	*Corbicula sp.*	Cellulase (–)	Farris et al. 1988
	In vitro 0.02 µM and 0.02 mM	*Sepia officinalis*	Trypsin (=)	Le Bihan et al. 2004
	In vivo 0.05 to 10 mg L^{-1} (24 h)	*Donax trunculus*	α-Amylase (–)	Mizrahi and Achituv 1989
	In vivo 6.11 mM	*Perna viridis*	Amylase (–)	Sabapathy and Teo 1992
	In vivo 0.01 to 0.5 mg L^{-1}	*Macrobrachium rosenbergii*	Pepsin (–), Cellulase (–), Amylase (–), Tryptase (–)	Li et al. 2008
Chrome	*In vivo* 0.136 mM/ 1.36 mM	*Salmo gairdneri*	Maltase (=)	Bogé et al. 1988
	In vivo 0.025 and 0.05 mg L^{-1}	*Cyprinus carpio*	Protease (–), Amylase (–), Lipase (–)	Sarita and Jain 2009
	In vivo 32 to 180 µg L^{-1} (48 h)	*Daphnia magna*	Amylase (+), β-Galactosidase (+), Trypsin (+); Cellulase (=), Esterase (=)	De Coen and Janssen 1997
	In vivo 33 to 180 µg L^{-1} (96 h)	*Daphnia magna*	Amylase (=), β-Galactosidase (=), Trypsin (=), Cellulase (=), Esterase (=)	
Lead	*In vitro* 0.4 to 1.6 µM	*Channa punctatus*	Lipase (–), Proteases (–)	Sastry and Gupta 1978b
	In vivo 0.05 to 10 mg L^{-1} (24 h)	*Donax trunculus*	α-Amylase (–)	Mizrahi and Achituv 1989
	In vivo 7.9 µM	*Perna viridis*	Sucrase (–)	Teo et al. 1990
	In vivo 17.1 mM	*Perna viridis*	Amylase (–)	Yan et al. 1996

(continued)

TABLE 11.1 (Continued)

Influence of Metals or Organic Contaminants on Digestive Enzyme Activities in Animals

Contaminant	Exposure Conditions	Species	Effects on Digestive Enzymes	Reference
	In vivo 1 mg L^{-1} (8 days)	*Perna viridis*	Amylase (–)	
	In vivo (contaminated food) 1 mg L^{-1} (8 d)	*Perna viridis*	Amylase (–)	
Nickel	*In vivo* 0.025 and 0.05 mg L^{-1}	*Cyprinus carpio*	Protease (–), Amylase (–), Lipase (–)	Sarita and Jain 2009
	In vivo 16.68 mM	*Perna viridis*	Amylase (–)	Sabapathy and Teo 1992
Mercury	*In vitro* 0.4 to 1.6 μM	*Channa punctatus*	Lipase (–), Proteases (–)	Sastry and Gupta 1978a
	In vivo 0.3 mg L^{-1} (15 or 30 days)	*Channa punctatus*	Peptidase (–), Lipase (–), Maltase (=), Lactase (=)	Sastry and Gupta 1980
	In vivo 0.025 and 0.05 mg L^{-1}	*Cyprinus carpio*	Protease (–), Amylase (–), Lipase (–)	Sarita and Jain 2009
	In vivo 0.3 mg L^{-1} (7, 15, and 30 days)	*Heteropneuste fossilis*	Amylase (–), Lactase (–), Maltase (–), Lipase (–), Trypsin (–), Pepsine (=)	Gupta and Sastry 1981
	In vivo 3 μg L^{-1} (7, 15, and 30 days)	*Channa punctatus*	Amylase (–), Lactase (–), Lipase (–)	Rana and Sharma 1982
	In vivo 0.05 to 10 mg L^{-1} (24 h)	*Donax trunculus*	α-Amylase (–)	Mizrahi and Achituv 1989
	In vivo 1.8 to 30 μg L^{-1} (48 h)	*Daphnia magna*	Amylase (+), β-Galactosidase (–), Trypsin (–), Cellulase (=), Esterase (=)	De Coen and Janssen 1997
	In vivo 1.8 to 30 μg L^{-1} (96 h)	*Daphnia magna*	Amylase (=), β-Galactosidase (=), Trypsin (=), Cellulase (=), Esterase (=)	
	In vivo 1.87 mM	*Perna viridis*	Amylase (–)	Sabapathy and Teo 1992
Silver	*In vitro* 0.02 μM/0.02 mM	*Sepia officinalis*	Trypsin (–)/ Trypsin (=)	Le Bihan et al. 2004
Zinc	*In vivo* 5 mg L^{-1} (11 days)	*Artemia salina*	Amylase (–), Trypsin (–)	Alayse-Danet et al. 1979

TABLE 11.1 (Continued)

Influence of Metals or Organic Contaminants on Digestive Enzyme Activities in Animals

Contaminant	Exposure Conditions	Species	Effects on Digestive Enzymes	Reference
	In vivo 87 to 975 μg L^{-1} (10, 20, and 30 days)	*Corbicula* sp.	Cellulase (–)	Farris et al. 1988
	In vitro 0.02 μM and 0.02 mM	*Sepia officinalis*	Trypsin (=)	Le Bihan et al. 2004
	In vivo 0.05 to 10 mg L^{-1} (24 h)	*Donax trunculus*	α-Amylase (–)	Mizrahi and Achituv 1989
	In vivo 13.3 μM	*Perna viridis*	Sucrase (–)	Teo et al. 1990
	In vivo 11.6 mM	*Perna viridis*	Amylase (–)	Yan et al. 1996
Alkyl sulfonic acid	*In vivo* 5 to 45 mg L^{-1}	*Daphnia magna*	β-Galactosidase (–), Esterase (–), Trypsin (–)	De Coen et al. 1998
2,4-dichlorophenoxy-acetic acid	*In vivo* 100 to 400 mg L^{-1} (50 min to 6 h)	*Daphnia magna*	Cellulase (=), Amylase (=), β-Galactosidase (–), Trypsin (– then +), Esterase (–)	De Coen and Janssen 1998
Deltamethrine (D) Methidation (M)	*In vivo* D and M: 10 to 100 μg mL^{-1}	*Cyprinus carpio*	D: Lipase (–), CpA (–), α-Chymotrypsin (+), Trypsin (=) M : α-Chymotrypsine (–), Lipase (–), CpA (–), Trypsin (=)	Simon et al. 1999
Dichlorvos	*In vitro* 0.2 to 100 mg L^{-1}	11 freshwater fish species	Amylase (=), Proteases (=)	Golovanova et al. 1999
	In vitro 0.2 to 100 mg L^{-1}	*Oreochromis mossambicus*	Carbohydrases (=)	Golovanova et al. 1994
	In vivo 0.46 mg L^{-1} 15 to 60 days	*Oreochromis mossambicus*	Carbohydrase (–)	
Fenitrothion	*In vitro* 2 to 10 mg L^{-1}	*Lampito mauritii*	Amylase (–), Invertase (–), Cellulase (–)	Patnaik and Dash 1993
	In vitro 2 to 10 mg L^{-1}	*Octochaetona surensis*	Amylase (–), Invertase (–), Cellulase (–)	
	In vitro 2 to 10 mg L^{-1}	*Drawida willsi*	Amylase (–), Invertase (–), Cellulase (–)	
Lindane	*In vivo* 0.5 to 6 mg L^{-1}	*Daphnia magna*	β-Galactosidase (–), Esterase (–), Trypsin (–)	De Coen et al. 1998

(continued)

TABLE 11.1 (Continued)

Influence of Metals or Organic Contaminants on Digestive Enzyme Activities in Animals

Contaminant	Exposure Conditions	Species	Effects on Digestive Enzymes	Reference
Permethrin (P) Cypermethrin (C)	*In vivo* P and C: 1 to 10 μg mL^{-1}	*Cyprinus carpio*	C and P: Trypsin (–), Lipase (–), CpA (–), α-Chymotrypsin (+)	Simon et al. 1999
Naphtalene	*In vitro* 0.3 to 15 mg L^{-1}	*Oreochromis mossambicus*	Carbohydrases (=)	Golovanova et al. 1994
	In vivo 1.5 mg L^{-1} (15 to 60 days)	*Oreochromis mossambicus*	Carbohydrases (=)	
Propiconazole	*In vitro* 500 μg L^{-1}	*Oncorhynchus mykiss*	Amylase (–)	Li et al. 2010
Sodium pentachlorophenolate	*In vivo* 0.1 to 2 mg L^{-1}	*Daphnia magna*	β-Galactosidase (–), Esterase (–), Trypsin (–)	De Coen et al. 1998
TBT	*In vivo* 5 to 45 μg L^{-1}	*Daphnia magna*	β-Galactosidase (–), Esterase (–), Trypsin (–)	De Coen et al. 1998
Mixture (acid mine drainage)	*In vivo*	*Gammarus fossarum*	β-Galactosidase (–), Esterase (–), Amylase (–), Cellulase (–), β-Glucosidase (–)	Dedourge-Geffard et al. 2009
Mixture (industrial wastewaters)	*In vivo*	*Orchestia gammarellus*	Esterase (–), α-Amylase (–)	Douhri and Sayah 2010
Mixture (three estuaries, pollution gradient)	*In vivo*	*Scrobicularia plana*	Amylase (–), Cellulase (–)	Boldina Cosqueric et al. 2010
Mixture (two estuaries, pollution gradient)	*In vivo*	*Nereis diversicolor*	Amylase (–), Cellulase (–)	Kalman et al. 2009

Notes: (–), inhibition; (+), induction; (=), no effect; TBT, tributyltin.

digestive fluids. Thus, Chen et al. (2002) observed in the annelid *Arenicola marina* induction followed by inhibition of the activity of different digestive enzymes (proteases, lipases, esterases, glucosidases) exposed to increasing concentrations of copper. These researchers also recorded in other invertebrate species (annelids, mollusks, echinoderms) a "bell-shaped" activity of digestive protease exposed to increasing concentrations of copper, the "transition" between inductor and inhibitor effects of the metal ion being observed at a variable concentration threshold according to the species exposed (e.g., 171–176 μM in clams *Mya arenaria* and oysters *Crassostrea virginica*, 868 μM in ribbed mussels *Modiolus demissus*). On the other hand, Hsieh et al. (2008) and Tsao et al. (2003), working on amylase isoforms isolated from two aquatic gastropods (*Haliotis sieboldii, Sulculus diversicolor aquatilis*), observed an induction of their activities when exposed to manganese, nickel, or silver, and a repression of their activities when exposed to copper, cadmium, or zinc. During intraluminal perfusion in rainbow trout, Bogé et al. (1988) did not observe any effect of chromium on maltase activity, even at high exposure concentrations. However, these potassium bichromate perfusions inhibited alkaline phosphatase activities, the intensity of inhibition being correlated positively with the concentration and length of exposure. The effect continued after the chromium had been withdrawn from perfusion (Bogé et al. 1988).

In addition to direct action on the hydrolytic properties of enzymes, pollutants may also exert indirect action on enzyme activity with toxic effects on their synthesis and/or secretion mechanisms. This intracellular toxicity may be especially observed in the digestive system of animals in which morphological changes (e.g., atrophy and necrosis of the digestive tissue) have been recorded during exposure to organic or metal pollutants (Auffret 1988; Lowe and Clarke 1989; Cajaraville et al. 1990). In addition to their implication in digestive processes, certain digestive tissues of vertebrates and invertebrates are also involved in the detoxification of pollutants accumulated at the level of the digestive system or the tegumentary and respiratory systems (and then transferred by the systemic route) (Najle et al. 2000). The digestive gland of bivalves is particularly implicated in the detoxification and sequestration of metals accumulated and stored in the lysosomes of the digestive cells (Janssen and Ertelt-Janssen 1983; Ballan-Dufrançais et al. 1985). However, this sequestration of metal is not always effective, so that some of the ions absorbed may remain biologically active within the cells and disturb the synthesis mechanisms and the secretion of enzymes, or their intracellular activity. For example, Zambare and Mahajan (2001) observed in the freshwater clam *Corbicula striatella* a decrease in the quantity of enzymes secreted by the digestive gland cells in clams exposed to copper or mercury. Likewise, Le Bihan et al. (2004), working on isolated digestive cells of cuttlefish, *Sepia officinalis*, recorded inhibition of protease secretions after exposure to silver. Furthermore, these researchers recorded induction of the intracellular and extracellular activities of cathepsin during exposure of these cells to a low concentration of cadmium and a repression of these same activities during exposure to a Cd concentration 1000 times higher.

In addition to exposures carried out at infra-individual levels of biological organization (i.e., on cells, tissues, organs, or isolated digestive fluids), exposures carried out on whole organisms have also shown modulation of digestive enzyme activities. For instance, Teo et al. (1990), Sabapathy and Teo (1992), and Yan et al. (1996) observed in green mussels, *Perna viridis*, a repression of amylolytic activity (digestive gland) in the presence of cadmium, copper, mercury, nickel, lead, or zinc. In addition to invertebrates, inhibition of enzymatic activities (amylase, proteases, lipases) has also been recorded in various species of fish exposed in their environment to mercury (Sastry and Gupta 1980; Gupta and Sastry 1981; Rana and Sharma 1982), cadmium (Golovanova et al. 1994), or lead (Teo et al. 1990; Yan et al. 1996). Finally, exposure of carp (*C. carpio*) to two doses (0.025 and 0.05 ppm) of arsenic, mercury, nickel, and chromium, individually or as a mixture, showed a decrease in all digestive enzymatic activities (proteases, amylase, lipases) regardless of the treatment used (Sarita and Jain 2009). Li et al. (2007) observed modulations of different digestive enzyme activity in *Oreochromis* sp. (Red Tilapia) during *in vitro* or *in vivo* exposure to copper, iron, and zinc. These differences may be due to (1) a dilution of ingested metal elements, by gastric liquid during intestinal transit leading to concentrations of metal too low to induce effects on digestive enzymes as opposed to exposure *in vitro* or (2) a secretion of digestive enzymes in close relationship with the alimentary intake of metal elements. Exposure to raised levels of metals may improve the nutritional condition of the organism and change the physicochemical characteristics of the intestinal membrane and thus affect digestive enzyme activities.

Examples of the inhibition of digestive enzyme activities have also been highlighted during exposures of organisms to organic contaminants (Table 11.1). In *Daphnia magna*, inhibitions of β-galactosidase and esterase activities have been observed upon exposure to lindane (0.5–6 mg L^{-1}) or 2.4-D (100–600 mg L^{-1}) (De Coen et al. 1998; De Coen and Janssen 1998). Furthermore, the exposure of fish to pyrethroid and organophosphorous pesticides has led to the inhibition of certain digestive enzymes (Golovanova et al. 1994; Simon et al. 1999). For example, in rainbow trout (Li et al. 2010), exposure to propiconazole (500 μg L^{-1}) for 30 days significantly inhibited the intestinal digestive enzymes (proteolytic enzymes and amylase) compared with lower concentrations of exposure (0.2 and 50 μg L^{-1}).

11.5 Responses of Digestive Enzyme Activities to Factors Other than Contaminants

In addition to their sensitivity to chemical stressors, digestive enzyme activities are also sensitive to various biotic or abiotic environmental factors likely to bring about the natural (and often seasonal) variations in significant activities. The life stage of the organism, the availability of food sources or the physicochemistry of the aquatic medium are just some of the many potentially confounding natural factors for which knowledge, and above all characterization, are needed for any perceptive interpretation of measured enzyme responses.

11.5.1 Modulation according to Quality/Quantity of Alimentary Resources

11.5.1.1 Aquatic Vertebrates: Fish Examples

According to the proportion of different components (carbohydrates, lipids, and proteins) in the food ingested, different enzymes will participate in digestion (Barnard 1973; Stevens 1988; Gilles et al. 2006). Thus, qualitative and quantitative differences in enzyme activities are generally observed between various species according to their alimentary regime (herbivorous, omnivorous, carnivorous).

Amylase activity in trout (*Oncorhynchus mykiss*) and eel (*Anguilla anguilla*), both omnivorous species, is much lower than in carp (*Cyprinus carpio*) or tench (*Tinca tinca*) (Hidalgo et al. 1999). Thus, measured amylase activities in herbivorous and omnivorous fish are higher than those found in carnivorous species, apparently to facilitate the digestion of the carbohydrates present in algae and other plants (Vonk and Western 1984; Sabapathy and Teo 1993; Hidalgo et al. 1999).

Similar differences have been found in lipase activities. Lipase activity (Chakrabarti et al. 1995) in carnivorous species, which generally have a diet richer in fats, is found to be higher than that in omnivorous or herbivorous fish (Opuszynski and Shireman 1995; Tengjaroenkul et al. 2000).

As regards proteins, carnivorous fish frequently show higher proteolytic enzyme activities, which encourage the digestion of food rich in animal proteins (Sabapathy and Teo 1993). Kuz'mina et al. (2002) thus observed protease activities that were higher in the carnivorous zander (*Stizostedion lucioperca*), moderate in the case of the perch (*Perca fluviatilis*), and lower in the omnivorous bream (*Abramis brama*). However, some results concerning proteolytic activities are contradictory. Indeed, Hidalgo et al. (1999), who studied protease activities in different species of fish, were not able to carry out a classification of species according to their feeding behavior. Rainbow trout (*O. mykiss*) and common carp (*C. carpio*) presented high values, whereas other carnivores such as the European eel (*A. anguilla*) and the gilthead sea bream (*Sparus aurata*) exhibited lower activities. However, when the phylogenetic parameter is taken into account and comparison is made between species of the same group, the hypothesis as to whether herbivores have higher amylase activities (α-amylase) than those of carnivores, and carnivores have higher protease activities (pepsin, trypsin) than those of herbivores, is less evident. Whereas Fernandez et al. (2001) found that differences of α-amylase activities in five species of Mediterranean sparids (family Sparidae) are primarily related to their feeding habits, other researchers (Chan et al. 2004; German et al. 2004) observed from studying four fish species of the family Stichaeidae, the shannies, that it was the phylogenetic link that predominantly affected the α-amylase activities. Chakrabarti et al. (1995), studying the activity of several digestive enzymes (α-amylase, cellulase, invertase, esterase, acid, and alkaline proteases) in 11 species of bony fish from one lake, also observed that there was no correlation between these enzyme activities and the feeding habits of the fish. Researchers attributed these results to fish habitats: life in a lake involves significant alimentary niche overlap of the different species within this system (Chakrabarti et al. 1995).

11.5.1.2 Aquatic Invertebrates

A relationship between the alimentary regime and the production of digestive enzymes is often found in invertebrates. Raised levels of enzyme activity according to the substrate most frequently present in the diet have been observed in a decapod crustacean, the shrimp *Litopenaeus vannamei* (Moss et al. 2001). Amylase and protease activities are thus modulated by the food composition in different species of shrimps and increase with the respective content of alimentary carbohydrates and proteins (Ceccaldi 1998). Mayer et al. (1997) studied relationships between different digestive enzymes in species of polychaetes, echinoderms, and holothurians. Lipase activities are generally higher in carnivorous polychaetes compared to other detritivorous species. The lipase/protease (L/P) ratio proves to be a food source indicator. In carnivorous polychaetes the L/P ratio is greater than 1, whereas in detritivorous species it is lower than 0.04.

In different species of crab, Johnston and Freeman (2005) also observed variations in digestive enzyme activities according to alimentary regime. *Nectocarcinus integrifons* eats marine higher plants and shows elevated cellulase activity in order to degrade cellulose. *Petrolisthes elongates,* which consumes phytoplankton and brown and green algae, shows raised laminarinase and β-glucosidase activities for digesting laminarin. Finally, *Leptograpsus variegates,* which has an omnivorous alimentary regime, presents increased activities of protease, α-glucosidase, and α-amylase, making best use of alimentary proteins and carbohydrates. Other research has also shown, according to species, elevated levels of enzyme activities in carnivores, moderate levels in omnivores, and low levels in herbivores (Jones et al. 1997; Kumlu and Jones 1997; Le Vay et al. 2001; Gutow et al. 2006). However, certain species do not show a relative importance of a particular class of enzyme. These species may have a very varied alimentary regime, as observed in the crab *Carcinus maenas* (Johnston and Freeman 2005), in turn, related to the large number of habitats occupied by this species.

Although the measurement of digestive enzymes is a commonly used tool for classification of alimentary regimes in many invertebrates, with the hypothesis that digestive enzymes activities are higher for the most common substrates in the diet (Gaudy and Boucher 1983; McClintock et al. 1991; Brêthes et al. 1994; Johnston and Freeman 2005; Johnston et al. 2005), the data are sometimes contradictory. Thus, Rodriguez et al. (1994)

unexpectedly noticed the highest protease activity (trypsin) in crustacean larvae (*Penaeus japonicus*) provided with food of plant origin. These responses are linked not only to the type of food, but also to the quantity of food available. By exposing copepods (*Calanus pacificus*) to different concentrations of the same food (diatoms), Hassett and Landry (1983) recorded higher activities of two carbohydrases (cellobiase and maltase) in individuals acclimatized to lower concentrations of food. These results suggest that the secretion of a large quantity of enzymes may maximize the use of the component present in low quantity in the food. Therefore, the maintenance of elevated enzyme activities when there is little food present may be energy efficient while retaining the potential to digest and make optimum use of high concentrations of food when these are present. Different studies support this hypothesis for different species of crustaceans and for different life stages (larva, juvenile, adult), particularly in the case of proteases (Jones et al. 1997; Kumlu and Jones 1997; Le Vay et al. 2001; Gutow et al. 2006). The same type of response has also been observed in mollusks. For example, in the common cockle, *Cerastoderma edule*, Ibarrola et al. (1998) recorded higher digestive enzyme activities in individuals subjected to nutritional stress (fasting for 3 days) than in nonstressed individuals. This type of response may be associated with the implementation of an energy optimization strategy, the decrease in dietary intake "encouraging" organisms to maintain, or even stimulate, their digestive enzymatic activities, to ensure efficient assimilation of the low quantity of food ingested (De Coen and Janssen 1998). This type of response has been observed by Palais et al. (2012) in the zebra mussel *Dreissena polymorpha* during a biomonitoring study in which induction of amylase and endocellulase digestive activities was observed in the digestive gland and crystalline style of organisms faced with reduced dietary intake. The induction observed was most marked during the period of gametogenesis, and indicated interactions between digestive enzyme activities, availability of food, and energy requirements (related to reproductive condition). In the same manner, the correlation between dietary intake and digestive activities in bivalves is not always obvious. Various studies have shown a decrease in the digestive enzymatic activities of bivalves in response to a reduction in their dietary intake (Teo et al. 1990; Yan et al. 1996; Ibarrola et al. 2000; Albentosa and Moyano 2008). Whatever the alimentary regime and the nutritional preferences of different species, they may be able to adapt to the available food, which may involve regulation of the synthesis of digestive enzymes (Labarta et al. 2002).

11.5.2 Modulation according to Ontogenesis, Life Cycle, or Sex of Organisms

11.5.2.1 Aquatic Vertebrates: Fish Examples

During larval development, changes in digestive enzyme activity profiles can be correlated with anatomical and physiological modifications, such as the opening of the mouth, the acquisition of a functional stomach, and also with changes in the type of food ingested (Buddington and Doroshov 1986; Kolkovski 2001; Cara et al. 2003; Chakrabarti et al. 2006; Zouiten et al. 2008; Galaviz et al. 2011). In the European sea bass *Dicentrarchus labrax*, amylase mRNA was higher in young larvae than in older ones (Péres et al. 1998). On the other hand, Douglas et al. (2000) noticed higher gene expression of amylase in the adult flounder *Pleuronectes americanus*. Bolasina et al. (2006) have monitored trypsin and lipase activities throughout the ontogenesis of Japanese flounder *Paralichthys olivaceus* [between 2 and 39 days after hatching (DAH)]. In addition to the increase in trypsin activity during feeding with *Artemia* (brine shrimp) (15 DAH), a decrease in lipase and trypsin activities at the end of metamorphosis (33 DAH) was noted. Similar results were observed for trypsin in Senegalese sole *Solea senegalensis* (Martinez et al. 1999). During the development of the carp (*Cirrhinus mrigala*), variations in digestive enzyme activities (amylase, protease, trypsin, chymotrypsin, lipase) were also detected in the course of ontogenesis (Chakrabarti and Rathore 2010). It appears that in the case of *Betta splendens* (Siamese fighting fish), acid proteases play a major role throughout the early life stages (10 days), whereas alkaline proteases take over in the later stages (Thongprajukaew et al. 2010).

Age-related modifications in the quality/quantity of the pool of individual digestive enzymes would signify evolutionary adaptations to different alimentary regimes and the different nutritional needs of the developmental stages (Kumar et al. 2000). Kuz'mina (1996) thus observed, at intestine level, considerable fluctuations of digestive enzyme activities according to age (0–7 years) in different species of fish. The general enzyme activity of the digestive tract increases with the age of the individual, in response to the increase in the size of the intestine and the weight of the mucus. On the other hand, changes concerning specific activities (adjusted in comparison to the weight of the individual) have shown that the higher digestive capacities observed during the early stages of development and also in other stages in some species of fish, are probably related to the period of sexual maturity of the organisms (Kuz'mina 1996). Recently, in fish endemic to Cuba (*Limia vittata* and *Gambusia punctata*), Falcón-Hidalgo et al. (2011) have shown that digestive enzyme activities can be used as an effective indicator of alimentary habits and of the development of the digestive tract in these organisms.

11.5.2.2 Aquatic Invertebrates

In invertebrates, variations in digestive enzyme activities likewise exist during their development cycle, notably in relation to changes in the alimentary regime (adaptation to an omnivorous or carnivorous regime) (Le Vay et al. 2001). Ontogenetic changes, under genetic control, such as physiological modifications (e.g., at the level of the gut diverticula of the anterior region of the tract in penaeid decapods during the development of the hepatopancreas), or the studied organ, have also had a major influence on enzyme activities (Lovett and Felder 1990; Kamarudin et al. 1994; Jones et al. 1997; Le Vay et al. 2001; Dai et al. 2009; Díaz et al. 2009). When different larval stages of different species of crustacean are fed with the same diet (phytoplankton or zooplankton), important variations of different digestive enzyme activities (protease, trypsin, amylase, lipase) are likewise observed according to development stage and individual size (Lovett and Felder 1990; Harms et al. 1991; Ceccaldi 1998; Le Vay et al. 2001; Johnston et al. 2004). In juvenile shrimps (*Litopenaeus vannamai*), Gamboa-Delgado et al. (2003) observed positive correlations between amylase, chymotrypsin, and lipase activities and the weight of the individuals. These increases in enzymatic activities are related to increases in the quantity of stomach content in the same individuals, in relation to their size (Gamboa-Delgado et al. 2003).

In the course of larval and postlarval development of invertebrates possessing an exoskeleton (arthropods), metamorphosis also seems to be an important factor in the variation of enzymatic activities (Van Wormhoudt 1983; Al-Mohanna and Nott 1989; Harms et al. 1991; Fernández et al. 1997; Ceccaldi 1998; Muhlia-Almazán and García-Carreño 2002). These variations according to the stage of metamorphosis are likewise linked to the sexual maturity cycle (Fernández et al. 1997). Indeed, variations of enzyme activity in relation to the process of vitellogenesis in mature female crustaceans have been observed (Van Wormhoudt and Ceccaldi 1973). Differences in amylase activities according to sex in copepod crustaceans have been proposed by Gaudy and Boucher (1983). These amylase activities would be higher in ovigerous females, in relation to a particular nutritional need at the time of oogenesis, notably for phosphorus. This would be obtained from the algal diet. Higher amylase activity in females may thus reflect a predominantly plant-based alimentary regime in comparison to that of the males.

11.5.3 Modulation according to Physicochemical Properties of Water: Light, Salinity, Temperature

During exposure in the laboratory, variations of enzymatic activity in juvenile European sea bass (*Dicentrachux labrax*) could be observed in relation to the light intensity (Cuvier-Péres et al. 2001). Trypsin and chymotrypsin activities were slightly affected at the time of weaning, whereas pepsin activities were considerably reduced at extreme values of light exposure (5 and 4000 lx). Variations in digestive enzyme activities in crustaceans have also been observed in relation to the time of day and following an internal circadian cycle (Harms et al. 1991; Ceccaldi 1998; Hernandez et al. 1999), principally with two peaks per day (Ceccaldi 1998). However, these diurnal variations are not always clearly established given the seasonal variation in the digestive enzyme activities (Baars and Oosterhuis 1984; Biesiot and Capuzzo 1990).

In a coastal bivalve (*Scrobicularia plana*), Fossi Tankoua et al. (2011) highlighted the effect of salinity on digestive enzyme activities notably concerning the amylase and cellulase activities of the crystalline style.

The rate of digestion and the quantity of enzymes produced is generally heavily dependent on the temperature (Vonk and Western 1984; Galgani 1985; Brock et al. 1986). General increases in activity are observed with a rise in temperature (Vonk and Western 1984; Brock et al. 1986; Ceccaldi 1998; Papoutsoglou and Lyndon 2005). However, differential effects of temperature on digestive enzyme activities have been observed, according to whether the fish concerned were hot or cold water species (Vonk and Western 1984; Hazel 1993; Savoie et al. 2008). Digestive enzyme activity response to temperature variation is thus specific to the species under consideration. In addition, it has been shown that temperature affects digestive enzyme activities in different ways according to the section of the digestive tract being examined. This suggests that fishes, which are poikilotherms, possess enzymatic digestion mechanisms capable of being active at different temperatures (Papoutsoglou and Lyndon 2005) as regards the expression of different isoenzymes with different temperature optima (literature cited by Papoutsoglou and Lyndon 2005). Temperature also plays an important role in the digestive enzyme activities of further invertebrates (Ceccaldi 1998). Han et al. (2002) observed in the crustacean copepod species *Calanus sinicus* a negative correlation between the level of protease or amylase activity and temperature. In fact, this species goes into diapause at higher temperatures (25°C) and is then characterized by low enzyme activities.

In direct or indirect relation to annual temperature variations, seasonal variations in the digestive enzyme activities (amylase, cellulase) of various bivalve and crustacean species have been observed (Stuart et al. 1985; Guarna and Borwsky 1995; Ibarrola et al. 1998; Wong and Cheung 2001; Le Pennec and Le Pennec 2002; Palais et al. 2012). However, these seasonal variations are not expressed in the whole digestive system, but, for example, in the crystalline style in bivalves (Wong and Cheung 2001; Palais et al. 2012). Unlike the digestive gland, the crystalline style forms a noncellular structure, not subjected to seasonal weight variation, probably explaining the lower variations in digestive enzyme activities observed in this structure. Thus variations in digestive organ weight (e.g., bivalve digestive gland linked to seasonal variations in food availability and/or the reproductive state of individuals) would explain the fluctuations in the measurement of some enzymatic activities rather more than the direct effect of temperature. This temperature/seasonal effect may represent one factor interacting with chemical stressors. Enzyme responses to contaminant exposure may vary according to season (Farris et al. 1988, 1994; Golovanova et al. 1999; Barfield et al. 2001). A more pronounced effect of cadmium on amylolytic activity has, for example, been observed in the bream (*Abramis brama*) during the period of heightened functional activity of its intestine (summer), in contrast to the winter period in which there was no alimentary activity (Golovanova et al. 1999).

11.6 Digestive Enzyme Activities in Environmental Biomonitoring

The direct relationship between the digestive activities of animals and their access to energy through food, underlines the relevance of digestive enzymes to indicate the exposure and/or the effect of one or more pollutants on an organism, but also possible implications for the entire energy metabolism and, consequently, major functions such as growth and reproduction. This relevance has led to different researchers suggesting that these enzymes should be measured as biomarkers (Lagadic et al. 1994; Hyne and Maher 2003). Hyne and Maher (2003) suggested using the decrease in cellulase, α-glucosidase, and β-glucosidase activities in macroinvertebrate larvae from the Namoi River in Australia as potential biomarkers for exposure to pesticides.

The measurement of digestive enzyme activities has shown the sensitivity of response to exposure of organisms to chemical stressors. Significant inhibitions of cellulolytic activity (cellulase) were observed in different species of mollusk (a bivalve: *Corbicula fulminea*; a gastropod: *Mudalia dilatata*) during exposure to copper and zinc in natural conditions or in the laboratory (Farris et al. 1988, 1994). Furthermore, inhibition of cellulolytic activity in bivalves (*Corbicula* sp.) placed in a river receiving factory effluent enabled the definition of an upstream–downstream gradient of contamination of the wastewater (Farris et al. 1988). Douhri and Sayah (2010) made a particular study of α-amylase activity in the amphipod crustacean *Orchestia gammarellus*, at several sites in Tangier Bay. The results showed a higher rate of amylolytic activity in individuals at sites furthest away from the contamination sources, in comparison to sites subjected to industrial wastewaters. During passive biosurveillance carried out on the polychaete annelid *Nereis diversicolor* at a lightly contaminated site (the Bay of Bourgneuf, France) and a multicontaminated site (the Loire estuary, France), the lowest enzyme activities (amylase, endocellulase) were measured in worms taken from the most highly contaminated site (Kalman et al. 2009). Similarly, during passive biomonitoring carried out with the bivalve *Scrobicularia plana* in the Goyen, Loire, and Seine estuaries in France (contamination gradient: Goyen < Loire < Seine), the lowest digestive activities (amylase, endocellulase in both digestive gland and crystalline style) were measured in clams taken from the most highly contaminated estuaries (apart from higher amylolytic activities in the Seine estuary) (Boldina-Cosqueric et al. 2010). Whereas the lowest digestive enzyme activities measured in the Loire and Seine estuaries probably resulted from exposure to elevated concentrations of contaminants, the authors did not exclude the associated effect of low food availability at the time of the study. During active biomonitoring carried out with the bivalve *Dreissena polymorpha*, the zebra mussel, in the catchment of the River Vesle (France) (Palais et al. 2012), upregulation of digestive activities of amylase (digestive gland, crystalline style) was observed in mussels exposed to various chemical contaminants and thus potentially facing an increase in their energy expenditure (General Adaptation Syndrome; Selye 1976). However, this upregulation has not been recorded at all contaminated sites, lower levels of pollution or higher levels of food availability probably explaining the low need for organisms to "invest" in their digestive activities (least energy requirement). These results highlight the adjustment of digestive enzymatic activities according to intake and energy requirements of the organism, factors which, if not taken into account, constitute potential sources of confoundment in the interpretation of responses measured.

The benefit of such biochemical biomarkers partly resides in their sensitivity in relation to the effects at higher levels of biological organization (individual, population, or community). Different studies have thus clarified the precocious character of the response of digestive enzymes in comparison to the emergence of effects at individual (growth, death) or community (biodiversity) levels. In the brine shrimp, *Artemia salina*, inhibitions of trypsin activity were observed after 24 and 72 h of exposure to copper (2 ppm) and zinc (5 ppm), respectively, and after 72 h (copper) and 24 h (zinc) in the case of amylase activity (Alayse-Danet et al. 1979). Inhibition of these enzyme activities (24 h for the two elements according to the enzyme under examination) occurred earlier than the effect on growth, which was only visible after 3 and 6 days of exposure to zinc and copper, respectively. During exposure carried out in the laboratory on the zebra mussel *D. polymorpha*, significant induction of digestive enzymatic activities (endocellulase in digestive gland and crystalline style) was recorded after 3 days at the highest copper exposure concentration (600 $\mu g\ L^{-1}$), this response having been accompanied by a high mortality rate in the exposed organisms (Palais 2011). During the exposure of another freshwater bivalve, the clam *Corbicula fluminea*, to different concentrations of cadmium, (3 to 25 ppb) over a period of 3 to 4 weeks, Barfield et al. (2001) observed significant reduction in cellulase activity before significant mortality. In these different examples, unrealistic exposure conditions (i.e., unrealistically high concentrations of toxic metals) were probably the origin of the induced or suppressed activities measures in the exposed organisms. These results bring to mind that an increase in digestive enzymatic activities is just as likely to take place as a decrease in these activities during exposure to trace metals. The response as such (increase or decrease) is nevertheless difficult to anticipate, since it varies relatively according to the enzyme, exposure (metal, exposure concentration, or duration), or the species under examination (Chen et al. 2002). The precocity of response was also highlighted during studies undertaken in natural conditions. Thus, during the transplantation of *C. fluminea* to different sites in a river receiving factory wastewater, cellulolytic activity emerged as a highly sensitive stress indicator (Farris et al. 1988). The decrease in activity was more sensitive than the reduction of macroinvertebrate diversity (Farris et al. 1988). During active biomonitoring carried out with the amphipod crustacean *Gammarus fossarum* in a water course strongly contaminated by various trace metals (L'Amous, France), significant repression of digestive enzymatic activities (amylase, endocellulase, esterase, galactosidase, glucosidase in the whole organism) was recorded after 7 days at the most contaminated site, this response having preceded a high mortality rate in the exposed amphipods (Dedourge-Geffard et al. 2009).

However, whatever the exposure conditions, multibiomarker approaches are necessary when attempting to define risk for an individual or its population. De Coen et al. (2001) have shown that the relationship between the enzyme activity parameters of carbohydrate metabolism, including digestive enzymes, and the observed effect at population level in *Daphnia magna*, was specific to the toxin being studied, but also that a single enzyme of carbohydrate metabolism cannot on its own predict changes at population level.

11.7 Perspectives

Intuitively, the local extinction or depletion of species used as prey by a consumer suggests cascading events on the energy metabolism, growth, physiological condition, and reproduction of the consumer. However, many species are highly adaptable and will shift to food resources other than those usually utilized, a change that may by transitorily accompanied by decreased assimilation efficiency (Labarta et al. 2002). From an operational point of view, establishing such an impairment of food resources comes down to studying alterations at the levels of the population and community, a practice that is not appropriate for a warning system, being too late in any ecotoxicological effect sequence.

Several studies have shown a link between feeding disturbances and consequences on the physiological condition and growth of organisms. Thus, effects on reproduction success and lastly on populations may be expected (Chapter 12). Feeding rate, which is a crucial parameter in an energy budget, may be relatively easily quantified, for instance, by using the method of postexposure depression described by McWilliam and Baird (2002) for *Daphnia* and adapted for various species (Chapter 10).

Responses linked to digestive enzyme activities are able to highlight exposure and/or the effect of single or multiple contaminants in aquatic organisms during exposure in laboratory conditions or *in situ*. The precocious nature of the response of digestive enzymes in relation to disturbances at individual level (condition index, growth, mortality) makes them of particular interest in ecotoxicology. Different lines of research remain to be developed: on the one hand to clarify the mechanisms of contaminant activity on enzymes or their harvesting/secretion sites; on the other hand, from the perspective of using digestive enzymes in biosurveillance, (1) to clarify the influence of biotic (age, sex, reproductive cycle, etc.) and abiotic (temperature, conductivity, salinity, etc.) parameters, and (2) to specify the link between the disturbance of the digestive capacity of an organism and the possible outcomes for their energy metabolism, and subsequently on their growth and reproduction.

References

Alayse-Danet, A.M. et al. 1979. Modèle de détection rapide des effets sublétaux des polluants: Modification des taux d'amylase et de trypsine d'*Artemia salina* contaminées par le cuivre ou le zinc. *Mar. Biol.* 51:41–6.

Albentosa, M., and F.J. Moyano. 2008. Influence of nutritional stress on digestive enzyme activities in juveniles of two marine clam species, *Ruditapes decussatus* and *Venerupis pullastra*. *J. Sea Res.* 59:249–58.

Al-Mohanna, S.Y., and J.A. Nott. 1989. Functional cytology of the hepatopancreas of *Penaeus semisulcatus* (Crustacean, Decapoda) during the molting cycle. *Mar. Biol.* 101:535–44.

Alvarez, A.O. et al. 2005. Responses to stress of *Caenorhabditis elegans* populations with different reproductive strategies. *Funct. Ecol.* 19:656–64.

Alvarez, M.d.C., and L.A. Fuiman. 2005. Environmental levels of atrazine and its degradation products impair survival skills and growth of red drum larvae. *Aquat. Toxicol.* 74:229–41.

Alvarez, O.A. et al. 2006. Physiological modes of action of toxic chemicals in the nematode *Acrobeloides nanus*. *Environ. Toxicol. Chem.* 25:3230–7.

Auffret, M. 1988. Histopathological changes related to chemical contamination in *Mytilus edulis* from field and experimental conditions. *Mar. Ecol. Prog. Ser.* 46:101–7.

Baars, M.A., and S.S. Oosterhuis. 1984. Diurnal feeding rhythms in North Sea copepods measured by gut fluorescence, digestive enzyme activity and grazing on labelled food. *Neth. J. Sea Res.* 18:97–119.
Baldwin, D.H. et al. 2009. A fish of many scales: Extrapolating sublethal pesticide exposures to the productivity of wild salmon populations. *Ecol. Appl.* 19:2004–15.
Ballan-Dufrançais, C. et al. 1985. Physiological features of heavy metal storage in bivalve digestive cells and amoebocytes: EPMA and factor analysis of correspondences. *Biol. Cell.* 53:283–92.
Barata, C. et al. 2002. Genetic variability in sublethal tolerance to mixtures of cadmium and zinc in clones of *Daphnia magna* Straus. *Aquat. Toxicol.* 60:85–99.
Barata, C. et al. 2007. Combined use of biomarkers and in situ bioassays in *Daphnia magna* to monitor environmental hazards of pesticides in the field. *Environ. Toxicol. Chem.* 26:370–9.
Barfield, M.L., J.L. Farris, and M.C. Black. 2001. Biomarker and bioaccumulation responses of Asian clams exposed to aqueous cadmium. *J. Toxicol. Environ. Health* 63A:495–510.
Barnard, E.A. 1973. Comparative biochemistry and physiology of digestion. In: *Comparative Animal Physiology*, ed. C.L. Prosser, 133–64. Philadelphia: Saunders Company.
Biesiot, P.M., and J.M. Capuzzo. 1990. Changes in the digestive enzyme activities during early development of the American lobster *Homarus americanus* Milne Edwards. *J. Exp. Mar. Biol. Ecol.* 136:107–22.
Björk, M. 1995. Bioavailability and uptake of hydrophobic organic contaminants in bivalve filter-feeders. *Ann. Zool. Fenn.* 32:237–45.
Bogé, G., D. Bussiere, and G. Pérès. 1988. Effects of hexavalent chromium on enzymatic activities and transport process of intestinal brush border membrane of rainbow trout (*Salmo gairdneri*). *Water Res.* 22:441–7.
Bolasina, S., A. Pérez, and Y. Yamashita. 2006. Digestive enzymes activity during ontogenetic development and effect of starvation in Japanese flounder, *Paralichthys olivaceus*. *Aquaculture* 252:503–15.
Boldina-Cosqueric, I. et al. 2010. Biochemical, physiological and behavioural markers in the endobenthic bivalve *Scrobicularia plana* as tools for the assessment of estuarine sediment quality. *Ecotoxicol. Environ. Saf.* 73:1733–41.
Brêthes, J.-C., B. Parent, and J. Pellerin. 1994. Enzymatic activity as an index of trophic resource utilization by the snow crab *Chionoecetes opilio* (O. Fabricius). *J. Crustacean Biol.* 14:220–5.
Brock, V., V.S. Kennedy, and A. Brock. 1986. Temperature dependency of carbohydrase activity in the hepatopancreas of thirteen and coastal bivalve species from the North American east coast. *J. Exp. Mar. Biol. Ecol.* 130:87–101.
Bryan, G.W., and P.E. Gibbs. 1991. Impact of low concentrations of tributyltin (TBT) on marine organisms: A review. In: *Metal Ecotoxicology, Concepts and Applications*, ed. M.C. Newman and A.W. McIntosh, 323–61. Chelsea, MI: Lewis Publishers.
Buddington, R.K., and S.I. Doroshov. 1986. Development of digestive secretions in white sturgeon juveniles (*Acipenser transmontanus*). *Comp. Biochem. Physiol.* 83A:233–8.
Cajaraville, M. et al. 1990. Responses of basophilic cells of the digestive gland of mussels to petroleum hydrocarbon exposure. *Dis. Aquat. Organ.* 9:221–8.
Cara, J.B. et al. 2003. Assessment of digestive enzyme activities during larval development of white bream. *J. Fish. Biol.* 63:48–58.
Ceccaldi, H. 1998. A synopsis of the morphology and physiology of the digestive system of some crustacean species studied in France. *Rev. Fish. Sci.* 6:13–39.
Chakrabarti, I. et al. 1995. Digestive enzymes in 11 freshwater teleost fish species in relation to food habit and niche segregation. *Comp. Biochem. Physiol.* 112A:167–77.
Chakrabarti, R., and R.M. Rathore. 2010. Ontogenic changes in the digestive enzyme patterns and characterization of proteases in Indian major carp *Cirrhinus mrigala*. *Aquacult. Nutr.* 16:569–81.
Chakrabarti, R. et al. 2006. Functional changes in digestive enzymes and characterization of proteases of silver carp (♂) and bighead carp (♀) hybrid, during early ontogeny. *Aquaculture* 253:694–702.
Chan, A.S. et al. 2004. Digestive enzyme activities in carnivores and herbivores: Comparisons among four closely related prickleback fishes (Teleostei: Stichaeidae) from a California rocky intertidal habitat. *J. Fish. Biol.* 65:848–58.

Chen, Z., and L.M. Mayer. 1998. Digestive proteases of the lugworm (*Arenicola marina*) inhibited by Cu from contaminated sediments. *Environ. Toxicol. Chem.* 17:433–8.
Chen, Z. et al. 2002. Inhibition of digestive enzyme activities by copper in the guts of various marine benthic invertebrates. *Environ. Toxicol. Chem.* 21:1243–8.
Correia, A.D. et al. 2007. Biochemical and behavioral responses in gilthead seabream (*Sparus aurata*) to phenanthrene. *J. Exp. Mar. Biol. Ecol.* 347:109–22.
Cuvier-Péres, A. et al. 2001. Effects of light intensity on animal husbandry and digestive enzyme activities in sea bass *Dicentrachus labrax* post-larvae. *Aquaculture* 202:317–28.
Dai, Y. et al. 2009. Activities of digestive enzymes during embryonic development in the crayfish *Procambarus clarkii* (Decapoda). *Aquat. Res.* 40:1394–9.
De Coen, W.M., and C.R. Janssen. 1997. The use of biomarkers in Daphnia magna toxicity testing: II. Digestive enzyme activity in *Daphnia magna* exposed to sublethal concentrations of cadmium, chromium and mercury. *Chemosphere* 35:1053–67.
De Coen, W.M., and C.R. Janssen. 1998. The use of biomarkers in *Daphnia magna* toxicity testing: I. The digestive physiology of daphnids exposed to toxic stress. *Hydrobiologia* 367:199–209.
De Coen, W.M., and C.R. Janssen. 2003. The missing biomarker link: Relationships between effects on the cellular energy allocation biomarker of toxicant-stressed *Daphnia magna* and corresponding population characteristics. *Environ. Toxicol. Chem.* 22:1632–41.
De Coen, W.M., M.L. Vangheluwe, and C.R. Janssen. 1998. The use of biomarkers in *Daphnia magna* toxicity testing: III. Rapid toxicity testing of pure chemicals and sediment pore waters using ingestion and digestive enzyme activity. *Chemosphere* 37:2677–94.
De Coen, W.M., C.R. Janssen, and H. Segner. 2001. The use of biomarkers in *Daphnia magna* toxicity testing: V. In vivo alterations in the carbohydrate metabolism of *Daphnia magna* exposed to sublethal concentrations of mercury and lindane. *Ecotoxicol. Environ. Saf.* 48:223–34.
Dedourge-Geffard, O. et al. 2009. Effects of metals on feeding rate and digestive enzymes in *Gammarus fossarum*: An in situ experiment. *Chemosphere* 77:1569–76.
Depledge, M.H., and M.C. Fossi. 1994. The role of biomarkers in environmental assessment. (2) Invertebrates. *Ecotoxicology* 3:161–72.
Díaz, A.C. et al. 2009. Ontogenetic changes in the digestive system of *Pleoticus muelleri* (Decapoda, Penaeoidea). *Invertebr. Reprod. Dev.* 52:1–12.
Douglas, S.E., S. Mandla, and J.W. Gallant. 2000. Molecular analysis of the amylase gene and its expression during development in the winter flounder, *Pleuronectes americanus*. *Aquaculture* 190:247–60.
Douhri, H., and F. Sayah. 2010. Validation of *Orchestia gammarellus* enzymatic activities in several sites of Tangier's bay (Morocco), *Afr. J. Environ. Sci. Technol.* 4:256–62.
Falcón-Hidalgo, B. et al. 2011. Digestive enzymes of two freshwater fishes (*Limia vittata* and *Gambusia punctata*) with different dietary preferences at three developmental stages. *Comp. Biochem. Physiol.* 158B:136–41.
Farris, J.L. et al. 1988. Application of cellulolytic activity of asiatic clams (*Corbicula sp.*) to in-stream monitoring of power plant effluents. *Environ. Toxicol. Chem.* 7:701–13.
Farris, J.L. et al. 1994. Molluscan cellulolytic activity responses to zinc exposure in laboratory and field stream comparisons. *Hydrobiologia* 287:161–78.
Fernández, I. et al. 1997. Digestive enzyme activities of *Penaeus notialis* during reproduction and moulting cycle. *Comp. Biochem. Physiol.* 118A:1267–71.
Fernández, I. et al. 2001. Characterization of α-amylase activity in five species of Mediterranean sparid fishes (Sparidae, Teleostei), *J. Exp. Mar. Biol. Ecol.* 262:1–12.
Fjeld, E., T.O. Haugen, and L.A. Vollestad. 1998. Permanent impairment in the feeding behavior of grayling (*Thymallus thymallus*) exposed to methylmercury during embryogenesis. *Sci. Total Environ.* 213:247–54.
Fleeger, J.W., K.R. Carman, and R.M. Nisbet. 2003. Indirect effects of contaminants in aquatic ecosystems. *Sci. Total Environ.* 317:207–33.
Fossi Tankoua, O. et al. 2011. Potential influence of confounding factors (size, salinity) on biomarkers in the sentinel species *Scrobicularia plana* used in programmes monitoring estuarine quality. *Environ. Sci. Pollut. Res.* 18:1253–63.

Galaviz, M.A. et al. 2011. Ontogeny of the digestive tract and enzymatic activity in white seabass, *Atractoscion nobilis*, larvae. *Aquaculture* 318:162–8.
Galgani, F. 1985. Régulation de l'activité des protéases digestives de *Penaeus japonicus* Bate en relation avec la température. *J. Exp. Mar. Biol. Ecol.* 94:11–8.
Gamboa-Delgado, J., C. Molina-Poveda, and C. Cahu. 2003. Digestive enzyme activity and food ingesta in juvenile shrimp *Litopenaeus vannamei* (Boone, 1931) as a function of body weight. *Aquat. Res.* 34:1403–11.
Gaudy, R., and J. Boucher. 1983. Relation between respiration, excretion (ammonia and inorganic phosphorus) and activity of amylase and trypsin in different species of pelagic copepods from an Indian Ocean equatorial area. *Mar. Biol.* 75:37–45.
German, D.P., M.H. Horn, and A. Gawlicka. 2004. Digestive enzyme activities in herbivorous and carnivorous prickleback fishes (Teleostei: Stichaeidae): Ontogenic, dietary, and phylogenetic effects. *Physiol. Biochem. Zool.* 77:789–804.
Gilles, R. et al. 2006. *Physiologie Animale*. Bruxelles: De Boeck.
Golovanova, I.L., G.M. Chuiko, and D.F. Pavlov. 1994. Effects of cadmium, naphthalene, and DDVP on gut carbohydrases activity in bream (*Abramis brama* L.) and Mozambique tilapia (*Oreochromis mossambicus* Peters). *Environ. Contam. Toxicol.* 52:338–45.
Golovanova, I.L. et al. 1999. in vitro effects of cadmium and DDVP (dichlorvos) on intestinal carbohydrase and protease activities in freshwater teleosts. *Comp. Biochem. Phys.* 122C:21–5.
Guarna, M.M., and R.L. Borowsky. 1995. Biochemical properties of amylase isozymes from *Gammarus palustris*. A comparative study. *Comp. Biochem. Phys.* 112B:619–28.
Gupta, P.H., and K.V. Sastry. 1981. Effect of mercuric chloride on enzyme activities in the digestive system and chemical composition of liver and muscles of the catfish, *Heteropneustes fossilis*. *Ecotoxicol. Environ. Saf.* 5:389–400.
Gutow, L. et al. 2006. Behavioural and metabolic adaptations of marine isopods to the rafting life style. *Mar. Biol.* 149:821–8.
Han, X., R. Wang, and J. Wang. 2002. Digestive gut structure and activity of protease, amylase, and alkaline phosphatase in *Calanus sinicus* during summer in the Yellow Sea and the East China Sea. *J. Exp. Mar. Biol. Ecol.* 270:131–46.
Hanazato, T. 2001. Pesticide effects on freshwater zooplankton: An ecological perspective. *Environ. Pollut.* 112:1–10.
Hansen, J.A. et al. 2004. Reduced growth of rainbow trout (*Oncorhynchus mykiss*) fed a live invertebrate diet pre-exposed to metal-contaminated sediments. *Environ. Toxicol. Chem.* 23:1902–11.
Harms, J. et al. 1991. Nutritional effects on digestion rate, digestive enzyme activity, growth, and biochemical composition of *Hyas araneus* L. (Decapoda: Majidae) larvae. *J. Exp. Mar. Biol. Ecol.* 145:233–65.
Hassett, R.P., and M.R. Landry. 1983. Effects of food-level acclimation on digestive enzyme activities and feeding behaviour of *Calanus pacificus*. *Mar. Biol.* 75:47–55.
Hazel, J.R. 1993. Thermal biology. In: *The Physiology of Fishes*, ed. D. Evans, 427–67. London: CRC Press.
Hernandez, P. et al. 1999. Rate of ingestion and proteolytic activity in digestive system during continuous feeding of juvenile shrimps. *J. Appl. Aquacult.* 9:35–45.
Hidalgo, M.C., E. Urea, and A. Sanz. 1999. Comparative study of digestive enzymes in fish with different nutritional habits: Proteolytic and amylase activities. *Aquaculture* 170:267–83.
Hsieh, M.S., L.J. Yin, and S.T. Jiang. 2008. Purification and characterization of the amylase from a small abalone *Haliotis sieboldii*. *Fish. Sci.* 74:425–32.
Hyne, R.V., and W.A. Maher. 2003. Review Invertebrate biomarkers: Link to toxicosis that predict population decline. *Ecotoxicol. Environ. Saf.* 54:366–74.
Ibarrola, I. et al. 1998. Seasonal variation of digestive enzyme activities in the digestive gland and the crystalline style of the common cockle *Cerastoderma edule*. *Comp. Biochem. Physiol.* 121A:25–34.
Ibarrola, I. et al. 2000. Acute and acclimated digestive responses of the cockle *Cerastoderma edule* (L.) to changes in the food quality and quantity – II. Enzymatic, cellular and tissular responses of the digestive gland. *J. Exp. Mar. Biol. Ecol.* 252:199–219.

Janssen, H.H., and U. Ertelt-Janssen. 1983. Cytochemical demonstration of cadmium and iron in experimental blue mussels (*Mytilus edulis*). *Mikroskopie* 40:329–40.

Jezierska, B. et al. 2009. Disturbances of early development of fish caused by heavy metals (a review). *Electron. J. Ichthyol.* 2:76–96.

Johnston, D., and J. Freeman. 2005. Dietary preference and digestive enzyme activities as indicators of trophic resource utilization by six species of crab. *Biol. Bull.* 208:36–46.

Johnston, D.J., A.J. Ritar, and C.W. Thomas. 2004. Digestive enzyme profiles reveal digestive capacity and potential energy sources in fed and starved spiny lobster (*Jasus edwardsii*) phyllosoma larvae. *Comp. Biochem. Phys.* 138B:137–44.

Johnston, M., D. Johnston, and A. Richardson. 2005. Digestive capabilities reflect the major food sources in three species of talitrid amphipods. *Comp. Biochem. Phys.* 140B:251–7.

Jones, D.A. et al. 1997. The digestive physiology of herbivorous, omnivorous and carnivorous crustacean larvae: A review. *Aquaculture* 155:285–95.

Kalman, J. et al. 2009. Assessment of the health status of populations of the ragworm *Nereis diversicolor* using biomarkers at different levels of biological organisation. *Mar. Ecol. Prog. Ser.* 393:55–67.

Kamarudin, M.S. et al. 1994. Ontogenetic change in digestive enzyme activity during larval development of *Macrobrachium rosenbergii*. *Aquaculture* 123:323–33.

Köhler, H.R. et al. 2000. Selection favours low hsp70 levels in chronically metal-stressed soil arthropods. *J. Evol. Biol.* 13:569–82.

Kolkovski, S. 2001. Digestive enzymes in fish larvae and juveniles – Implications and applications to formulated diets. *Aquaculture* 200:181–201.

Krång, A.S., and G. Rosenqvist. 2006. Effects of manganese on chemically induced food search behaviour of the Norway lobster, *Nephrops norvegicus* (L.). *Aquat. Toxicol.* 78:284–91.

Kumar, S., J.G. Sharma, and R. Chakrabarti. 2000. Quantitative estimation of proteolytic enzyme and structural study of anterior part of intestine of Indian major carp (*Catla catla*) larvae during ontogenesis. *Curr. Sci.* 79:1007–101.

Kumlu, M., and D.A. Jones. 1997. Digestive protease activity in planktonic crustaceans feeding at different trophic levels. *J. Mar. Biol. Assoc. U.K.* 77:159–65.

Kuz'mina, V.V. 1996. Influence of age on digestive enzyme activity in some freshwater teleosts. *Aquaculture* 148:25–37.

Kuz'mina, V.V., I.L. Golovanova, and E. Kovalenko. 2002. Separate and combined effects of cadmium, temperature, and pH on digestive enzymes in three freshwater teleosts. *Bull. Environ. Contam. Toxicol.* 69:302–8.

Labarta, U. et al. 2002. Enzymatic digestive activity in epifaunal (*Mytilus chilensis*) and infaunal (*Mulinia edulis*) bivalves in response to changes in food regimes in a natural environment. *Mar. Biol.* 140:669–76.

Lagadic, L., T. Caquet, and F. Ramade. 1994. The role of biomarkers in environmental assessment. (5) Invertebrate populations and communities. *Ecotoxicology* 3:193–208.

Lagadic, L. et al. 2000. *Use of Biomarkers for Environmental Quality Assessment*. Enfield: NH Science Publisher.

Le Bihan, E., A. Perrin, and N. Koueta. 2004. Development of a bioassay from isolated digestive gland cells of the cuttlefish *Sepia officinalis* L. (Mollusca Cephalopoda): Effect of Cu, Zn and Ag on enzyme activities and cell viability. *J. Exp. Mar. Biol. Ecol.* 309:47–66.

Le Pennec, G., and M. Le Pennec. 2002. Molecular analysis of the seasonal expression of genes coding for different functional markers of the digestive gland of the bivalve mollusk *Pecten maximus* (L.). *Comp. Biochem. Physiol.* 133B:417–26.

Le Vay, L. et al. 2001. Digestion in relation to feeding strategies exhibited by crustacean larvae. *Comp. Biochem. Physiol.* 128A:623–30.

Li, N., Y. Zhao, and J. Yang. 2008. Effects of water-borne copper on digestive and metabolic enzymes of the giant freshwater prawn *Macrobrachium rosenbergii*. *Arch. Environ. Contam. Toxicol.* 55:86–93.

Li, S.J., J.L. Li, and T.T. Wu. 2007. The effects of copper, iron and zinc on digestive enzyme activities in the hybrid tilapia *Oreochromis niloticus* (L.) × *Oreochromis aureus* (Steindachner). *J. Fish. Biol.* 71:1788–98.

Li, Z.-H. et al. 2010. Effects of exposure to sublethal propiconazole on intestine-related biochemical responses in rainbow trout, *Oncorhynchus mykiss*. *Chem.-Biol. Interact.* 185:241–6.
Lovett, D.L., and D.L. Felder. 1990. Ontogenetic change in digestive enzyme activity of larval and postlarval white shrimp *Penaeus setiferus* (Crustacea, Decapoda, Penaeidae). *Biol. Bull.* 178:144–59.
Lowe, D.M., and K.R. Clarke. 1989. Contaminant-induced changes in the structure of the digestive epithelium of *Mytilus edulis*. *Aquat. Toxicol.* 15:345–58.
Lucia, M. et al. 2010. Effect of dietary cadmium on lipid metabolism and storage of aquatic bird *Cairina moschata*. *Ecotoxicology* 19:163–70.
Martinez, I. et al. 1999. Digestive enzyme activity during larval development of the Senegal sole (*Solea senegalensis*). *Fish. Physiol. Biochem.* 21:317–23.
Mayer, L.M. et al. 1997. Digestive environments of benthic macroinvertebrate guts: Enzymes, surfactants and dissolved organic matter. *J. Mar. Res.* 55:785–812.
McClintock, J.B. et al. 1991. Digestive carbohydrases of the blue crab *Callinectes sapidus* (Rathbun): Implications in utilization of plant-derived detritus as a trophic resource. *J. Exp. Mar. Biol. Ecol.* 148:233–9.
McGeer, J.C. et al. 2000. Effects of chronic sublethal exposure to waterborne Cu, Cd or Zn in rainbow trout: 1. Iono-regulatory disturbance and metabolic costs. *Aquat. Toxicol.* 50:231–43.
McWilliam, R.A., and D.J. Baird. 2002. Postexposure feeding depression: A new toxicity endpoint for use in laboratory studies with *Daphnia magna*. *Environ. Toxicol. Chem.* 21:1198–205.
Mizrahi, L., and Y. Achituv. 1989. Effect of heavy metals ions on enzyme activity in the mediterranean mussel, *Donax trunculus*. *Bull. Environ. Contam. Toxicol.* 42:854–9.
Morrow, M.D., D. Higgs, and C.J. Kennedy. 2004. The effects of diet composition and ration on biotransformation enzymes and stress parameters in rainbow trout, *Oncorhynchus mykiss*. *Comp. Biochem. Physiol.* 137C:143–54.
Moss, S.M., S. Divakaran, and B.G. Kim. 2001. Stimulating effects of pond water on digestive enzyme activity in the Pacific white shrimp, *Litopenaeus vannamei* (Boone). *Aquacult. Res.* 32:125–31.
Muhlia-Almazán, A., and F.L. García-Carreño. 2002. Influence of molting and starvation on the synthesis of proteolytic enzymes in the midgut of the white shrimp *Penaeus vannamei*. *Comp. Biochem. Physiol.* 133B:383–94.
Munkittrick, K.R. et al. 1991. Altered performance of white sucker populations in the Manitouwadge chain of lakes is associated with changes in benthic macroinvertebrate communities as a result of copper and zinc contamination. *Ecotoxicol. Environ. Saf.* 21:318–26.
Najle, R. et al. 2000. Histopathology of the digestive gland of an Antarctic limpet exposed to cadmium. *Sci. Total Environ.* 247:263–8.
Opuszynski, K., and J.V. Shireman. 1995. Digestive mechanisms. In: *Herbivorous Fishes: Culture and Use for Weed Management*, ed. K. Opuszynski and J.V. Shireman, 21–31. Boca Raton, FL: CRC Press.
Palais, F. 2011. Etude du potentiel informative des activités enzymatiques digestives d'un bivalve d'eau douce (*Dreissena polymorpha*) dans le suivi de la qualité des eaux d'un hydrosystème régional (La Vesle, Champagne Ardenne) selon une stratégie de biosurveillance intégrée. Thèse de Doctorat, Université Reims Champagne Ardenne.
Palais, F. et al. 2012. One year monitoring of core biomarker and digestive enzyme responses in transplanted zebra mussels (*Dreissena polymorpha*). *Ecotoxicology* 21:888–905.
Papoutsoglou, E.S., and A.R. Lyndon. 2005. Effect of incubation temperature on carbohydrate digestion in important teleosts for aquaculture. *Aquat. Res.* 36:1252–64.
Patnaik, H.K., and M.C. Dash. 1993. Effect of fenitrothion on gut enzyme activity of tropical earthworms. *Fresenius Environ. Bull.* 2:540–6.
Péres, A., J.L. Zambonino-Infante, and C.L. Cahu. 1998. Dietary regulation of activities and mRNA levels of trypsin and amylase in sea bass (*Dicentrarchus labrax*) larvae. *Fish. Physiol. Biochem.* 19:145–52.
Rainbow, P.S. 2002. Trace metal concentrations in aquatic invertebrates: Why and so what? *Environ. Pollut.* 120:497–507.

Rana, S.V.S., and R. Sharma. 1982. Co-enzyme effects of inorganic mercury in the liver of a freshwater fish *Channa punctatus*. *J. Appl. Toxicol.* 2:275–7.
Reynaldi, S. et al. 2006. Linking feeding activity and maturation of *Daphnia magna* following short-term exposure to fenvalerate. *Environ. Toxicol. Chem.* 25:1826–30.
Riddell, D.J., J.M. Culp, and D.J. Baird. 2005. Behavioral responses to sublethal cadmium exposure within an experimental aquatic food web. *Environ. Toxicol. Chem.* 24:431–41.
Rodriguez, A. et al. 1994. Biochemical composition and digestive enzyme activity in larvae and postlarvae of *Penaeus japonicus* during herbivorous and carnivorous feeding. *Mar. Biol.* 118: 45–51.
Sabapathy, U., and L.H. Teo. 1992. A kinetic study of the α-amylase from the digestive gland of *Perna viridis* L. *Comp Biochem Physiol.* 101B:73–7.
Sabapathy, U., and L.H. Teo. 1993. A quantitative study of some digestive enzymes in the rabbitfish, *Siganus canaliculatus*, and the sea bass, *Lates calcarifer*. *J. Fish. Biol.* 42:595–602.
Sánchez, M., E. Andreu-Moliner, and M.D. Ferrando. 2004. Laboratory investigation into the development of resistance of *Daphnia magna* to the herbicide molinate. *Ecotoxicol. Environ. Saf.* 59:316–23.
Sarita, and K.L. Jain. 2009. Chronic effects of heavy metals on the activity of some digestive and metabolic enzymes in *Cyprinus carpio*. *Ann. Biol.* 25:63–7.
Sastry, K.V., and P.K. Gupta. 1978a. In vitro inhibition of digestive enzymes by heavy metals and their reversal by chelating agent: Part I. Mercuric chloride intoxication. *Bull. Environ. Contam. Toxicol.* 20:729–35.
Sastry, K.V., and P.K. Gupta. 1978b. In vitro inhibition of digestive enzymes by heavy metals and their reversal by chelating agent: Part II. Lead nitrate intoxication. *Bull. Environ. Contam. Toxicol.* 20:736–42.
Sastry, K.V., and P.K. Gupta. 1980. Changes in the activities of some digestive enzymes of *Channa punctatus*, exposed chronically to mercuric chloride. *J. Environ. Sci. Health* 15B:109–19.
Savoie, A. et al. 2008. Metabolic and digestive activity profiles of newly hatched spotted wolffish (*Anarhichas minor* Olafsen): Effect of temperature. *Aquat. Res.* 39:382–9.
Selye, H. 1976. *Stress in Health and Disease*. Boston: Butterworth Publishers, 1286 pp.
Shipp, E., and A. Grant. 2006. *Hydrobia ulvae* feeding rates: A novel way to assess sediment toxicity. *Environ. Toxicol. Chem.* 25:3246–52.
Simon, L.M. et al. 1999. Effects of synthetic pyrethroids and methidation on activities of some digestive enzymes in carp (*Cyprinus carpio* L.). *J. Environ. Sci. Health* 34B:819–28.
Sonawane, S.M. 2004. Effect of heavy metal ($CuSO_4$) on digestive enzymes amylase on fresh water bivalve *Lamellidens marginalis*. *Ecol. Environ. Conserv.* 10:97–9.
Stevens, C.E. 1988. Comparative physiology of the vertebrate digestive system. In: *Comparative Nutrition*, ed. K. Blaxter and I. MacDonald, 21–36. London: Libbey.
Stuart, V., E.J.H. Head, and K.H. Mann. 1985. Seasonal changes in the digestive enzyme levels of the amphipod *Corophium volutator* (Pallas) in relation to diet. *J. Exp. Mar. Biol. Ecol.* 88:243–56.
Swaileh, K.M., and A. Ezzughayyar. 2000. Effects of dietary Cd and Cu on Feeding and Growth Rates of the Landsnail *Helix engaddensis*. *Ecotoxicol. Environ. Saf.* 47:253–60.
Tengjaroenkul, B. et al. 2000. Distribution of intestinal enzyme activities along the intestinal tract of cultured Nile tilapia, *Oreochromis niloticus* L. *Aquaculture* 182:317–27.
Teo, L.H., Z. Lateef, and Y.K. Ip. 1990. Some properties of the sucrase from the digestive gland of the green mussel *Perna viridis* L. *Comp. Biochem. Physiol.* 96B:47–51.
Thongprajukaew, K. et al. 2010. Characterization and expression levels of protease enzymes at different developmental stages of Siamese fighting fish (*Betta splendens* Regan, 1910). *Kasetsart. J. Nat. Sci.* 44:411–23.
Tsao, C.Y., Y.Z. Pan, and S.T. Jiang. 2003. Purification and characterization of amylases from small abalone (*Sulculus diversicolor aquatilis*). *J. Agric. Food Chem.* 51:1064–70.
Van der Oost, R., J. Beyer, and N.P.E. Vermeulen. 2003. Fish bioaccumulation and biomarkers in environmental risk assessment: A review. *Environ. Toxicol. Pharmacol.* 13:57–149.

Van Wormhoudt, A. 1983. Variations immunoquantitatives de l'alpha-amylase au cours du cycle d'intermue à différentes saisons chez *Palaemon serratus* (Crustacea: Decapoda: natantia). *Mar. Biol.* 74:127–32.

Van Wormhoudt, A., and H.J. Ceccaldi. 1973. Activités des enzymes au cours de la vitellogenèse chez *Palaemon serratus*. *C. R. Séances Soc. Biol.* 167:679–84.

Vonk, H.J., and J.R.H. Western. 1984. *Comparative Biochemistry and Physiology of Enzymatic Digestion*. London: Academic Press.

Wang, W.X., and N.S. Fisher. 1999. Delineating metal accumulation pathways for marine invertebrates. *Sci. Total Environ.* 237–238:459–72.

Weis, J.S. et al. 2011. Living in a contaminated estuary: Behavioral changes and ecological consequences for five species. *Biosciences* 61:375–85.

Wicklum, D., and R.W. Davies. 1996. The effects of chronic cadmium stress on energy acquisition and allocation in a freshwater benthic invertebrate predator. *Aquat. Toxicol.* 35:237–52.

Wilding, J., and L. Maltby. 2006. Relative toxicological importance of aqueous and dietary metal exposure to a freshwater crustacean: Implications for risk assessment. *Environ. Toxicol. Chem.* 25:1795–801.

Wong, W.H., and S.G. Cheung. 2001. Feeding rates and scope for growth of green mussels Perna viridis (L.) and their relationship with food availability in Kat O, Hong Kong. *Aquaculture* 193:123–37.

Yan, T., L.H. Teo and Y.M. Sin. 1996. Effects of metals on α-amylase activity in the digestive gland of the green mussel, *Perna viridis* L. *Bull. Environ. Contam. Toxicol.* 56:677–82.

Zambare, S.P., and A.Y. Mahajan. 2001. Heavy metal (copper and mercury) induced alterations in the enzyme secretory activity of hepatopancreas of a freshwater bivalve *Corbicula Striatella*. *Pollut. Res.* 20:143–6.

Zouiten, D. et al. 2008. Ontogeny of the digestive tract of thick lipped grey mullet (*Chelon labrosus*) larvae reared in "mesocosms." *Aquaculture* 279:166–72.

12

Consequences of Energy Metabolism Impairments

Catherine Mouneyrac, Cyril Durou, and Alexandre Péry

CONTENTS

Today, early warning and sensitive diagnosis tools with an ecological relevance seem to be promising for the early assessment of the ecosystem degradation resulting from damaging contaminants.

Among these tools called "biomarkers," the most appropriate ones seem to be suborganismal responses revealing the health status of various organisms, and more particularly those related to energy metabolism and bioenergetics linked to growth and reproduction. There are different types of biomarkers that can be considered biomarkers by themselves, or when in association with others. In this chapter, we will first evoke the currently available biomarkers related to metabolism, and their responses to exposure of organisms to contaminants. Then, we will analyze how these individual or subindividual responses related to energy metabolism and bioenergetics may be used for anticipating the consequences on populations. In addition, mathematical models of effects based on the impact of energy have been developed over the past 10 years.

12.1 Biomarkers Linked to Energy Metabolism and Bioenergetics

A very large number of biological responses to stress linked to the metabolism and energy of living organisms may be investigated (Le Gal et al. 1997; Mayer et al. 2002). Several parameters related to energy metabolism and characteristic of a certain level of biological organization are available (Figure 12.1):

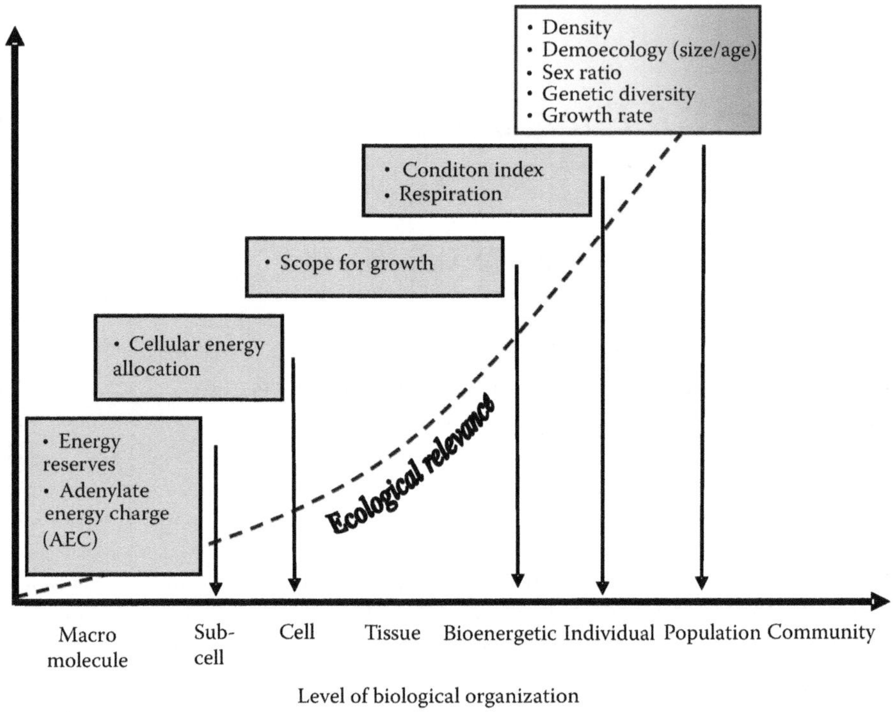

FIGURE 12.1
Biomarkers linked to energy metabolism and bioenergetics, according to the level of biological organization.

- Adenylate energy charge (AEC) assesses the chemical energy necessary for intermediary metabolism.
- Both the energy reserves (glycogen, lipids, proteins) and caloric content indicate the energy available.
- The cellular energy allocation (CEA) determines the net cellular energy budget.
- The scope for growth (SfG) estimates the net energy budget allocated to the growth and reproduction of organisms.
- Condition indices provide information on the general condition of organisms.

12.1.1 Adenylate Energy Charge

The AEC is an ubiquitous system in animals and is usually used for assessing the metabolic energy potentially available in a living organism. This endpoint is based on the measurement of adenosine triphosphate (ATP), adenosine diphosphate, and adenosine monophosphate (AMP) concentrations (Le Gal et al. 1997). By definition, AEC may vary from 0 (if all adenylates are stored in the form of AMP) to 1 (if all adenylates are stored in the form of ATP). In optimal conditions, a steady state between the degradation and regeneration of ATP is reached, which seems to maintain an AEC value of 0.8 approximately. The AEC values have been studied in connection with the ability of organisms to survive, grow, and reproduce. Therefore, in the absence of stress ($0.75 < AEC < 0.90$), the growth rate is high, and survival and reproduction are ensured. Under moderate stress conditions (0.50 < AEC < 0.75), survival is ensured, but growth and reproduction rates tend to decrease or to stop. However, the situation remains reversible. Severe stress conditions (AEC < 0.50) prevent any ability of growth and reproduction and result in decreased viability, even after the return to optimal conditions. Strict correlations between variations in AEC and the ability of organisms to survive are not always recorded. A decrease in AEC indicates that a living organism responds to stress by consumption of energy. On the other hand, when an organism has gone beyond its ability to cope with stress, AEC values are extremely varied. At this point, stress is the cause of serious disturbances in the physiology of organisms, and mortality is recorded in the concerned population. Besides the presence of chemical stresses in the environment, natural stress factors such as temperature, salinity, pH, and oxygen concentration modify the AEC (Le Gal et al. 1997). Abiotic factors of the environment are so many sources of natural stress for living organisms, and their potential influence on energy metabolism must be taken into account.

Surveys based on AEC are mainly carried out on aquatic organisms from various taxonomic groups (annelids, mollusks, crustaceans, and in fish species). Usually, the AEC decreases after exposure to metals and lipophilic contaminants [pesticides, polychlorinated biphenyls (PCBs), hydrocarbons] whether *in situ* or under laboratory conditions (Le Gal et al. 1997; Mayer et al. 2002).

12.1.2 Energy Reserves

Numerous stresses (physical and/or chemical) may result in the mobilization of energy reserves. When the energy intake is higher than the energy needed, energy is stored in the form of macromolecules (glycogen and/or lipids). Glycogen and lipids are always used first; in the case of severe stress, proteins may be catabolized. From the data available in the literature, energy reserves are measured in relevant organs (muscle, gonad, mantle, hepatopancreas, digestive gland) or in the whole body from very varied taxonomic groups (mollusks, arthropods, isopods, annelids, and fish species) from aquatic or terrestrial environments. The use as biomarkers of key molecules or enzyme activities linked to energy metabolism has been the subject of several bibliographical reports (Le Gal et al. 1997; Mayer et al. 2002).

Usually, the response tends to be a decrease in energy reserves after exposure to different types of contaminants, either under laboratory conditions or in the field: copper, zinc, mercury, cadmium, lead, organochlorines (lindane, endosulfan), organophosphates (parathion), PCBs, domestic and/or industrial wastewater effluents or contaminated sediments (Table 12.1).

This response can be assessed within a few hours to a few days after initial exposure. This delay in the response has the advantage of not taking into account the stress linked to the handling of organisms. Besides, the detected variations persist over several days (Gimeno et al. 1995). In the collembolan (*Folsomia candida*, soil arthropods) exposed to the active substance dinoseb (herbicide), during 6 days at 15–30 mg g^{-1} dry soil, an increase in energy reserves and weight of treated groups has been reported compared to control groups. This result seems to indicate that collembolan adopt a strategy of increasing their growth in order to improve their reproduction since the number of eggs laid was greater in exposed organisms. But after 21 days, all measured parameters decreased, indicating that after they have produced an effort to increase growth and reproduction, lethality increases (Staempfli et al. 2007).

TABLE 12.1

Examples of Studies Investigating Influence of Chemical Exposure on Energy Reserves, under Laboratory Conditions

Chemical	Species	Exposure Conditions	Variable	Material Used for Measurements	Energy Reserves Response	Reference
Zn	*Daphnia magna*	Laboratory Concentrations: 0.1–1.0 μM	Proteins, carbohydrates, lipids	Whole body	Increase after preexposure	Canli 2005
Zn, Cd	*Poecilus cupreus*	Laboratory, contaminated food	Calorilic content	Whole body	Decrease	Maryanski et al. 2002
Cd	*Macoma balthica*	Laboratory for 5 weeks Concentrations: 0, 10, 30, 100, 300 ppb	Glycogen	Whole body	Decrease	Duquesne et al. 2004
Cd, Cu, Zn, Pb, Ni	*Dreissena polymorpha*	Active biomonitoring study	Glycogen Lipids	Digestive gland, gonad	Decrease	Palais et al. 2011
Cd	*Nucella lapillus*	Laboratory for 80 days Concentrations: 0.01 and 400 μg L^{-1}	Glycogen	Digestive gland, gonad	No impact	Leung and Furness 2001
Cd	*Nereis virens*	Laboratory for 2–80 days Concentrations: 0.5–40 mg L^{-1}	Glycogen	Rear segments	Decrease in the long term	Carr and Neff 1982
Pb, Cd, Cu Al, Ni, Zn	*Dendrobaena octaedra*	*In situ*	Glycogen	Whole body	No change Decrease	Holmstrup et al. 2011
Cd	*Anguilla anguilla*	Laboratory for 1 month Concentrations: 0–5 μg L^{-1}	Lipids	Muscle	Decrease	Pierron et al. 2007
Cu	*Daphnia magna*	Laboratory Concentrations: 0.5, 1, 5, 12, 35, 100 μg L^{-1}	Energy reserves (proteins, glycogen, lipids)	Whole body	Significant decrease in energy reserves in first generation specimens, compared with third generation specimens	Bossuyt and Janssen 2003

Pb	*Porcellio scaber*	Laboratory, for 7–80 days; Concentrations: 1–100 mg L^{-1}	Protein, Glycogen	Whole body	Decrease	Knigge and Köhler 2000
Hg Lindane	*Daphnia magna*	Laboratory, for 96 h Hg: 1.8, 3.2, 5.6, 10, 18, 24, 32 μg L^{-1} Lindane: 0.056, 0.10, 0.18, 0.32, 0.56, 0.75 mg L^{-1}	Polysaccharides	Whole body	Decrease Decrease	De Coen and Janssen 1997
	id.		Lipids	id.	Decrease Decrease	id.
	id.		Proteins	id.	Decrease Decrease	id.
Tebuconazole	*Daphnia magna*	Laboratory for 5 days: 0.41, 0.52, 0.71, and 1.14 mg L^{-1}	Proteins Lipids Glycogen	Whole body	Decrease	Sancho et al. 2009
Azadirachtin	*Plodia interpunctella* (bdc larvae)	Diet exposure for 7 days: 2 and 4 mg L^{-1}	Proteins Lipids Glycogen	Whole body	Decrease	Rharrabe et al. 2008
Endosulfan	*Anguilla anguilla*	Laboratory for 12, 24, 48, 72, 96 h; Concentrations: 4.1, 8.2 μg L^{-1}	Glycogen	Muscle	Decrease	Gimeno et al. 1995
Carbofuran	*Hyalella castroi*	Laboratory exposure for 7 days: 1 or 10 μg L^{-1}	Proteins Lipids Glycogen	Whole body	Decrease	Dutra et al. 2009
Parathion	*Porcellio dilatatus*	Laboratory for 21 days; Concentrations: 0.1, 1, 10, 25, 50, 100, 250, 500 μg g^{-1}	Glycogen Lipids Proteins	Whole body	Decrease Decrease Decrease	Ribeiro et al. 2001
Endosulfan	id.	id	id.	id.	Decrease Decrease No impact	
Fenhexamide	*Tubifex tubifex*	Laboratory, for 2, 4, and 7 days 0.1, 1, 10 mg L^{-1}	Glycogen, Proteins	Whole body	Significant decrease 2–4 days after exposure	Mosleh et al. 2005
Dinoseb	*Folsomia candida*	Laboratory for 6 and 21 days 15–30 mg g^{-1} dry soil	Lipids Proteins	Whole body	Increase (6 days) Decrease (21 days)	Staempfli et al. 2007
Chitosan	*Tubifex tubifex*	Laboratory for 2, 4, and 7 days 25, 62, 125 mg L^{-1}	Proteins	Whole body	Decrease	Mosleh et al. 2007
PCBs	*Crassostrea virginica*	Laboratory for 8 weeks 0.35–3.5 μg day^{-1} $individual^{-1}$ (algae)	Glycogen	Adductor muscle Mantle Gonad	Decrease No impact No impact	Encomio and Chu 2000

Moreover, the activation of any reserve constituents depends on the type of contaminant. For example, mobilization of glycogen and lipids can be observed in *Porcellio dilatatus* as a result of an exposure to endosulfan (0.1 to 500 μg g^{-1}) in the laboratory, without changing the protein levels, whereas the latter decrease after exposure to parathion (0.1 to 500 μg g^{-1}) (Ribeiro et al. 2001).

Among vertebrates, *Anguilla anguilla* (the European eel) exposed to cadmium showed lower lipid storage capacity owing to an increased use of triglycerides and lower body weight growth, all likely to compromise successful reproduction (Pierron et al. 2007).

Besides variations resulting from exposure to toxicants, the varying concentrations of energy reserves in organisms depend on numerous biotic and/or abiotic factors. The most significant natural variations in energy reserves, and notably in glycogen, are unquestionably linked to seasonal variations (Le Gal et al. 1997). Seasons, through variations in temperature and photoperiod, have a significant influence on the reproductive status of organisms in their natural habitats and lead them to accumulate energy reserves to meet the energy requirements of reproduction (Pellerin-Massicotte et al. 1994; Gauthier-Clerc et al. 2002; Mouneyrac et al. 2006; Durou and Mouneyrac 2007). The influence of parasitism can also lead to a depletion of energy reserves as observed in zebra mussels, *Dreissena polymorpha* (Minguez et al. 2009).

In *Daphnia magna*, physiological acclimatization to copper (from 0.5 to 100 μg L^{-1}) determines the high levels of energy reserves, when organisms are exposed for several successive generations in the laboratory (Bossuyt and Janssen 2003). In the third generation, individuals show an increased tolerance to copper and higher levels of energy reserves compared with first-generation individuals. In *Daphnia magna,* animals preexposed to zinc and having acquired a tolerance to this metal do not mobilize their energy reserves after exposure to zinc (0.1 and 1.0 μM) in the laboratory, compared with organisms that have not been preexposed (Canli 2005).

Regarding the influence of diet on energy reserve levels in *Nucella lapillus* (predatory mollusk), nutrition state and more particularly the prey type help organisms tolerate the toxic effects of cadmium and maintain glycogen reserves at appropriate levels (Leung and Furness 2001). The authors did not record any decrease in glycogen concentrations following exposure to cadmium. However, long periods of fasting in organisms exposed to cadmium led to mass loss, with high metal concentrations resulting in increased mortality. The authors insist on the necessity of taking into account growth (weight dilution of toxicants), condition, and food quality of organisms for a sound interpretation of results. On the other hand, in order to prevent toxicant intrusion [e.g., polycyclic aromatic hydrocarbons (PAHs)], bivalve mollusks such as oysters can close their shells, leading to a reduced food and thus energy intake and consequently accelerate the lipid and carbohydrate metabolism in order to meet the energy demand (Jeong and Cho 2007).

Therefore, it is not surprising that the utilization of energy reserves as an accurate biomarker of pollution is still being debated. Actually, even though analytical techniques are now perfected, the influence of natural factors (age, sex, reproductive state, etc.) or environmental ones (temperature, oxygen levels, season, etc.) on those biomarkers is still being studied (Verslycke and Janssen 2002). However, in *Dreissena polymorpha*, energy reserves of lipids are directly connected to the weight condition of organisms and their reproductive status (Voets et al. 2006). Recently, in the freshwater gammarid *Gammarus roeseli*, higher levels of lipids, proteins, and antioxidant activities measured in females suggest that females could have better defense capacities against oxidative stress than males (Sroda and Cossu-Leguille 2011).

Specific measurements of energy reserves are of little use—when considered alone—for assessing the health condition of individuals exposed to pollutants in the field. However,

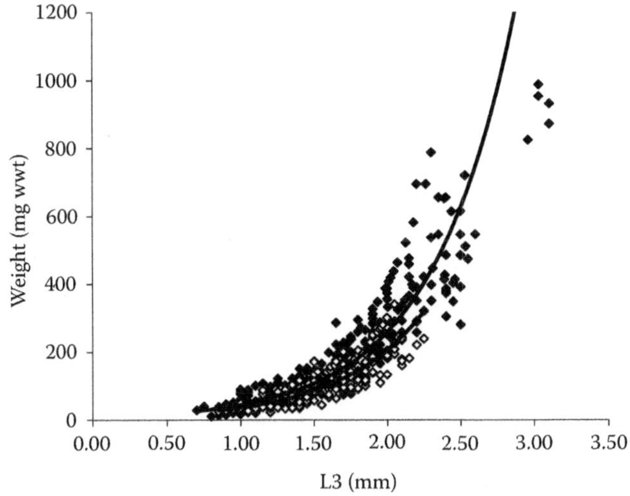

FIGURE 12.2
Intersite difference of size–weight index in *Nereis diversicolor* originating from Authie (◆) and Seine (◇) estuaries (Durou 2006).

within the framework of a survey on the condition of populations from one species, energy reserves could provide essential information on the individuals' ability to survive, grow, and reproduce, and consequently they yield long-term information about the fitness of a population. For example, in *Nereis diversicolor,* originating from both clean and contaminated sites, the measured energy reserve levels of organisms from the contaminated site generally decrease and are associated with mass loss (see Section 12.2 and Figure 12.2). And yet, these results are not universal since the effects vary depending on the studied species and involved contaminants. In *Biomphalaria glabrata,* a freshwater gastropod mollusk, exposure in the laboratory to 0.1 and 0.05 mg Cd L^{-1}; 0.5, 0.1, and 0.05 mg Pb L^{-1}; and 0.5, 0.1, and 0.05 mg As L^{-1}, showed that arsenic has a stronger impact on the glycogen in the gonad than cadmium (Ansaldo et al. 2006). The cadmium toxic effect is counterbalanced by metallothionein, which explains the latter result (Amiard et al. 2006). In a field study, the earthworm, *Dendrobaena octaedra,* was collected from polluted and unpolluted reference areas. Pb, Cd, and Cu accumulation did not correlate with glycogen reserves of worms, whereas Al, Ni, and Zn were negatively correlated with glycogen levels. These results suggest that, in these earthworms, coping with different metals is associated with differential energy demands depending on the detoxification strategy involved (Holmstrup et al. 2011).

12.1.3 Cellular Energy Allocation

If the adenylate charge or energy reserves provide evidence of the available energy in organisms, it is also necessary to consider integration parameters of both the availability and consumption of energy by organisms. In this regard, the CEA accounts for the net energy budget at cellular level (De Coen and Janssen 1997):

$$\mathrm{CEA} = \frac{\int_0^t E_\mathrm{a}\,\mathrm{d}t - \int_0^t E_\mathrm{c}\,\mathrm{d}t}{t}. \tag{12.1}$$

TABLE 12.2

Examples of Studies Investigating Influence of Chemical Exposure on Cellular Energy Allocation under *in situ* or Laboratory Conditions

Exposure	Species	Material Used for Measurements	Effect	Reference
Tributyltin chloride	*Neomysis integer*	Whole body	Decrease	Verslycke et al. 2003
Chlorpyrifos	*Neomysis integer*	Whole body	Decrease	Verslycke et al. 2004
Hg; Lindane	*Daphnia magna*	Whole body	Decrease	De Coen and Janssen 1997
Cd Cr TBT 2,4-D* Lindane PCP	*Daphnia magna*	Whole body	Decrease Decrease Decrease Decrease Decrease Decrease	De Coen and Janssen 2003
In situ (transplantation) pollution gradient	*Dreissena polymorpha*	Whole body	Decrease	Smolders et al. 2004
Oil related compounds	*Gammarus setosus* *Onisimus litoralis* *Liocyma fluctuosa*	Whole body	Decrease Increase No change	Olsen et al. 2007
Oil related compounds	*Gammarus wilkitzkii*	Whole body	Decrease	Olsen et al. 2008
Cd, Zn, Cd+Zn Cd, Zn, Cd+Zn	*Melanoides tuberculata* *Helisoma duryi*	Soft tissues	Decrease	Moolman et al. 2007

*2,4 dichloro phenoxyacetic (2,4-D).

Equation 12.1 shows the measurement of the CEA, where E_a represents the available energy in the organism (the sum of the energy contributions of the three energy constituents: carbohydrates, lipids, and proteins), E_c is the consumed energy, t is the exposure duration, and dt is the time derivate.

CEA has been successfully applied in laboratory and field studies (Table 12.2) in freshwater environments using daphnia and zebra mussel (Smolders et al. 2004) and gastropods (Moolman et al. 2007); in estuarine environments using *Neomysis integer* (Verslycke et al. 2004; Erk et al. 2008) and *Mytilus galloprovincialis* (Erk et al. 2011); and in polar marine environments using Arctic benthic amphipods and bivalves (Olsen et al. 2007, 2008). The CEA approach has been shown to be ecologically relevant as cellular effects have been linked to higher levels of biological organization (De Coen and Janssen 2003; Smolders et al. 2004). Moreover, a significant influence of natural factors such as salinity—a key factor involved in osmoregulation processes—on CEA has been demonstrated in mussels (*M. galloprovincialis*). The decrease in CEA value observed in mussels from the estuarine site compared to those from the coastal site indicates a lower amount of energy available for growth, reproduction, or defense against pollution and may implicate their susceptibility to stress (Erk et al. 2011).

12.1.4 Scope for Growth

To cope with the suffered stress, an increase in the energy consumption of the basal metabolic rate may result in a decrease in the energy invested into growth and/or reproduction.

In a freshwater invertebrate (*Nephelopsis obscura*), changes in energy allocation (i.e., increased mucus production) have been observed in animals exposed to cadmium under laboratory conditions, thus contributing to a decreased growth rate compared with the control animals (Wicklum and Davies 1996). Actually, the net available energy budget for growth and reproduction may be considered a biomarker of the general condition of health. An interesting and convenient application of this principle is the energy allocation to growth and reproduction, that is, SfG. The SfG notably takes into account the assimilation of energy through food, the energy released in respiration and lost in excretion—and more particularly, nitrogen derivatives (Le Gal et al. 1997).

In fact, the SfG has been used for a long time to measure the impact of contaminants on biological organisms in their natural habitat (Smaal and Widdows 1994). Such a method is largely used in mussels, because of the extensive knowledge of their characteristics and behavior, and their use as sentinel species in numerous biomonitoring programs (Smaal and Widdows 1994; Widdows et al. 1997, 2002). Significant negative correlations between PCBs, DDT, and organochlorine concentrations have been measured in tissues of exposed mussels in the Venice lagoon, but not in the case of metals (Widdows et al. 1997). Some surveys have shown that the significant decreases in SfG are essentially attributable to the presence of PAHs (Widdows et al. 2002), or low values of SfG were attributed to exposure to organotins and PAHs (Halldórsson et al. 2005).

Besides the use of SfG in environmental monitoring, variations of SfG are also studied under laboratory experiments. In the laboratory, Baillieul et al. (2005) investigated the influence of food availability and cadmium exposure on daphnids' growth and reproduction. Both types of stress affected the body size of daphnids. The SfG was determined after 14 days of exposure and calculated either per individual or weight corrected. Decreasing body weight was associated with decreasing SfG, regardless of whether SfG was calculated on individual basis or weight corrected. Instead, reproduction was positively correlated with SfG calculated on individual basis and negatively with SfG weight corrected.

Verslycke et al. (2004) studied the responses of *N. integer* in terms of SfG and CEA following exposure to the active substance chlorpyrifos. The treated groups were exposed—under laboratory conditions—to environmentally relevant test concentrations of 0.038, 0.056, 0.072, and 0.100 μg L^{-1} during 48, 96, and 168 h. Under the conditions of experiments, chlorpyrifos exposure significantly decreases SfG of *N. integer*, the effect being much stronger at the highest test concentration of chlorpyrifos. Comparison of SfG versus CEA allows the authors to conclude that the CEA would be a more sensitive biomarker than SfG. Numerous factors have an impact on the SfG: temperature, dissolved oxygen concentration, food availability, sizes/weights of organisms, reproductive cycle (Mayer et al. 2002). Natural factors can also act as stressors and affect SfG. In the estuarine *Atrina maura* (a bivalve mollusk), SfG was significantly correlated with food availability, i.e., microalgae (Arrieche et al. 2011). Respiration measurement has proved to be integrative of metabolism since it is characteristic of the intensity of the physiological processes at the time of measurement. Increases in the standard metabolic rate (SMR) assessed through the measurement of oxygen consumption were observed in various species (crustaceans, amphibians, reptiles) exposed to metals in the field or under laboratory conditions (Rowe 1998; Hopkins et al. 1999). In oysters (*Crassostrea virginica*) exposed to cadmium (50 μg L^{-1}) under laboratory conditions and acclimatized to various temperatures (20, 24, and 28°C), increases in SMR were higher at 20°C, intermediate at 24°C, and lacking at 28°C, in comparison with the control groups. Therefore, the efficiency of oxygen supply mechanisms could be cancelled by the combination of high temperatures and exposure to cadmium (Lannig et al. 2006). At cell level, significant respiratory costs were highlighted in rainbow trout

(*Oncorhynchus mykiss*) hepatocytes during the biotransformation of pyrene, an organic contaminant (Bains and Kennedy 2004). Impaired aerobic capacities were observed in the muscle of metal-contaminated wild yellow perch (*Perca flavescens*) and were also reflected in aerobic capacities of the whole fish (Couture and Kumar 2003). However, interpretation of results must consider the fact that the response does not take into account anaerobic energy production—representing approximately 3%—(Abele et al. 1998). Moreover, the interpretations of available energy variations must also take into account the tolerance limits of benthic invertebrates to temperature increase in intertidal areas, which varies highly according to species. These tolerance variations involve complex interactions in which responses to stress activate the pro-oxidant potential of catabolic activities resulting in membrane lipid peroxidation. The resulting accumulation of neutral lipids is often associated with the destabilization of the lysosomal membrane, the latter being probably an indicator of the degradation of the lipid metabolism and development of autophagic processes (see Chapter 5).

12.1.5 Condition Indices

Morphological and/or biometric measurements can be used to assess condition indices in various organisms such as bivalve mollusks, fish, birds, and mammals (Mayer et al. 2002). Hepatosomatic and gonadosomatic indices are the ratio between the organ mass (liver, gonad) and the whole body mass. Body condition indices (condition index/factor) based on the measurements of the masses and lengths of organisms are used to determine their general condition of health.

In *Mya arenaria*, an intrasedimentary bivalve mollusk, decreases in gonadosomatic condition and index are observed in the samples gathered from the areas impacted by effluents from harbor activities (Blaise et al. 2002; Gagné et al. 2002; Gauthier-Clerc et al. 2002; Gagné et al. 2003; Siah et al. 2003). In freshwater mussels (*Dreissena polymorpha*), a decrease in the condition index of organisms from the multipolluted estuary of the Seine is assessed (Minier et al. 2006). In *Platichthys flesus*, flatfish taken from several impacted estuaries (Loire, Seine, or Gironde), an overall decrease in the condition factor is observed, compared with samples taken from the reference Ster estuary (Marchand et al. 2004). In the gudgeon (*Gobio gobio*), no correlation could be established between the condition factor and the metal contamination level. Nevertheless, within the pollution gradient, a concentration threshold was found in the liver, gill, and kidney, above which the condition factor is always low (Bervoets and Blust 2003). In the wild yellow perch (*P. flavescens*) inhabiting clean and contaminated areas, the condition factor was correlated with indicators of aerobic metabolism (citrate synthase, cytochrome *c* oxidase), protein anabolism (nucleoside diphosphokinase), and indicators of lipid accumulation such as visceral lipid index (Gauthier et al. 2011). These types of indices used in biomonitoring programs are not currently available in annelids, and more particularly in *N. diversicolor*, an intrasedimentary annelid usually recommended in biomonitoring programs on environmental quality (Chapter 7). Within the framework of the population monitoring from the Seine and the Authie (site of reference), a size–weight index with size "L3" (measurement of the first three segments) has been developed (Durou et al. 2008). This index, tested in the field over more than a 1-year cycle, clearly highlights a difference in the condition of organisms depending on sites, with a particularly deteriorated condition in samples from the Seine estuary (Figure 12.2).

This type of index can be used on a larger scale, notably on sites with more varied food availability and contamination levels. Considering the significant level of contaminants stored in the sediment compartment, the determination of benthic bioindicators is now relevant.

12.2 Energy Biomarkers: Predictors for a Change of Scale?

The life history of organisms depends on internal and environmental factors. It may be described using life history traits such as life expectancy, age or size at sexual maturity, and fecundity. Any modification of these characteristics has rapid and direct consequences on a population, such as decreases in biomass and structural changes in the population (size/age relationships, recruitment dynamics). In this chapter, we have discussed the different bioenergy biomarkers likely to detect a change in energy allocation. These biomarkers are expected to provide a forecast of the variations at the population level, since populations are among the major concerns of ecological risk assessment. Today, it is necessary to establish a link between the changes in energy metabolism and variations in life history traits.

AEC was one of the first bioenergy biomarkers used. Values are compared with different life history traits such as survival and an organism's ability to grow and reproduce (Le Gal et al. 1997). However, in case of severe stress, these values may vary significantly, very likely because of an increasing number of adverse effects on the physiology of organisms.

Decreases in energy reserves are usually observed in organisms following exposure to contaminants (Le Gal et al. 1997; Mayer et al. 2002). Energy reserves may be mobilized to trigger mechanisms of defense (storage, excretion, detoxification of contaminants) (Chapter 3).

In the case of field surveys, a large number of parameters should be taken into account. First, it is generally assumed that organisms living in conditions close to their threshold of tolerance are more sensitive to additional stress (Heugens et al. 2001). The ecological factors to be especially taken into account are temperature, salinity, and the availability of food in the natural habitat (Heugens et al. 2001). In case of insufficient food availability, organisms reallocate their resources so as to favor survival to the detriment of reproduction (Smolders et al. 2005). High levels of energy reserves often coincide with a suitable weight condition and abundant food supplies. Consequently, the nutritional status of organisms greatly influences the response of energy reserves as biomarkers. In addition, numerous surveys combine the measurement of energy reserves with variables highlighting food intake such as the filtering rate in microphages (Muyssen et al. 2006). Various authors consider that exposure to contaminants even disturbs the nutrition of organisms, and make nutrition assessment within the framework of ecotoxicity surveys (Reynaldi et al. 2006; Shipp and Grant 2006). From this perspective, we wish to report the experience we gained during the "Programme National d'Écotoxicologie" (National Program in Ecotoxicology) (2002–2004), which developed a multiparameter approach (chemical, biochemical biomarkers, ecological) to *Nereis diversicolor*, a key species of the sediment compartment in estuaries. This survey in the field was carried out on annelid populations of the Authie and Seine estuaries, during a 3-year monitoring period. Biomarkers usually used in ecotoxicology [acetylcholinesterase (AChE), catalase (CAT), glutathione *S*-transferase (GST) activities, and thiobarbituric acid reactive substance concentrations; Chapter 2], energy reserves (glycogen, lipids, and proteins), and contaminants (PCBs, PAHs, and metals) were determined in the whole organism. Relationships between biomarkers, energy reserves, and contaminants were examined, using a principal component analysis (Durou et al. 2007). The distribution of data in the principal plane (Figure 12.3) clearly contrasts the Seine observations located on the positive axis 1—*positively correlated with the GST activity and chemical*

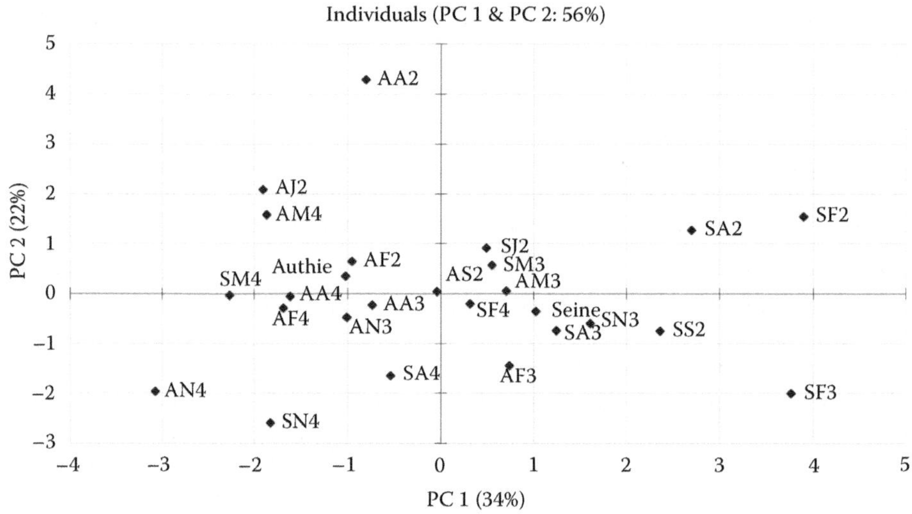

FIGURE 12.3
Principal component analysis representing on plane 1–2, the observations carried out in the Authie and Seine estuaries. Initials of each data set include the collection site (A = Authie, S = Seine), the sampling month (F = February, J = July, M = May, N = November, S = September, A2 = April 2002, A3 = August 2003) and the year (2 for 2002, 3 for 2003 and 4 for 2004).

contaminant concentrations—and the results of Authie on the negative axis—*correlated with high energy reserves*. Besides showing induction of GST resulting from higher contaminant contents, organisms from the Seine generally had lower energy reserve levels. Organisms from both sites were also contrasted in the principal plane 1–3 (not shown) with high levels of PCBs and a lower AChE activity in the Seine compared to the Authie (Durou 2006).

Concomitantly, a survey into the reproduction of these organisms was conducted to assess more particularly the potential fecundity of females originating from both sites. In this species, gametes are produced in the coelom and consequently, fecundity largely depends on body volume. Organisms from the Authie generally had higher energy reserve levels than those from the Seine, and also had bigger size and mass. Observations revealed that females from the Seine (impacted site) had lower fecundity than females from the Authie, in accordance with their smaller sizes (Figure 12.4).

At the population level, structures clearly varied (Durou et al. 2005). During a 3-year monitoring period, six cohorts (cohorts of individuals) from the Seine (CSx) and eight cohorts from the Authie (CAx) were observed (Figure 12.5). It is also probable that life expectancy in the Seine was shorter; the recruitment of a new generation of individuals appeared concomitantly with the disappearance of the previous generation.

Thus, in the field, responses of biochemical, bioenergy, and reproduction biomarkers were compared with the population status in *N. diversicolor* originating from the Authie and Seine estuaries. The difficulty of using energy reserves as pollution biomarkers mainly comes from the fact that they provide a "snapshot" of the situation and do not help to effectively understand the energy investment already made for growth or reproduction functions. For example, the Seine samples had high levels of energy reserves in the spring, a period when females reach sexual maturity and can reproduce (Durou and Mouneyrac 2007). Nevertheless, in the spring, animals had smaller size and mass and lower fecundity than those from the Authie.

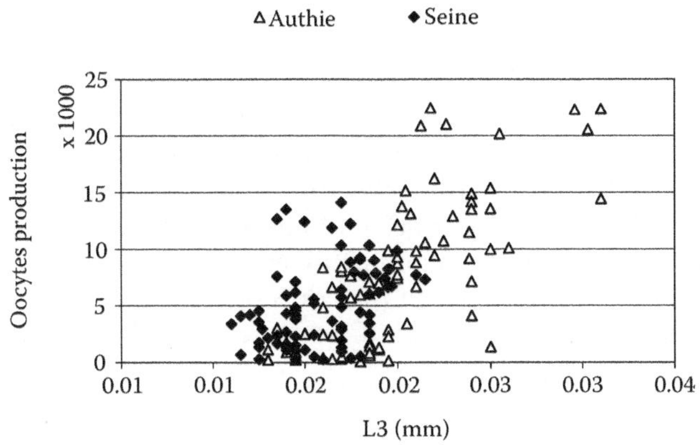

FIGURE 12.4
Fecundity in *N. diversicolor* females from the Authie and Seine estuaries, according to the length of first three segments.

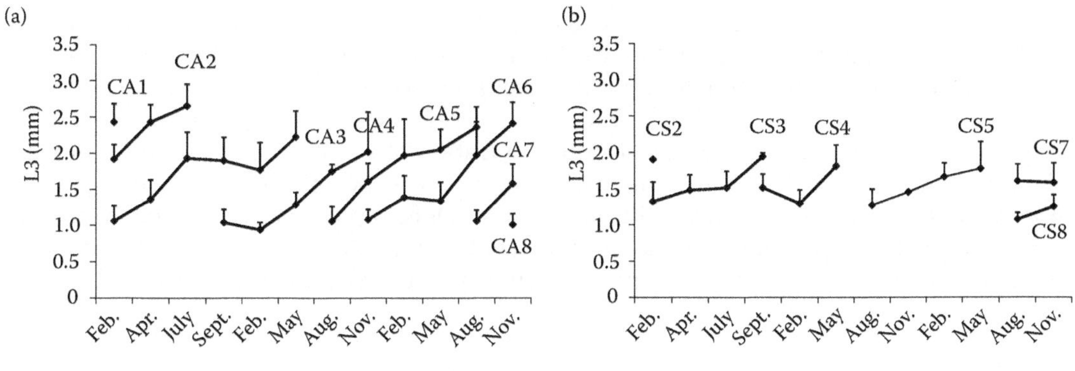

FIGURE 12.5
Population variations in *N. diversicolor* originating from Authie (a) and Seine (b) estuaries, over a 3-year monitoring period.

Could we use the variations in CEA to forecast effects at population level? In *Daphnia magna*, which has a very short life cycle, the ecological relevance of such an energy biomarker was examined. In this species, exposure to various toxic agents under laboratory conditions (Table 12.2) resulted in a decrease in the cellular energy budget, along with a decrease in the available energy (energy reserves) and an increase in the energy used (respiration). At the same time, decreases in the size of organisms and offspring number were noted (De Coen and Janssen 1997, 2003). These research works established a link between CEA variations identified at the cell level and individual responses having direct consequences on the population.

Over the past 10 years, mathematical models of the effects based on the impact of energy have been set up, using the DEBtox software package developed in the Netherlands (Kooijman and Bedaux 1996). These models aim to explain the disruption of dynamic energy in organisms, in connection with the accumulation of toxic compounds.

12.3 Modeling of Effects Based on Impact of Energy

To model the effects of a chemical on energy, the first point is to examine the relationships between its accumulation in the organisms and its effects to obtain a certain amount of toxicological relevance. In general, only exposure concentration is available, whereas the effect is mainly linked to the concentration at target level. Usually, a simple but realistic toxicokinetic model, also called one-compartment linear model, is used. The absorption rate of the compound is assumed to be proportional to exposure concentration, and its elimination rate proportional to its internal concentration. The following equation is then obtained, for nongrowing organisms:

$$\frac{dC_i}{dt}(t) = k_a \times c_e(t) - k_e \times C_i(t) \tag{12.2}$$

where C_i is the compound internal concentration, c_e is its exposure concentration, and k_a and k_e are its absorption and elimination rates, respectively. Taking compound kinetics into account has the advantage of highlighting the temporal aspects of the exposure. In effect models, the effect is directly linked to the internal concentration, the only time factor involved being the determination of the concentration using the kinetic model. As a consequence, parameter estimates do not depend on test duration.

Energy-based models presuppose the existence of a no-effect concentration, i.e., that organisms can tolerate a certain level of concentration of xenobiotics, either by inhibiting them through various trap mechanisms, or because a sufficient number of functions are operational in a normally functioning individual. Once organisms are unable to tolerate the accumulated toxicant, proportionality exists between the observed effect and excess toxin; we will call "toxicity rate" the proportionality factor k_i between effects and threshold excess.

$$Effect = k_i(C_i - NEC) \tag{12.3}$$

The effect is a disruption of the parameters controlling energy in organisms and is represented by the Dynamic Energy Budget theory. This theory (Kooijman 2010) aims to provide a mathematical representation of the use of energy from food intake, shared among four major physiological functions: growth, maintenance of vital functions, maintenance of reproductive capacity, and reproduction. This theory was tested on roughly 270 vertebrate and invertebrate species. The physiological action mechanism of the toxicant corresponds to the affected parameter. More specifically, regarding survival, the effect is an increase in mortality rate. Regarding growth, the effect involves either a maintenance cost increase, an increase in the synthesis cost of new tissues, or a decrease in the nutrition rate. Regarding reproduction, the effect could either disturb growth and thus reduce reproduction, or involve a direct reduction in fecundity.

The accuracy of this approach to assess the dose–response relationships was verified with several species of invertebrates and different types of toxicants (pesticides, metals, PAHs, etc.), and the results obtained were often significantly different from those expected with a more standard statistical approach (Kooijman and Bedaux 1996; Péry et al. 2003; Ducrot et al. 2004; Alvarez et al. 2005). Similarly, the accuracy of the assessed toxicity threshold was demonstrated in *Chironomus riparius* (Péry et al. 2008) for cadmium, copper, and zinc. The quality of data description can, in theory, help identify the mechanism of toxic action if it is not known at the time of test. This identification is feasible but is rather tricky and generally requires more toxicity tests (Péry et al. 2003).

One of the major advantages of the energy-based models is their relevance when shifting from the individual to the population level. In fact, these models enable the effects to be extrapolated to any temporal profile of exposure. They allow a dynamic approach to the occurrence and degree of effects over time, which corroborates the relevance of studying population dynamics (Péry et al. 2006). In addition, by focusing on energy, these models help to assess the influence of compensation mechanisms (e.g., an increase in mortality involves more food for survivors, and increasing reproduction and growth) and assess the effects on biomass (Péry et al. 2004) or on the capacity for domination in interspecies competition (Hülsmann et al. 2005).

12.4 Recommendations, Conclusions, and Perspectives

The current use of biomarkers connected with energy metabolism, as contamination biomarkers, is rather limited for multiple reasons.

First, whatever the considered biomarker, it is quite obvious that irrespective of any contamination effect, intrinsic factors such as age or reproductive status as well as environmental factors such as temperature or oxygen level, influence the response of these biomarkers. Consequently, an in-depth knowledge of the biological model is a prerequisite. In addition, tolerance levels of the studied organisms toward natural factors should be established precisely, notably through laboratory experiments. Given the current changes in climatic conditions, the influence of temperature—an essential ecological factor—should be looked at as a matter of priority, more particularly within the framework of long-term environmental monitoring programs.

Second, these biomarkers are sometimes associated with other general stress markers such as decrease in growth rate (Ribeiro et al. 2001; Mosleh et al. 2005), decrease in organisms' size and weight (Maryanski et al. 2002), or in condition index. Therefore, we recommend that in future surveys including bioenergy indicators, general condition indices should be assessed as well as the reproductive status of organisms.

Responses of bioenergy parameters to chemical stressors differ, depending on the delay between the actual exposure and the observed effect. The effects at subcellular level (energy reserves) are quicker than effects at cellular level (CEA), and even quicker at individual or population levels. Although variations in energy reserves arise before CEA variations, the delay of the latter ranges from a few hours to a few days (De Coen and Janssen 1997). This quick response time requires meticulous sampling and handling of organisms.

Finally, few surveys have aimed to establish links between energy-based models and energy metabolism biomarkers. First, it is obvious that the modeling approach described in this article includes the generalization and formalization of the SfG. And yet, there are no ecotoxicological experiments establishing a link between these two tools. Some authors sought to connect energy reserve measurements with the effects of copper on the growth and reproduction of *Chironomus riparius* (Servia et al. 2005). It seemed that energy reserve measurements could not provide an appraisal of the effects on growth. In fact, one animal with 2-day growth delay cannot be distinguished from an animal that is measured

2 days before, and energy measurements appear totally redundant with development measurements. However, there is a close correlation between lipid reserves and reproduction measurements. The most valuable measurement provided by energy-based models is the internal concentration measurement (Péry et al. 2008), which does not require examining bioavailability issues and enables a measurement related to effects that is relatively independent from experimental conditions.

References

Abele, D. et al. 1998. Exposure to elevated temperatures and hydrogen peroxide elicits oxidative stress and antioxidant response in the Antarctic intertidal limpet *Nacella concinna*. *Comp. Biochem. Physiol.* 120B:425–35.

Alvarez, O. A. et al. 2005. Responses to stress of *Caenorhabditis elegans* populations with different reproductive strategies. *Funct. Ecol.* 19:656–64.

Amiard, J. C. et al. 2006. Metallothioneins in aquatic invertebrates: Their role in metal detoxification and their use as biomarkers. *Aquat. Toxicol.* 76:160–202.

Ansaldo, M. et al. 2006. Potential use of glycogen level as biomarker of chemical stress in *Biomphalaria glabrata*. *Toxicology* 224:119–27.

Arrieche, D. et al. 2011. Scope for growth, biochemical composition, and antioxidant immune responses of the penshell *Atrina maura* to flow velocity and concentration of microalgae. *Aquaculture* 319:211–20.

Baillieul, M., R. Smolders, and R. Blust. 2005. The effect of environmental stress on absolute and mass-specific scope for growth in *Daphnia magna* Strauss. *Comp. Biochem. Physiol.* 140C:364–73.

Bains, O. S., and C. J. Kennedy. 2004. Energetic costs of pyrene metabolism in isolated hepatocytes of rainbow trout, *Oncorhynchus mykiss*. *Aquat. Toxicol.* 67:217–26.

Bervoets, L., and R. Blust. 2003. Metal concentrations in water, sediment and gudgeon (*Gobio gobio*) from a pollution gradient: Relationship with fish condition factor. *Environ. Pollut.* 126:9–19.

Blaise, C. et al. 2002. Molluscan shellfish biomarker study of the Quebec, Canada, Saguenay Fjord with the soft-shell clam, *Mya arenaria*. *Environ. Toxicol.* 17:170–86.

Bossuyt, B. T. A., and C. R. Janssen. 2003. Acclimation of *Daphnia magna* to environmentally realistic copper concentrations. *Comp. Biochem. Physiol.* 136C:253–64.

Canli, M. 2005. Dietary and water-borne Zn exposures affect energy reserves and subsequent Zn tolerance of *Daphnia magna*. *Comp. Biochem. Physiol.* 141C:110–6.

Carr, R. S., and J. M. Neff. 1982. Biochemical indices of stress in the sandworm *Neanthes virens* (Sars): II. Sublethal responses to cadmium. *Aquat. Toxicol.* 2:319–33.

Couture, P., and P. R. Kumar. 2003. Impairment of metabolic capacities in copper and cadmium contaminated wild yellow perch (*Perca flavescens*). *Aquat. Toxicol.* 64:107–20.

De Coen, W. M., and C. R. Janssen. 1997. The use of biomarkers in *Daphnia magna* toxicity testing: IV. Cellular Energy Allocation: A new methodology to assess the energy budget of toxicant-stressed *Daphnia* populations. *J. Aquat. Ecosyst. Stress Recov.* 6:43–55.

De Coen, W. M., and C. R. Janssen. 2003. The missing biomarker link: Relationship between effects on the cellular energy allocation biomarker of toxicant-stressed *Daphnia magna* and corresponding population characteristics. *Environ. Toxicol. Chem.* 22:1632–41.

Ducrot, V. et al. 2004. Energy based modelling as a basis for the analysis of reproduction data with the midge *Chironomus riparius*. *Environ. Toxicol. Chem.* 23:225–31.

Duquesne, S., M. Liess, and D. J. Bird. 2004. Sub-lethal effects of metal exposure: Physiological and behavioural responses of the estuarine bivalve *Macoma balthica*. *Mar. Environ. Res.* 58:245–50.

Durou, C. 2006. Recherche d'indicateurs de l'état physiologique de l'annélide polychète endogée *Nereis diversicolor* en relation avec la qualité du milieu. PhD thesis, Université de Nantes, France.

Durou, C., and C. Mouneyrac. 2007. Linking steroid hormone levels to sexual maturity index and energy reserves in *Nereis diversicolor* from clean and polluted estuaries. *Gen. Comp. Endocrinol.* 150:106–13.
Durou, C. et al. 2005. Linking the physiological and population status of worms *Nereis diversicolor* from a clean and a multi-polluted estuary. In SETAC E (ed) *15th Annual Meeting, The Raison d'être of Environmental Toxicology and Chemistry*, Lille, France.
Durou, C. et al. 2007. Biomonitoring in a clean and a multi-contaminated estuary based on biomarkers and chemical analyses in the endobenthic worm *Nereis diversicolor. Environ. Pollut.* 148:445–58.
Durou, C., C. Mouneyrac, and C. Amiard-Triquet. 2008. Environmental quality assessment in estuarine ecosystems: Use of biometric measurements and fecundity of the ragworm *Nereis diversicolor* (Polychaeta, Nereididae). *Water Res.* 42:2157–65.
Dutra, B. K. et al. 2009. Carbofuran-induced alterations in the energy metabolism and reproductive behaviors of *Hyalella castroi* (Crustacea, Amphipoda). *Comp. Biochem. Physiol.* 149C:640–6.
Encomio, V., and F. L. Chu. 2000. The effect of PCBs on glycogen reserves in the eastern oyster *Crassostrea virginica. Mar. Environ. Res.* 50:45–9.
Erk, M. et al. 2008. Metallothionein and cellular energy allocation in the estuarine mysid shrimp *Neomysis integer* exposed to cadmium at different salinities. *J. Exp. Mar. Biol. Ecol.* 357:172–80.
Erk, M., D. Ivanković, and Z. Strižak. 2011. Cellular energy allocation in mussels (*Mytilus galloprovincialis*) from the stratified estuary as a physiological biomarker. *Mar. Pollut. Bull.* 62:1124–9.
Gagné, F. et al. 2002. Alteration of the biochemical properties of female gonads and vitellins in the clam *Mya arenaria* at contaminated sites in the Saguenay Fjord. *Mar. Environ. Res.* 53:295–310.
Gagné, F. et al. 2003. Sex alteration in soft-shell clams (*Mya arenaria*) in an intertidal zone of the Saint Lawrence River (Quebec, Canada). *Comp. Biochem. Physiol.* 134C:189–98.
Gauthier, C. P., C. Campbell, and P. Couture. 2011. Enzymatic correlates of energy status in wild yellow perch inhabiting clean and contaminated environments. *Environ. Toxicol. Chem.* 30:2148–56.
Gauthier-Clerc, S. et al. 2002. Delayed gametogenesis of *Mya arenaria* in the Saguenay fjord (Canada): A consequence of endocrine disruptors? *Comp. Biochem. Physiol.* 131C:457–67.
Gimeno, L. et al. 1995. Pesticide effects on eel metabolism. *Ecotoxicol. Environ. Saf.* 31:153–7.
Halldórsson, H. P., J. Svavarsson, and Å. Granmo. 2005. The effect of pollution on scope for growth of the mussel (*Mytilus edulis* L.) in Iceland. *Mar. Environ. Res.* 59:47–64.
Heugens, E. H. W. et al. 2001. A review of the effects of multiple stressors on aquatic organisms and analysis of uncertainty factors for use in risk assessment. *Crit. Rev. Toxicol.* 31:247–84.
Holmstrup, M. et al. 2011. Body metal concentrations and glycogen reserves in earthworms (*Dendrobaena octaedra*) from contaminated and uncontaminated forest soil *Environ. Pollut.* 159:190–7.
Hopkins, W. A., C. L. Rowe, and J. D. Congdon. 1999. Elevated trace element concentrations and standard metabolic rate in banded water snakes (*Nerodia fasciata*) exposed to coal combustion wastes. *Environ. Toxicol. Chem.* 18:1258–63.
Hülsmann, S., K. Rinke, and W. M. Mooij. 2005. A quantitative test of the size efficiency hypothesis by means of a physiologically structured model. *Oikos* 110:43–54.
Jeong, W. G., and S. M. Cho. 2007. Long-term effect of polycyclic aromatic hydrocarbon on physiological metabolisms of the Pacific oyster, *Crassostrea gigas. Aquaculture* 265:343–50.
Knigge, T., and H.R. Köhler. 2000. Lead impact on nutrition, energy reserves, respiration and stress protein (hsp 70) level in *Porcellio scaber* (Isopoda) populations differently preconditioned in their habitats. *Environ. Pollut.* 108:209–17.
Kooijman, S. A. L. M. 2010. *Dynamic Energy Budget Theory for Metabolic Organisation*. Cambridge: Cambridge University Press.
Kooijman, S. A. L. M., and J. J. M. Bedaux. 1996. *The Analysis of Aquatic Toxicity Data*. Amsterdam: VU University Press.
Lannig, G., J. F. Flores, and I. M. Sokolova. 2006. Temperature-dependent stress response in oysters, *Crassostrea virginica*: Pollution reduces temperature tolerance in oysters. *Aquat. Toxicol.* 79:278–87.

Le Gal, Y. et al. 1997. Charge énergétique en adénylates (CEA) et autres biomarqueurs associés au métabolisme énergétique. In Lagadic, L., Caquet, T., Amiard, J. C., Ramade, F. *Biomarqueurs en Écotoxicologie: Aspects Fondamentaux*. Paris: Masson, 241–85.

Leung, K. M., and R. W. Furness. 2001. Survival, growth, metallothionein and glycogen levels of *Nucella lapillus* (L.) exposed to sub-chronic cadmium stress: The influence of nutritional state and prey type. *Mar. Environ. Res.* 52:173–94.

Marchand, J. et al. 2004. Physiological cost of tolerance to toxicants in the European flounder *Platichthys flesus*, along the French Atlantic Coast. *Aquat. Toxicol.* 70:327–43.

Maryanski, M. et al. 2002. Decreased energetic reserves, morphological changes and accumulation of metals in carabid beetles (*Poecilus cupreus* L.). *Ecotoxicology* 11:127–39.

Mayer, F. L. et al. 2002. Physiological and non specific biomarkers. In Huggett, R. J., Kimerle, R. A., Mehrle, P. M. J., Bergman, H. L. *Biomarkers: Biochemical, Physiological and Histological Markers of Anthropogenic Stress*. SETAC Special Publication Series. Boca Raton, FL: Lewis Publishers, 5–85.

Minguez, L. et al. 2009. Interactions between parasitism and biological responses in zebra mussels (*Dreissena polymorpha*): Importance in ecotoxicological studies. *Environ. Res.* 109:803–50.

Minier, C. et al. 2006. A pollution-monitoring pilot study involving contaminant and biomarker measurements in the Seine estuary, France, using zebra mussels (*Dreissena polymorpha*). *Environ. Toxicol. Chem.* 25:112–9.

Moolman, L., J. H. J. Van Vuren, and V. Wepener. 2007. Comparative studies on the uptake and effects of cadmium and zinc on the cellular energy allocation of two freshwater gastropods. *Ecotoxicol. Environ. Saf.* 68:443–50.

Mosleh, Y. Y. et al. 2005. Metallothionein induction, antioxidative responses, glycogen and growth changes in *Tubifex tubifex* (Oligochaete) exposed to the fungicide, fenhexamid. *Environ. Pollut.* 135:73–82.

Mosleh, Y. Y. et al. 2007. Effects of chitosan on oxidative stress and metallothioneins in aquatic worm *Tubifex tubifex* (Oligochaeta, Tubificidae). *Chemosphere* 67:167–75.

Mouneyrac, C. et al. 2006. *In situ* relationship between energy reserves and steroid hormone levels in *Nereis diversicolor* (O.F. Müller) from clean and contaminated sites. *Ecotoxicol. Environ. Saf.* 65:181–7.

Muyssen, B., K. A. C. De Schamphelaere, and C. R. Janssen. 2006. Mechanisms of chronic waterborne Zn toxicity in *Daphnia magna*. *Aquat. Toxicol.* 77:393–401.

Olsen, G. H. et al. 2007. Alterations in the energy budget of Arctic benthic species exposed to oil-related compounds. *Aquat. Toxicol.* 83:85–92.

Olsen, G. H. et al. 2008. Cellular energy allocation in the Arctic sea ice amphipod *Gammarus wilkitzkii* exposed to the water soluble fractions of oil. *Mar. Environ. Res.* 66:213–4.

Palais, F. et al. 2011. One-year monitoring of reproductive and energy reserve cycles in transplanted zebra mussels (*Dreissena polymorpha*). *Chemosphere* 83:1062–73.

Pellerin-Massicotte, J. et al. 1994. Seasonal variability in biochemical composition of the polychaete *Nereis virens* (Sars) in two tidal flats with different geographic orientations. *Comp. Biochem. Physiol.* 107A:509–16

Péry, A. R. R. et al. 2003. Modelling toxicity and mode of action of chemicals to analyse growth and emergence tests with the midge *Chironomus riparius*. *Aquat. Toxicol.* 65:281–92.

Péry, A. R. R., R. Mons, and J. Garric. 2004. Energy-based modelling to study population growth rate and production for the midge *Chironomus riparius* in ecotoxicological risk assessment. *Ecotoxicology* 13:647–56.

Péry, A. R. R., A. Geffard, and J. Garric. 2006. Mechanistic models to perform population risk assessment with the midge *Chironomus riparius*: Application to heavy metals. *Environ. Sci. Technol.* 40:6026–31.

Péry, A. R. R. et al. 2008. Assessing the risk of metal mixtures in contaminated sediments on *Chironomus riparius* based on cytosolic accumulation. *Ecotoxicol. Environ. Saf.* 71:869–73.

Pierron, F. et al. 2007. Impairment of lipid storage by cadmium in the European eel (*Anguilla anguilla*) *Aquat. Toxicol.* 81:304–11.

Reynaldi, S. et al. 2006. Linking feeding activity and maturation of *Daphnia magna* following short-term exposure to fenvalerate. *Environ. Toxicol. Chem.* 25:1826–30.
Rharrabe, K. et al. 2008. Effects of azadirachtin on post-embryonic development, energy reserves and α-amylase activity of *Plodia interpunctella* Hübner (Lepidoptera: Pyralidae). *J. Stored Prod. Res.* 44:290–4.
Ribeiro, S. et al. 2001. Effect of endosulfan and parathion on energy reserves and physiological parameters of the terrestrial isopod *Porcellio dilatatus*. *Ecotoxicol. Environ. Saf.* 49:131–8.
Rowe, C. L. 1998. Elevated standard metabolic rate in a freshwater shrimp (*Palaemonetes paludosus*) exposed to trace element-rich coal combustion waste. *Comp. Biochem. Physiol.* 121A:299–304.
Sancho, E. et al. 2009. Disturbances in energy metabolism of *Daphnia magna* after exposure to tebuconazole. *Chemosphere* 74:1171–8.
Servia, M. J. et al. 2005. Effects of copper on energy metabolism and larval development in the midge *Chironomus riparius*. *Ecotoxicology* 15:229–40.
Shipp, E., and A. Grant. 2006. *Hydrobia ulvae* feeding rates: A novel way to assess sediment toxicity. *Environ. Toxicol. Chem.* 25:3246–52.
Siah, A. et al. 2003. Delayed gametogenesis and progesterone levels in soft-shell clams (*Mya arenaria*) in relation to *in situ* contamination to organotins and heavy metals in the St. Lawrence River (Canada). *Comp. Biochem. Physiol.* 135C:145–56.
Smaal, A. C., and J. Widdows. 1994. The scope for growth of bivalves as an integrated response parameter in biological monitoring. In Kramer, K. J. M. *Biomonitoring of Coastal Waters and Estuaries*. Boca Raton, FL: CRC Press, 247–61.
Smolders, R., M. Baillieul, and R. Blust. 2005. Relationship between the energy status of *Daphnia magna* and its sensitivity to environmental stress. *Aquat. Toxicol.* 73:155–70.
Smolders, R. et al. 2004. Cellular energy allocation in zebra mussels exposed along a pollution gradient: Linking cellular effects to higher levels of biological organization. *Environ. Pollut.* 129:99–112.
Sroda, S., and C. Cossu-Leguille. 2011. Seasonal variability of antioxidant biomarkers and energy reserves in the freshwater gammarid *Gammarus roeseli*. *Chemosphere* 83:538–44.
Staempfli, C., J. Tarradellas, and K. Becker-van Slooten. 2007. Effects of dinoseb on energy reserves in the soil arthropod *Folsomia candida*. *Ecotoxicol. Environ. Saf.* 68:263–71.
Verslycke, T., and C. R. Janssen. 2002. Effects of a changing abiotic environment on the energy metabolism in the estuarine mysid shrimp *Neomysis integer* (Crustacea: Mysidacea). *J. Exp. Mar. Biol. Ecol.* 279:61–72.
Verslycke, T. et al. 2003. Cellular energy allocation in the estuarine mysid shrimp *Neomysis integer* (Crustacea: Mysidacea) following tributyltin exposure. *J. Exp. Mar. Biol. Ecol.* 288:167–79.
Verslycke, T. et al. 2004. Cellular energy allocation and scope for growth in the estuarine mysid *Neomysis integer* (Crustacea: Mysidacea) following chlorpyrifos exposure: A method comparison. *J. Exp. Mar. Biol. Ecol.* 306:1–16.
Voets, J. et al. 2006. Microcontaminant accumulation, physiological condition and bilateral asymmetry in zebra mussels (*Dreissena polymorpha*) from clean and contaminated surface waters. *Aquat. Toxicol.* 79:213–25.
Wicklum, D., and R. W. Davies. 1996. The effects of chronic cadmium stress on energy acquisition and allocation in a freshwater benthic invertebrate predator. *Aquat. Toxicol.* 35:237–52.
Widdows, J. et al. 2002. Measurement of stress effects (scope for growth) and contaminant levels in mussels (*Mytilus edulis*) collected from the Irish Sea. *Mar. Environ. Res.* 53:327–56.
Widdows, J., C. Nasci, and V. Fossato. 1997. Effects of pollution on the scope for growth of mussels (*Mytilus galloprovincialis*) from the Venice Lagoon, Italy. *Mar. Environ. Res.* 43:69–79.

13

Biomarkers of Genotoxicity for In Situ Studies at Individual and Population Levels

Paule Vasseur, Franck Atienzar, Carole Cossu-Leguille, François Rodius, and Sébastien Lemière

CONTENTS

13.1 Introduction

In every living organism, the genome governs cell functioning in response to signals from its environment. Any physical or chemical agent capable of interfering with these signals through genetic and epigenetic mechanisms may be toxic for cell cycle, growth, division, and differentiation.

For decades, genomic disturbances have been considered to be a result of alteration in the primary structure of DNA. The genetic material is known to encode all the information necessary for an organism to grow, divide or reproduce, and die. DNA structural integrity has been seen as essential in life. Nowadays, epigenetic phenomena have been discovered to be fundamental for proper development and cellular differentiation within normal tissues. The term epigenetics refers to functionally relevant modifications to the genome that do not involve a change in the nucleotide sequence. Examples of such changes are DNA methylation and histone deacetylation, both of which suppress gene expression without altering the sequence of the silenced genes. Despite these advances and active research in cancer development, there is still little knowledge on the pathology of epigenetic changes beyond the field of human health. Pollutant-induced changes in genome expression in natural biota are still being studied through structural alteration of DNA, resulting in mutations affecting gene sequence and in chromosomal alterations (chromosomal aberrations, DNA breaks, crossing-over, etc.).

Key objectives defined at the onset of genetic ecotoxicology are (1) to explore the extent to which ecosystems are contaminated with genotoxicants and (2) to identify adverse effects in individuals and more importantly in populations and communities. In human risk assessment, cancer is a major fear of exposure to genotoxic agents. As far as environmental health is concerned, carcinogenic effects in individuals are not analyzed as a threat, compared to adverse effects on populations and communities. On the other hand, mutations are considered major threats, since they can be transmitted to subsequent generations when gametes are mutated. Altered fertility, development, and embryonic survival will reduce reproductive success and thus impair population size or structure. Heritable mutations, gamete loss due to cell death, embryo mortality resulting from lethal mutations, and abnormal development are environmentally significant. Therefore, linkages between genotoxic responses, reproduction, and developmental effects justify research on genotoxic responses as ecologically relevant biomarkers.

The challenge regarding exposure to genotoxic pollutants is to establish thresholds that will preserve environmental and human health in the long term. Strategies to achieve this goal will be presented in this chapter. We shall first provide an overview of the main biomarkers of exposure to genotoxic pollutants, and then the relationships between genotoxicity and populational effects established from environmentally relevant experiments and *in situ* studies.

13.2 *In Situ* Biomarkers of Exposure to Genotoxic Pollutants

In their review on genotoxicity biomarkers for environmental biomonitoring published 15 years ago, Rether et al. (1997) underlined the wide use of mutagenicity assays carried out on environmental samples in the laboratory, as opposed to the few studies focusing on living species in contaminated environments. Nowadays, the situation has changed with the development and widespread use of the comet assay for *in situ* investigations and analyses of exposed populations. Owing to its breadth, this assay has advantages over the classical methods in use for the survey of populations at risk, namely, measurement of DNA adducts, cytogenetic assays, and the micronucleus (MN) test. The alkaline unwinding assay has now been replaced by the comet assay for screening clastogens.

13.2.1 DNA Adducts

The ^{32}P-postlabeling test described by Randerath et al. (1981) is the reference method to measure the covalent binding of electrophilic intermediates to DNA, *in vitro, in vivo,* and *in situ*. Detection of adducts provides a way of documenting exposure to genotoxic chemicals without any knowledge of their identity. The method of quantification is highly sensitive for detecting bulky hydrophobic xenobiotic–DNA adducts. The latter are expressed as nanomoles adducts per mole nucleotides, or as the number of adducts per 10^9 nucleotides.

Alkylating agents, active metabolites of xenobiotics such as polycyclic aromatic hydrocarbons (PAHs), and endogenous active radicals may bind covalently to DNA and form adducts. Lipid peroxidation products, such as aldehydes, covalently modify lipids, proteins, and DNA, whose nucleophilic groups present in purine and pyrimidine bases are targets for electrophilic moieties (Nair 1999). Etheno-adducts are biomarkers of DNA damage arising from the reaction of lipid peroxidation end products, and recognized as a typical signature of DNA damage in white blood cells (Arab et al. 2009). Despite the biological relevance of these adducts, shown to be mutagenic in bacteria, there are few studies measuring DNA adducts generated by oxidative stress in aquatic model organisms. The first analyses of etheno-adducts in tissues of the mussel *Perna perna* were conducted by Almeida et al. (2006), who checked for effects of seasonality in the mantle and in the digestive gland. The method of analysis of these adducts using high-performance liquid chromatography coupled to mass spectrometry (HPLC-MS) has been described by Machado Garcia et al. (2012).

DNA adducts have been used for two decades as a molecular dosimetric index by the National Oceanic and Atmospheric Administration (NOAA) for studying factors of toxicopathic liver lesions in fish, including neoplasms. Populations at risk include flatfish, which are bottom-feeding fish species living in close contact with sediments. They accumulate a variety of contaminants from sediments, which are a known repository for a large number of inorganic and organic hydrophobic pollutants. A large monitoring program of sediment and bottom fish for chlorinated hydrocarbon pesticides, polychlorobiphenyls (PCBs), and PAHs was launched along the west coast of the United States, from 1984 to 1993, in order to identify factors of fish disease and establish remediation. The studied species on U.S. coasts included *Pleuronectes vetulus* (English sole), *Platichthys stellatus* (starry flounder), *Genyonemus lineatus* (white croaker) on the Pacific coast, and *Pleuronectes americanus* (winter flounder) on the northeast coast (Myers et al. 1998a).

The goals were to identify pollutants responsible for the adverse effects recorded and to validate biomarkers of neoplastic and nonneoplastic lesions (Chapter 15). Hepatic lesions included proliferation, specific degeneration/necrosis, foci of cellular alterations, and neoplasms (Myers et al. 2003). PAH pollutants were identified among risk factors, whereas PCBs were only detected as potentially PAH-interacting agents (Myers et al. 1990). Liver lesions could be related with hepatic DNA adducts or biliary fluorescent aromatic compounds (FACs) representing PAH metabolites released into the bile. Liver lesions, however, did not correlate well with levels of sediment contamination.

DNA adducts have been validated as biomarkers in more than 28 teleost fish species (Reichert et al. 1998). Exposure to a mixture of polyaromatic compounds leads to typical autoradiographic patterns of hepatic DNA digests that distribute within a diagonal on thin layer chromatograph maps. The "diagonal radioactive zone" results from spots of DNA adducts of different molecular weights produced by PAH metabolites (Varanasi et al. 1989).

Hepatic DNA adducts are rapidly formed within 24–48 h of exposure (Collier and Varanasi 1991). They also have slow repair rates (French et al. 1996) with estimated half-lives of 40–100 days in English sole. Therefore, the biomarker is considered a reliable and useful dose-responsive indicator of cumulative and chronic exposure to PAHs in wild fish. On the other hand, FACs, which are rapidly eliminated from fish, reflect recent exposure to PAH-contaminated sediment. This FAC biomarker of short-term PAH exposure responds within hours, is dose-responsive, reaches a maximum within 2 weeks in conditions of continuous exposure, and has a half-life of 14–16 days in English sole (Collier and Varanasi 1991; French et al. 1996).

In a heavily polluted site (Duwamish Waterway) in Puget Sound, Washington, USA, on the Pacific coast, the level of hepatic DNA adducts in English sole could reach 140 ± 27 nmol adducts mol^{-1} nucleotides, whereas biliary fluorescent metabolites were present at 250 ng mg^{-1} protein and sediment contamination was 3600 ng ΣPAHs (18 congeners) g^{-1} sediment (fresh weight) and 570 ng ΣPCBs g^{-1}. Higher levels of hepatic DNA adducts (275 ± 50 nmol adducts mol^{-1} nucleotides) were found in the black croaker (*Cheilotrema saturnum*), in San Diego Bay, California (Reichert et al. 1998); this species is very sensitive to genotoxic pollutants and to neoplasia, and at the same site, other fish species such as the white croaker (*G. lineatus*) present much fewer DNA adducts and are less prone to liver lesions.

In addition to ecological factors, metabolic and toxicokinetic specificities of PAH metabolites play a role in species sensitivity differences. For instance, the starry flounder (*P. stellatus*), a sympatric species to the English sole (*P. vetulus*), rarely develops neoplasia. Yet, levels of hepatic DNA adducts are identical in the two species sampled at the same site (Reichert et al. 1998). This example explains why a combined use of several indicators, including histopathological examination of liver in this case, is an optimal approach in environmental biomonitoring.

In Puget Sound, the site of Eagle Harbor was designated as a superfund site in 1987, because of its high level of sediment contamination and the highest recorded prevalence of hepatic lesions in the resident epibenthic flatfish, such as the English sole (*Parophrys vetulus*). PAH sediment contamination originated from a facility that used creosote as a wood preservative. The closure of the facility and remediation of sediments from 1993 to 2002 contributed to reduced PAH exposure in English sole (Myers et al. 2008). The results of the biological survey up to 2004 confirmed that xenobiotic–DNA adduct level was a highly significant risk factor for toxicopathic hepatic lesions and neoplasia in English sole, with each additional unit of adduct increasing the risk of initiating neoplastic lesions by 1.036 times. The level of biliary FACs was also a significant risk factor for several of the toxicopathic liver lesions, but not of the occurrence of hepatic neoplasms. On the other hand, hepatic CYP1A levels, as estimated by aryl hydrocarbon hydroxylase activities, were not significant risk factors of any of the hepatic lesions categories and correlated poorly with biliary FACs and hepatic DNA adducts. Therefore, their use as a monitoring tool in Eagle Harbor has been discontinued (Myers et al. 2008; Chapter 15).

Hepatic neoplasia in flounder *Platichthys flesus* have been recorded in different areas in the Baltic Sea, especially in the southern Baltic Sea off the German and Polish coasts, during fish disease surveys in the late 1980s (Lang and Dethlefsen 1994) and in more recent studies (Lang et al. 2006). DNA adducts as molecular indicators of PAH exposure were screened in the flounder in a short retrospective study over 1995–1996, but bulky hydrophobic DNA adducts were not found as a major factor in the etiology of preneoplastic and neoplastic liver lesions during these sampling years (Malmström et al. 2009).

DNA adducts were also studied after oil spill accidents that occurred during the past two decades to evaluate their impacts on marine species. In Europe, this concerns the wrecks of the *Sea Empress* near the South coast of Wales (UK) in February 1996 and of the *Erika* near the Atlantic coast of France in December 1999. No specific biomarker for genotoxicity was used during the *Exxon Valdez* disaster (March 1989) in South Alaska. The ecosystem effects of this oil spill, and of the recent oil spill produced by the explosion of the Deepwater Horizon (DWH) drilling platform in the Gulf of Mexico (April 2010) will be discussed in Section 13.2.1.

Two months after the *Sea Empress* disaster, DNA adducts in the gills of the shanny *Lipophrys pholis* sampled at the spillage zone were still at a level of 470/10^9 nucleotides, whereas the total hydrocarbons in sediments detected at the same site was 22 ± 4 μg g^{-1} dry weight (d.w.). The high levels of DNA adducts in ovaries, testis, and liver augured adverse effects on reproduction and progeny output. High levels of DNA adducts were also quantified in the sole (*Pleuronectes platessa*) and the dab (*Limanda limanda*), even 16 months after the wreck (Table 13.1). Surprisingly, DNA adducts were absent in soft tissues of mussels and sponges.

On the other hand, after the *Erika* wreck, high levels of DNA adducts were measured in the digestive gland of mussels (*Mytilus edulis*), although these were three times lower than in the liver of fish (*Solea solea*) (Table 13.1). The highest values were recorded at Pointe Chemoulin and Baie de Bourgneuf, the sites most impacted by the oil spill (Amat et al. 2006). The range of PAH concentrations (16 PAH congeners) in mussels was 200–1500 ng g^{-1} (d.w. soft tissues) at impacted sites, versus 20–200 ng g^{-1} d.w. in apparently nonimpacted sites.

Generally speaking, DNA adducts in fish livers are also biomarkers used in the biomonitoring of estuaries and rivers (Table 13.1). High values have been recorded at Southampton, in the Thames, the Tyne, and the Clyde, in comparison with the rivers Mersey and Alde in England, the Forth in Scotland, and Belfast in Northern Ireland (Lyons et al. 2004). In France, the Baie de Seine was compared to the reference site of the Baie de Somme (Akcha et al. 2003, 2004). The levels of DNA adducts in fish livers sampled in 2001 in Baie de Seine were rather low (about 10 adducts/10^9 nucleotides). Levels were slightly higher in adults than in juveniles, and in males compared to females. On the other hand, records in freshwater mussels (*Dreissena polymorpha*) sampled in the Seine river (15–40 adducts/10^9 nucleotides) in 2002 indicated exposure to genotoxic pollutants, possibly PAHs. The latter were quantified in mussel tissues at high concentrations at polluted sites compared to the reference site (Le Goff et al. 2006). Temporal variations were registered, and 1 year later (2003) DNA adduct levels were lower (<10 adducts/10^9 nucleotides), as were the PAH concentrations in the mussels (Rocher et al. 2006).

Levels of DNA adducts in gills of *M. edulis* were found in the same range at polluted sites on the coasts of Iceland, Norway and Sweden (10–20 adducts/10^9 nucleotides), for instance, in Reykjavik harbor (Ericson et al. 2002; Skarphédinsdóttir et al. 2007). Gill DNA adduct levels were found to be higher in mussels in the intertidal zone compared to the subtidal zone, the difference being largest in winter (Skarphédinsdóttir et al. 2005). This was explained by a higher pollution level of the surface microlayer, an additional stress for intertidal mussels under the effect of the tidal cycle, and photoactivation of hydrocarbons, such as PAHs, known to be rendered more toxic by sunlight and ultraviolet (UV) radiation. DNA adduct levels in mussels were higher in gills than in the digestive gland (Skarphédinsdottir et al. 2007) as already shown in *Mytilus galloprovincialis* on Mediterranean coasts by Pisoni et al. (2004). In fish, the difference was not so clear, with higher adduct levels in the liver than in the gills at different sites of the Baltic Sea (Ericson et al. 1998), whereas the inverse was found in the Venice lagoon (Venier and Zampieron 2005).

The comparison of adduct levels between studies requires consideration of the adduct enrichment method used before chromatographic separation of ^{32}P-nucleotides. Indeed, the comparison of the same samples indicated that results differed according to whether enrichment was achieved by digestion with nuclease P1 or by extraction in pure butanol (Harvey et al. 1999; Venier and Zampieron 2005).

As a whole, these data validate the measurement of DNA adducts in fish as a dosimetry biomarker of exposure to genotoxic pollutants. The levels of hepatic DNA adducts correlate

TABLE 13.1

DNA Adducts Measured in Vertebrates (Fish) or Invertebrates Collected *In Situ* in Rivers or in Estuaries and Exposed to Hydrocarbons Released by Oil Spills (Shaded) or Produced by Anthropogenic Activities

Sites	Organisms	Station Site or Point (pt)	Sampling + *x* months (m.) *Period*	Adducts[a] Nb/10^9 Nucleotides		ΣHAP ng/g d.w. Mussel Tissues or in Sediments	Other Measurement Biliary Metabolites	Authors
Oil spill accident								
Aegean sea. Dec. 1992 Galicia (ES)	*Mytilus edulis*	Reference oiled (3 pts)	+6 m.	2.0 ± 1.6 *110 ± 18; 699 ± 196; 1280 ± 373*	ΣHAP_{13}	27 307; 131; 216	ng/g dry weight	Solé et al. 1996
Sea Empress Feb. 1996 South Wales (UK)	*Lipophrys pholis*	oiled pt 2 oiled pt 2 oiled pt 2	+2 m.	*477 ± 24 gills* *437 ± 81 liver* *254 ± 13 ovary/testis*	*THC µg/g d.w.*	*22.0 ± 4.1*	+11 m., 17 m.: ND	Harvey et al. 1999
	Pleuronectes platessa	non oiled pt oiled pt 4	+3 m.	*22 ± 19* *155 ± 50*	*THC µg/g d.w.*	*ref: ND* *point 4: 1.9*	No adducts in invertebrates:	
	Limanda limanda	non oiled pt oiled pt 7 oiled pt 7	+5 and 16 m. +5 m. +16 m.	*ND* *12 ± 4* *71 ± 20*	*THC µg/g d.w.*	*point 7: ND* *ND*	*Mytilus edulis Halichondria panicea*	
Erika Dec. 1999 South Britain Sud (F)	*Mytilus edulis*	Chemoulin Chemoulin Tresson Tresson other sites	+1m. +4 m. +4 m. +13 m. +13 m.	110 39 21 22 < 10		Ref: 50–200 Polluted: 200–1500		Bocquené et al. 2004
	Solea solea Juveniles	Pertuis, Bourgneuf	+2 m. +9 m.	$90_{Pertuis}$ – $290_{Bourgneuf\ Bay}$ $20_{Pertuis}$ – $50_{Bourgneuf\ Bay}$				Amat et al. 2006
Puget Sound (Washington) (USA)	*Pleuronectes vetulus*	Duwamish Waterway Polnell Pt		140 ± 27 35 ± 1	ΣPCB_8 ΣPAH_{18} ΣPCB_8 ΣPAH_{18}	*570 ng/g f.w. sed.* *3600 ng/g f.w.* *28 ng/g f.w. sed.* *21 ng/g f.w. sed.*	*biliary FACs* *250 ng/mg prot.* *41 ng/mg prot.*	Myers et al. 1998b Stein et al. 1992

Pacific coast (USA)	*Cheilotrema saturnum*	North San Diego Bay		275 ± 50			*neoplastic lesions preneoplastic lesions*	Reichert et al. 1998
		Reference		0–10			*nd*	
	Genyonemus lineatus	North San Diego Bay		10 ± 10			*nd*	
		Ref		0–5			*nd*	
Tyne estuary (UK)	*Platichthys flesus*	Tyne at Newcastle	*Oct 1997*	*344 ± 37*	*ΣPAH*$_{15}$	*µg/kg d.w. sediment 12,548–43,470*	*Biliary metabolites 14572 ± 1887*[b]	Lyons et al. 1999
		Alde (ref)		*101 ± 48*		*n.d. in Alde*	*632 ± 56*[b]	
Estuaries (UK) Sept.–Dec. 2002	*Platichthys flesus*	Tyne		137 ± 49		*10,790–43,470*		Lyons et al. 2004
		Alde		52 ± 14				
		Southampton		939 ± 370*		*750*		
		Thames		511 ± 192*		*597–5350*		
		Forth		93 ± 27				
		Mersey		174 ± 59		*1811–5740*		
		Clyde		362 ± 158*				
		Belfast		42 ± 5				
Seine/Somme (France)	*Limanda limanda*	Baie Seine 4 stations	*March 2001*	7 ± 7 (mean 4 st.)	ΣPAH$_{16}$ ΣPCB$_8$	354 ± 47 1036 ± 132	Comet: higher in Baie de Seine	Akcha et al. 2003
		Baie Somme 2 stations		1 ± 1; 14 ± 12	ΣPAH$_{16}$ ΣPCB$_8$	113 ± 19 99 ± 14	than in Baie de Somme	
Seine/Somme (France)	*Limanda limanda*	Baie Seine	*Sept 2001*	8.3 ± 2.7 m. juv. 15.5 ± 10.5 m. ad. 7.2 ± 3.2 fem. juv. 12.3 ± 5.1 fem. ad.			Comet, adducts: No significant difference between sites	Akcha et al. 2004
		Baie Somme		6.2 ± 2.7 m. ad. 2.2 ± 0.8 fem. ad.				
Seine (France)	*Dreissena polymorpha*	Yville ref	*June 2002*	3.9 ± 4.4	ΣPAH$_{21}$[c]	53 ± 7[c]		Le Goff et al. 2006
		Oissel		40.1 ± 24.5		1426 ± 1226[c]		
		La Bouille		15.4 ± 15.6		1604 ± 1053[c]		

(*continued*)

TABLE 13.1 (Continued)

DNA Adducts Measured in Vertebrates (Fish) or Invertebrates Collected *In Situ* in Rivers or in Estuaries and Exposed to Hydrocarbons Released by Oil Spills (Shaded) or Produced by Anthropogenic Activities

Sites	Organisms	Station Site or Point (pt)	Sampling + *x* months (m.) *Period*	Adducts[a] Nb/10^9 nucleotides		ΣHAP ng/g d.w. Mussel Tissues or in Sediments	Other Measurement Biliary Metabolites	Authors
Seine (France)	*Dreissena polymorpha*	Yville ref	*June 2003*		ΣPAH_{21}	49	47 ΣPCB_7	Rocher et al. 2006
		Oissel		<5		652	490 ng/g d.w.	
		La Bouille				990	580 soft tissues	
Seine Estuary (France)	*Mytilus edulis*	Villerville		<10		447	583	
		Le Havre		<10		404	591	
		Antifer		<10		147	324	
		Moulard ref		170 ± 100 (!)		15	120	
Baltic Sea Swedish coasts (SW)	*Perca fluviatilis*	reference	*Oct 1994*	<1 liver <1 gills	ΣPAH_{15}	*1 μg/g d.w. spm*		Ericson et al. 1998
		site 1 (poll)		29 ± 20 liver 15 gills		*5–41 μg/g d.w. spm* *14–211 μg/g d.w. sediments*		
	Esox lucius	reference		4.7 ± 3 liver 0.3 ± 0.2 gills				
		site 1(poll)		13.5 liver 6.2 gills				
Iceland (IS)	*Mytilus edulis*	Reykjavik	*March 2000*	18.9 ± 2.2 gills 1.8 digest. gl.				Ericson et al. 2002
			August 2000	10.3 ± 1.2 gills 3 digest. gl.				
		reference	*March 2000*	0.74 ± 0.18 gills 0.43 ± 0.1 digest. gl.				
			August 2000	2.1 ± 0.26 gills 1.2 ± 0.7 digest. gl.				

Nordic coasts	*Mytilus* sp.	references	*Oct 2000*	0.5–1.2 gills	ΣPAH_{32}	40–240	Skarphédinsdottir et al. 2007
Iceland (IS)		+ poll sites					
Norway (NO)		Reykjavik		10		11,670	
Baltic Sea		Norway		10		5120	
		Sweden *SW*		2.8		925	
		Baltic		2.6		1330	
		Remote site		3.8		100	
Mediterranean Sea	*Mytilus galloprovincialis*	Carteau		gills	ΣPAH_{13}	101 ± 60 Σtissues	Bodin et al. 2004
Riviera (F)			*1999*	9 –33 spring	ΣPCB_9	90 ± 27	
				19–28 summer	Σmetals	202 ± 78	
			2000	3–45 winter	As, Cd, Cr,		
				17–44 spring	Cu, Hg, Pb		
				6–54 summer	Ni, Zn		
Ionian Sea	*M. galloprovincialis*		*2001*	3–66 gills			Pisoni et al. 2004
				0.1–33 digest. gl.			
			2002	8–53 gills			
				0.8–45 digest. gl.			
Venice lagoon (IT)	*Zosterisessor ophiocephalus*	site1 –poll	*1998 summer*	*5.9 ± 4 gills*		Site1/site 7	Venier and Zampieron 2005
				5.1 ± 4 liver	ΣPAH_{16}	101/1058 ng/g d.w. lipids	
		site7 +poll		*20.6 ± 14 gills*			
				4.7 ± 4 liver	Aroclor 1254		
						153/1176 ng/g	
	M. galloprovincialis	site1 –poll	*1999*	*4.3 ± 0.6 gills*	+ 1260	d.w. lipids	
		site7 +poll	*summer*	*37 ± 1.2 gills*			

Note: PAH concentrations in soft tissues in mussels or in *sediments* (*in italic*), PAH biliary metabolites in fish or other measured parameters. NS, not significant; ref, reference; ctm, contaminated; f.w., fresh weight; d.w., dry weight; ND, not detected; nd, no data; prot., protein; pt, point; spm, settling particulate matter; m., males; fem., females; juv., juveniles; ad., adults; Poll., polluted; –poll, less polluted; +poll, the most polluted.

[a] Adduct enrichment method by butanol extraction (numbers in italic) or by using P1 nuclease.

[b] µg/kg fresh tissue expressed in equivalent 1-hydroxypyrene (biliary FACs, biliary fluorescent aromatic compounds measured in the bile; THC, total hydrocarbons).

[c] Mean results (in ng/g d.w. soft tissues) of three samples of zebra mussels, collected in June 2001, June 2003, and June 2004.

* Significantly different from control site.

well with biliary fluorescent metabolites in fish exposed to PAHs. Mussels are also useful sentinel species in PAH-contaminated areas, provided that adducts are measured in specific tissues (gills or digestive gland) instead of whole tissues, which may dilute the adduct pool and weaken the signal. Yet, abnormal DNA adducts detected by the ^{32}P-postlabeling method remain generally nonidentified, which may be considered a limitation, but a limitation that is not specific to this biomarker. Another confusing result is the presence of abnormal DNA adducts on autoradiograms of controls or specimens from reference sites.

13.2.2 Cytogenetic Assays and MN Test

A wide range of *in vitro* and *in vivo* cytogenetic assays have been standardized to detect genotoxic agents in mammals for regulatory and sanitary purposes. Criteria are structural or numerical chromosomal aberrations, sister chromatid exchanges, and micronuclei, measured in cells or in dividing tissues, using microscopic examination during mitosis, or using flow cytometry. These criteria have rapidly been applied to environmental biomonitoring. Anaphase aberrations were proposed for use to survey fish populations (Kocan et al. 1982). Their measurement was applied to study pollution and developmental abnormalities of Atlantic fish (Longwell et al. 1992). Planktonic eggs provided evidence that pollution was associated with mortality, malformation, and abnormal chromosomal division of fish embryos developing near the surface water off Atlantic coasts. Correlations were obtained between mitotic abnormalities, lethality and pollution by PAHs, metals and embryo abnormalities, chlorinated hydrocarbons and defects in late stages of embryogenesis in mackerel (*Scomber scombrus*) on the east coast of New Jersey, USA (Hose 1994).

These markers have also been applied to invertebrates, especially to sea urchins (*Strongylocentrus purpuratus*) to assess the effects of polluted sediments in San Francisco Bay (Long et al. 1990).

In the Baltic Sea, karyotypic changes in gill epithelial cells were described in the bivalve *Macoma balthica* from the Gulf of Gdansk, and have been associated with gill neoplasia (Thiriot-Quiévreux and Wolowicz 2001; Smolarz et al. 2003). Cytogenetic damage in somatic cells and in gonads of the gastropod *Lymnaea ovata*, the bivalve *Mytilus edulis* (micronuclei), and the crustacean *Balanus improvisus* were described on Lithuanian coasts from 1995 to 1999 (Baršienė 2002).

Today, the MN test is one of the most widely used bioassays for genotoxicity in field studies. It was first carried out on hemocytes in invertebrates or erythrocytes in fish and is now applied to a wide range of aquatic organisms including sea urchins, oysters, crabs, and worms, and in both wild and transplanted species (Bolognesi and Hayashi 2011). It was applied to larvae of amphibians, the newt *Pleurodeles waltl* (Gauthier et al. 1989) and the toad *Xenopus laevis*. The amphibian MN assay has been standardized (ISO 2006) and used for the biomonitoring of surface waters, sampled, and tested in the laboratory (Gauthier et al. 2004) or with caged amphibians *in situ* (Wirz et al. 2005). Besides assessing chemicals and recently nanomaterials (Mouchet et al. 2011), micronuclei measurement is also used for offshore biomonitoring using mussels (Gorbi et al. 2008; Sundt et al. 2011) or fish (Rybakovas et al. 2009).

The measurement of MN is easier and more rapid than karyotype analysis (Vasseur et al. 1995). Indeed, cytogenetic studies are difficult because of the relatively large number of chromosomes of most fish species and the small chromosome size of many aquatic organisms. The MN assay targets interphase cells of any proliferating cell population regardless of its karyotype, which is an advantage in environmental monitoring programs. Micronuclei (MN) originate from chromosome fragments, or whole chromosomes that lag behind at anaphase during nuclear division. In environmental studies, micronuclei are counted after cell harvesting, fixation, and staining with chromatin reagents, and expressed per 1000 scored cells. Nuclear abnormalities, which include nuclear buds, binucleated cells, nucleoplasmic bridges, and apoptotic nuclei, may be also counted and reported, but separately to MN (Fenech 2007). An MN requires cell division after damage induction to be observed in the cytoplasm of the daughter cells. Therefore, the division kinetics of the tissue studied governs the sensitivity of the response and needs to be known for correct interpretation of MN frequencies. A low mitotic index of the cell population may explain negative results, even at polluted sites, either as a result of exposure to cytotoxic pollutants that impair growth and cell division, or of physiologically low division rates of the analyzed tissue. The age of the animals or their developmental stage is not a trivial factor in the response.

The results of the micronuclei assays in a number of field studies carried out using wild fish and mussels have been reviewed by Bolognesi and Hayashi (2011), and by Mouchet and Gauthier (2012) regarding MN responses in amphibians.

In fish, micronuclei are measured in erythrocytes, gills, and fins of native species in the field (Baršienė et al. 2012) or in laboratory studies (Goanvec et al. 2008; Hoshina et al. 2008) to assess the impact of anthropogenic contaminants such as petroleum refinery effluents. In the Baltic Sea, a large survey of genotoxicity in native species, flounder *P. flesus*, herring *Clupea harengus*, and eelpout *Zoarces viviparus*, over the past decade was carried out using the MN assay in erythrocytes. Background responses of MN incidences determined using data from reference sites obtained in 2001–2011 were below 0.39 MN/1000 erythrocytes in the native species, with offshore values lower than those in coastal zones. Long-term environmental genotoxicity studies (2001–2011) in different zones of the Baltic Sea showed clear increases in the level of environmental genotoxicity in 2009–2011, compared to lower MN levels in fish collected in 2001–2007 (Hylland et al. 2008; Baršienė et al. 2012). Contamination in zones close to estuaries or industrial activities that were measured within the HELCOM Baltic Sea Action Plan (HELCOM 2010), and accidental spills of contaminants have been suggested as possible causes for this increase.

In bivalves, hemolymph is easily collectable by noninvasive sampling, but gills are preferred for MN examination, although their preparation is not adequately standardized. Gill cells are proliferative and more sensitive than hemocytes. Persistence of MN frequencies in gill cells of *Mytilus* sp. was shown to be >1 month in short-term experiments using genotoxic agents. Under laboratory conditions, higher baseline and induced MN frequencies have been found in gill cells compared to hemocytes. Because of the influence of temperature on the mitotic rate studied in controlled experimental conditions, seasonal effects have often been taken into consideration in field studies (Baršienė et al. 2006a). This also explains the higher background levels in warm southern countries compared to northern ones, for instance, at Mediterranean sites compared to the Baltic Sea (Fernandez et al. 2011). At a reference site in the Baltic Sea, the lowest MN incidence (0.37 MN/1000 gill cells) occurred in *M. edulis* collected at a temperature of 5°C in autumn, whereas a higher level (1.16 MN/1000 cells) was detected at 13.6°C (Baršienė et al. 2006b).

The MN frequency in caged mussels requires several weeks of deployment, and comparison shows lower induction levels than in native mussels (Bolognesi and Hayashi 2011). In field studies, the MN test is generally applied in a multibiomarker approach among other genotoxicity markers and indicators of cytotoxicity, oxidative stress, and bioaccumulation of contaminants. Correlations between response of the comet assay, apoptosis, and micronuclei measurements have been found in *Dreissena polymorpha* exposed *ex situ* (in the laboratory) to river samples from the Lambro/Po confluence (Binelli et al. 2010).

Yet, a higher sensitivity of the MN test compared to the comet assay has been found in *D. polymorpha* (gill cells) after 2 months' caging in the Orge river in the Seine Basin (Bourgeault et al. 2010) and in *Mytilus galloprovincialis* caged for 4 weeks for offshore monitoring in the Adriatic sea (Gorbi et al. 2008). Higher sensitivity of the MN response has also been reported by Bombail et al. (2001) in the fish *Pholis gunnellus* from the Firth of Forth, Scotland.

A MN represents irreversible damage that lasts until cell death (Kirsch-Volders et al. 2003). In contrast, increased DNA fragmentation may occur rapidly but transiently, being subjected to repair (Collins 2004). The MN level is an index of accumulated genetic damage during the life span of the cells, and is therefore one of the most suitable techniques to identify integrated responses to complex mixtures of contaminants. In addition to clastogens, the MN assay detects aneugens, responsible for aneuploïdy (loss or gain of one chromosome), which cannot be identified using the comet assay. Aneugens can be discriminated from clastogens by visualizing the centromere on a whole chromosome in the fluorescence *in situ* hybridization test (Eastmond et al. 1995), whereas an acentric chromosome fragment (a fragment without a centromere) will not be probed. The technique is widely developed in medical diagnostics, but not yet in aquatic ecotoxicology.

13.2.3 Comet Assay

The single cell gel electrophoresis assay, also known as the comet assay, is based on the fact that the DNA organization and its association with matrix protein in the nucleus are disrupted when damaged. The individual strands of DNA lose their compact supercoiled structure and relax, allowing DNA extension. This occurs regardless of the pH, neutral or alkaline (Collins et al. 2008). The assay involves embedding of cells in agarose, membrane lysis with detergents, and high salt concentrations to solubilize histones. Under an electric field, damaged DNA is drawn to the anode, whereas undamaged DNA does not move. The halo of relaxed DNA looks like a comet and is visualized after staining, generally using fluorescent dyes. The level of damage is related to the intensity of DNA staining in the tail of the comet as stressed by Collins et al. (2008). Fragments in the tail will result if two breaks occur within one strand or if DNA detaches from the matrix at high pH.

The assay originated from Ostling and Johänson (1984), who first used microelectrophoresis, lysis with detergents, and pH 9.5 to detect breaks induced by ionizing radiation. A few years later, Singh et al. (1988) proposed a similar method, but using alkaline conditions (pH >13), to enlarge detection of damage, especially to alkaline labile sites. The latter include apurinic and apyrimidic (AP) sites, which arise from the loss of a damaged base, leaving a base-less sugar in the backbone. AP sites occur as intermediates during base excision repair (BER) and may also arise spontaneously owing to altered chemical stability resulting from changes in bases or sugars (Collins et al. 2008). Alkaline pH > 13 (usual protocol) converts AP sites into breaks, whereas pH ca. 12 (less commonly used) does not. Performing the assay in parallel at these two pH levels distinguishes agents that induce just strand breaks from those that produce AP sites. Using the alkaline version at pH > 13 alone integrates breaks resulting from both exposure and repair.

Protocols have been optimized and applications defined by several authors successively (Mitchelmore and Chipman 1998; Tice et al. 2000; Devaux and Bony 2012). Results are preferably expressed as the percentage of DNA in the tail that covers the widest range of damage and is linearly related to break frequency (Collins et al. 2008).

The assay provides the opportunity to study DNA damage, including oxidative damage, repair, and cell death in different cell types, without prior knowledge of karyotype and cell turnover (Jha 2008). Ecotoxicological applications and significance of the comet assay have recently been reviewed by Jha (2008) and Frenzilli et al. (2009), who underlined the potential ecological relevance of comet assay–derived genotoxicity data. A wide range of phylogenetically disparate groups of organisms from oligochaetes to fishes, amphibians, and mammals have been successfully investigated in aquatic toxicology (Jha 2008; Dhawan et al. 2009). A number of advantages explain the popularity of the comet assay in environmental biomonitoring. The assay is relatively simple and sensitive, and requires a small number of cells. It is applicable to cell suspensions and dissociated tissues of many eukaryotes. Hemocytes and erythrocytes are often analyzed given their practical and noninvasive sampling. Yet, cells of solid tissues such as gills, liver, and digestive gland are also examined, although tissue dissociation is a critical step that requires careful handling to avoid artifactual DNA breakage. Although mechanical dissociation was preferred for the bivalve *Unio tumidus* (Lemière et al. 2005a), enzymatic digestion of gill epithelium with dispase or collagenase was more efficient in another freshwater mussel, *D. polymorpha* (Vincent-Hubert et al. 2011). Comparison of cell type sensitivity indicates that circulating cells appear less sensitive than gill cells (Rank et al. 2005; Bourgeault et al. 2010; Vincent-Hubert et al. 2011) or hepatocytes (Frenzilli et al. 2009). Gill cells and hepatocytes are involved in the uptake of environmental contaminants from water and food, and they are at the first line of exposure to pollutants and their active metabolites. Approaches for the preservation and storage of cell samples from remote marine sites and from large experiments have also been developed (Hartl et al. 2010).

The comet assay is used to assess the genotoxic potential of chemicals, environmental samples (Lee and Steinert 2003; Kilemade et al. 2004), and more recently, nanomaterials (Galloway et al. 2010; Bernardeschi et al. 2010). Aquatic organisms or cells (Vincent-Hubert et al. 2012) are used in laboratory testing facilities.

The assay has been widely applied to evaluate the impact of urban, agricultural, and industrial activities on native fish species in river and estuarine biomonitoring. Generally, DNA damage in erythrocytes parallels the degree of pollution (Larno et al. 2001; Flammarion et al. 2002; Roy et al. 2003; Akcha et al. 2003; Bony et al. 2008; Nahrgang et al. 2010). The comet assay in fish embryos is now being developed to assess the genotoxicity of chemicals (Pereira et al. 2012) and environmental samples, to study the long-term consequences of DNA damage and embryo abnormalities (Vicquelin et al. 2011).

In the context of linking genotoxicity in individuals to population effects, the impact of paternal genotoxin exposure was explored on the reproductive biology of two marine invertebrates: the polychaete *Arenicola marina* and the mussel *M. edulis* (Lewis and Galloway 2009). A 72-h *in vivo* exposure of males to methylmethane sulfonate (MMS) and benzo[*a*]pyrene at high concentrations resulted in DNA damage in somatic cells and sperm. The fertilization success of DNA-damaged sperm was unaffected, but abnormal development was recorded in the two species, suggesting potential long-term consequences for population success. In *M. edulis* allowed to recover for 72 h after exposure, hemocytes showed rapid and better recovery from DNA damage than sperm.

An analogous experiment, but a long-term survey, was performed on two fish species, the brown trout *Salmo trutta* and the Arctic charr *Salvelinus alpinus* injected with a high dose of MMS (50 mg kg^{-1} body weight) (Devaux et al. 2011). DNA damage increased in sperm collected 3 weeks after the injection. It had no impact on fertilization and hatching survival, but induced embryo abnormalities (that persisted 1 year after breeding in brown trout). The results of these experiments were in line with those reported in earlier studies (Longwell et al. 1992; Hose 1994; Anderson and Wild 1994) that described links between exposure to genotoxic agents and deleterious consequences on progeny.

Gametes were also analyzed in the amphipod *Gammarus fossarum*, males and females, encaged on both sides of a pollution source. DNA damage was higher in spermatozoa and ova than in hemocytes at the polluted sites (Lacaze et al. 2011). Application of the comet assay to spermatozoa of this species required the time of lysis to be extended to 18 h before electrophoresis, as a consequence of a high degree of compaction of sperm chromatin (Lacaze et al. 2010). Such a high compaction rate is disturbing and possibly constitutes one mechanism of protection of the genetic material.

The comet assay has been often used in multibiomarker approaches in field studies to determine:

- The relationships between genotoxicity effects and pollution gradient as pioneered by Large et al. (2002) and Shaw et al. (2002)
- The influence of seasonal parameters, duration of exposure, delay for occurrence of lesions, and sensitivity of transplanted and native species
- The sensitivity of the comet assay compared to other genotoxicity tests

Mollusks, including bivalves such as *Mytilus* sp. and *D. polymorpha*, have often been used in biomonitoring marine/estuarine and freshwater, respectively. In transfer experiments using bivalves, fluctuations in DNA breaks were observed in the first 10 days (Steinert et al. 1998a; Regoli et al. 2004). Therefore, a very short transplantation experiment is not recommended. DNA breaks in gill cells of transplanted *M. galloprovincialis* reached the levels of native species within 30 days of transfer, whereas micronuclei measures remained lower than in native species (Nigro et al. 2006). Seasonal, temporal, and intrasite variations had been registered (Nigro et al. 2006; Le Goff et al. 2006; Rocher et al. 2006; Rank et al. 2007; Baussant et al. 2009), which underlined the importance of standardizing and defining precise deployment and sampling protocols in field studies. DNA damage was generally lower in winter than in summer (Shaw et al. 2000) and in subtidal mussels compared to intertidal mussels from the same site. This may be due to sunlight and photoactivation of organic pollutants, such as PAHs (Steinert et al. 1998b).

The occurrence of adaptation mechanisms in chronically exposed populations through up-regulation of DNA repair mechanisms has been suggested. This was supported by results in native (Shaw et al. 2002; Noventa et al. 2011) and transplanted invertebrates (Large et al. 2002), and also in fish (McFarland et al. 1999).

Among other prospects of the assay, it is worth mentioning its usefulness to follow DNA repair (Lemière et al. 2005b), and to evaluate the time-course variations in DNA damage in the field and transplantation experiments (Frenzilli et al. 2009). It has been used to explore the genetic mechanisms of damage involving alterations of oncogenes and antioncogenes induced by oxidative stress (Mai et al. 2010). Specific damage, such as oxidative injury and pyrimidine dimerization, can be highlighted using DNA repair enzymes (Figure 13.1). The comet assay offers the possibility of analyzing epigenetic mechanisms of toxicity in relation to gene methylation. DNA methylation plays an important role in gene regulation and maintaining genome stability. Hypermethylation of cytosine may lead to aberrant silencing of crucial genes, such as DNA repair genes and genes involved in cell cycle control. Wentzel et al. (2010) coupled the comet assay with use of different restriction endonucleases to assess the general DNA methylation status. The endonucleases used are

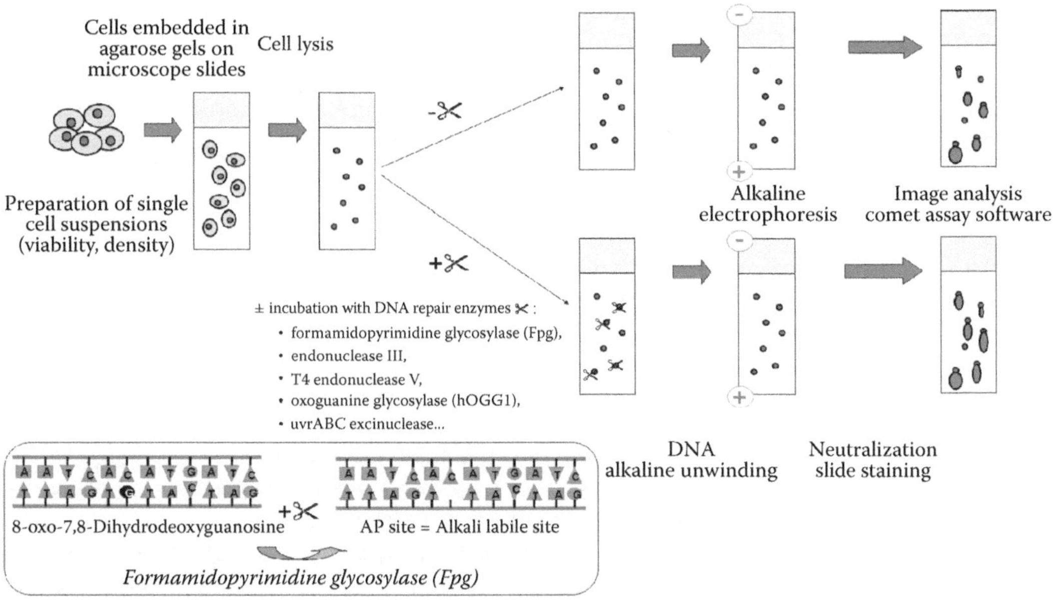

FIGURE 13.1
Scheme of the main steps of the comet assay protocol, coupled to the use of DNA repair enzymes.

able to recognize CpG islands, depending on the presence of methylated or unmethylated cytosine, and induce strand breaks at these sites.

For apoptosis assessment, the DNA diffusion assay has been developed and is now widely used (Singh 2005), including in environmental studies (Frenzilli et al. 2004, 2008, 2009). The DNA diffusion assay developed by Singh (2000) is a visualization method in which cells are embedded in agarose and lysed as usual, but electrophoresis is omitted and replaced by precipitation with ethanol. Small DNA fragments diffuse in the agarose and form a halo and a hazy outer boundary. Indeed, the apoptotic-like comets, the so-called "hedgehog" comets, with all DNA in the tail and a small head as they appear in a standard alkaline comet assay, are contested as a specific indicator of apoptosis. In fact, they may disappear over time after repair, in contradiction with the fact that apoptosis is irreversible (Collins et al. 2008). Vasquez and Tice (1997) underlined that the very low molecular weight (LMW) DNA in terminal apoptotic and necrotic cells would be lost from the gels under typical electrophoretic conditions and proposed a LMW diffusion assay to detect apoptotic/necrotic cells (Frenzilli et al. 2009). Because of the controversy, apoptotic cells are not counted in a standard alkaline comet assay.

13.2.4 Oxidative DNA Damage

DNA can be subjected to several types of oxidative damage, including gross alterations, e.g., single- and double-strand breaks, minor helix distorting DNA base modifications that can be generated after reactive oxygen (ROS) exposure. 8-Oxodeoxyguanosine (8-oxodG) is an abundant lesion formed during oxidative stress and possesses strong miscoding properties responsible for G:C → T:A transversions in DNA. This lesion has been measured for years as a marker of DNA oxidative damage, using the HPLC separation method coupled with electrochemical detection described by Shigenaga et al. (1994). In spite of its sensitivity, this method suffers from several drawbacks: oxidative artifacts during DNA extraction and sample preparation. To avoid artifactual oxidation, extraction using chaotropic reagents such as NaI, and addition of desferrioxamine as an iron chelator to inhibit ROS production were recommended (Helbock et al. 1998; Ravanat et al. 2002). Basal levels measured in aquatic species were in the range 0.6–5 8-oxodGuo/10^5 dGuo in bivalves (Lemière et al. 2005a; Almeida et al. 2007) and fish (Rodríguez-Ariza et al. 1999). Higher values have been found in the digestive glands of *M. galloprovincialis* (Akcha et al. 2000a, 2000b) and in the gills of *Unio tumidus* (Charissou et al. 2004). This method is now replaced by the comet assay with enzymatic detection.

The modified comet assay relies on the addition of endonucleases after cell lysis and DNA digestion with the lesion-specific endocucleases (Figure 13.1). The method is detailed by Friedman-Angeli et al. (2012). Two bacterial DNA repair endonucleases were first used to recognize oxidized purines and pyrimidines and convert them into strand breaks: formamidopyrimidine DNA glycosylase (Fpg), which recognizes the oxidized purine 8-oxoGua, and also ring-opened purines, or formamidopyrimidines; endonuclease III, which recognizes oxidized pyrimidines (Collins 2004; Collins et al. 2008). Fpg has been used to study oxidative damage in *Limanda limanda* from the Seine and Somme estuaries in France (Akcha et al. 2003), in isolated hemocytes of oyster and clam exposed to oxidant (Gielazyn et al. 2003) and in gill cells of *M. edulis* from the Mersey estuary in the United Kingdom (Emmanouil et al. 2008).

The human hOGG1, as 8-oxoguanine DNA glycosylase involved in the BER pathway of 8-oxodG, has been recommended by Smart et al. (2006) and Smith et al. (2006) owing to its better sensitivity and specificity. Its better specificity was confirmed by Michel and Vincent-Hubert (2012), who compared the performances of Fpg and hOGG1 in gill cells of *D. polymorpha* exposed to contaminants.

13.2.5 Mutations

DNA damage generally induces DNA repair mechanisms to protect genome integrity. It is of primary importance that DNA repair occurs before DNA replication takes place; otherwise, mutations will be fixed. The rapid DNA repair mechanisms primarily target genes that are being transcribed. Despite the high efficiency of DNA repair, DNA lesions may persist particularly in case of many lesions in both DNA strands (coding and noncoding regions). Genotoxins can also directly or indirectly inhibit DNA repair enzymes. Mutated DNA is not stable and is more prone to rearrangements such as the loss, migration, and translocation of DNA sequences into another chromosome. This may affect somatic cells as well as germinal cells, and consequently, mutations could be transmitted to future generations.

Genes controlling cell cycle and differentiation are widely studied in medicine because their mutation may initiate cancer. Mutations activating the protooncogene ras have been found in flounder *P. flesus*, in relation to hepatic neoplasms in contaminated sites (Peck-Miller et al. 1998; Vincent-Hubert 2000). High frequency at mutational hotspots in K-ras genes was measured in pink salmon *Oncorhynchus gorbuscha* experimentally exposed to *Exxon Valdez* oil (Roy et al. 1999). The so-called "tumor suppressor" *P53* gene has been sequenced in a number of fish species including the flounder, since its human homologue is often inactivated by mutation in mammals and human carcinogenesis (Cachot et al. 2004). In contrast to human, the fish *P53* gene is infrequently mutated in tumors, but a p53 polymorphism has been found at contaminated sites; its functional role has yet to be explored (Marchand et al. 2010).

Studies of germinal mutations are of great interest in ecotoxicology because they are heritable. Analysis of tandem repeat DNA loci (TRDL), which showed much higher rates of mutation than unique sequence DNA, appeared promising. These TR sequences consist of short repeat series of 2 to 100 bp, located in noncoding regions of the genome and called minisatellites or microsatellites depending on their structure. They are generally GC-rich, and constitute hotspots for mutations. Increased rates of TR mutation have been observed in wildlife as a result of environmental exposure. Thus, the relationship between germline TR instability and exposure to pollutants has been explored (Yauk 1998, 2006; Bouffler et al. 2006).

Yauk and Quinn (1996) have used multilocus DNA fingerprinting methodology to study mutations *in situ* in herring gulls (*Larus argentatus*) nesting in industrialized urban sites (the Great Lakes in north America). They found the mutation rate in herring gulls inhabiting a heavily industrialized urban harbor to be more than twice as high as at three rural sites. Such lesions, which were transmitted to future generations (Yauk 1998), could be important biomarkers for heritable genetic changes resulting from *in situ* exposure to environmental genotoxins in a free-living vertebrate species. Yauk et al. (2000) further studied germinal mutations in herring gulls and demonstrated the impact of environmental pollutants. They reported that gulls nesting near integrated steel mills showed significantly higher mutation rates than gulls from rural locations. Distance of the nesting location of herring gulls from the steel industry's coking ovens was negatively correlated with minisatellite mutation rate, demonstrating significant risk for induced germline mutations in cities with steel operations. For review, please refer to Yauk (2004, 2006). Mutations in minisatellites have also been widely studied in mammals (Dubrova et al. 1993) and fish (the medaka *Oryzias latipes*) (Shimada and Shima 1998, 2004) after irradiation. The studies have shown that some mutations were transmissible to following generations. Instability of minisatellites observed after exposure to nongenotoxic agents raises the question of the influence of epigenetic events on genomic functions.

Another fundamental question concerns the impact of mutations at the population level. Bickham and Smolen (1994) have suggested that mutations can reduce survival and reproductive capacities, which could lead to the extinction of a species. On the other hand, mutations were also shown to be important for evolution and life on earth (Wurgler and Kramers 1992). Only, neutral mutations or mutations that bring a selective advantage have persisted, and adaptive mutations are necessary for individuals to cope with environmental changes (Foster 1998). For the time being, the consequences of *in situ* mutations are not well understood and should be better studied. We hope that in coming years, more studies will be launched to increase our understanding.

13.2.6 Biomolecular Techniques for Genomic and Transcriptomic Studies

Molecular biology techniques have been developed to detect an array of lesions such as mutations, DNA breakages, and recombinations with potential consequences on the genome and its activity. A very limited number of plants and non-mammalian animal species have been fully sequenced. This significantly restricts the development of DNA and RNA microarray techniques in ecotoxicology for genomic studies.

Several techniques for genomic analysis, initially used for species characterization, can also be used for nonsequenced species. Such techniques include the random amplification of DNA or randomly amplified polymorphic DNA (RAPD; Williams et al. 1990), arbitrary primed polymerase chain reaction (Welsh and McClelland 1990), DNA amplification fingerprinting (Caetano-Anollés et al. 1991), and amplified fragment length polymorphism (AFLP; Vos et al. 1995). All these techniques are based on DNA amplification after random primer annealing. Each primer generates amplimers of different size that can be separated by electrophoresis. Patterns from control and treated samples are then compared. The appearance and disappearance of bands, and changes in band intensity could be due to alterations at DNA level such as mutations, DNA breakages, and DNA adducts that could interfere with the *Taq* (*Thermus aquaticus*) polymerase. An example of control and treated RAPD profiles is presented in Figure 13.2. Patterns have been generated from *M. galloprovincialis* gill DNA exposed to benzo(*a*)pyrene diol epoxyde or solvent. Figure 13.2 reveals, for instance, the disappearance of bands 6-1 and 8-2, the appearance of amplimers 8-3 and 8-4, as well as the overexpression of band 5-1 after amplification with primers OPB8, OPB6, and OPB5 in treated samples (Atienzar et al. 2002). In ecotoxicology, RAPD technology has been used to study the effects of genotoxins in environmental species (for review, please refer to Atienzar and Jha 2006). Atienzar and collaborators originally optimized and validated the technique to finally use it to detect DNA alterations. They studied the effects of UV on macroalgae (Atienzar et al. 2000) as well as the consequences of benzo(*a*)pyrene exposure on *Daphnia* and on following generations (Atienzar et al. 1999; Atienzar and Jha 2004). *Daphnia* species are ideal for RAPD studies because all individuals are identical at DNA level (reproduction by parthenogenesis). This significantly reduces the variability in control RAPD profiles. *Daphnia* is also an excellent model to study the transgenerational effects of genotoxins. After exposure to B(*a*)P, changes in RAPD profiles appeared after 3 and 6 days. Nevertheless,

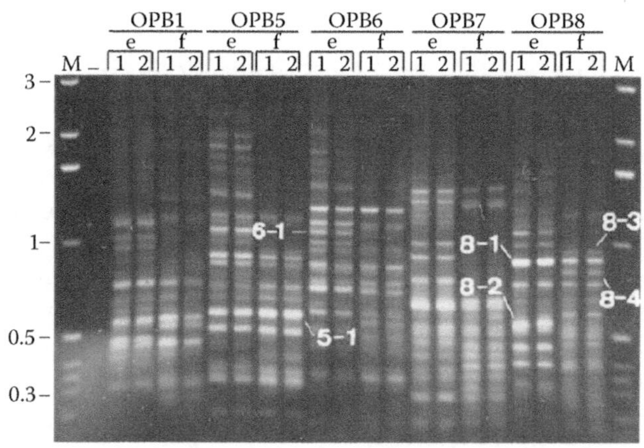

FIGURE 13.2
RAPD profiles of genomic DNA from gill cells of *Mytilus galloprovincialis* nonexposed (e) or exposed (f) to benzo[*a*]pyrene diol epoxyde (677 adducts/10^8 nucleotides). This figure shows disappearance of bands 8-2 and 6-1; appearance of bands 8-3 and 8-4, and an increased intensity of band 5-1. (Well 1 and 2: 20 and 5 ng DNA; (–): control without DNA; *M* = 1 kb, the molecular size (in kilobases) of selected bands is indicated on the left of the gel; OPB 1 to 8 10-mer primers (5′→3′) of different sizes). (From Atienzar, F.A. et al., *Mutat. Res.*, 521, 151–163, 2002. With permission.)

nearly all changes in RAPD profiles were reversible after 8 or 12 days of recovery, which suggests that *Daphnia* can repair most of the DNA lesions. The changes in RAPD profiles were more pronounced in moribund individuals compared to active swimming *Daphnia* exposed to B(*a*)P. Some of the changes were transmitted from the B(*a*)P-exposed adult *Daphnia* to the neonates (first two generations). It is noteworthy that none of the neonates was exposed directly to B(*a*)P. This suggests that genomic alterations occurred in germinal cells in adult *Daphnia* and were transmitted to the next generations (Atienzar and Jha 2004). These results illustrate well the power of the RAPD approach to detect genomic changes after careful optimization.

Similar approaches have been used to study the effect of genotoxins on gene expression. The techniques developed without any knowledge of the genome are: RNA-arbitrarily primed polymerase chain reaction (RAP-PCR; Welsh et al. 1992), subtractive hybridization techniques (Sargent and Dawid 1983), and differential display (Liang and Pardee 1992). For instance, Rodius et al. (2002, 2004) performed RAP-PCR to study the impact of aquatic pollutants on the transcriptome of freshwater bivalves. This approach uses random primers to amplify cDNA, then the amplimers are separated by electrophoresis on agarose gel, and the expression patterns of organisms from polluted sites are compared to controls. Freshwater mussels (*U. tumidus*) from a control area were transferred for 21 days to two sites in the Moselle River, upstream and downstream of urban and industrial areas (Rodius et al. 2004). An additional amplimer was observed in the pattern of the bivalves transplanted to the downstream site compared to controls (Figure 13.3).

This amplimer has been cloned and sequenced. PCR performed with specific primers showed that this product is not the result of a polymorphism but corresponds to gene induction in the transplanted mussels. However, the analysis of about 40 mineral and organic pollutants in the sediments did not allow the identification of pollutants which could explain the modification of the expression pattern. Moreover, the amplimer coding and translated sequences did not show any homology with nucleotide or protein databases at the National Center for Biotechnology Information site, so no function concerning this gene could be suggested.

Considering that the sequencing of the genome of aquatic species is becoming increasingly accessible, the field of ecotoxicogenomics is expected to progress in the coming years (Lettieri 2006; Piña and Barata 2011). This will help us understand the observed responses to environmental stressors in contaminated environments (Calzolai et al. 2007).

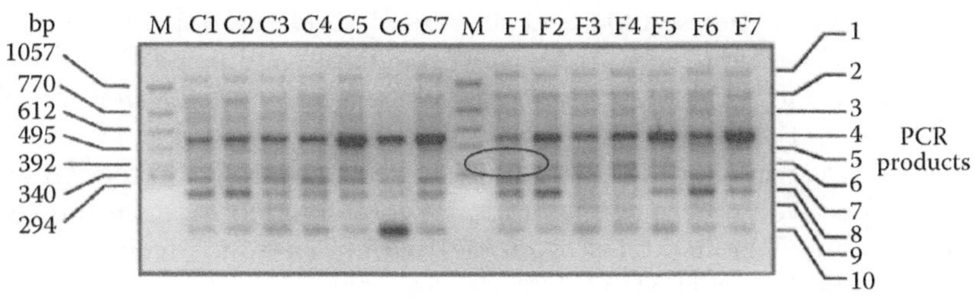

FIGURE 13.3
Comparison of RAP-PCR amplification patterns of mRNA from digestive glands of *Unio tumidus* using the AP1 primer. PCR products (amplimers) are numbered to facilitate pattern description. Mussels have been transplanted at C (control) and F (polluted) sites. M, DNA size marker.

13.3 Genotoxicity, Physiological, and Populational Effects

The impact of genotoxicity on populations is a matter of debate in ecotoxicology. However, there are very few *in situ* studies allowing the establishment of links between genomic alterations, physiological effects in individuals and populational changes. Two case studies will be discussed here:

- The impact of the *Exxon Valdez* oil (EVO) spill on fish, especially herring populations
- The relationships between genomic changes and health status of populations exposed to contaminants, such as mosquito fish *Gambusia affinis* exposed to radioactive elements and *Daphnia longispina* exposed to chemical pollutants

13.3.1 *Exxon Valdez* Accident

On March 24, 1989, 40,000 metric tons of crude oil were spilled into Prince William Sound (PWS), Alaska, from the tanker *Exxon Valdez*. The accident occurred coincidently with the spawning of Pacific herring (*Clupea pallasi*) stocks in PWS. The spill trajectory overlapped the route of the adult herring spawning migration and spawning locations. The herring did not spawn on the most heavily oiled beaches but did spawn on shorelines considered to be lightly to moderately oiled.

No mortality of adult fish was reported, but eggs and larvae were severely impacted. Larvae hatched from contaminated sites had high rates of abnormalities and mortality (Brown et al. 1996; Hose et al. 1996; Norcross et al. 1996; Marty et al. 1997). A high level of mitotic aberrations and micronuclei was measured in larvae that had been collected between 6 and 10 weeks after the grounding (Hose and Brown 1998): 40–51% of aberrant anaphases and telophases were recorded in impacted areas, versus <10% in non-impacted zones. Major oil-associated effects in larvae taken from oiled sites in spring 1989 included small size, ascites, pericardial edema, delayed development, and genetic damage.

Studies focused on the Pacific herring because of their economic value and the coincidence of the EVO spill with spawning, but similar effects have been observed with pink salmon (*O. gorbuscha*). Although pink salmon did not spawn until months after the spill, they deposited their eggs in intertidal stretches of stream deltas. Oiled gravel persisting in some streams was a continuous source of exposure for pink salmon embryos, resulting in elevated mortality.

Toxicity was related to the soluble and dispersed PAH fraction in the water column; sediment concentrations were low in the few specimens collected. Naphthalenes, fluorenes, phenanthrenes, and dibenzothiophenes were found in seawater samples at oiled sites. Total PAH (TPAH) concentrations in water in spring peaked at 1.9 and 2.6 $\mu g\ L^{-1}$ in oiled areas, which was considered weak pollution, because it never exceeded water quality standards. Retrospectively, these considerations were erroneous, since the concentration thresholds for genomic lesions and morphological abnormalities were estimated later to range between 0.4 and 0.7 μg TPAHs L^{-1} (Carls et al. 2002).

Significant effects on fish populations were not reported in the 2 years following the oil spill, until a collapse of more than 75% in the adult herring population was recorded during the winter 1992–1993, responsible for an economic crash never before seen. A link with the EVO accident could not be demonstrated, because monitoring studies had stopped in 1989 for the reason that TPAH concentrations in water were considered too low to be hazardous to wildlife. The herring collapse was explained by a disease caused by a viral hemorrhagic septicemia virus, possibly favored by immunological deficiency consecutive to PAH contamination (Chapter 6). Other factors have been put forward, such as insufficient food supply, population at carrying capacity, and poor overwintering conditions for the fish. The tools were not in place at the time of the oil spill and the collapse to validate any hypothesis. Therefore, the link between the EVO spill and disappearance of the herring population has remained a contentious subject.

Recent papers reanalyzed the question in the light of documented data demonstrating that the herring decline began earlier than 1992 and was associated with the EVO spill.

Thorne and Thomas (2008) produced comprehensive data on the behavior of adult herring, showing them to be vulnerable to damage from oil spills, which was unknown or misunderstood at the time of the spill: for instance, the surfacing behavior of herring that undergo vertical migration and gulp air at the surface, contributing to their exposure to oil via ingestion and gill coating.

Hulson et al. (2008) confirmed that the population model used overestimated herring biomass from 1990 to 1992, and that back-calculated estimates from hydroacoustic abundance showed a decline actually starting in 1989. For these authors, there were two declines, one in 1989 due to the oil spill, and the main one in 1992 when disease affected a large population that was in weakened condition.

Incardona et al. (2005, 2009) and Hicken et al. (2011) gave mechanistic explanations to the decline of the Pacific herring in PWS. Fish embryos exposed to low levels of low molecular tricyclic PAHs in weathered crude oil were shown to develop a syndrome of pericardial edema, and craniofacial and body axis defects (Incardona et al. 2005). Cardiac syndrome was the primary response of herring embryos to ANSCO crude oil during weathering (Incardona et al. 2009). Transient embryonic exposure to very low concentrations of oil causes sublethal and delayed toxicity: nearly a year after oil embryonic exposure, adult fishes showed subtle changes in performance, indicative of reduced cardiac output sufficient to compromise survival under stress conditions. These delayed physiological impacts on cardiovascular performance at later stages provide a potential mechanism linking reduced individual survival to population-level ecosystem responses of fish species to chronic, low-level oil pollution.

The population effects on fish populations after the EVO spill are cause for concern of the effects of any major spill, including the recent Deepwater Horizon (DWH) event. The long-term biological consequences of this oil spill are still unknown, especially for resident organisms, with the exception of evident toxicity on avian species and mortality in wildlife. Embryotoxicity studies are being conducted. Genomic and physiological footprints of the DWH oil spill on resident marsh fishes attested to exposure to hydrocarbon-like chemicals and indicate physiological and reproductive impairment (Whitehead et al. 2011).

13.3.2 Genotoxicity, Population Genetics, and Environmental Contamination

We have not found any experimental ecoepidemiological studies directly answering the question of the involvement of genomic alterations induced by environmental stressors on population dynamics. But there are a few studies retrospectively relating genomic changes and good health of living populations in contaminated environments.

Theodorakis and Shugart (1997) studied these relationships in populations of mosquito fish (*G. affinis*) living in four ponds—two noncontaminated (A, B) and two radioactively-contaminated (C, D)—on or near the Department of Energy Oak Ridge Reservation (US-TN). Site C was colonized in the late 1970s by the transplantation of approximately

250 individuals from pond A, so that any differences between the two phylogenetically related populations A and C could be explained by environmental factors.

The study, which used RAPD and allozyme analyses of the genomic structure of the four populations, showed:

- Identical patterns between the two noncontaminated populations A and B, which differed from that of the radionuclide-contaminated populations.
- An increased genetic diversity with a higher percentage of polymorphism and heterozygosity in the radionuclide-exposed populations C and D.
- Patterns of population C were much closer to those of population D than to patterns of population A, from which it drifted phylogenetically. Observed changes in banding patterns (lower or higher frequencies) conferred selective advantages to the contaminated populations.

The influence of radionuclides on the genetic structure of the population was clearly demonstrated. Site contamination resulted in severe stress affecting the viability and reproductive success of transplanted individuals, which exhibited a high number of embryonic abnormalities and a lower fertility than that observed at reference sites (Theodorakis 2003).

Contaminant-indicative bands (CIBs) were present at higher frequencies in the descendants of the exposed populations and were also observed in control mosquito fish that had been encaged *in situ* in pond C (Theodorakis et al. 1999). These changes in banding patterns of contaminated fish were associated with increased fertility, and a decrease in embryonic abnormalities and in DNA damage.

The hypothesis of the selection of resistant populations from surviving individuals has been advanced to explain the observed changes in genetic structure. Thus, the contaminant-indicative RAPD markers could be generated from, or linked to, loci involved in the response to radiation, or even a mechanism of radioresistance. The induction of an identified band was associated with that of a locus (NP nucleoside phosphorylase) involved in nucleoside synthesis and DNA repair. The authors suggested two links, between the induction of this NP locus and DNA breaks, and between increased genetic diversity and a mechanism of adaptation or radioresistance. The characterization of other contaminant-indicative RAPD markers and the identification of the function of the corresponding genes have yet to be carried out (Theodorakis and Bickham 2004). Very similar results were subsequently obtained by the same authors with redbreast sunfishes (*Lepomis auritus*), collected along a contamination gradient in a pulp mill-contaminated river (the Pigeon River located in North Carolina and Tennessee, USA) (Theodorakis et al. 2006).

These studies support the existence of a relationship between genotoxicity and population effects.

One putative CIB (perhaps related to a locus involved in metal tolerance) was also found in AFLP patterns (1) of a historically metal-exposed population and (2) of the more copper-tolerant individuals from populations of *D. longispina* studied by Martins et al. (2009). These authors studied *D. longispina* collected in or near the aquatic system of the abandoned cupric-pyrite mine of Sao Domingos (Portugal). Three populations were sampled [two from control sites and one from a historically acid mine drainage (AMD) impacted station]. The authors expected a loss of genetic diversity as a result of increased resistance in the impacted population. If they could discriminate (genetically) the exposed AMD population from the reference populations based on the AFLP patterns, they could not conclude to a reduction of genetic diversity in this AMD-impacted population. They consider that the complete molecular characterization of the CIB found would be essential to assess its possible relationship with the increased tolerance.

A qualitative relationship linking genomic and population changes has been highlighted in these examples, reflecting accidental situations or severe chronic stress levels. It would be worthwhile to investigate example situations of lower stress levels to determine thresholds of contamination acceptable for populations in the long term.

13.4 Quality Threshold for Health Safety

Determination of quality thresholds of aquatic pollutants is one of the outcomes of ecotoxicity studies. Usually, quality thresholds are determined for every pollutant separately, on the basis of the results of controlled experiments. Ideally, concentration–effects relationships should be established from *in situ* studies, in order to integrate pollutant interactions within complex environments. The latter approach still requires much research. Such an innovative approach has been used by the National Marine Fisheries Service (NMFS) of the NOAA to determine sediment quality thresholds for PAHs. This approach will be highlighted here, because it answers our dilemma/questions on how to take into account multipollution, environmental complexity, and management constraints in environmental regulation (Johnson et al. 2002). In the NMFS approach, robust data based on actual responses of living species to PAH polluted sites have been collected over decades. The responses analyzed consisted of genotoxicity effects, hepatic lesions, disturbances of reproduction, development, growth, and immunological effects in the English sole *P. vetulus*.

The response of the English sole to PAHs has been widely documented, especially in Puget Sound, as discussed in Section 13.1.1 and Chapter 15. Relationships between sediment TPAH concentrations and biological responses have allowed an estimation of a threshold with confidence limits, for each type of effect measured. The list of 18 PAHs considered (10 congeners with 2–3 rings, 8 congeners with 4–6 rings) was close to the list of the 16 PAHs defined by the EPA. BaP equivalents were not used because of the high emphasis given to high molecular weight carcinogenic congeners, whereas LMW PAHs contributed to reproduction and growth alterations (Johnson et al. 2002).

In order to assess the likely degree of injury experienced by English sole exposed to sediments with different sediment PAH concentrations, expected prevalences of liver lesions and reproductive abnormalities were calculated using a regression model. The sediment TPAH threshold concentration below which no significant carcinogenic or adverse reproductive effects in English sole are observed has been estimated as 1000 $\mu g\ kg^{-1}$ dry weight (or 1000 $ng\ g^{-1}$ dry weight).

This threshold was well below the sediment quality criteria for PAHs set by the Washington State Department of Ecology in 1995 of 7400 $\mu g\ kg^{-1}$ dry weight for LMW aromatic hydrocarbons and 19,200 $\mu g\ kg^{-1}$ dry weight for high molecular weight aromatic hydrocarbons, assuming a typical value of 2% total organic carbon for Puget Sound sediment. This estimated threshold was not based on the lowest threshold values (DNA damage and nonneoplastic liver lesions in English sole) and was therefore not overly conservative; indeed, the estimated thresholds for DNA damage were 290 $ng\ g^{-1}$ d.w. sediment (with upper confidence limit of 1318 $ng\ g^{-1}$ d.w. sediment), and 630 $ng\ g^{-1}$ d.w. sediment for several reproductive effects.

This guideline of 1000 μg kg^{-1} dry weight (1000 ng g^{-1} dry weight) did not incorporate a safety factor to account for uncertainty in this threshold estimate, as risk analyses often do. It was a practical value for management decisions that should be protective of estuarine fish populations and workable from the perspective of sediment remediation and management (Johnson et al. 2002).

It would be worth developing a similar approach to protect invertebrates and consumers at higher levels in the food chain. Indeed, invertebrates bioaccumulate pollutants that may be transferred to their predators via the food chain. After the *Erika* oil spill (13.2.1), high levels of PAHs in soft tissues of marine mussels were measured, and a threshold of total PAHs in seafood had to be defined to protect human health. Lemière et al. (2005b) showed that ingestion by mammals of mussels contaminated with the *Erika* oil induced DNA damage in their hepatocytes, blood cells, and bone marrow. The authors established that a threshold level of BaP toxic equivalents in food, i.e., mussel tissues, is in the range of 6–34 μg kg^{-1} d.w., which would not induce any DNA damage and be safe to consumers.

13.5 Conclusions

Links between genotoxicity at the level of the individual and population effects have been established retrospectively in several cases and in some situations of severe stress. It would be useful if future studies explored genomic changes associated not only with contaminant exposure, but also with tolerance and adaptation (Chapter 14). There is a real need for integrated studies to better understand mechanisms developed at the population level, for species to survive and reproduce in unsafe environments, in the field, and over several generations. Stress conditions that are not too stringent would be worth studying in order to help determine pollution thresholds that can be accepted in the long term and be compatible with individual/population fitness.

The present review has shown that several assays are efficient tools to detect genotoxic impacts in exposed species *in situ*. In some cases, genotoxicity in somatic cells could be associated with embryo mortality, decreased viability, and reproduction impairment.

Measuring genotoxicity in early life stages and juveniles may be envisaged to link more closely genotoxicity in individuals and population dynamics. The comet assay would be a useful tool to investigate genotoxicity in juveniles, with careful interpretation of the responses and time-course studies of repair efficiency.

Bearing in mind that the gametes of a number of invertebrates are emitted into the surrounding water and that larvae are in direct contact with pollutants in contaminated environments, these critical life stages would be worth studying as targets for genotoxic aquatic pollutants. Prospects for research on gametes have emerged recently. The significance of such studies in an ecological perspective is obvious, provided they mimic conditions of exposure representative of the environment. In invertebrates and vertebrates, it would be worth analyzing the sensitivities of gametes and zygotes at environmental concentrations of contaminants.

In the case of LMW PAHs, it has been demonstrated that exposure at low concentrations to these congeners may induce cardiac and skeleton abnormalities in embryos that will have consequences on adult performance. These congeners are considered nongenotoxic in mammals, but their genotoxicity in embryos is not well known. Whether or not their mode of action refers to genotoxicity in embryos, it is necessary to have a better knowledge of the underlying mechanisms of toxicity at the gene level and to understand the network of impacted genomic functions. Basic fundamental research is needed, not only genotoxicity studies for DNA structure integrity, but also transcriptomic studies to identify sets of genes subjected to alterations and dysfunctioning. No doubt the development of the "omics" technologies will be useful to investigate the mechanisms underlying pollutant toxicity. Furthermore, they have the potential to provide new high-throughput tools for the detection of genomic disturbance and pollution fingerprints.

Acknowledgment

The authors warmly thank Mireille Ollivier, CNRS, for a literature survey and help in providing documentation.

References

Akcha, F. et al. 2000a. Benzo[*a*]pyrene-induced DNA damage in *Mytilus galloprovincialis*: Measurement of bulky DNA adducts and DNA oxidative damage in terms of 8-oxo-7,8-dihydro-2′-deoxyguanosine formation. *Biomarkers* 5:355–61.

Akcha, F. et al. 2000b. Induction and elimination of bulky benzo[*a*]pyrene-related DNA adducts and 8-oxodGuo in mussels *Mytilus galloprovincialis* exposed *in vivo* to B[*a*]P-contaminated feed. *Mar. Ecol. Progr. Ser.* 205:195–206.

Akcha, F. et al. 2003. Potential value of the comet assay and DNA adduct measurement in dab (*Limanda limanda*) for assessment of *in situ* exposure to genotoxic compounds. *Mutat. Res.* 534:21–32.

Akcha, F., G. Leday, and A. Pfohl-Leszkowicz. 2004. Measurement of DNA adducts and strand breaks in dab (*Limanda limanda*) collected in the field: Effects of biotic (age, sex) and abiotic (sampling site and period) factors on the extent of DNA damage. *Mutat. Res.* 552:197–207.

Almeida, E.A. et al. 2006. Oxidative stress in *Perna perna* and other bivalves as indicators of environmental stress in the Brazilian marine environment: Antioxidants, lipid peroxidation and DNA damage. *Comp. Biochem. Physiol.* 135C:295–303.

Almeida, E.A. et al. 2007. Oxidative stress in *Perna perna* and other bivalves as indicators of environmental stress in the Brazilian marine environment: Antioxidants, lipid peroxidation and DNA damage. *Comp. Biochem. Physiol.* 146A:588–600.

Amat, A. et al. 2006. DNA adducts in fish following an oil spill exposure. *Environ. Chem. Lett.* 4:93–9.

Anderson, S.L., and G.C. Wild. 1994. Linking genotoxic responses and reproductive success in ecotoxicology. *Environ Health Perspect.* 102(Suppl 12):9–12.

Arab, K. et al. 2009. Typical signature of DNA damage in white blood cells: A pilot study on etheno adducts in Danish mother–newborn child pairs. *Carcinogenesis* 30:282–5.

Atienzar, F.A., and A.N. Jha. 2004. The random amplified polymorphic DNA (RAPD) assay to determine DNA alterations, repair and transgenerational effects in B(*a*)P exposed *Daphnia magna. Mutat. Res.* 552:125–40.

Atienzar, F.A., and A.N. Jha. 2006. The random amplified polymorphic DNA (RAPD) assay and related techniques applied to genotoxicity and carcinogenesis studies: A critical review. *Mutat. Res.* 613:76–102.

Atienzar, F.A. et al. 1999. Qualitative assessment of genotoxicity using random amplified polymorphic DNA: Comparison of genomic template stability with key fitness parameters in *Daphnia magna* exposed to benzo[*a*]pyrene. *Environ. Toxicol. Chem.* 18:2275–82.

Atienzar, F.A. et al. 2000. Comparison of ultraviolet-induced genotoxicity detected by random amplified polymorphic DNA with chlorophyll fluorescence and growth in a marine macroalgae, *Palmaria palmata*. *Aquat. Toxicol.* 50:1–12.

Atienzar, F.A. et al. 2002. Evaluation of the random amplified polymorphic DNA (RAPD) assay for the detection of DNA damage and mutations. *Mutat. Res.* 521:151–63.

Baršienė, J. 2002. Genotoxic impacts in Klaipeda Marine port and Butinge oil terminal areas (Baltic Sea). *Mar. Environ. Res.* 54:475–9.

Baršienė, J. et al. 2006a. Biomarker responses in flounder (*Platichthys flesus*) and mussel (*Mytilus edulis*) in the Klaipėda–Būtingė area (Baltic Sea). *Mar. Pollut. Bull.* 53:422–36.

Baršienė, J. et al. 2006b. Cytogenetic and cytotoxic effects in gill cells of the blue mussel *Mytilus* spp. from different zones of the Baltic Sea. *Mar. Pollut. Bull.* 53:469–78.

Baršienė, J. et al. 2012. Risk of environmental genotoxicity in the Baltic Sea over the period of 2009–2011 assessed by micronuclei frequencies in blood erythrocytes of flounder (*Platichthys flesus*), herring (*Clupea harengus*) and eelpout (*Zoarces viviparus*) *Mar. Environ. Res.* 77:35–42.

Baussant, T. et al. 2009. Enzymatic and cellular responses in relation to body burden of PAHs in bivalve molluscs: A case study with chronic levels of North Sea and Barents Sea dispersed oil. *Mar. Pollut. Bull.* 58:1796–807.

Bernardeschi, M. et al. 2010. Genotoxic potential of TiO_2 on bottlenose dolphin leukocytes. *Anal. Bioanal. Chem.* 396:619–23.

Bickham, J.W., and M.J. Smolen. 1994. Somatic and heritable effects of environmental genotoxins and the emergence of evolutionary toxicology. *Environ. Health Perspect.* 102:25–8.

Binelli, A. et al. 2010. Multi-biomarker approach to investigate the state of contamination of the Lambro/R. Po confluence (Italy) by zebra mussels (*Dreissena polymorpha*). *Chemosphere* 79: 518–28.

Bocquené, G. et al. 2004. Biological effects of the "Erika" oil spill on the common mussel (*Mytilus edulis*). *Aquat. Liv. Res.* 17:309–16.

Bodin, N. et al. 2004. Seasonal variations of a battery of biomarkers and physiological indices for the mussel *Mytilus galloprovincialis* transplanted into the northwest Mediterranean Sea. *Comp. Biochem. Physiol.* 138C:411–27.

Bolognesi, C., and M. Hayashi. 2011. Micronucleus assay in aquatic animals. *Mutagenesis* 26:205–13.

Bombail, V. et al. 2001. Application of the comet and micronucleus assays to butterfish (*Pholis gunnellus*) erythrocytes from the Firth of Forth, Scotland. *Chemosphere* 44:383–92.

Bony, S. et al., 2008. Genotoxic pressure of vineyard pesticides in fish: Field and mesocosm surveys. *Aquat. Toxicol.* 89:197–203.

Bouffler, S.D. et al. 2006. Assessing radiation-associated mutational risk to the germline: Repetitive DNA sequences as mutational targets and biomarkers. *Radiat. Res.* 165:249–68.

Bourgeault, A. et al. 2010. Lessons from a transplantation of zebra mussels into a small urban river: An integrated ecotoxicological assessment. *Environ. Toxicol.* 25:468–78.

Brown E.D., B.L. Norcross, and J.W. Short. 1996. An introduction to studies on the effects of the *Exxon Valdez* oil spill on early life history stages of Pacific herring, *Clupea pallasi*, in Prince William Sound, Alaska. *Can. J. Fish. Aquat. Sci.* 53:2337–42.

Cachot, J. et al. 2004. Functional analysis of chemically-induced mutations at the flounder TP53 locus, the FACIM assay. *Mutat. Res.* 552:51–60.

Caetano-Anollés, G., B.J. Bassam, and P.M. Gresshoff. 1991. DNA amplification fingerprinting using very short arbitrary oligonucleotide primers. *Biotechnology* 9:553–7.

Calzolai, L. et al. 2007. Transcriptomics and proteomics. Applications to ecotoxicology. *Comp. Biochem. Physiol.* 2D:245–9.

Carls, M.G., G.D. Marty, and J.E. Hose. 2002. Synthesis of the toxicological impacts of the *Exxon Valdez* oil spill on Pacific herring (*Clupea pallasi*) in Prince William Sound, Alaska, U.S.A. *Can. J. Fish. Aquat. Sci.* 59:153–72.

Charissou, A.M., C. Cossu-Leguille, and P. Vasseur. 2004. Relationship between two oxidative stress biomarkers, malondialdehyde and 8-oxo-7,8-dihydro-2′-deoxyguanosine, in the freshwater bivalve *Unio tumidus*. *Sci. Total Environ.* 322:109–22.

Collier, T.K., and U. Varanasi. 1991. Hepatic activities of xenobiotic metabolizing enzymes and biliary levels of xenobiotics in English sole (*Parophrys vetulus*) exposed to environmental contaminants. *Arch. Environ. Contam. Toxicol.*20:462–73.

Collins, A.R. 2004. The comet assay for DNA damage and repair. *Mol. Biotechnol.* 26:249–61.

Collins, A.R. et al. 2008. The comet assay: Topical issues. *Mutagenesis* 23:143–51.

Devaux, A., and S. Bony. 2012. Genotoxicity of contaminants: Comet assay. In: *Encyclopedia of Aquatic Ecotoxicology* (Volumes 1 and 2), ed. J.F. Férard and C. Blaise. Dordrecht, The Netherlands: Springer (in press).

Devaux, A. et al. 2011. Reproduction impairment following paternal genotoxin exposure in brown trout (*Salmo trutta*) and Arctic charr (*Salvelinus alpinus*). *Aquat. Toxicol* 101:405–11.

Dhawan, A., M. Bajpayee, and D. Parmar. 2009. Comet assay: A reliable tool for the assessment of DNA damage in different models. *Cell. Biol. Toxicol* 25:5–32.

Dubrova, Y.E., A.J. Jeffreys, and A.M. Malashenko. 1993. Mouse minisatellite mutations induced by ionizing radiation. *Nat. Genet.* 5:92–4.

Eastmond, D.A., M. Schuler, and D.S. Rupa. 1995. Advantages and limitations of using fluorescence *in situ* hybridization for the detection of aneuploidy in interphase human cells. *Mutat. Res. Lett.* 348:153–62.

Emmanouil, C. et al. 2008. Oxidative damage in gill of *Mytilus edulis* from Merseyside, UK, and reversibility after depuration. *Environ. Pollut.* 151:663–8.

Ericson, G., E. Lindesjöö, and L. Balk. 1998. DNA adducts and histopathological lesions in perch (*Perca fluviatilis*) and northern pike (*Esox lucius*) along a polycyclic aromatic hydrocarbon gradient on the Swedish coastline of the Baltic Sea. *Can. J. Fish. Aquat. Sci.* 55:815–24.

Ericson, G. et al. 2002. DNA adducts as indicators of genotoxic exposure in indigenous and transplanted mussels, *Mytilus edulis* L. from Icelandic coastal sites. *Mutat. Res.* 516:91–9.

Fenech, M. 2007. Cytokinesis-block micronucleus cytome assay. *Nat. Protoc.* 2:1084–104.

Fernandez, B. et al. 2011. Micronuclei and other nuclear abnormalities in mussels (*Mytilus galloprovincialis*) as biomarkers of cyto-genotoxic pollution in Mediterranean waters. *Environ. Mol. Mutagen.* 52:479–91.

Flammarion, P. et al. 2002. Multibiomarker responses in fish from the Moselle River (France). *Ecotoxicol. Environ. Saf.* 51:145–53.

Foster, P.L. 1998. Adaptive mutation: Has the unicorn landed? *Genetics* 148:1453–9.

French, B. et al. 1996. Accumulation and dose-response of hepatic DNA adducts in English sole (*Pleuronectes vetulus*) exposed to a gradient of contaminated sediments. *Aquat. Toxicol.* 36:1–16.

Frenzilli G. et al. 2004. DNA damage in eelpout (*Zoarces viviparus*) from Göteborg harbour. *Mutat. Res./Fund. Mol. Mech. Mutagen.* 552:187–95.

Frenzilli, G. et al. 2008. Cellular responses in the cyprinid *Leuciscus cephalus* from a contaminated freshwater ecosystem. *Aquat. Toxicol.* 89:188–96.

Frenzilli, G., M. Nigro, and B.P. Lyons. 2009. The Comet assay for the evaluation of genotoxic impact in aquatic environments. *Mutat. Res./Rev. Mutat. Res.* 681:80–92.

Friedmann-Angeli, J.P. et al. 2012. Damage in aquatic animal: Comet assays and 8-oxo-7,8-dihidro-2′-deoxyguanosine levels. In: *Oxidative Stress in Aquatic Ecosystems*, ed. D. Abele, J.P. Vazquez-Medina, and T. Zenteno-Savin. Chichester, UK: Wiley-Blackwell.

Galloway, T. et al. 2010. Sublethal toxicity of nano-titanium dioxide and carbon nanotubes in a sediment dwelling marine polychaete. *Environ. Pollut.* 158:1748–55.

Gauthier, L., Y. Levi, and A. Jaylet. 1989. Evaluation of the clastogenicity of water treated with sodium hypochlorite or monochioramine using a micronucleus test in newt larvae (*Pleurodeles waltl*). *Mutagenesis* 4:170–3.

Gauthier, L. et al. 2004. Biomonitoring of the genotoxic potential (micronucleus assay) and detoxifying activity (EROD induction) in the River Dadou (France), using the amphibian *Xenopus laevis*. *Sci. Total Environ.* 323:47–61.

Gielazyn, M.L. et al. 2003. Detection of oxidative DNA damage in isolated marine bivalve hemocytes using the comet assay and formamidopyrimidine glycosylase (Fpg). *Mutat. Res.* 542:15–22.

Goanvec, C. et al. 2008. Flow cytometry for the evaluation of chromosomal damage in turbot *Psetta maxima* (L.) exposed to the dissolved fraction of heavy fuel oil in sea water: A comparison with classical biomarkers. *J. Fish Biol.* 73:395–413.

Gorbi, S. et al. 2008. An ecotoxicological protocol with caged mussels, *Mytilus galloprovincialis*, for monitoring the impact of an offshore platform in the Adriatic sea. *Mar. Environ. Res.* 65:34–49.

Hartl, M.G.J., S.J.W. Grigson, and E. Sinet. 2010. Maintenance of bivalve hemocytes for the purpose of delayed DNA strand break assessment using the comet assay. *Environ. Mol. Mutagen.* 51:64–8.

Harvey, J.S. et al. 1999. An assessment of the genotoxic impact of the *Sea Empress* oil spill by the measurement of DNA adduct levels in selected invertebrate and vertebrate species. *Mutat. Res.* 441:103–14.

Helbock, H.J. et al. 1998. DNA oxidation matters: The HPLC-electrochemical detection assay of 8-oxo-deoxyguanosine and 8-oxo-guanine. *Proc. Natl. Acad. Sci. U. S. A.* 95:288–93.

HELCOM. 2010. Hazardous substances in the Baltic Sea–An intergated thematic assessment of hazardous substances in the Baltic Sea. Balt. Sea Environ. Proc. No. 120B.

Hicken, C.E. et al. 2011. Sublethal exposure to crude oil during embryonic development alters cardiac morphology and reduces aerobic capacity in adult fish. *Proc. Natl. Acad. Sci. U. S. A.* 108:7086–90.

Hose, J.E. 1994. Large-scale genotoxicity assessments in the marine environment. *Environ. Health Perspect.* 102(Suppl 12):29–32.

Hose, J.E., and E.D. Brown. 1998. Field applications of the piscine anaphase aberration test: Lessons from the *Exxon Valdez* oil spill. *Mutat. Res.* 399:167–78.

Hose, J.E. et al. 1996. Sublethal effects of the (*Exxon Valdez*) oil spill on herring embryos and larvae: Morphological, cytogenetic, and histopathological assessments, 1989–1991. *Can. J. Fish Aquat. Sci.* 53:2355–65.

Hoshina M.M., D. de Franceschi de Angelis, and M.A. Marin-Morales. 2008. Induction of micronucleus and nuclear alterations in fish (*Oreochromis niloticus*) by a petroleum refinery effluent. *Mutat. Res./Gen. Toxicol. Environ. Mutagen.* 656:44–8.

Hulson, P.-J.F. et al. 2008. Data conflicts in fishery models: Incorporating hydroacoustic data into the Prince William Sound Pacific herring assessment model. *ICES J. Mar. Sci.* 65:25–43.

Hylland, K. et al. 2008. Water column monitoring near oil installations in the North Sea 2001–2004. *Mar. Pollut. Bull.* 56:414–29.

Incardona, J.P. et al. 2005. Aryl hydrocarbon receptor-independent toxicity of weathered crude oil during fish development. *Environ. Health Perspect.* 113:1755–62.

Incardona, J.P. et al. 2009. Cardiac arrhythmia is the primary response of embryonic Pacific herring (*Clupea pallasi*) exposed to crude oil during weathering. *Environ. Sci. Technol.* 43:201–7.

ISO. 2006. International Standard. Water quality—Evaluation of genotoxicity by measurement of the induction of micronuclei—Part 1: Evaluation of genotoxicity using amphibian larvae. ISO 21427-1, ICS:13.060.70, Geneva, CH, August 2006, 15 pp.

Jha, A.N. 2008. Ecotoxicological applications and significance of the comet assay. *Mutagenesis* 23:207–21.

Johnson, L.L., T.K. Collier, and J.E. Stein. 2002. An analysis in support of sediment quality thresholds for polycyclic aromatic hydrocarbons (PAHs) to protect estuarine fish. *Aquat. Conserv. Fresh. Ecosyst.* 12:517–38.

Kilemade, M.F. et al. 2004. Genotoxicity of field-collected inter-tidal sediments from Cork Harbor, to juvenile turbot (*Scophthalmus maximus* L.) as measured by the Comet assay. *Environ. Mol. Mutagen.* 44:54–64.

Kirsch-Volders, M. et al. 2003. Report from the *in vitro* micronucleus assay working group, *Mutat. Res.* 540:153–63.

Kocan, R.M., M.L. Landholt, and K.M. Sabo. 1982. Anaphase aberrations: A measure of genotoxicity in mutagen-treated fish cells. *Environ. Mutagen.* 4:181–9.

Lacaze, E. et al. 2010. Genotoxicity assessment in the amphipod *Gammarus fossarum* by use of the alkaline Comet assay. *Mutat. Res./Gen. Toxicol. Environ. Mutagen.* 700:32–8.

Lacaze, E. et al. 2011. DNA damage in caged *Gammarus fossarum* amphipods: A tool for freshwater genotoxicity assessment. *Environ. Pollut.* 159:1682–91.

Lang, T., and V. Dethelfsen. 1994. Externally visible diseases and liver neoplasms in flounder (*Platichtys flesus*) in the southern Baltic Sea. In: *Diseases and Parasites of Flounder Platichtys flesus in the Baltic Sea*, vol. 15, ed. G. Byland and L.-G. Lönnström, 39–58. BMB Publications: Baltic Marine Biologists, Åbo.

Lang, T. et al. 2006. Liver histopathology in Baltic flounder (*Platichtys flesus)* as indicator of biological effects of contaminants. *Mar. Pollut. Bull.* 53:488–96.

Large, T. et al. 2002. Different levels of mussel (*Mytilus edulis*) DNA strand breaks following chronic field and acute laboratory exposure to polycyclic aromatic hydrocarbons. *Mar. Environ. Res.* 54:493–7.

Larno, V. et al. 2001. Responses of chub (*Leuciscus cephalus*) populations to chemical stress, assessed by genetic markers, DNA damage and cytochrome P4501A induction. *Ecotoxicology* 10:145–58.

Le Goff, J. et al. 2006. DNA adduct measurements in zebra mussels, *Dreissena polymorpha*, Pallas: Potential use for genotoxicant biomonitoring of fresh water ecosystems. *Aquat. Toxicol.* 79:55–64.

Lee, R.F., and S. Steinert, 2003. Use of the single cell gel electrophoresis/comet assay for detecting DNA damage in aquatic (marine and freshwater) animals. *Mutat. Res./Rev. Mutat. Res.* 544: 43–64.

Lemière, S. et al. 2005a. DNA damage (comet assay) and 8-oxodGuo (HPLC-EC) in relation to oxidative stress in the freshwater bivalve *Unio tumidus*. *Biomarkers* 10:41–57.

Lemière, S. et al. 2005b. DNA damage measured by the single-cell gel electrophoresis (Comet) assay in mammals fed with mussels contaminated by the 'Erika' oil-spill. *Mutat. Res.* 581:11–21.

Lettieri, T. 2006. Recent applications of DNA microarray technology to toxicology and ecotoxicology. *Environ. Health Perspect.* 114:4–9.

Lewis, C., and T. Galloway T. 2009. Reproductive consequences of paternal genotoxin exposure in marine invertebrates. *Environ. Sci. Technol.* 43:928–33.

Liang, P., and A.B. Pardee. 1992. Differential display of eukaryotic messenger RNA by means of the polymerase chain reaction. *Science* 257:967–71.

Long, E.R. et al. 1990. Comparative evaluation of five toxicity tests with sediments from San Francisco Bay and Tomales Bay, California. *Environ. Toxicol. Chem.* 9:1193–214.

Longwell, A.C. et al. 1992. Pollution and developmental abnormalities of Atlantic fishes. *Environ. Biol. Fish* 35:1–21.

Lyons, B.P., C. Stewart, and M.F. Kirby. 1999. The detection of biomarkers of genotoxin exposure in the European flounder (*Platichthys flesus*) collected from the River Tyne Estuary. *Mutat. Res.* 446:111–9.

Lyons, B.P. et al. 2004. DNA adduct analysis and histopathological biomarkers in European flounder (*Platichthys flesus*) sampled from UK estuaries. *Mutat. Res.* 552:177–86.

Machado Garcia, C.C. et al. 2012. Evaluation of DNA adducts formed by lipid peroxidation by-products. In: *Oxidative Stress in Aquatic Ecosystems*, ed. D. Abele, J.P. Vázquez-Medina, and T. Zenteno-Savín, 475–86. Chichester, UK: Wiley-Blackwell.

Mai, W.J. et al. 2010. Acute acidic exposure induces p53-mediated oxidative stress and DNA damage in tilapia (*Oreochromis niloticus*) blood cells. *Aquat. Toxicol.* 100:271–81.

Malmström, C. et al. 2009. Screening of hydrophobic DNA adducts in flounder (*Platichtys flesus*) from the Baltic Sea. *Chemosphere* 77:1514–9.

Marchand, J. et al. 2010. Genetic polymorphism and its potential relation to environmental stress in five populations of the European flounder, *Platichthys flesus*, along the French Atlantic Coast. *Mar. Environ. Res.* 70:201–9.

Martins, N. et al. 2009. Effects of acid mine drainage on the genetic diversity and structure of a natural population of *Daphnia longispina*. *Aquat. Toxicol.* 92:104–12.

Marty, G.D. et al. 1997. Histopathology and cytogenetic evaluation of Pacific herring larvae exposed to petroleum hydrocarbons in the laboratory or in Prince William Sound, Alaska, after the *Exxon Valdez* oil spill. *Can. J. Fish. Aquat. Sci.* 54:1846–57.

McFarland, V.A. et al. 1999. Biomarkers of oxidative stress and genotoxicity in livers of field-collected brown bullhead, *Ameriurus nebulosus*. *Arch. Environ. Contam. Toxicol.* 37:236–41.

Michel, C., and F. Vincent-Hubert. 2012. Detection of 8-oxodG in *Dreissena polymorpha* gill cells exposed to model contaminants. *Mutat. Res./Gen. Toxicol. Environ. Mutagen.* 741:1–6.

Mitchelmore, C.L., and J.K. Chipman. 1998. DNA strand breakage in aquatic organisms and the potential value of the comet assay in environmental monitoring. *Mutat. Res.* 399:135–47.

Mouchet, F., and L. Gauthier. 2012. Genotoxicity of contaminants: Amphibian micronucleus assays. In: *Encyclopedia of Aquatic Ecotoxicology* (Volumes 1 and 2), ed. J.F. Férard and C. Blaise. Dordrecht, The Netherlands: Springer (in press).

Mouchet, F. et al. 2011. International amphibian micronucleus standardized procedure (ISO 21427-1) for *in vivo* evaluation of double-walled carbon nanotubes toxicity and genotoxicity in water. *Environ Toxicol.* 26:136–45.

Myers, M.S. et al. 1990. Overview of studies on liver carcinogenesis in English sole from Puget Sound; evidence for a xenobiotic chemical etiology: I. Pathology and epizootiology. *Sci. Total Environ.* 94:33–50.

Myers, M.S. et al. 1998a. Toxicopathic hepatic lesions as biomarkers of chemical contaminant exposure and effects in marine bottomfish species from the Northeast and Pacific Coasts, USA. *Mar. Pollut. Bull.* 37:92–113.

Myers, M.S. et al. 1998b. Toxicopathic hepatic lesions in subadult English sole (*Pleuronectes vetulus*) from Puget Sound, Washington, USA: Relationships with other biomarkers of contaminant exposure. *Mar. Environ. Res.* 45:47–67.

Myers, M.S., L.L. Johnson, and T.K. Collier. 2003. Establishing the causal relationship between polycyclic aromatic hydrocarbon (PAH) exposure and hepatic neoplasms and neoplasia-related liver lesions in English sole (*Pleuronectes vetulus*). *Hum. Ecol. Risk Assess.* 9:67–94.

Myers, M.S. et al. 2008. Improved flatfish health following remediation of a PAH-contaminated site in Eagle Harbor, Washington. *Aquat. Toxicol.* 88:277–88.

Nahrgang, J. et al. 2010. Biomarker responses in polar cod (*Boreogadus saida*) exposed to the water soluble fraction of crude oil. *Aquat. Toxicol.* 97:234–42.

Nair, J. 1999. Lipid peroxidation-induced etheno-adducts in humans. In: *Exocyclic DNA adducts in Mutagenesis and Carcinogenesis*, ed. B. Singer and H. Bartsch, 55–61. IARC Scientific Publications No. 150. Lyon: International Agency for Research on Cancer, 361 pp.

Nigro, M. et al. 2006. Cellular biomarkers for monitoring estuarine environments: Transplanted versus native mussels. *Aquat. Toxicol.* 77:339–47.

Norcross, B.L. et al. 1996. Distribution, abundance, morphological condition, and cytogenetic abnormalities of larval herring in Prince William Sound, Alaska, following the (*Exxon Valdez*) oil spill. *Can. J. Fish Aquat. Sci.* 53:2376–87.

Noventa, S., B. Pavoni, and T.S. Galloway. 2011. Periwinkle (*Littorina littorea*) as a sentinel species: A field study integrating chemical and biological analyses. *Environ. Sci. Technol.* 45:2634–40.

Ostling, O., and K.J. Johanson. 1984. Microelectrophoretic study of radiation induced DNA damages in individual mammalian cells. *Biochem. Biophys. Res. Commun* 123:291–8.

Peck-Miller, K. et al. 1998. Complete cDNA sequence of the Ki-ras proto-oncogene in the liver of wild English sole (*Pleuronectes vetulus*) and mutation analysis of hepatic neoplasms and other toxicopathic lesions. *Mol. Carcinog.* 23:207–16.

Pereira., S. et al. 2012. Genotoxicity of uranium contamination in embryonic zebrafish cells. *Aquat. Toxicol.* 109:11–6.

Piña, B., and C. Barata. 2011. A genomic and ecotoxicological perspective of DNA array studies in aquatic environmental risk assessment. *Aquat. Toxicol.* 105S:40–9.

Pisoni, M. et al. 2004. DNA adducts, benzo(*a*)pyrene monooxygenase activity, and lysosomal membrane stability in *Mytilus galloprovincialis* from different areas in Taranto coastal waters (Italy). *Environ. Res.* 96:163–75.

Randerath, K., M.V. Reddy, and R.C. Gupta. 1981. ^{32}P-postlabelling tests for DNA damage. *Proc. Natl. Acad. Sci. U. S. A.* 78:6126–9.

Rank, J., K. Jensen, and P.H. Jespersen. 2005. Monitoring DNA damage in indigenous blue mussels (*Mytilus edulis*) sampled from coastal sites in Denmark. *Mutat. Res./Gen. Toxicol. Environ. Mutagen.* 585:33–42.

Rank, J. et al. 2007. DNA damage, acetylcholinesterase activity and lysosomal stability in native and transplanted mussels (*Mytilus edulis*) in areas close to coastal chemical dumping sites in Denmark. *Aquat. Toxicol.* 84:50–61.

Ravanat, J.L. et al. 2002. Cellular background level of 8-oxo-7,8-dihydro-2′-deoxyguanosine: An isotope based method to evaluate artefactual oxidation of DNA during its extraction and subsequent work-up. *Carcinogenesis* 23:1911–8.

Regoli, F. et al. 2004. Time-course variations of oxyradical metabolism, DNA integrity and lysosomal stability in mussels, *Mytilus galloprovincialis*, during a field translocation experiment. *Aquat. Toxicol.* 68:167–78.

Reichert, W.L. et al. 1998. Molecular epizootiology of genotoxic events in marine fish: Linking contaminant exposure, DNA damage, and tissue-level alterations. *Mutat. Res.* 411:215–25.

Rether, B. et al. 1997. Biomarqueurs de génotoxicité chez les végétaux et les animaux. In: *Biomarqueurs en Ecotoxicologie: Aspects Fondamentaux*, ed. L. Lagadic, T. Caquet, J.C. Amiard, F. Ramade, 186–208. Paris: Masson.

Rocher, B. et al. 2006. Genotoxicant accumulation and cellular defence activation in bivalves chronically exposed to waterborne contaminants from the Seine River. *Aquat. Toxicol.* 79:65–77.

Rodius, F., C. Hammer, and P. Vasseur. 2002. Use of RNA arbitrarily primed PCR to identify genomic alterations in the digestive gland of the freshwater bivalve *Unio tumidus* at a contaminated site. *Environ. Toxicol.* 17:538–46.

Rodius, F., C. Hammer, and P. Vasseur. 2004. Use of RNA-arbitrarily-primed PCR to detect the induction of gene expression in freshwater bivalves (*Unio tumidus*) transplanted into the Moselle River. *Environ. Mol. Mutagen.* 43:209–16.

Rodríguez-Ariza, A. et al. 1999. Content of 8-oxodG in chromosomal DNA of *Sparus aurata* fish as biomarker of oxidative stress and environmental pollution. *Mutat. Res.* 438:97–107.

Roy, L.A. et al. 2003. The relationships of biochemical endpoints to histopathology and population metrics in feral flatfish species collected near the municipal wastewater outfall of Orange County, California, USA. *Environ. Toxicol. Chem.* 22:1309–17.

Roy, N.K. et al. 1999. High frequency of K-*ras* mutations in pink salmon embryos experimentally exposed to *Exxon Valdez* oil. *Environ. Toxicol. Chem.* 18:1521–8.

Rybakovas, A., J. Baršienė, and T. Lang. 2009. Environmental genotoxicity and cytotoxicity in the offshore zones of the Baltic and the North Seas. *Mar. Environ. Res.* 68:246–56.

Sargent, T.D., and I.B. Dawid. 1983. Differential gene expression in the gastrula of *Xenopus laevis*. *Science* 222:135–9.

Shaw, J.P. et al. 2000. Seasonal variation in mussel *Mytilus edulis* digestive gland cytochrome P4501A- and 2E-immunoidentified protein levels and DNA strand breaks (Comet assay). *Mar. Environ. Res.* 50:405–9.

Shaw, J.P. et al. 2002. Elevation of cytochrome P450-immunopositive protein and DNA damage in mussels (*Mytilus edulis*) transplanted to a contaminated site. *Mar. Environ. Res.* 54:505–9.

Shigenaga, M.K. et al. 1994. Assays of oxidative damage biomarkers 8-OhdG and 8-oxoguanine in nuclear DNA and biological fluids by HPLC with ED. *Methods Enzymol.* 234:16–33.

Shimada, A., and A. Shima. 1998. Combination of genomic DNA fingerprinting into the medaka specific-locus test system for studying environmental germ-line mutagenesis. *Mutat. Res.* 399:149–65.

Shimada, A., and A. Shima. 2004. Transgenerational genomic instability as revealed by a somatic mutation assay using the medaka fish. *Mutat. Res.* 552:119–24.

Singh, N.P. 2000. A simple method for accurate estimation of apoptotic cells. *Exp. Cell. Res.* 256:328–37.

Singh, N.P. 2005. Apoptosis assessment by the DNA diffusion assay. *Methods Mol. Med.* 111:55–67.

Singh, N.P. et al. 1988. A simple technique for quantitation of low levels of DNA damage in individual cells. *Exp. Cell Res.* 175:184–91.

Skarphédinsdóttir, H. et al. 2005. Seasonal and intertidal impact on DNA adduct levels in gills of blue mussels (*Mytilus edulis* L.). *Environ. Pollut.* 136:1–9.

Skarphédinsdóttir, H. et al. 2007. DNA adducts and polycyclic aromatic hydrocarbon (PAH) tissue levels in blue mussels (*Mytilus spp.*) from Nordic coastal sites. *Mar. Environ. Res.* 64:479–91.

Smart, D.J., J.K. Chipman, and N.J. Hodges. 2006. Activity of OGG1 variants in the repair of pro-oxidant-induced 8-oxo-2′-deoxyguanosine. *DNA Repair* 5:1337–45.

Smith, C.C. et al. 2006. hOGG1 recognizes oxidative damage using the comet assay with greater specificity than FPG or ENDOIII. *Mutagenesis* 21:185–90.

Smolarz, K., M. Wolowicz, and C. Thiriot-Quiévreux. 2003. Argyrophilic nucleolar organizer regions (AgNORs) in interphases and metaphases of normal and neoplastic gill cells of *Macoma balthica* (Bivalvia: Tellinidae) from the Gulf of Gdansk, Baltic Sea. *Dis. Aquat. Org.* 56:269–74.

Solé, M. et al. 1996. Effects of the "Aegean Sea" oil spill on biotransformation enzymes, oxidative stress and DNA-adducts in digestive gland of the mussel (*Mytilus edulis* L.). *Comp. Biochem. Physiol.* 113C:257–65.

Stein, J.E. et al. 1992. Bioindicators of contaminant exposure and sublethal effects: Studies with benthic fish in Puget Sound, Washington. *Environ. Toxicol. Chem.* 11:701–14.

Steinert, S.A. et al. 1998a. DNA damage in mussels at sites in San Diego Bay. *Mutat. Res.* 399:65–85.

Steinert, S.A., R. Streib Montee, and M.P. Sastre. 1998b. Influence of sunlight on DNA damage in mussels exposed to polycyclic aromatic hydrocarbons. *Mar. Environ. Res.* 46:355–8.

Sundt, R.C. et al. 2011. PAH body burden and biomarker responses in mussels (*Mytilus edulis*) exposed to produced water from a North Sea oil field: Laboratory and field assessments. *Mar. Pollut. Bull.* 62:1498–505.

Theodorakis, C.W. 2003. Establishing causality between population genetic alterations and environmental contamination in aquatic organisms. *Hum. Ecol. Risk Assess.* 9:37–58.

Theodorakis, C.W., and J.W. Bickham. 2004. Molecular characterization of contaminant-indicative RAPD markers. *Ecotoxicology* 13:303–9.

Theodorakis, C.W., and L.R. Shugart. 1997. Genetic ecotoxicology: II. Population genetic structure in mosquitofish exposed *in situ* to radionuclides. *Ecotoxicology* 6:335–54.

Theodorakis, C.W., T. Elbl, and L.R. Shugart. 1999. Genetic ecotoxicology: IV. Survival and DNA strand breakage is dependent on genotype in radionuclide-exposed mosquitofish. *Aquat. Toxicol.* 45:279–91.

Theodorakis, C.W. et al. 2006. Evidence of altered gene flow, mutation rate, and genetic diversity in redbreast sunfish from a pulp-mill-contaminated river. *Environ. Sci. Technol.* 40:377–86.

Thiriot-Quiévreux, C., and M. Wolowicz. 2001. Chromosomal study of spatial variation of the prevalence of a gill neoplasia in *Macoma balthica* (L.) from the Gulf of Gdansk (Baltic Sea). *Ophelia* 54:75–81.

Thorne, R.E., and G.L. Thomas. 2008. Herring and the "Exxon Valdez" oil spill: An investigation into historical data conflicts. *ICES J. Mar. Sci.* 65:44–50.

Tice, R.R. et al. 2000. The single cell gel/comet assay: Guidelines for *in vitro* and *in vivo* genetic toxicology testing. *Environ. Mol. Mutagen.* 35:206–21.

Varanasi, U., W.L. Reichert, and J.E. Stein. 1989. ^{32}P-postlabeling analysis of DNA adducts in liver of wild English sole (*Parophrys vetulus*) and winter flounder (*Pseudopleuronectes americanus*). *Cancer Res.* 49:1171–7.

Vasquez, M., and R.R. Tice. 1997. Detecting genotoxic activity against high molecular weight DNA using the alkaline single cell gel (SCG) assay. *Environ. Mol. Mutagen.* 29(S28):53.

Vasseur, P. et al. 1995. Indices for carcinogenicity in aquatic ecosystems: Significance and development. NATO ASI Series, vol. 128. *Evaluating and Monitoring the Health of Large Scale Ecosystems*, ed. D.J. Rapport, C.I. Gaudet, and P. Calow, 179–93. Berlin: Springer-Verlag.

Venier, P., and S. Zampieron. 2005. Evidence of genetic damage in grass gobies and mussels from the Venice lagoon. *Environ. Int.* 31:1053–64.

Vicquelin, L. et al. 2011. A new spiked sediment assay using embryos of the Japanese medaka specifically designed for a reliable toxicity assessment of hydrophobic chemicals. *Aquat. Toxicol.* 105:235–45.

Vincent-Hubert, F. 2000. cDNA cloning and expression of two Ki-ras genes in the flounder, *Platichthys flesus*, and analysis of hepatic neoplasms. *Comp. Biochem. Physiol.* 126B:17–27.

Vincent-Hubert, F., A. Arini, and C. Gourlay-Francé. 2011. Early genotoxic effects in gill cells and haemocytes of *Dreissena polymorpha* exposed to cadmium, B[*a*]P and a combination of B[*a*]P and Cd. *Mutat. Res./Gen. Toxicol. Environ. Mutagen.* 723:26–35.

Vincent-Hubert, F. et al. 2012. Mutagenicity and genotoxicity of suspended particulate matter in the Seine river estuary. *Mutat. Res./Gen. Toxicol. Environ. Mutagen.* 741:7–12.
Vos, P. et al. 1995. AFLP: A new technique for DNA fingerprinting. *Nucleic Acids Res.* 23:4407–14.
Welsh, J., and M. McClelland. 1990. Fingerprinting genomes using PCR with arbitrary primers. *Nucleic Acids Res.* 18:7213–8.
Welsh, J. et al. 1992. Arbitrarily primed PCR fingerprinting of RNA. *Nucleic Acids Res.* 20:4965–70.
Wentzel, J.F. et al. 2010. Assessing the DNA methylation status of single cells with the comet assay. *Anal. Biochem.* 400:190–4.
Whitehead, A. et al., 2011. Genomic and physiological footprint of the Deepwater Horizon oil spill on resident marsh fishes. *PNAS* doi/10.1073/pnas.1109545108.
Williams, J.G. et al. 1990. DNA polymorphisms amplified by arbitrary primers are useful as genetic markers. *Nucleic Acids Res.* 18:6531–5.
Wirz, M.V.M.A., P.H. Saldiva, and D.V. Freire-Maia. 2005. Micronucleus test for monitoring genotoxicity of polluted river water in *Rana catesbeiana* tadpoles. *Bull. Environ. Contam. Toxicol.* 75:1220–7.
Wurgler, F.E., and P.G.N. Kramers. 1992. Environmental effects of genotoxins (ecogenotoxicology). *Mutagenesis* 7:321–7.
Yauk, C.L. 1998. Monitoring for induced heritable mutations in natural populations: Application of minisatellite DNA screening. *Mutat. Res.* 411:1–10.
Yauk, C.L. 2004. Advances in the application of germline tandem repeat instability for *in situ* monitoring. *Mutat. Res.* 566:169–82.
Yauk, C.L. 2006. Special issue on induced tandem repeat instability. *Mutat. Res.* 598:1–5.
Yauk, C.L., and J.S. Quinn. 1996. Multilocus DNA fingerprinting reveals high rate of heritable genetic mutation in herring gulls nesting in an industrialized urban site. *Proc. Natl. Acad. Sci. U. S. A.* 93:12137–41.
Yauk, C.L. et al. 2000. Induced minisatellite germline mutations in herring gulls (*Larus argentatus*) living near steel mills. *Mutat. Res.* 452:211–8.

14

Evolutionary Toxicology and Transcriptomic Approaches

Justine Marchand, Françoise Denis, and Jean Laroche

CONTENTS

14.1 Introduction

Biomarkers can reflect the actual state of health of organisms or populations, but do not provide information on the potential of a population to survive in a polluted environment. Genetic approaches can, however, produce relevant knowledge in an ecotoxicology study. Indeed, the genetic variation in populations results from their evolutionary histories and represents the raw material that has, or has not, enabled the evolutionary adaptation of these populations faced with environmental challenges (global change, hypoxia, emerging contaminants, multistress, etc.).

Genetic tools can be used particularly to analyze the differential capacity of organisms (i.e., of genotypes) in a population to cope with a chemical stress, although the between-individuals variability in biochemical or physiological responses is still being widely ignored in ecotoxicology, or is considered background noise (Forbes and Forbes 1997).

Genetic investigations in ecotoxicology can actually lead to:

1. The assessment of within-population genetic diversity, a key factor allowing species survival; any alteration of this diversity could lead to a possible loss of adaptive capacity to meet any additional stress (one of the first principles of conservation genetics; Soulé 1987).
2. The estimation of the potential connectivity between populations by the analysis of the between-populations genetic diversity. Are the populations more or less isolated and thus more or less genetically differentiated? What is the level of gene flow between populations? Does this level allow a possible recolonization of the habitat by a species, after a massive demographic bottleneck?
3. The identification of relationships between genetic diversity and physiology by searching for possible couplings between genotypes and phenotypes, considering fitness-relevant biological traits.
4. The detection of modifications at gene expression level potentially leading to cellular damage, or to physiological acclimation or even to resistance to chemical stress.

Thus, we set out in this chapter to investigate the possible cause-and-effect relationships between chemical stress and genetic diversity/gene expression/biological traits. Finally, we will analyze the potential of genetic markers as diagnostic tools for a better knowledge of the impact of contaminants on populations in the field.

14.2 Major Evolutionary Forces That Change Genetic Variability

Across the whole distribution area of a particular species, random mating between individuals is indeed impossible, the dispersion–migration capacity of the species being limited. Thus, one observes generally in the field different local populations (or demes) considered as reproduction units displaying spatial and temporal stability.

Within-population genetic variability is subjected to four major evolutionary forces (mutation, genetic drift, migration, and selection) (Figure 14.1) that can lead to changes in allele frequencies (Belfiore and Anderson 1998, 2001; Bagley et al. 2002).

14.2.1 Mutation

Mutation is a process that creates genetic diversity, producing a new allele different from the ancestral allele. The different alleles observed for a particular gene are derived from mutational modifications frequently linked to errors during DNA replication. Mutations are rare events (single locus mutation rate per generation: 10^{-4} to 10^{-6}); thus, they can change allele frequencies in a population, but these changes are usually very slow. Mutations in noncoding regions generally have no impact on the phenotypes and thus are not submitted

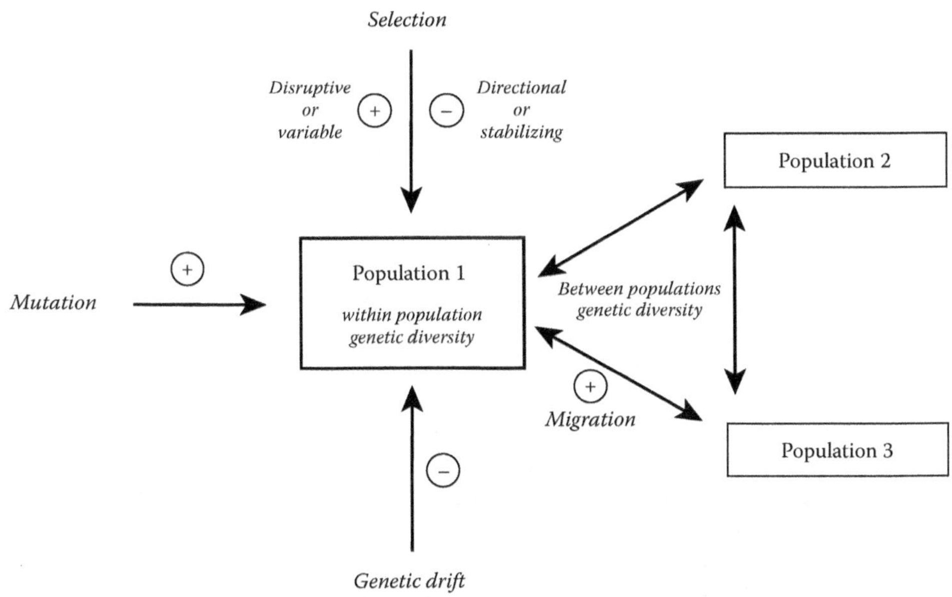

FIGURE 14.1
Effects of the four main evolutionary forces on within-population genetic diversity: a process leading to conservation or increase ⊕, or to reduction ⊖ of the within-population genetic diversity.

to selection; this kind of polymorphism is particularly analyzed in population genetics to characterize within- and between-population genetic diversity. On the other hand, mutations in coding sectors for a functional protein or for a transcription factor can considerably affect individual physiological performance or organism fitness.

Particular noncoding regions of the genome consist of many copies of the same sequence (two to nine nucleotides) and are called microsatellites. These markers display a high natural mutation rate, and thus are particularly targeted in explorations of the genetic structure of populations in the field and to analyze mutation load in populations.

14.2.2 Genetic Drift

Genetic drift refers to fluctuations in allele frequency that occur by chance, especially in small populations, as a result of sampling error. In small populations, in particular, genetic drift effects are frequently relatively dominant to those of selection, high-performance genotypes and genotypes sensitive to stress displaying the same extinction probability (Hauser et al. 2003). Genetic drift leads to an increase in the genetic differentiation between populations.

14.2.3 Migration

Migration is characterized by the movement of individuals between populations, whereas gene flow is the movement of genes between populations, i.e., is linked to the successful reproduction of an immigrant in its new habitat. Generally, gene flow leads to an increase of within-population genetic diversity and to a decrease of between-populations genetic diversity, i.e., to a lower level of genetic differentiation.

Gene flow between moderately contaminated populations and heavily polluted populations can induce a reduction of the adaptation speed in the polluted systems (Groenendijk et al. 1999). In a contaminated population, the identification of immigrants by genetic tools can be particularly pertinent, allowing the exploration of the effects of contaminants, primarily on residents and secondly on immigrants (Theodorakis 2001).

14.2.4 Selection

Selection preferentially retains in a particular environment the individuals displaying the higher fitness. Selection can oppose genetic drift in extreme environments where selection can maintain a particular polymorphism over thousands of generations, even for isolated populations of a few dozens of individuals (Nevo 2001). The demonstration of the effects of selection on molecular variants (i.e., the estimation of their values in terms of physiological performance) is generally very difficult to carry out, DNA diversity being very complex to analyze. Thus, selective pressure is sometimes more easily detected by measuring the heritability of a phenotypic trait in a population, considering that the trait is integrating the physiological effects of several genes (Hauser et al. 2003).

The selection modes induced by contamination on populations are very diversified and complex (Van Straalen and Timmermans 2002). If a polymorphism is linked to a pattern of resistance, one should detect an increase in the tolerant genotype frequency, possibly by different modes:

- Directional selection, without a heterosis effect (i.e., without heterozygote superiority), where a reduced genetic variability is observed over the population.
- Stabilizing selection, fitness being highest at some intermediate phenotypes and lower at the phenotypic extremes (decrease of genetic variability).
- Variable selection in space and time (alternately toxic and nontoxic periods); if fitness is highest for the tolerant genotype in the toxic period but lower in the nontoxic period, genetic variability will thus be maintained.
- Disruptive selection where fitness is lowest at some intermediate phenotype and is higher at the phenotypic extremes; genetic diversity will be also maintained.

14.3 Common Markers of Genetic Diversity in Ecotoxicology

Effects of contaminants on genetic diversity can be considered:

- Indirectly through the measure of biometric characters linked to fitness with a heritable component (quantitative genetics approach)
- Directly through the study of allozymes or DNA markers (population genetics approach)

Biometric characters, potentially under control of several genes, may generate differential physiological performances and thus lead to population effects in the face of environmental stress (Woods and Hoffman 2000). These biometric approaches are based on the hypothesis that selective pressures determine the genetic characteristics of a population and thus its adaptive potential. For these approaches, large laboratory experiments performed on successive generations are essential and usually the genetic component of the physiological trait associated with tolerance (i.e., its heritability) has to be studied (see Section 14.4.2 for examples).

The second family of markers associated with genetic diversity concerns molecular markers. Allozymes (enzymatic proteins) were the first markers to be used in populations facing environmental stress. Allozyme loci may have several allelic variants with different biochemical characteristics, in particular, molecular variations that induce variations in charge or structure, leading therefore to differences in migration speed during electrophoresis. Considering several individuals and several allozyme loci, the number and the frequency of the different alleles of the whole population may be estimated and its genetic diversity characterized. Enzymatic polymorphism reveals about 30% of the DNA genetic variability since many molecular mutations do not lead to variations in charge or structure of the subsequent protein. The "tolerance to contaminants/genotype" association of allozyme loci has been encountered in many aquatic organisms since the 1980s, both in freshwater and marine organisms. For some authors, allelic variants are important for the determination of tolerance (Schluter 2000). For others, heterozygosity of particular loci (Benton and Gutman 1992) or multiloci heterozygosity (Peles et al. 2003) are determinant characters for tolerance. Allozyme markers can still be considered pertinent tools for the evaluation of genetic variability in ecotoxicology (see Section 14.4.1 for examples), but more often, current research considers genetic variation directly at the DNA level (genomic approaches).

At the DNA level, two types of markers—neutral markers and markers under selection—are important to consider in ecotoxicology. Neutral markers are loci not evolving directly in response to selection, the dynamics of which are controlled mainly by mutation, genetic drift, and migration. These markers are usually noncoding regions of the nuclear DNA (such as microsatellites, introns, and intergenic regions) and mitochondrial loci (the control region called the D-loop, the different subunits of NADH, etc.) and are useful in ecotoxicology:

- For identifying genetic structure in natural populations and, in particular, potential gene flow between chronically contaminated populations and reference populations
- For exploring the genetic history of populations in different environmental contexts
- For detecting the impact of drastic reduction in population size that can lead to the loss of genetic variation through genetic drift

Their level of variability may, however, be affected by mutagen toxicants (see Section 14.4.1.1 for examples) or influenced by selection on nearby (linked) loci (through the phenomenon of hitchhiking).

In an ecotoxicological context, markers under selection are fragments of gene that code for a protein conferring a selective advantage or disadvantage on the individual subjected to a chemical stress. Identifying the genes that potentially affect particular traits is the major difficulty. Studies were initially focused on individual candidate genes such as the gene coding for xenobiotic metabolizing enzymes (i.e., cytochrome P450s, glutathione-*S*-transferases), for a transcription factor AhR (aryl hydrocarbon receptor), a well-known chemical sensor, or for proteins involved in energy production (Hahn 1998; Hulla et al. 1999; Burnett et al. 2007). The identification of their polymorphism relied in particular on techniques such as restriction fragment length polymorphism (RFLP), single strand conformation polymorphism, and denaturating gel gradient electrophoresis.

Recent advances in molecular technologies have made genome-wide sampling in populations feasible through genome typing (Luikart et al. 2003). Genome typing is the simultaneous genotyping of hundreds of loci from across the genome [microsatellites, amplified fragment length polymorphism (AFLPs), or nucleotide polymorphisms], combining both neutral markers and markers potentially under selection. The classical AFLP protocol uses a polymerase chain reaction PCR-based DNA fingerprinting technique to produce dozens of polymorphic markers (presence or absence of a restriction enzyme site) that cover the entire genome. AFLPs are increasingly used to identify markers associated with traits under selection in nonmodel organisms (Skøt et al. 2002; Wang et al. 2003); these markers were successfully used to detect signatures of selection in natural populations adapted to chronic pollution (Williams and Oleksiak 2008).

The Diversity Array technology (DArt) approach uses a microarray, in which each spot contains a DNA fragment that has been amplified from a library of polymorphic markers that were identified during an initial screening phase (Jaccoud et al. 2001). DArt is attractive because a single PCR can amplify hundreds of polymorphic markers and because automation is easier using images rather than gel electrophoresis. However, this technique can only be used on species for which large data sets and effective genomic tools have been developed.

Single nucleotide polymorphisms (SNPs) have recently attracted the interest of geneticists and ecotoxicologists; they require species-specific marker development but they have potential for automation and high throughput (Bozinovic and Oleksiak 2011). The rapid development of numerous SNPs (including nonsynonymous and functional SNPs) is becoming feasible in some nonmodel species, owing to the rapid growth of expressed sequence tag (EST) databases, data-mining software, and primer-design strategies such as Comparative Anchor-Tag Sequences and Exon-Primed Intron-Crossing PCR (Luikart et al. 2003; Moen et al. 2008a, 2008b).

14.4 Links between Chemical Stress, Genetic Diversity, and Phenotypic Responses

14.4.1 Chemical Contamination and Impact on Genetic Diversity

Interest in evolutionary ecotoxicology (i.e., the study of effects of chemical pollutants on the genetics of natural populations) is rapidly growing among ecotoxicologists; the major population genetic impacts in the context of contamination studies include (1) changes in allelic or genotypic frequencies caused by increased mutation rates, (2) genome-wide changes in genetic diversity linked to demographic bottlenecks, (3) changes in allelic or genotypic frequencies caused by contaminant-induced selection, and (4) changes in dispersal patterns of gene flow that alter the genetic relationship among populations (Bickham 2011). In this part 14.4.1, we will focus mainly on the points (1) to (3).

14.4.1.1 Direct Interactions between Pollutants and DNA through Mutations

Evidence of direct effects of pollution on genomes has been highlighted in terrestrial organisms and especially in the case of radionuclide contamination. For example, Dubrova et al. (1996) found that germinal mutations were two times higher at three minisatellite loci in a human population living around Chernobyl compared to a reference population living in England. Ellegren et al. (1997) highlighted higher germinal mutations in swallows (*Hirundo rustica*) whose parents had reproduced in the radionuclide-contaminated zones of the Chernobyl region compared to reference swallows; they noted higher germinal mutations not only in noncoding regions of the DNA (especially microsatellites) but also in coding regions such as mutations causing albinism. Herring gulls (*Larus argentatus*) nesting near steel mills on the Great Lakes were shown to have higher germline mutation rates at minisatellite DNA loci than those at rural sites (Yauk and Quinn 1996), and mutation frequency increased with colony proximity to integrated steel mills (Yauk et al. 2000). The authors postulated that inhaled airborne contaminants emitted from steel mills, such as polycyclic aromatic compounds, were largely responsible for mutation induction (Yauk et al. 2000). To address this issue, mice were exposed *in situ* to ambient air in a polluted industrial area near steel mills. Heritable mutation frequency at tandem-repeat DNA loci in mice exposed 1 km downwind from two integrated steel mills was 1.5- to 2-fold elevated compared with those at a reference site 30 km away (Somers et al. 2002).

14.4.1.2 Indirect Effects

Indirect effects of pollutants on genomes may be due to:

- A massive reduction of the size of the population due to mortality
- A selection of alleles or genotypes associated with "tolerance" in contaminated areas and an elimination of the most "sensitive" genotypes

Only a few studies have highlighted the existence of genetic erosion due to a massive reduction of the size of populations in contaminated sites. For example, Demarais et al. (1993) concluded that the modifications of genetic structure of populations of the freshwater fish (*Gila seminuda*) observed after an acute accidental exposure to the insecticide rotenone, were probably due to a genetic bottleneck caused by a high mortality. Murdoch and Herbert (1994) suggested the same hypothesis to explain the weak mitochondrial genetic diversity observed in populations of catfish (*Ameiurus nebulosus*) from contaminated sites. Facemire et al. (1995) have also hypothesized that the mixture of contaminants [mercury, polychlorobiphenyls (PCBs), etc.] occurring in Florida induced mortalities in Florida panther *Puma concolor* populations, leading to a loss of genetic diversity in these populations. In a more recent study, Matson et al. (2006) observed a general reduction of the haplotype and nucleotide diversities in marsh frog (*Rana ridibunda)* populations in the highly contaminated industrial sector of Sumgayit, Azerbaijan, compared to uncontaminated reference sites.

Numerous studies have been conducted over the last three decades trying to identify associations or correlations between the level of contaminant exposure and genotypic and/or allelic frequency variations. The first studies deal essentially with allozyme markers and field populations, considering simple contamination, with metals (Nevo et al. 1984), radionuclides (Gillepsie and Guttman 1989), and pesticides (Tanguy et al. 1999). For example, Gillepsie and Guttman (1989) detected significant changes in allelic frequencies at the PGM (phosphoglucomutase) locus between fish populations (*Campostoma anomalum*) from upstream sites (less contaminated) and fish from downstream sites (with high radionuclide concentrations). These changes occurred within 500 meters from the original site of the contamination with the genotype PGM-BB increasing significantly, suggesting a higher tolerance to radionuclides for individuals carrying this genotype compared to the others. Phosphoglucomutase plays a critical role in glycolysis and gluconeogenesis, and the efficiency of this enzyme to produce energy may potentially constitute a major target for selective pressures induced by stress, either natural or anthropogenic. These selective pressures may lead to changes in the distribution of allelic and genotypic frequencies in exposed populations. Benton et al. (1994) have also highlighted distinct genotypic frequencies between populations of caddisfly larvae *Nectopsyche albida* from sites highly contaminated by trace metals (arsenic, cadmium, chromium, copper, and zinc) and caddisflies from reference sites. Several other studies have found selective pressures acting on allozyme markers in natural populations exposed to complex mixtures of contaminants (Heithaus and Laushman 1997; Mulvey et al. 2002; Marchand et al. 2003; Virgilio et al. 2003). Allozyme markers studied were particularly enzymes involved in energy production and/or general metabolism and for which the polymorphism detected may lead to differential physiological performances for organisms exposed to stress (Lawrence et al. 2003). In a recent study, Yap et al. (2011) found correlations between the genetic diversity of allozymes and the levels of heavy metals in the horseshoe crab *Carcinoscorpius rotundicauda* populations in polluted and unpolluted intertidal areas in Peninsular Malaysia, higher heterozygosities being detected in the contaminated populations.

At the DNA level, strong modifications of allelic frequencies at the oncogene c-abl (Wirgin et al. 1990) and at the cytochrome P4501A1 (Roy et al. 1995) loci were detected in populations of fish (*Microgadus tomcod*) from the Hudson River compared to reference sites. Kim et al. (2003) found that a particular haplotype at the mitochondrial ND6 locus in the intertidal gastropod *Littorina brevicula* had a weaker frequency in contaminated sites compared to several reference sites. Timmermans et al. (2007) suggested that the metallothionein gene (which codes for a metal-binding protein) and its surrounding DNA region were affected by environmental metal contamination in the collembolan *Orchesella cincta*. Nadig et al. (1998) and Krane et al. (1999) have used a more general approach for identifying modifications: they worked on RAPD (Random Amplified Polymorphic DNA) markers in perch (*Lepomis auritius*) and crayfish (*Orconectes rusticus*) exposed to complex mixtures of contaminants (HAP, PCB, trace metals). They both found higher genetic similarities between individuals from exposed populations than between individuals from exposed and reference populations. Nadig et al. (1998) have also highlighted a significant correlation between a particular genotype frequency in redbreast sunfish *L. auritus* and a pollution gradient. The analysis of persistent fish populations living in chronically contaminated estuaries for more than 60 years in the United States and Europe, compared to populations living in "pristine" estuaries, have provided new insights into the identification of candidate genes potentially involved in the response of fish to chemical stress in the field. The polymorphism of AhR 1 and AhR 2 loci in the killifish *Fundulus heteroclitus* (Hahn et al. 2004, 2005), and the polymorphism of p53, BHMT3, and PGDS in the European flounder *Platichthys flesus* (Marchand et al. 2010), could be related to the possible tolerance of populations faced with complex mixtures of pollutants. Hemmer-Hansen et al. (2007) confirmed the great capacity of the European flounder to adapt to contrasted environments in estuaries; they highlighted the genetic basis of the adaptation of this species to the particular environment of the Baltic Sea compared to the North Sea (gradient of salinity, chemical stress, hypoxia, etc.), considering a candidate gene (*Hsc 70*), and neutral loci (microsatellites).

However, the examples cited above, either with allozymes or DNA markers, observed in different contexts (simple contamination or mixtures) showed only simple correlations between allelic or genotypic frequencies and chemical contamination but did not explore how chemical stress can select different genetic variants. Examples cited below integrate individual physiological performances (phenotypes) in order to establish "genotype–phenotype" relations, for example, differential genotype sensitivity to pollutants. The first studies performed in this way were carried out essentially in the laboratory. For example, differential mortalities associated with particular alleles or genotypes at the loci PGM and PGI (phosphoglucose isomerase) were observed in fish exposed to experimental metal contamination (see the synthesis of Gillepsie and Guttman 1999). More recently, Virgilio et al. (2005) highlighted robust relations between genetic variability at several allozyme loci and differential mortality during exposure to copper, whatever the site of origin of the population of the estuarine polychaete *Hediste diversicolor* examined. More directly, Chagnon and Guttman (1989) found in the mosquitofish *Gambusia affinis* different enzyme activities between allelic variants at the PGM locus, suggesting the existence of a tolerant allele to copper and a more sensitive one. In the same species and studying exposure to mercury, Kramer et al. (1992) have highlighted differential metabolic activities between individuals carrying different genotypes at the PGI locus. In a recent analysis, a genotype-dependent recovery was detected from acute exposure to trace metal contamination in the freshwater clam *Sphaerium novaezelandiae* (Philips and Hickey 2010); no significant genotype-specific differences in mortality were detected, considering particularly the allozyme PGM, but the authors did identify differences in reburial rates across all genotypes. In this experiment, the populations display genetically based variation in their ability to recover from zinc exposure in the laboratory and that such variation is linked to a physiological trait (reburial).

Other studies coupling genotypes and phenotypes have been performed but this time with individuals collected in the field. Their objectives were to detect differential physiological responses measured in the organisms directly after collection. Theodorakis and Shugart (1997) identified an increase in the frequency of particular RAPD bands in individuals from two natural populations of *Gambusia affinis* exposed to radionuclides, compared to a reference population; individuals carrying these particular bands also had better fecundity, a lower level of DNA damage and better survival compared to the others (Theodorakis and Shugart 1997, 1999). The authors hypothesized that these bands may be linked to loci involved in the protection of macromolecules or in the reparation of cellular and molecular lesions (Theodorakis and Shugart 1999). Larno et al. (2001) showed in highly polluted populations of chub (*Leuciscus cephalus*) from different rivers of the Rhône basin, that particular alleles or genotypes at the PGM and EST (esterase) loci were correlated to higher DNA integrities. They also found an increase in frequency of these alleles or genotypes with the intensity of contamination, leading thus to the description of alleles as potentially "tolerant" to contamination (Larno et al. 2001). Different studies performed on estuarine populations of flounder (*Platichthys flesus*) have confirmed previous findings (Laroche et al. 2002; Marchand et al. 2003, 2004). The authors found a better ability to maintain DNA integrity in individual flounders carrying the PGM-85 allele in different contaminated estuaries along the Atlantic coast of France (Loire, Seine, and Vilaine estuaries), compared to a reference estuary (Ster); a higher frequency of the PGM-85 allele was detected in populations from contaminated estuaries and the hypothesis of a selective pressure acting on the PGM locus was formulated. Such associations (genotype–phenotype) have also been observed for several aquatic species exposed to various contaminants (Gillepsie and Guttman 1999; Kammenga et al. 2000; Van Straalen and Hoffman 2000; Weis 2002).

Correlations between individual heterozygosity (essentially estimated with allozyme markers) and particular fitness components (survival, development stability, growth rate, fecundity, metabolism, etc.) have been observed in different aquatic species exposed to contamination (Benton and Guttman 1992; Kopp et al. 1992; Gillepsie and Guttman 1999; Harper-Arabie et al. 2004). Marchand et al. (2003, 2004) observed that the most genetically variable (the most heterozygous) individual flounders were also the most able to maintain their DNA integrity in contaminated estuaries, confirming the findings of several studies performed on natural populations of fish exposed to complex mixtures of pollutants (Shugart and Theodorakis 1998; Larno et al. 2001; Laroche et al. 2002). Maes et al. (2005) observed reduced genetic variability in strongly polluted eels *Anguilla anguilla*, as well as a negative correlation between level of contaminant metal bioaccumulation and allozymatic multilocus heterozygosity. Microsatellite genetic variability did not show any pollution-related differences, suggesting a differential response of metabolic enzymes and possibly direct overdominance of heterozygous individuals (Maes et al. 2005). These correlations may be in part explained by a slower basal metabolism in heterozygous individuals (compared to homozygous), leading thus to a weaker need for energy for vital functions and finally to a better ability to cope with environmental stress (Depledge 1996; Hauser et al. 2003).

Considering the studies cited in this part 14.4.1, many studies conducted in the past three decades have suggested that contamination may constitute a strong selective pressure, able to lead to adaptive changes in natural populations. However, the specific identification of the target on which selection acts remains very explorative, even in the case of the mummichog *Fundulus heteroclitus*, the premier teleost fish model in environmental biology, for which the genetic basis for resistance to pollution still needs to be elucidated (Burnett et al. 2007; Williams and Oleksiak 2008; Bozinovic and Oleksiak 2011). Thorough analyses of the molecular mechanisms of resistance to toxicants were first performed during experiments conducted with antibiotics, medicine, and pesticides (synthesis of Feyereisen 1995; Taylor and Feyereisen 1996). Several modifications of the target were detected:

- Punctual mutations in the coding regions of the enzyme targeted, leading to structural alterations of the active site and thus to a weaker sensitivity of the enzyme
- Duplications of the gene, thus leading to more efficient expression
- Mutations in the regulatory region of a gene (promoter) or in a gene coding for a protein that binds to a promoter

In the case of complex contaminations, such as those observed in aquatic systems, mechanisms of selection remain, for the most part, unknown. Selection may possibly act on loci physically close to those studied. Proximity on a chromosome may allow genes to be dragged along, through a mechanism of "hitchhiking," with a selective sweep experienced by the advantageous gene nearby (Hebert and Luiker 1996).

Several studies have not observed differences in genetic diversity between exposed populations and reference populations. For example, Mulvey et al. (2002) did not observe genetic alteration in allozyme markers in populations of *Fundulus heteroclitus* from a site highly contaminated by polycyclic aromatic hydrocarbons (PAHs). Neither did Whitehead et al. (2003), considering AFLP and microsatellites markers in populations of the freshwater fish *Catostomus occidentalis* exposed to pesticides, compared to unpolluted populations. Several factors may explain their findings:

- An insufficient contaminant stress
- The weak bioavailability of contaminants, their heterogeneous distribution, or their complex mixture

- A high gene flow that tended to homogenize the distribution of genotypic frequencies in populations, thus masking potential local adaptation
- A lack of sensitivity to pollutants of the genetic markers used

The latter factor is particularly fundamental. Indeed, a selective pressure does not affect the entire genome equally. The combination of different classes of molecular markers (neutral markers vs. markers potentially under selection) is particularly interesting, and, as described in Section 14.3, strong advances in genome typing has made this feasible for many species at a large scale, called genome-wide sampling. Genome-wide sampling (microsatellites, AFLPs, nucleotide polymorphisms, etc.) is used to identify and separate locus-specific effects (such as selection, mutation, assortive mating, and recombination) from genome-wide effects (such as drift or bottlenecks, gene flow, and inbreeding). Genome-wide effects inform us reliably about population demography and phylogenetic history, whereas locus-specific effects help identify genes that are important for fitness and adaptation (Luikart et al. 2003). This approach, called "population genomics" (Black et al. 2001), relies on the principle that neutral loci across the genome will be similarly affected by demography and the evolutionary history of populations, and loci under selection will often behave differently and therefore reveal "outlier" patterns of variation (Luikart et al. 2003; Storz 2005). Population-genomics studies of adaptive molecular variation will improve our understanding of the genetic mechanisms of speciation and will speed up the discovery of genes that are important for health and human medicine (Black et al. 2001; Jorde et al. 2001; Gibson and Mackay 2002) but also for biodiversity conservation (Luikart et al. 2003). This approach has been applied to identify candidate loci responsible for local adaptations among populations of the common frog *Rana temporaria* (Bonin et al. 2006) and among populations of deer mice *Peromyscus maniculatus* distributed across altitudinal gradients (Storz and Dubach 2004), and among ecotypes of lake whitefish *Coregonus clupeaformis* (Campbell and Bernatchez 2004) and populations of intertidal snails (*Littorina saxatilis*) (Wilding et al. 2001). The biggest hurdle encountered by this approach is the determination of the relevance of specific polymorphisms to protein function. This field is called "functional genomics" and requires (1) characterization of polymorphisms in both the coding and regulatory regions, (2) studies associating performances to protein variants, and (3) studies combining genetic polymorphism and particular environments. For the moment, this part is well documented for human diseases (cancer, Alzheimer, etc.), but evidence remains scarce for aquatic organisms. The most significant example concerns environmental adaptations in populations of the fish *F. heteroclitus* in the gene coding the lactate dehydrogenase-B enzyme (LDH), a locus that has not, however, been discovered by the genome typing approach. Both phenotypic (increased LDH activity) and genotypic (changes in Ldh-B regulatory sequences) differences between populations distributed along a steep thermal gradient have been found to be affected by natural selection, rather than genetic drift (see the synthesis of Schulte 2001).

14.4.2 Transmission of Resistance to Contaminants

Resistance (or tolerance) to contaminants has been identified in many taxa, from microorganisms to vertebrates or plants (Amiard-Triquet 2011), considering (1) a physiological acclimation of individuals or (2) heritable adaptive responses in a population where contaminants can be considered selective factors. However, demonstrations of the resistance (adaptive responses) of particular populations exposed to complex mixtures of pollutants, in freshwater or marine waters, are still rare in the literature. Crossing of individuals from contaminated systems can produce F1 and F2 generations that display a resistance to pollutants similar to the one observed for their parents; thus, this tolerance can be considered a genetic adaptation of the population (Belfiore and Anderson 2001; Johnston 2011).

Using this approach, Klerks and Moreau (2001) investigated the potential resistance of the fish *Cyprinodon variegatus* to chemical stress under experimental conditions; they highlighted a decreasing heritability of resistance to chemical stress with the growing number of pollutants in the mixture. For this fish species, the complexity of the mixture considerably reduced the possibility of resistance or slowed down the development of this adaptation. On the other hand, artificial selection carried out in the laboratory on another fish (*Heterandria formosa*) showed a fast response to cadmium selection; after only one generation of selection, a proportion of two selected lineages out of three displayed an increasing resistance to cadmium (Xie and Klerks 2003).

After the remediation of a site highly contaminated by cadmium, a complete and a rapid loss (between 9 and 18 generations) of genetically based cadmium resistance was observed for a population of a freshwater oligochaete *Limnodrilus hoffmeisteri*; this result was linked to the dominance of one pollutant and to a probable selective pressure on one locus (Levinton et al. 2003).

Several studies have underlined that *F. heteroclitus* populations from three chronically polluted estuarine systems in the United States are resistant to the aromatic hydrocarbons in their environment, as compared to nearby fish from relatively clean habitats; resistance in first- and sometimes second-generation embryos suggests that differential survival could be due to genetic adaptation rather than to physiological acclimation (synthesis in Burnett et al. 2007; Van Veld and Nacci 2008).

Recent work has also been conducted on the effects of multigenerational exposure to pollutants on arthropods. The long-term effects of nine trace metals were studied in insect larvae, those of the midges *Chironomus plumosus* and *Culicoides furens*; the sensitivity to trace metals decreased with LC_{50} values becoming larger between first and third generations (Vedamanikam and Shazilli 2008). *Daphnia magna* were exposed to a pesticide mixture consisting of a pyrethroid insecticide and a pre-emergent herbicide, and significant differences were detected between F0 and F1 *D. magna* for survival, in which F1 waterfleas were less sensitive to pesticide mixtures than F0 (Brausch and Salice 2011).

14.4.3 Cost of Resistance to Toxicants

The energy allocated by an organism to stress resistance increases its survival probability, but will reduce the energy allocated to essential functions such as growth or reproduction, perhaps significantly. Trade-offs are widely detected in the literature between maintenance (survival) and production (measured by developmental rate, fecundity, etc.), particularly in stressed populations (Mouneyrac et al. 2011).

In their synthesis on the cost of tolerance to toxins measured in experimental approaches, Van Straalen and Hoffman (2000) underlined the cost of tolerance to metals for plants, invertebrates, and fishes, generally identified by a very low frequency of tolerant individuals in "pristine" conditions and by different alterations of life history traits. Arthropods resistant to insecticides also displayed a cost that is characterized in "pristine" habitats by a loss of physiological performance in terms of fecundity, developmental rate, and fertility (Roush and McKenzie 1987). The resistance of the mosquito *Culex pipiens* to organophosphorous insecticides evolved in the Montpellier region of France from 1972 to 1984, by the substitution of a costly allele in terms of physiological performance by a less costly one (Guillemaud et al. 1998). The different molecular mechanisms of resistance to toxins (antibiotics, herbicides, insecticides) showed differential physiological costs (Taylor and Feyereisen 1996), and a major question was not elucidated: what is the validity of the cost estimated in the laboratory, relatively to the real cost in the field?

In natural populations, a smaller number of studies have explored the cost of resistance to chemical stress that has remained very difficult to estimate because numerous confounding factors, not linked to toxins, can dramatically modify estimations of this physiological cost; furthermore, we must consider that the heritability of numerous biological traits decreases from experimental conditions to natural environments (Woods and Hoffman 2000).

Several studies involving successive generations of fish selected in the laboratory for their resistance to contamination have characterized the cost of resistance, showing:

- Reduced size of offspring, lower fecundity, increased age at first maturity, for cadmium-exposed fish versus control fish in experimental exposures (Xie and Klerks 2004)
- Higher susceptibility of fish exposed to contaminated estuarine sediments compared to control fish when faced with other stressors such as UV radiation and hypoxia (Meyer et al. 2003)

In the case of the European flounder (*Platichthys flesus*), resistant genotypes characterized by their high capacity to maintain their DNA integrity in polluted estuaries, showed lower fecundity and condition index, compared to other genetic variants considered sensitive (Marchand et al. 2004). A number of fitness disadvantages were also observed in *F. heteroclitus* resident in polluted estuaries, including poor prey-capture capacity and reduced salinity range tolerance for egg fertilization (reviewed by Burnett et al. 2007). A recent study also highlighted a fitness cost of copper resistance in the marine copepod *Tigriopus japonicus*, as the intrinsic population growth rate of the copper-resistant lineage was significantly lower than that of the control (Kwok et al. 2009).

14.5 Toxicant-Altered Gene Expression

The action of environmental stress on an organism can affect the intracellular transfer of genetic information at sublethal levels. Toxicogenomic approaches using the tools of molecular biology provide an understanding of a new aspect of the impacts of chemical contaminants on living organisms. The level of genetic information transfer is assessed by the measurement of the rate of transcription of those products, which can affect—directly or indirectly—the processes of detoxification and/or resistance. At present, studies are performed in controlled experiments in which organisms are exposed to one contaminant or to a mixture of contaminants in order to evaluate the potential effects of synergy or antagonism between different molecules. The objective is to identify transcripts whose level of expression varies significantly between exposed and unexposed organisms. The information given by the entire sequenced genomes of particular species allows the recognition of a large number of sequences, and these are accessible in databanks. Moreover, functional genomic techniques are becoming available for a growing number of aquatic species (Pierron et al. 2009; Hook 2010; Scown et al. 2010; Pierron et al. 2011; Xuereb et al. 2011; Martini et al. 2012).

Overall, two approaches can be defined to explore the alteration of gene expression by pollutants: an *a priori* approach and a global one.

14.5.1 *A Priori* Approach

When genes linked to the response of an organism to contaminants are identified and their sequences known, their expression levels can be assessed. This approach (called *a priori*) must anticipate variations in the transcripts in order to be effective, and has been mainly developed on species that are of economic interest and/or have been subjected to many prior biochemical and genetic studies. In this frame, the semiquantitative estimation of gene expression can be conducted first by the assessment of band intensities of reverse transcription RT-PCR products on electrophoresis gels. This approach was carried out in the case of glutathione-*S*-transferase (GST) genes, a family encoding multifunctional proteins that catalyze the conjugation of glutathione to a wide variety of hydrophobic compounds leading to their elimination. Among GST studies, Boutet et al. (2004) evaluated the level of the transcription of GST genes in the oyster *Crassostrea gigas*; this work showed that two classes of GST (mu and omega) should be useful biomarkers of exposure to hydrocarbons, whereas two other classes (pi and sigma) should be relevant as biomarkers of exposure to pesticides. Li et al. (2012) showed that two genes encoding different GST proteins (sigma and pi) present different expression patterns in the case of exposure of the clam *Venerupis philippinarum* to bacterial stress, confirming different roles for these two families of proteins.

Among *a priori* methods, quantitative measurement of transcription processes is widely carried out using quantitative PCR or real-time PCR. Thus, Zhou et al. (2011) measured the impact of bisphenol A on the expression of three genes involved in the larval development of the gastropod *Haliotis diversicolor*, demonstrating an embryotoxic effect at cellular and molecular levels. In another study on the oyster *C. gigas* for which previous knowledge was available on the effect of cadmium on energy metabolism, Achard-Joris et al. (2006) assessed the transcription level of the mitochondrial gene encoding the first subunit of the cytochrome C oxidase protein complex CoxI, involved in electron transfer, in oysters exposed experimentally to cadmium contamination. They showed an induction of the transcription of CoxI, but no relationship was detected between the level of gene expression and the concentration of metal in the tissues. This induction was maintained but at a lower level in the case of contamination with both cadmium and zinc, but disappeared in the case of zinc contamination alone; these results confirmed that zinc could limit cadmium toxicity. In this experiment, the processes of gene overexpression could be considered compensation mechanisms to restore the level of metabolism activities necessary to cope with cadmium contamination.

14.5.2 "Global" Approach (Open System)

When no data are available on the nature of pollutants or on the genes involved in the responses to stress, a "global" approach can be carried out, exploring the patterns of gene expression over several hundred or thousand genes, by different genomic tools (Ghosh et al. 2011).

In this "global" approach, the techniques most frequently used recently are differential display (Liang and Pardee 1992) and suppression subtractive hybridization (SSH) (Diatchenko et al. 1996; Gurskaya et al. 1996). Differential display is a relatively rapid and not very expensive technique, but showed a major inconvenience: the detection of a large number of false positives (Martin and Pardee 2000). So, it must be accompanied by necessary steps of verification. Using this technique successfully, Bultelle et al. (2002) detected differentially expressed genes in the freshwater zebra mussel *Dreissena polymorpha* exposed to contaminants. In the case of the fish *Fundulus heteroclitus* exposed to a polyaromatic hydrocarbon (anthracene), Peterson and Bain (2004) revealed 26 DNA fragments displaying differential expression; among these sequences, 8 showing homologies with data available in international databanks were identified, the other 18 sequences being unknown. The 26 differential inductions were verified with macroarray technology. This technique consists of leaving PCR products of genes whose expression might be disturbed on a solid support (membrane). Membranes are then placed in contact with complementary DNAs (cDNAs) under hybridization conditions, cDNA reflecting the transcription of exposed and unexposed animals. By hybridization processes, fixed fragments of DNA keep the homologous cDNA. The comparison of the membranes obtained with the two transcriptomic extracts allows the determination of different levels of expression between contaminated and control organisms. With this technical approach, Peterson and Bain (2004) identified three sequences as markers of contamination by anthracene. A parallel study conducted on natural populations submitted to complex contaminations confirmed the previous inductions but revealed other overexpressions; these results underline the difficulties of extrapolating observations from controlled environments to natural environments.

The aim of SSH is to obtain a picture of the induction (positive regulation) or the repression (negative regulation) of several genes, the quantification of the differences between the expression levels in two experimental or environmental conditions not being possible. The SSH technique considers the study of two banks of cDNA to identify separately the two kinds of regulation. This approach was used to analyze differential expression in the oyster *C. gigas* subjected to pesticides (Tanguy et al. 2005). The authors revealed 137 sequences differentially expressed between contaminated versus uncontaminated oysters; among them, 81 were new markers. The 56 identified sequences were linked to different processes: detoxification, energy metabolism. Brown et al. (2006), analyzing the responses of the mussel *Mytilus edulis* to PAHs, by SSH, showed different biomarkers compared to those observed for vertebrates. Sheader et al. (2004) explored by SSH the transcriptomic responses of the flatfish *Platichtys flesus* to intraperitoneal contamination by benzo[*a*]pyrene and cadmium; they revealed 184 differentially expressed genetic fragments. The same SSH method allowed the identification of new candidate genes involved in the responses of *P. flesus* to experimental cocktails of pesticides; the induction or the inhibition of particular genes observed in the previous experiments was confirmed by comparison of fish from pesticide-impacted versus "pristine" estuaries (Marchand et al. 2006).

The target genes revealed by these global approaches could be potentially subjected to selective pressure induced by contaminants; they could constitute good candidate genes:

1. To analyze their differential expression levels in relation to the nature of the chemical stress
2. To explore the possible functional role of their polymorphism in the ability of organisms or populations to resist contamination

14.5.3 Combining the Two Approaches

Denslow et al. (2004) efficiently combined the two previous approaches, (1) revealing (using SSH) that the gene CYP 1A was overexpressed in the fish *Micropterus salmonides* subjected to the effluent of a paper mill, and (2) using real-time PCR method to quantify the CYP 1A expression level, as a tool for environmental monitoring.

The SSH methodology was carried out to assess the effects of ethinylestradiol on the transcription of males of the plaice *Pleuronectes platessa* (Brown et al. 2004). Thus, the first step was to generate banks of cDNA containing the genes with modified expression in the previous experimental contamination; then, the selected clones were used in DNA macroarray tests. Since 2004, numerous trancriptomic studies in ecotoxicology have been developed considering different combinations of the *a priori* and "global" approaches (Snape et al. 2004; Buckley et al. 2006; Hook et al. 2010; Milan et al. 2011; Vandersteen 2011).

14.5.4 New Insights Exploring Gene Regulation and Genomics

Over the past 10 years, numerous studies conducted in aquatic ecotoxicology have widely contributed to our understanding that the most prominent environmental contaminants cause changes in gene expression through activation of receptors or other ligand-activated transcription, particularly for fish models. In a review considering the basic processes of cell signaling and gene expression explored in *Fundulus heteroclitus*, Burnett et al. (2007) underlined the major roles played, for example: (1) by AhR, a family of transcription factors regulating the expression of cytochrome P450s and other genes in response to organic pollutants, and (2) by the hypoxia-inducible factor (HIFα) that could be toxicologically relevant.

In the past decade, sequence data have facilitated research in aquatic toxicology. Genomic information can be used to design and develop microarrays for some selected genes to thousands of genes, microarrays being constructed by cDNAs or oligonucleotides, and spotted arrays are commercially available for a few fish species (Fent and Sumpter 2011). Microarrays can be used to determine upregulation and downregulation of gene transcripts (mRNA) in exposed organisms. It is widely considered that transcriptomics aids in the identification of new toxicological effects as well as molecular, cellular, and biochemical pathways, but the link between transcript expression changes and effects at higher biological levels has rarely been addressed (Fent and Sumpter 2011). Furthermore, Nikinmaa and Rykönen (2011) have emphasized the need to relate mRNA and protein levels, and protein activity of individual genes in aquatic toxicology, as studies of mRNA expression do not necessarily reflect protein expression.

The comparison of transcriptome responses between natural populations in the field from polluted and reference sites could be particularly effective to detect contemporary evolution in heavily contaminated systems (Whitehead et al. 2010; Bozinovic and Oleksiak 2011); in addition to revealing mechanisms of toxicity and tolerance to pollutants, these studies underline the value of comparative transcriptomics to explore adaptive differences among wild populations.

14.6 Conclusions

The two complementary genomic approaches (population genomics and transcriptomics) are becoming particularly pertinent to thoroughly explore the impact of contaminants on natural populations (Bozinovic and Oleksiak 2011). However, although a growing number of studies provide descriptions of how genes or proteins are altered by chemical stress, only a few have tried to link these changes to physiological effects (Fent and Sumpter 2011).

Recent analyses conducted on wild fish populations resident in heavily contaminated estuaries appeared very efficient in exploring the mechanistic basis of resistance to pollutants (Whitehead et al. 2010; Wirgin et al. 2011). These two studies have underlined the major role played by the AhR in the tolerance of fish to chemical stress. The second study (Wirgin et al. 2011) showed that the evolved resistance to PCBs that occurs in some populations is linked to a particular allele of AhR2 encoding an AhR protein characterized by reduced binding affinity for dioxin-like compounds. Thus, research has clearly shown that a richer understanding of genome–environment interactions can be derived from studies of natural populations of outbred species that are routinely exposed to stressors in their environment.

Considering unicellular models, experimental manipulations in microcosms on microalgae subjected to petroleum contamination underlined the existence of resistant cells arising from rare spontaneous mutations that occur randomly before crude oil exposure (Carrera-Martinez et al. 2010); an adaptation of microalgae to a gradient of continuous petroleum contamination being clearly detected in a chronically contaminated river (Carrera-Martinez et al. 2011).

In the general context of global change, aquatic populations are increasingly subjected to multistress in their natural habitats (including the impacts of toxins, thermal stress, hypoxia, pathogens). Several studies have suggested that populations subjected to elevated temperatures may become more susceptible to pollution (Lapointe et al. 2011), and that the thermal tolerance window is narrowed in ectotherms inhabiting polluted areas compared with their counterparts in "pristine" environments (Lannig et al. 2008). Furthermore, the combined effects of parasites and contaminants can reduce either resistance or tolerance to infection (Marcogliese and Pietrock 2011). The cumulative effects of multiple stressors could become a priority concern for ecotoxicologists, particularly in their efforts to explore how adaptation to chronically contaminated habitats could affect the ability of populations to cope with other stressors. Thus, chemical exposure combined with global warming may result in evolutionary responses of populations currently not accounted for in any environmental risk assessment procedure (Messiaen et al. 2010).

Acknowledgment

The authors are much indebted to the excellent field and laboratory assistance of Louis Quiniou, over 12 years, in the context of several ecotoxicological programs.

References

Achard-Joris, M. et al. 2006. Cytochrome *c* oxidase subunit I gene is up-regulated by cadmium in freshwater and marine bivalves. *BioMetals* 19:237–44.

Amiard-Triquet, C. 2011. Pollution tolerance: From fundamental biological mechanisms to ecological consequences. In: *Tolerance to Environmental Contaminants*, ed. C. Amiard-Triquet, P.S. Rainbow, and M. Roméo, 1–24. Boca Raton, FL: CRC Press.

Bagley, M.J. et al. 2002. *Genetic Diversity as an Indicator of Ecosystem Condition and Sustainability: Utility for Regional Assessments of Stream Condition in the Eastern United States.* Cincinnati, OH: US Environmental Protection Agency.

Belfiore, N.M., and S.L. Anderson. 1998. Genetic patterns as a tool for monitoring and assessment of environmental impacts: The example of genetic ecotoxicology. *Environ. Monit. Assess.* 51:465–79.

Belfiore, N.M., and S.L. Anderson. 2001. Effects of contaminants on genetic patterns in aquatic organisms: A review. *Mutat. Res.* 489:97–122.

Benton, M.J., and S.I. Guttman. 1992. Allozyme genotype and differential resistance to mercury pollution in the caddisfly, *Nectopsyche albida*: II. Mutilocus genotypes. *Can. J. Fish. Aquat. Sci.* 49:147–9.

Benton, M.J., S.A. Diamond, and S.I. Guttman. 1994. A genetic and morphometric comparison of *Helisoma trivolvis* and *Gambusia holbrooki* from clean and contamined habitats. *Ecotoxicol. Environ. Saf.* 29:20–37.

Bickham, J.W. 2011. The four cornerstones of evolutionary toxicology. *Ecotoxicology* 20:497–502.

Black, W.C. et al. 2001. Population genomics: Genome-wide sampling of insect populations. *Annu. Rev. Entomol.* 46:441–69.

Bonin, A. et al. 2006. Explorative genome scan to detect candidate loci for adaptation along a gradient of altitude in the common frog (*Rana temporaria*). *Mol. Biol. Evol.* 23:773–83.

Boutet, I., A. Tanguy, and D. Moraga. 2004. Characterisation and expression of four mRNA sequences encoding glutathione *S*-transferases pi, mu, omega and sigma classes in the Pacific oyster *Crassostrea gigas* exposed to hydrocarbons and pesticides. *Mar. Biol.* 146:53–64.

Bozinovic, G., and M.J. Oleksiak. 2011. Genomic approaches with natural fish populations from polluted environments. *Environ. Toxicol. Chem.* 30:283–9.

Brausch, J.M., and C.J. Salice. 2011. Effects of an environmentally realistic pesticide mixture on *Daphnia magna* exposed for two generations. *Arch. Environ. Contam. Toxicol.* 61:272–9.

Brown, M. et al. 2004. Temporal changes in gene expression in the liver of male plaice (*Pleuronectes platessa*) in response to exposure to ethinylestradiol analysed by macroarray and real-time PCR. *Mutat. Res.* 552:35–49.

Brown, M. et al. 2006. Application of SSH and a macroarray to investigate altered gene expression in *Mytilus edulis* in response to exposure to benzo[*a*]pyrene. *Mar. Environ. Res.* 62:S128–35.

Buckley, B.A., A.Y. Gracey, and G.N. Somero. 2006. The cellular response to heat stress in the goby *Gillichthys mirabilis*: A cDNA microarray and protein-level analysis. *J. Exp. Biol.* 209:1660–77.

Bultelle, F. et al. 2002. Identification of differentially expressed genes in *Dreissena polymorpha* exposed to contaminants. *Mar. Environ. Res.* 54:385–9.

Burnett, K.G. et al. 2007. *Fundulus* as the premier teleost model in environmental biology: Opportunities for new insights using genomics. *Comp. Biochem. Physiol.* 2D:257–86.

Campbell, D., and L. Bernatchez. 2004. Generic scan using AFLP markers as a means to assess the role of directional selection in the divergence of sympatric whitefish ecotypes. *Mol. Biol. Evol.* 21:945–56.

Carrera-Martinez, D. et al. 2010. Microalgae response to petroleum spill: An experimental model analysing physiological and genetic response of *Dunaliella tertiolecta* (Chlorophycae) to oil samples from the tanker *Prestige*. *Aquat. Toxicol.* 97:151–9.

Carrera-Martinez, D. et al. 2011. Adaptation of microalgae to a gradient of continuous petroleum contamination. *Aquat. Toxicol.* 101:342–50.

Chagnon, N.L., and S.I. Guttman. 1989. Differential survivorship of allozyme genotypes in mosquitofish populations exposed to copper or cadmium. *Environ. Toxicol. Chem.* 8:319–26.

Demarais, B.D., T.E. Dowling, and W.L. Monckley. 1993. Post-perturbation genetic changes in populations of endangered Virgin River chubs. *Conserv. Biol.* 7:334–41.

Denslow, N.D. et al. 2004. Gene expression fingerprints of largemouth bass (*Micropterus salmoides*) exposed to pulp and paper mill effluents. *Mutat. Res.* 552:19–34.

Depledge, M.H. 1996. Genetic ecotoxicology: An overview. *J. Exp. Mar. Biol. Ecol.* 200:57–66.

Diatchenko, L. et al. 1996. Suppression subtractive hybridization: A method for generating differentially regulated or tissue-specific cDNA probes and libraries. *Proc. Nat. Acad. Sci. U. S. A.* 93:6025–30.
Dubrova, Y.E. et al. 1996. Human minisatellite mutation rate after the Chernobyl accident. *Nature,* 380:683–6.
Ellegren, H. et al. 1997. Fitness loss and germline mutations in barn swallows breeding in Chernobyl. *Nature* 389:593–6.
Facemire, C.F., T.S. Gross, and L.J. Guillette Jr. 1995. Reproductive impairment in the Florida panther: Nature or nurture? *Environ. Health Perspect.* 103:79–86.
Fent, K., and J.P. Sumpter. 2011. Progress and promises in toxicogenomics in aquatic toxicology: Is technical innovation driving scientific innovation? *Aquat. Toxicol.* 105:25–39.
Feyereisen, R. 1995. Molecular biology of insecticide resistance. *Toxicol. Lett.* 82–83:83–90.
Forbes, V.E., and T.L. Forbes. 1997. *Ecotoxicologie, Théorie et Applications.* Paris: Institut National de la Recherche Agronomique.
Ghosh, S. et al. 2011. Global gene expression and Ingenuity biological functions analysis on PCBs 153 and 138 induced human PBMC *in vitro* reveals differential mode(s) of action in developing toxicities. *Environ. Int.* 37:838–57.
Gibson, G., and T. Mackay. 2002. Enabling population and quantitative genomics. *Genet. Res.* 80:1–6.
Gillepsie, R.B., and S.I. Guttman. 1989. Effects of contaminants on the frequencies of allozymes in populations of the central stoneroller. *Environ. Toxicol. Chem.* 8:309–17.
Gillepsie, R.B., and S.I. Guttman. 1999. Chemical-induced changes in the genetic structure of populations: Effects on allozymes. In: *Genetics and Ecotoxicology*, ed. V.E. Forbes, 55–77. Philadelphia: Taylor & Francis.
Groenendijk, D. et al. 1999. Fluctuating life-history parameters indicating temporal variability in metal adaptation in riverine chironimids. *Arch. Environ. Contam. Toxicol.* 37:175–81.
Guillemaud, T. et al. 1998. Evolution of resistance in *Culex pipiens*: Allele replacement and changing environment. *Evolution* 52:443–53.
Gurskaya, N.G. et al. 1996. Equalizing cDNA subtraction based on selective suppression of polymerase chain reaction: Cloning of Jurkat cell transcripts induced by phytohemaglutinin and phorbol 12-myristate 13-acetate. *Anal. Biochem.* 240:90–7.
Hahn, M.E. 1998. Mechanisms of innate and acquired resistance to dioxin-like compounds. *Rev. Toxicol.* 2:395–443.
Hahn, M.E. et al. 2004. Aryl hydrocarbon receptor polymorphisms and dioxin resistance in Atlantic killifish (*Fundulus heteroclitus*). *Pharmacogenetics* 14:131–43.
Hahn, M.E. et al. 2005. Mechanism of PCB- and dioxin-resistance in fish in the Hudson River estuary: Role of receptor polymorphisms. In: *Final Report*. New York: Hudson River Foundation Grant. Available at http://www.hudsonriver.org/ls/.
Harper-Arabie, R.M. et al. 2004. Protective effects of allozyme genotype during chemical exposure in the grass shrimp, *Palaemonetes pugio*. *Aquat. Toxicol.* 70:41–54.
Hauser, L. et al. 2003. Molecular/cellular processes and the population genetics of a species. In: *Effects of Pollution on Fish, Molecular Effects and Population Responses*, ed. A.J. Lawrence and K.L. Hemingway, 256–88. Oxford, UK: Blackwell Science.
Hebert, P.D.N., and M. Luiker. 1996. Genetic effects of contaminant exposure—Towards an assessment of impacts on animal populations. *Sci. Total Environ.* 191:23–58.
Heithaus, M.R., and R.H. Laushman. 1997. Genetic variation and conservation of stream fishes: Influence of ecology, life history, and water quality. *Can. J. Fish. Aquat. Sci.* 54:1822–36.
Hemmer-Hansen, J. et al. 2007. Adaptive divergence in a high gene flow environment: Hsc70 variation in the European flounder (*Platichthys flesus* L.). *Heredity* 99:592–600.
Hook, S.E. 2010. Promise and progress in environmental genomics: A status report on the applications of gene expression-based microarray studies in ecologically relevant fish species. *J. Fish Biol.* 77:1999–2022.
Hook, S.E. et al. 2010. Temporal patterns in the transcriptomic response of rainbow trout, *Oncorhynchus mykiss*, to crude oil. *Aquat. Toxicol.* 99:320–9.

Hulla, J. et al. 1999. Symposium overview: The role of genetic polymorphism and repair deficiencies in environmental disease. *Toxicol. Sci.* 47:135–43.

Jaccoud, D. et al. 2001. Diversity arrays: A solid state technology for sequence information independent genotyping. *Nucleic Acids Res.* 29:e25.

Johnston, E. 2011. Tolerance to contaminants: Evidence from chronically exposed populations of aquatic organisms. In: *Tolerance to Environmental Contaminants*, ed. C. Amiard-Triquet, P.S. Rainbow, and M. Roméo, 25–48. Boca Raton, FL: CRC Press.

Jorde, L., W. Watkins, and M. Banshad. 2001. Population genomics: A bridge from evolutionary history to genetic medicine. *Hum. Mol. Genet.* 10:2199–207.

Kammenga, J., M.J. Arts, and A. Doroszuk. 2000. Multi-generation effects at the population level: Fitness maximisation and optimal allocation in a nematode. In: *Demography in Ecotoxicology*, ed. J. Kammenga and J. Laskowski, 163–77. Chicester, UK: John Wiley and Sons Ltd.

Kim, S.J. et al. 2003. Emergent effects of heavy metal pollution at a population level: *Littorina brevicula* a study case. *Mar. Pollut. Bull.* 46:74–80.

Klerks, P., and C. Moreau. 2001. Heritability of resistance to individual contaminants and to contaminant mixtures in the sheepshead minnow (*Cyprinodon variegatus*). *Environ. Toxicol. Chem.* 20:1746–51.

Kopp, R.L., S.I. Guttman, and T.E. Wissing. 1992. Genetic indicators of environmental stress in central mudminnows (*Umbra limi*) populations exposed to acid deposition in the Adirondack Mountains. *Environ. Toxicol. Chem.* 11:665–76.

Kramer, V.J. et al. 1992. Glycolysis and Krebs cycle metabolites in mosquitofish, *Gambusia holbrooki*, Girard 1859, exposed to mercuric chloride: Allozyme genotype effects. *Environ. Toxicol. Chem.* 11:357–64.

Krane, D.E., D.C. Sternberg, and G.A. Burton. 1999. Randomly amplified polymorphic DNA profile-based measures of genetic diversity in crayfish correlated with environmental impacts. *Environ. Toxicol. Chem.* 18:504–8.

Kwok, K.W.H., E.P.M. Grist, and K.M.Y. Leung. 2009. Acclimation effect and fitness cost of copper resistance in the marine copepod *Tigriopus japonicus*. *Ecotoxicol. Environ. Saf.* 72:358–64.

Lannig, G. et al. 2008. Cadmium-dependent oxygen limitation affects temperature tolerance in eastern oysters (*Crassostrea gigas* Gmelin). *Am. J. Physiol. Reg. Integr. Comp. Physiol.* 294: R1338–46.

Lapointe, D., F. Pierron, and P. Couture. 2011. Individual and combined effects of heat stress and aqueous or dietary copper exposure in fathead minnows (*Pimephales promelas*). *Aquat. Toxicol.* 104:80–5.

Laroche, J. et al. 2002. Genetic and physiologic responses of flounder (*Platichthys flesus*) populations to chemical contamination in estuaries. *Environ. Toxicol. Chem.* 21:2705–12.

Larno, V. et al. 2001. Responses of chub (*Leuciscus cephalus*) populations to chemical stress, assessed by genetic markers, DNA damage and cytochrome P450 1A induction. *Ecotoxicology* 10:145–58.

Lawrence, A.J. et al. 2003. Molecular/cellular processes and the physiological response to pollution. In: *Effects of Pollution on Fish, Molecular Effects and Population Responses,* ed. A.J. Lawrence and K.L. Hemmingway, 83–133. Oxford, UK: Blackwell Sciences.

Levinton, J.S. et al. 2003. Rapid loss of genetically based resistance to metals after the cleanup of a Duperfund site. *Proc. Natl. Acad. Sci. U. S. A.* 100:9889–91.

Li, C. et al. 2012. Two classes of glutathione *S*-transferase genes with different response profiles to bacterial challenge in *Venerupis philippinarum*. *Fish. Shellfish Immunol.* 32:219–22.

Liang, P., and A.B. Pardee. 1992. Differential display of eukaryotic messenger RNA by means of polymerase chain reaction. *Science* 257:967–71.

Luikart, G. et al. 2003. The power and promise of population genomics: From genotyping to genome typing. *Nat. Rev. Genet.* 4:981–94.

Maes, G.E. et al. 2005. The catadromous European eel *Anguilla anguilla* (L.) as a model for freshwater evolutionary ecotoxicology: Relationship between heavy metal bioaccumulation, condition and genetic variability. *Aquat. Toxicol.* 73:99–114.

Marchand, J. et al. 2003. Responses of European flounder *Platichthys flesus* populations to contamination in different estuaries along the Atlantic coast of France. *Mar. Ecol. Progr. Ser.* 260:273–84.
Marchand, J. et al. 2004. Physiological cost of tolerance to toxicants in the European flounder *Platichthys flesus*, along the French Atlantic Coast. *Aquat. Toxicol.* 70:327–43.
Marchand, J. et al. 2006. Molecular identification and expression of differentially regulated genes of the European flounder, *Platichthys flesus*, submitted to pesticide exposure. *Mar. Biotechnol.* 8:275–94.
Marchand, J. et al. 2010. Genetic polymorphism and its potential relation to environmental stress in five populations of the European flounder *Platichthys flesus*, along the French Atlantic coast. *Mar. Environ. Res.* 70:201–9.
Marcogliese, D.J., and M. Pietrock. 2011. Combined effects of parasites and contaminants on animal health: Parasites do matter. *Trends Parasitol.* 27:123–30.
Martin, K.J., and A.B. Pardee. 2000. Identifying expressed genes. *Proc. Natl. Acad. Sci. U. S. A.* 97:3789–91.
Martini, F. et al. 2012. Gene expression of heat shock protein 70, interleukin-1β and tumor necrosis factor α as tools to identify immunotoxic effects on *Xenopus laevis*: A dose-response study with benzo[*a*]pyrene and its degradation products. *Environ. Pollut.* 160:28–33.
Matson, C.W. et al. 2006. Evolutionary toxicology and population genetic effects of chronic contaminant exposure on marsh frogs (*Rana ridibunda*) in Sumgayit, Azerbaijan. *Environ. Health Perspect.* 114:547–52.
Messiaen, M. et al. 2010. The micro-evolutionary potential of *Daphnia magna* population exposed to temperature and cadmium stress. *Ecotoxicol. Environ. Saf.* 73:1114–22.
Meyer, J.N. et al. 2003. Antioxidant defenses in killifish (*Fundulus heteroclitus*) exposed to contaminated sediments and model prooxidants: Short term and heritable responses. *Aquat. Toxicol.* 65:377–95.
Milan, M. et al. 2011. Transcriptome sequencing and microarray development for the Manila clam, *Ruditapes philippinarum*: Genomic tools for environmental monitoring. *BMC Genomics* 12:234–51.
Moen, T. et al. 2008a. Identification and characterization of novel SNP markers in Atlantic cod: Evidence for directional selection. *BMC Genet.* 9:18.
Moen, T. et al. 2008b. A linkage map of the Atlantic salmon (*Salmo salar*) based on EST-derived SNP markers. *BMC Genomics* 9:223–36.
Mouneyrac, C., P.T.Y. Leung, and K.M.Y. Leung. 2011. Cost of tolerance. In: *Tolerance to Environmental Contaminants*, ed. C. Amiard-Triquet, P.S. Rainbow, and M. Roméo, 265–97. Boca Raton, FL: CRC Press.
Mulvey, M. et al. 2002. Genetic structure of *Fundulus heteroclitus* from PAH-contaminated and neighboring sites in the Elizabeth and York Rivers. *Aquat. Toxicol.* 61:195–209.
Murdoch, M.H., and P.D.N. Hebert. 1994. Mitochondrial DNA diversity of brown bullhead from contaminated and relatively pristine sites in the Great Lakes. *Environ. Toxicol. Chem.* 13:1281–9.
Nadig, S.G., K.L. Lee, and S.M. Adams. 1998. Evaluating alterations of genetic diversity in sunfish populations exposed to contaminants using RAPD assay. *Aquat. Toxicol.* 43:163–78.
Nevo, E. 2001. Evolution of genome–phenome diversity under environmental stress. *Proc. Natl. Acad. Sci. U. S. A.* 98:6233–40.
Nevo, E., R. Ben-Shlomo, and B. Lavie. 1984. Mercury selection of allozymes in marine organisms: Predictions and verification in nature. *Proc. Natl. Acad. Sci. U. S. A.* 81:1258–9.
Nikinmaa, M., and K.T. Rytkönen. 2011. Functional genomics in aquatic toxicology—Do not forget the function. *Aquat. Toxicol.* 105:16–24.
Peles, J.D., W.I. Towler, and S.I. Guttman. 2003. Population genetic structure of earthworms (*Lumbricus rubellus*) in soils contaminated by heavy metals. *Ecotoxicology* 12:379–86.
Peterson, J.S.K., and L.J. Bain. 2004. Differential gene expression in anthracene-exposed mummichogs (*Fundulus heteroclitus*). *Aquat. Toxicol.* 66:345–55.
Philips, N.R., and C.W. Hickey. 2010. Genotype-dependent recovery from acute exposure to heavy contamination in the freshwater clam *Sphaerium novaezelandiae*. *Aquat. Toxicol.* 99:507–13.

Pierron, F. et al. 2009. Transcriptional responses to environmental metal exposure in wild yellow perch (*Perca flavescens*) collected in lakes with differing environmental metal concentrations (Cd, Cu, Ni). *Ecotoxicology* 18:620–31.

Pierron, F. et al. 2011. Effects of chronic metal exposure on wild fish populations revealed by high-throughput cDNA sequencing. *Ecotoxicology* 20:1388–99.

Roush, R.T., and J.A. McKenzie. 1987. Ecological genetics of insecticide and acaricide resistance. *Annu. Rev. Entomol.* 32:361–80.

Roy, N.K. et al. 1995. Characterization and prevalence of a polymorphism in the 3′ untranslated region of cytochrome P4501A1 in cancer-prone Atlantic tomcod. *Arch. Biochem. Biophys.* 322:204–13.

Schluter, D. 2000. *The Ecology of Adaptative Radiation*. New York: Oxford University Press.

Schulte, P. 2001. Environmental adaptations as windows on molecular evolution. *Comp. Biochem. Physiol.* 128B:597–611.

Scown, T.M. et al. 2010. Effects of aqueous exposure to silver nanoparticles of different sizes in rainbow trout. *Toxicol. Sci.* 115:521–34.

Sheader, D.L. et al. 2004. Isolation of differentially expressed genes from contaminant exposed European flounder by suppressive, subtractive hybridisation. *Mar. Environ. Res.* 58:553–7.

Shugart, L., and C. Theodorakis. 1998. New trends in biological monitoring: Application of biomarkers to genetic ecotoxicology. *Biotherapy* 11:119–27.

Skøt, L. et al. 2002. Molecular genecology of temperature response in *Lolium perenne*: 2. Association of AFLP markers with ecogeography. *Mol. Ecol.* 11:1865–76.

Snape, J.R. et al. 2004. Ecotoxicogenomics: The challenge of integrating genomics into aquatic and terrestrial ecotoxicology. *Aquat. Toxicol.* 67:143–54.

Somers, C.M. et al. 2002. Air pollution induces heritable DNA mutations. *Proc. Natl. Acad. Sci. U. S. A.* 99:15904–7.

Soulé, M.E. 1987. *Viable Populations for Conservation*. Cambridge, UK: Cambridge University Press.

Storz, J.F. 2005. Using genome scans of DNA polymorphism to infer adaptive population divergence. *Mol. Ecol.* 14:671–88.

Storz, J., and J. Dubach. 2004. Natural selection drives altitudinal divergence at the albumin locus in deer mice *Peromyscus maniculatus*. *Evolution* 58:1342–52.

Tanguy, A. et al. 1999. Effects of an organic pollutant (tributyltin) on genetic structure in the Pacific oyster *Crassostrea gigas*. *Mar. Pollut. Bull.* 38:550–9.

Tanguy, A. et al. 2005. Molecular identification and expression study of differentially regulated genes in the Pacific oyster *Crassostrea gigas* in response to pesticide exposure. *FEBS J.* 272:390–403.

Taylor, M., and R. Feyereisen. 1996. Molecular biology and evolution of resistance to toxicants. *Mol. Biol. Evol.* 13:719–34.

Theodorakis, C.W. 2001. Integration of genotoxic and population genetic endpoints in biomonitoring and risk assessment. *Ecotoxicology* 10:245–56.

Theodorakis, C.W., and L.R. Shugart. 1997. Genetic ecotoxicology: II. Population genetic structure in mosquitofish exposed *in situ* to radionuclides. *Ecotoxicology* 6:335–54.

Theodorakis, C.W., and L.R. Shugart. 1999. Natural selection in contaminated environments: A case study using RAPD genotypes. In: *Genetics and Ecotoxicology*, ed. V.E. Forbes, 123–50. Philadelphia: Taylor & Francis.

Timmermans, M.J., J. Ellers, and N.M. Van Straalen. 2007. Allelic diversity of metallothionein in *Orchesella cincta* (L.): Traces of natural selection by environmental pollution. *Heredity* 98:311–9.

Vandersteen, W. 2011. Detecting gene expression profiles associated with environmental stressors within an ecological context. *Mol. Ecol.* 20:1322–3.

Van Straalen, N.M., and A.A. Hoffman. 2000. Review of experimental evidence for physiological costs of tolerance to toxicants. In: *Demography in Ecotoxicology*, ed. J. Kammenga and R. Laskowski, 147–61. New York: Wiley.

Van Straalen, N., and M. Timmermans. 2002. Genetic variation in toxicant-stressed populations: An evaluation of the "genetic erosion" hypothesis. *Hum. Ecol. Risk Assess.* 8:983–1002.

Van Veld, P.A., and D.E. Nacci. 2008. Toxicity resistance. In *The Toxicology of Fishes*, ed. R.T. Di Giulio and D.E. Hinton DE, 597–641. Boca Raton, FL: Taylor & Francis.

Vedamanikam, V.J., and N.A.M. Shazili. 2008. The effect of multi-generational exposure to metals and resultant change in median lethal toxicity tests values over subsequent generations. *Bull. Environ. Contam. Toxicol.* 80:63–7.

Virgilio, M. et al. 2003. Relationships between sediments and tissue contamination and allozymic patterns in *Hediste diversicolor* (Polychaeta Nereididae) in the Pialassa lagoons (north Adriatic Sea). *Oceanol. Acta* 26:85–92.

Virgilio, M., S. Maci, and M. Abbiati. 2005. Comparisons of genotype-tolerance responses in populations of *Hediste diversicolor* (Polychaeta: Nereididae) exposed to copper stress. *Mar. Biol.* 147:1305–12.

Wang, Z. et al. 2003. Reconciling actual and inferred population histories in the house finch (*Carpodacus mexicanus*) by AFLP analysis. *Evolution* 57:2852–64.

Weis, J.S. 2002. Tolerance to environmental contaminants in the mummichog, *Fundulus heteroclitus. Hum. Ecol. Risk Assess.* 8:933–53.

Whitehead, A. et al. 2003. Genetic variation among interconnected populations of *Catostomus occidentalis*: Implications for distinguishing impacts of contaminants from biogeographical structuring. *Mol. Ecol.* 12:2817–33.

Whitehead, A. et al. 2010. Comparative transcriptomics implicates mechanisms of evolved pollution tolerance in a killifish population. *Mol. Ecol.* 19:5186–203.

Wilding, C., R. Butlin, and J. Grahame. 2001. Differential gene exchange between parapatric morphs of *Littorina saxatilis* detected using AFLP markers. *J. Evol. Biol.* 14:611–9.

Williams, L.M., and M.J. Oleksiak. 2008. Signatures of selection in natural populations adapted to chronic pollution. *BMC Evol. Genet.* 8:282–93.

Wirgin, I. et al. 1990. Genetic diversity at an oncogene locus and in mitochondrial DNA between populations of cancer-prone Atlantic tomcod. *Biochem. Gen.* 28:459–75.

Wirgin, I. et al. 2011. Mechanistic basis of resistance to PCBs in Atlantic tomcod from the Hudson river. *Science* 331:1322–5.

Woods, R.E., and A.A. Hoffman. 2000. Evolution in toxic environments: Quantitative versus major gene approaches. In: *Demography in Ecotoxicology*, ed. J. Kammenga and R. Laskowski. Chichester, UK: Wiley.

Xie, L., and P. Klerks. 2003. Responses to selection for cadmium resistance in the least killifish, *Heterandria formosa. Environ. Toxicol. Chem.* 22:313–20.

Xie, L., and P. Klerks. 2004. Fitness cost of resistance to cadmium in the least killifish (*Heterandria formosa*). *Environ. Toxicol. Chem.* 23:1499–503.

Xuereb, B. et al. 2011. Vitellogenin-like gene expression in freshwater amphipod *Gammarus fossarum* (Koch, 1835): Functional characterization in females and potential for use as an endocrine disruption biomarker in males. *Ecotoxicology* 20:1286–99.

Yap, C.K., C.M. Chong, and S.G. Tan. 2011. Allozyme polymorphisms in horseshoe crabs, *Carcinoscorpius rotundicauda,* collected from polluted and unpolluted intertidal areas in Peninsular Malaysia. *Environ. Monit. Assess.* 174:389–400.

Yauk, C.L., and J.S. Quinn. 1996. Multilocus DNA fingerprinting reveals high rates of heritable genetic mutations in herring gulls in an industrialized urban site. *Proc. Natl. Acad. Sci. U. S. A.* 93:12137–41.

Yauk, C.L. et al. 2000. Induced minisatellite germline mutations in herring gulls (*Larus argenteus)* living near steel mills. *Mutat. Res.* 452:211–8.

Zhou, J., X.S. Zhu, and Z.H. Cai. 2011. The impacts of bisphenol A (BPA) on abalone (*Haliotis diversicolor supertexta*) embryonic development. *Chemosphere* 82:443–50.

15

Biomarkers Currently Used in Environmental Monitoring

Tracy K. Collier, Michael W.L. Chiang, Doris W.T. Au, and Philip S. Rainbow

CONTENTS

15.1 Introduction

It is clear from the preceding chapters in this book that biomarkers have an important contribution to make to environmental assessment. It is relevant then to ask whether and which biomarkers have made the full transition from the research arena to use by regulatory bodies with responsibility for assessing and ultimately controlling the release of toxic contaminants into the ecosystem.

It is impossible to be comprehensive in any such survey, and we have concentrated on three regions of the world to provide examples—Europe, the United States of America, and Asia.

15.2 European Approach

In Europe, the European Union (EU) is addressing the harmonization of the management of water quality among member nations. One aspect of the EU approach is delivered by REACH (Registration, Evaluation, Authorization, and Restriction of Chemical Substances), a regulation of June 2007, with the aim to improve the protection of human and environmental health through better and earlier identification of the properties, particularly toxic properties, of chemicals produced by industry. REACH puts greater responsibility on industry to manage the risks from chemicals and provide safety information on the potential hazard or toxicity of the chemical concerned, and information on amounts emitted.

The second aspect of the EU management of water quality concerns the Water Framework Directive (WFD), which addresses cumulative ecological risks from all stressors for each watershed in member nations. The WFD is an EU Directive of 2000 that commits EU Member States to achieve good qualitative and quantitative status of all water bodies (including coastal marine waters up to 1 nautical mile from the shore) by 2015. Both ecological status and chemical status need to be at least good to achieve this good water status. The WFD relies on ambient environmental standards to classify water bodies on the basis of chemical contamination. However, such directives can be more narrative than numerical (Luoma and Rainbow 2008). The WFD, for example, sets a target of "progressive reduction" for emissions of the 33 priority substances [including the trace metals As, Cd, Cr, Cu, Hg, Ni, Pb, Zn, and tributyltin (TBT)], but legally binding numerical criteria are retained in member states. The WFD is in the process of implementation with details varying among member states. It represents a watershed-by-watershed approach to risk assessment and risk management as compared to a chemical-by-chemical approach, although chemical loading data from REACH could ultimately be beneficial to evaluate potential loadings for individual watersheds.

Complementarily, the Marine Strategy Framework Directive (2008/56/EC; MSFD) has the objective of taking necessary measures to achieve or maintain good environmental status in the marine environment by the year 2020 at the latest, not only in coastal waters but also in marine waters defined as the waters, seabed, and subsoil on the seaward side of the baseline from which the extent of territorial waters is measured extending to the outmost reach of the area where a Member State has and/or exercises jurisdictional rights.

15.2.1 OSPAR and the Marine Environment

The OSPAR Convention for the Protection of the Marine Environment of the northeast Atlantic is the mechanism by which 15 western European nations cooperate to protect the waters of the northeast Atlantic. The convention was opened for signature in 1992 and entered into force in 1998. By 2007, the OSPAR Convention had been ratified by Belgium, Denmark, Finland, France, Germany, Iceland, Luxembourg, Netherlands, Norway, Portugal, Sweden, Switzerland, and the United Kingdom, and approved by the European Union and Spain. The resulting OSPAR Commission is guided by an ecosystem approach to an integrated management of human activities in the marine environment through the application of best available techniques and best environmental practice.

The OSPAR Convention requires Contracting Parties to "cooperate in carrying out monitoring programs," to develop quality assurance methods, and assessment tools and to "carry out...research which is considered necessary...to increase knowledge and understanding of the marine environment," "take into account scientific progress which is considered to be useful for...[such]...assessment purposes and which has been made elsewhere," and imposes on the OSPAR Commission duties to "define and implement collaborative monitoring programs," to "approve the presentation and interpretation of their results" and to "carry out [quality status] assessments," including in such assessments "both an evaluation of the effectiveness of the measures taken and planned for the protection of the marine environment and the identification of priorities for action."

Key to the fulfillment of these obligations is the series of OSPAR Joint Assessment and Monitoring Programmes (JAMP) that set out the basis on which the OSPAR Contracting Parties will work together. JAMP 2003–2010 delivered the holistic OSPAR Quality Status Report 2010 (http://ospar2010.squizmix.com/content/themes/joint-assessment-and-monitoring-programme-jamp). JAMP 2010–2014 is primarily orientated to support the activities of Contracting Parties in respect of the EU MSFD, and has, as its end point, the establishment of monitoring programs under that Directive. Part of JAMP is CEMP (OSPAR Co-ordinated Environmental Monitoring Programme http://www.ospar.org/content/content.asp?menu=00120000000116_000000_000000) addressing coordination across OSPAR, with the aim of delivering comparable data from across the OSPAR maritime area that can be used to address specific questions raised in the JAMP. JAMP regularly produces Guidelines for Contaminant-specific Biological Effects, in turn, supported by CEMP reports.

A final acronym needs to appear here, that of ICES—the International Council for the Exploration of the Sea. ICES provides OSPAR with scientific advice and information, and the ICES Secretariat serves as a data center for JAMP and CEMP. Such data will include reports of contaminant concentrations in water, sediment, and biota, and the results of biological effects monitoring.

It is in the guidelines and reports of the likes of JAMP and CEMP that biomarkers appear (OSPAR Commission 2007; ICES 2011). OSPAR has embraced biological effects monitoring (OSPAR Commission 2007), addressing such JAMP questions as:

1. What are the concentrations in the marine environment, and the effects of the substances on the OSPAR list of chemicals for priority action?
2. Are there any problems emerging related to the presence of hazardous substances in the marine environment? In particular, are there any unintended/unacceptable biological responses, or unintended/unacceptable levels of such responses, being caused by exposure to hazardous substances?

To address these questions, biological effects techniques can be deployed (OSPAR Commission 2007):

1. For investigative monitoring or screening
2. As an exploratory tool for identifying both the effects of contaminants not included in the monitoring program and combined effects of contaminants
3. As part of an integrated chemical and biological effects monitoring

In short, biomarkers have a role to play, while recognizing that the current deployment of biological effects techniques in general remains limited. Quality assurance is recognized as an essential component of the biological effects monitoring programs, but is currently only partially in place for most of the recommended techniques (Table 15.1).

TABLE 15.1

Current Status of Biomarker Quality Assurance

Biomarkers	Quality Assurance
PAH metabolites	QUASIMEME
Cytochrome P4501A activity (EROD)	BEQUALM, MEDPOL
Vitellogenin	WGBEC
Acetylcholinesterase	WGBEC
Comet assay	WGBEC
Micronucleus formation	MEDPOL
DNA adducts	BEQUALM
Metallothionein	BEQUALM (fish) MEDPOL (mussels)
Lysosomal stability (Cytochemical and Neutral Red)	MEDPOL (mussels)
Liver histopathology	BEQUALM
Macroscopic liver neoplasms	BEQUALM
Imposex/Intersex in gastropods	QUASIMEME
Externally visible fish diseases	BEQUALM
Reproductive success in eelpout	BEQUALM

Source: ICES 2010, 2011.

Note: BEQUALM, Biological Effects Quality Assurance in Monitoring Programmes; MEDPOL, Mediterranean Action Plan for the Barcelona Convention; QUASIMEME, Quality Assurance of Information for Marine Environmental Monitoring in Europe; WGBEC, Working Group of Biological Effect of Contaminants (ICES).

Thus, in the OSPAR 2007 document on Biological Effects Monitoring Techniques (OSPAR Commission 2007), the following biomarkers are recognized—hepatic metallothionein in fish, δ-aminolevulinic acid dehydratase in fish blood (for lead exposure), ethoxyresorufin *O*-deethylase (EROD) in fish liver, lysosomal stability, polycyclic aromatic hydrocarbon (PAH) metabolites in fish bile, and externally visible diseases and liver pathology in fish. Baseline values for the first five of these have been determined from reference sites. In the JAMP Guidelines for Contaminant-specific Biological Effects (OSPAR Agreement 2008–2009), further biomarkers are included, such as those relevant to (1) oxidative stress, (2) genotoxicity (DNA adducts), and (3) endocrine effects including imposex (TBT) and intersex, and estrogen-specific effects including plasma vitellogenin concentrations. Furthermore, these 2008–2009 JAMP guidelines separate out biological effects monitoring techniques that are specific, in turn, to metals, PAHs, TBT, and estrogens. Consideration of biomarkers is also strongly emphasized in the CEMP assessment report 2008–2009: assessment of trends and concentrations of selected hazardous substances in sediments and biota, with detailed distributional data, for example, on TBT-induced imposex in coastal gastropod mollusks and on EROD activities in livers of several fish including cod *Gadus morhua*, flounder *Platichthys flesus*, and dab *Limanda limanda*.

ICES advice to OSPAR on integrated monitoring and assessment of chemicals and biological effects in June 2011 (ICES 2011) recommends that groups of biological effects methods be deployed to address specific questions. Biomarkers figure strongly, including intersex (ovotestis) measurement in marine and estuarine fish; acetylcholinesterase activities to assess neurotoxic effects in aquatic organisms; histopathology of mussels (*Mytilus* spp.); micronucleus and Comet assays for assessing cytogenetic and DNA damage in aquatic organisms; metallothioneins in mussels; and externally visible fish diseases, macroscopic liver neoplasms, and liver histopathology in fish.

It can be concluded, therefore, that a role for biomarkers is being increasingly recognized in investigations of the ecotoxicology of northwestern European coastal environments, but what about freshwater systems?

15.2.2 Freshwater

The application of biomarkers in the ecotoxicological assessment of contaminants in freshwater lags behind that in coastal environments, but there are specific projects actively pursuing this topic.

In France, the National Institute for Industrial Environment and Risks (INERIS), set up in 1990, is a public research institute under the aegis of the French Ministry of Environment. Its mission is to carry out or commission studies and research programs aimed at preventing the risks associated with commercial activities on health, the safety of persons and property, and on the environment. Since 2000, the ecotoxicological unit of INERIS has been developing biomarkers in several freshwater fish species that can assess the effects of contaminants with the potential to be used in environmental regulation within the WFD (Sanchez and Porcher 2009). These biomarkers have included biomarkers of oxidative stress, neurotoxicity, and reproductive and immunological disruption. Of particular interest recently have been biomarkers of the effects of androgenic endocrine disruptors in freshwater fish, for example, in gudgeon *Gobio gobio* downstream from the discharge of a pharmaceutical manufacturing plant (Sanchez et al. 2011).

The NORMAN network (Network of Reference Laboratories for Monitoring of Emerging Environmental Pollutants), initially funded under Framework Programme 6 of the European Commission in 2005, has been established since 2009 as a permanent self-sustaining network of (now 45) reference laboratories, research centers, and related organizations for the monitoring and biomonitoring of emerging environmental substances, including in freshwater systems. Its mission includes the encouragement of the validation and harmonization of common measurement methods and monitoring tools so that the demands of risk assessors can be better met. The application of biomarkers is of central importance. The NORMAN network is of international significance with one member (from Canada) beyond the boundaries of Europe. The countries of origin of European members make up an impressive list—France, Germany, Denmark, Netherlands, United Kingdom, Switzerland, Italy, Slovakia, Portugal, Spain, Belgium, Slovenia, Sweden, Cyprus, Serbia, and the Czech Republic.

As a further example, the FISHBIO project is a river monitoring project in France that seeks to apply a multibiomarker approach to complement chemical analyses. Thus, established biomarkers (acetylcholinesterase, catalase, glutathione *S*-transferase, malondialdehyde (MDA), EROD) are analyzed in three fish species (barbel *Barbus barbus*, chub *Squalius cephalus*, and trout *Salmo trutta*) from 11 sites including a reference site.

Thus, biomarkers are being increasingly used to assess the biological effects of contaminants in European freshwater systems. Knowledge to be disseminated widely through the NORMAN network will catalyze the further incorporation of biomarkers in the environmental assessment of these freshwaters with the potential for them to be written into environmental regulations.

15.3 United States of America

In the United States, three federal agencies have recently used biomarkers of ecotoxicological effects in nationwide monitoring programs designed to examine the effects of chemical contaminants on aquatic habitats. These agencies, and the respective programs, include:

- The National Oceanic and Atmospheric Administration (NOAA) has run the National Status and Trends (NS&T) Program since approximately 1984, and biomarkers were measured in one component of the NS&T from 1984 through 1994.
- The U.S. Environmental Protection Agency (USEPA) collected field data under the Environmental Monitoring and Assessment Program (EMAP) from 1990 to 2006.
- The US Geological Survey (USGS) ran the Biomonitoring of Environmental Status and Trends (BEST) program from 1995 until approximately 2006.

Building on lessons learned from these national programs, a number of regional monitoring programs that use biomarkers to better understand the effects of chemical contaminants on aquatic ecosystems have been initiated in the United States. One of the longest-standing and most developed of these is the Puget Sound Assessment and Monitoring Program (PSAMP), which has run continuously since 1989. In this section, we briefly review how biomarkers are, or were, used in each of these four U.S. programs, and end with a prospectus for the use of biomarkers in the United States.

15.3.1 National Status and Trends Program

NOAA began the NS&T Program in 1984 to provide a way to monitor the health of the nation's coastal ecosystems. The program included Mussel Watch,* focused on measuring levels of persistent chemical contaminants in bivalves, and the National Benthic Surveillance Project (NBSP), which focused on the health of marine fish. The primary biomarker used in NBSP was the determination of toxicopathic lesions, determined by histopathological examination of liver, kidney, gonads, gills, and fins of demersal (i.e., bottom-dwelling) fish. This biomarker suite was evaluated in relation to a wide range of chemical contaminants in sediments at the site of capture, as well as contaminants in fish tissues and in stomach contents. Results from this monitoring program clearly established that prevalences of liver disease in several demersal fish species were strongly correlated with sediment levels of PAHs and, to a lesser extent, polychlorobiphenyls (PCBs), DDTs, and chlordanes (Myers et al. 1998). Moreover, results from the NBSP and more contemporary work led to the conclusion that PAHs are causally linked to several forms of liver disease, including neoplasia, in certain demersal fish (Myers et al. 2003). Because toxicopathic lesions represent an integration of the effects of contaminant exposure over space and time, and are considerably less expensive to measure compared to analyses of chemical contaminants in sediments and tissues, they have become widely used as a tool for assessing ecosystem health (Myers et al. 1998; Myers et al. 2008).

* NOAA's Mussel Watch was patterned closely after EPA's Mussel Watch Program, which had ended in 1979. Today, it is the longest running continuous contaminant monitoring program in U.S. coastal waters. Although some bivalve tissue samples have been archived (by freezing or tissue fixation) that could be used for biomarker measurements, to date Mussel Watch is focused on measurements of tissue burdens of contaminants.

The NBSP also included the measurement of fluorescent aromatic compounds (FACs) in bile of demersal fish, as a way to estimate exposure to PAHs, because PAHs are readily metabolized and excreted (Krahn et al. 1984). The monitoring data on bile FACs that were generated during implementation of the NBSP has proven to be valuable for use in rapidly assessing situations such as the emergency response following Hurricane Katrina (Krahn et al. 2005; Hom et al. 2008). In 1988, the NBSP added measures of CYP1A to other biomarkers already in use. This biomarker, hepatic activity of aryl hydrocarbon hydroxylase (AHH), was surveyed by a standardized method (Collier et al. 1995) for several years in demersal fish from the Atlantic, Gulf, and Pacific coasts of the United States.

Trends analysis throughout the 10 years of the NBSP did not show any monotonic trends in toxicopathic lesions in any tissues, nor in levels of FACs in fish bile. Interestingly, trends of increasing hepatic AHH activities were reported for the period from 1988 through 1994, in winter flounder (*Pleuronectes americanus*) collected at several urbanized sites along the Northeastern United States (Collier et al. 1998). Although the NBSP was discontinued after 1994, its paradigm (assessing the health of fish from sites showing a range of degradation by chemical contaminants) was continued in a number of regional monitoring efforts. Notable among these are the large rivers monitoring network under the U.S. Geological Survey Biomonitoring of Environmental Status and Trends (BEST) program, and the PSAMP, which are discussed later in this section.

15.3.2 Environmental Monitoring and Assessment Program

The USEPA's EMAP was one of the first national monitoring programs to use a probabilistic sampling design to monitor the conditions of the nation's waterways, both marine and freshwater. Field data were collected from 1990 through 2006, generally through a series of regional efforts conducted in collaboration with local (e.g., state) agencies. Although EMAP largely focused on chemical analyses and large-scale measures of biological organization, such as benthic community indices, a biomarker tested in this program was the prevalence of grossly visible external abnormalities in fish and shellfish. Generally, there was a positive correlation between increased levels of sediment contamination and these gross pathologies. In one report from EMAP, almost 25,000 fish were examined, and background prevalences of abnormalities ranged between 0.5% and 0.7%, whereas prevalences ranged as high as 2–3% at contaminated sites (Fournie et al. 1996). This is in contrast to histopathological examination of internal organs of fish, where prevalences of liver lesions have been reported as high as ~80% for fish from some highly contaminated areas (Malins et al. 1985; Myers et al. 1987). External pathologies, because of their low resolving power, were dropped from EMAP in the mid-1990s, and no statistically significant temporal trends in this biomarker appear to have been reported. On a much smaller scale, EMAP investigators also evaluated splenic macrophage aggregates (SMAs), measured by image analysis, for discriminating degraded versus nondegraded estuaries. Although it was concluded that the number and mean size of SMAs were effective indicators of chemical contaminants, as well as in some cases of low dissolved oxygen (Fournie et al. 2001), this biomarker is not currently in widespread use.

15.3.3 Biomonitoring of Environmental Status and Trends

In 1995, the USGS initiated an ambitious effort to determine the effects of environmental contaminants on fish health in major U.S. river systems, and to use a suite of biomarkers to do this. Biomarkers included histopathological examination of several tissues and hepatic CYP1A (similar to NBSP), but also included measures of reproductive status such as vitellogenin and a variety of steroid hormones. Macrophage aggregates, and a health assessment index, were also included in the suite of analyses conducted. The systems covered included the Mississippi, Columbia, Rio Grande, Yukon, and Colorado Rivers, as well as an investigation of rivers in the Southeastern United States. River systems were intensively sampled over a 1- to 2-year period, but individual sites were only sampled once in that period, with the intention of returning to the same system several years later. However, the BEST program ended in approximately 2006. Nonetheless, this biomarker-focused approach, for the first time, gave a broad view of fish health in rivers of the United States. One of the more notable findings from this effort was documentation of widespread intersex in certain fish species from U.S. rivers. The most common form of intersex was the presence of testicular oocytes, but the specific etiology of this condition remains to be determined (Hinck et al. 2009). To look at temporal trends, fish tissue contaminant data were compared to similar data collected 10–20 years earlier (Schmitt et al. 2005; Hinck et al. 2006); however, there were no similar earlier measures of biomarkers, so trends in these measures could not be assessed. Despite not being currently funded, the BEST program provides a rich legacy of data, archived tissue samples, and methods that can be used in the future to determine if there are, in fact, temporal trends in fish health in response to chemical contaminant exposure. Most impressively, the methods used in BEST are well documented and described (Schmitt and Dethloff 2000) and are available online (http://www.cerc.usgs.gov/pubs/BEST/methods.pdf), and the data collected are likewise available online, in an easily searchable form (http://www.cerc.usgs.gov/data/best/search/).

15.3.4 Puget Sound Assessment and Monitoring Program

Washington State, in the northwestern corner of the contiguous United States, contains within its boundaries a large estuary that is unique among U.S. estuaries, being that it is a deep fjord, formed by glacial scouring, and home to a growing and urbanizing human population. Beginning in the 1970s, Puget Sound has served as an "outdoor laboratory" to study the effects of chemical contaminants on marine fish species, and many of the findings from Puget Sound were used to design the NBSP component of the NS&T (Pierce et al. 1978; Malins et al. 1985; Myers et al. 1987). In 1989, a state-funded monitoring program, PSAMP, was founded, and part of its mission was to monitor for trends in fish health and chemical contaminant exposure in Puget Sound.

Central to this trends investigation has been the continued histopathological examination of tissues of English sole (*Pleuronectes vetulus*), because it was in this species that the strongest link between PAH exposure and liver disease, including neoplasia, had been demonstrated. Results from PSAMP show convincingly that the prevalences of toxicopathic liver disease have declined sharply along the Seattle waterfront, starting in 1999, through at least 2007, and that this trend is accompanied by reductions in PAH exposure as measured by FACs in bile (Figure 15.1; State of the Sound 2009). This improvement in a biomarker of fish health is very likely a reflection of the efforts to achieve source control and clean up contaminated sediments in this area. An additional example of biomarkers being used to demonstrate the success of management actions was recently reported in Puget Sound as well. Eagle Harbor, one of the most severely contaminated sites in this basin, was remediated by placing a cap of clean sediment over areas where total sediment PAHs exceeded 100,000 ppb on a dry weight basis. Before capping, almost 80% of the English sole from this site had toxicopathic liver disease. After capping, however, liver disease in this population of fish declined over the course of several years, and recently was reported to be at levels consistent with other nonurban sites in Puget Sound (Myers et

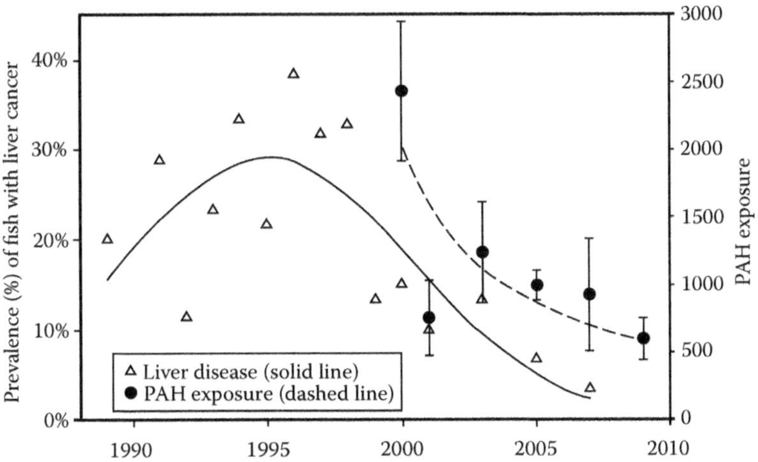

FIGURE 15.1
Twenty-year trend in liver disease and PAH exposure in English sole from Elliott Bay, Puget Sound, Washington, USA. Prevalence of liver disease was calculated using the number of 60 fish per year exhibiting any of four lesion types; preneoplasm, neoplasm, specific degenerative necrosis, or nonneoplastic proliferation. PAH exposure was estimated as the concentration of benzo(*a*)pyrene (BaP) equivalents in fish bile. No trend in mean fish age indicates that age was not a factor in the decline of liver disease. (Adapted from State of the Sound 2009, and personal communication to T.K.C. from J.E West, S.M. O'Neill, and M.S. Myers.)

al. 2008). Other biomarkers measured in this study included biliary FACs, hepatic CYP1A, and hepatic DNA adducts (measured by ^{32}P-postlabeling). Similar decreasing trends were seen in FACs and DNA adducts, but not in CYP1A, suggesting that this biomarker was not specific to PAH exposure in fish from this site. These successful demonstrations of how environmental improvement can be measured by using biomarkers of fish health have led to a recent decision by the state of Washington to incorporate liver disease as an indicator of environmental quality, for determining progress in protecting and restoring the Puget Sound ecosystem (State of the Sound 2009). PSAMP, in collaboration with other programs, continues to develop and test additional biomarkers, such as the measurement of vitellogenin in marine and anadromous fish (Peck et al. 2011).

Because of the large number of studies of fish health and contaminant exposure in Puget Sound and along the west coast of the United States, investigators are now able to construct quantitative relationships between levels of sediment contaminants and different biological endpoints, such as liver disease, reproductive function, and growth. This linking of biomarker responses and sediment contamination has been useful for managers facing decisions about cleaning up contaminated sites, because they can estimate the improvements in biological function that will result from costly remedial actions. An example of information generated from such an analysis is shown in Table 15.2, which reports the estimated levels of biological impacts associated with increasing levels of PAHs in sediments (Johnson et al. 2002). This is especially important for contaminants, such as PAHs, that are not readily measured in fish species because of rapid metabolism and excretion.

15.3.5 U.S. Summary and Prospectus

A number of national monitoring programs in the United States have included one or more biomarkers as part of their approach, but none of these programs has been sustainable. As in other parts of the world, there is an increasing recognition that effects-based monitoring

TABLE 15.2

Estimated Effects Levels Associated with Increasing Sediment PAH Concentration for Selected Liver Lesions, Indicators of Reproductive Dysfunction, and DNA Damage in English Sole

PAHs (ppb dry wt)	Any Toxicopathic Liver Lesion (%)	Liver Cancer (%)	Inhibited Gonadal Development (%)	Inhibition of Spawning (%)	Infertile Eggs, % of Eggs Spawned	DNA Damage (nmol Adducts/mol bases)
50	0	0	15	12	38	5
100	0	0	15	12	38	5
1000	9	0	15	17	42	25
2000	18	0	15	25	48	36
3000	24	1	15	30	51	43
5000	31	3	18	35	55	51
10,000	40	6	27	43	61	63
100,000	71	16	58	69	80	100

Source: Adapted from Johnson, L.L. et al., *Aquat. Conserv.: Mar. Fresh. Ecosyst.*, 12, 517–538, 2002.

is needed (Handy et al. 2002; Lam and Gray 2003), but such approaches probably cannot be standardized across a large geographic scale, such as the entire United States, because of differences in the biological communities across such distances. However, regional approaches to effects-based monitoring are proving to be effective, especially when linked to management actions. In Puget Sound, an approach was recently recommended for addressing the complex issue of toxic chemical contaminants in that aquatic ecosystem (Collier et al. 2007). The recommended approach consisted of three parts: (1) identification of sources and quantification of loadings to the estuary; (2) development and implementation of strategies for source reductions and source control; and (3) biologically based monitoring and assessment, to determine the effectiveness of source control strategies. Biomarkers, appropriately applied, will be important tools in regional efforts such as these, and future monitoring programs should be sustainable if they are able to make linkages between biological outcomes and management actions.

15.4 Approaches in Asia

Rapid urban and industrial development and associated pollution in Asia have seen growing public awareness and concern on environmental problems and related human health issues. In the Asia Pacific region in general, and Hong Kong and China in particular, environmental sustainability is among the top government priorities and policies. It has now been recognized that proper management of the environment, in particular, water resources, is vital to protect public health and sustain socioeconomic development.

The history and development of biomarkers for water quality monitoring have been fully discussed in other chapters. It is recognized that the environmental fate and bioavailability of contaminants can be accounted for when biomarkers are used for water quality monitoring. There is therefore a worldwide trend to supplement chemical-based monitoring data with effects-related biomonitoring data (Handy et al. 2002; Lam and Gray 2003). Although many biomarkers are still in the experimental stage of development and are not yet ready for use in practical monitoring purposes, some biomarkers (e.g., body burden of trace metals/trace organics in mussels and barnacles; lysosomal integrity in mussels) have become well established and have already been adopted in major national monitoring programs by regulatory authorities in Europe and North America, as discussed above. Owing to large differences in physical environments and marine/freshwater biota, biomarkers or biological indicators proven effective in the temperate regions of Europe and the United States may not necessarily be applicable to Asia (a mainly tropical and subtropical environment). It is therefore necessary to design appropriate laboratory and field studies to validate and quantify responses of selected biomarkers to pollution in the local environment.

In the past decade, there have been increasing numbers of research studies led by tertiary institutions/research centers/nongovernment organizations on the development and validation of biomarkers for monitoring water pollution in Asia. Table 15.3 provides a selection of large-scale biomarker-based monitoring programs conducted in some Asian countries/regions using local sentinel species (such as mussels and fish). The Asia-Pacific Mussel Watch, a comprehensive monitoring program jointly participated in by many tertiary institutions and government agencies from 10 Asian countries, was conducted to monitor the levels of organic pollutants [e.g., TBT, polybrominated diphenyl ethers (PBDE), and organochlorines] in mussels (*Perna viridis, Mytilus edulis*) sampled along coastal areas (Table 15.3). Until now, the use of biomarker-based monitoring has been limited only to experimental trials largely on an *ad hoc* basis. Biomarkers have not been used in Asian countries for routine environmental monitoring and regulatory purposes. There is no statutory requirement on the use of biomarkers in formulating water quality objectives/guidelines and/or environmental risk assessment, and respective governments have no specific programs for these initiatives.

Table 15.4 summarizes current water quality monitoring practices adopted by Asian countries/regions, including Hong Kong, China, South Korea, Japan, India, Taiwan, Singapore, Malaysia, the Philippines, Thailand, Indonesia, Vietnam, Lao People's Democratic Republic (PDR), Myanmar, and Cambodia. In general, national-level monitoring programs have been well established for environmental management by the respective governments in most Asian countries, except for a few developing ones. However, monitoring and assessments of environmental quality are largely based on physicochemical data and the levels of hazardous chemicals in water and sediment samples. The application of biomarkers for water quality monitoring is significantly lagging behind their counterparts in European Countries and the North American continent.

15.4.1 Hong Kong

Since the 1980s, water monitoring programs have been regularly conducted by the Environmental Protection Department (EPD), mainly focused on physicochemical parameters of marine waters and sediments. Biological monitoring was restricted to the monitoring of sewage bacteria and phytoplankton as part of the general marine water monitoring. Between 2001 and 2003, the EPD led a consultant study on the "Development of a Biological Indicator System for Monitoring Marine Pollution," with the objectives of choosing a set of biomarkers/bioindicators that can be used to track the biological effects of pollution in Hong Kong waters. In 2004–2010, the EPD launched a biomonitoring program using a battery of selected biomarkers including EROD activity, histopathology, and condition factor (i.e., weight/length ratio) for fish (*Siganus oramin* and *Leiognathus brevirostris*), lysosomal integrity for the mussel *Perna viridis*, and relative penis size index (TBT biomarker) for the whelk *Thais clavigera* (Table 15.3; HKEPD 2007), to monitor the pollution

TABLE 15.3

Selected Examples of Biomarker-Based Monitoring Programs in Asian Countries Since 2000

Country/Region	Monitoring Programs/Studies	Biomarkers/Details of the Techniques or Related Technologies	Status	References
Cambodia, China, Hong Kong, India, Indonesia, Japan, South Korea, Malaysia, the Philippines, Singapore and Vietnam	Biomonitoring of contaminants: Asia-Pacific Mussel Watch	Green-lipped mussels *Perna viridis* and/or blue mussels *Mytilus edulis*: • Butyltin compounds (BTs) including mono-, di-, and tributyltin, and total tin (∑Sn) • Polybrominated diphenyl ethers (PBDEs) including BDE-3, BDE-15, BDE-28, BDE-47, BDE-99, BDE-100, BDE-153, BDE-154, BDE-183, BDE-196, BDE-197, BDE-206, BDE-207, and BDE-209 • Organochlorines (OCs) including polychlorinated biphenyls (PCBs), DDTs, HCHs, chlordane related compounds (CHLs), and hexachlorobenzene (HCB)	R&D	Sudaryanto et al. 2002; Monirith et al. 2003; Ramu et al. 2007
Hong Kong SAR, China	Monitoring of biomarkers and biological indicators in coastal waters	(1) Rabbit fish *Siganus oramin*: PAHs and PCBs in muscle; Condition Factor (CF), Hepatosomatic Index (HSI), and Gonadosomatic Index (GSI); ethoxyresorufin-*O*-deethylase (EROD) enzyme activities and accumulation of secondary/tertiary lysosomes	R&D	Au 2004; Fang et al. 2009a, b, 2010
		(2) *Siganus oramin* and pony fish *Leiognathus brevirostris*: CF, HIS, and GSI; fin erosion and epidermal hyperplasia; EROD activities in liver	A	HKEPD 2007
		(3) Green-lipped mussel *Perna viridis*: PAH and PCB body burdens.	R&D	Fang et al. 2009c
		Lysosomal integrity.	A	HKEPD 2007
		(4) Gastropod *Thais clavigera*: relative penis size index (RPSI) (imposex) as a biomarker of tributyltin (TBT) contamination	A	HKEPD 2007
South Korea	Biomarker responses in coastal waters	(1) Rockfish *Sebastes schleegeli* and marbled flounder *Pseudopleuronectes yokohamae*: Biliary PAH metabolites and hepatic biomarkers after the *Hebei Spirit* oil spill incident	R&D	Jung et al. 2011

TABLE 15.3 (Continued)

Selected Example of Biomarker-Based Monitoring Programs in Asian Countries Since 2000

Country/ Region	Monitoring Programs/Studies	Biomarkers/Details of the Techniques or Related Technologies	Status	References
		(2) Gastropod *Thais clavigera*: Imposex and organotin contamination	R&D	Shim et al. 2000
		(3) Mussels *Mytilus coruscus, Mytilus edulis*, and oyster *Crassostrea gigas*: butyltin and phenyltin before and after the regulation of banning TBT-based antifouling paints in Korea.	R&D	Shim et al. 2005
India	Biomarker responses in coastal waters	(1) Gastropod *Cronia contracta*: Acetylcholinesterase (AChE) activities as neurotoxin biomarker of neurotoxic contaminants, Goa	R&D	Gaitonde et al. 2006
		(2) *Cronia contracta*: impairment of DNA integrity (alkaline unwinding assay) as genotoxin biomarker, Goa	R&D	Sarkar et al. 2008
		(3) *Perna viridis*: oxidative stress indices (lipid peroxidation, superoxide dismutase, catalase, glutathione peroxidase, glutathione reductase, glutathione *S*-transferase, ascorbic acid, reduced glutathione) in gills and digestive glands, Goa	R&D	Jena et al. 2009

Note: A, application, R&D, research and development.

status of Hong Kong waters. However, as yet, there is no mandate on a biomarker-based monitoring policy in Hong Kong waters.

15.4.2 China

In China, other than the Special Administrative Region (SAR) of Hong Kong, nationwide marine environmental pollution monitoring was started in 1984. The State Environmental Protection Administration is responsible for all surface water (lakes, reservoirs, and rivers), underground water, coastal and nearshore seawater, and wastewater discharge. It monitors water quality, biology, sediments, and discharge volumes. In addition to typical physicochemical analyses, the levels of nutrients, trace metals, oils, and persistent organic pollutants (POPs) are becoming the key concerns in water monitoring programs (Wen 2005). The State Oceanic Administration has carried out monitoring of trace metals in biota including tissues of fish, shrimps, and mussels (*Mytilus* and *Perna* spp.) (Huang Zongguo, personal communication to D.W.T.A.). However, the biomonitoring program is only at the experimental trial stage. Moreover, most biomarker development studies have been

TABLE 15.4

Water Quality Monitoring: Current Practices in Asia

Country/ Region	Nationwide Monitoring of Surface Water (Rivers and Lakes)	Nationwide Marine Environmental Monitoring
Hong Kong	Environmental Protection Department (EPD) regularly monitors water quality at more than 80 sampling stations in more than 30 rivers and streams, mostly on a monthly basis. The monitoring covers more than 40 parameters including physicochemical characteristics, organic compounds, nutrients, toxic metals and coliform bacteria (HKEPD 2005a).	EPD marine monitoring program covers about 90 water and 60 sediment sampling stations for physical and chemical parameters, including temperature, pH, salinity, turbidity, and dissolved oxygen measured *in situ*. Plus laboratory analysis of some 40 parameters such as nutrients, metals, organics, and coliform bacteria. Water quality monitoring is generally conducted once a month, whereas sediment quality monitoring is done twice a year (HKEPD 2005b).
China	State Environmental Protection Administration (SEPA) monitors water quality, biology, sediments, and discharge volumes. Monthly—479 monitoring stations for parameters including pH, conductivity, DO, COD, BOD, NH_4-N, oil, volatile phenols, Hg, Pb Weekly—43 automatic monitoring stations report temperature, pH, DO, COD, conductivity, turbidity, TOC, NH_4-N (POMRAC 2009).	SEPA: (1) Seawater quality monitoring of parameters including depth, transparency, pH, salinity, DO, COD, nutrients, oil, chlorophyll *a* (2) Mussel watch: bivalve (mussels, oysters, and clams) body burdens oil, total Hg, Cd, Pb, Cu, As, DDT, PCB, etc. (Wen 2005).
South Korea	Local Environmental Offices affiliated with the Ministry of Environment (MOE) are responsible for monitoring lakes and rivers. (1) River water quality parameters: • 48 times per year for 16 physical, chemical, and biological parameters including water level, flow rate, pH, Biological Oxygen Demand (BOD), COD, nitrate, phytoplankton • 12 times per year for 11 parameters including phosphate, chlorophyll *a*, metals, and CN • Once per year for 4 organic compounds including PCBs. (2) Lake water quality parameters: • 12 times per year for 21 physical, chemical, and biological parameters including water level, flow rate, pH, BOD, COD, nitrate, phosphate, phytoplankton, suspended solids (SS) • 4 times per year for 7 parameters including metals and CN • Once per year for 4 organic compounds including PCBs. (POMRAC 2009)	Coastal environmental monitoring of water quality is implemented by National Fisheries Research & Development Institute (NFRDI) with intensive environmental monitoring at 296 stations: • 4 times per year: general physical and chemical parameters items including salinity, pH, DO, COD, SS, chlorophyll *a* • 2 times per year: selected trace metals, CN and organic compounds (PCB, TBT) Plus selective monitoring of metals and organic compounds (PCB, TBT, PAH, PCDD/DFs) in organisms (fish, bivalves) and sediments (Wen 2005; POMRAC 2009).

TABLE 15.4 (Continued)

Water Quality Monitoring: Current Practices in Asia

Country/ Region	Nationwide Monitoring of Surface Water (Rivers and Lakes)	Nationwide Marine Environmental Monitoring
Japan	River/lake water quality monitoring parameters are similar to those for marine coastal water monitoring: 8 parameters listed in list I and 26 chemicals (23 of list II + fluoride, boron, and nitrate- and nitrite-nitrogen). There are an additional 27 precautionary monitoring parameters consisting of further metals and organic compounds (POMRAC 2009; MEJ 2011).	The water quality survey of coastal areas covers: I. 8 physical/chemical/biochemical parameters: pH, chemical oxygen demand (COD), dissolved oxygen (DO), *n*-hexane extract (oil, etc.), total nitrogen (TN), total phosphorous (TP), total coliforms II. 23 chemicals including 8 metals and 15 organic compounds.
India	National Water Quality Monitoring Programme (NWMP) (CPCB 2008a) (1) Water samples analyzed for 28 parameters: • 9 core parameters including pH, conductivity, DO, BOD, nitrate-N, nitrite-N, fecal and total coliforms • 19 other physicochemical and bacteriological parameters including turbidity, alkalinity, COD, total N, hardness, SS, phosphate, sulfate, Na, Ca, Mg, K, F, Cl, ammonia, boron. Plus another 9 trace metals and 15 pesticides in selected samples. (2) Biological monitoring in specific locations: saprobity index, diversity index, P/R ratio.	Parameters monitored include (CPCB 2008b): (1) Water quality: salinity, DO, pH, SS, BOD, phosphate, nitrate, nitrite, ammonia, petroleum hydrocarbons, pathogenic *Vibrio*, pathogenic Enterobacteriaceae. (2) Sediment quality: TOC, bacteria, Cd, Pb, Hg. (3) Biological characteristics: • primary productivity, chlorophyll *a*, pheophytin, phytoplankton, zooplankton, benthos, fish • biomonitoring analysis of selected organisms (e.g., mussels) for Cd, Hg, Pb as well as pesticide residues (DFDT, BHC)
Taiwan	Department of Environmental Monitoring and Information Management responsible for all national environmental water monitoring. Rivers: monthly for mainly physicochemical parameters including conductivity, pH, DO, BOD, COD, SS, nitrogen (ammonia, nitrite, nitrate), but also coliforms; seasonally—including phosphate and some trace metals (EPA Taiwan 2010a, b).	19 regions and 104 sampling points: seasonal monitoring of physicochemical parameters only, including salinity, pH, DO, SS, chlorophyll *a*, nitrogen (ammonia, nitrite, nitrate), phosphate, silicate, and selected trace metals (Cd, Cr, Cu, Hg, Pb, Zn) (EPA Taiwan 2010a, b).
Singapore	National Environment Agency (NEA): regular monitoring of pH, DO, BOD, SS, ammonia, sulfide (NEA 2002).	NEA: monitoring of water samples for trace metals, TOC, and other physical, chemical, and bacteriological parameters (NEA 2002).
Malaysia	Malaysian Department of Environment (DOE): water monitoring at regular intervals (WEPA 2009). Six main traditional physicochemical measurements: BOD, CO, NH_3N, pH, DO, SS. Other parameters such as trace metals and bacteria measured according to site requirement (Morse et al. 2007; DOE 2010a).	DOE monitoring program: • *In situ* water quality parameters include pH, DO, salinity, turbidity, tar balls • laboratory analyses of parameters including fecal coliforms, SS, ammonia, nitrite, nitrate, selected metals, CN, phenol, TBT, PAH (DOE 2010b).

(*continued*)

TABLE 15.4 (Continued)

Water Quality Monitoring: Current Practices in Asia

Country/ Region	Nationwide Monitoring of Surface Water (Rivers and Lakes)	Nationwide Marine Environmental Monitoring
The Philippines	Environmental Management Bureau (EMB) conducts regular water quality monitoring (2001 to 2005—238 water bodies monitored for classification or regular water quality monitoring). Monthly or quarterly monitoring (WEPA 2009) of parameters that include: • Primary parameters—pH, temperature, SS, BOD, DO, fecal coliforms, nitrate, phosphate, chloride • Secondary parameters—ammonia, barium, boron, fluoride, iron, sulfate, CN, selected metals (As, Cd, Cr, Cu, Hg, Ni, Pb), organics (animal–vegetable fat and oil, benzene, toluene, ethylbenzene, xylene) (DENR-EMB 2008).	EMB: 39 bays and coasts (WEPA 2009) for parameters including DO, coliforms, trace metals. Manila Bay has its own monitoring program: • Water quality parameters including DO, pH, salinity, fecal and total coliforms, nutrients, oil and grease in seawater • Sediment, shellfish, and other benthos—trace metals and persistent organic pollutants (POPs) (DENR-EMB 2007).
Thailand	Pollution Control Department (PCD) monitors water quality in rivers 3 to 4 times per year (WEPA 2009; Boonsoong et al. 2010). Parameters (Morse et al. 2007) include: pH, DO, BOD, fecal and total coliforms, nitrate, ammonia, phenols, CN, selected trace metals, radioactivity, organochlorine pesticides (DDT, alpha-BHC, dieldrin, aldrin, heptachlor, endrin) (PCD 2004a).	PCD monitors coastal water quality twice a year, in 23 seaside provinces both rainy and dry seasons at 100 and 500 m offshore. Mainly physicochemical parameters include floatable solids, floatable oil and grease, pH, salinity, DO, fecal and total coliforms, nitrate, phosphate, phenols, ammonia, sulfide, selected trace metals, radioactivity, PCBs, total organochlorine pesticides (PCD 2004b).
Indonesia	In 2006, 30 Regional Environmental Impact Control Agencies monitored 35 rivers (WEPA 2009) for mainly physicochemical parameters including pH, SS, DO, BOD, COD, ammonia, nitrite, nitrate, P, chlorine, phenol, oil and grease, detergent, fecal and total coliforms (Syafrul 2009).	Ministry of the Environment conducts regular monitoring for parameters including turbidity, SS, pH, salinity, DO, BOD, COD, ammonia, nitrite, nitrate, CN, H_2S, oil, phenol, organochlorine pesticides, PCB, detergent, trace metals, coliforms, pathogens, plankton, radionuclides (Hutomo 1993).
Vietnam	Ministry of Environment and Natural Resources (MoNRE): Surface water monitored 4 times per year (Son 2008) for parameters including pH, SS, turbidity, conductivity, DO, BOD, COD, NH_4, NO_3, PO_4. Cl^-, SiO_2, Ca, Mg, Na, K, SO_4, HCO_3, total Fe, total coliforms. At particular sites trace metals and pesticides may also be monitored (World Bank 2006; Son 2008).	Ministry of Science, Technology, and Environment of Vietnam (MOSTE): coastal water monitoring addresses 29 parameters for 3 categories: beaches, aquaculture, and other areas (WEPA 2009), parameters including BOD, COD, nutrients, trace metals, pesticides, coliforms. Phytoplankton and harmful microalgae are also sampled at times (Dien and Thuoc 2003).

TABLE 15.4 (Continued)

Water Quality Monitoring: Current Practices in Asia

Country/ Region	Nationwide Monitoring of Surface Water (Rivers and Lakes)	Nationwide Marine Environmental Monitoring
Lao PDR	Public agencies (WEPA 2009) monitor 23 sites monthly (Komany 2008) for DO, salts (Ca, Mg, Na, K, Cl, SO_4), nutrients (NO_2, NO_3, NH_4, PO_4, TP, SiO_2), organic matter: COD. No metals but plans for As, Hg, Cd, Pb. • *Ad hoc* analyses may be carried out (Komany 2008) for: pentachlorobenzene, hexachlorobenzene, heptachlor, endrin, dieldrin, α-HCH, β-HCH, γ-HCH, *p,p'*-DDE, *p,p'*-DDD, *p,p'*-DDT, *o,p'*-DDT, α-endosulfan, α-chlordane, γ-chlordane, pentachloroaniline, quintozene, and diazinon, plus 3 organophosphorous pesticides (mevenphos, dichlorvos, and parathion).	Not applicable.
Cambodia	Ministry of Environment: annual monitoring program with 12 monitoring stations (WEPA 2009). Parameters include water depth, pH, transparency, SS, DO, BOD, TN, TP (EMCZ 2006).	Annual monitoring program with 8 stations. Parameters include water depth, pH, transparency, SS, DO, BOD, TN, TP (EMCZ 2006).

focused on the assessment of the biological effects induced by single chemical/compound and limited to laboratory exposures (Liu and Wang 2007; Ji and Lu 2010). Comprehensive field studies for validation of selected biomarkers are needed. Financial support, laws, and policy related to biomonitoring are lagging behind chemical monitoring. Legislative measures and policy on chemical effluents have been implemented and enforced by authorities at different levels. Regulatory authorities rarely apply biological data (biomarkers or bioindicators) in environmental management.

15.4.3 South Korea

In South Korea, the National Fisheries Research and Development Institute has conducted nationwide regular environmental monitoring since 1972, to assess coastal water quality and its impact on marine flora and fauna. Under this project, trace metals, nutrients, POPs, and other physical/chemical/biological parameters in water, sediment, or biota are being monitored (Wen 2005; Kim 2010). The program does not involve any biomarker-based monitoring, although, since 2002, the government has supported research projects to develop biomarkers, such as physiological and molecular biomarkers for specific target pollutants (e.g., POPs and endocrine disruptors). The Korea Ocean Research and Development Institute has conducted many short-term monitoring and surveys on pollution levels and effects of pollutants on marine organisms such as marine gastropods and bivalves. These research projects were funded by the Ministry of Maritime Affairs and Fisheries, the Ministry of Environment, the Ministry of Science and Technology, and by other government agencies and the private sector (Wen 2005).

15.4.4 Japan

Information on biological monitoring in Japan is scarce. The National Institute of Environmental Studies (NIES), an independent administrative institution closely affiliated with the Ministry of the Environment, Japan (MEJ), has been leading research in global environment monitoring including monitoring and risk assessment of endocrine disruptors and dioxin. Monitoring of water quality is the responsibility of the prefectural governments and local offices of national administrative agencies. Moreover, companies that discharge effluents from specified factories are mandated to measure pollutant levels in their effluents and keep records of the measured findings for government inspection. A total of 271 stations report monthly on environmental standard and supplementary parameters for the first-order rivers. In 1984, the MEJ initiated and conducted a nationwide biomonitoring program for organic pollution of freshwater environments, named the Nationwide Survey of Aquatic Organisms, and based on biological index of macroinvertebrates (Morse et al. 2007). Currently, MEJ is active in developing the CALUX (Chemical Activated Luciferase Gene Expression) assay to monitor water quality. The MEJ also launched a marine monitoring program (the Marine Environment Monitoring Survey) in 1998–2007 based on the "Marine Environment Monitoring Review Guideline" (Environment Agency of Japan 1998). This was a comprehensive survey performed on seawater, sediments, marine organisms, benthic communities, floating plastic wastes, and so forth, not only with respect to coastal waters but also offshore sea areas. The pollution levels of hazardous chemical substances (trace metals, PCBs, dioxins, organotin compounds, and substances covered in the POPs Treaty) were monitored. The levels of PCBs and dioxins, etc., within the bodies of five types of marine organisms (mussels, benthic sharks in coastal areas, squid, cod, and crustaceans in offshore areas) were also monitored (MEJ 2009). The use of census data for biological indicators is still under development.

15.4.5 India

Taking into consideration the mounting pollution pressure along the coast of India, the Government of India has initiated national-level monitoring programs such as Coastal Ocean Monitoring and Prediction System (COMAPS) and Integrated Coastal Mapping and Management (ICMAM) to document the concentrations of chemical constituents in coastal waters and sediments, and changes in biological communities in the coastal area over the years. However, there has been lack of efforts to establish the causal relationship between pollutants and biological effects in the resident species. The Government of India is not involved in any of the international collaboration programs dealing with biomarker research. The National Institute of Oceanography, one of the constituent laboratories of the Council of Scientific and Industrial Research (CSIR) of India, has on its own initiative carried out studies on biomarker-based monitoring (Verlecar et al. 2006). The Central Pollution Control Board (CPCB) is the chief regulatory agency on water quality issues in India. It collaborates with State Pollution Control Boards (SPCBs) and Pollution Control Committees (PCCs) to implement the various policies and regulations that govern water quality management and to monitor the nationwide network of water quality monitoring stations in India. With the assistance of the SPCBs and PCCs, the CPCB has set up a network of freshwater monitoring stations under the National Water Quality Monitoring Programme. At present, the inland water quality-monitoring network is operated under a three-tier program, i.e., Global Environment Monitoring System, Monitoring of Indian National Aquatic Resources System, and Yamuna Action Plan. The Department of Ocean

Development has been managing the COMAPS in collaboration with the Ministry of Environment and Forests since 1991.

15.4.6 Other Asian Countries

In other Asian countries and regions, such as Taiwan, Singapore, Malaysia, the Philippines, Thailand, Indonesia, Vietnam, Lao PDR, Myanmar, and Cambodia, the official guidelines for monitoring water quality in rivers, lakes and reservoirs, and coastal waters mainly rely on the analysis of physicochemical analyses such as dissolved oxygen, biological oxygen demand, pH, electrical conductivity, nutrients, oil, and trace metals, etc. Total coliform and fecal coliform bacteria are often the only biological parameters included (NEA 2002; WEPA 2009; EPA Taiwan 2010a, b). There is no information on the use of biomarker-based monitoring techniques.

In Taiwan, the Department of Environmental Monitoring and Information Management has the responsibility for all national environmental water monitoring for river and reservoir water. Since 2002, coastal waters have been divided into 19 regions and 104 sampling points, which are monitored once a season.

In Malaysia, the Department of Environment (DOE) started its monitoring program in 1978 to establish baselines and detect water quality changes in river water quality. Water samples are collected at regular intervals from designated stations for *in situ* and laboratory analyses (WEPA 2009). The DOE started a marine monitoring program in 1978 for Peninsular Malaysia and in 1985 for Sabah and Sarawak, followed by the Island Marine Water Quality Monitoring Programme in 1998.

In the Philippines, the Environmental Management Bureau (EMB) and its regional offices conduct regular water quality monitoring throughout the country. The EMB has monitored a total of 39 bays and coasts in the Philippines regularly since 1996 (WEPA 2009). Biological parameters such as phytoplankton, zooplankton, and benthos have also been monitored in selected water bodies (DENR-EMB 2007), but biomarkers are not used for monitoring water quality.

In Thailand, the Pollution Control Department, under the direction of Ministry of Natural Resources and Environment, has been monitoring water quality in rivers since 1983 (Boonsoong et al. 2010), and preliminary rapid bioassessment studies on the water quality of streams (using benthos) have been conducted in northern and northeastern Thailand (Morse et al. 2007). However, there is no information on the use of biomarker based monitoring techniques in Thailand.

In Vietnam, water monitoring of surface waters (rivers, ponds, and lakes) at the national level is undertaken by the Ministry of Environment and Natural Resources (MoNRE) and other ministries/departments. The monitoring locations are concentrated mainly in environmental hot spots and sensitive areas. A monitoring station system along the coast of Vietnam in general, and north Vietnam in particular, has been established and is managed by the Ministry of Science, Technology, and Environment to monitor coastal water quality. However, owing to limited budgets, monitoring activities have not covered all of the designated stations and the intensity designed into the program has not been met; a regular and routine water quality monitoring program is only now being established (World Bank 2006).

In Lao PDR, public agencies carry out water quality monitoring, laboratory analysis, and other functions (WEPA 2009). In Myanmar, there is a lack of appropriate monitoring facilities, proper and systematic keeping of records, regular monitoring, and surveillance data for water quality control and basic standards of water quality (WEPA 2009).

15.4.7 Ways Forward

Listed below are the recommended ways forward to implement biomarker-based monitoring in Asia:

1. Educate the general public, government officials, and policy makers in understanding the usefulness and importance of supplementing chemical-based monitoring data with effects-based biomonitoring data for environmental monitoring.
2. Government funding is needed to support research programs led by universities and relevant research institutes/centers on the development and validation of biomarkers for water quality assessment.
3. Local fauna/flora (sentinel species) relevant to subtropical/tropical environment should be identified, selected, and inventoried (database).
4. Overseas protocols for biomonitoring should be optimized for local sentinel species; tolerance values for invertebrates/vertebrate species of different ecoregions should be determined.
5. Standardization and calibration of selected biomarkers specific for each country's freshwater/marine ecosystems.
6. Training for personnel must be provided to overcome the shortage of skilled technicians in government agencies.
7. Provision of equipment/literature resources/protocols with other Asian countries.
8. Establishment of collaborative regional biomarker-based programs for technology transfer and capacity building.

15.5 Conclusions

Although our coverage here of regulatory bodies and their procedures has necessarily been patchy, it can be concluded that the use of biomarkers is starting to appear with some prominence in environmental assessment programs across the world. Furthermore, the use of biomarkers in such programs is being increasingly recommended by official bodies, or at least by quasi-official organizations taking national or international responsibility for the acquisition of biologically relevant data to supplement (or occasionally offer an alternative to) chemical data. Thus, the OSPAR Commission of Western Europe has increasingly over time endorsed the utilization of biomarkers by the Contracting Parties to the OSPAR Convention in environmental assessments of the quality of NW European coastal waters.

The presence of tumors and the like in fish, both marine and freshwater, was among the earliest biological effects of contaminants noted with records now from all three continents addressed here. Ease of observation was clearly an early driving factor. In time, recommended biomarkers have been extended essentially to the lower levels of biological organization in a range of animals including invertebrates, not least because biological effects at these lowest molecular, biochemical, and cytological levels are considered the most sensitive and therefore the most appropriate candidates to be part of any early warning system. These and further biomarkers can now be labeled as indicative, for example, of oxidative stress, genotoxicity, and endocrine disruption, as our knowledge of the biology of biomarkers continues to increase. Similarly the specificity, or indeed generality, of the factors inducing biomarker changes is being increasingly understood, and the application of a suite of biomarkers in a suite of model organisms of known suitability for each biomarker represents a very strong tool in understanding, or even predicting, which toxic contaminant may be present at a high enough bioavailability to cause concern. Perhaps, even more interesting against this background of relevant interpretive knowledge, is the potential for the use of a suite of biomarkers in an environmental assessment program to betray the presence of a high bioavailability of an, as yet, unknown toxic contaminant.

It appears then that the use of biomarkers in environmental assessment internationally will only increase, and rightly so. Although Western nations appear to be leading in the application of biomarkers, the technology is being rapidly exported and developing countries around the world are beginning to follow, as they address their own local contamination problems associated with industrial and economic development.

References

Au, D.W.T. 2004. The application of histo-cytopathological biomarkers in marine pollution monitoring: A review. *Mar. Pollut. Bull.* 48:817–34.

Boonsoong, B. et al. 2010. An implementation plan for using biological indicators to improve assessment of water quality in Thailand. *Environ. Monit. Assess.* 165:205–15.

Collier, T.K. et al. 1995. A field evaluation of cytochrome P4501A as a biomarker of contaminant exposure in three species of flatfish. *Environ. Toxicol. Chem.* 14:154–62.

Collier, T.K., B.F. Anulacion, and B.D. Bill. 1998. Hepatic CYP1A in winter flounder (*Pleuronectes americanus*) along the Northeast coast: Results from the National Benthic Surveillance Project. *Mar. Pollut. Bull.* 37:86–91.

Collier, T.K. et al. 2007. Toxic chemical contaminants and Puget Sound. In: *Proc. 2007 Georgia Basin Puget Sound Research Conference.* Available online at: http://www.engr.washington.edu/epp/psgb/2007psgb/2007proceedings/oral_abstracts.html#12E.

CPCB. 2008a. Central Pollution Control Board, Ministry of Environment & Forests, India. *Water.* Available at http://www.cpcb.nic.in/water.php. Accessed January 10, 2012.

CPCB. 2008b. Central Pollution Control Board, Ministry of Environment & Forests, India. *Coastal Ocean Monitoring and Prediction Systems.* Available at http://www.cpcb.nic.in/comaps.php. Accessed January 10, 2012.

DENR-EMB. 2007. Department of Environment and Natural Resources, Environmental Management Bureau, the Philippines. *National Water Quality Status Report (2001–2005).* Available at http://www.emb.gov.ph/wqms/2001-2005%20NWQSR/NWQSR%20-%20Body.pdf. Accessed January 10, 2012.

DENR-EMB. 2008. Department of Environment and Natural Resources, Environmental Management Bureau, the Philippines. *Water Quality Monitoring Manual. Volume I—Manual on Ambient Water Quality Monitoring.* Available at http://www.emb.gov.ph/wqms/policies/ambient_14aug08.pdf. Accessed January 10, 2012.

Dien, C.V., and T.V. Thuoc. 2003. An overview of current activities relating to phytoplankton monitoring in coastal waters of North Vietnam. Workshop on Red Tide Monitoring in Asian Coastal Waters. University of Tokyo, Tokyo, Japan, 10–12 March 2003. pp. 69–73. Available at http://fol.fs.a.u-tokyo.ac.jp/rtw/TOP/EXabst/020ChuVanThuoc.pdf. Accessed January 10, 2012.

DOE. 2010a. Department of Environment, Ministry of Natural Resources and Environment, Malaysia. *River Water Quality Monitoring.* Available at http://www.doe.gov.my/portal/water-marine-river-water/river-water-quality-monitoring/. Accessed January 10, 2012.

DOE. 2010b. Department of Environment, Ministry of Natural Resources and Environment, Malaysia. *Marine and Water Quality Monitoring*. Available at http://www.doe.gov.my/portal/water-marine-marine-island-marine-water/marine-and-water-quality-monitoring/. Accessed January 10, 2012.

EMCZ. 2006. Environmental Management in the Coastal Zone, Ministry of Environment, Cambodia. Water Quality in Coastal Rivers and Near Shore Coastal Waters of Cambodia, Second Annual Monitoring Report, May 2005–April 2006, 102 pp. Available at http://www.czmcam.org/content/documents/Report_WQM_CZM_Final%20_ENG-KH_.pdf. Accessed January 10, 2012.

EPA Taiwan. 2010a. Environmental Protection Administration Executive Yuan, Taiwan. *Environmental Water Quality Information, Monitoring Plan*. Available at http://wq.epa.gov.tw/WQEPA/Code/Business/Plan.aspx?Languages=en. Accessed December 28, 2011.

EPA Taiwan. 2010b. Environmental Protection Administration Executive Yuan, Taiwan. *Environmental Water Quality Information: Water Monitoring Introduction, Monitoring Plan*. Available at http://wq.epa.gov.tw/WQEPA/Code/Business/Plan.aspx?Languages=en. Accessed January 10, 2012.

Fang, J.K.H. et al. 2009a. The use of muscle burden in rabbitfish *Siganus oramin* for monitoring polycyclic aromatic hydrocarbons and polychlorinated biphenyls in Victoria Harbour, Hong Kong and potential human health risk. *Sci. Total Environ.* 407:4327–32.

Fang, J.K.H. et al. 2009b. The use of physiological indices in rabbitfish *Siganus oramin* for monitoring of coastal pollution. *Mar. Pollut. Bull.* 58:1229–44.

Fang, J.K.H. et al. 2009c. Concentrations of polycyclic aromatic hydrocarbons and polychlorinated biphenyls in green-lipped mussel *Perna viridis* from Victoria Harbour, Hong Kong and possible human health risk. *Mar. Pollut. Bull.* 58:601–34.

Fang, J.K.H. et al. 2010. Ethoxyresorufin-*O*-deethylase enzyme activities and accumulation of secondary/tertiary lysosomes in rabbitfish *Siganus oramin* as biomarkers for xenobiotic exposures. *Sci. Total Environ.* 408:4833–40.

Fournie, J.W., J.K. Summers, and S.G. Weisberg. 1996. Prevalence of gross pathological abnormalities in estuarine fishes. *Trans. Am. Fish. Soc.* 125:581–90.

Fournie, J.W. et al. 2001. Utility of splenic macrophage aggregates as an indicator of fish exposed to degraded environments. *J. Aquat. Anim. Health* 13:105–16.

Gaitonde, D. et al. 2006. Acetylcholinesterase activities in marine snail (*Cronia contracta*) as a biomarker of neurotoxic contaminants along the Goa coast, West coast of India. *Ecotoxicology* 15:353–8.

Handy, R., A. Jha, and M. Depledge. 2002. Biomarker approaches for ecotoxicological biomonitoring at different levels of biological organization. In: Burden, F.R., McKelvie, I., Förstner, U., Guenther, A. (Eds.), *Environmental Monitoring Handbook*. New York: McGraw-Hill, 1100 pp.

Hinck, J.E. et al. 2006. Environmental contaminants and biomarker responses in fish from the Columbia River and its tributaries: Spatial and temporal trends. *Sci. Total Environ.* 366:549–78.

Hinck, J.E. et al. 2009. Widespread occurrence of intersex in black bass (*Micropterus* spp.) from U.S. rivers, 1995–2004. *Aquat. Toxicol.* 95:60–70.

HKEPD. 2005a. Hong Kong Environmental Protection Department. *River Water Quality*. Available at http://www.epd.gov.hk/epd/english/environmentinhk/water/river_quality/rwq_monitoring.html. Accessed January 10, 2012.

HKEPD. 2005b. Hong Kong Environmental Protection Department. *Marine Water Quality*. Available at http://www.epd.gov.hk/epd/english/environmentinhk/water/marine_quality/mwq_monitoring.html. Accessed January 10, 2012.

HKEPD. 2007. Hong Kong Environmental Protection Department. *Marine Water Quality, Biological Indicator Monitoring in Hong Kong*. Available at http://www.epd.gov.hk/epd/english/environmentinhk/water/marine_quality/bio_detail.html. Acessed December 28, 2011.

Hom, T. et al. 2008. Assessing seafood safety in the aftermath of Hurricane Katrina. In: McLaughlin, K.D. (Ed.), *Mitigating Impacts of Natural Hazards on Fishery Ecosystems*, 73–94. American Fisheries Society Symposium 64, 446 pp.

Hutomo, M. 1993. Development of marine environmental quality criteria in Indonesia. In: Watson, D., Setiapermana, S., and Vigers, G. (Eds.). *Marine Environmental Quality: Perspectives on ASEAN Criteria and Monitoring. Proceedings of the first ASEAN–Canada technical Planning Workshop on Marine Science*, Jakarta, Indonesia, 16–21 November 1992, Volume 1. Available at http://www.coremap.or.id/downloads/1334.pdf. Accessed January 16, 2012.

ICES Advice. 2010. Further development of guidance on integrated monitoring and assessment of chemicals and biological effects. Book 1.

ICES Advice. 2011. Further development of guidance on integrated monitoring and assessment of chemicals and biological effects. Book 1.

Jena, K.B., X.N. Verlecar, and G.B.N. Chainy. 2009. Application of oxidative stress indices in natural populations of *Perna viridis* as biomarker of environmental pollution. *Mar. Pollut. Bull.* 58:107–13.

Ji, Y., and G.H. Lu. 2010. Biomarker technology for early warning of water pollution risks. *J. Hohai Univ. (Nat. Sci.)* 38:258–62 (in Chinese).

Johnson, L.L., T.K. Collier, and J.E. Stein. 2002. An analysis in support of sediment quality thresholds for polycyclic aromatic hydrocarbons (PAHs) to protect estuarine fish. *Aquat. Conserv.: Mar. Fresh. Ecosyst.* 12:517–38.

Jung, J.-H. et al. 2011. Biomarker responses in pelagic and benthic fish over 1 year following the *Hebei Spirit* oil spill (Taean, Korea). *Mar. Pollut. Bull.* 62:1859–66.

Kim, H.G. 2010. An overview on the occurrences of Harmful Algal Blooms (HABs) and mitigation strategies in Korean coastal waters. In: Ishimatsu, A., and Lie, H.J. (Eds.). *Coastal Environmental and Ecosystem Issues of the East China Sea, Nagasaki University Major Research Project Restoration of Marine Environment and Resources in East Asia*, 121–131. Tokyo, Japan: TERRAPUB and Nagasaki University. Available at http://www.terrapub.co.jp/onlineproceedings/fs/nu/pdf/nu2010121.pdf. Accessed December 28, 2011.

Komany, S. 2008. *Environmental Water Quality Standards and Water Monitoring System in Lao PDR*. Available at http://www.wepa-db.net/pdf/0810forum/paper18.pdf. Accessed January 10, 2012.

Krahn, M.M. et al. 1984. Determination of metabolites of xenobiotics in the bile of fish from polluted waterways. *Xenobiotica* 14:633–46.

Krahn, M.M. et al. 2005. Analysis of bile of fish collected in coastal waters of the Gulf of Mexico potentially affected by Hurricane Katrina to determine recent exposure to polycyclic aromatic compounds (PACs). NOAA Fisheries, Northwest Fisheries Science Center, Seattle, WA. Report submitted to Steven Murawski of NOAA Fisheries on 9/26/05.

Lam, P.K.S., and J.S. Gray. 2003. The use of biomarkers in environmental monitoring programmes. *Mar. Pollut. Bull.* 46:182–6.

Liu, H.F., and F. Wang. 2007. Progress of research on molecular biomarkers for monitoring contamination level of water system. *J. Anhui Agric. Sci.* 35:5502–6 (in Chinese).

Luoma, S.N., and P.S. Rainbow. 2008. *Metal Contamination in Aquatic Environments. Science and Lateral Management*. Cambridge: Cambridge University Press.

Malins, D.C. et al. 1985. Toxic chemicals in sediments and biota from a creosote-polluted harbor: Relationships with hepatic neoplasms and other hepatic lesions in English sole (*Parophrys vetulus*). *Carcinogenesis* 6:1463–9.

MEJ. 2009. Ministry of the Environment, Japan. *Present Status of Marine Pollution in the Sea around Japan as based on data from Marine Environment Monitoring Survey results Fiscal Years 1998–2007*. Tokyo, Japan: Global Environmental Issues Division, Global Environmental Bureau, Ministry of the Environment, Japan. 23 pp. Available at http://www.env.go.jp/water/kaiyo/monitoring/status_report/en-1.pdf and http://www.env.go.jp/water/kaiyo/monitoring/status_report/en-2.pdf. Accessed January 10, 2012.

MEJ. 2011. Ministry of the Environment, Japan. *Environmental Quality Standards of Japan-Water.* Available at http://www.env.go.jp/en/water/wq/wp.pdf. Accessed January 12, 2012.

Monirith, I. et al. 2003. Asia-Pacific mussel watch: Monitoring contamination of persistent organochlorine compounds in coastal waters of Asian countries. *Mar. Pollut. Bull.* 46:281–300.

Morse, J.C. et al. 2007. Freshwater biomonitoring with macroinvertebrates in East Asia. *Front. Ecol. Environ.* 5:33–42.

Myers, M.S., L.D. Rhodes, and B.B. McCain. 1987. Pathologic anatomy and patterns of occurrence of hepatic neoplasms, putative preneoplastic lesions, and other idiopathic hepatic conditions in English sole (*Parophrys vetulus*) from Puget Sound, Washington. *J. Natl. Cancer Inst.* 78:333–63.

Myers, M.S. et al. 1998. Toxicopathic hepatic lesions as biomarkers of chemical contaminant exposure and effects in marine bottomfish species from the Northeast and Pacific coasts, USA. *Mar. Pollut. Bull.* 37:92–113.

Myers, M.S., L.L. Johnson, and T.K. Collier. 2003. Establishing the causal relationship between polycyclic aromatic hydrocarbon (PAH) exposure and hepatic neoplasms and neoplasia-related liver lesions in English sole (*Pleuronectes vetulus*). *Hum. Ecol. Risk Assess.* 9:67–94.

Myers, M.S. et al. 2008. Improved flatfish health following remediation of a PAH-contaminated site in Eagle Harbor, Washington. *Aquat. Toxicol.* 88:277–88.

NEA. 2002. National Environment Agency, Singapore. *Water Pollution.* Available at http://app2.nea.gov.sg/topics_waterpollution.aspx. Accessed December 28, 2011.

OSPAR Agreement 2008–09. JAMP guidelines for contaminant-specific biological effects. OSPAR Commission, Monitoring guidelines Ref N° 2008–9.

OSPAR Commission 2007. Background document on biological effects monitoring techniques. Assessment and Monitoring Series.

PCD. 2004a. Pollution Control Department, Ministry of Natural Resources and Environment, Thailand. *Water Quality Standard-Surface Water Quality Standards.* Available at http://www.pcd.go.th/info_serv/en_reg_std_water05.html#s3. Accessed January 16, 2012.

PCD. 2004b. Pollution Control Department, Ministry of Natural Resources and Environment, Thailand. *Water Quality Standard-Coastal Water Quality Standards.* Available at http://www.pcd.go.th/info_serv/en_reg_std_water02.html#s2. Accessed January 16, 2012.

Peck, K.A. et al. 2011. Development of an enzyme-linked immunosorbent assay for quantifying vitellogenin in Pacific salmon and assessment of field exposure to environmental estrogens. *Environ. Toxicol. Chem.* 30:477–86.

Pierce, K.V., B.B. McCain, and S.R. Wellings. 1978. Pathology of hepatomas and other liver abnormalities in English sole (*Parophrys vetulus*) from the Duwamish River estuary, Seattle, Washington. *J. Natl. Cancer Inst.* 60:1445–53.

POMRAC. 2009. Pollution Monitoring Regional Activity Centre, Northwest Pacific Action Plan (NOWPAP). *Regional Overview on River and Direct Inputs of Contaminants into the Marine and Coastal Environment in NOWPAP Region with Special Focus on the Land Based Sources of Pollution.* POMRAC Technical Report No 7. Vladivostok, Russian Federation: POMRAC. 80 pp. Available at ftp://pomrac.dvo.ru/Publ/POMRAC_RDI_2009.pdf. Accessed January 10, 2012.

Ramu, K. et al. 2007. Asian mussel watch program: Contamination status of polybrominated diphenyl ethers and organochlorines in coastal waters of Asian countries. *Environ. Sci. Technol.* 41:4580–6.

Sanchez, W., and J.M. Porcher. 2009. Fish biomarkers for environmental monitoring within the Water Framework Directive. *Trends Anal. Chem.* 28:150–8.

Sanchez, W. et al. 2011. Adverse effects in wild fish living downstream from pharmaceutical manufacture discharges. *Environ. Int.* 37:1342–8.

Sarkar, A. et al. 2008. Evaluation of impairment of DNA integrity in marine gastropods (*Cronia contracta*) as a biomarker of genotoxic contaminants in coastal water around Goa, West coast of India. *Ecotox. Environ. Saf.* 71:473–82.

Schmitt, C.J., and G.M. Dethloff. 2000. Eds. *Biomonitoring of Environmental Status and Trends (BEST) Program: Selected Methods for Monitoring Chemical Contaminants and Their Effects in Aquatic Ecosystems.* U.S. Geological Survey, Biological Resources Division, Columbia, MO: Information and Technology Report USGS/BRD-2000— 0005. 81 pp.

Schmitt, C.J. et al. 2005. Environmental contaminants and biomarker responses in fish from the Rio Grande and its US tributaries: Spatial and temporal trends. *Sci. Total Environ.* 350:161–93.

Shim, J.W. et al. 2000. Imposex in the rock shell, *Thais clavigera*, as evidence of organotin contamination in the marine environment of Korea. *Mar. Environ. Res.* 49:435–51.

Shim, J.W. et al. 2005. Assessment of butyl- and phenyltin pollution in the coastal environment of Korea using mussels and oysters. *Mar. Pollut. Bull.* 51:922–31.

Son, N.M. 2008. *Report on Water quality Component.* Available at http://www.vnwatersectorreview.com/files/Water_quality_EN.pdf. Accessed January 11, 2012.

State of the Sound 2009. 2010. Puget Sound Partnership, Olympia, WA, USA. Available online at: http://www.psp.wa.gov/sos2009.php. Accessed August 10, 2012

Sudaryanto, A. et al. 2002. Asia-Pacific mussel watch: Monitoring of butyltin contamination in coastal waters of Asian developing countries. *Environ. Toxicol. Chem.* 21:2119–30.

Syafrul, H. 2009. *Water Quality Monitoring in Indonesia.* Available at http://www.wepa-db.net/pdf/0910indonesia/f.pdf. Accessed January 10, 2012.

Verlecar, X.N. et al. 2006. Marine pollution detection through biomarkers in marine bivalves. *Curr. Sci.* 91:1153–7.

Wen, Q. 2005. *First Progress Report on Drafting Regional Pollution Monitoring Guidelines.* Second Meeting of the Regional Working Group for the Pollution Component, UNDP/GEF Project entitled "Reducing Environmental Stress in the Yellow Sea Large Marine Ecosystem." Busan, Republic of Korea, 7–10 November 2005. UNDP/GEF/YS/RWG-P.2/6 rev. 1. Available at http://www.yslme.org/doc/rwg2p/6-monit%20guideline.pdf. Accessed December 29, 2011.

WEPA. 2009. Water Environment Partnership in Asia. *Outlook of Water Environmental Management Strategies in Asia.* Kanagawa, Japan: Institute for Global Environmental Strategies (IGES), Ministry of the Environment, Japan. 53 pp. Available at http://www.wepa-db.net/pdf/outlook.pdf. Accessed December 26, 2011.

World Bank. 2006. *Vietnam Environment Monitor 2006: Water Quality in Vietnam with a Focus on the Cau, Nhue-Day and Dong Nai River Basins.* Available at http://www.sxsh.vn/en-US/Files/Details.aspx?ID=167. Accessed January 10, 2012.

16

Conclusions: Biomarkers in Environmental Risk Assessment

Claude Amiard-Triquet, Jean-Claude Amiard, and Philip S. Rainbow

CONTENTS

During the second half of the 20th century, a number of dramatic events (major oil spills associated with the wrecks of large tankers, contamination of the food chain by industrial effluents with consequences on human health as experienced in Minamata Bay, Japan, depletion of populations of emblematic species such as the bald eagle attributed to the uncontrolled use of the organochlorine pesticide DDT, etc.) occurred, which led to an increased awareness and concern in society at large about environmental pollution. To answer the newly emerging demands of society, many laws and regulations were adopted, and regulatory bodies were created in order to ensure their implementation. Environmental chemistry and ecotoxicology played major roles in the development of tools aimed at the assessment of environmental quality, itself a precondition for environmental management. Retrospective risk assessments are performed when sites have potentially been impacted in the past. When they show a degradation of environmental quality, the restoration of degraded habitats and ecosystems must be addressed. Prospective or predictive risk assessments aim to assess future risks from releases of chemicals into the environment.

In the United States, the federal Toxic Substances Control Act gives the Environmental Protection Agency (USEPA) the authority to regulate, and even ban, the manufacture, use, and distribution of both new and existing chemicals. In Europe, significant improvement has occurred recently with the adoption of a new chemical policy, REACH (Registration, Evaluation, and Authorization of CHemicals).

16.1 How Can We Assess Environmental Quality?

Environmental assessment is classically based on a triad of analyses including (1) chemical analyses in environmental matrices; (2) bioassay methods for the evaluation of toxic effects of contaminants typically in the laboratory; (3) biological responses in the field at the level of community structure. This approach was particularly well developed as the Sediment Quality Triad (Chapman 1990) since, in the aquatic environment, this compartment is the main store for most of the contaminants entering ecosystems. It is extremely difficult to establish a cause–effect relationship when effects result from exposure to mixtures, a situation, however, which is that most commonly encountered in the field.

16.1.1 Chemical Monitoring

The pros and cons of measurements of chemicals in environmental matrices (water, sediment, biota) have been recently reviewed (Amiard-Triquet and Rainbow 2009; Knoery and Claisse 2010). Chemical analyses mainly provide information on chemicals that are perceived to be relatively easy to analyze such as trace metals, DDT and its metabolites, γHCH, αHCH, some congeners of polychlorinated biphenyls (PCBs), and some individual polycyclic aromatic hydrocarbons (PAHs). Among the roughly 100,000 chemicals that have at least limited toxicity information available, only 1150 unique compounds are measured in waters (Egeghy et al. 2012). However, complex mixtures are present in certain environments (such as major estuaries across the world), including many classes of compounds (many persistent organic pollutants and emerging contaminants such as pharmaceuticals and nanoparticles) that are not yet accessible to analysis or are extremely expensive to analyze. Moreover, the physicochemical parameters of the medium interfere with contaminants, thus modifying their characteristics and their bioavailability (Figure 16.1). Analyzing contaminants in biota rather than in water or sediment gives access to the bioavailable fraction of the pollutant load in the medium that is the component that has the potential to induce toxic effects.

16.1.2 Bioindicator Indices

One of the main objectives of aquatic ecotoxicological research is the assessment of contaminant impacts on populations, communities, and ecosystems in support of legislation that has been adopted on a worldwide scale to determine the ecological integrity of surface waters [United States' Clean Water Act, 1972; European Community Water Framework Directive (WFD), 2000]. Many bioindicator indices (at the level of populations and communities, see Chapter 7) have been proposed, but it is difficult to attribute effects at these high levels of biological organization to the presence of contaminants rather than to other stress factors (e.g., habitat alteration, overfishing), particularly in highly dynamic and complex systems

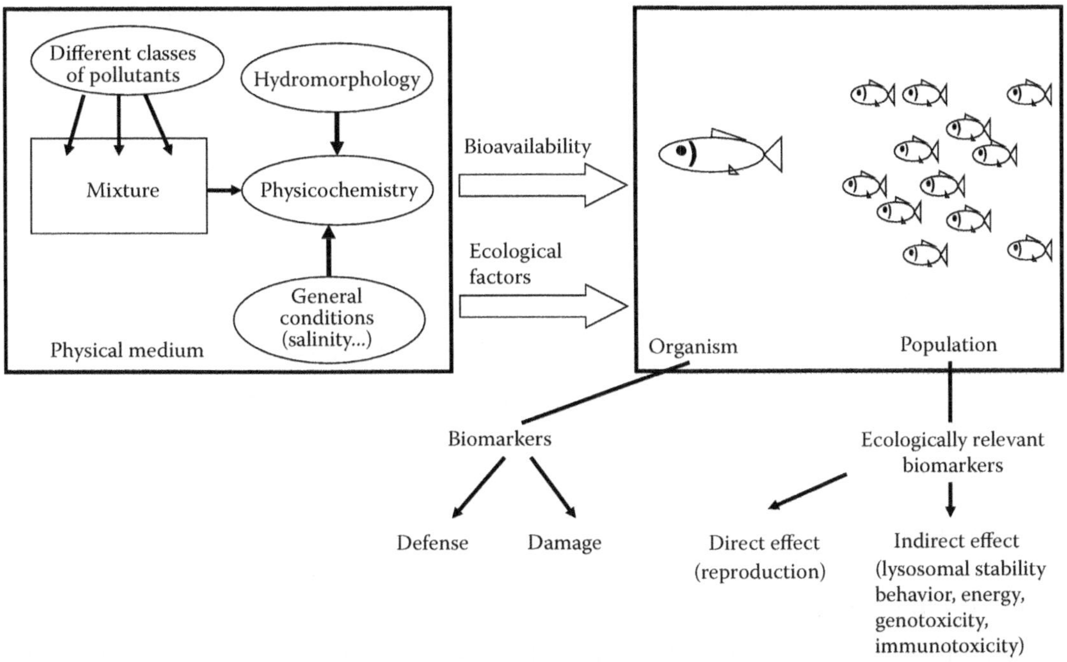

FIGURE 16.1
Methods for quality assessment of aquatic environments.

characterized by important spatiotemporal changes in temperature, pH, dissolved O_2, redox potential, hydrodynamics, and sedimentary processes, plus salinity in the case of estuaries. Bioindicators at supra-individual levels of biological organization are long-term responses that often appear when the pollution impact has been at work for years (Figure 1.3). This lack of both specificity and precociousness is a serious restraint to their usefulness for environmental protection even though their use is imposed in regulations such as the WFD.

16.1.3 Biomarkers

In contrast to these high-level bioindicators, individual and infra-individual responses used as biomarkers have attracted considerable interest as they are able to constitute early and sensitive tools for the assessment of environmental health. The methodology of biomarkers has given rise to a rocketing number of publications, from about 100 in 1991 to about 3000 in 2007 (Jemec et al. 2010). These authors have reviewed the criteria for the use of biomarkers in environmental pollution studies:

- The relationship between the level and the duration of exposure and the biomarker response must be established.
- The effect of confounding factors, both intrinsic and extrinsic, must be well understood. Infra-individual biomarkers should be detected at lower concentrations of exposure than individual responses.
- The changes that occur at a given level of organization must be indicative of changes at the next level (infra-individual toward individual, individual toward supra-individual).

These authors also list practical considerations (economic, easy to perform, reliable). However, biomarkers have been criticized for their lack of specificity since they can respond to several classes of contaminants and to nonchemical stressors (confounding factors) and their lack of ecological relevance (Forbes et al. 2006). In order to mitigate these disadvantages, different strategies are proposed in the next sections for a sound use of biomarkers for environmental quality assessment as well as predictive risk assessment.

16.2 Using Biomarkers to Assess Environmental Quality

Although biomarkers are being increasingly studied (Jemec et al. 2010) and used in biomonitoring (Chapter 15), a number of serious problems still need to be solved. An important one is our ability to distinguish between normal fluctuations of biological responses used as biomarkers and fluctuations attributable to stress, particularly chemical stress induced by pollutants (Figure 1.6).

16.2.1 Confounding Factors

Confounding factors include both extrinsic factors (temperature, pH, redox status, hardness in freshwaters, salinity in estuarine waters, food availability, parasitism) and intrinsic factors (weight/size/age, sex, reproductive status, tolerance acquired by populations chronically exposed to contaminants in their environment). Seasonal fluctuations that have been evoked as influencing most of the biomarkers examined in the different chapters of this book include both extrinsic and intrinsic factors, among which the most important are certainly temperature, food availability, and reproductive status, but also the pollutant inputs associated with different river flows and applications of pesticide treatments.

Despite this limitation, most of the existing information on the influence of temperature on biomarker responses has been collected as part of field surveys covering temporal variations. However, according to the fundamentals of biochemistry, temperature is recognized as influencing the activity of enzymes by changing the protein conformation and their catalytic efficiency or binding capacity. This is important for the methodology of biomarkers since many subindividual markers are based on enzymatic activities. In estuarine and coastal environments, salinity is also an important source of variability for numerous biomarkers, as exemplified in the endobenthic invertebrates *Scrobicularia plana* and *Nereis diversicolor* (Kalman et al. 2010; Fossi Tankoua et al. 2011 and references quoted by these authors). The influence of the size/weight/age complex on biomarker responses to toxicants is at least partly the consequence of differences in uptake of contaminants, which are well documented after many previous studies on contamination biomonitors such as the bivalves used in Mussel Watch programs (NAS 1980). It is also evident that biomarkers of contaminants that interfere specifically with development or reproduction (Chapters 8 and 9) are strongly influenced by the life history stage (larvae, embryos, juveniles, adults) or sex and sexual status of the relevant organism. Molting, which is a key period in the life span of arthropods, must be taken into account for a sound interpretation of biological responses in species belonging to this zoological group (Chapter 8).

Other factors have been less documented. However, the influence of pH and oxygenation on biomarker responses was shown for several enzyme activities in the freshwater bivalves *Corbicula fluminea* and *Anodonta cygnea* (Vidal et al. 2002; Robillard et al. 2003), whereas reproduction biomarkers (% females carrying a dead brood) in the amphipod *Monoporeia affinis* were related to oxygen concentration of the bottom water (Eriksson Wiklund and Sundelin 2004). The importance of parasitism in ecotoxicological studies is highlighted by changes affecting biomarkers of oxidative stress in freshwater mussels *Dreissena polymorpha* and marine mussels *Mytilus galloprovincialis* (Minguez et al. 2009; Canesi et al. 2010) or the impact of the parasitic acanthocephalan worm *Echinorhynchus truttae* on the feeding rate of the amphipod *Gammarus pulex* (Fielding et al. 2003). Infection with another acanthocephalan parasite *Acanthocephalus tehlequahensis* causes a significant and direct energetic cost to its freshwater isopod intermediate host *Caecidotea communis* at both individual and population levels (Lettini and Sukhdeo 2010).

The sampling matrix must also be carefully chosen. For instance, Bechshøft et al. (2011) have shown that retinol was not uniformly distributed in the kidney of polar bears *Ursus maritimus*. Many different tissues have been used for the determination of acetylcholinesterase (AChE) activity in different species (Chapter 4). However, after a comparison between eyes, brain, gills, digestive tract, and muscles of the common prawn *Palaemon serratus*, Frasco et al. (2010) recommend the use of eyes since this tissue expresses the highest levels of ChE while possessing an enzyme that shows the properties of a vertebrate AChE. In the case of genotoxic biomarkers, mussels are useful sentinel species in PAH-contaminated areas, provided DNA adducts are measured in specific tissues (gills or digestive gland), as opposed to whole tissues that may dilute the adduct pool and weaken the signal (Chapter 13). In the gammarid amphipod crustacean *Gammarus fossarum* caged upstream and downstream of wastewater treatment plant effluent output, spermatozoa appeared to be more susceptible to accumulating DNA damage than hemocytes and oocytes (Lacaze et al. 2011a). Studying genotoxic damage in polychaetes, Lewis and Galloway (2008) demonstrated that different cell types within the same species and different species within the polychaetes show significantly different responses.

It is important to consider the confounding roles of all these factors before using any biomarker in any organ or cell of any species in the medium under investigation. However, it must be duly noted that studies highlighting the influence of certain biological or ecological conditions on particular biomarkers also reveal that other biomarkers are robust, not being influenced or only marginally by confounding factors.

Important differences in responsiveness to both contamination factors and natural factors appear when comparing different species. Ethoxyresorufin O-deethylase (EROD), and more recently following the development of molecular biology, CYP genes are frequently used as core biomarkers of exposure to organic xenobiotics. However, a comparison of the responses of tomcod *Microgadus tomcod* and killifish *Fundulus heteroclitus* submitted to AH exposure in North American estuaries, underlines that, depending on species and site, responses may be different with the consequent possibility of false negatives or positives when interpreting the results of a biomonitoring program (Berthet et al. 2011). In a comparison of biomarker responses in two endobenthic invertebrates, it appears that the annelid *Nereis diversicolor* is less sensitive than the bivalve *Scrobicularia plana* to the effects of size and salinity differences (Table 16.1). In addition, this annelid is more responsive for a number of enzyme activities, as also underlined by Solé et al. (2009), whereas it is not efficient at revealing metal exposure by use of metallothionein (MT) concentrations (Poirier et al. 2006). For behavioral tests, in the case of petroleum hydrocarbons, bivalves (*Mercenaria mercenaria*, Olla et al. 1983; *Protothaca staminea*, Pearson et al. 1981) seem more responsive than annelids (*Nereis virens*, Olla et al. 1984), but, at a high degree of mixed pollution such as in the Seine estuary, France, delayed burrowing was also observed in *N. diversicolor* (Mouneyrac et al. 2009). Even within the same taxon, Hagger et al. (2010) report

TABLE 16.1

Responsiveness of Different Classes of Biomarkers Determined in the Bivalve *Scrobicularia plana* (Fossi Tankoua et al. 2011) and the Annelid *Nereis diversicolor* (Kalman et al. 2010) to Estuarine Contamination and to Confounding Factors

	Scrobicularia plana			*Nereis diversicolor*		
	Contamination	**Size Class**	**Salinity**	**Contamination**	**Size Class**	**Salinity**
Biochemical Markers						
MT	Increase (46–68%)	NS	23% > at lower S‰	Irrelevant[a]	ND	ND
LDH	Unclear	NS	133% > at higher S‰	Increased	NS	> at higher S‰
GST	Increase (25–94%)	56% > in small specimens	76% > at higher S‰	Increased	NS	NS
AChE	Decrease (38–53%)	40% > in big specimens	53% > at lower S‰	Decreased	NS	NS
CMCase[b]	Unclear	NS	NS	Decreased	NS	> at higher S‰
CMCase *Crystalline style*	Unclear	NS	100% > at higher S‰			
Amylase[b]	Unclear	NS	NS	Decreased	NS	NS
Amylase *Crystalline style*	Unclear	NS	133% > at higher S‰			
Physiological Markers						
Condition index	NS	> in small specimens	NS	ND	ND	ND
Gonadosomatic index	NS	NS	NS	ND	ND	ND
Water content	NS	ND	2% > at lower S‰	ND	NS	NS
Glycogen *Gonad*	77% decrease	192% > in small specimens	NS			
Glycogen[b]	NS	550% > in small specimens	NS	Decreased	> in larger specimens	NS
Lipids	NS	NS	NS	Decreased	> in larger specimens	NS
Proteins	NS	NS	NS	Decreased	Unclear	NS or > at higher S‰
Behavioral Markers						
Burrowing speed	Decreased	> in small specimens	> at higher S‰	Decreased	NS	NS
Egestion rate	ND	ND	ND	Decreased	> in larger specimens	> at higher S‰
Post-feeding rate	ND	ND	ND	Decreased	NS	Moderate effect

Note: ND, not documented; NS, no significant difference.

[a] Poirier et al. (2006).

[b] In the digestive gland of *S. plana*; in the whole tissues of *N. diversicolor*.

that the feeding rate exhibits different seasonal patterns in different bivalve species, the mussel *Mytilus edulis*, the cockle *Cerastoderma edule*, and the Venus clam *Chamelea gallina*.

16.2.2 Active Biomonitoring

Active biomonitoring (Chapter 2) provides an opportunity for controlling biotic factors, allowing the selection of standard organisms (homogeneous size, sex, reproductive status) with similar previous histories (specimens produced by aquaculture or collected from a reference site). Such experimental procedures are particularly applicable to filter-feeders that access their food in the water column, especially bivalves, which are commonly used for biomonitoring in Mussel Watch–type programs. However, it is important to know how caging itself affects biological responses to chemical pollution, particularly through its effects on food availability as discussed in Chapter 2, and also, in the case of fish, the effect of caging densities (Oikari 2006). In cases where invertebrates have been caged, the question of food availability has received a number of relevant responses such as the provision of rocks covered with biofilm introduced into containers with the mudsnail *Potamopyrgus antipodarum* (Gust et al. 2010), an *ad libitum* supply of alder leaves for amphipods *Gammarus fossarum* (Forrow and Maltby 2000; Coulaud et al. 2011), or the presence of a significant volume of sediment from the site of origin for polychaetes *Arenicola marina* (Ramos-Gómez et al. 2011).

In addition, certain biomarkers can respond to stress associated with the transplantation process (handling stress, anoxia, oxidative burst), as for total oxyradical scavenging capacity responses in mussels *Mytilus galloprovincialis* (Camus et al. 2004). In fish, the induction of plasma cortisol has been observed in response to general stress such as that encountered in handling and capture (Hontela 2000). In addition, stress induction during sampling requires particular attention as it has been demonstrated to strongly affect immune functions (Chapter 6).

Despite these limitations, caging has been widely used in recent years for biomonitoring based on biomarkers determined in many species belonging to different taxa such as gastropods (e.g., Gust et al. 2010), bivalves (Chapter 2), crustaceans (e.g., Lacaze et al. 2011a), and fish (e.g., Brammell et al. 2010; Ji et al. 2010; Klobučar et al. 2010; Kerambrun et al. 2011).

16.2.3 Reference Site

As stated by Amiard-Triquet and Rainbow (2009), in our present state of knowledge, it is unacceptable to study the ecotoxicological status of an aquatic medium without comparison against a reference site. Such a reference site must be chosen to be as close as possible to the site under investigation, in terms of temperature, granulometry, organic content of sediment, etc., in order to mitigate the importance of confounding factors. In freshwater environments, an appropriate strategy is often based on an upstream/downstream procedure. In estuarine areas, the reference site needs to be chosen to be as close as possible to the first estuary in terms of salinity regime. In practice, it will probably be impossible to reach perfect agreement between estuaries, but every attempt must be made to assess the effects of any such differences where possible. Even this may prove impossible, for example, in the case of differences in hydrodynamic conditions, but it remains necessary to make a careful assessment of these differences. Even paired comparisons in freshwater environments between stations upstream/downstream from identified point-sources of pollution, which are based on the assumption that physicochemical conditions are similar between stations except for levels of bioavailable toxic compounds, are not totally satisfactory. The addition of more than one reference site into any comparative study (Hanson et al. 2010), however superficially attractive, has significant resource implications (Amiard-Triquet and Rainbow 2009). An alternative to the use of multiple reference sites may be to use historical data (Hanson et al. 2010). However, special attention must be paid to the fact that historical data may include temporal variations such as those due to global warming or increasing input of emerging contaminants.

16.2.4 Use of Modeling to Establish Baseline Values of Biomarkers

Is it possible to be free from the need for a reference site(s) by determining a baseline for comparison, as for example, for chemical concentrations in bivalves used in Mussel Watch program?

OSPAR documents prepared with assistance of the International Council for the Exploration of the Sea provide a number of baseline values determined in samples from reference sites such as for MT and EROD in fish liver, δ-aminolevulinic acid dehydratase in fish blood, and lysosomal stability (OSPAR Commission 2007). For vitellogenin (VTG) in male fish or Scope for Growth (SfG), provisional background values are indicated. More recently (OSPAR 2011), the question of baseline values has been examined for AChE activity, frequency of micronuclei, and DNA damage determined by using the Comet assay.

It does seem preferable to assess the normal physiological ranges of biomarkers at reference sites. Accordingly, Xuereb et al. (2009a) determined a AChE basal activity of 8.4 nmol min^{-1} for the amphipod *Gammarus fossarum* (male with a weight ranging from 15 to 20 mg), corresponding to the AChE annual mean value obtained for the Morcille River, France, a station unimpacted by anthropogenic activities. In the absence of any significant seasonal effect on *G. fossarum* AChE activity, the maximal and minimal threshold values were 7.4 and 9.5 nmol min^{-1}, corresponding to the 95% confidence intervals. Thus, a mean AChE activity lower than the minimal threshold value could be interpreted as resulting from a neurotoxic effect. A similar approach was used to assess the value of the base level of DNA damage measured using the Comet assay in adult male amphipods originating from four sites located upstream of contaminant sources. From all experiments, a mean value of 3.1% (95% confidence interval, 2.7–3.6% tail DNA) was determined (Lacaze et al. 2011a). This low range results from the lack of significant spatial and temporal variations in the DNA damage level. Deviation from the mean of four different sites on the Swedish Baltic coast with low anthropogenic impact was established for a large number of biomarkers (blood variables, VTG in male plasma, EROD, and glutathione reductase in liver) determined in perch *Perca fluviatilis* (Hanson et al. 2010).

Hagger et al. (2010) investigated the seasonal responses of a battery of biomarkers Neutral Red Retention (NRR, heart rate, feeding rate) at different levels of biological organization in mussels *Mytilus edulis* collected from a reference estuary in Devon, UK. By integrating biomarker responses in a biomarker response index and creating an index of health, these authors consider that it is possible to limit the natural variability of individual responses.

To improve water quality biomonitoring by using *in situ* feeding assays, attempts at modeling the influence of confounding factors were recently made following on from the previous work of Maltby et al. (2002) or Moreira et al. (2006). The influence of natural factors (body size, temperature, and conductivity for *G. fossarum* studied by Coulaud et al. 2011; temperature, salinity, and sediment grain size for *Hydrobia ulvae* tested by Krell et al. 2011) were first quantified under laboratory conditions. Then, it is possible to test the influence of each factor by using classical statistical models (multiple regression model, analysis of variance), and to evaluate the suitability of the model by comparing feeding activity values predicted from the model and those measured in the field. This strategy was successfully applied to an *in situ* assay with *H. ulvae* at reference (Mira River) and contaminated (Sado River) Portuguese estuaries. In the case of *G. fossarum*, Coulaud et al. (2011) have determined a feeding inhibition index that allows comparison of a given observed feeding rate with the expected value of feeding activity in a noncontaminated context at the same temperature.

Experience acquired in the framework of Mussel Watch type programs has demonstrated that chemical contamination levels are strongly influenced by the condition of the organisms analyzed, linked to the trophic characteristics of the ecosystem where the specimens were collected. Models linking a biometric parameter (the condition index) to pollutant levels in mussels allow an adjustment of raw data on contamination to predicted data for a reference individual, thus making clear the distinction between physiological condition and environmental factors (Andral et al. 2004). These models may be adapted for concentrations of biological macromolecules used as biomarkers, as proposed by Mouneyrac et al. (2008) for steroid hormones in the bivalve *Scrobicularia plana*.

It thus appears possible to establish baseline values for a number of biomarkers of interest and in addition, by using models such as the feeding inhibition index (Coulaud et al. 2011) or the correction factor derived from the condition index (Andral et al. 2004), to be able to distinguish between variations due to contamination and those attributable to natural factors. A next step is the use of models of classification such as those proposed by Burgeot et al. (2010) for a spatial and temporal evaluation of neurotoxic effects and by Hagger et al. (2010) for an integrated index of health. Nevertheless, the interpretation of the ecological significance of many biomarkers remains a major challenge for ecotoxicology.

16.3 Using Biomarkers for Predictive Risk Assessment

For many authors, searching for the "missing biomarker link" that allows prediction of ecological effects from biomarker responses is of utmost importance (e.g., Weis et al. 2001; De Coen and Janssen 2003a; Amiard-Triquet 2009; Canti et al. 2009; Almeida et al. 2010; Connon et al. 2011; Lacaze et al. 2011; Sánchez-Argüello et al. 2012; Bergek et al. 2012). As discussed above (Section 16.1.3), ideally biomarkers that reveal a change at a given level of organization must be indicative of changes at the next level (Jemec et al. 2010). Even if this criterion is rarely fulfilled, from a conceptual point of view, it is clear that a number of individual responses have a great potential in relating adverse outcome pathways (AOPs) to ecologically relevant levels of biological organization (Ankley et al. 2010; Kramer et al. 2011). In the introductory chapter, we asked the question: "Is it possible to reconcile the benefits of biochemical markers and ecological responses?" In the light of the reviews devoted to different biomarkers in this book, it appears that the conceptual links described by Adams et al. (1989) have been well documented. This is partly attributable to methodological and technical progress allowing improvement of mechanistic information (transcriptomics, proteomics, metabolomics) at infra-individual levels, thus supporting the identification of AOPs, whereas life history, ecosystem, and other stressors provide context to the population-level models built on the individual organism response (Kramer et al. 2011).

However, linking different levels of biological organization is not a simple task as illustrated by discrepancies occurring between effects expected from individual responses in the laboratory and the fate of a population in the field. Studies on PCBs and fish from contaminated sites have shown that effects observed experimentally on individual organisms do not necessarily imply that measurable or ecologically significant effects on wild populations of fish will occur at the same or similar exposure levels, as documented in the Hudson River for striped bass *Morone saxatilis* and white perch *Morone americana* (Barnthouse et al. 2003, 2009). Physiological acclimation or genetic adaptation in populations chronically exposed to contaminants in their medium is one of the possible explanations (Amiard-Triquet et al. 2011), but ecological processes involving interactions between tolerant and sensitive species may be another part of the story (Figure 3.2).

16.3.1 Biomarkers of Ecological Relevance

Individual responses to chemical stress such as physiological and behavioral impairments or life history traits are ecologically relevant since survival, growth, and reproduction are fundamental to organism fitness (Figure 16.1). In addition, they can be also linked to infra-individual biomarkers as exemplified by Xuereb et al. (2009b) for AChE inhibition and behavioral alteration, or for endocrine disrupting chemicals (Figure 8.5).

There are no clear links from induction of defenses to individual health or reproduction. These mechanisms help an organism to cope with the presence of hazardous compounds in the environment. This situation is well documented in areas where organisms chronically exposed to contaminants in their environment have become tolerant, allowing them to survive and reproduce (Chapters 3 and 14). It is only when defenses are overwhelmed that damage can occur even if indirect effects, namely, the cost of tolerance, can negatively affect population dynamics, biomass, and densities (Chapter 12). Certain mechanisms of defense can turn out to be less protective than initially expected, such as the production of reactive metabolites resulting from xenobiotic biotransformation by phase I and II enzymes (Chapter 3). However, biomarkers of ecological relevance must be preferably searched for among biomarkers of damage.

The ecological significance of oxidative stress—which occurs when the rate of production of reactive oxygen species (ROS) exceeds the capacity of the antioxidant defense and repair mechanisms leading to oxidative damage to biological macromolecules—has recently generated an abundant literature, with a special issue of *Functional Ecology* (vol. 24) devoted to *The Ecology of Antioxidants and Oxidative Stress in Animals* in 2010, whereas a whole book was edited by Abele et al. (2012) on *Oxidative Stress in Aquatic Ecosystems* with several chapters devoted to marine animal models for aging, development, and disease. Mammalian and avian studies have provided evidence that oxidative stress has an important physiological cost, affecting longevity, reproduction, immune responses, and intense physical activity (Costantini et al. 2010). Whether and how increased ROS impinges on fitness in the wild is a judgment difficult to make. The gaps in our knowledge are larger when considering fish, and even more in the case of invertebrates. The latter are generally characterized by low metabolic rates and ROS formation, associated with lower antioxidant enzyme activities compared to vertebrates (Buttemer et al. 2010). In aquatic organisms, Pašková et al. (2011) report that only few experimental studies with pesticides have directly linked developmental toxicity with key oxidative stress endpoints, such as lipid peroxidation, oxidative DNA damage, or modulation of antioxidant mechanisms, but they also mention that pesticide-related oxidative damage occurs in exposed adult fish, amphibians, and invertebrates. They conclude that the available experimental data provide support to the concept that oxidative stress is a highly important mechanism in pesticide-induced reproductive or developmental toxicity, and that other stressors may also act by oxidative mechanisms. Maintaining and up-regulating antioxidant defenses and repair of the damaged molecules require resources, thus generating a cost of tolerance that can have repercussions on other functions, including life-history and signal traits (Hõrak and Cohen 2010). Germline cells may be especially vulnerable to oxidative stress with a risk of transgenerational effects on offspring viability (Metcalfe and Alonso-Alvarez 2010). However, Costantini et al. (2010) recommend that ecologists should be careful to avoid the oversimplification of pronouncing all ROS as "bad," since they are also essential components of cell signaling and regulation (i.e., redox messengers). In addition, the exposure of an organism to a low dose of a chemical agent or environmental stressor that is damaging at higher doses induces an adaptive beneficial effect on the organism itself, a process termed hormesis (Mattson and Calabrese 2010 quoted by Costantini et al. 2010), whereas enhancing photoautotrophs' own protective systems may also improve the nutritional quality of foods for heterotrophic organisms (Snoeijs et al. 2012). The quality of aquatic primary producers as food for other organisms can both increase or decrease as a result of increased oxidative stress through accumulation of antioxidants or through lower growth rates caused by oxidative damage, respectively (Snoeijs et al. 2012).

Links between genomic changes, either deleterious or adaptive, and population effects have been documented in Chapters 13 and 14. Measuring genotoxicity in early life stages and juveniles may be envisaged to link more closely genotoxicity in individuals and population dynamics. Gametes and larvae emitted by invertebrates into the water column are particularly at risk from genotoxicity effects of contaminants entering the aquatic environment. The determining of genotoxic effects on these life stages at realistic environmental doses is crucial from an ecological perspective. No doubt the development of the "omics" technologies will be useful to investigate mechanisms underlying pollutant toxicity and to provide new tools for the detection of genomic disturbance and pollution fingerprints. [See special issue of *Environmental Toxicology and Chemistry* (Vol. 30, n°2) devoted to *Omics and Environmental Science*].

Currently, several assays reviewed in Chapter 13 are efficient tools to detect genotoxic impacts in species exposed in the field. The micronucleus and Comet assays are recommended methods for assessing DNA damage in aquatic organisms under the auspices of OSPAR (2011).

Moore et al. demonstrate in Chapter 5 that the lysosomal–vacuolar system is very strongly linked in a mechanistic manner to higher level physiological processes of protection, protein and cellular turnover, SfG, and is directly correlated with larval viability. Developing an expert system for the integration of biomarker responses in mussels into an animal health index, Dagnino et al. (2007) considered lysosomal membrane stability to be the most representative biomarker of the development of stress syndrome. The link with SfG in mussels and other bivalves is of particular interest since provisional values of health status thresholds have been proposed (OSPAR 2007):

- Animals may be considered healthy if the SFG is greater than +5 J g^{-1} h^{-1}.
- Animals may be considered stressed but compensating if the SFG is between +5 and –2 J g^{-1} h^{-1}.
- Animals may be considered severely stressed if the SFG is less than –2 J g^{-1} h^{-1}.

Similarly, Crowe et al. (2004) were able to link SfG values of mussels from sites on the west coast of the United Kingdom differentially affected by hydrocarbon contamination with the diversity of associated macrofaunal communities. Diversity was reduced at sites with mussels showing low SfG (<10 J g^{-1} h^{-1}), compared with sites where mussels had an SfG greater than 15 J g^{-1} h^{-1}. The possibility of lysosomal biomarker reactions having an interpretative capacity in terms of ecosystem level effects is still tentative, but recent results showing a significant correlation between lysosomal stability and macrofaunal diversity in the Langesund Fjord in Norway (Figure 5.5) are exciting.

It is well established that sublethal exposure affecting AChE can alter an animal's behavior and locomotory abilities, potentially affecting reproduction success, fitness, and survival (Chapters 4 and 10; OSPAR 2011), and ultimately population effects may be expected (Chapter 12) as shown in the multipolluted Seine estuary, France, for populations of the ragworm *Nereis diversicolor* (Gillet et al. 2008; Bessineton 2009). Behavioral alterations result in changes in feeding rate—a crucial parameter in an energy budget—that may be relatively easily quantified, for instance by using the method of postexposure depression described by McWilliam and Baird (2002) for *Daphnia* and adapted for various species (Chapter 10). In addition to changes in food consumption, the acquisition of energy may be affected by impairment of assimilation, linked to pollutant effects on digestive enzyme activities (Chapter 11). Therefore, both AChE and digestive enzyme activities should be considered ecologically relevant parameters. From an operational point of view, AChE as an "old" biomarker benefits from an important background of scientific knowledge, allowing the assessment of baseline levels of AChE activity in different aquatic species and the establishment of relevant environmental assessment criteria (Burgeot et al. 2010; OSPAR 2011).

Reviewing the application of histocytopathological biomarkers in marine pollution monitoring, Au (2004) has highlighted the symptoms that are the most ecologically relevant: internal fish disease (liver, gill and kidney histopathology, oocyte atresia, embryonic effects) and histopathology of mollusks (Table 4.3). Methodologies and diagnostic criteria involved in the monitoring of contaminant-specific macroscopic liver neoplasms and liver histopathology have largely been developed based on experience with flatfish species such as dab *Limanda limanda* and flounder *Platichthys flesus* (OSPAR 2010). Abnormalities of the byssus organ in bivalves and disruption of sensory system function in fish are able to induce behavioral disturbance (Chapter 4). Alterations of liver ultrastructure in fish or histological lesions of the digestive gland in gastropods may be associated with changes in energy metabolism, which in turn could explain decreased swimming velocity in the case of fish (Chapter 4). Effects on germinal cells, impairments of embryo–larval development, histopathological impairments that occur as a consequence of pollution by endocrine disruptors such as imposex in gastropods and intersex in fish and bivalves, all have the potential to predict ecological disturbance through altered reproductive success.

The literature provides evidence of consistent links between infra- and supra-individual effects of endocrine disrupting compounds (Chapters 8 and 9). Historically, this was first demonstrated in the case of TBT-based antifouling paints, the use of which has led to the presence of imposex in many gastropod species, leading in turn to local extinctions. An initial UK ban on the use of TBT on small vessels in 1987, followed by the International Maritime Organisation ban in 2008, allowed the recovery of gastropod populations, particularly the well-studied neogastropod *Nucella lapillus* (Chapter 9), but macroinfaunal recovery was also observed within 3 to 5 years of the initial ban (Smith et al. 2008). However, at historically heavily affected sites, TBT levels remain above environmental standards (e.g., EQS equivalent levels, EcoQO), and monitoring will continue until TBT concentrations fall below those levels. In the meantime, the source of TBT has shifted from ships' hulls to historical sediment burdens.

Intersex in bivalves and fish is also widely present in many geographical areas, but it remains impossible today to firmly establish a link between the phenomenon of intersex and reproductive success with potential consequences at the population level (Chapter 9; Fossi Tankoua et al. 2012). On the other hand, dramatic freshwater fish population consequences of endocrine disruption effects associated with the presence of EE2 have been established in the field (Kidd et al. 2007) or forecast through modeling of the effects of five chemicals (17β-trenbolone, 17α-trenbolone, prochloraz, fenarimol, and fadrozole) that inhibit VTG production through different mechanisms (Miller et al. 2007; see Figure 16.2). A recent paper by Sanchez et al. (2011) supports the hypothesis that pharmaceutical compounds discharged into streams are involved in endocrine disruption effects and fish population disturbances, and threaten the disappearance of resident fish species. Recently, Lewis and Ford (2012 have reviewed the current knowledge regarding pollutants impacting male infertility across several aquatic invertebrate phyla. If the number of studies in invertebrates remains low, biomarkers developed for human infertility clinics (sperm tracking, measurements of DNA strand breaks and sperm viability) provide useful tools for the study of infertility in male aquatic invertebrates. So these authors conclude that it represents a good opportunity for a better understanding of reproductive abnormalities, providing a sound basis to link sperm toxicity and male infertility effects to higher level impacts on the reproductive biology and dynamics of populations.

All the evidence for an endocrine effect in amphibians is reliant on the correct interpretation of biomarker data. In Chapter 9, it is concluded that there is an effect of atrazine

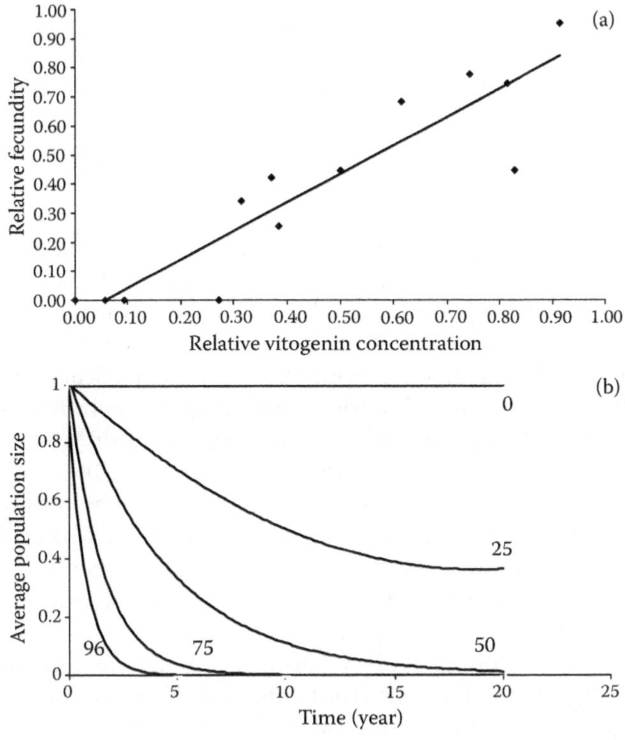

FIGURE 16.2
Linking biochemical responses to population-level effects in the freshwater fish *Pimephales promelas*. (a) Relationship between vitellogenin (VTG) concentration in plasma and fecundity. (b) Effect of reduction in VTG concentration in plasma and population size—0: no VTG reduction; 25, 50, 75, 96: VTG reduced by 25%, 50%, 75%, or nearly 96%.

exposure on amphibians, probably through altered steroid balance, leading potentially to complete sex reversal but more commonly to abnormal gonadal development and reduced reproductive success. Knowledge of various potential morphological, behavioral, and biochemical biomarkers in amphibians is slowly building. We must remain cautious about interpreting data about endocrine disruption in amphibians as a cause for amphibian population decline, but, nevertheless, the biomarkers observed in amphibians are useful signposts to the fate of those amphibians unable to adapt to chemical pollution.

In Chapter 6, Brousseau et al. highlight that several biomarkers of the immune response (immunomarkers) can be affected by xenobiotics such as drugs and pollutant chemicals. These effects can result in immunodepression or immunostimulation, leading to hypersensitivity and autoimmunity disorders. However, effects on immune functions do not necessarily imply an impact on host resistance to pathogens or an increased susceptibility to cancer, thanks to detoxification and compensation processes that protect higher levels of organization (tissue and organism). In addition, the degree of exposure, the physiological status of individuals, and the complexity of the immune system can modulate the immunotoxic effects of toxicants. Brousseau et al. (Chapter 6) conclude that immunomarkers do indicate an impairment of the immune response, allowing the identification of populations at risk among marine mammals. Establishing a sound relationship between exposure to persistent toxic substances, the impacts on the immune system, and the host's reduced resistance to infection and cancer, is not a simple task, being complicated by numerous confounding factors. However, these authors present a very convincing body of evidence indicating that environmental pollution can significantly contribute to stress on the immune response resulting in major consequences for the health of marine mammal populations.

16.3.2 Modeling

Historically, ecological risk assessments of contaminants have focused at the level of the individual, for example, based on estimations of toxicological endpoints from laboratory exposure experiments, yet the protection goal is usually a field population rather than an individual organism (Schmolke et al. 2010). At the population level, other factors come into play such as life history characteristics, population structure, seasonal effects, and habitat heterogeneity in the field. Ecological models can integrate such factors and so offer great potential as tools to predict population-level effects of contaminant exposure, for example, when extrapolating from laboratory to field (Schmolke et al. 2010).

Interest in the use of ecological modeling to link individual biomarkers to consequences at the population level may be illustrated by some examples. Miller et al. (2007) demonstrated a relationship between VTG concentration in the plasma of the freshwater fish *Pimephales promelas* and fecundity, and an associated effect of reduction in VTG concentration in plasma and population size (Figure 16.2).

In the realm of behavioral biomarkers, Chaumot et al. (2003) modeled how toxin avoidance behavior of a hypothetical brown trout *Salmo trutta* population living in a river network might alter the impact of pollution, concluding that this behavior could weakly raise the asymptotic population growth rate, while, in the presence of density-dependent effects on recruitment, leading to a substantial drop in the equilibrium size of the population. Baird et al. (2007) used a simulated model to predict that a 50% impairment in the feeding rate of the amphipod *Gammarus pulex* would cause population growth rate to fall below the replacement level, leading to population decline and ultimately local extinction. De Coen et al. (2003b) used a multivariate model based on short-term biomarker measurements to successfully predict population-level responses of *Daphnia magna*. These biomarkers included activities of digestive enzymes, enzymes of intermediary metabolism, cellular energy allocation, DNA damage, and antioxidative stress activity. Rose et al. (2003) modeled the effects of PCB exposure on populations of fish, integrating experimental results on fecundity, egg mortality, and larval behavior.

Such studies merit being developed because they not only link effects at different levels of biological organization, but also link the results of laboratory experiments with field studies. The experimental results are in effect being used to predict the order of magnitude of the population effects in the field. This predicted response can be compared against the variability in population abundance and population dynamics observed in the field. The cumulative effect of multiple stressors can thus be investigated through the modeling. Moreover, Rose et al. (2003) propose that their model can be used to simulate the growth of fish populations with and without the effects of PCB exposure, or indeed in the presence or absence of other stress factors such as increased fishing.

16.3.3 Regulatory Acceptability

Because it is difficult to study directly the fate of a population at a contaminated site, the extrapolation from individual to population level in an ecological model appears an attractive substitute in order to meet the demands of regulatory bodies (Grimm et al. 2009; Preuss et al. 2009; Galic et al. 2010; Hommen et al. 2010). However, the use of ecological models to support regulatory risk assessments of contaminants is patchy, being rarely used, for example, for pesticides (Schmolke et al. 2010). These authors reviewed relevant models available in an assessment of their suitability for use in such regulatory risk assessments and found that, in spite of meeting high scientific standards, most currently available ecological models would need modifications before being considered suitable for regulatory risk assessments. Similar conclusions have been reached by Galic et al. (2010) and Forbes et al. (2011). Galic et al. (2010) highlighted the advantages of the use of ecological models as the relatively straightforward integration of the sensitivity of species to chemicals, the mode of action and fate in the environment of toxicants, life history traits of the species of concern, and landscape features. These authors reviewed 148 publications in an attempt to establish whether existing published ecological modeling studies have addressed (or have the potential to address) the protection aims and requirements of the chemical directives of the EU. Most models had been developed to establish population-level responses on the basis of individual effects, but the lack of clarity about protection goals in legislative documents made it impossible to establish a direct link between modeling studies and protection goals (Galic et al. 2010).

Ecological risk assessments within regulatory frameworks are tailored to fit decision-making needs, the quality of available data, and legally mandated protection goals (Kramer et al. 2011). Although the protection and associated risk assessment for an individual organism might be appropriate in the management of an endangered species, it is clear from the above discussion that ecological risk assessments of environmental contaminants need to address population-level effects. Population models, especially those informed by AOPs, can be used to translate individual-level study results to population-level responses (Kramer et al. 2011). The challenge is to effectively translate quantitative mechanistic AOP data to higher-level responses meaningful to population modeling and realistic ecological risk assessment; in this context, model validation is a critical requirement (Kramer et al. 2011), but a criterion rarely met (Schmolke et al. 2010).

16.4 Multidisciplinary Approach for Classifying Polluted Environments

16.4.1 Implementation of Biomarkers

The biomarkers of ecological relevance that have been reviewed in this book have provided significant progress in the field of environmental assessment compared to the "core biomarkers" (mainly biochemical biomarkers) described in Chapter 2. They can be very useful for biomonitoring purposes, since they are able to reveal environmental health impairments before massive effects occur at the level of populations or communities. They can contribute to a prognostic risk assessment, contrary to biotic indices based on the community structure that allow only a retrospective assessment when degradation has already occurred.

Biomarkers of ecological relevance will also allow a classification of contaminated sites based on the potential risk of complex mixtures of pollutants for ecosystem health, integrating across total contaminant availabilities rather than relying on the concentrations of a limited number of chemical pollutants, in effect those perceived as easy to analyze (*a priori* analysis). So biomarkers are potentially of great assistance to stakeholders, for instance, when there is a need to establish priorities for remediation campaigns (Chapter 15).

Compared to the classical biomarkers of defense or damage, they are situated at a higher level of biological organization, but consequently they are often less sensitive. However, for each of them, the different chapters of this book provide many examples highlighting significant responses observed in contaminated natural media or—in the case of laboratory experiments—at environmentally realistic doses.

Biomarkers of ecological relevance (Figure 16.1) are generally nonspecific biomarkers, interestingly providing a general assessment of environmental quality, but on the other hand, responding to a large number of chemical contaminants and even to nonchemical stresses. However, in order to decide upon actions aimed to improve environmental quality, environmental managers need to know the substances responsible for the impaired health status of the local biota. For this purpose, toxicity identification evaluation studies may be designed, combining fractionation and chemical analysis of fractions obtained from an effluent or a contaminated sediment with bioassays for the characterization of the toxicity associated with each fraction (Amiard et al. 2009). At this stage, core biomarkers provide a relevant complementary approach (Figure 16.1), even though they are less specific than initially supposed when the methodology of biomarkers was first developed in the early 1990s; this applies even in the cases of AChE initially labeled as a specific biomarker for organophosphorous pesticides and carbamates, and MT, a biomarker of exposure to metal contaminants. At any rate, this deficiency is now well recognized, and it is widely admitted that the solution lies in the use of multibiomarker approaches (Chapter 2).

For certain authors, except when it has been firmly established that there is no risk present, any detection of a biomarker response should be considered a sign of a potential, and therefore an unacceptable, risk for organisms (the Precautionary Principle). Biomarkers of defense are activated by their (more or less specific) inducers below a certain threshold above which the toxicity is so high that defense mechanisms are inhibited. In the absence of data on concentrations of contaminant chemicals in the environment or in the organisms themselves, it is impossible to decide if the biomarker response lies in the increase phase (when the organisms are able to tolerate that level of contamination) or in the decrease phase (when the health of the organisms is compromised) of a bell-shaped curve as shown in Figure 16.3 (e.g., Correia et al. 2007; Ortiz-Delgado et al. 2008; Amiard-Triquet and Roméo

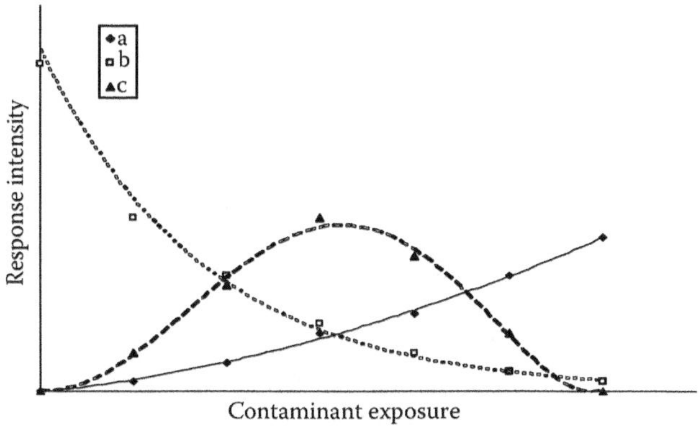

FIGURE 16.3
Changes in the response of biomarkers of damage (a, b) and of defense (c) as a function of the level of contaminant exposure. (a) Increasing response (e.g., lipofuscin lysosomal content, DNA damage). (b) Decreasing response (e.g., lysosomal membrane stability, AChE activity). (c) Bell-shaped curve (EROD, GST, MT). (Adapted from Dagnino, A. et al., *Biomarkers*, 12, 155–172, 2007.)

in Amiard-Triquet et al. 2011; Ji and Lu 2011). In the case of biomarkers of damage, the interpretation is easier, even in the absence of chemical data, since the response trend has no reversal of direction with increasing level of contaminant exposure (Figure 16.3). Even for core biomarkers the present degree of validation is still insufficient to avoid the parallel use of chemical analysis, an outcome that would be beneficial, particularly in developing countries, since it would allow reduction of the cost of environmental monitoring.

In order to reinforce the ecological value of biomarkers, the choice of the biological matrix in which they are to be determined is crucial. The selection of biological models or sentinel species has to cope with a major problem that is the importance of any inherent variability, between individuals as well as between species. It can be recommended to focus (1) on the most sensitive species or life stages, the protection of which will probably ensure the security of less sensitive species; (2) on species that have a key role in the structure and functioning of the ecosystem; or, even better, (3) on a set of species able to represent the different functional aspects of the ecosystem (Chapter 7).

As highlighted in Chapter 14, the cumulative effects of multiple stressors should become a priority concern for ecotoxicologists, particularly to explore how adaptation to chronically contaminated habitats could affect the ability of populations to cope with other stressors, particularly from the perspective of global warming.

16.4.2 Integrating Biomarkers in a Weight of Evidence Approach

Environmental quality assessments inevitably now call upon data from a variety of multidisciplinary sources, and the weight of evidence (WOE) approach has been developed to combine information from these different lines of evidence (LOEs). The WOE process can, for example, help to determine the extent of contamination or its ecological significance, and can contribute to defining an optimal remediation program (Luoma and Rainbow 2008). Thus, the WOE approach is being increasingly adopted within the context of ecological risk assessments, including the quality assessment of polluted sediments, and biomarkers have a key role to play as a contributory LOE to such ends (Chapman and Hollert 2006; Morales-Caselles et al. 2009; Benedetti et al. 2012). For assessment of sediment quality, Chapman and Hollert (2006) proposed an additional LOE to the original Sediment Quality Triad of Long and Chapman (1985), including biomarkers of genotoxicity (micronucleus assay) and endocrine disruption, EROD, and stress protein concentrations. For quality assessment of a sediment impacted by an oil spill derived from the tanker *Prestige*, Morales-Caselles et al. (2009) proposed a set of biomarkers as a LOE: for example, mixed functional oxidase activity (EROD), glutathione-*S*-transferase (GST) activity, glutathione reductase activity, and glutathione peroxidase activity in the crab *Carcinus maenas* and the clam *Ruditapes philippinarum* in the laboratory and the field. The selected biomarkers provide information about the initial biological responses to the presence of contaminants with the potential then to predict effects to the ecosystem (Morales-Caselles et al. 2009). Benedetti et al. (2012) used a WOE approach to classify polluted sediments from the Venice Lagoon, involving a wide battery of biomarkers based on the European eel *Anguilla anguilla*: for example, enzymes of the cytochrome P450 biotransformation pathway, antioxidant responses and the onset of oxidative damage, lysosomal membrane stability, and genotoxic effects. As expected, responses measured at organism level were in general accordance with, but less marked than, the onset of sublethal effects measured by the biomarkers. The WOE model proposed by the authors allowed a good quantitative characterization of sediment hazard and good discrimination on the basis of the different chemical and biological data lines of evidence (Benedetti et al. 2012).

16.5 Conclusions

To date, scientists have been able to provide a large number of complementary tools (chemical, biochemical, physiological, behavioral, statistical) to assess the environmental risks induced by anthropogenic inputs of contaminants and to establish environmental quality standards favoring biological conservation. Very simplistically, risk assessment is based on a causal relationship between a contaminant and its target that may be represented by a curve such as that shown in Figure 16.4. If the level of exposure of a given contaminant remains lower than the No Observed Effect Concentration (NOEC), both the scientist and the stakeholder will agree that it is not necessary to envisage any action aiming at the reduction of contamination since the presence *per se* of a contaminant is not a danger. As stated by Paracelsus in the sixteenth century, "the dose makes the poison", at least for many chemical contaminants. If the exposure, however, is responsible for a biological response (effect) in 50% of the exposed population (EC_{50}), then it might also be easy to reach agreement, in this case to consider that the risk is not acceptable and that remediation programs are needed to reduce and finally eliminate this risk. Then, it will be necessary to monitor environmental quality in order to verify the efficiency of any remediation, whereas, in less exposed areas, the objective of monitoring will be to verify that the situation, initially acceptable, does not worsen. At exposures intermediate between the NOEC and EC_{50} exposures (Figure 16.4), a political decision is needed to determine whether the risk is acceptable or not. Chemists, biologists, and modelers are able to assess a risk with an associated uncertainty, but not to decide whether that risk is "acceptable." The advice of scientific experts must contribute to the debate but the environmental management strategy must take into account the opinion of the general public and particularly the users of relevant environmental goods and services. Their contribution is called for in international regulations such as the European directive 2008/105/EC on environmental quality standards

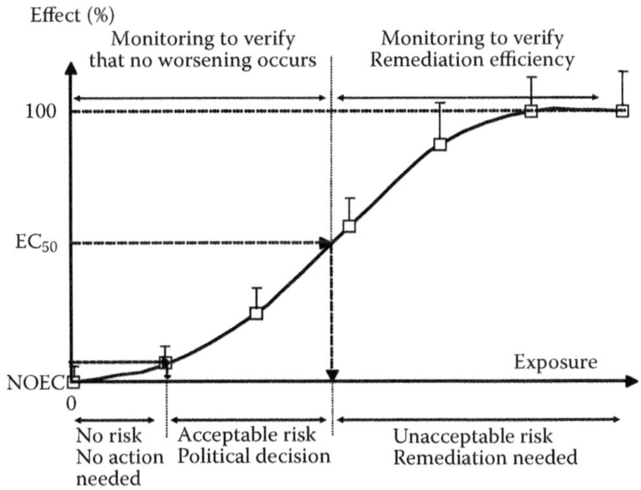

FIGURE 16.4
Relationship between the level of exposure, expected effects and actions needed. EC_{50}, mean effective concentration; NOEC, no observed effect concentration.

in the field of water policy, which explicitly mentions the contribution of the European Economic and Social Committee. At the interface between experts and the general public, the contribution of sociologists would be useful to improve the environmental governance of the ecosociosystem.

References

Abele, D., J.P. Vázquez-Medina, and T. Zenteno-Savín. 2012. *Oxidative Stress in Aquatic Ecosystems*. Chichester, UK: Wiley-Blackwell.

Adams, S.M. et al. 1989. The use of bioindicators for assessing the effects of pollutant stress on fish. *Mar. Environ. Res.* 28:459–64.

Almeida, J.R. et al. 2010. Linking behavioural alterations with biomarkers responses in the European seabass *Dicentrarchus labrax* L. exposed to the organophospate pesticide fenitrothion. *Ecotoxicology* 19:1369–81.

Amiard, J.C. et al. 2009. Quantification of contaminants. In: *Environmental Assessment of Estuarine Ecosysqtems. A Case Study*, ed. C. Amiard-Triquet and P.S. Rainbow, 31–57. Boca Raton, FL: CRC Press.

Amiard-Triquet, C. 2009. Behavioral disturbances: The missing link between sub-organismal and supra-organismal responses to stress? Prospects based on aquatic research. *Hum. Ecol. Risk Assess.* 15:87–110.

Amiard-Triquet, C., and P.S. Rainbow. 2009. Conclusions. In: *Environmental Assessment of Estuarine Ecosystems. A Case Study*, ed. C. Amiard-Triquet and P.S. Rainbow, Chapter 16, 323–48. Boca Raton, FL: CRC Press.

Amiard-Triquet, C., P.S. Rainbow, and M. Roméo. 2011. *Tolerance to Environmental Contaminants*. Boca Raton, FL: CRC Press.

Andral, B. et al. 2004. Monitoring chemical contamination levels in the Mediterranean based on the use of mussel caging. *Mar. Pollut. Bull.* 49:704–12.

Ankley, G.T. et al. 2010. Adverse outcome pathways: A conceptual framework to support ecotoxicological research and risk assessment. *Environ. Toxicol. Chem.* 29:730–41.

Au, D.W.T. 2004. The application of histo-cytopathological biomarkers in marine pollution monitoring: A review. *Mar. Pollut. Bull.* 48:817–34.

Baird, D.J. et al. 2007. *In situ*-based effects measures: Determining the ecological relevance of measured responses. *Integr. Environ. Assess. Manag.* 3:259–67.

Barnthouse, L.W., D. Glaser, and J. Young. 2003. Effects of historic PCB exposures on the reproductive success of the Hudson River striped bass population. *Environ. Sci. Technol.* 37:223–8.

Barnthouse, L.W., D. Glaser, and L. Desantis. 2009. Polychlorinated biphenyls and Hudson River white perch: Implications for population-level ecological risk assessment and risk management. *Integr. Environ. Assess. Manag.* 5:435–44.

Bechshøft, T.Ø. et al. 2011. Distribution of vitamins A (retinol) and E (α-tocopherol) in polar bear kidney: Implications for biomarker studies. *Sci. Total Environ.* 409:3508–11.

Benedetti, M. et al. 2012. A multidisciplinary weight of evidence approach for classifying polluted sediments: Integrating sediment chemistry, bioavailability, biomarkers responses and bioassays. *Environ. Int.* 38:17–28.

Bergek, S. et al. 2012. From individuals to populations: Impacts of environmental pollution on natural eelpout populations. *Ecotox. Environ. Saf.* 79:1–12.

Berthet, B., K.M.Y. Leung, and C. Amiard-Triquet. 2011. Inter- and intra-specific variability of tolerance: Implications for bioassays and biomonitoring. In: *Tolerance to Environmental Contaminants*, ed. C. Amiard-Triquet, P.S. Rainbow, and M. Roméo, Chapter 3, 49–84. Boca Raton, FL: CRC Press.

Bessineton, C. 2009. Historical Records of the *Nereis diversicolor* populations in the Seine Estuary. In: *Environmental Assessment of Estuarine Ecosystems. A Case Study*, ed. C. Amiard-Triquet and P.S. Rainbow, 183–97. Boca Raton, FL: CRC Press.

Brammell, B.F. et al. 2010. CYP1A expression in caged rainbow trout discriminates among sites with various degrees of polychlorinated biphenyl contamination. *Arch. Environ. Contam. Toxicol.* 58:772–82.

Burgeot, T. et al. 2010. Acetylcholinesterase: Methodology development of a biomarker and challenges of its application for biomonitoring. ICES CM Code: F-25. International Council for the Exploration of the Sea, Annual Science Conference, 20–24 September 2010, Nantes, France. e-paper: http://www.ices.dk/products/CMdocs/CM-2010/F/F2510.pdf.

Buttemer, W.A., D. Abele, and D. Costantini. 2010. From bivalves to birds: Oxidative stress and longevity. *Funct. Ecol.* 24:971–83.

Camus, L. et al. 2004. Total oxyradical scavenging capacity responses in *Mytilus galloprovincialis* transplanted into the Venice lagoon (Italy) to measure the biological impact of anthropogenic activities. *Mar. Pollut. Bull.* 49:801–8.

Canesi, L. et al. 2010. Effects of vibrio challenge on digestive gland biomarkers and antioxidant gene expression in *Mytilus galloprovincialis*. *Comp. Biochem. Physiol.* 152C:399–406.

Canti, M.N. et al. 2009. Linking genotoxic responses with cytotoxic and behavioural or physiological consequences: Differential sensitivity of echinoderms (*Asterias rubens*) and marine molluscs (*Mytilus edulis*). *Aquat. Toxicol.* 94:68–76.

Chapman, P.M. 1990. The sediment quality triad approach to determining pollution-induced degradation. *Sci. Total Environ.* 97–98:815–25.

Chapman, P.M., and H. Hollert. 2006. Should the sediment quality triad become a tetrad, a pentad, or possibly even a hexad? *J. Soils Sediments* 6:4–8.

Chaumot, A. et al. 2003. Do migratory or demographic disruptions rule the population impact of pollution in spatial networks? *Theor. Popul. Biol.* 64:473–80.

Connon, R.E. et al. 2011. Linking molecular biomarkers with higher level condition indicators to identify effects of copper exposures on the endangered delta smelt (*Hypomesus transpacificus*). *Environ. Toxicol. Chem.* 30:290–300.

Correia, A.D. et al. 2007. Biochemical and behavioral responses in gilthead seabream (*Sparus aurata*) to phenanthrene. *J. Exp. Mar. Biol. Ecol.* 347:109–22.

Costantini, D. et al. 2010. From molecules to living systems: Historical and contemporary issues in oxidative stress and antioxidant ecology. *Funct. Ecol.* 24:950–9.

Coulaud, R. et al. 2011. *In situ* feeding assay with *Gammarus fossarum* (Crustacea): Modelling the influence of confounding factors to improve water quality biomonitoring. *Water Res.* 45:6417–29.

Crowe, T.P. et al. 2004. Measurement of sublethal stress effects on individual organisms indicate community-level impacts of pollution. *J. Appl. Ecol.* 41:114–23.

Dagnino, A. et al. 2007. Integration of biomarker data into an organism health index: Development of an expert system and its validation with field and laboratory data in mussels. *Biomarkers* 12:155–72.

De Coen, W.M., and C.R. Janssen. 2003a. The missing biomarker link: Relationship between effects on the cellular energy allocation biomarker of toxicant-stressed *Daphnia magna* and corresponding population characteristics. *Environ. Toxicol. Chem.* 22:1632–41.

De Coen, W.M., and C.R. Janssen. 2003b. A multivariate biomarker-based model predicting population-level responses of *Daphnia magna. Environ. Toxicol. Chem.* 22:2195–201.

Egeghy, P.P. et al. 2012. The exposure data landscape for manufactured chemicals. *Sci. Total Environ.* 414:159–66.

Eriksson Wiklund, A.K., and B. Sundelin. 2004. Biomarker sensitivity to temperature and hypoxia—A seven year field study. *Mar. Ecol. Prog. Ser.* 274: 209–14.

Fielding, N.J. et al. 2003. Effects of the acanthocephalan parasite *Echinorhynchus truttae* on the feeding ecology of *Gammarus pulex* (Crustacea: Amphipoda). *J. Zool.* 261:321–5.

Forbes, V.E., A. Palmqvist, and L. Bach. 2006. The use and misuse of biomarkers in ecotoxicology. *Environ. Toxicol. Chem.* 25:272–80.

Forbes, V.E. et al. 2011. Adding value to ecological risk assessment with population modeling. *Hum. Ecol. Risk Assess.* 17:287–99.

Forrow, D.M., and L. Maltby. 2000. Toward a mechanistic understanding of contaminant- it would be difficult to make an accurate judgement on whether pollution might be masking the natural variability of the biomarker response induced changes in detritus processing in streams: Direct and indirect effects on detritivore feeding. *Environ. Toxicol. Chem.* 19:2100–6.

Fossi Tankoua, O. et al. 2011. Potential influence of confounding factors (size, salinity) on biomarker tools in the sentinel species *Scrobicularia plana* used in monitoring programmes of estuarine quality. *Environ. Sci. Pollut. Res.* 18:1253–63.

Fossi Tankoua, O. et al. 2012. Physiological status and intersex in the endobenthic bivalve *Scrobicularia plana* from thirteen estuaries in Northwest France. *Environ. Pollut.* 167:70–7.

Frasco, M.F. et al. 2010. Localization and properties of cholinesterases in the common prawn (*Palaemon serratus*): A kinetic–histochemical study. *Biol. Bull.* 218:1–5.

Galic, N. et al. 2010. Potential application of population models in the European ecological risk assessment of chemicals: II. Review of models and their potential to address environmental protection aims. *Integr. Environ. Assess. Manag.* 6:338–60.

Gillet, P. et al. 2008. Response of *Nereis diversicolor* population (Polychaeta, Nereididae) to the pollution impact—Authie and Seine estuaries (France). *Estuar. Coast. Shelf Sci.* 76:201–10.

Grimm, V. et al. 2009. CREAM: A European project on mechanistic effect models for ecological risk assessment of chemicals. *Environ. Sci. Pollut. Res.* 16:614–17.

Gust, M. et al. 2010. *In situ* biomonitoring of freshwater quality using the New Zealand mudsnail *Potamopyrgus antipodarum* (Gray) exposed to waste water treatment plant (WWTP) effluent discharges. *Water Res.* 44:4517–28.

Hagger, J.A. et al. 2010. The influence of seasonality on biomarker responses in *Mytilus edulis. Ecotoxicology* 19:953–62.

Hanson, N., L. Förlin, and A. Larsson. 2010. Spatial and annual variation to define the normal range of biological endpoints: An example with biomarkers in perch. *Environ. Toxicol. Chem.* 29: 2616–24.

Hommen, U. et al. 2010. Potential application of ecological models in the European environmental risk assessment of chemicals: I. Review of protection goals in EU directives and regulations. *Integr. Environ. Assess. Manag.* 6:325–37.

Hontela, A. 2000. Endocrine biomarkers: Hormonal indicators of sublethal toxicity in fishes. In: *Use of Biomarkers for Environmental Quality Assessment*, ed. L. Lagadic et al., 187–204. Enfield, NH: Science Publishers.

Hõrak, P., and A. Cohen. 2010. How to measure oxidative stress in an ecological context: Methodological and statistical issues. *Funct. Ecol.* 24:960–70.

Jemec, A. et al. 2010. Biochemical biomarkers in environmental studies—lessons learnt from enzymes catalase, glutathione *S*-transferase and cholinesterase in two crustacean species. *Environ. Sci. Pollut. Res.* 17:571–81.

Ji, Y., and G. Lu, 2011. Sublethal effects of PAHs on selected biomarkers of freshwater fish *Carassius auratus*. In: *Remote Sensing, Environment and Transportation Engineering (RSETE)*, 2011 International Conference, Nanjing, China, 24–26 June 2011, pp. 6392–5.

Ji, Y. et al. 2010. Fish transplantation and stress-related biomarkers as useful tools for assessing water quality. *J. Environ. Sci.* 22:1826–32.

Kalman, J. et al. 2010. Assessment of the influence of confounding factors (weight, salinity) on the response of biomarkers in the estuarine polychaete *Nereis diversicolor*. *Biomarkers* 15:461–9.

Kerambrun, E. et al. 2011. Are biochemical biomarker responses related to physiological performance of juvenile sea bass (*Dicentrarchus labrax*) and turbot (*Scophthalmus maximus*) caged in a polluted harbour? *Comp. Biochem. Physiol.* 154C:187–95.

Kidd, K.A. et al. 2007. Collapse of a fish population after exposure to a synthetic estrogen. *Proc. Natl. Acad. Sci.* 104:8897–901.

Klobučar, G.I.V. et al. 2010. Genotoxicity monitoring of freshwater environments using caged carp (*Cyprinus carpio*). *Ecotoxicology* 19:77–84.

Knoery, J., and D. Claisse. 2010. Insights from 30 years of experience in running the French marine chemical monitoring network. International Council for the Exploration of the Sea, Nantes, pp. 1–9. http://www.ices.dk/products/CMdocs/CM-2010/F/F0810.pdf.

Kramer, V.J. et al. 2011. Adverse outcome pathways and ecological risk assessment: Bridging to population-level effects. *Environ. Toxicol. Chem.* 30:64–76.

Krell, B. et al. 2011. An estuarine mudsnail *in situ* toxicity assay based on postexposure feeding. *Environ. Toxicol. Chem.* 30:1935–42.

Lacaze, E. et al. 2011a. DNA damage in caged *Gammarus fossarum* amphipods: A tool for freshwater genotoxicity assessment. *Environ. Pollut.* 159:1682–91.

Lacaze, E. et al. 2011b. Linking genotoxic responses in *Gammarus fossarum* germ cells with reproduction impairment, using the Comet assay. *Environ. Res.* 111:626–34.

Lettini, S.E., and M.V.K. Sukhdeo. 2010. The energetic cost of parasitism in isopods. *Ecoscience* 17:1–8.

Lewis, C., and A.T. Ford. 2012. Infertility in male aquatic invertebrates: A review. *Aquat. Toxicol.* 120–121:79–89.

Lewis, C., and T. Galloway. 2008. Genotoxic damage in polychaetes: A study of species and cell-type sensitivities. *Mutat. Res.* 654:69–75.

Long, E.R., and P.M. Chapman. 1985. A sediment quality triad: Measures of sediment contamination, toxicity and infaunal community composition in Puget Sound. *Mar. Pollut. Bull.* 16:405–15.

Luoma, S.N., and P.S. Rainbow. 2008. *Metal Contamination in Aquatic Environments. Science and Lateral Management*. Cambridge: Cambridge Press.

Maltby, L. et al. 2002. Evaluation of the *Gammarus pulex in situ* feeding assay as a biomonitor of water quality: Robustness, responsiveness, and relevance. *Environ. Toxicol. Chem.* 21:361–8.

McWilliam, R.A., and D.J. Baird. 2002. Postexposure feeding depression: A new toxicity endpoint for use in laboratory studies with *Daphnia magna*. *Environ. Toxicol. Chem.* 21:1198–205.

Metcalfe, N.B., and C. Alonso-Alvarez. 2010. Oxidative stress as a life-history constraint: The role of reactive oxygen species in shaping phenotypes from conception to death. *Funct. Ecol.* 24: 984–96.

Miller, D.H. et al. 2007. Linkage of biochemical responses to population-level effects: A case study with vitellogenine in the fathead minnow (*Pimephales promelas*). *Environ. Toxicol. Chem.* 26: 521–7.

Minguez, L. et al. 2009. Interactions between parasitism and biological responses in zebra mussels (*Dreissena polymorpha*): Importance in ecotoxicological studies. *Environ. Res.* 109:843–50.
Morales-Caselles, C., I. Riba, and T. Ángel DelValls. 2009. A weight of evidence approach for quality assessment of sediments impacted by an oil spill: The role of a set of biomarkers as a line of evidence. *Mar. Environ. Res.* 67:31–7.
Moreira, S.M. et al. 2006. An *in situ* postexposure feeding assay with *Carcinus maenas* for estuarine sediment-overlying water toxicity evaluations. *Environ. Pollut.* 139:318–29.
Mouneyrac, C. et al. 2008. Biological indices, energy reserves, steroid hormones and sexual maturity in the infaunal bivalve *Scrobicularia plana* from three sites differing by their level of contamination. *Gen. Comp. Endocrinol.* 157:133–41.
Mouneyrac, C. et al. 2009. Influence of anthropogenic stress on fitness and behaviour of a key-species of estuarine ecosystems, the ragworm *Nereis diversicolor*. *Environ. Pollut.* 158: 121–8.
NAS. 1980. The International Mussel Watch. Report of a workshop sponsored by the environment studies board. Vol. Natural Resources Commission National Research Council National Academy of Sciences, Washington, DC.
Oikari, A. 2006. Caging techniques for field exposures of fish to chemical contaminants. *Aquat. Toxicol.* 78:370–81.
Olla B.L., A.J. Bedja, and W.H. Pearson. 1983. Effects of oiled sediment on the burrowing behaviour of the hard clam, *Mercenaria mercenaria*. *Mar. Environ. Res.* 9: 183–93.
Olla, B.L. et al. 1984. Sublethal effects of oiled sediment on the sand worm, *Nereis (Neanthes) virens*: Induced changes in burrowing and emergence. *Mar. Environ. Res.* 13: 121–39.
Ortiz-Delgado, J.B. et al. 2008. Tissue-specific induction of EROD activity and CYP1A protein in *Sparus aurata* exposed to B(*a*)P and TCDD. *Ecotoxicol. Environ. Saf.* 69:80–8.
OSPAR. 2010. Further development of guidance on integrated monitoring and assessment of chemicals and biological effects. Annex 5: Externally visible fish diseases, macroscopic liver neoplasms and liver histopathology.
OSPAR. 2011. Further development of guidance on integrated monitoring and assessment of chemicals and biological effects.
OSPAR Commission. 2007. Background document on biological effects monitoring techniques. Assessment and Monitoring Series.
Pašková, V., K. Hilscherová, and L. Bláha. 2011. Teratogenicity and embryotoxicity in aquatic organisms after pesticide exposure and the role of oxidative stress. *Rev. Environ. Contam. Toxicol.* 211:25–61.
Pearson, W.H. et al. 1981. Effects of oiled sediment on predation on the littleneck clam, *Protothaca staminea*, by the Dungeness crab, *Cancer magister*. *Estuar. Coast. Shelf Sci.* 13: 445–54.
Poirier, L. et al. 2006. A suitable model for the biomonitoring of trace metal bioavailabilities in estuarine sediment: The annelid polychaete *Nereis diversicolor*. *J. Mar. Biol. Assoc. U.K.* 86:71–82.
Preuss, T.G. et al. 2009. Mechanistic effect models for ecological risk assessment of chemicals (MEMoRisk)—A new SETAC–Europe Advisory Group. *Environ. Sci. Pollut. Res.* 16:250–2.
Ramos-Gómez, J. et al. 2011. Validation of *Arenicola marina* in field toxicity bioassays using benthic cages: Biomarkers as tools for assessing sediment quality. *Mar. Pollut. Bull.* 62:1538–49.
Robillard, S., G. Beauchamp, and M. Laulier. 2003. The role of abiotic factors and pesticide levels on enzymatic activity in the freshwater mussel *Anodonta cygnea* at three different exposure sites. *Comp. Biochem. Physiol.* 135C:49–59.
Rose, K.A. et al. 2003. Using nested models and laboratory data for predicting population effects of contaminants on fish: A step toward a bottom-up approach for establishing causality in field studies. *Hum. Ecol. Risk Assess.* 9:231–57.
Sanchez, W. et al. 2011. Adverse effects in wild fish living downstream from pharmaceutical manufacture discharges. *Environ. Int.* 37:1342–8.
Sánchez-Argüello, P., N. Aparicio, and C. Fernández. 2012. Linking embryo toxicity with genotoxic responses in the freshwater snail *Physa acuta*: Single exposure to benzo(a)pyrene, fluoxetine, bisphenol A, vinclozin and exposure to binary mixtures with benzo(a)pyrene. *Ecotox. Environ. Saf.* 80:152–60.

Schmolke, A. et al. 2010. Ecological models and pesticide risk assessment: Current modelling practice. *Environ. Toxicol. Chem.* 29:1006–12.

Smith, R. et al. 2008. Macrofaunal recovery following TBT ban—Long-term recovery of subtidal macrofaunal communities in relation to declining levels of TBT contamination. *Environ. Monit. Assess.* 136:245–56.

Snoeijs, P., P. Sylvander, and N. Häubner. 2012. Oxidative stress and trophic interactions. In: *Oxidative Stress in Aquatic Ecosystems,* ed. D. Abele, J.P. Vázquez-Medina, and T. Zenteno-Savín, 72–88. Chichester, UK: Wiley-Blackwell.

Solé, M., J. Kopecka-Pilarczyk, and J. Blasco. 2009. Pollution biomarkers in two estuarine invertebrates, *Nereis diversicolor* and *Scrobicularia plana,* from a marsh ecosystem in SW Spain. *Environ. Int.* 35: 523–31.

Vidal, M.L., A. Bassères, and J.F. Narbonne. 2002. Influence of temperature, pH, oxygenation, water-type and substrate on biomarker responses in the freshwater clam *Corbicula fluminea* (Müller). *Comp. Biochem. Physiol.* 132C:93–104.

Weis, J.S. et al. 2001. Effects of contaminants on behavior. Biochemical mechanisms and ecological consequences. *Biosciences* 51:209–17.

Xuereb, B. et al. 2009a. Acetylcholinesterase activity in *Gammarus fossarum* (Crustacea Amphipoda). Intrinsic variability, reference levels, and a reliable tool for field surveys. *Aquat. Toxicol.* 93:225–33.

Xuereb, B. et al. 2009b. Acetylcholinesterase activity in *Gammarus fossarum* (Crustacea Amphipoda): Linking AChE inhibition and behavioural alteration. *Aquat. Toxicol.* 94:114–22.

Index

Page numbers followed by f and t indicate figures and tables, respectively.

D

F

H

I

U

V

W

X

Y

Z

For Product Safety Concerns and Information please contact our EU representative GPSR@taylorandfrancis.com Taylor & Francis Verlag GmbH, Kaufingerstraße 24, 80331 München, Germany

Batch number: 10731069

Printed by Printforce, the Netherlands